COST, EFFECTIVENESS, AND DEPLOYMENT OF FUEL ECONOMY TECHNOLOGIES FOR LIGHT-DUTY VEHICLES

Committee on the Assessment of Technologies for Improving
Fuel Economy of Light-Duty Vehicles, Phase 2

Board on Energy and Environmental Systems

Division on Engineering and Physical Sciences

NATIONAL RESEARCH COUNCIL
OF THE NATIONAL ACADEMIES

THE NATIONAL ACADEMIES PRESS
Washington, D.C.
www.nap.edu

THE NATIONAL ACADEMIES PRESS **500 Fifth Street, NW** **Washington, DC 20001**

NOTICE: The project that is the subject of this report was approved by the Governing Board of the National Research Council, whose members are drawn from the councils of the National Academy of Sciences, the National Academy of Engineering, and the Institute of Medicine. The members of the committee responsible for the report were chosen for their special competences and with regard for appropriate balance.

This study was supported by Contract No. DTNH22-11-H-00352 between the National Academy of Sciences and the National Highway Traffic Safety Administration. Any opinions, findings, conclusions, or recommendations expressed in this publication are those of the author(s) and do not necessarily reflect the views of the organizations or agencies that provided support for the project.

International Standard Book Number 13: 987-0-309-37388-3
International Standard Book Number 10: 0-309-37388-3
Library of Congress Control Number: 2015947372

Additional copies of this report are available for sale from the National Academies Press, 500 Fifth Street, NW, Keck 360, Washington, DC 20001; (800) 624-6242 or (202) 334-3313; http://www.nap.edu.

Printed in the United States of America

THE NATIONAL ACADEMIES

Advisers to the Nation on Science, Engineering, and Medicine

The **National Academy of Sciences** is a private, nonprofit, self-perpetuating society of distinguished scholars engaged in scientific and engineering research, dedicated to the furtherance of science and technology and to their use for the general welfare. Upon the authority of the charter granted to it by the Congress in 1863, the Academy has a mandate that requires it to advise the federal government on scientific and technical matters. Dr. Ralph J. Cicerone is president of the National Academy of Sciences.

The **National Academy of Engineering** was established in 1964, under the charter of the National Academy of Sciences, as a parallel organization of outstanding engineers. It is autonomous in its administration and in the selection of its members, sharing with the National Academy of Sciences the responsibility for advising the federal government. The National Academy of Engineering also sponsors engineering programs aimed at meeting national needs, encourages education and research, and recognizes the superior achievements of engineers. Dr. C.D. Mote Jr. is president of the National Academy of Engineering.

The **Institute of Medicine** was established in 1970 by the National Academy of Sciences to secure the services of eminent members of appropriate professions in the examination of policy matters pertaining to the health of the public. The Institute acts under the responsibility given to the National Academy of Sciences by its congressional charter to be an adviser to the federal government and, upon its own initiative, to identify issues of medical care, research, and education. Dr. Victor J. Dzau is president of the Institute of Medicine.

The **National Research Council** was organized by the National Academy of Sciences in 1916 to associate the broad community of science and technology with the Academy's purposes of furthering knowledge and advising the federal government. Functioning in accordance with general policies determined by the Academy, the Council has become the principal operating agency of both the National Academy of Sciences and the National Academy of Engineering in providing services to the government, the public, and the scientific and engineering communities. The Council is administered jointly by both Academies and the Institute of Medicine. Dr. Ralph J. Cicerone and Dr. C.D. Mote Jr. are chair and vice chair, respectively, of the National Research Council.

www.national-academies.org

COMMITTEE ON THE ASSESSMENT OF TECHNOLOGIES FOR IMPROVING FUEL ECONOMY OF LIGHT-DUTY VEHICLES, PHASE 2

JARED COHON, *Chair*, NAE,[1] Carnegie Mellon University, Pittsburgh, Pennsylvania
KHALIL AMINE, Argonne National Laboratory, Chicago, Illinois
CHRIS BAILLIE, AxleTech International, Troy, Michigan
JAY BARON, Center for Automotive Research, Ann Arbor, Michigan
R. STEPHEN BERRY, NAS,[2] University of Chicago, Chicago, Illinois
L. CATHERINE BRINSON, Northwestern University, Evanston, Illinois
MATT FRONK, Matt Fronk & Associates, LLC, Honeoye Falls, New York
DAVID GREENE, University of Tennessee-Knoxville, Knoxville, Tennessee
ROLAND HWANG, Natural Resources Defense Council, San Francisco, California
LINOS JACOVIDES, NAE, Michigan State University, East Lansing, Michigan
THERESE LANGER, American Council for Energy Efficient Economy, Washington, D.C.
REBECCA LINDLAND, King Abdullah Petroleum Studies and Research Center, Riyadh, Saudi Arabia
VIRGINIA McCONNELL, Resources for the Future, Washington, D.C.
DAVID MERRION, Merrion Expert Consulting, LLC, Brighton, Michigan
CLEMENS SCHMITZ-JUSTEN, CSJ Schmitz-Justen & Company, Greenville, South Carolina
ANNA STEFANOPOULOU, University of Michigan Automotive Research Center, Ann Arbor, Michigan
WALLACE WADE, NAE, Ford Motor Company (retired), Novi, Michigan
WILLIAM WALSH, Automotive Safety Consultant, McLean, Virginia

Staff

K. JOHN HOLMES, Study Director
DANA CAINES, Financial Manager
LINDA CASOLA, Senior Program Assistant
ELIZABETH EULLER, Program Assistant
STEVE GODWIN, Director, Studies and Special Programs, Transportation Research Board
LaNITA JONES, Administrative Coordinator
MICHELLE SCHWALBE, Program Officer
E. JONATHAN YANGER, Research Associate
ELIZABETH ZEITLER, Associate Program Officer
JAMES J. ZUCCHETTO, Director, Board on Energy and Environmental Systems

[1] NAE, National Academy of Engineering.
[2] NAS, National Academy of Science.

BOARD ON ENERGY AND ENVIRONMENTAL SYSTEMS

Staff

[1] NAE, National Academy of Engineering.
[2] NAS, National Academy of Sciences.

Preface

In 2012, the U.S. Department of Transportation's National Highway Traffic Safety Administration (NHTSA) and the U.S. Environmental Protection Agency (EPA) proposed significant new Corporate Average Fuel Economy (CAFE)/greenhouse gas (GHG) emission standards for light-duty vehicles. These standards will require the new vehicle fleet to double in fuel economy by 2025. Importantly, the vehicle manufacturers and suppliers by and large supported these new regulations. However, the manufacturers understandably had reservations in light of the aggressive nature of the standards. In order to address such concerns and meet statutory regulations, the Agencies proposed a mid-term review of the fuel economy standards. This review is to be completed by April 2018 in order to finalize the 2022-2025 standards.

The Committee on Assessment of Technologies for Improving the Fuel Economy of Light-Duty Vehicles, Phase 2, was established upon the request of NHTSA to help inform the mid-term review. Our committee was asked to assess the CAFE standard program and the analysis leading to the setting of the standards, as well as to provide its opinion on costs and fuel consumption improvements of a variety of technologies likely to be implemented in the light-duty fleet between now and 2030. The committee took the implications of our work very seriously, given the large potential impacts of the CAFE/GHG rules on the environment, consumers and vehicle manufacturers.

The committee comprised a wide array of backgrounds and sought input from agency analysts, vehicle manufacturers, equipment suppliers, consultants, academicians and many other experts. In addition to regular committee meetings, committee members held workshops on several critical topics, visited agency laboratories for extended discussions with their experts, and conducted numerous information-gathering site visits to automobile manufacturers and suppliers. The committee put great effort into thorough preparation for these meetings, asked probing questions and requested follow-up information in order to understand the perspectives of the many stakeholders. In addition, the committee commissioned a vehicle simulation modeling study from the University of Michigan in order to better understand the impacts of technology interactions. I greatly appreciate the considerable time and effort contributed by the committee's individual members throughout our information-gathering process, report writing and deliberations, and the committee extends its gratitude to the highly qualified experts who provided us with excellent presentations and rigorous discussions and graciously hosted us on our many excursions.

The committee operated under the auspices of the National Research Council Board on Energy and Environmental Systems (BEES). I would like to recognize the BEES staff for organizing and planning meetings, and assisting with information gathering and report development. The efforts of K. John Holmes, Elizabeth Euller, LaNita Jones, Michelle Schwalbe, Jonathan Yanger, Elizabeth Zeitler, James Zucchetto, and Steve Godwin were invaluable to the committee's ability to deliver its final report. I would also like to recognize David Cooke and Dharik Mallapragada for their early input. Thanks also to the many presenters, too numerous to name individually, who contributed to the committee's data-gathering process. Their contributions were invaluable and are listed in Appendix C.

This report has been reviewed in draft form by individuals chosen for their diverse perspectives and technical expertise, in accordance with procedures approved by the NRC's Report Review Committee. The purpose of this independent review is to provide candid and critical comments that will assist the institution in making its published report as sound as possible and to ensure that the report meets institutional standards for objectivity, evidence, and responsiveness to the study charge. The review comments and draft manuscript remain confidential to protect the integrity of the deliberative process. We wish to thank the following individuals for their review of this report:

Alexis Bell, NAS, University of California, Berkeley,
Andrew Brown Jr., Delphi Corporation,
John German, International Council for Clean Transportation,

Kenneth Gillingham, Yale University,
Imtiaz Haque, Clemson University,
Roger Krieger, University of Wisconsin, Madison,
Robert Lindeman, Northrop Grumman/Mission Systems (retired),
Shaun Mepham, Drive System Design, Inc.,
Margo Oge, U.S. Environmental Protection Agency (retired),
Gary Rogers, Roush Industries, Inc.,
Robert Sawyer, University of California, Berkeley,
Alan Taub, University of Michigan,
Thomas Wenzel, Lawrence Berkeley National Laboratory,
Ron Zarowitz, AutoPacific, and
Martin Zimmerman, University of Michigan.

Although the reviewers listed above have provided many constructive comments and suggestions, they were not asked to endorse the conclusions or recommendations nor did they see the final draft of the report before its release. The review of this report was overseen by Elisabeth M. Drake, Massachusetts Institute of Technology, and Elsa Garmire, Dartmouth College. Appointed by the NRC, they were responsible for making certain that an independent examination of this report was carried out in accordance with institutional procedures and that all review comments were carefully considered. Responsibility for the final content of this report rests entirely with the authoring committee and institution.

Jared Cohon, *Chair*
Committee on Assessment of Technologies for Improving the Fuel Economy of Light-Duty Vehicles, Phase 2

Contents

APPENDIXES

Boxes, Figures, and Tables

BOXES

FIGURES

TABLES

Summary

The light-duty vehicle fleet is expected to undergo substantial technological changes in the coming decades. New powertrain designs, alternative fuels, advanced materials and significant changes to the vehicle body are being driven by increasingly stringent fuel economy and greenhouse gas emission standards. By the end of the next decade, new vehicles will be more fuel efficient, lighter, emitting less air pollutants, safer, and more expensive to purchase relative to current vehicles. Given their increased efficiency, these vehicles will be less expensive to fuel than in the absence of such standards. Though the gasoline-fueled spark ignition (SI) engine will continue to be the dominant powertrain configuration even through 2030, such vehicles will be equipped with advanced technologies, materials, electronics and controls, and aerodynamics. And by 2030, the deployment of alternative methods to propel and fuel vehicles and alternative modes of transportation, including autonomous vehicles, will be well underway. In this context, the US Department of Transportation's National Highway Traffic Safety Administration (NHTSA) requested that the National Research Council (NRC) study the costs, benefits, and issues related to the implementation of new light-duty vehicle technologies.

NHTSA is responsible for fuel economy standards and, together with the US Environmental Protection Agency (EPA), has been progressively tightening Corporate Average Fuel Economy (CAFE) and greenhouse gas (GHG) emission standards. The most recent set of CAFE/GHG standards, termed the National Program,[1] cover model years (MY) 2017-2025 and call for an average light-duty vehicle fleet fuel economy of 40.3-41.0 miles per gallon (mpg) by 2021 and 48.7-49.7 mpg by 2025.[2] The carbon dioxide (CO_2) emission standard by 2025 is 163 grams/mile, which is equivalent to 54.5 mpg if vehicles were to meet this CO_2 level entirely through fuel economy improvements.[3]

Recognizing the uncertainties for setting standards out to 2025, NHTSA is committed to a mid-term review to evaluate progress. Furthermore, under the provisions set out by the Energy Independence and Security Act of 2007, NHTSA does not have the statutory authority for setting CAFE standards for greater than five years at a time; thus, only the MY 2017-2021 CAFE standards can be considered final. Therefore, beginning in 2016, NHTSA and EPA will coordinate a mid-term review that must be finalized by April 2018. In conjunction with EPA and the California Air Resources Board, NHTSA will perform a joint technical analysis as part of its rulemaking for the MY 2022-2025 standards. This NRC study is designed to feed into the mid-term review and provide an independent review of technologies and the CAFE program more generally. The full task statement is in Appendix A.

TECHNOLOGIES FOR REDUCING FUEL CONSUMPTION

The committee's report discusses a wide range of technologies and opportunities for reducing fuel consumption in light-duty vehicles. Figure S.1 demonstrates some of the progress to date and remaining needs for meeting these stan-

[1] The National Program is the combination of NHTSA's CAFE program and EPA's light-duty vehicle GHG emissions program. The first phase of the National Program covered MY 2012-2016 vehicles.

[2] As discussed later in the summary and in the body of the report, the National Program uses a footprint-based standard that means that compliance for each manufacturer is not a pre-determined target but dependent on the mix of vehicles sold each year. For these purposes, vehicle footprint is defined as the average track width of a vehicle multiplied by its wheelbase. The higher mpg values are based on the agencies' projections using a MY 2008 baseline; the lower mpg values are projections using a MY 2010 baseline.

[3] Reducing air conditioning leakage and deploying low greenhouse warming potential refrigerants are included as compliance options in the GHG standards but not the CAFE program, since they reduce GHGs but not fuel consumption.

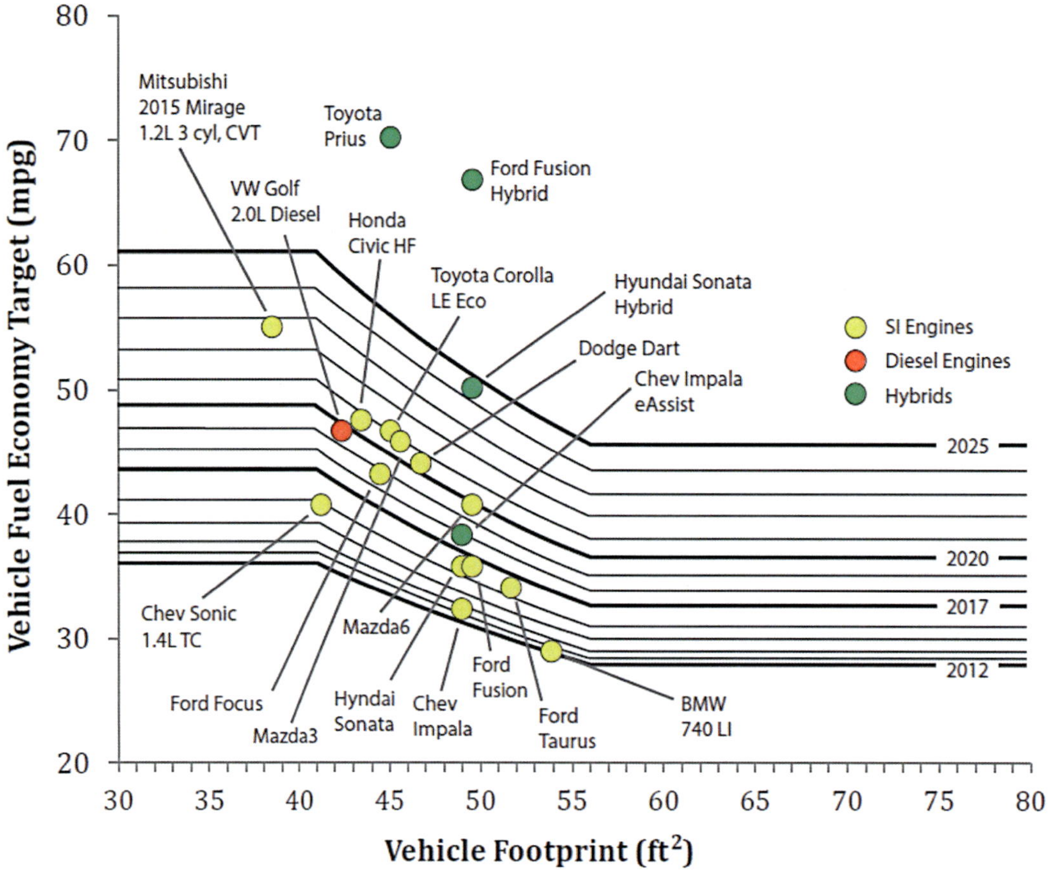

FIGURE S.1 Certification fuel economy values of selected 2013 and 2014 MY cars plotted on NHTSA CAFE target curves.

dards. It shows the official EPA certification fuel economy values compared to CAFE targets for particular vehicles given their respective footprints. These vehicles have already incorporated some of the technologies identified by NHTSA and represent a range of technologies and powertrains in the passenger car segment. It includes many high volume models. Most of the vehicles use spark-ignition gasoline engines and demonstrate the potential for conventional technologies to meet these standards. This figure shows fuel economy values for SI engines clustered around the 2016 MY targets and between the 2019 and 2021 MY targets. The outliers on this figure are also significant. The hybrids high fuel economy show why auto manufacturers might pursue these technologies as part of a strategy to meet the new standards. Also notable is the largest footprint vehicle on this figure that has many of the technologies for improving fuel economy but only meets the 2012-2013 targets. It demonstrates that implementing these technologies can be used to improve performance rather than fuel economy.

ESTIMATED FUEL CONSUMPTION REDUCTIONS AND COST

A central task for the committee was to develop estimates of the cost and potential fuel consumption reductions for technologies that might be employed from 2020 to 2030. This is a challenging task given the inherent difficulty in projecting technology outcomes so far into the future. In order to accomplish this task, the committee focused its efforts on projecting technology effectiveness and costs in years 2017, 2020, and 2025. The committee found the analysis conducted by NHTSA and EPA in their development of the 2017-2025 standards to be thorough and of high caliber on the whole. In particular, the committee notes that the use of full vehicle simulation modeling in combination with lumped parameter modeling has improved the Agencies' estimation of fuel economy impacts. Increased vehicle testing has also provided input and calibration data for these models. Similarly, the use of teardown studies has improved the Agencies' estimates of costs. The committee assessed the methodologies and assumptions used to develop these estimates and then conducted its own evaluation, giving greatest attention to those technologies that have large potential benefits. This work relied on committee expertise and analysis; presentations from and discussions with agency experts, auto manufacturers, suppliers, and others; and information contained in regulatory documents, academic literature, and press accounts. Using these same sources, the committee also considered technologies not included in the Agencies' analysis but in many cases had difficulty obtaining sufficient information, especially on costs. Therefore, in some cases specific conclusions especially on costs cannot be made for technologies not considered by the Agencies.

Tables S.1 and S.2 show the committee's estimates of fuel consumption benefits and direct manufacturing costs for technologies. The committee provides the benefit estimates for the individual technologies in Table S.1, which remain constant through the time frame, and the costs for 2025 in Table S.2. The committee focused on estimates of direct costs rather than total costs due to the uncertainties around mark-up factors discussed in Chapter 7. Chapter 8 and its appendices contain the complete tabulation of the committee's cost and benefit estimates and the comparison to the Agency estimates. The estimates are based on the technological assessments in Chapters 2-6.

Based on its analysis, the committee concurred with the Agencies' costs and effectiveness values for many technologies. In other cases the committee developed estimates that differed from the Agencies' values. For some technologies, committee members held different views on the best estimate of cost and effectiveness; these are represented in the different values in Tables S.1 and S.2. In such cases, the committee feels that both values are supported by the available data and analysis. It is important to note that these values are not meant to represent the full range of possible values for technology cost and effectiveness, but rather the different possible most likely values based on expert views represented on the committee.

Although the focus of the committee was individual technologies, it is valuable to consider vehicles of the future and the technology packages they might include. In doing selected technology pathway analyses, the committee considered how the order of technology application might impact the effectiveness of an individual technology; how the application of multiple technologies in a pathway prioritized by cost effectiveness might produce overall cost and fuel consumption reductions; and how alternative approaches and technologies could provide additional reductions in fuel consumption at lower cost increments. The committee found that understanding the base or null vehicle, the order of technology application, and the interactions among technologies are critical for assessing the costs and effectiveness for meeting the standards. It also noted that flexibilities contained within the standards in the form of air conditioning efficiency credits and off-cycle credits can reduce compliance costs.

Due to the committee's limited ability to model fleet and vehicle models in a more detailed manner, it did not estimate compliance costs for the MY2017- 2025 standards. Furthermore, the committee notes that a simple "roll up" of the its cost and effectiveness estimates, examples of which are provided in Chapters 2 and 8 for a specific example vehicle, cannot be used to estimate future compliance costs without similar roll-ups for all representative vehicles together with the consideration of compliance flexibilities discussed in Chapter 10.

Recommendations: While the committee concurred with the Agencies' costs and effectiveness values for a wide array of technologies, in some cases the committee developed estimates that significantly differed from the Agencies' values. Therefore, the committee recommends that the Agencies pay particular attention to the reanalysis of these technologies in the mid-term review (Recommendation 8.2). The committee also recommends that the Agencies establish a new definition of a "null" vehicle, representative of the most basic vehicle in the 2016 MY time frame as well as a baseline 2016 MY fleet reflecting actual technology penetration rates (Recommendation 8.1). This will assist in distinguishing between technologies available for the MY 2017-2025 CAFE targets and technologies that have already been applied to reach the MY 2016 CAFE targets. Further, the committee notes that the use of full vehicle simulation modeling in combination with lumped parameter modeling and teardown studies contributed substantially to the value of the Agencies' estimates of fuel consumption and costs, and it recommends they continue to increase the use of these methods to improve their analysis. The committee recognizes that such methods are expensive but believes that the added cost is well justified because it produces more reliable assessments (Recommendation 8.3).

SPARK-IGNITION ENGINES

The spark-ignition (SI) engine fueled with gasoline is by far the primary powertrain configuration in the United States for light-duty vehicles and will continue this dominance through the 2025 timeframe and likely beyond. The committee applied technologies in the order of cost effectiveness and engineering requirements to estimate the effectiveness and costs of the technologies in the 2017 to 2025 time frame. The committee focused on turbocharged, downsized engines in conjunction with other SI technologies to replace larger displacement, naturally aspirated engines. This is the pathway considered by the Agencies in the NHTSA/EPA compliance demonstration path[4] to be a major option for reducing fuel consumption to meet the standards. The committee estimated

[4] The Agencies developed what the committee termed the "compliance demonstration path" representing a cost effective set of model-specific technology packages that auto manufacturers could adopt to meet the standards. Although the Agencies' analysis demonstrates a possible technology path to compliance, auto manufacturers will plot their own future course to compliance.

TABLE S.1 NRC Committee's Estimated Fuel Consumption Reduction Effectiveness of Technologies

Percent Incremental Fuel Consumption Reductions: NRC Estimates					
		Midsize Car I4 DOHC	Large Car V6 DOHC	Large Light Truck V8 OHV	
Spark Ignition Engine Technologies	Abbreviation	Most Likely	Most Likely	Most Likely	Relative To
NHTSA Technologies					
Low Friction Lubricants - Level 1	LUB1	0.7	0.8	0.7	Baseline
Engine Friction Reduction - Level 1	EFR1	2.6	2.7	2.4	Baseline
Low Friction Lubricants and Engine Friction Reduction - Level 2	LUB2_EFR2	1.3	1.4	1.2	Previous Tech
VVT- Intake Cam Phasing (CCP - Coupled Cam Phasing - OHV)	ICP	2.6	2.7	2.5	Baseline for DOHC
VVT- Dual Cam Phasing	DCP	2.5	2.7	2.4	Previous Tech
Discrete Variable Valve Lift	DVVL	3.6	3.9	3.4	Previous Tech
Continuously Variable Valve Lift	CVVL	1.0	1.0	0.9	Previous Tech
Cylinder Deactivation	DEACD	N/A	0.7	5.5	Previous Tech
Variable Valve Actuation (CCP + DVVL)	VVA	N/A	N/A	3.2	Baseline for OHV
Stoichiometric Gasoline Direct Injection	SGDI	1.5	1.5	1.5	Previous Tech
Turbocharging and Downsizing Level 1 - 18 bar BMEP 33%DS	TRBDS1	7.7 - 8.3	7.3 - 7.8	6.8 - 7.3	Previous Tech
Turbocharging and Downsizing Level 2 - 24 bar BMEP 50%DS	TRBDS2	3.2 - 3.5	3.3 - 3.7	3.1 - 3.4	Previous Tech
Cooled EGR Level 1 - 24 bar BMEP, 50% DS	CEGR1	3.0 - 3.5	3.1 - 3.5	3.1 - 3.6	Previous Tech
Cooled EGR Level 2 - 27 bar BMEP, 56% DS	CEGR2	1.4	1.4	1.2	Previous Tech
Other Technologies					
By 2025:					
Compression Ratio Increase (with regular fuel)	CRI-REG	3.0	3.0	3.0	Baseline
Compression Ratio Increase (with higher octane regular fuel)	CRI-HO	5.0	5.0	5.0	Baseline
Compression Ratio Increase (CR~13:1, exh. scavenging, DI (aka Skyactiv, Atkinson Cycle))	CRI-EXS	10.0	10.0	10.0	Baseline
Electrically Assisted Variable Speed Supercharger[a]	EAVS-SC	26.0	26.0	26.0	Baseline
Lean Burn (with low sulfur fuel)	LBRN	5.0	5.0	5.0	Baseline
After 2025:					
Variable Compression Ratio	VCR	Up to 5.0	Up to 5.0	Up to 5.0	Baseline
D-EGR	DEGR	10.0	10.0	10.0	TRBDS1
Homogeneous Charge Compression Ignition (HCCI) + Spark Assisted CI[b]	SA-HCCI	Up to 5.0	Up to 5.0	Up to 5.0	TRBDS1
Gasoline Direct Injection Compression Ignition (GDCI)	GDCI	Up to 5.0	Up to 5.0	Up to 5.0	TRBDS1
Waste Heat Recovery	WHR	Up to 3.0	Up to 3.0	Up to 3.0	Baseline
Alternative Fuels[c]:					
CNG-Gasoline Bi-Fuel Vehicle (default UF = 0.5)	BCNG	Up to 5 Incr [42]	Up to 5 Incr [42]	Up to 5 Incr [42]	Baseline
Flexible Fuel Vehicle (UF dependent, UF = 0.5 thru 2019)	FFV	0 [40 thru 2019, then UF TBD]	0 [40 thru 2019, then UF TBD]	0 [40 thru 2019, then UF TBD]	Baseline
Ethanol Boosted Direct Injection (CR = 14:1, 43% downsizing) (UF~0.05)	EBDI	20 [24]	20 [24]	20 [24]	Baseline

Percent Incremental Fuel Consumption Reductions: NRC Estimates

		Midsize Car I4 DOHC	Large Car V6 DOHC	Large Light Truck V8 OHV	
Diesel Engine Technologies	Abbreviation	Most Likely	Most Likely	Most Likely	Relative To
NHTSA Technologies					
Advanced Diesel	ADSL	29.4	30.5	29.0	Baseline
Other Technologies					
Low Pressure EGR	LPEGR	3.5	3.5	3.5	ADSL
Closed Loop Combustion Control	CLCC	2.5	2.5	2.5	ADSL
Injection Pressures Increased to 2,500 to 3,000 bar	INJ	2.5	2.5	2.5	ADSL
Downspeeding with Increased Boost Pressure	DS	2.5	2.5	2.5	ADSL
Friction Reduction	FR	2.5	2.5	2.5	ADSL
Waste Heat Recovery	WHR	2.5	2.5	2.5	ADSL
Transmission Technologies	Abbreviation	Most Likely	Most Likely	Most Likely	Relative To
NHTSA Technologies					
Improved Auto. Trans. Controls/Externals (ASL-1 & Early TC Lockup)	IATC	2.5 - 3.0	2.5 - 3.0	2.5 - 3.0	4 sp AT
6-speed AT with Improved Internals - Lepelletier (Rel to 4 sp AT)	NUATO-L	2.0 - 2.5	2.0 - 2.5	2.0 - 2.5	IATC
6-speed AT with Improved Internals - Non-Lepelletier (Rel to 4 sp AT)	NUATO-NL	2.0 - 2.5	2.0 - 2.5	2.0 - 2.5	IATC
6-speed Dry DCT (Rel to 6 sp AT - Lepelletier)	6DCT-D	3.5 - 4.5	3.5 - 4.5	N/A	6 sp AT
6-speed Wet DCT (Rel to 6 sp AT - Lepelletier) (0.5% less than Dry Clutch)	6DCT-W	3.0 - 4.0	3.0 - 4.0	3.0 - 4.0	6 sp AT
8-speed AT (Rel to 6 sp AT - Lepelletier)	8AT	1.5 - 2.0	1.5 - 2.0	1.5 - 2.0	Previous Tech
8-speed DCT (Rel to 6 sp DCT)	8DCT	1.5 - 2.0	1.5 - 2.0	1.5 - 2.0	Previous Tech
High Efficiency Gearbox Level 1 (Auto) (HETRANS)	HEG1	2.3 - 2.7	2.3 - 2.7	2.3 - 2.7	Previous Tech
High Efficiency Gearbox Level 2 (Auto, 2017 and Beyond)	HEG2	2.6 - 2.7	2.6 - 2.7	2.6 - 2.7	Previous Tech
Shift Optimizer (ASL-2)	SHFTOPT	0.5 - 1.0	0.5 - 1.0	0.5 - 1.0	Previous Tech
Secondary Axle Disconnect	SAX	1.4 - 3.0	1.4 - 3.0	1.4 - 3.0	Baseline
Other Technologies					
Continuously Variable Transmission with Improved internals (Rel to 6 sp AT)	CVT	3.5 - 4.5	3.5 - 4.5	N/A	Previous Tech
High Efficiency Gearbox (CVT)	CVT-HEG	3.0	3.0	N/A	Previous Tech
High Efficiency Gearbox (DCT)	DCT-HEG	2.0	2.0	2.0	Previous Tech
High Efficiency Gearbox Level 3 (Auto, 2020 and beyond)	HEG3	1.6	1.6	1.6	Previous Tech
9-10 speed Transmission (Auto, Rel to 8 sp AT)	10SPD	0.3	0.3	0.3	Previous Tech
Electrified Accessories Technologies	Abbreviation	Most Likely	Most Likely	Most Likely	Relative To
NHTSA Technologies					
Electric Power Steering	EPS	1.3	1.1	0.8	Baseline
Improved Accessories - Level 1 (70% Eff Alt, Elec. Water Pump and Fan)	IACC1	1.2	1.0	1.6	Baseline
Improved Accessories - Level 2 (Mild regen alt strategy, Intelligent cooling)	IACC2	2.4	2.6	2.2	Previous Tech
Hybrid Technologies	Abbreviation	Most Likely	Most Likely	Most Likely	Relative To
NHTSA Technologies					
Stop-Start (12V Micro-Hybrid) (Retain NHTSA Estimates)	SS	2.1	2.2	2.1	Baseline
Integrated Starter Generator	MHEV	6.5	6.4	3.0	Previous Tech
Strong Hybrid - P2 - Level 2 (Parallel 2 Clutch System)	SHEV2-P2	28.9 - 33.6	29.4 - 34.5	26.9 - 30.1	Baseline

Percent Incremental Fuel Consumption Reductions: NRC Estimates					
		Midsize Car I4 DOHC	Large Car V6 DOHC	Large Light Truck V8 OHV	
Strong Hybrid - PS - Level 2 (Power Split System)	SHEV2-PS	33.0 - 33.5	32.0 - 34.1	N/A	Baseline
Plug-in Hybrid - 40 mile range	PHEV40	N/A	N/A	N/A	Baseline
Electric Vehicle - 75 mile	EV75	N/A	N/A	N/A	Baseline
Electric Vehicle - 100 mile	EV100	N/A	N/A	N/a	Baseline
Electric Vehicle - 150 mile	EV150	N/A	N/A	N/A	Baseline
Other Technologies					
Fuel Cell Electric Vehicle	FCEV	N/A	N/A	N/A	Baseline
Vehicle Technologies	Abbreviation	Most Likely	Most Likely	Most Likely	Relative To
NHTSA Technologies					
Without Engine Downsizing[d]					
0 - 2.5% Mass Reduction (Design Optimization)	MR2.5	0.80	0.80	0.85	Baseline
2.5 - 5% Mass Reduction		0.81	0.81	0.85	Previous MR
0 - 5% Mass Reduction (Material Substitution)	MR5	1.60	1.60	1.69	Baseline
With Engine Downsizing (Same Architecture)[d]					
5 - 10% Mass Reduction		4.57	4.57	2.85	Previous MR
0 - 10% Mass Reduction (HSLA Steel and Aluminum Closures)	MR10	6.10	6.10	4.49	Baseline
10 - 15% Mass Reduction (Aluminum Body)		3.25	3.25	2.35	Previous MR
0 - 15% Mass Reduction (Aluminum Body)	MR15	9.15	9.15	6.73	Baseline
15 - 20% Mass Reduction		3.37	3.37	2.41	Previous MR
0 - 20% Mass Reduction (Aluminum Body, Magnesium, Composites)	MR20	12.21	12.21	8.98	Baseline
20 - 25% Mass Reduction		3.47	3.47	2.46	Previous MR
0 - 25% Mass Reduction (Carbon Fiber Composite Body)	MR25	15.26	15.26	11.22	Baseline
Summary - Mass Reduction Relative to Baseline					
0 - 2.5% Mass Reduction	MR2.5	0.80	0.80	0.85	Baseline
0 - 5% Mass Reduction	MR5	1.60	1.60	1.69	Baseline
0 - 10% Mass Reduction	MR10	6.10	6.10	4.49	Baseline
0 - 15% Mass Reduction	MR15	9.15	9.15	6.73	Baseline
0 - 20% Mass Reduction	MR20	12.21	12.21	8.98	Baseline
0 - 25% Mass Reduction	MR25	15.26	15.26	11.22	Baseline
Low Rolling Resistance Tires - Level 1 (10% Reduction)	ROLL1	1.9	1.9	1.9	Baseline
Low Rolling Resistance Tires - Level 2 (20% Reduction)	ROLL2	2.0	2.0	2.0	Previous Tech
Low Drag Brakes	LDB	0.8	0.8	0.8	Baseline
Aerodynamic Drag Reduction - Level 1 (10% Reduction)	AERO1	2.3	2.3	2.3	Baseline
Aerodynamic Drag Reduction - Level 2 (20% Reduction)	AERO2	2.5	2.5	2.5	Previous Tech

[a] Comparable to TRBDS1, TRBDS2, SS, MHEV, IACC1, IACC2.
[b] With TWC aftertreatment. Costs will increase with lean NO_x aftertreatment.
[c] Fuel consumption reduction in gge (gasoline gallons equivalent) [CAFE fuel consumption reduction].
[d] FC Reductions – Ricardo 2007. Car without engine downsizing: +3.3% mpg/10% MR = -3.2% FC/10% MR. Car with engine downsizing (for MR > 10%): +6.5% mpg/10%MR = -6.1% FC/10% MR. Truck without engine downsizing: +3.5% mpg/10% MR = -3.4% FC/10% MR. Truck with engine downsizing (for MR > 10%): +4.7% mpg/10%MR = 4.5% FC/10% MR.
NOTE: Midsize car: 3,500 lbs, large car: 4,500 lbs, large light truck: 5,500 lbs.

TABLE S.2 NRC Committee's Estimated 2025 Direct Manufacturing Costs of Technologies

2025 MY Incremental Direct Manufacturing Costs (2010$): NRC Estimates					
		Midsize Car I4 DOHC	Large Car V6 DOHC	Large Light Truck V8 OHV	
Spark Ignition Engine Technologies	Abbreviation	Most Likely	Most Likely	Most Likely	Relative To
NHTSA Technologies					
Low Friction Lubricants - Level 1	LUB1	3	3	3	Baseline
Engine Friction Reduction - Level 1	EFR1	48	71	95	Baseline
Low Friction Lubricants and Engine Friction Reduction - Level 2	LUB2_EFR2	51	75	99	Previous Tech
VVT- Intake Cam Phasing (CCP - Coupled Cam Phasing - OHV)	ICP	31 - 36	63 - 73	31 - 36	Baseline for DOHC
VVT- Dual Cam Phasing	DCP	27 - 31	61 - 69	31 - 36	Previous Tech
Discrete Variable Valve Lift	DVVL	99 - 114	143 - 164	N/A	Previous Tech
Continuously Variable Valve Lift	CVVL	49 - 56	128 - 147	N/A	Previous Tech
Cylinder Deactivation	DEACD	N/A	118	133	Previous Tech
Variable Valve Actuation (CCP + DVVL)	VVA	N/A	N/A	235 - 271	Baseline for OHV
Stoichiometric Gasoline Direct Injection	SGDI	164	246	296	Previous Tech
Turbocharging and Downsizing Level 1 - 18 bar BMEP 33%DS	TRBDS1	245 - 282	-110 to -73	788 - 862	Previous Tech
V6 to I4 and V8 to V6			-396* to -316*	700* - 800*	
Turbocharging and Downsizing Level 2 - 24 bar BMEP 50%DS	TRBDS2	155	155	261	Previous Tech
I4 to I3		-82* to -86*			
Cooled EGR Level 1 - 24 bar BMEP, 50% DS	CEGR1	180	180	180	Previous Tech
Cooled EGR Level 2 - 27 bar BMEP, 56% DS	CEGR2	310	310	523	Previous Tech
V6 to I4				-453* to -469*	
Other Technologies					
By 2025:					
Compression Ratio Increase (with regular fuel)	CRI-REG	50	75	100	Baseline
Compression Ratio Increase (with higher octane regular fuel)	CRI-HO	75	113	150	Baseline
Compression Ratio Increase (CR~13:1, exh. scavenging, DI (aka Skyactiv, Atkinson Cycle))	CRI-EXS	250	375	500	Baseline
Electrically Assisted Variable Speed Supercharger	EAVS-SC	1,302	998	N/A	Baseline
Lean Burn (with low sulfur fuel)	LBRN	800	920	1,040	Baseline
After 2025:					
Variable Compression Ratio	VCR	597	687	896	Baseline
D-EGR	DEGR	667	667	667	TRBDS1
Homogeneous Charge Compression Ignition (HCCI) + Spark Assisted CI[a]	SA-HCCI	450	500	550	TRBDS1
Gasoline Direct Injection Compression Ignition	GDCI	2,500	2,875	3,750	Baseline
Waste Heat Recovery	WHR	700	805	1,050	Baseline
Alternative Fuels:					
CNG-Gasoline Bi-Fuel Vehicle	BCNG	6,000	6,900	7,800	Baseline
Flexible Fuel Vehicle	FFV	75	100	125	Baseline
Ethanol Boosted Direct Injection (incr CR to 14:1, 43% downsizing)	EBDI	740	870	1,000	Baseline

2025 MY Incremental Direct Manufacturing Costs (2010$): NRC Estimates

		Midsize Car I4 DOHC	Large Car V6 DOHC	Large Light Truck V8 OHV	
Diesel Engine Technologies	Abbreviation	Most Likely	Most Likely	Most Likely	Relative To
NHTSA Technologies					
Advanced Diesel	ADSL	2,572	3,034	3,228	Baseline
Other Technologies					
Low Pressure EGR	LPEGR	113	141	141	ADSL
Closed Loop Combustion Control	CLCC	58	87	87	ADSL
Injection Pressures Increased to 2,500 to 3,000 bar	INJ	20	22	22	ADSL
Downspeeding with Increased Boost Pressure	DS	24	24	24	ADSL
Friction Reduction	FR	54	82	82	ADSL
Waste Heat Recovery	WHR	700	805	1,050	ADSL
Transmission Technologies	Abbreviation	Most Likely	Most Likely	Most Likely	Relative To
NHTSA Technologies					
Improved Auto. Trans. Controls/Externals (ASL-1 & Early TC Lockup)	IATC	42	42	42	Baseline 4 sp AT
6-speed AT with Improved Internals - Lepelletier (Rel to 4 sp AT)	NUATO-L	-11	-11	-11	IATC
6-speed AT with Improved Internals - Non-Lepelletier (Rel to 4 sp AT)	NUATO-NL	165	165	165	IATC
6-speed Dry DCT (Rel to 6 sp AT - Lepelletier)	6DCT-D	-127 to 26	-127 to 26	N/A	6 sp AT
6-speed Wet DCT (Rel to 6 sp AT - Lepelletier)	6DCT-W	-75 to 75	-75 to 75	-75 to 75	6 sp AT
8-speed AT (Rel to 6 sp AT - Lepelletier)	8AT	47 - 115	47 - 115	47 - 115	Previous Tech
8-speed DCT (Rel to 6 sp DCT)	8DCT	152	152	152	Previous Tech
High Efficiency Gearbox Level 1 (Auto) (HETRANS)	HEG1	102	102	102	Previous Tech
High Efficiency Gearbox Level 2 (Auto, 2017 and Beyond)	HEG2	165	165	165	Previous Tech
Shift Optimizer (ASL-2)	SHFTOPT	22	22	22	Previous Tech
Secondary Axle Disconnect	SAX	86	86	86	Baseline
Other Technologies					
Continuously Variable Transmission with Improved internals (Rel to 6 sp AT)	CVT	154	154	NA	Baseline
High Efficiency Gearbox (CVT)	CVT-HEG	107	107	NA	Baseline
High Efficiency Gearbox (DCT)	DCT-HEG	127	127	127	Baseline
High Efficiency Gearbox Level 3 (Auto, 2020 and beyond)	HEG3	128	128	128	Baseline
9-10 speed Transmission (Auto, Rel to 8 sp AT)	10SPD	65	65	65	Baseline
Electrified Accessories Technologies	Abbreviation	Most Likely	Most Likely	Most Likely	Relative To
NHTSA Technologies					
Electric Power Steering	EPS	74	74	74	Baseline
Improved Accessories - Level 1 (70% Eff Alt, Elec. Water Pump and Fan)	IACC1	60	60	60	Baseline
Improved Accessories - Level 2 (Mild regen alt strategy, Intelligent cooling)	IACC2	37	37	37	Previous Tech
Hybrid Technologies	Abbreviation	Most Likely	Most Likely	Most Likely	Relative To
NHTSA Technologies					
Stop-Start (12V Micro-Hybrid)	SS	225 - 275	255 - 305	279 - 329	Baseline
Integrated Starter Generator	MHEV	888 - 1,018	888 - 1,115	888 - 1,164	Previous Tech
Strong Hybrid - P2 - Level 2 (Parallel 2 Clutch System)	SHEV2-P2	2,041 - 2,588	2,410 - 3,086	2,438 - 3,111	Baseline

2025 MY Incremental Direct Manufacturing Costs (2010$): NRC Estimates					
		Midsize Car I4 DOHC	Large Car V6 DOHC	Large Light Truck V8 OHV	
Strong Hybrid - PS - Level 2 (Power Split System)	SHEV2-PS	2,671	2,889	N/A	Baseline
Plug-in Hybrid - 40 mile range	PHEV40	8,325 - 9,672	11,189 - 13,135	N/A	Baseline
Electric Vehicle - 75 mile	EV75	8,451 - 8,963	11,025 - 11,929	N/A	Baseline
Electric Vehicle - 100 mile	EV100	9,486	11,971	N/A	Baseline
Electric Vehicle - 150 mile	EV150	12,264	14,567	N/A	Baseline
Other Technologies					
Fuel Cell Electric Vehicle	FCEV	N/A	N/A	N/A	
Vehicle Technologies	Abbreviation	Most Likely	Most Likely	Most Likely	Relative To
NHTSA Technologies					
Without Engine Downsizing					
0 - 2.5% Mass Reduction (Design Optimization)	MR2.5	0 - 22	0 - 28	0 - 39	Baseline
2.5 - 5% Mass Reduction		0 - 66	0 - 85	0 - 112	Previous MR
0 - 5% Mass Reduction (Material Substitution)	MR5	0 - 88	0 - 113	0 - 151	Baseline
With Engine Downsizing (Same Architecture)[b]					
5 - 10% Mass Reduction		151 - 315	194 - 405	264 - 558	Previous MR
0 - 10% Mass Reduction (HSLA Steel and Aluminum Closures)	MR10	151 - 403	194 - 518	264 - 710	Baseline
10 - 15% Mass Reduction (Aluminum Body)		431 - 730	554 - 938	751 - 1,279	Baseline
0 - 15% Mass Reduction (Aluminum Body)	MR15	431 - 730	554 - 938	751 - 1,279	Baseline
15 - 20% Mass Reduction		486 - 600	626 - 772	866 - 1,064	Previous MR
0 - 20% Mass Reduction (Aluminum Body, Magnesium, Composites)	MR20	917 - 1,330	1,179 - 1,710	1,617 - 2,343	Baseline
20 - 25% Mass Reduction		1,026 - 1,260	1,319 - 1,620	1,807 - 1,947	Previous MR
0 - 25% Mass Reduction (Carbon Fiber Composite Body)	MR25	1,943 - 2,590	2,498 - 3,330	3,424 - 4,290	Baseline
Mass Reduction Cost ($ per lb.)					
0 - 2.5% Mass Reduction	MR2.5	0.00 - 0.25	0.00 - 0.25	0.00 - 0.28	Baseline
0 - 5% Mass Reduction	MR5	0.00 - 0.49	0.00 - 0.49	0.00 - 0.55	Baseline
0 - 10% Mass Reduction	MR10	0.43 - 1.15	0.43 - 1.15	0.48 - 1.29	Baseline
0 - 15% Mass Reduction	MR15	0.82 - 1.39	0.82 - 1.39	0.91 - 1.55	Baseline
0 - 20% Mass Reduction	MR20	1.31 - 1.90	1.31 - 1.90	1.47 - 2.13	Baseline
0 - 25% Mass Reduction	MR25	2.22 - 2.96	2.22 - 2.96	2.49 - 3.12	Baseline
Low Rolling Resistance Tires - Level 1 (10% reduction in rolling resistance)	ROLL1	5	5	5	Baseline
Low Rolling Resistance Tires - Level 2 (20% reduction in rolling resistance)	ROLL2	31	31	31	Previous Tech
Low Drag Brakes	LDB	59	59	59	Baseline
Aerodynamic Drag Reduction - Level 1	AERO1	33	33	33	Baseline
Aerodynamic Drag Reduction - Level 2	AERO2	100	100	100	Previous Tech

* Costs with reduced number of cylinders, adjusted for previously added technologies – see Appendix T for the derivation of the turbocharged, downsized engine costs.
[a] With TWC aftertreatment. Costs will increase with lean NO_x aftertreatment.
[b] Includes mass decompounding: 40% for cars, 25% for trucks.
NOTE: Midsize car: 3,500 lbs, large car: 4,500 lbs, large light truck: 5,500 lbs.

that the combined SI engine technologies would provide overall fuel consumption reduction close to that estimated by NHTSA but with as much as 15 percent higher direct manufacturing costs for several of the technologies. The committee notes that the cost and effectiveness of SI engine technologies are complicated by several factors: the limitations on compression ratio due to currently available octane levels; the tradeoffs between drivability and fuel economy; and the impact of power-to-weight ratio on effectiveness. A separate concern is the degree to which the technologies used to meet the fuel economy standards will increase the relative deviation of real-world fuel economy from CAFE compliance values.

There are also new technologies not considered by EPA and NHTSA that might provide additional fuel consumption reductions for SI engines, or provide alternative approaches by 2025 and beyond. These technologies include higher compression ratio, exhaust scavenging, lean burn, and electrically assisted supercharger approaches and alternative fuels such as compressed natural gas-gasoline bi-fuel engines and ethanol-boosted direct injection engines.

Recommendations: Since spark-ignition engines are expected to be dominant beyond 2025, updated effectiveness and cost estimates of the most effective spark-ignition engine technologies should be developed for the mid-term review of the CAFE standards. Updated effectiveness estimates should be derived from full system simulations using engine maps based on measured data or generated engine model maps derived from validated baselines and include models for fuel octane requirements and drivability. Updated cost estimates using teardown cost studies of recently introduced spark-ignition engine technologies, including all vehicle integration costs, should be developed to support the mid-term review (Recommendation 2.1; Recommendation 2.2).

COMPRESSION-IGNITION ENGINES

Compression-ignition (CI) engines fueled by diesel have the highest thermodynamic efficiency of all internal combustion engine types. Compared to SI engines, CI engines possess three major benefits: lean-burn fuel mixtures, lack of air intake throttling, and higher compression ratios. In addition, diesel fuel has a higher energy content, due to a higher carbon content, than comparable volumes of gasoline so fewer gallons of fuel are used to provide identical work. Diesel's higher carbon content results in 15 percent more CO_2 per gallon burned than gasoline. This relative carbon emissions increase is one reason why manufacturers might not invest significantly in diesel technologies to comply with the National Program standards. The EPA/NHTSA compliance demonstration path included a vehicle fleet that had less than one percent diesel vehicles in 2025. The committee notes, however, that diesel vehicle product offerings and sales have increased recently.

CI engines provide large reductions in fuel consumption relative to baseline naturally aspirated SI engines with a higher cost and price, but with a lower total cost of ownership. While the committee agreed with the Agencies' fuel consumption reduction estimates, it found that the current EPA fuel economy certification data did not bear this out. However, in the future, CI engines will have even lower fuel consumption, most likely from applying higher levels of turbocharging with downsizing-downspeeding, improvements in closed-loop combustion control, and higher fuel injection pressures. Challenges for diesel vehicles to meet new emission standards for particulate matter, nitrogen oxides, and volatile organic compounds have prompted the development of improved aftertreatment systems that may reduce both the costs and size of the systems.

Recommendation: EPA and NHTSA should expand their full system simulations supported by mapping the latest diesel engines that incorporates as many of the latest technologies discussed in Chapter 3 as possible. EPA and NHTSA should conduct a teardown cost study of a modern diesel engine with the latest technologies to provide an up-to-date estimate of diesel engine costs. The teardown study should evaluate all costs, including vehicle integration, which includes the cooling system; torsional vibration damper; electrical systems, which include starter motors, batteries, and alternators; noise, vibration, and harshness control technologies; and vehicle costs resulting from the increased weight of the diesel engine. The study should also include an analysis of the increased residual value of a diesel-powered vehicle (Recommendation 3.1).

HYBRID AND ELECTRIFIED POWERTRAINS

Electrification of the powertrain is a powerful method to reduce fuel consumption and GHG emissions. Electrification ranges from hybrid electric vehicles (HEVs) to plug-in electric vehicles (PEVs)[5] to fuel cell electric vehicles (FCEVs). Fully realizing the GHG emission benefits from increased electrification in PEVs or FCEVs requires concomitant changes in the fuel mix for electricity generation or a low carbon source for hydrogen. The committee generally agrees with the Agencies' estimates of fuel consumption benefits for HEVs, but has concerns about the regulatory treatment of the GHG emissions from the generation of electricity for BEVs and PHEVs.

The committee finds that the battery cost estimates used by the Agencies are broadly accurate while the cost of the non-battery elements may be too low. The Agencies determined that an emerging HEV design, the P2, is likely to be the dominant strong hybrid technology based on projected cost and effectiveness versus the currently-dominant

[5] PEVs include both plug-in hybrid vehicles (PHEVs) and pure battery electric vehicles (BEVs).

power split design.[6] However, the committee found that the Agencies made critical assumptions about implementation of the P2 design that need to be revised, particularly the size of the motor and technologies required for launch performance and consumer comfort.

Battery cost is the dominant cost for PEVs and is a function of energy and power requirements, battery chemistry and required battery life. The committee finds that real-world battery life validation data do not exist beyond simulations and accelerated aging tests, so the appropriate state of charge swing to meet the prevalent electric powertrain warranty of 8 years and 100,000 miles is unknown. Meeting and exceeding this battery lifetime is important because the cost for a replacement battery pack is large.

Although current penetration is low, the use of electrification is likely to increase. In the opinion of the committee, the penetration of strong HEVs, PHEVs, and BEVs by MY 2025 will be larger than the respective 5%, 0% and 2% that the Agencies included in their compliance demonstration path. California Zero Emissions Vehicle mandates may require a higher penetration of the PEVs than included in the Agencies' compliance demonstration path. The committee believes that fuel cell electric vehicles will have minimal impact on CAFE compliance to 2025, but may play a bigger role in 2030 and beyond.

Recommendations: For their mid-term review, the Agencies should examine auto manufacturers' experiences of battery life to determine the appropriate state of charge swing for PHEVs and BEVs so that they can assign costs appropriately (Recommendation 4.2). The Agencies should undertake a teardown study of the next generation PS and P2 architectures to update cost and full system simulation of P2 and PS architectures to update effectiveness (Recommendation 4.3). Further, at the time of the mid-term review, there will be several vehicles with electrified powertrains in the market. The Agencies should commission teardown studies of the most successful examples of (1) stop-start, (2) strong hybrids (PS, P2, and two motor architectures), (3) PHEV20 and PHEV40, and (4) BEV100. At that time there will be better estimates of volumes for each type in the 2020 to 2025 time frame so that a better estimate of cost can be calculated (Recommendation 4.4).

TRANSMISSIONS

Transmission design reduces fuel consumption through increasing the number of gears, allowing the engine to operate closer to its best efficiency condition, and by minimizing parasitic losses within transmission architecture. Some technologies, such as turbocharged, downsized engines, require transmission design changes to maintain launch behavior and overcome the acceleration lag that accompanies such engines. The most popular transmission design is the planetary automatic transmission (AT), which is expected to remain the dominant architecture in the US in the 2025 timeframe; however, the market will transition from today's typical 6-speed AT to 8- to 10-speed AT designs. Dual-clutch transmissions (DCTs) can reduce parasitic losses by 40 to 60 percent relative to current automatic transmissions, but may have a lower market penetration than estimated by the Agencies' compliance demonstration path. Continuously variable transmissions (CVTs) provide continuously variable gear ratios to improve efficiency. The committee believes CVTs will experience higher market penetration than assumed in the Agencies' compliance demonstration path.

Negative synergies[7] between engine and transmission technologies could reduce the expected effectiveness of certain transmission technologies. For example, the efficiency gains from turbocharged and downsized engines reduce some efficiency gains from increased transmission gears since both technologies focus on reducing engine pumping and friction losses. The committee finds that achieving the fuel consumption reduction the Agencies attribute to moving to 8- and 9-speed transmissions would require a combination of high efficiency gearbox technologies, torque converter lockup, and aggressive shift logic. Current analyses of increasing the number of gear ratios have found minimal gains beyond 7 gear ratios; however, reducing parasitic losses within a transmission can offer an approximate 5 percent reduction in fuel consumption in the 2025 timeframe.

Recommendations: NHTSA and EPA should add the CVT to the list of technologies applicable for the 2017-2025 CAFE standards. (Recommendation 5.2). NHTSA and EPA should update the analyses of technology penetration rates for the midterm review to reflect the anticipated low DCT penetration rate in the U.S. market (Recommendation 5.1).

NON-POWERTRAIN TECHNOLOGIES

There are many opportunities outside of the vehicle's powertrain to adopt fuel-saving technologies. The committee considered mass reduction, aerodynamics, tires, vehicle accessories (such as power steering and heating/air conditioning systems) and the rapidly developing area of vehicle automation systems.

[6] The P2 hybrid uses a clutch connecting a single electrical machine and engine crankshaft, and incorporates a conventional transmission. As discussed in Chapter 4, the P2 architecture has an inherent design advantage over the power split in that there is no double energy conversion during certain operating conditions that occurs in the power split design.

[7] The committee uses synergies to mean that the effect of two or more technologies applied together may produce a result not obtainable by simply adding the effect of the individual technologies. Thus, the combination of technologies may produce either negative synergies, meaning the sum of the effects of the technologies is less than the impact of the individual technologies, or positive synergies, meaning the sum of the effects of the technologies is greater than the impact of the individual technologies.

The committee finds the levels of mass reduction assumptions identified in the EPA/NHTSA compliance demonstration path to be overly conservative for midsize and large cars. Even with additional mass required for safety improvements, the industry is likely to implement mass reduction of 10 to 20 percent in passenger vehicles, which is higher than what is in the Agencies' compliance demonstration path. In a few cases with specialty vehicles, greater than 20 percent could be achieved using advanced materials and new design strategies. For light-trucks, 15 to 20 percent reductions are expected. By taking mass reduction into account early in vehicle design, even more benefits can be achieved by drivetrain optimization and decompounding.[8] The committee found that the costs of mass reduction range from the Agencies estimates to much higher costs, and it concurs with the Agencies' assertion that incremental costs are likely to increase as more mass is removed from a vehicle design.

There have been several excellent teardown studies to help assess the opportunities and cost for reducing mass in vehicles, but there has been little attention given to interpreting how best to use the results. The committee feels that these studies are hard to generalize and future teardowns would benefit from careful selection of vehicles that are representative of their class.

It is the committee's view that mass will be reduced across all vehicle sizes with proportionately more mass removed from heavier vehicles. The most current studies that analyze the relationships among vehicle footprint, mass, and safety support the argument that removing mass across the fleet in this manner while keeping vehicle footprints constant will have a beneficial effect on safety for society as a whole.[9] Additionally, with the introduction of improved crash simulation and vehicle design techniques, new materials, and crash avoidance technology (such as lane change warning and autonomous frontal braking), crashworthiness and crash avoidance should be improved. During the transition period when vehicle masses are being reduced, there could be a negative impact on safety due to variance in the distribution of the mass across the vehicle fleet.

Recommendation: The committee recommends that the Agencies augment their current work with a materials-based approach that looks across the fleet to better define opportunities and costs for implementing lightweighting techniques, especially in the area of decompounding. A characterization of current vehicles in terms of materials content is a prerequisite for such a materials-based approach and for quantifying the opportunities to incorporate different lightweighting materials in the fleet (Recommendation 6.3).

COST AND MANUFACTURING CONSIDERATIONS

The committee notes that technology and design changes will impact the indirect costs of firms. In theory, the committee agrees with the Indirect Cost Multiplier (ICM) method as it attempts to assign indirect costs to products based on the activities they require, as opposed to assuming identical indirect cost multipliers for all technologies as with the Retail Price Equivalent. However, attribution for these indirect costs can be ambiguous, especially for future costs, and it was not possible for this committee to validate the Agencies' ICMs. The committee also notes that the method used by the agencies to estimate how costs change with increasing production volume is unconventional in that it is strictly a function of time rather than cumulative production. Such an approach allows a technology to accomplish significant cost reductions even if its production volumes remain very low.

The product development process of auto manufacturers is accelerating for several reasons, including the need to implement new technologies faster to meet the steadily increasing CAFE/GHG standards. More rapid deployment, although better for meeting regulations and responding to consumer demands, will increase stranded capital and incur higher product deployment costs. Further complicating the deployment of new technologies is the growth of global platforms. The movement by manufacturers towards global platforms can be considered a constraint, especially in the short term where supply chains are not fully developed, as well as an opportunity, especially in the long term where scale economies can provide cost reductions.

Recommendation: The Agencies should continue research on indirect cost multipliers with the goal of developing a sound empirical basis for their estimation (Recommendation 7.1). The committee also recommends that the Agencies continue to conduct and review empirical evidence for the cost reductions that occur in the automobile industry with volume, especially for large volume technologies that will be relied on to meet the CAFE/GHG standards (Recommendation 7.2).

CONSUMER IMPACTS AND ACCEPTANCE ISSUES

A critical element for the success of the National Program is how consumers respond to the more fuel efficient vehicles of the future. This requires understanding trends in the new vehicle market as well as consumer reaction to new technology for fuel economy. There have been substantial improvements in vehicle technology over time, with increases in mass, horsepower, and acceleration during the period of flat fuel economy standards from 1985 to 2005, and increases

[8] Mass decompounding is the opportunity for additional, or secondary, mass reduction in a vehicle's design based on the new specifications of the newly designed vehicle following the initial, or primary, implemented mass reductions.

[9] In this report, societal risk is used to describe the statistical probability of a fatality occurring for the occupants of the subject vehicle, the occupants of any involved vehicle(s), and any pedestrians or cyclists involved in a given crash. Personal risk, or occupant risk, is the statistical probability of a fatality occurring for only the occupants of the subject vehicle.

in average fuel economy and performance since 2006. The committee noted that vehicle segments are changing in response to consumer preferences, such as a shift away from truck-based sport utility vehicles towards more fuel-efficient car-based crossover utility vehicles. Other changes may be more driven by regulations to improve fuel economy. There is evidence that consumers will not widely adopt technologies that interfere with driver experience, comfort or perceived utility even for large improvements in fuel economy.

Regulatory efficiency and consumer benefit are further related to how consumers value fuel economy and other vehicle attributes. The extent to which consumers undervalue fuel economy (the energy paradox) in new car purchases remains a subject of debate with empirical evidence mixed on the overall magnitude of consumers' unwillingness to make energy-efficient investments even when those investments appear to have short payback periods. Consumer response is likely diverse, but the committee finds that manufacturers perceive that consumers require relatively short payback periods of one to four years for fuel economy improvements.

Recommendations: The committee recommends that the Agencies do more research on the existence and extent of the energy paradox in fuel economy, the reasons for consumers' undervaluation of fuel economy relative to its discounted expected present value, and differences in consumers' perceptions across the population (Recommendation 9.1). The Agencies should study the value of vehicle attributes to consumers, consumer willingness to trade off other attributes for fuel economy, and the likelihood of consumer adoption of new, unfamiliar technologies in the vehicle market (Recommendation 9.3). The Agencies should also conduct more research on the existence and extent of supply-side barriers to long-term investments in fuel economy technologies (Recommendation 9.2).

ASSESSMENT OF CAFE PROGRAM METHODOLOGY AND DESIGN

The committee found that the National Program standards adopted for the MY 2017-2021 and proposed through 2025 are different from the earlier CAFE standards in a number of important ways, including the development of combined fuel economy and GHG standards and added flexibility for manufacturer compliance through credit markets. These standards also continue the use of the footprint-based metric that began with the MY 2008-2011 fuel economy standards for light trucks. The committee appreciates the difficulty of developing a single national program for reducing LDV petroleum consumption and GHG emissions and commends the Agencies for their combined efforts. An important motivation for adopting a standard based on vehicle footprint (the vehicle's wheelbase times the average track width) is to be safety-neutral. The committee found the empirical evidence from historical data appears to support the argument that the new footprint-based standards are likely to have little effect on vehicle and overall highway safety.

There has long been recognition that the existing two-cycle certification tests used for fuel economy compliance and GHG emissions are not accurate representations of real-world driving behavior. The 5-cycle test procedure appears to provide a better representation of the range of real-world driving conditions. Additionally, there is no comprehensive source of information on the real-world fuel economy of light-duty vehicles to assess the gap between real-world fuel economy and the certification values, and this relationship may change in the future as vehicle technologies change.

Other important elements of the current standards are the treatment of alternative fuel vehicles and assessment of technology improvements that would occur in the absence of the standards. The 2017-2025 CAFE/GHG standards use a variety of incentives to spur production of alternative fuel vehicles including natural gas and electric vehicles, which have potential for greatly reduced oil use. These incentives include the CAFE program use of a 0.15 divisor for the fuel economy of alternative fuel vehicles and the GHG program use of sales multipliers and temporary zero emissions treatment. These incentives are more consistent with the reduced petroleum use of alternative fuel vehicles and less consistent with GHG benefits of all alternative fuels. Additionally, the Agencies' analyses assume a reference case for which no fuel economy is added after the 2016 MY. Assuming there is continued technology improvement after 2016, and it does not go to fuel economy in the reference (no-additional standards) case, then the improvements would go to enhance other vehicle attributes.

Recommendations: The Agencies should monitor the effects of the CAFE/GHG standards by collecting data on fuel efficiency, vehicle footprint, fleet size mix, and price of new vehicles to understand the impact of the rules on consumers' choices and manufacturers' products offered (Recommendation 10.1). The Agencies, perhaps in collaboration with other federal agencies, should conduct an on-going, scientifically-designed survey of the real-world fuel economy of light-duty vehicles. The survey should also collect information on real-world driving behavior and driving cycles. This information will be useful in determining the adequacy of the current test cycle and could inform the establishment of improved, future (post 2025) test cycles, if necessary (Recommendation 10.2). The Agencies should consider how to develop a reference case for the analysis of societal costs and benefits that includes accounting for the potential opportunity costs of the standards in terms of alternative vehicle attributes forgone (Recommendation 10.7). The midterm review is also a time when the Agencies should consider how the credit markets are different between the CAFE and GHG rules, and what the implications of these differences are for the auto manufacturers (Recommendation 10.8).

Further, the committee recommends that the Agencies study the potential benefits, costs and risks of establishing a standard based on a single metric that achieves both GHG and petroleum reductions in addition to continued efforts to harmonize the two regulations (Recommendation 10.12). Permanent regulatory treatment of alternative fuel vehicles should be commensurate with their well-to-wheels GHG and petroleum reduction benefits (Recommendation 10.6).

1

Introduction

Light-duty vehicles (LDVs) are the primary mode of transportation for the majority of the American public, being used for over 80% of all personal trips and comprising almost 90% of all passenger miles traveled (FHWA 2011). These vehicles consume about 130 billion gallons of gasoline each year, or approximately 18% of the total annual energy consumption in the United States, a significant fraction of which is produced from imported petroleum (EIA 2013; Davis et al. 2013). The impacts of this consumption are great, influencing economic prosperity, national security, and the environment. Gasoline consumed by the LDV fleet results in the production of air pollutants including carbon dioxide, the largest contributor to anthropogenic climate change. Political and economic instability in the Middle East, from which the United States receives about one eighth of its total petroleum supply, has contributed to concern over the nation's dependence on foreign energy sources (EIA 2014). In the past, this dependence has affected the United States through supply shocks, price uncertainty, and foreign policy commitments.

In response to these issues, Congress and executive branch Agencies have moved to reduce the petroleum consumption of LDVs, primarily through the tightening of fuel economy and greenhouse gas emissions (GHG) standards. Meeting these standards, as with earlier fuel economy, emissions and safety regulations, will require the introduction of new technologies and increased deployment of existing ones, resulting in a future LDV fleet that will be lighter, use smaller, boosted engines, and be powered by a greater diversity of fuels and powertrains compared to the current vehicle fleet.

STUDY BACKGROUND AND SETTING

The Nation's official pursuit of increased vehicle efficiency started 40 years ago when price shocks resulting from the 1973 Arab oil embargo led to the first regulations of fuel economy as part of the Environmental Policy and Conservation Act of 1975 (EPCA). Under the EPCA, Congress set the Corporate Average Fuel Economy (CAFE) standard for new passenger cars with the expectation of doubling the fuel efficiency of new passenger vehicles to 27.5 miles per gallon (mpg) by 1985. This legislation gave the National Highway Traffic Safety Administration (NHTSA), within the Department of Transportation (DOT), the authority to regulate the fuel economy of LDVs beginning with the 1978 model year and the authority to set fuel economy standards for other classes of vehicles, including light trucks.

After the CAFE standard set out in the EPCA was achieved in 1985, Congress relaxed the standard for passenger cars from 27.5 mpg to 26.0 mpg for a few years (1986-1988), responding to low oil prices, consumer demand, and pressure from the automotive industry (Bamberger 2002). After the standard returned to 27.5 mpg in 1990, passenger car fuel economy standards were not adjusted again for over two decades, until model year (MY) 2011. However, in 2003, NHTSA issued regulations increasing the fuel economy standards for light trucks for MY 2005-2007. In 2006, NHTSA again increased the fuel economy standards for light trucks for MY 2008-2011 and introduced the footprint standard, which is based on the product of a vehicle's wheelbase and its track width. Figure 1.1 demonstrates the use of this footprint standard in the calculation of target curves for passenger cars for MY2012-2025.

In 2007, Congress passed the Energy Independence and Security Act (EISA), which authorized NHTSA, in consultation with the Environmental Protection Agency (EPA) and the Department of Energy (DOE), to establish regulations that would maintain vehicle performance while raising the fuel economy of the new LDV fleet to a minimum of 35 mpg by 2020 and to a "maximum feasible average fuel economy standard" thereafter until 2030. Also in 2007, the outcome of Supreme Court case *Massachusetts v. EPA* established that the EPA was obligated to determine whether to regulate the emission of carbon dioxide (CO_2) and other GHGs from motor vehicles under Section 202 of the Clean Air Act. The Supreme Court decision and EPA's finding on the need to regulate LDV GHG emissions also enabled the state of California, through the California Air Resources Board (CARB), to set its own regulations for GHG emissions under Section

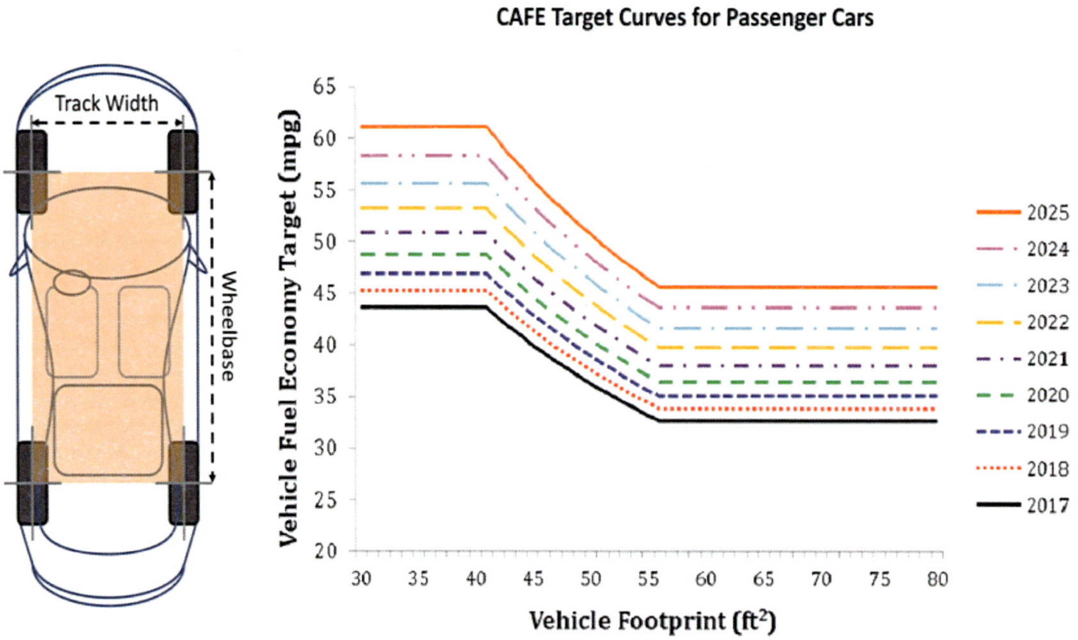

FIGURE 1.1 (Left) Vehicle footprint (track width × wheelbase) shown in orange. (Right) CAFE target curves for passenger cars as a function of footprint for MY 2017-2025.

209 of the Clean Air Act. The NHTSA and EPA standards are based on vehicle efficiency on a miles per gallon or GHG emission per mile basis.

Model Year 2012-2016 National Standards

Recognizing the problem of having as many as three separate regulatory Agencies setting standards governing the fuel economy of the light-duty fleet, the Obama Administration requested in 2009 that NHTSA, EPA, and the CARB work together to produce a single set of fuel economy and GHG emissions standards. The fuel economy and GHG emissions standards for MY 2012-2016 (EPA/NHTSA 2010) represent the first standard under this new National Program.[1] The standard required that fleet-averaged fuel economy reach an equivalent of 35.4 mpg by MY 2016.

The National Program also continued its use of the attribute-based standard that was first used for the MY 2008-2011 light truck standards, in which a vehicle's fuel economy target is related to its footprint (see Figure 1.1). The reasons given by NHTSA for a footprint standard are threefold: (1) the regulations encourage manufacturers with fleets that consist primarily of small passenger cars to continue to reduce the fuel consumption of even the smallest vehicles; (2) the footprint standard discourages manufacturers from downsizing vehicles to meet the standard, which could have led to safety concerns; and (3) the footprint standard is intended to be more equitable since it will require all manufacturers to make improvements. The footprint standard also means that compliance for each manufacturer is not a predetermined target but dependent on the mix of vehicle sizes it sells each year.

The National Program also made significant changes in the flexibility of manufacturer compliance. Credits for overcompliance could now be shared not just within a single manufacturer's fleet between cars and trucks, as with the earlier CAFE program, but also traded or sold among manufacturers. Credits can also be "banked" and used in later years, which can help a manufacturer achieve compliance within its own refresh and redesign product cycles. Credits are available in MY2012-2016 under both the CAFE and CO_2 programs for dual-fueled vehicles, which can run on gasoline and an alternative fuel such as electricity or a fuel mixture of 85% ethanol/15% gasoline (E85). There are also credits available under the EPA's CO_2 program for CO_2 reductions from switching to alternative air conditioning refrigerants, applying advanced technologies such as hybridization and electrification in the 2012-2016 timeframe, and applying off-cycle technologies such as active aerodynamics, whose efficiency benefits are not captured by the test cycles. Finally, in the MY2012-2016 regulations, EPA created an opportunity for banking "early credits" generated for MY2009-2011 vehicles for overcompliance with California or CAFE regulations over that 3-year time span.

[1] Throughout this report the combined GHG and fuel economy standards developed jointly by the NHTSA and EPA are referred to as the "National Program" or the "CAFE/GHG standards."

Model Year 2017-2025 National Standards

The standards for MY2017-2025 were adopted in 2012 (EPA/NHTSA 2012). The footprint-based standards are meant to achieve an approximately 4.2 percent annual increase in fuel economy for passenger cars and 3 percent annual increase for light trucks over this period. The target curves for passenger cars are shown in Figure 1.1, and Table 1.1 shows the fleetwide efficiencies that are estimated to result from the application of the footprint standard curves. As a point for comparison with the estimates shown in Table 1.1, the achieved fleetwide fuel economy averages for all manufacturers in MY2012 were 35.3 mpg for cars and 25.0 mpg for trucks (NHTSA 2014). Key changes to the MY2017-2025 standards include new accounting for alternative fuels and changes to the crediting program meant to align more closely the CAFE and greenhouse gas standards as well as reflect the current levels of technology. For example, for MY2017-2025, credits for off-cycle technologies and air conditioner efficiency will be made available for CAFE as well as for the CO_2 program. Chapter 10 contains a more detailed discussion of the 2017-2025 CAFE/GHG standards.

An important feature of the standards is their treatment of alternative-fuel vehicles. Dual-fueled vehicles—that is, vehicles that can be propelled using two different fuels (e.g., plug-in hybrid electric vehicles powered by either electricity or gasoline and flex-fuel vehicles powered by E85 or conventional gasoline)—have historically been credited equally for each fuel regardless of whether drivers use the alternative fuel. Beginning in 2020, credit for dual-fueled vehicles will consider how much of the alternative fuel an average consumer will use. This is especially important for plug-in hybrid electric vehicles, where battery size strongly dictates the fraction of electric miles. Vehicles using a single alternative fuel, such as battery electric vehicles or vehicles that run solely on compressed natural gas, also receive incentives for petroleum and GHG emissions reductions. Under the CAFE program, such vehicles are given a "petroleum equivalence factor," which increases the compliance fuel economy. Under the EPA's program, alternative-fuel vehicles will be considered based on "upstream accounting" of their emissions, in which net emissions including fuel production and distribution–related upstream GHG emissions are considered after an individual automaker exceeds its cumulative production cap. However, fuel cell, battery electric, and plug-in hybrid electric vehicles are currently credited only by their tailpipe emissions (0 g/mile) for an initial volume of vehicles in order to accelerate adoption. To further accelerate the production of alternative-fuel vehicles, those powered by natural gas, electricity, and hydrogen also receive credit multiplication under EPA's program.

In addition to alternative-fuel vehicles, the Agencies consider a number of other advanced technologies to be "game changers" that offer significant petroleum displacement and GHG emissions reductions. In order to accelerate the production of these technologies, the Agencies offer additional credits to manufacturers that introduce these technologies at significant volumes in their fleet. For the MY2017-2025 rulemaking, the Agencies considered the hybridization of full-size trucks to qualify as a game-changing technology and established credits under both programs for manufacturers who apply strong or mild hybrid technology to a minimum fraction of their truck fleet (10+ percent for strong hybrids and 20+ percent for mild hybrids with a goal of reaching 80+ percent penetration by 2021).

One final important note about the MY2017-2025 National Program comes as a result of the differing statutory authorities granted to EPA and NHTSA: EISA 2007 precludes NHTSA from setting CAFE standards for more than five years at a time. Therefore, only the MY2017-2021 CAFE standards can be considered binding. In 2017, NHTSA and EPA will coordinate a mid-term review of the standard as part of NHTSA's rulemaking for MY2022-2025, which must be finalized by April 2018.

Other Vehicle Regulations

There are other categories of vehicle regulations that will impact the deployment of new fuel economy technologies.

TABLE 1.1 Estimated Required Fleetwide Average Efficiencies under the National Program

	2012	2013	2014	2015	2016	2017	2018	2019	2020	2021	2022	2023	2024	2025
Passenger Car														
CAFE (mpg)	33.3	34.2	34.9	36.2	38.7	40.1	41.6	43.1	44.8	46.8	49.0	51.2	53.6	56.2
EPA (g CO_2/mi)	263	256	247	236	225	212	202	191	182	172	164	157	150	143
Light Truck														
CAFE (mpg)	25.4	26.0	26.6	27.5	29.2	29.4	30.0	30.6	31.2	33.3	34.9	36.6	38.5	40.3
EPA (g CO_2/mi)	346	337	326	312	298	295	285	277	269	249	237	225	214	203

NOTE: Values are shown in miles per gallon for NHTSA's CAFE standard and in grams CO_2 per mile for EPA's GHG regulations, with averages reflecting the MY 2008 baseline.
SOURCE: (EPA/NHTSA 2010, 2012).

These include the California Zero Emissions Vehicle (ZEV) mandates adopted by ten states, national criteria pollutant emissions standards, global fuel economy and emissions standards, and safety requirements (C2ES 2013). The most recent amendments to the California ZEV program (CARB 2013) stipulate that starting in 2017 zero tailpipe emissions vehicles (battery and fuel cell electric vehicles) and plug-in hybrid electric vehicles must together account for 15.4 percent of new sales in California and nine other states by 2025. Implementation of the ZEV mandate will increase the deployment of such vehicles above what NHTSA and EPA estimated might be required for compliance with the national CAFE/GHG standards.

Criteria pollutant standards, particularly the recently enacted Tier 3 standards for particulates, oxides of nitrogen, and hydrocarbon emissions, will add emission control costs and may create implementation issues for some fuel economy technologies (EPA 2014). The criteria pollutant standards are particularly important for diesel-fueled vehicles and some technologies that could be used for gasoline-powered vehicles.

Given that vehicle manufacturers, which are often referred to as Original Equipment Manufacturers (OEMs), operate in multiple markets around the world, their vehicle fleets must demonstrate compliance with the standards on the test cycles relevant for a given market. However, different technology options may yield varied benefits across the different test cycles used in various markets. For example, the lower speeds and accelerations and longer idling times of the test cycle used in Europe will confer greater advantages to technologies that reduce fuel consumption at lower speeds or idle compared to the test cycle used in this country (EPA n.d.; SAE 2012; VTA 2011). However, Europe will convert to the Worldwide Harmonized Light Vehicles Test Procedure in the 2020 time period, which will lessen the difference between the European and United States test cycles. Safety requirements will also impose additional equipment requirements and may limit some potential approaches to fuel economy improvement.

Fuel Consumption versus Fuel Economy

In understanding how the standards are defined and perceived, it is important to distinguish fuel consumption, the volume of fuel consumed divided by the distance traveled, from its inverse, the more commonly reported fuel economy, which is distance traveled divided by the volume of fuel used, usually reported in mpg. Fuel economy in mpg is familiar to consumers because it is the commonly used metric for vehicle efficiency, and it also tells the consumer something about the utility of the vehicle—how far one can travel on a given quantity of fuel. Fuel consumption, on the other hand, is proportional to the money a consumer would save or the benefits from reduced petroleum use that the nation may accrue from fuel economy standards. While fuel economy is regularly quoted and familiar to the public, it is not linear with fuel use because it is a distance/volume ratio. This non-linear relationship is not well understood by consumers and leads to misinterpretation of the benefits of improved fuel economy for vehicles of different efficiencies (Larrick and Soll 2008). Compared to less efficient vehicles, more efficient vehicles show a smaller improvement in fuel consumption for a given incremental improvement in fuel economy, see Figure 1.2.

Fuel consumption scales linearly with fuel volume and also accordingly with fuel cost and CO_2 emissions, which are the metrics of interest for the CAFE/GHG standards. If the goal of a consumer or policy maker is to minimize fuel use and maximize energy efficiency of passenger vehicle travel, then fuel consumption, rather than fuel economy, is the more straight-forward metric to use for accurate comparisons of the efficiency of different vehicles or vehicle technologies. In this report, the fuel efficiency of passenger vehicles or vehicle technologies will typically be reported in terms of fuel consumption, in units of gallons/100 miles. The fuel economy of vehicles will also be reported at times because the CAFE standards are defined in terms of fuel economy and the mpg metric is so generally familiar.

All fuel economy or fuel consumption values used in the report for actual vehicles fueled exclusively with gasoline or diesel will be compliance numbers as recorded by the EPA unless otherwise noted. This is to distinguish them from label-reported fuel economy values, which are adjusted to reduce mpg from the city and highway compliance numbers to mimic more closely what consumers will experience in real world driving. The label value will be occasionally reported when discussing consumer perceptions and will be identified as the fuel economy label value.

APPROACH TO TECHNOLOGY COST AND FUEL CONSUMPTION REDUCTION ESTIMATES

A central task for the committee was to develop estimates of the cost[2] and potential fuel consumption reductions for technologies that might be employed from 2020 to 2030. This is challenging given the inherent difficulty in projecting technology outcomes so far into the future. In order to accomplish this task, the committee focused its efforts on projecting technology effectiveness and costs in years 2017, 2020, and 2025. This allowed the committee to collect and assess information developed by the Agencies for the current and upcoming

[2] In general, the committee reports direct manufacturing costs (DMC) and does not estimate the indirect costs (IC) or total costs (TC=DMC+IC). As discussed in Chapter 7, DMCs are defined alternatively as the price an OEM would pay a supplier for a fully manufactured part ready for assembly in a vehicle, or the OEM's total cost of internally manufacturing the same part. Indirect costs (IC) include expenditures not directly required for manufacturing a component technology but necessary for the operation of an automobile manufacturing firm. Chapter 7 discusses methods and issues associated with estimating DMC, IC, and TC.

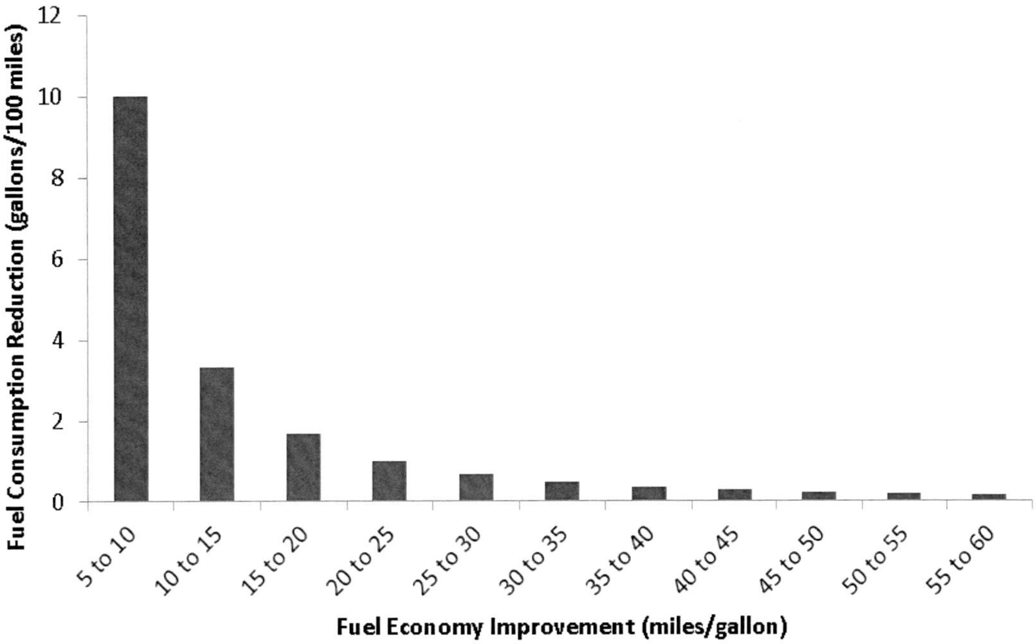

FIGURE 1.2 Diminishing fuel consumption reduction from 5 mpg fuel economy improvement at low to high fuel economy. Fuel consumption reductions are the fundamental metric of vehicle efficiency as they are directly proportional to fuel cost as well as CO_2 emissions reductions. The graph highlights the diminishing fuel consumption reduction benefit to marginal increases in fuel economy for increasingly efficient vehicles.

standards from 2017-2025. The committee also used input from OEMs, suppliers and other analysts whose predictions are often geared towards near-term implementations. Recognizing both the importance of looking at fuel consumption reduction technologies beyond 2025 and the high degree of uncertainty associated with making projections so far into the future, the committee felt it best to discuss only qualitatively the prospects for most technologies for the time frame beyond 2025. The committee has included effectiveness and cost estimates for SI technologies that might be available beyond 2025 in Tables S.1 and S.2, with the recognition that there are many hurdles associated with these technologies, as discussed in Chapter 2.

The committee's conclusions on cost and effectiveness are summarized in Tables S.1 and S.2. The committee found the analysis conducted by NHTSA and EPA in their development of the 2017-2025 standards to be thorough and of high caliber on the whole, and recognized in particular the value of their use of teardown studies and full-system simulation for estimating cost and effectiveness of many technologies. Consequently, for technologies considered by the Agencies, the committee used the Agencies' effectiveness and cost findings as a starting point and assessed the methodologies and assumptions used to develop them. It then conducted its own analysis, giving greatest attention to those technologies that have large potential benefits and/or that raised particular methodological or technical concerns. This analysis, which is presented in the following chapters, relied on committee expertise; presentations from Agency experts, OEMs, suppliers, and others; and information contained in regulatory documents, academic literature, and press accounts. As part of this effort, the committee reviewed the extensive modeling and analysis done by the Agencies during the rulemaking, as documented in the supporting studies for the final rule. In addition, the University of Michigan was engaged to perform full system simulation of a hypothetical gasoline-powered spark ignition vehicle to further the committee's understanding of the interactions of multiple technologies.

Based on its analysis, the committee concurred with the Agencies' costs and effectiveness values for many technologies. In other cases, however, the committee developed cost and effectiveness values using its expert judgement that differed from the Agencies' reported cost and effectiveness values. As documented in the report, updates or adjustments to the Agencies' estimates were based on the committee's expertise and judgment that incorporated a review of available information. This included reviews of information contained in published studies; input from experts in the automobile field including consultants, OEMs, and suppliers; and discussions with experts from NHTSA and EPA. Each of the committee's chapters contains an extensive reference list; Appendix C provides a list of all the public presentations the committee received (over fifty presentations, including three workshops); and the committee's public access file contains additional input (over 130 items) it received. The committee spent much time reviewing the regulatory documents produced by NHTSA and EPA, and followed up with the Agencies on particular issues through exchanges of questions from the committee and responses from the Agencies (contained in our Public Access File) and open session presentations/discussions with the Agencies' experts. The committee recommends that the Agencies revisit these areas of difference in depth in the midterm evaluation.

For some technologies, committee members held different views on the best estimate of cost and effectiveness; these are represented in the different values in Tables S.1 and S.2. In such cases, the committee feels that both values are supported by the available data and analysis. It is important to note that these values are not meant to represent the full range of possible values for technology cost and effectiveness, but rather the different possible most likely values based on expert views represented on the committee. The committee has included a general discussion of uncertainty in Chapter 10.

Other Considerations for Study

An important consideration for this study is whether the committee should explicitly consider the "unknown unknowns" (technologies and design beyond what is known today) in its estimates of fuel economy effectiveness and costs. The committee based its estimates on those fuel efficiency opportunities for which it could foresee a technology pathway to deployment over the 2017-2030 time period. The committee realizes that there will be unanticipated technological innovations and market trends that will produce vehicles with technologies not fully considered in the committee's analysis. The committee acknowledges the possibility that these unanticipated innovations may permit the industry to meet emission standards at lower than predicted cost. Though the committee could not describe such unpredictable technologies nor could it quantify their effectiveness and cost, it did look forward as far as possible to evaluate technologies that are on the development horizon but which do not have a clear technology pathway to deployment, such as advanced vehicle batteries and highly turbocharged engines. The committee does not believe that the automobile industry has reached the end of innovation, but quantifying possible improvements for unknown innovations was beyond the scope of the committee's study.

Understanding the impacts of LDVs on petroleum consumption and GHG emissions requires considering the fuels used to power such vehicles. The committee notes that it was not tasked with addressing fuels for light-duty vehicles independent of the interaction with technologies. This meant that the committee was not tasked with how different fuels, such as natural gas or biofuels, might independently reduce GHG emissions or contribute to energy security. Further, although the committee recognized the impact of fuel prices on consumer adoption of fuel economy technologies in Chapter 9, it was not tasked with providing a detailed assessment of the relationship between fuel prices and fuel economy technology deployment. This relationship has been emphasized due to the fluctuations in fuel costs that have occurred over the timeframe of this study. For example, during 2014 alone, average U.S. gasoline prices ranged from a high of over $3.60 per gallon to a low of under $2.25 per gallon. As discussed in Chapter 9, it is unknown how consumers will alter purchase decisions and assess total cost of ownership in the face of short-term fluctuations of fuel costs. Also, it is difficult to predict the effect lower fuel costs will have on marketing efforts for fuel-efficient vehicles. The committee was not tasked with predicting how fuel prices will change in the future beyond noting that the short-term fluctuations observed recently do not provide any insights into fuel prices that will occur during the time period (2017-2030) that is the focus of the committee's assessment.

STUDY ORIGIN AND ORGANIZATION OF REPORT

The National Research Council (NRC) has long had a role in helping to inform NHTSA on issues related to fuel economy. In 1992, the NRC released *Automotive Fuel Economy: How Far Should We Go?*, which considered the feasibility and the desirability of a variety of efforts to improve the fuel economy of the light-duty vehicle fleet. Subsequently, the NRC released *Effectiveness and Impact of Corporate Average Fuel Economy (CAFE) Standards* (NRC 2002), which studied the impact and effectiveness of CAFE standards originally mandated in the Energy Policy and Conservation Act of 1975.

More recently, Section 107 of EISA instructed NHTSA to contract with the NRC to "develop a report evaluating vehicle fuel economy standards, including an assessment of automotive technologies and costs to reflect developments since the [NRC]'s 2002 report (NRC 2002) evaluating the corporate average fuel economy standards was conducted and an assessment of how such technologies may be used to meet the new fuel economy standards." Section 107 also noted that the report should be updated at 5-year intervals through 2025. In 2011, the first such report in response to this mandate was released, *Assessment of Fuel Economy Technologies for Light-Duty Vehicles* (NRC 2011). This will often be referred to as the Phase 1 report throughout this report. That report examined categories of near-term technologies important for reducing fuel consumption, their costs, issues associated with estimating costs and price impacts of these technologies, and approaches for estimating the fuel consumption benefits from combinations of these technologies.

The current report is the second in the series and is timed to inform the mid-term review mentioned above by considering technologies applicable in the 2020 to 2030 timeframe. In particular, the committee was asked to include the following in its assessment:

- Methodologies and programs used to develop standards for passenger cars and light trucks under current and proposed CAFE programs;
- Potential for reducing mass by up to 20%, including materials substitution and downsizing of existing vehicle designs, systems or components;
- Other vehicle technologies whose benefits may not be captured fully through the federal test procedure, including aerodynamic drag reduction and improved efficiency of accessories;
- Electric powertrain technologies, including the capabilities of hybrids, plug-in hybrids, battery electric vehicles, and fuel cell vehicles;
- Advanced gasoline and diesel engine technologies that will increase fuel economy;
- Assumptions, concepts, and methods used in estimating the costs of fuel economy improvements, including the degree to which time-based cost learning for well-developed existing technologies and/or volume-based cost learning for newer technologies should apply and the differences between Retail Price Equivalent and Indirect Cost Multipliers;
- Analysis of how fuel economy technologies may be practically integrated into automotive manufacturing processes and how such technologies are likely to be applied;
- Costs and benefits in vehicle value that could accompany the introduction of advanced vehicle technologies;
- Test procedures and calculations used to determine fuel economy values for purposes of determining compliance with CAFE standards;
- Consumer responses to factors that may affect changes in vehicle use.

The complete statement of task for this study is given in Appendix A and biographical information of the committee members can be found in Appendix B.

This report is organized to correspond with the statement of task. Powertrain technologies are discussed first (Chapters 2-5), followed by non-powertrain technologies (Chapter 6). All cost and effectiveness tables for these sections, unless otherwise specified, represent the committee's estimates. Cost and manufacturing issues are discussed in Chapter 7. Chapter 8 then provides the committee's estimates of the costs and fuel consumption reduction benefits of employing these technologies. That is followed by the impacts on future consumers (Chapter 9) and a general assessment of the regulatory process (Chapter 10). A central result of the Phase I committee was a pair of tables summarizing the committee's estimate of each technology's costs and fuel economy benefits. This same format was followed, and updated tables are provided in Chapter 8.

REFERENCES

Bamberger, R. 2002. Automobile and Light Truck Fuel Economy: The CAFE Standards. Almanac of Policy Issues. Congressional Research Services, September 25. http://www.policyalmanac.org/environment/archive/crs_cafe_standards.shtml.

C2ES (Center for Climate and Energy Solutions). 2013. ZEV Program. http://www.c2es.org/us-states-regions/policy-maps/zev-program.

CARB (California Air Resource Board). 2013. Zero Emission Vehicle Program. http://www.arb.ca.gov/msprog/zevprog/zevprog.htm.

Davis, S.C., S.W. Diegel, and R.G. Boundy. 2013. Transportation Energy Data Book: Edition 32. Oak Ridge National Laboratory, Oak Ridge, TN.

De Carvalho, R., A. Villela, and S. Botero. 2012. Fuel Economy and CO_2 Emission – A Comparison between Test Procedures and Driving Cycles. SAE International Technical Paper 2012-36-0479. doi:10.4271/2012-36-0479.

EIA (Energy Information Agency). 2013. Annual Energy Review 2012. Washington, D.C.

EIA. 2014. Frequently Asked Questions: How much petroleum does the United States import and from where? U.S. Energy Information Administration, Washington, D.C. http://www.eia.gov/tools/faqs/faq.cfm?id=727&t=6.

EPA (Environmental Protection Agency). n.d. Emission Test Cycles: EPA Highway Fuel Economy Test Cycle. DieselNet. https://www.dieselnet.com/standards/cycles/hwfet.php.

EPA. 2013. EPA Proposes Tier 3 Motor Vehicle Emission and Fuel Standards. Regulatory Announcement. http://www.epa.gov/otaq/documents/tier3/420f13016a.pdf.

EPA. 2014. Control of Air Pollution From Motor Vehicles: Tier 3 Motor Vehicle Emission and Fuel Standards; Final Rule. Federal Register 79(81): 23414-23886.

EPA/NHTSA (National Highway Traffic Safety Administration). 2010. Light-Duty Vehicle Greenhouse Gas Emission Standards and Corporate Average Fuel Economy Standards; Final Rule. Federal Register 75(88):25323-25728. May 7.

EPA/NHTSA. 2012. 2017 and Later Model Year Light-Duty Vehicle Greenhouse Gas Emissions and Corporate Average Fuel Economy Standards. Federal Register 77(199): 62623–3200. October 15.

FHWA (Federal Highway Administration). 2011. Summary of Travel Trends: 2009 National Household Travel Survey. FHWA-PL-11-02. U.S. Department of Transportation, Federal Highway Administration. Accessed March 12, 2013. http://nhts.ornl.gov/2009/pub/stt.pdf.

Larrick, R., and J. Soll. 2008. The Mpg Illusion. Science 320(5883):1593-1594.

NHTSA. 2009. Average Fuel Economy Standards Passenger Cars and Light Trucks Model Year 2011. Federal Register 74(59): 14196-14456. March 30.

NHTSA. 2014. Summary of Fuel Economy Performance (Public Version). U.S. Department of Transportation, NHTSA, NVS-220, June 26.

NRC (National Research Council). 1992. Automotive Fuel Economy: How Far Should We Go? Washington, D.C.: National Academy Press.

NRC. 2002. Effectiveness and Impact of Corporate Average Fuel Economy (CAFE) Standards. Washington, D.C.: The National Academies Press.

NRC. 2011. Assessment of Fuel Economy Technologies for Light-Duty Vehicles.Washington, D.C.: The National Academies Press.

VCA (United Kingdom Vehicle Type Approval Agency). 2011. Exhaust Air Quality Pollutant Emissions Testing. http://www.dft.gov.uk/vca//fcb/exhaust-emissions-testing.asp.

2

Technologies for Reducing Fuel Consumption in Spark-Ignition Engines

INTRODUCTION

The spark-ignition (SI) engine, fueled with gasoline, has long been the dominant engine for the light-duty fleet in the United States. This dominance is expected to continue through the 2025 time frame and beyond. EPA and NHTSA, in their analysis for the MY 2017-2025 standards, have projected potential compliance paths for each company and for the industry fleet as a whole using the Environmental Protection Agency's (EPA) OMEGA model and the National Highway Traffic Safety Administration's (NHTSA) Corporate Average Fuel Economy (CAFE) model, also known as the Volpe model (EPA/NHTSA 2012a).[1] The EPA/NHTSA projected compliance demonstration path for the industry fleet as a whole, shown in Table 2.1, indicates that SI engines are projected to be used in 98 percent of the 2025 MY fleet, with 2 percent projected to be battery electric vehicles. Of the 98 percent of gasoline engines, 15 percent are projected to be in stop-start (SS), 26 percent are projected to be used in mild hybrid electric drivetrains (MHEVs), and 5 percent will be used in hybrid electric drivetrains (HEVs). With this continuing dominance projected for spark-ignition gasoline engines, technologies for reducing the fuel consumption of these engines will be essential for achieving the future CAFE standards.

This chapter considers technologies and associated costs for reducing fuel consumption in SI gasoline engines. The fundamentals of SI engine efficiency will be reviewed first to provide a context for examining the potential of individual technologies. With this background, the individual technologies for reducing fuel consumption will be reviewed within the following categories:

- Technologies EPA/NHTSA included in the final CAFE rule analysis;
- Technologies EPA/NHTSA considered for but did not include in the final CAFE rule analysis;
- Technologies EPA/NHTSA neither considered for nor included in the final CAFE rule analysis;
- Control systems, models and simulation techniques; and
- Emission control systems for meeting future criteria pollutant emission standards.

Estimates of the potential effectiveness of each of the technologies are presented and expressed in terms of percent reduction in fuel consumption. The fundamental means by which each technology achieves the reduction in fuel consumption—such as through reductions in friction, reductions in pumping loss, or improvements in thermodynamic efficiency—are identified. Potential interactions with other technologies, whereby the effectiveness of an individual technology might be enhanced but more likely would be diminished, are discussed. For each technology that EPA/NHTSA considered applicable in complying with the final CAFE rule, EPA/NHTSA provided estimates of the technology's effectiveness and cost. These estimates are reviewed in this chapter, and, where appropriate, modifications to the effectiveness and/or cost are suggested.

SI ENGINE EFFICIENCY FUNDAMENTALS

SI engines are often referred to as Otto cycle engines to describe the idealized thermodynamic processes of the engine. The idealized thermodynamic cycle for the Otto cycle engine is shown on a pressure versus volume (P-V) diagram in Appendix D, together with other thermodynamic cycles for several other engines discussed later. An energy balance for an SI gasoline engine operating at a condition representative of the Federal Test Procedure (FTP)[2] cycle,

[1] Although the final rule illustrates possible compliance paths, each company is expected to plot its own future course to compliance.

[2] The FTP represents the city driving portion of the test cycles used to estimate fuel economy and compliance with the CAFE/GHG standards. Chapter 10 discusses these test cycles and issues associated with them.

TABLE 2.1 EPA/NHTSA Technology Penetration for the MY 2025 Control Case with the 2017-2025 CAFE/GHG Standards in Effect for the Combined Light-Duty Truck and Car Fleet (percent)[a]

	Mass Reduction[b]	Turbocharged / Downsized 18-27 BMEP[c]	8-speed Automatic Transmission	8-speed Dual Clutch Transmission	Mild Hybrid	Strong Hybrid	Electric Vehicle	Diesel
Fleet	–7	93	35	56	26	5	2	0

[a]Technology penetrations for Aston Martin, Lotus and Tesla are not included here but can be found in EPA's Regulatory Impact Analysis (RIA).
[b]Negative values for mass reduction represent percentage of mass removed.
[c]BMEP, brake mean effective pressure.
SOURCE: EPA/NHTSA (2012a, Table III-29).

shown in Figure 2.1, is useful for identifying potential fuel consumption reduction opportunities. Technologies that improve thermodynamic efficiency or reduce losses and result in an increase in brake work as a percentage of the total fuel energy are effective in reducing fuel consumption. Factors affecting the various components of the energy balance are discussed in this section, while definitions and efficiency fundamentals are discussed in Appendix E.

The energy balance in Figure 2.1 illustrates current typical efficiencies and opportunities for reducing fuel consumption in SI gasoline engines. Energy input into a vehicle in the form of fuel produces energy output in the form of heat and work. Energy output goes into three areas: exhaust enthalpy, heat to coolants, and indicated work, where the indicated work is work done on the piston and exhaust enthalpy and heat to coolants are thermodynamic efficiency losses. A portion of the indicated work on the piston is further categorized as accessory work (required for the engine-driven pumps, cooling fan and alternator), rubbing friction work, and pumping work to move air into and exhaust out of the cylinders. The remaining portion of indicated work is the work that goes into the driveline (through the transmission, final drive, axles and tires) to propel the vehicle. That portion of the indicated work is termed brake work.[3] The energy output from the input fuel energy is described further below:

- Approximately one-third of the fuel energy is lost as exhaust enthalpy and another one-third is lost as heat rejected to coolant. Friction losses are generally manifested as additional heat transferred to coolant or oil.
- Brake work is nearly 40 percent lower than indicated work on the FTP cycle due to pumping losses, rubbing friction losses and accessory drive requirements.
- Improvements in thermodynamic efficiency increase the fraction of fuel energy that goes into indicated work.
- Pumping losses comprise approximately 5 percent of the total fuel energy. If pumping losses could be reduced by 20 percent, or 1 percent of the fuel energy, fuel consumption could be reduced by 2.8 percent (applying an indicated thermal efficiency (ITE) of 36 percent).[4]
- By increasing indicated work (work done on the piston) through improvements to thermodynamic efficiency by 1 percentage point, from 36 percent to 37 percent, fuel consumption could be reduced by 2.7 percent.[5]

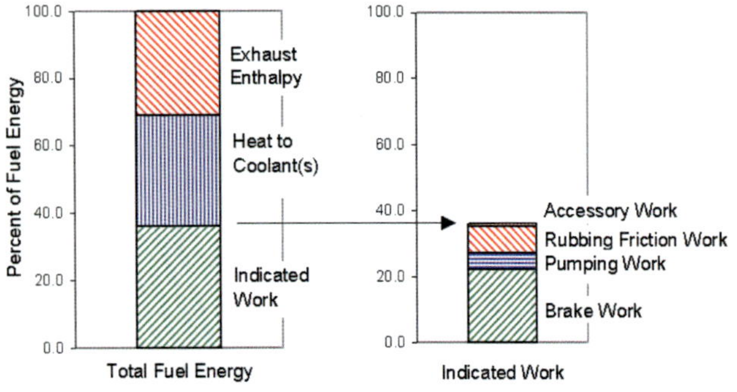

FIGURE 2.1 Energy balance for SI gasoline engine for an operating condition representative of the FTP cycle.
SOURCE: derived from data in Heywood (1988).

[3] Engine torque is measured with the engine connected to a dynamometer. The power delivered by the engine, which is absorbed by the dynamometer, is the product of torque and speed (Ameri 2010). The value of engine power measured in this manner is called brake power. This power is the usable power delivered by the engine to the load, in this case a brake (Heywood 1988).

[4] Indicated Thermal Efficiency (ITE) = Indicated work (energy)/fuel consumption (energy).

Reducing work by an amount equal to 1 percent of fuel energy would reduce fuel consumption by 2.8 percent, which is the fuel required to produce this work as governed by the cycle efficiency, so that fuel consumption (FC) = Indicated Work/ITE = 1.0/.36 = 2.8 percent.

[5] The baseline case assumes 36 percent ITE, as shown in Figure 2.1. The indicated work is calculated as follows:

100 percent fuel energy × 36% ITE/100 = 36 percent indicated work.

The amount of fuel energy required to produce the same indicated work of 36 percent (on the original fuel energy basis) is then calculated as follows:

fuel energy × 37% ITE/100 = 36 percent indicated work, or

fuel energy = 36/37 × 100 = 97.3 percent fuel energy.

With 37 percent ITE (a 1 percentage point increase), the fuel energy required is reduced by 2.7 percent relative to the baseline case.

- Rubbing friction losses are approximately 8 percent of the total fuel energy. If rubbing friction losses could be reduced by 25 percent, or 2 percent of the fuel energy, fuel consumption could be reduced by 5.6 percent (applying 36 percent ITE).
- Engine accessories (oil pump, water pump, fan and alternator) require work that consumes approximately 1 percent of the fuel energy. If engine accessory power requirements could be reduced by 50 percent, or 0.5 percent of the fuel energy, fuel consumption could be reduced by 1.4 percent (applying 36 percent ITE).
- Although a typical SI engine may have 22 percent brake thermal efficiency at representative FTP operating conditions (where brake work is shown as a percent of fuel energy input), brake thermal efficiencies significantly greater than 30 percent are typical at optimum operating conditions.

Approaches to increasing the brake work output are summarized below, following which specific technologies to implement these approaches are discussed in the remainder of this chapter.

Thermodynamic Factors

Thermodynamic factors affect indicated thermal efficiency. Thermodynamic factors include combustion timing and duration, compression and expansion ratios, working fluid properties, and heat transfer. Improvements in ITE can be achieved by modifying these thermodynamic factors as follows:

- *Reduced combustion duration with optimum timing.* Reducing the combustion duration while maintaining optimum timing releases more of the fuel energy closer to the optimum piston location (top dead center), thereby allowing a longer expansion to yield an increase in cycle work. Fast burn combustion systems that meet manufacturers' combustion pressure rise rates for acceptable noise, vibration and harshness (NVH) have been developed over the past several decades and are generally incorporated in current vehicles (NRC 2011). The final CAFE rule does not specifically propose technologies that would further reduce combustion duration, probably because of concerns about NVH.
- *Increased compression ratios.* Increases in the mechanical compression ratio can provide an increase in cycle efficiency. Variable valve timing can also be used to modify the effective expansion ratio and compression ratio. Late exhaust valve opening increases the effective expansion ratio to increase cycle work. Early or late intake valve closing decreases the effective compression ratio, thereby reducing the compression work while maintaining the same expansion ratio, resulting in a further increase in cycle efficiency for a given compression ratio.
- *High specific heat ratio of the working fluid.* For an idealized Otto cycle, the thermodynamic efficiency increases with increased specific heat ratio (Heywood 1988).[6] Air is preferred over exhaust gas as a diluent due to the higher specific heat ratio of air, but exhaust emission requirements using three-way catalysts (TWC) currently preclude the use of air as a diluent. Exhaust gas recirculation (EGR) is an option instead of air. However, some manufacturers are considering lean burn combustion systems using air as a diluent, if fuel changes are sufficient to reduce sulfur content in gasoline to facilitate the application of suitable emission control systems.
- *Reduced heat transfer from the working fluid.* Approximately one-third of the fuel energy is lost to the combustion chamber walls, which lowers the average combustion gas temperature and pressure, in turn reducing the work transferred to the piston. This heat transfer is generally required to protect engine materials, limit oil degradation, and preclude the onset of combustion knock.[7] Although reduced cooling might be considered in engine locations beyond the combustion chamber, such as around the exhaust ports, thermodynamic efficiency will not be improved with such reductions. Split cooling systems are used by some manufacturers to independently optimize the cooling of the cylinder head and the block to achieve friction reductions and faster warm-up during cold starting.
- *More efficient operating conditions.* As noted earlier, the 22 percent brake thermal efficiency shown in Figure 2.1 at representative FTP operating conditions is significantly lower than the >30 percent brake thermal efficiency typically achieved at an optimum operating condition. The significant technologies that directly address this potential improvement include

---

[6] The efficiency of the ideal Otto cycle is defined as follows:
$\eta = 1 - 1/CR^{\gamma-1}$ where:
η = efficiency
CR = compression ratio
$\gamma = cp/cv$ = ratio of specific heats
(= specific heat at constant pressure/specific heat at constant volume)
This equation indicates that larger values of γ result in higher values of efficiencies.

[7] Knock is an abnormal combustion phenomenon characterized by noise resulting from the autoignition of a portion of the fuel-air mixture ahead of the advancing flame. As the flame propagates across the combustion chamber, the unburned mixture ahead of the flame, called the end gas, is compressed, causing its pressure, temperatures and density to increase. The end gas may autoignite, thereby spontaneously and rapidly releasing a large part of the chemical energy. This causes high-frequency pressure oscillations inside the cylinder that produce the sharp metallic noise called knock. The knock phenomena are governed by engine variables and the anti-knock quality of the fuel, defined by the fuel's octane number (Heywood 1988, pp. 375 and 470).

turbocharging and downsizing, cylinder deactivation, and hybridization. Also addressing this opportunity are transmissions with a higher number of ratios.

Pumping Work

Reductions in pumping work can be achieved with systems such as variable valve timing and variable valve lift, turbocharged and downsized engines, and cylinder deactivation. Transmissions with a higher number of gears also provide the opportunity to reduce pumping work of the engine.

Friction Work

Approaches for reducing engine friction include low-friction lubricants, reduction of engine friction through design modifications, turbocharged and downsized engines, cylinder deactivation, and hybridization. Transmissions with a higher number of gears also provide the opportunity to reduce engine speed to reduce friction work.

Accessory Work

Electrically-driven water and oil pumps controlled to meet demands, rather than belt-driven pumps that operate at a fixed ratio of engine speed, also will reduce fuel consumption.

FUEL CONSUMPTION REDUCTION TECHNOLOGIES – IDENTIFIED IN FINAL CAFE RULE ANALYSIS

Specific technologies to implement the approaches previously identified for reducing fuel consumption are discussed in this section in the order presented in the final CAFE rule. Table 2.2 lists some of the fuel consumption reduction technologies directly applicable to SI gasoline engines from the Final Regulatory Impact Analysis (FRIA) (EPA 2012; NHTSA 2012). The table shows the specific categories of improvements in thermal efficiency or reduction in losses that are impacted by each technology. The fuel consumption reductions for each technology listed in the table are estimates by NHTSA (2012), and the distributions of the reductions in losses and improvements in ITE for each technology are from the EPA Lumped Parameter Model.

In the first part of this section, overviews of each technology are provided and the fuel consumption reduction principles are described. The committee's estimates of fuel consumption reductions and 2025 costs (2010 dollars) are presented and compared to NHTSA's estimates. The second part of this section discusses costs estimated by the committee that differed from those of NHTSA. Fuel consumption reduction effectiveness and costs are generally presented for a midsize car with an I4 engine for simplicity. However, a complete set of estimates for a midsize car with an I4 dual overhead cam (DOHC) engine, a large car with a V6 DOHC engine, and a large light truck with a V8 overhead valve (OHV) engine are provided in Table 2A.1 for effectiveness and Tables 2A.2a, b, and c for 2017, 2020, and 2025 direct manufacturing costs, respectively (Annex tables at end of chapter).

Rubbing Friction Reduction

Engine friction losses comprise approximately 8 percent of the fuel energy, as shown in Figure 2.1 and Table 2.2. As discussed earlier, if friction could be reduced by 25 percent, a 5.6 percent reduction in fuel consumption could be achieved. This section will describe technologies that can be applied to reduce engine friction.

Low Friction Lubricants - Level 1 (LUB1)

Lower viscosity engine lubricants are capable of reducing engine rubbing friction. The final CAFE/GHG TSD proposes that shifting to lower viscosity lubricants—in particular, changing from a 5W-30 motor oil to 5W-20 or 0W-20—would reduce friction through reductions in high and/or low and high temperature viscosities (EPA/NHTSA 2012b). The TSD recognizes that testing would be needed in order to ensure that durability is maintained. Since some manufacturers currently specify 5W-20 motor oil, the fuel consumption benefit is already incorporated in some current vehicles. However, 5W-30 may need to be retained for turbocharged engines. Low friction lubricants were projected in the TSD to provide a 0.5 to 0.8 percent reduction in fuel consumption at a cost of $4.02, which is consistent with the estimates provided in the Phase 1 study (NRC 2011).

Reducing the viscosity of motor oils to improve fuel economy can be accomplished with (1) better base stocks and/or (2) more friction modifiers in the additive package. The quality and service classifications of motor oil, as well as an indication of their fuel economy improvement potential, are provided or certified by the following organizations (Carley 2007):

- SAE provides a numerical code system for grading motor oils according to their viscosity characteristics. Taking 10W-30 motor oil as an example, the first number (10W) refers to the viscosity grade at low temperatures (W for winter) and the second number (30) refers to the viscosity grade at high temperatures. The relationship of SAE numerical codes and kinematic viscosity, which directly affect fuel economy, is shown for several examples in Table 2.3.
- The International Lubricant Standardization and Approval Committee (ILSAC) consolidates and coordinates standards for motor oil testing. ILSAC developed minimum performance standards for gasoline-powered passenger car and light truck oils, which became known as gasoline-fueled (GF) motor oil standards.

TABLE 2.2 Analysis of Improvements in Thermal Efficiency or Reductions in Losses for SI Engine Technologies Based on Fuel Consumption Reduction Estimates by EPA/NHTSA and Distributions of Reductions in Losses and Improvements in ITE from the EPA Lumped Parameter Model (percent)

Midsize Car Technologies[a,b]	Overall % Reduction in FC[c]	Indicated Efficiency %[d]	Indicated Work as a % of Baseline Fuel	Friction Loss as a % of Baseline Fuel	Pumping Loss as a % of Baseline Fuel	Accessory Loss as a % of Baseline Fuel	Brake Work as a % of Baseline Fuel[e]
Baseline Engine - Initial Values		36.00	36.00	8.00	5.00	1.00	22.0
Low Friction Lubricants - Level 1	0.70	36.00	35.75	7.75	5.00	1.00	22.0
Incremental Changes:				(0.25)			
Engine Friction Reduction - Level 1	2.60	36.00	34.82	6.81	5.00	1.00	22.0
Incremental Changes:				(0.94)			
Low Friction Lubricants and Engine Friction Reduction - Level 2	1.26	36.00	34.38	6.36	5.00	1.00	22.0
Incremental Changes:				(0.45)			
Variable Valve Timing - Dual Cam Phasing - DOHC	5.10	36.10	32.63	6.17	3.53	1.00	22.0
Incremental Changes:			(0.09)	(0.19)	(1.47)		
			32.71				
Continuously Variable Valve Life	4.6	36.10	31.21	5.97	2.22	1.00	22.0
Incremental Changes:				(0.20)	(1.31)		
Cylinder Deactivation	0.7	36.10	30.99	5.94	2.03	1.00	22.0
Incremental Changes:				(0.03)	(0.19)		
SGDI	1.50	36.65	30.53	5.94	2.03	1.00	22.0
Incremental Changes:			0.46				
			30.99				
18 bar BMEP Turbocharging and Downsizing	8.30	37.48	28.42	5.04	1.00	1.00	22.0
Incremental Changes:			0.64	(0.90)	(1.03)		
			29.06				
Cumulative (Multiplicative) Reduction in Fuel Consumption	-22.50			-3.0	-4.0		
Remaining				5.0	1.0		
Percent Reductions in Losses/Improvements in Indicated Efficiency		4.1		-37.0	-80.0		

[a] Reductions in fuel consumption (FC) for specific technologies are from NHTSA RIA, Table V-126 (2012).
[b] Distributions of reductions in losses for each technology are from EPA Lumped Parameter Model.
[c] Fuel consumption reductions for technologies listed are multiplicatively combined to provide overall reductions using the factor (100-%FC)/100.
[d] Indicated Thermal Efficiency = Indicated Work/Fuel Consumed (where fuel consumed is reduced by % reduction in fuel consumption for each technology).
[e] Brake Work = Indicated Work - Friction Loss - Pumping Loss - Accessory Loss.

These standards cover all aspects of oil performance in engines together with emission system durability and fuel economy. In 1997, ILSAC introduced an "Energy Conserving-EC" rating for motor oils that demonstrated improved fuel economy. Since that time, a number of GF oil ratings have been introduced, each one providing a target level improvement in fuel economy. However, ILSAC test procedures do not correspond to the EPA fuel economy test procedure. The latest rating, GF-5, introduced in late 2010, was expected to improve fuel economy by 0.5 percent over the previous GF-4 rated motor oil (Lubrizol 2010), which is in the range expected with the first level of low friction lubricants.

- The American Petroleum Institute (API) provides motor oil specifications. The latest API specification for gasoline engines is "SN" which matches the ILSAC GF-5 rating.

TABLE 2.3 Viscosity Grades of Engine Motor Oils

Automotive Lubricant Viscosity Grades (Engine Oils – SAE J 300, Dec. 1999)

SAE	Low Temperature Viscosities		High-Temperature Viscosities		
			Kinematic at 100°C (mm^2/s)		High Shear Rate at 150°C, 10/s (mPa.s)
ViscosityGrade	Cranking max at temp °C (mPa.s)	Pumping max at temp °C (mPa.s)	min	max	min
0W	6200 at –35	60 000 at –40	3.8	—	—
5W	6600 at –30	60 000 at –35	3.8	—	—
10W	7000 at –25	60 000 at –30	4.1	—	—
20	—	—	5.6	<9.3	2.6
30	—	—	9.3	<12.5	2.9

SOURCE: www.tribology-abc.com.

Lubricants are also important enablers of some technologies. A GF-6 oil rating is under development, with a target release in late 2016, to introduce a new, lower viscosity grade oil (Miller et al. 2012). The GF-6 rating will address several needs specific to turbocharged, downsized engines. The rating will ensure increased fuel economy throughout the oil drain interval. Perhaps more important, it will protect against engine-oil-caused, low-speed pre-ignition (LSPI), which has become a concern for turbocharged, downsized engines, as discussed later in this chapter. The GF-6 rating will also provide adequate wear protection for stop-start engines, which experience frequent starts and stops after extended periods of downtime.

As described in Appendix F, by changing from 5W-30 to 5W-20 oil, the committee estimated that low-friction lubricants – level 1 could provide approximately a 0.5 percent reduction in fuel consumption, which is within the range estimated by EPA/NHTSA in the final CAFE rule. EPA/NHTSA estimated that the incremental direct manufacturing cost of $3 for changing lubricants is due to the incremental cost of the oil. The overall cost, however, may be offset because fewer oil changes will be required. The amortized durability testing costs by the vehicle manufacturers would be reflected in the indirect cost.

The wide range of engine motor oils specified for 2013 MY vehicles certified by EPA are listed in Table 2.4. Not all of the vehicles certified by EPA specified 5W-20 or lower viscosity motor oil, suggesting that some vehicles may have the opportunity of using the lower friction lubricants after completing adequate testing. However, other vehicles may be limited in changing to lower viscosity oils due to operating loads and temperature concerns.

TABLE 2.4 Engine Motor Oils Specified for 2013 MY Light-Duty Vehicles (LDVs)

10W Low-Temperature Viscosity Oils	5W Low-Temperature Viscosity Oils	0W Low-Temperature Viscosity Oils
10W-60 (Aston Martin)	5W-40	0W-40
10W-40	5W-30 GF4	0W-30
	5W-20 GF4	0W-20 (Toyota)
	5W-20 GF5	0W-20 GF4 (Mazda, Kia)
		0W-20 GF5 (Honda)

SOURCE: EPA Fuel Economy Data MY 2013.

Low-Friction Lubricants - Level 2 (LUB2)

Several years ago, a 0W-20 synthetic motor oil with lower viscosity during cold-start and warm-up operation was introduced in some high-end cars. Recently, Japanese automakers approved the use of 0W-20 motor oils in some of their mainstream vehicles. The 0W-20 motor oil improves fuel economy during cold-start and warm-up operation and has been reported to improve fuel economy by 0.5 to 1.0 percent on the EPA test procedure.

In 2013, SAE released a new standard for SAE viscosity grade 16 that is likely to appear as 5W-16 and 0W-16 oils. SAE is currently working on a specification for 0W-12 motor oils. These lower viscosity oils at operating temperatures are intended to improve the fuel economy of engines specifically designed for these oils. Use of these oils in other engines could result in premature wear. One automaker is reported to be specifying 0W-16 oil in several vehicles, but these vehicles have not yet been certified by EPA in the United States, and the extent of the engine design modifications to ensure adequate durability is unknown (Swedberg 2013).

The combined effects of low-temperature viscosity reduction and 100°C viscosity reduction are estimated in Appendix F to provide an overall 1.0 percent reduction in fuel consumption for low friction lubricants - level 2, which is similar to the level of effectiveness estimated by EPA/NHTSA in the TSD.

Synthetic 0W-20 motor oil costs $7.17 to $8.79 per quart compared to $3.99 to $6.29 per quart for nonsynthetic 5W-20 motor oil. Therefore, oil changes for a car requiring 0W-20 motor oil would cost $4.40 to $24.00 more than a car using conventional oil. Since oil change intervals may be nearly twice as long compared to cars using conventional motor oils, an owner would expect to have the same or even lower annual maintenance costs. However, the higher cost of the initial oil fill for the 0W-20 motor oil is assumed to be included under Low Friction Lubricants (LUB2). Engine design changes are expected to be required to provide compatibility with these low-viscosity oils. These changes may include changes in oil pressure, bearing materials, and clearances, and other changes in specifications for wear surfaces in the engine.

Engine Friction Reduction - Level 1 and Level 2 (EFR1 and LUB2_EFR2)

Engine design changes are capable of reducing engine rubbing friction. The design of engine components, including low-tension piston rings, piston skirt design, roller cam followers, crankshaft design and bearings, material coatings, material substitution, optimal thermal management, and piston and cylinder surface treatments are projected in the final TSD to provide reductions in fuel consumption (EPA/NHTSA 2012a). For engine friction reduction through design of engine components (EFR1), NHTSA projected a 2.0 to 2.7 percent reduction in fuel consumption.

In addition to the first level of engine friction reduction, the final CAFE TSD added a second level of incremental reductions in engine friction, which may be required when a second level of low-friction lubricants is applied. For this second level of reductions in engine friction and low-friction lubricants, referred to as LUB2_EFR2, NHTSA projected an incremental 1.04 to 1.37 percent reduction in fuel consumption.

Examples of the main engine components on which vehicle manufacturers and suppliers are working to reduce friction are smaller, low-friction bearings; pistons with smaller skirts with coatings and low tension piston rings; diamond-like coatings on valve lifters; low-friction crankshaft seals; and the elimination of balance shafts (Truett 2013). Further discussion of these opportunities is provided in Appendix F.

By applying the design changes described in Appendix F, consisting of 50 percent reduction in bearing losses, 50 percent reduction in piston ring pressure, 10 percent reduction in valvetrain losses, and 50 percent reduction in seal losses, to the baseline overall engine friction, a 10 percent reduction in overall friction would be expected (Ricardo Inc. 2012). A 10 percent reduction in friction could reduce fuel consumption by 2.2 percent, based on the relationship developed earlier in this chapter and in Appendix G. An engine with balance shafts having roller bearings instead of journal bearings could realize an additional 0.4 percent reduction in fuel consumption.

The application of these technologies to reduce engine mechanical friction is illustrated in Figure 2.2 for a Nissan 1.2L three-cylinder gasoline engine (Kobayashi et al. 2012). This engine has the first known application of diamond-like carbon (DLC) coated piston rings for reduced friction. A variable displacement oil pump is used to supply the additional oil pressure for this high-output engine without increasing the oil pump work at moderate loads. Mirror-finished bearing surfaces and bore circularity are applied to further reduce engine friction.

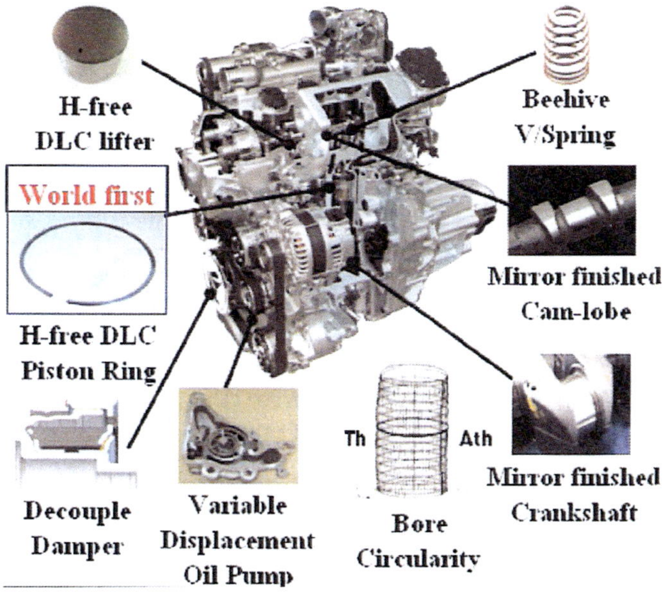

FIGURE 2.2 Low-friction technologies in a Nissan 1.2L three-cylinder gasoline engine.
SOURCE: Kobayashi (2012). Reprinted with permission from SAE paper 2012-01-0415. Copyright © 2012 SAE International.

To provide the same performance with a smaller displacement engine, higher brake mean effective pressure (BMEP)[8] is required and can be achieved through turbocharging and downsizing. A high BMEP engine is likely to have higher friction than a naturally aspirated engine with the same displacement, even after applying the modifications described above to reduce engine friction (Truett 2013). This is because the higher cylinder pressures and temperatures exert greater loads on the rubbing surfaces. The directional impact of friction reduction on fuel consumption after engine downsizing is discussed in Appendix G. The friction reduction required in a 50 percent downsized engine would be approximately double that required in the baseline, naturally aspirated engine to achieve the same reduction in fuel consumption. However, the 50 percent downsizing of the engine would provide up to 50 percent reduction in friction, resulting in approximately an 11 percent reduction in fuel consumption at a typical FTP operating condition, which is a significant portion of the fuel consumption benefit of the turbocharged, downsized engine, as discussed in Appendix G.

Engine modifications required to accommodate the low-viscosity synthetic oils beginning with SAE0W-16 are assumed to be included in the EFR2 technology. Specific modifications that would be required have not been described by NHTSA or the original equipment manufacturers (OEMs).

Thermal Management

Thermal management offers an opportunity for additional engine friction reduction. Several thermal management methods being investigated are described in this section.

- *Dual Cooling Circuit.* A dual circuit cooling system with separate cooling circuits for the cylinder head and cylinder block, together with reduced coolant volumes, allows the block to warm up faster for reduced friction during cold-start and warm-up operation. Tests on thermal management systems using split cooling with an electric water pump revealed nearly a 3 percent reduction in fuel consumption (Lodi 2008). However, these experimental tests found that there was little change in oil sump temperatures, so only a portion of the reduction in fuel consumption could be attributed to friction, while the remainder would be attributed to a reduction in heat losses from the combustion process. Schaeffler has developed an advanced thermal management system to better control drivetrain temperatures and is claimed to improve fuel economy by as much as 4 percent through shortened warm-up times (Green Car Congress 2012).
- *Waste Heat Utilization.* Several studies of waste heat utilization to reduce engine friction are under way or have been recently completed, with mixed results. A joint team from Chrysler and the Center for Automotive Research at The Ohio State University recently investigated an approach to capture the waste heat energy and distribute it to the transmission and engine oils (Sniderman 2012a, 2012b). Since higher temperature oil is less viscous, less torque is required to overcome friction, allowing the transmission and engine to operate at higher mechanical efficiencies. Fuel economy improvements of almost 4 percent were projected. The largest efficiency gains were obtained while heating the oil during a cold start, and approximately half of the improvement came from the engine and half from the transmission.

Dana Holding Corporation is marketing an Active Warm-Up (AWU) heat exchanger, which uses otherwise wasted thermal energy, such as heat lost through cooling systems or engine exhaust, to warm the engine and transmission oils (Dana n.d.).

Delphi, in a DOE research program, is investigating exhaust heat recovery as a technology for friction reduction (Confer et al. 2013). Delphi's exhaust heat recovery system (EHRS) employs a heat exchanger in the exhaust downstream of the catalytic converter to provide captured waste exhaust heat to the engine lubricating oil. Delphi concluded that only a marginal benefit could be attributed to exhaust heat recovery.

The effectiveness and direct manufacturing cost estimates for engine friction reduction technologies in naturally aspirated engines are shown in Table 2.5. The committee concurred with NHTSA's estimates of the overall fuel consumption reductions and direct manufacturing costs (DMC) for low-friction lubricants and engine friction reductions. An extensive number of modified engine components, including bearings, pistons and rings, cylinders, valve train components, timing chains, seals, and the oil pump and cooling system, are required to achieve the estimated fuel consumption reductions, and these actions can only be applied during a major engine redesign.

In addition to the technologies listed in Table 2.5, the potential fuel consumption reductions for engine friction reduction resulting from engine thermal management ranged from marginal to 4 percent. NHTSA included an unspecified friction reduction resulting from thermal management in the estimated reductions shown in Table 2.5. The committee assumed that thermal management was limited to a dual cooling circuit, while waste heat utilization technologies were considered under waste heat recovery technologies, as discussed later in this chapter. The estimated reductions in fuel consumption shown in Table 2.5 are valid for naturally aspirated engines only, as discussed previously and in Appendix G.

[8] BMEP is the theoretical constant pressure exerted during each power stroke of the engine to produce power equal to brake power. Current naturally aspirated production engines typically average 10-12 bar BMEP, while turbocharged engines average 18- 20 bar BMEP (Lawal and Garba 2013; NHTSA/EPA 2012b).

TABLE 2.5 Estimated Fuel Consumption Reductions (percent) and 2025 MY Direct Manufacturing Costs (2010 dollars) for Friction Reduction Technologies in a Midsize Car with a Naturally Aspirated I4 Engine

Friction Reduction Technology	NRC Estimated Most Likely Fuel Consumption Reduction (%)[a]	NHTSA Estimated Fuel Consumption Reduction (%)[a]	NRC Estimated Most Likely 2025 MY DMC Costs (2010$)[a]	NHTSA Estimated 2025 MY DMC Costs (2010$)[a]
LUB1	0.5 - 0.8	0.5 - 0.8	3	3
LUB2_EFR2 (Incremental)	1.0 - 1.4	1.04 - 1.37	48	48
EFR1				
Friction	2.0 - 2.2	2.0 - 2.7	51	51
Thermal Mgmt.	0.0 - 0.5	Incl. thermal mgmt.		
Total	3.5 - 4.9	3.5 - 4.9	102	102

[a] Relative to baseline except as noted.
SOURCE: EPA/NHTSA (2012b); additional references cited in section on rubbing friction.

Variable Valve Timing

Variable valve timing (VVT) was discussed extensively in the Phase 1 study (NRC 2011), so highlights from the Phase 1 study are summarized in this section. Valve timing influences volumetric efficiency, and therefore torque and power, over the engine speed range. At moderate speeds and light loads, valve timing influences pumping losses, effective compression and expansion ratios, and residual exhaust gas retention. Valve overlap can be minimized at idle for good combustion stability. A summary of these effects is presented in Table 2.6.

Dual Overhead Cam Engines

Many current VVT systems employ a cam phaser that rotates the position of the camshaft relative to the timing chain sprocket driven by the crankshaft. Oil-pressure-activated systems (OPA) use engine oil pressure to rotate the camshaft relative to the timing chain. BorgWarner has a cam-torque-actuated (CTA) system, which differs from the OPA system. The CTA system does not require engine oil pressure for actuation but uses instead the reaction force from the valve springs. The operation of both systems is described in Appendix H.

Manufacturers use many different names to describe their implementation of the various types of VVT systems. Some of the dominant names include, besides VVT, variable cam timing (VCT), VANOS (BMW), variable cam phasing (VCP), intake cam phasing (ICP), dual cam phasing (DCP), twin independent variable camshaft timing (Ti-VCT) and variable valve timing and lift electronic control (VTEC). EPA reports that 97.5 percent of 2014 vehicles have some form of VVT (EPA 2014a).

Single Overhead Cam Engines

Single overhead cam engines (SOHCs) have the intake and exhaust cams on the same camshaft. Applying a camshaft phaser to the single overhead cam provide variable valve timing, but on SOHC engines, this feature is often referred to as coupled cam phasing (CCP) or VCT. Since the intake and exhaust cam lobes are on the same camshaft, a VVT mechanism advances or retards the entire camshaft (intake and exhaust) equally. The lobe centerlines change in relation to top dead center, but the lobe-separation angle (the distance

TABLE 2.6 Predominant Effects with VVT

Operating Condition	Intake Valve Timing	Exhaust Valve Timing	Valve Overlap
Wide-open throttle - low speed • Maximize torque	Early closing	Late opening	Decreased
Wide-open throttle - high speed • Maximize power	Late closing	Early opening	Increased
Light load • Reduced pumping losses • Maximize expansion work	Late closing (compression ratio lower than expansion ratio), or early closing (intake valve throttling)	Late opening	
Light load • Internal EGR gas retention for lean gas/fuel ratio		Late closing	Increased
Idle stability			Minimized

SOURCE: NRC (2011).

between the intake and exhaust lobe centerlines) stays the same. Generally, the camshaft would be advanced to improve low-speed torque and for better idle characteristics. Retarding the camshaft would improve high-speed power. A typical production cam optimized for a SOHC advance/retard VVT system is generally designed with less overlap.

Effectiveness and Cost

Fuel consumption reductions for a VVT system were estimated by analyzing the fundamental effects of VVT, which include (1) the thermodynamic advantage of a lower effective compression ratio relative to the expansion ratio and (2) the reduced pumping losses and heat losses resulting from the increased internal EGR. By estimating these effects on the Otto cycle efficiency, fuel consumption reductions comparable to EPA and NHTSA's estimates were obtained, as shown in Table 2.7. NHTSA's estimated fuel consumption reductions for CCP are also shown in Table 2.7. NHTSA has estimated that CCP for SOHC engines can provide reductions in fuel consumption nearly equal to DCP on DOHC engines, which appears to be overly optimistic.

The direct manufacturing costs for intake and exhaust VVT systems—ICP and DCP—applied to DOHC engines are shown in Table 2.7 for an I4 engine and discussed in detail in a later section of this chapter. The committee's estimates of incremental direct manufacturing costs are approximately 15 percent higher than NHTSA's estimates due to the inclusion of all system components, including the cam phaser, an up-sized oil pump, an oil control valve, drivers for engine control unit (ECU), oil drillings, position feedback sensor and trigger wheel, wiring, and connectors.

Variable Valve Lift (DVVL and CVVL)

A variety of both discrete variable valve lift (DVVL) and continuously variable valve lift (CVVL) mechanisms have recently been incorporated in production vehicles. VVL systems reduce pumping losses by transferring a significant portion of airflow control from the throttle to the engine valves. The resulting higher manifold pressures (reduced manifold vacuum levels) reduce the negative work done on the piston to reduce pumping losses. Appendix I reviews several systems that have been introduced with the objective of reducing fuel consumption. DVVL for SOHC engines is generally being implemented with one of the mechanisms described in Appendix I for DOHC engines, since both types of engines apply VVA only to the intake valves.

The committee's estimates of fuel consumption reductions for DVVL and CVVL agree with NHTSA's, as shown in Table 2.8. NHTSA's estimated fuel consumption reductions for DVVL applied to SOHC engines are the same as for DVVL applied to DOHC engines since, as noted above, VVL is only applied to the intake valves. Although NHTSA identifies DVVL for OHV engines, the system mechanism was not described in the support documents for the final rulemaking. NHTSA's fuel consumption reduction estimate for DVVL applied to OHV engines is shown in Table 2.8. NHTSA has applied coupled cam phasing (CCP) together with DVVL to OHV engines, which have only one camshaft, and labeled the combination variable valve actuation (VVA). NHTSA has estimated the fuel consumption savings for this combination in OHV engines to be slightly less than DVVL alone applied to SOHC and DOHC engines. NHTSA assumes that cylinder deactivation will be applied to OHV engines prior to applying VVT.

Estimates of direct manufacturing costs for DVVL and CVVL systems are shown in Table 2.8 for an I4 engine and discussed later in this chapter. Direct manufacturing costs are estimated to be approximately 15 percent higher than NHTSA's estimates due to inclusion of the total system, including an additional intermediate shaft with additional cam lobes and roller elements for the CVVL systems, cylinder head modifications, hydraulic or electric actuation, drivers for the engine control unit (ECU), wiring, and connectors.

Multiair Electrohydraulic Valve-timing System

Multiair is an electrohydraulic valve-timing system developed by Fiat that provides both VVT and VVL. It provides dynamic and direct control of air and combustion, cylinder-by-cylinder and stroke-by-stroke. With Multiair, direct

TABLE 2.7 Estimated Fuel Consumption Reductions (percent) and 2025 MY Direct Manufacturing Costs (2010 dollars) for VVT Technologies in a Midsize Car with an I4 Engine

Variable Valve Timing Technology	NRC Estimated Most Likely Fuel Consumption Reduction (%)[a]	NHTSA Estimated Fuel Consumption Reduction (%)[a]	NRC Estimated Most Likely 2025 MY DMC Costs (2010$)[a]	NHTSA Estimated 2025 MY DMC Costs (2010$)[a]
DOHC				
ICP	2.6	2.6	31 – 36	31
DCP (Relative to ICP)	2.5	2.5	27 – 31	27
DCP (Relative to base)	5.0	5.0	58 – 67	58
SOHC	3.5	5.0	31 – 36	31

[a] Relative to baseline except as noted.

TABLE 2.8 Estimated Fuel Consumption Reductions (percent) and 2025 MY Direct Manufacturing Costs (2010 dollars) for VVL Technologies in a Midsize Car with an I4 Engine (except as noted)

Variable Valve Lift Technology	NRC Estimated Most Likely Fuel Consumption Reduction (%)[a]	NHTSA Estimated Fuel Consumption Reduction (%)[a]	NRC Estimated Most Likely 2025 MY DMC Costs (2010$)[a]	NHTSA Estimated 2025 MY DMC Costs (2010$)[a]
DOHC				
DVVL	3.6	3.6	99 - 114	99
CVVL (Incremental)	1.0	1.0	49 - 56	49
SOHC - DVVL	3.6	3.6	99 - 114	99
OHV - V8				
VVA (CCP + DVVL)	3.2	3.2	235 - 271	235[b]

[a] Relative to baseline except as noted.
[b] $31 for CCP + $204 for DVVL.

control of the air is provided by the intake engine valves without using the throttle (Green Car Congress 2009c). The operation of the Multiair system is described in Appendix I. Through solenoid valve opening and closing time control, a wide range of optimum intake valve opening and closing schedules can be obtained to improve maximum power, low-speed torque, and partial valve opening to control trapped air mass in the cylinders. Although the Multiair system could theoretically provide fuel consumption reductions similar to a mechanical VVT and VVL system, its lower mechanical efficiency (since mechanical energy is not recovered as in a conventional cam follower and spring system), is expected to provide lower benefits than the mechanical systems. Multiair systems are in production on the 2014 MY Fiat 500 and the Dodge Dart in the United States.

Cylinder Deactivation

Cylinder deactivation, which shuts off multiple cylinders and results in higher loads on the remaining operating cylinders, can be utilized during part load operation to reduce pumping losses and friction losses. Pumping losses are reduced due to the higher loads of the operating cylinders, which require less throttling. Friction losses are reduced due to the lower piston loads of the deactivated cylinders, which have near-zero mean cylinder pressures. Cylinder deactivation has been applied to six- and eight-cylinder engines. Recently, Volkswagen introduced cylinder deactivation, known as active cylinder management, on a 1.4L four-cylinder engine in Europe.

In order to deactivate a cylinder, the intake and exhaust valves are held closed. This creates an "air spring" in the combustion chamber, in which the preceding cycle's exhaust gases are trapped and compressed in the upstroke and expanded in the downstroke. This compression and expansion result in reduced engine friction losses for the deactivated cylinders. In cylinder deactivation systems, the engine management system stops fuel from being delivered to the deactivated cylinders. Ignition and cam timing, as well as throttle position, are adjusted to ensure that switching from full cylinder operation to cylinder deactivation is nearly imperceptible. Until recently, cylinder deactivation primarily has been employed in engines with high displacement, which have low efficiency at light loads.

There are two primary categories of cylinder deactivation. The first, used in pushrod engines, employs solenoids to spill the oil supplied to the hydraulic tappet. As a result, the lifters are collapsed and cannot activate their respective pushrods, thereby deactivating the valves.

The second category of cylinder deactivation is employed in overhead cam engines. In this type of cylinder deactivation, two interlocked rocker arms on the same fulcrum are used for each valve that can be deactivated. The first rocker arm follows the cam, and the second is used as a valve actuator. On cylinder deactivation, the oil pressure (controlled by a solenoid) causes a pin to be released between the rocker arms. The arm that has been unlocked by the release of the pin cannot activate the valve. A variation of this system achieves cylinder deactivation by adding a second lobe with zero lift to a sleeve on the camshaft, which is hydraulically shifted to position the normal lift lobe or the zero-lift cam lobe at the location of the cam follower.

After an early commercial failure with cylinder deactivation in the 1980s, Mercedes-Benz revived the idea of cylinder deactivation. In 1999, an Active Cylinder Control (ACC) system was included in full-size Mercedes-Benz models that were sold in Europe. For the V8 and V12 engines, the ACC system deactivated half of the engine's cylinders (J.D. Power 2012). Cylinder deactivation now is being extensively applied to V8 and V6 engines with a variety of different names. Some examples include General Motors' Active Fuel Management (AFM), used on many V8 and V6 engines; Chrysler's Multi-Displacement System (MDS) on its V8 engines; and Honda's Variable Cylinder Management (VCM) on its V6 engines. In addition, some of the VVL systems, discussed in the previous section, include the capability of cylinder deactivation.

The first OEM to implement a cylinder deactivation

system in order to reduce fuel usage in small four cylinder engines was Volkswagen. The system, which is called active cylinder management, has been implemented on a 1.4L turbocharged, gasoline-fueled engine in the Polo Blue GT in Europe. In this engine, two of the four cylinders are deactivated and fuel to these cylinders is shut off. By shutting down the second and third cylinders under low and medium loads, Volkswagen has reported an 8.5 percent reduction in fuel consumption on the EU driving cycle.

Tula Technology Inc. is developing a different approach for cylinder deactivation (Tula n.d.). Its system controls each cylinder individually and fires only enough of them at any moment to deliver the torque required. Tula claims its system can boost fuel efficiency of a V8 engine 18 percent, which is claimed to be more than twice the gain possible with a conventional deactivation system. No engineering details of the system, or engineering test data, are available to confirm the company's claims. The company has said that it is working with several automakers to commercialize the technology.

The fuel consumption reductions and direct manufacturing costs for cylinder deactivation estimated by NHTSA are shown in Table 2.9 and compared to the committee's estimates. NHTSA estimated, and the committee agrees, that cylinder deactivation for OHV engines can provide up to a 5.5 percent reduction in fuel consumption, assuming that it is applied before VVT and VVL. However, for SOHC and DOHC engines, NHTSA assumed that cylinder deactivation would be applied after DCP and VVL, resulting in a less than 1 percent reduction in fuel consumption. In contrast to NHTSA's estimates of up to 5.5 percent reduction in fuel consumption, the Department of Energy has estimated cylinder deactivation can increase efficiency by 7.5 percent over VVT (DOE 2013).

The committee agrees with NHTSA's estimated direct manufacturing costs.

Stoichiometric Gasoline Direct Injection

Stoichiometric, gasoline direct injection (SGDI) engines inject fuel directly into the combustion chamber instead of the intake port, as in many current engines with port fuel injection. Direct injection requires a new injector design; an engine-driven, high-pressure fuel pump; new fuel rails; and changes to the cylinder head and piston (Confer et al. 2013). Injecting the fuel directly into the cylinder cools the air/fuel charge within the cylinder due to fuel evaporation, which produces two beneficial results. First, since the cooler charge is less prone to detonation, compression ratios can be increased to achieve higher thermodynamic efficiency without combustion knock. Second, since the cooled mixture is denser, the engine will produce more power. With higher power density, direct injection is an enabler for higher BMEP engines.

The committee estimated that SGDI can provide a fuel consumption reduction of 1.5 percent, which is in agreement with NHTSA's estimates. This reduction is achieved by the following means: The compression ratio can be increased due to the evaporative cooling of the air/fuel charge in the cylinder. As discussed later in the section High Compression Ratio with High Octane Gasoline, an increase of 1.0 compression ratio facilitated by direct injection would provide an estimated 1.5 percent reduction in fuel consumption A modest increase in power accompanies the application of SGDI. With this increased power, an engine with SGDI could be downsized to provide power equivalent to a port-fuel-injected (PFI) engine. This modest downsizing could provide a small additional reduction in fuel consumption.

The application of SGDI has increased significantly over the past few years, often in conjunction with turbocharging and downsizing. Most major light-duty vehicle manufacturers have SGDI in production in at least some MY 2014 vehicles. *Automotive News* recently published the percentage of light duty vehicles with gasoline direct injection, which is shown in Table 2.10.

As shown in Table 2.11, the committee agrees with NHTSA's estimate that SDGI is expected to provide up to a 1.5 percent reduction in fuel consumption. It also concurs with NHTSA's estimates of incremental direct manufacturing costs for SGDI.

Turbocharged, Downsized Engines and Cooled EGR

Turbocharging increases the engine airflow and specific power output, which allows engine size to be reduced while maintaining performance. As a result, friction and pump-

TABLE 2.9 Estimated Fuel Consumption Reductions (percent) and 2025 MY Direct Manufacturing Costs (2010 dollars) for Cylinder Deactivation Technologies in V6 and V8 Engines

Cylinder Deactivation Technology	NRC Estimated Most Likely Fuel Consumption Reduction (%)[a]	NHTSA Estimated Fuel Consumption Reduction (%)[a]	NRC Estimated Most Likely 2025 MY DMC Costs (2010$s)[a]	NHTSA Estimated 2025 MY DMC Costs (2010$s)[a]
DOHC[a]	0.7	0.7	118	118
SOHC[a]	0.7	0.7	118	118
OHV[b]	5.5	5.5	133	133

[a] V6 – Applied after DCP and VVL.
[b] V8 – Applied before VVT and VVL.

TABLE 2.10 Percent of LDVs with Gasoline Direct Injection

Year	Percent
2008	2.3
2009	4.2
2010	8.3
2011	15.4
2012	22.7
2013	30.8

SOURCE: Automotive News (2014).

ing losses are reduced at lighter loads relative to a larger, naturally aspirated engine. Downsizing facilitates operating closer to the minimum fuel consumption region of the engine map than is possible with a larger, naturally aspirated engine. Higher levels of brake mean effective pressure (BMEP) may require cooled, exhaust gas recirculation (CEGR) to reduce susceptibility to knocking at high loads and provide additional charge dilution at part loads for further reductions in fuel consumption.

The fuel consumption benefits of turbocharging and downsizing are illustrated in Figure 2.3. At low to moderate torque levels, the turbocharged, downsized engine provides significant reductions in brake specific fuel consumption (BSFC) relative to a naturally aspirated, port-fuel-injected, SI engine and allows the engine to operate closer to its minimum BSFC over a wider range of speeds and loads. These reductions result from reductions in friction, due to fewer or smaller cylinders and associated moving components, (although partially offset by the higher friction due to higher cylinder pressures and temperatures) and pumping losses, due to the reduction or elimination of throttling at light loads. An analysis of the friction reduction that results from downsizing an engine is provided in Appendix G.

The final TSD describes CEGR, also called the "boosted EGR combustion concept," as a charge diluent for reducing combustion temperatures. At full load, the additional charge dilution provided by cooled EGR reduces the need for fuel enrichment by reducing the susceptibility to knocking combustion. The reduced susceptibility to knock facilitates higher boost pressure and/or compression ratio, which may enable further reductions in engine displacement with accompanying reductions in pumping and friction losses. High BMEP engines are anticipated by EPA/NHTSA to use gasoline direct injection, DCP, and discrete or continuously VVL. For the higher BMEP levels, the final CAFE rule suggests a dual-loop EGR system consisting of both high and low pressure EGR loops and dual EGR coolers. The final CAFE rule indicates that the 27 bar BMEP engine would require cooled EGR while the 24 bar BMEP engine could optionally use EGR for additional fuel consumption reduction.

The final CAFE rule considers four different levels of turbocharged, downsized, high BMEP engines. The terminology applied to these engines by NHTSA is shown in Table 2.12 together with the BMEP levels, percent downsizing, cooled EGR usage, boost pressure required, and the boost system that may be applied. Boost systems that NHTSA anticipates being applied for reaching 18, 24, 27 bar BMEP are described in Table 2.12. Each incremental increase in BMEP requires increasingly complex boost systems, which begin with turbochargers with wastegates for 18 bar BMEP and move up to variable geometry turbine turbochargers for 24 bar BMEP with absolute boost pressure of 2 bar, and two stage turbochargers for 27 bar BMEP. Ricardo has indicated that advanced boosting systems with 3 bar absolute boost pressure are required for BMEP levels exceeding 27 bar that may be applied in the 2020-2025 time frame (EPA/NHTSA 2012b).

Most vehicle manufacturers have introduced turbocharged, downsized engines as replacements or as options for larger displacement, naturally aspirated engines with the objective of reducing fuel consumption instead of improving the performance of the vehicle, as had been the practice previously. As an example, one vehicle manufacturer has planned and implemented turbocharged, downsized engines for most applications in its vehicle product lines, including replacements for V8 engines, V6 engines and I4 engines with smaller, turbocharged engines. In 2009, Ford introduced a 3.5L V6 turbocharged engine, called an EcoBoost engine, which had power output comparable to a V8 engine. This engine was applied in several vehicle lines. Ford subsequently applied a 3.5L turbocharged V6 engine to the F150 pickup truck, where V8 engines had been dominant. Recent sales data indicate that the 3.5L V6 EcoBoost engine had

TABLE 2.11 Estimated Fuel Consumption Reductions (percent) and 2025 MY Direct Manufacturing Costs (2010 dollars) for Stoichiometric Gasoline Direct Injection Technology in a Midsize Car with an I4 Engine

Stoichiometric Gasoline Direct Injection Technology	NRC Estimated Most Likely Fuel Consumption Reduction (%)	NHTSA Estimated Fuel Consumption Reduction (%)	NRC Estimated Most Likely 2025 MY DMC Costs (2010$)	NHTSA Estimated 2025 MY DMC Costs (2010$)
I4	1.5	1.5	164	164
V6	1.5	1.5	246	246
V8	1.5	1.5	296	296

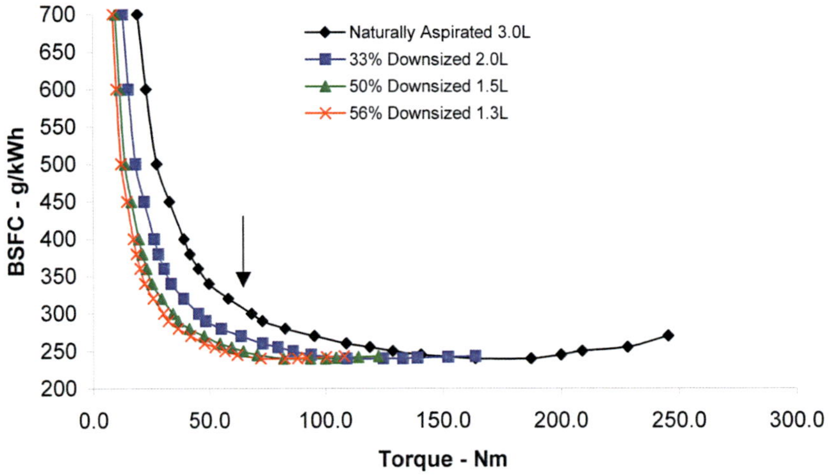

FIGURE 2.3 Effect of turbocharging and downsizing on BSFC versus torque.
Graph developed by progressively scaling a generic brake specific fuel island map from Dick et al. (2013).

TABLE 2.12 Boost Systems for Turbocharged, Downsized Engines

System	BMEP (bar)	Downsizing (%)	Cooled EGR	Absolute Boost Pressure (bar)	Boost System
Turbocharging and downsizing-Level 1	18	33	No	~1.7	Single turbocharger for I engines with wastegate Dual turbocharger for V engines with wastegate
Turbocharging and downsizing-Level 2	24	50	No	2.0	Variable geometry turbocharger
Cooled EGR-Level 1	24	50	Yes	2.0	Variable geometry turbocharger
Cooled EGR-Level 2	27	56	Yes	2.3	Two stage turbocharger

SOURCE: EPA/NHTSA (2012b).

been installed in nearly half of the F150 vehicles sold. Ford's next step was to develop four-cylinder turbocharged engines as replacements for V6 engines. A 2.0L turbocharged engine was recently applied to a number of vehicles as options or replacements for V6 engines. In addition, a 1.6L turbocharged engine was introduced for the 2013 MY as a replacement for V6 engines or larger I4 engines in several additional vehicle lines.

In the most extreme case of downsizing to date, Ford introduced a 1.0L three-cylinder turbocharged engine in the 2014 MY Fiesta. This engine has direct injection, turbocharging, and variable timing for the intake and exhaust camshafts and produces 123 hp, with a specific power output of 123 hp/L. The naturally aspirated, four-cylinder 1.6L engine in the Fiesta has the same power output. However, the 1.0L EcoBoost engine produces more torque (125 lb-ft) at lower rpm (1,400) and has an overboost feature (which allows increased boost for short periods of time) that increases torque to 148 lb-ft. The 45 mpg EPA highway rating for the Fiesta is the highest of any non-hybrid or non-diesel vehicle currently sold in the United States. Table 2.13 lists the three-cylinder turbocharged engines that are in production or under consideration for applications in the United States.

Ford recently announced another significant step in turbocharged, downsized engines. Following the announcement that the 2015 MY F150 pickup truck would have a body and cargo bed made of aluminum instead of steel for a weight savings of up to 700 lb, the company announced a new 2.7L V6 turbocharged engine for this vehicle (Truett 2014). This engine produces 315 hp resulting in a 15 percent increase in power to weight ratio over the 5.0L V8 engine in the 2014 MY F150 (Ford Media Center 2014). The 2.7L V6 engine, which would have 46 percent less displacement than the 5.0L V8 engine, will have a two-piece cylinder block. The upper section contains the cylinder bores and is made of compacted graphite iron (CGI) to enhance strength. To save weight, the lower section is die-cast aluminum (Truett 2014). The compacted graphite iron upper section also helps to reduce noise as combustion temperatures and pressures increase.

The implementation of turbocharged engines in production vehicles has been increasing since 2008. As shown in Table 2.14, the percentage of LDVs with turbocharged engines increased to 14.8 percent in the 2013 MY. This trend is

TABLE 2.13 Three-Cylinder Gasoline Engines in Production or Under Consideration for U.S. Applications

Manufacturer	Engine	Power (hp)	Reference
Current production			
Ford (2014 MY Fiesta, 2015 MY Focus)	1.0L 3 cylinder, TC	123 hp	
Mercedes-Benz Smart for Two/Mitsubishi Engine (Since 2008 MY)	1.0L 3 cylinder, NA		
Under consideration			
BMW (Mini, i8 Hybrid, 1 Series, 3 Series)	1.5L 3 cylinder, TC	120-222 hp	carscoops.com, Jan 4, 2014
VW	1.0L 3 cylinder, TC	110 hp	Automotive News, July 1, 2013
GM/Opel	1.0L 3 cylinder, TC	115 hp	greencarcongress.com, Jan 4, 2014
Mercedes/Renault (Smart for Two)	3 cylinder, TC	N/A	Autonews.com, May 24, 2013
Honda	1.0L 3 cylinder, TC	N/A	Honda.com, November 19, 2013
Kia (Currently in Europe)	1.0L 3 cylinder, NA (MFI)	69 PS	Kia-buzz.com, March 24, 2011

NOTE: MFI, multiport fuel injection; NA, naturally aspirated; N/A, not applicable; TC, turbocharged.

TABLE 2.14 Percent of Light-Duty Vehicles with Turbochargers

Year	Percent
2008	3.0
2009	3.3
2010	3.3
2011	6.8
2012	8.4
2013	14.8

SOURCE: Automotive News (2014).

expected to continue. Honda announced in November 2013 that it is developing a new family of engines that includes a 1.0L three-cylinder and two four-cylinder engines with 1.5 and 2.0L displacements (Autoweek 2013). In March 2014, GM announced that it is developing a new family of small 1.0L to 1.5L gasoline engines that will include turbocharging (Saporito 2014). In May 2014, Chrysler announced that it will launch a new line of small gasoline engines that are turbocharged (Zoia 2014). In July 2014, Toyota announced that it is embarking on a "massive engine overhaul" that will include the development of turbocharged engines with EGR (Greimel 2014b).

Effectiveness of Turbocharged, Downsized Engines

The committee used several methods to estimate the fuel consumption reduction effectiveness of turbocharged, downsized engines. First, the committee reviewed the basis of NHTSA's estimates. NHTSA's estimate of the effectiveness of a 27 bar BMEP engine was based on an analytical study described in the Ricardo (2011) report. The results from this analytical study were subsequently used to estimate the effectiveness of the 18 bar and 24 bar BMEP engines. The starting point for the analytical study of the 27 bar BMEP engine was test data from an experimental 3.2L V6 ethanol boosted direct injection (EBDI) engine. Ricardo tested this engine using E85 and indolene (98 RON) fuels (Cruff et al. 2010). When tested with indolene, the engine produced 5 bar lower BMEP, indicating that significant spark retard was required with indolene to avoid knock, in contrast to the higher octane E85 fuel.

Starting with the BSFC map for the 3.2L V6 EBDI engine, Ricardo added the following features: cam profile switching (CPS); 2 stage boosting, replacing the single stage boosting system; and a compression ratio increase of 0.5 (from 10:1 to 10.5:1). A 3.5 percent improvement in friction was also added, but was not included in the BSFC map for the 27 bar BMEP engine. The method used for developing the resulting BSFC map for this engine with these added features was not described in the Ricardo (2011) report. The committee concluded that there is ambiguity concerning the fuel for the 27 bar BMEP engine. Specifically, the 3.2L V6 EBDI engine was knock-limited when tested with indolene (98 RON), and features were added that further increased, rather than decreased, the knock susceptibility of the engine (see Fuel Octane Issues section for a definition of RON).

The EPA "ground rules" stated that the engine should operate on 87 AKI (91 RON) fuel (see Fuel Octane Issues section for a definition of AKI). Although the engine may operate on 87 AKI fuel, the knock control system likely would retard the spark timing from the best efficiency timing under more conditions than was the case with the original EBDI engine. Even though the tendency to knock occurs at high loads, controlling knock at these conditions is essential for engine integrity. Controlling knock with spark retard in a turbocharged engine can be problematic due to the likelihood of exceeding the temperature capability of the turbocharger. Effective control of knock generally requires a reduction in compression ratio, which would also have a detrimental effect on fuel consumption under the CAFE driving cycle conditions. Based on the foregoing considerations, the committee determined that reductions in compression ratio of turbocharged, downsized engines could be needed to provide satisfactory operation on 87 AKI fuel. The impact of reductions in compression ratio on effectiveness is discussed at the conclusion of this section.

The second method to estimate fuel consumption reduction effectiveness consisted of a review of EPA certification fuel economy test data for the 2014 and 2015 model years for similar vehicles equipped with a turbocharged, downsized engine or a naturally aspirated engine. To provide information at comparable performance levels, the EPA fuel economy data were adjusted to equal power-to-weight ratio for each set of comparable vehicles using the technique described in the TSD (EPA/NHTSA 2012b). The turbocharged, downsized engines, at equal power to the naturally aspirated engines, were found to have nearly comparable peak torque levels within less than +/- 8 percent so that further adjustments for torque differences were not applied to these comparisons.

The following empirical expression developed by NHTSA was used to adjust the fuel economy comparisons to equal power to weight ratio (NHTSA 2012):

$$CO_{2_i} \text{ or } GPM = \beta_{\frac{hp}{wt}} + \left(\frac{Horsepower}{Weight}\right)_i + \beta_{weight} Weight_i + C$$

where GPM (gal/mi) = CO_2 (g/mi)/8,887 g CO_2/gal gasoline,
hp/weight= the rated horsepower of the vehicle divided by the curb weight,
Weight = the curb weight of the vehicle in pounds,
C, $\beta_{hp/wt,}$, β_{weight} = constants, and
i = individual vehicle.

Values for the constants in the above equation are listed in Table 2.15, as described in the NHTSA RIA (2012).

TABLE 2.15 Values for Constants in the Empirical Equation of NHTSA

	Cars	Trucks
$\beta_{hp/wt}$ =	1.09×10^3	1.13×10^3
β_{weight} =	3.29×10^{-2}	3.45×10^{-2}
C =	–3.29	2.73

A further adjustment to equal performance, as measured by 0 to 60 mph acceleration time, would have required a full system simulation using complete torque curves for each engine in the vehicles listed in the table, but this was beyond the scope of the committee.

EPA certification test vehicles with different engines often have other powertrain and vehicle differences. The Lumped Parameter Model (LPM) was used to adjust the certification data to account for these features so that only the effectiveness of the turbocharged, downsized engine could be determined. These adjusted fuel consumption data were compared with the LPM predictions of the effectiveness of turbocharged, downsized engines after accounting for the other technologies on the certification vehicles. The LPM was chosen since EPA and NHTSA used it in the final rulemaking process and it is a reasonably accurate method for this purpose.

Annex Table 2A.5 (at end of this chapter) shows the adjusted fuel consumption data compared with the LPM predictions for turbocharged, downsized engines. Also shown in Table 2A.5 for reference are the EPA label fuel economy data, the CAFE unadjusted fuel economy data, and the fuel economy data adjusted for power to weight ratio.

The comparisons of adjusted fuel consumption data with LPM predictions generally indicate the actual fuel consumption data show less of a reduction than the LPM predictions. The normalized certification vehicle fuel consumption reductions ranged from 1 to 13 percentage points below the fuel consumption reductions estimated by the LPM for turbocharged, downsized engines. Assuming some of the vehicles with large deficits relative to the LPM estimates were early implementations, the committee estimated that the representative fuel consumption reduction potential for turbocharged, downsized engines may be in the range of 1 to 2 percentage points lower than the EPA and NHTSA estimates, as embodied in the LPM. The normalized certification vehicle fuel consumption reduction for two vehicles exceeded the LPM estimated fuel consumption reduction for turbocharged, downsized engines.

The third method to estimate fuel consumption reduction effectiveness consisted of contracting with University of Michigan (U of M) to conduct a full system simulation of a midsize car starting with a baseline I4 engine. The details of that simulation are discussed in Chapter 8. Several of the technologies evaluated in the full systems simulation were turbocharging and downsizing to 33 percent and 50 percent with cooled EGR. These technologies were applied to the engine after applying reduced friction, dual cam phasing (variable valve timing), discrete variable valve lift, and stoichiometric gasoline direct injection. Table 2.16 compares the results from this modeling with the estimates contained in NHTSA's Regulatory Impact Analysis (RIA) (2012) and modeling results from the EPA's LPM. The LPM is described in EPA's RIA (2012a) and the TSD (EPA/NHTSA 2012b). All of the estimates shown in the table are relative to the previous technologies already applied to the engine, as described in Chapter 8, and they are significantly less than the estimates relative to a baseline I4 engine, as shown in other tables in this chapter, due to negative synergies. The U of M full system simulation modeled the interactive effects of the engine technologies listed in Table 2.16. Likewise, negative synergies were included in the NHTSA RIA estimates for the engine technologies and in the LPM estimates.

The fuel consumption reduction result from the full system simulation for the turbocharged, 50 percent downsized, 24 bar BMEP engine with cooled EGR was within 2 percent-

TABLE 2.16 Comparisons of Full System Simulation Results with NHTSA Estimates for Turbocharged, Downsized Engines (percent fuel consumption reduction)

Technology	U of M Full System Simulation	NHTSA Estimates (based on RIA)	Estimates Based on EPA's LPM
18 bar BMEP (33% downsizing) (rel. to NA baseline)	9.6	8.3	6.4
24 bar BMEP (50% downsizing) with cooled EGR (incremental)	4.6	6.9	6.1
24 bar BMEP (50% downsizing) with cooled EGR (rel. to NA baseline)	13.8	14.6	12.1

Note: All estimates are relative to the previous technologies already applied to the engine (previous technologies include low friction lubricants, engine friction reduction, dual cam phasing, discrete variable valve lift, and direct injection), as described in Chapter 8. NA, naturally aspirated engine.

age points of EPA's and NHTSA's estimate, although the results for the two steps used to achieve the 24 bar BMEP engine showed some differences from EPA's and NHTSA's estimates. The results for the 18 bar BMEP engine showed a reduction in fuel consumption of 1 to 3 percentage points more than EPA's and NHTSA's estimates, while the incremental fuel consumption reduction result from the full system simulation for the 24 bar BMEP engine with cooled EGR (relative to 18 bar BMEP) was up to 1.5 percentage points lower than EPA's and NHTSA's estimates. As discussed in the cooled EGR section later in this chapter and shown in Figure 2.1, the pumping losses are already very low in the 18 bar BMEP engine, which included many fuel consumption reduction features before adding EGR, so the effectiveness of cooled EGR in further reducing pumping losses is significantly diminished.

The U of M full system simulations are within the range of the Agencies' effectiveness estimates. The simulation study selected the optimum compression ratio for the CAFE test cycles but did not address the control of high load knock and drivability concerns. However, addressing these concerns could reduce the effectiveness of the turbocharged, downsized engine in a production vehicle. In the U of M modeling study, the trade-off between borderline knock and compression ratio was optimized within the CAFE test cycles, but controlling knock at full load without exceeding the turbocharger temperature limits might require the application of spark timing retard and/or air/fuel ratio enrichment. Likewise, driveability was not part of the full system simulation but likely would require changes to the torque converter and/or final drive ratio to ensure driveability comparable to the naturally aspirated engine. Similarly, the modeling that served to calibrate the LPM may not have fully addressed these issues.

Taking into account all three methods considered for estimating the fuel consumption effectiveness of turbocharging and downsizing technologies, and factoring in the knock and driveability concerns, the committee recommends expanding the range of effectiveness for these technologies, as shown in Table 2.17. In contrast to Table 2.16, the fuel consumption reductions shown in this table are relative to a baseline naturally aspirated engine with fixed valve timing and lift, except as noted.

Reduced Compression Ratio for 87 AKI (91 RON) Gasoline

The foregoing review of NHTSA's analysis from the Ricardo (2011) report indicated that reductions in compression ratio of turbocharged, downsized engines are likely to be needed to provide satisfactory operation on 87 AKI fuel. In addition, other references in the TSD related to experimental, turbocharged, downsized engines (the Sabre engine from Lotus Engineering and the 30 bar BMEP engine from MAHLE Powertrain) were developed in Europe and used European "regular" 95-98 RON gasoline. If U.S. regular gasoline instead of European "regular" gasoline were used in the 24 bar BMEP turbocharged, downsized engine, then approximately a 1 ratio reduction in compression ratio may be required to avoid knocking at high load conditions, as described in Appendix J. This reduction in compression ratio would result in up to a 1.5 percent loss in fuel consumption reduction effectiveness.

TABLE 2.17 Recommended Expanded Most Likely Range of Effectiveness for Turbocharging and Downsizing Technologies

Technology	High Most Likely Range - NHTSA Estimate of Fuel Consumption Reductions (Relative to baseline NA Engine with Fixed Valve Timing and Lift) (%)	Low Most Likely Range - NRC Adjustment (Relative to baseline)
18 bar BMEP	12.1 – 14.9	Reduce by 1 pct point
24 bar BMEP	16.4 – 20.1	Reduce by 2 pct points
24 bar BMEP with CEGR	19.3 – 23.0	Reduce by 3 pct points
27 bar BMEP with CEGR	17.6 – 24.6	Reduce by 3 pct point

Note: All estimates are relative to a baseline naturally aspirated engine with fixed valve timing and lift, except as noted.

Spark Retard at Some Higher Load Regions

The elevated intake pressures of a turbocharged, downsized engine increases the knock susceptibility of an engine. Intake pressures on the CAFE drive cycles could be 1.5 times the levels in naturally aspirated engines. With the likely onset of knock within the CAFE drive cycles for turbocharged, downsized engines, spark retard would be required to prevent knocking conditions. Spark retard to avoid knock was estimated to result in an increase in fuel consumption of approximately a 6 percent at the high load conditions susceptible to knock, as described in Appendix J.

Wider Ratio Transmissions and/or Modified Torque Converters to Compensate for Turbocharger Lag during Launch

Drivability was not characterized for the 27 bar BMEP engine in the Ricardo (2011) report, or for the MAHLE engine and other experimental turbocharged, downsized engines that were referenced by EPA and NHTSA. However, for a vehicle launch from an idle condition, a turbocharged, downsized engine cannot develop the torque of the comparable naturally aspirated engine due to turbocharger lag. A higher transmission ratio or a modified torque converter may be required to provide higher torque multiplication at launch. These changes would result in higher engine speeds, which could increase fuel consumption by up to 6 percent during launch conditions, as described in Appendix J. This condition is important since there are 18 launch conditions (idle periods followed by an acceleration mode) in the FTP75 cycle.

MAHLE Turbocharged, Downsized Engine

There are no production examples of light-duty SI engines at the upper end of turbocharging to 27 bar BMEP and downsizing to a 56 percent reduction in displacement. As noted later in this section, several vehicle manufacturers commented that they considered the limitations for turbocharging and downsizing to be about 50 percent downsizing and 25 bar BMEP. However, given the long time frame for this rule and the committee's mandate to consider fuel economy technologies out to 2030, it is important to consider 27 bar BMEP engines. MAHLE Powertrain has explored the capability of achieving 30 bar BMEP in an experimental, downsized 1.2L 3 cylinder engine that would replace a 2.4L naturally aspirated engine (Blaxill 2012). MAHLE concluded that 50 percent downsizing is feasible, although driveability in launch modes due to turbocharger lag was acknowledged as an issue. As illustrated in Figure 2.4, MAHLE showed that 50 percent downsizing provided a 26 percent reduction in fuel consumption, which compares to EPA/NHTSA projections of 20.6 to 24.6 percent for a lower 27 bar BMEP engine with cooled EGR.

Although MAHLE has demonstrated the power and fuel consumption capability of an experimental, highly turbo-

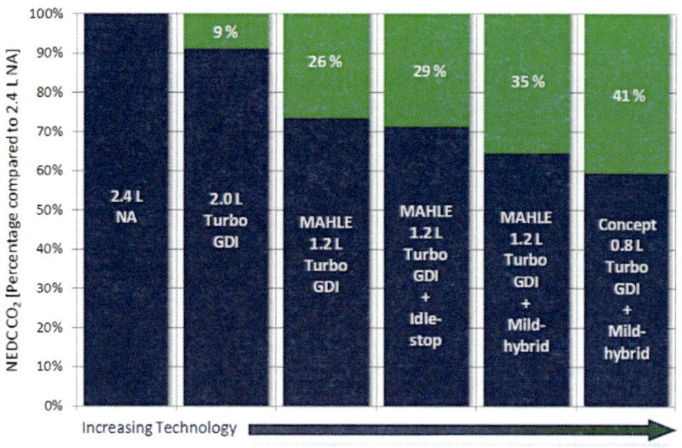

FIGURE 2.4 Fuel consumption reduction of MAHLE's 30 bar BMEP, turbocharged and downsized engine.
SOURCE: MAHLE (2012). Used with permission of MAHLE Powertrain LLC.

charged and downsized engine, some aspects of MAHLE's development require clarification. The fuel consumption reduction data shown in Figure 2.4 are for the New European Driving Circle (NEDC) rather than for the U.S. urban and highway cycles used for CAFE compliance. MAHLE's engine requires 95 RON gasoline, whereas mainstream vehicles in the United States today use regular gasoline with 91 RON (87 AKI). Although the engine was not tested with 91 RON regular grade fuel, the achievement of 27 bar BMEP with this fuel may be an issue. Failure to achieve 27 bar BMEP would result in the need for a larger engine to maintain performance of the baseline vehicle, which would provide less than the expected reduction in fuel consumption. MAHLE has not evaluated turbocharger lag at altitudes much above sea level, although turbocharged engines typically experience exaggerated turbocharger lag at altitude because of the reduced exhaust mass flow available to accelerate the turbocharger.

MAHLE is also considering further improvements to the highly turbocharged and downsized engine: further reduction in displacement to 0.8L (67 percent downsizing), exhaust gas recirculation, lean combustion, variable valve trains, and friction reduction. The fuel consumption reductions estimated by MAHLE for these technologies, shown in Figure 2.4, are expected to be significantly less when applied in combination with an already highly downsized and boosted engine. MAHLE did not provide its plans for exploring the benefits of these additional technologies.

In addition to the research program conducted by the MAHLE Powertrain Group, which was directed toward reaching 30 bar BMEP, research programs have also been pursued by other organizations. One example is the experimental Sabre research engine developed by Lotus Engineering, which reached 20 bar BMEP with a 32 percent downsized engine (Coltman 2008). Another example is a General Motors experimental turbocharged engine, which reached

26.4 bar BMEP (Schmuck-Soldan 2011). The Ultraboost project, which had a target of 32.4 bar BMEP with 60 percent downsizing, is discussed separately in the next section.

Ultraboost

A recent paper entitled "Ultraboost: Investigations into the Limits of Extreme Engine Downsizing" provided insights into downsizing from a U.K collaborative project (Turner et al. 2014). In that project, a 60 percent downsized engine (from 5.0L V8 to 2.0L I4) provided a projected 15 percent reduction in NEDC fuel consumption based on steady-state mapping data. An analytical adjustment was made to reduce the measured "high friction" in the engine to match the friction of a "typical boosted engine," but this adjustment is probably optimistic since this engine operates at much higher boost pressures than "typical boosted engines." With this analytical adjustment, a 22.6 percent reduction in fuel consumption was estimated. Depreciating this "warm" value by approximately 2.5 percent for the CAFE drive cycle, this engine is estimated to provide approximately a 20 percent reduction in fuel consumption (Ricardo Inc. 2011). In contrast, NHTSA projects 20.6 to 24.6 percent reduction in fuel consumption for the 27 bar BMEP, 56 percent downsized engine.

Several characteristics of this engine are significant. The engine "required" 95 RON gasoline. The compression ratio was reduced from 11.5:1 for the naturally aspirated engine to 9:1 for the Ultraboost engine requiring 95 RON gasoline. The engine had variable cam phasing and cam profile switching. Cooled EGR was used. An engine-driven supercharger (to fill the gap in boost pressure at lower speeds) and a turbocharger were used to obtain the 3.5 bar absolute boost pressure (2.5:1 pressure ratio). Two charge air coolers were used.

Issues with Turbocharged, Downsized Engine

Several remaining technical issues for turbocharged and downsized engines are described below.

Turbocharger Lag

In a turbocharged engine when an increase in torque is commanded, due to the inertia of the turbocharger, the time required to increase boost pressure depends on the increase in rotational speed of turbocharger. Figure 2.5 illustrates the effect of turbocharger lag on the ability of an engine to respond to an increase in torque demand. The curve labeled "Ricardo-assumed" was used by Ricardo for its full system simulation study. Based on EPA's concern with Ricardo's assumption, EPA provided Ricardo with its proposed time constants for naturally aspirated and turbocharged engines based on test data. Ricardo subsequently recalculated the acceleration times using EPA's time constants. As noted, a turbocharged engine may take between 2.5 and 5 seconds to generate the full value of the demanded torque, which can be a source of customer complaint. This issue is being addressed with increasingly smaller turbochargers with reduced rotational inertia.

Reducing turbocharger response times to achieve maxi-

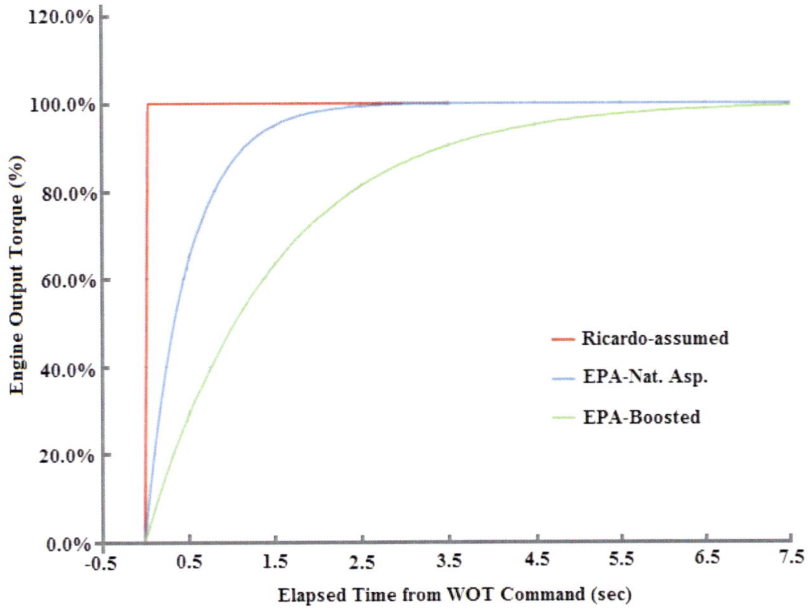

FIGURE 2.5 EPA-proposed time constants and resulting effect on torque rise time for turbocharging.
SOURCE: EPA/NHTSA (2012b).

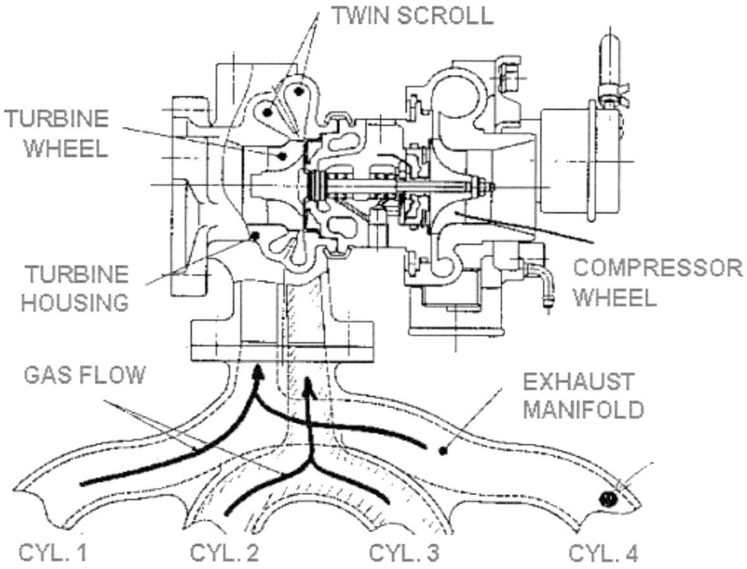

FIGURE 2.6 Twin-scroll turbocharger.
SOURCE: Bundy (2009).

mum torque is an enabler for achieving maximum feasible downsizing of a turbocharged engine while maintaining 0-60 mph acceleration times equal to those of a naturally aspirated engine. Control strategies can also affect transient performance of turbocharged engines. An analytical study of the trade-off between fuel economy and transient performance in turbocharged engines has shown that an engine control strategy optimized for best fuel economy could result in a loss in transient performance (Eriksson 2002). The fuel-optimized strategy keeps the wastegate open to maintain low pressures before and after the engine, whereas the transient performance strategy tries to keep the turbocharged speed as high as possible by closing the wastegate. The typical calibration in production turbocharged gasoline vehicles will strike a balance between the two extreme calibrations based on the analysis of the trade-off between fuel economy and transient response (Gorzelic 2012).

A twin scroll turbocharger has been introduced on some turbocharged, downsized engines to provide higher boost pressures and reduced turbocharger lag times during transients (Bundy 2009). A twin-scroll turbocharger separates the cylinders whose exhaust pulses interfere with each other. The result is superior scavenging of the engine's cylinders and more efficient delivery of exhaust gas energy to the air charge entering each cylinder. The twin-scroll turbocharger includes not only the complex twin-scroll exhaust gas collectors from the turbine of the turbocharger but also a bifurcated exhaust manifold for the separation of exhaust flowing from the engine, as shown in Figure 2.6. With the increased complexity, the twin-scroll system increases the cost of the turbocharger.

Another approach to eliminating turbocharger lag is to use an engine-driven supercharger in place of the turbocharger. However, the power consumption of the supercharger will diminish the fuel consumption reduction obtained with downsizing unless measures are taken to reduce or eliminate the power consumption, such as with a bypass valve arrangement and/or a clutch mechanism to disengage the supercharger when it is not required at light loads. Several manufacturers have applied superchargers. Audi produces a 3.0L supercharged engine installed in the A6 Quattro, A8, and Q5 vehicles. However, these vehicles do not have larger displacement, naturally aspirated engine counterparts to provide a comparison of the potential of supercharging to reduce fuel consumption relative to turbocharging. Nissan has applied a supercharger with a bypass valve and electromagnetic clutch to a 1.2L three-cylinder gasoline engine with the objective of achieving the lowest fuel consumption in the European B segment market. This engine also benefited from a high compression ratio (13:1), direct injection, and low friction (Kobayashi 2012).

Another approach to eliminating turbocharger lag is to use an electrically assisted turbocharger or supercharger, which is discussed later in this chapter.

Limits of Downsizing - Octane Requirement

Fuel octane requirements for high BMEP engines remain a concern. EPA and NHTSA have proposed the use of cooled EGR to reduce the octane requirement of 24 and 27 bar BMEP engines. Limited results on the ability of cooled EGR to reduce the octane requirements of engines are available. Southwest Research Institute (SwRI) has found that in a modern GDI engine every 10 percent increase in EGR can provide about 1 compression ratio increase in knock-limited BMEP (Alger et al. 2012). Instead of reducing compression ratio, EGR could be used to reduce the octane requirement, with estimates ranging from approximately 2.5 RON (Leone 2014) to 5 RON (Heywood). Unlike using higher octane fuel to control knock, using EGR to control knock slows the combustion process, resulting in less complete combustion and higher exhaust temperatures, which may present durability concerns.

Some vehicle manufacturers and suppliers are conducting research on the effectiveness of cooled EGR for reducing octane requirements in high BMEP engines. However, results from their research were not available to demonstrate that EGR could reduce the octane requirement to 91 RON in a high BMEP engine. Many manufacturers are members of the SwRI High-Efficiency, Dilute, Gasoline Engine (HEDGE) consortium, which is conducting research on cooled EGR. One manufacturer plans to specify premium fuel for its turbocharged, downsized engines, since it found that the use of cooled EGR is not adequate to facilitate operation on 91 RON fuel. Some European manufacturers also specify premium fuel for turbocharged engines. Specifying premium fuel for turbocharged downsized engines will raise the cost of operation for the consumer.

Several vehicle manufacturers commented on the limitations for turbocharging and downsizing. They said that 50 percent downsizing and 25 bar BMEP were the limits due to NVH and knock limits, assuming the use of regular grade gasoline. However, a few manufacturers indicated that higher BMEP levels would require 100 RON gasoline, which is not currently available in the United States. These manufacturers were doubtful that EGR was a sufficient enabler to reach higher BMEP levels. Another vehicle manufacturer indicated that further fuel consumption reductions could not be obtained with downsizing beyond approximately 50 percent.

Limits of Downsizing - Preignition

Preignition and detonation or knock are concerns with downsized, turbocharged engines. MAHLE illustrated these limits at high BMEP levels in Figure 2.7, which shows that the spark timing range for acceptable operation between preignition and detonation limits is significantly reduced at higher BMEP levels. SwRI has identified low-speed preignition (LSPI), which can seriously damage engine parts or cause complete engine failure, as a major impediment to aggressive engine downsizing and downspeeding to reduce fuel consumption (Alger 2013). SwRI has demonstrated that LSPI can be suppressed in turbocharged engines by using cooled EGR and advanced ignition timing. SwRI launched a Preignition Prevention Program (P3) consortium in 2010 that is looking at the root causes and at fuels and lubricants to discover ways to suppress LSPI.

LSPI is abnormal combustion at low engine speeds and high loads. It is characterized by preignition that leads to high cylinder pressure and heavy knock. LSPI often occurs in multiple cycles and usually oscillates between preignition and spark ignition. LSPI is typically measured in the range of fewer than six preignition events per 30,000 engine cycles. In the case of the SwRI engine, 15 percent cooled EGR was found to eliminate LSPI completely (Alger 2010). SwRI has hypothesized that low-speed preignition results from the oil and fuel mixture being ejected from the crevice volumes of the piston and igniting the main charge. To address this cause,

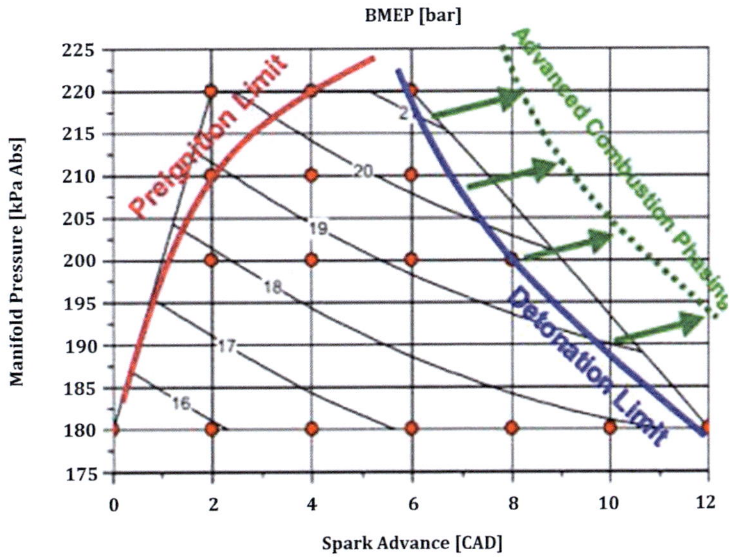

FIGURE 2.7 Preignition and detonation limits for a turbocharged, downsized engine.
SOURCE: Blaxill (2012). Used with permission of MAHLE Powertrain LLC.

the new GF-6 oil rating is being developed to protect against LSPI (see earlier section on Low Friction Lubricants).

Higher-Temperature Turbochargers

Turbochargers with a 950°C temperature limit, as assumed by EPA and NHTSA, may not be sufficient for achieving the full potential fuel consumption reductions or the largest amount of downsizing (NHTSA/EPA 2014). Engine exhaust temperatures increase with load and can easily exceed 950°C before full load is reached. To protect the turbocharger, fuel enrichment is often used, which can deteriorate the fuel consumption of the vehicle. To extend the load range at a stoichiometric air/fuel ratio, higher-temperature turbochargers with a capability of 1050°C are being applied (Merkelbach 2009; Bickerstaff 2012). Achieving this temperature capability requires expensive alloys (MAR M246 nickel-cobalt-tungsten superalloy) that significantly increase the cost of the turbocharged engine. As increased levels of downsizing are applied, increasing turbocharger temperature capabilities are expected to be required.

Transmissions

Torsional dampers are required between the engine and transmission to decouple the engine rotational irregularities and reduce vibration and noise levels in the transmission. The task and complexity, and therefore cost, increase as one downsizes from V8 to V6, V6 to I4, and I4 to I3 engines. I3 engines will require the most expensive damper. Increasingly complex damping systems could include single- or two-stage dampers, a dual-mass flywheel, and/or a torque converter damper.

Noise, Vibration, and Harshness

Vehicle modifications will be required to isolate downsized engines from the passenger compartments. These modifications may consist of complex engine mounting systems and engine and turbocharger noise isolation.

Cooled Exhaust Gas Recirculation

EGR can increase the efficiency of gasoline engines through several mechanisms:

- Reduced throttling losses with the increased flow of air and EGR into the cylinders;
- Reduced heat rejection due to the lowered peak combustion temperatures;
- Reduced chemical dissociation, with the lower peak temperatures resulting in more of the released energy near top dead-center; and
- Higher specific heat ratio (gamma), which increases the work done on the piston.

The potential fuel consumption reduction provided by cooled EGR was estimated for each of these mechanisms. The introduction of 20 percent EGR at a part load condition in a conventional engine would increase manifold pressure by 20 percent, which would reduce pumping losses by approximately 10 percent. However, by adding EGR to an engine with VVT, continuously variable valve lift, and turbocharging and downsizing, the pumping losses will already be very low, so adding EGR is not expected to provide significant additional reductions in pumping losses. Pumping losses could possibly increase due to the requirement for higher exhaust pressure to achieve the required EGR flow. EGR will increase the specific heat ratio, which is estimated to provide a 1.5 reduction in fuel consumption. Adding in benefits from reduced heat rejection, reduced dissociation losses and minor reductions in pumping losses would result in about 2.5 percent reduction in fuel consumption, which is 1 percentage point lower compared to the 3.5 percent effectiveness estimated by NHTSA, as shown previously in Table 2.17. MAHLE Behr recently reported that cooled EGR could provide about a 2 to 4 percent reduction in fuel consumption at light to moderate loads (Morey 2014).

A supplier confirmed that NHTSA's estimate of $305 total cost, or $212 direct manufacturing costs, for the dual-loop, high- and low-pressure, cooled EGR system is in the appropriate range. Water condensation problems, which would require a sophisticated trap and drain system, would increase this cost. However, this supplier felt that single-loop EGR systems would likely be the preferred approach.

A supplier suggested that high dilution rates with EGR might require upgraded ignition systems to achieve acceptable combustion stability (low coefficient of variation of IMEP). Today's ignition systems produce approximately 40 mJ of energy, but high rates of EGR may require more than double the energy, which would necessitate a new ignition system with an unknown incremental cost.

Summary of Fuel Consumption Reductions and Costs of Turbocharged, Downsized Engines

A summary of the estimated fuel consumption reductions and associated direct manufacturing costs for turbocharging and downsizing (TRBDS) is shown in Table 2.18. The committee's high effectiveness estimates of turbocharged, downsized engines agree with NHTSA's estimates, while the committee's low effectiveness estimates are lower than NHTSA's estimates by the amounts shown previously in Table 2.17, which were relative to the baseline engine. For the incremental estimates relative to the previously applied SI engine technologies, the ratio of NHTSA's incremental to baseline effectiveness was applied to the committee's baseline estimates to provide the committee's incremental estimates shown in Table 2.18.

The committee's most likely low estimate of incremental direct manufacturing cost (DMC) for turbocharged, down-

TABLE 2.18 Fuel Consumption Reductions and 2025 MY Direct Manufacturing Costs (2010$s) for Turbocharged, Downsized I4 Engines in a Midsize Car with an I4 Engine (not including cost of SGDI, which is considered an enabler for TRBDS)

Turbocharged, Downsized Engine Technology	NRC Estimated Most Likely Fuel Consumption Reduction (%)[a,b]	NHTSA Estimated Fuel Consumption Reduction (%)[a]	NRC Estimated Most Likely 2025 MY DMC Costs (2010$)[a]	NHTSA Estimated 2025 MY DMC Costs (2010$)[a]
18 bar BMEP[b]	11.1 - 14.9	12.1 - 14.9	245 - 282	245
Incremental[c]	7.7 - 8.3	8.3	245 - 282	245
24 bar BMEP[b]	14.4 - 20.1	16.4 - 20.1	400 - 437	400
Incremental[c]	3.2 - 3.5	3.5	155	155
24 bar BMEP w/ CEGR[b]	16.3 - 23.0	19.3 - 23.0	580 - 617	580
Incremental[c]	3.0 - 3.5	3.5	180	180
27 bar BMEP w/ CEGR[b]	17.6 - 24.6	20.6 - 24.6	890 - 927	890
Incremental[c]	1.4	1.4	310	310
Other Possible Costs				
Turbocharger (Upgrade to 1050 °C)			25 - 75	
Ignition Upgrade (for EGR)			20 - 70	
Transmission (Upgrades for 3 cyl)			0 - 50	
Vehicle Integration (NVH, Thermal Mgmt.)			0 - 25	

[a] Baseline is 12 bar BMEP natural aspirated engine.
[b] Relative to baseline with fixed valve timing and lift, PFI.
[c] Incremental to all previous SI technologies (LUB, EFR, DCP, CVVL, SGDI, TRBDS as applicable).
[d] Ranges are shown for all vehicle classifications.
SOURCE: EPA/NHTSA (2012b) and committee.

sized engines match NHTSA's projections, while the committee's most likely high costs are higher than NHTSA's projections due to increased estimated costs for some of the system components, including the turbocharger and charge air cooler. Other possible costs are noted in Table 2.18, which do not appear to have been considered by NHTSA; they include upgrades for higher temperature capability turbochargers, ignition system upgrades to provide adequate ignition energy with cooled EGR, transmission upgrades, particularly with three-cylinder engines, and vehicle integration components for NVH reduction and thermal management.

DOE Research Projects on Turbocharged and Downsized Engines

DOE currently has programs with Ford, General Motors, and Chrysler to demonstrate a 25 percent improvement in fuel economy while achieving Tier 2 Bin 2 emissions requirements with downsized, boosted engines and a variety of other technologies, including lean combustion, cooled EGR, advanced ignition systems, and friction reduction technologies. These programs are described in Appendix K. Final results from these programs are not yet available.

Accessories

Approximately 2.8 percent of the fuel consumption (equal to 1 percent of the fuel energy) is required to drive the accessories, most of which are required by the engine when tested on the CAFE drive cycle. This estimate can be derived from Figure 2.1 by dividing the 1 percent of the fuel energy shown for accessory loads by the ITE of 36 percent. Additional discussion of vehicle accessories such as air conditioning is contained Chapter 6, "Non-Powertrain Technologies." NHTSA accounts for engine-required accessories and vehicle accessories in the combined category of Improved Accessories, Levels 1 and 2 (IACC1 and IACC2). NHTSA has defined the improved accessories as follows:

- IACC1: electric water pump, electric cooling fan, high efficiency alternator and
- IACC2: mild alternator regenerative braking (specifically excluded are an electric oil pump and electrically driven air conditioner compressor).

NHTSA estimated the following fuel consumption reductions for the improved accessories (EPA/NHTSA 2012):

- IACC1: 0.91-1.61 percent (relative to EPS) and
- IACC2: 1.74-2.55 percent (relative to IACC1).

The estimated fuel consumption reductions with improved accessories for each category are shown in Table 2.19. The following steps are shown in the table to estimate the fuel consumption reduction for the improved accessories: (1) estimate the engine BMEP required for each accessory, (2) replace mechanically driven accessories with electrically driven accessories (which increases power requirements due to the electric motor and alternator inefficiencies relative to the mechanical drives), (3) apply on-demand operation of the electrically driven accessories, and (4) replace DC brush motors with brushless motors for improved efficiency. The alternator efficiency was improved from 65 percent to 70 percent as specified by NHTSA. The fuel consumption reduction for the improved accessories level 1 was estimated to be 1.1 percent, which is at the low end of the range of NHTSA's estimates. For improved accessories level 2, the fuel consumption reduction was estimated to be 2.0 percent, which was also within the range of NHTSA's estimates.

Water Pump

An electric water pump can be controlled to provide the flow of coolant through the engine to maintain required engine temperatures. An electric water pump will be more efficient than one that is belt driven at a fixed ratio of engine speed, which is independent of the coolant flow required. A turbocharged engine with an electrically-driven water pump can continue to run the water pump to cool the turbocharger even if the engine is shut off. As an example, BMW uses electrically-driven water pumps on most of its mainstream turbocharged engines. Concern about failure modes with an electric water pump was reported.

Cooling Fan

Most front-wheel-drive cars and many rear-wheel-drive vehicles currently use electrically driven cooling fans. Direct current (DC) motor-driven cooling fans have a wide range of maximum wattages in light-duty vehicles. For a 400 W cooling fan, assuming 70 percent alternator efficiency, a load of 571 W, or 0.75 horsepower, would be applied to the engine at maximum cooling conditions. Cooling fan loads can be reduced with two-speed or infinitely variable speed operation, in addition to shutting them off when not needed. Two speeds were often achieved by using a resistor to reduce voltage to the motor. Infinitely variable speeds are provided by a pulsewidth modulated controller, which reduces the amount of energy wasted.

Oil Pump

Fixed-displacement oil pumps are used on most vehicles today. Typically, these pumps are oversized in order to operate under harsh engine operating conditions. They typically consume more power and deliver significantly higher oil pressures and flow rates than needed. They contain pressure-relief valves to avert excessively high oil pressures. Since they consume significant amounts of energy at high oil flow rates, these designs are inefficient.

Variable-displacement oil pumps help keep these energy losses to a minimum. Active control matches the oil flow and pressure to the engine needs. It eliminates excess oil flow, substantially reduces the parasitic load on the engine, and ultimately saves fuel. In variable-displacement pumps, changing the displacement volume controls the flow rate. Vane-pump designs have hydraulic and electrical controls and actuators that move the pump housing and vary the eccentricity of the rotor. Electronic controls vary the pressure set points as dictated by operating conditions. Some vehicle manufacturers adopted these kinds of pumps starting in 2011, using them in engines for high-end vehicles in Europe. Recently, Chrysler introduced a variable-displacement oil pump on its 3.6L DOHC V6 engine, which is used in about a third of Chrysler products (Witzenburg 2013).

Although not considered by NHTSA, electrified engine oil pumps provide further opportunities to reduce fuel consumption.

Brushless Motors

Brushless motors are replacing brushed DC motors. Brushless motors are typically 85-90 percent efficient, whereas brushed DC motors are 75-80 percent efficient (Quantum Devices 2013). The higher efficiency of a brushless motor would result in a 12 percent reduction in electrical power required. However, brushless motors are more expensive than brushed motors, partly due to their control requirements. One vehicle manufacturer is planning to apply brushless motors to its entire product line.

Summary of Effectiveness of Spark Ignition Engine Technologies

EPA and NHTSA expended significant effort and resources in estimating the fuel consumption reductions for a range of SI engine technologies. To accomplish this task, they used full system simulations, response surface modeling, and the LPM together with literature reviews, data from vehicle manufacturers and suppliers, and expert opinions. From this input, EPA and NHTSA developed the estimates of fuel consumption reductions shown in the TSD. Since the committee did not have the resources to develop similar estimated fuel consumption reductions, the following approaches were used to examine the Agencies' estimates and to develop the committee's estimates of most likely effectiveness values: fundamental technical analysis, literature reviews, full system simulations, EPA certification data, expert input from vehicle manufacturers and others, and the committee's expertise.

A summary of the committee's low and high most likely

TABLE 2.19 Estimated Fuel Consumption Reductions with Improved Engine Required Accessories Included in NHTSA's IACC Categories

Values for 1500 RPM – Average Engine Speed on CAFE Test Cycle

Engine Required Accessories Included in IACC Category		Mechanical Load MEP (kPa)[a]	Electrical Load (watts)	Baseline percent of CAFE Load IMEP (%)[b]	Modified percent of CAFE Load IMEP (%)[b]	Percent of CAFE load IMEP with Electrification (%)[f]	Percent of CAFE Load IMEP with 25% On-Demand Operation (%)	Percent of CAFE Load IMEP with Brushless Motors (%)[g]	Modified percent of Total FC[d]	Reduction in FC (from baseline to modified) (%)
IACC 1										
Water Pump	Mechanical	7.0		1.4						
	Electrical - 70% Alt/Motor Eff					2.2	0.55	0.48	0.48	0.92
Cooling Fan[c]	Electrical		400							
	Mechanical - 65% Alt Eff, 25% Duty Cycle	1.04		0.21			0.21			
	Mechanical - 70% Alt/ Motor Eff, 25% Duty Cycle	0.9	371		0.18		0.18	0.16	0.16	0.05
Alternator[d]	Mechanical	10.0		2.0						
	Electrical Output - 65% Alt Eff				1.3					
	Mechanical - 70% Alt Eff				1.9					0.14
Sub-Total				3.6						1.1
IACC2										
Regen Braking	80% of Remaining Electric Power from Regeneration			2.5					0.5	2.0
Total										3.1

[a] Heywood (1988), pp. 739-740.
[b] Heywood (1988), p. 825. Full load IMEP = 1000 kPa or 14.5 bar, CAFE cycle IMEP = 500 kPa (7.25 bar).
[c] 400 watt electric fan, 25% on-demand, 0.4 kW/(150 kW engine x 1500 rpm/6000 rpm) x 500kPa x .25 =2.7 kPa.
[d] Alternator for engine electrical power (ignition, controls) only.
[e] Assuming 100% of fuel produces 100% of CAFE cycle IMEP.
[f] Assuming 70% motor efficiency, 70% alternator efficiency.
[g] Assuming 12% efficiency improvement for brushless motors.
NOTE: Grey color indicates input for the calculations in the table.

effectiveness estimates for SI engine technologies are compared to NHTSA's estimates for an I4 engine in Table 2A.3 (Annex at end of this chapter). The committee's estimated effectiveness values for I4 DOHC, V6 DOHC, and V8 OHV engines in midsize cars, large cars, and large light trucks, respectively, are provided in Table 2A.1 (Annex). The committee's estimates of effectiveness agreed with many of NHTSA's estimates. For several technologies, the committee's high estimates agreed with NHTSA's estimates, while the low estimate for the 18 bar BMEP engine was 1 percentage point lower than the high estimate of 14.9 percent; for the 24 bar BMEP engine, the low estimate was 2 percentage points lower than the high estimate of 20.1 percent; and, for cooled EGR, the low estimate was 1 percentage point lower than the high estimate of 3.5 percent. To achieve these estimates, some of the technologies will require new developments, such as a new low-friction lubricant meeting a new specification for 0W-12 oil, which is under development, and higher BMEP engines with cooled EGR.

Costs of Spark Ignition Engine Technologies

Teardown Cost Studies of Turbocharged, Downsized Engines

Of all the SI engine technologies identified by NHTSA, teardown cost studies by FEV were conducted only for turbocharging and downsizing (TRBDS), which included SGDI technologies. These teardown cost studies assessed the direct manufacturing costs of vehicle technologies. They were conducted by FEV for the following three turbocharging and downsizing cases:

1. SGDI and turbocharging with engine downsizing from a DOHC four-cylinder engine to a small DOHC four-cylinder engine.
2. SGDI and turbocharging with engine downsizing for a DOHC V6 engine to a DOHC four-cylinder engine.
3. SGDI and turbocharging with engine downsizing from a SOHC three-valve/cylinder V8 engine to a DOHC V6 engine.

EPA extrapolated the results of these studies to several other downsizing scenarios.

Each of the FEV teardown cost studies developed incremental costs associated with 17 subsystems. These subsystem costs were assigned to three major technologies under consideration: SGDI, turbocharging, and engine downsizing. In some cases, a portion of the overall cost result was distributed over several technologies (Olechiw 2009). Table 2.20 summarizes the results of binning the costs.

Teardown costs of the three most costly subsystems provide insight into these overall costs. The three most costly subsystems are the following (FEV Inc. 2009):

1. Induction air charging subsystem,
2. Fuel induction subsystem, and
3. Engine management, engine electronic and electrical subsystem.

A further breakdown of these subsystem costs in Table 2.21 illustrates the most costly components in these subsystems. Using the 2.4L I4 NA to 1.6L I4 TC as an example, these three subsystems comprise nearly 80 percent of the total cost of turbocharging and downsizing technology.

The vehicle manufacturers and suppliers that the NRC committee met with were asked to comment on all of the technology costs, with particular attention to the teardown costs. The following comments were received:

- The cost of a turbocharger assembly could be up to twice the cost shown in Table 2.21 depending on the materials required to achieve a specified temperature capability and the boost control system (wastegate, variable geometry turbine). EPA and NHTSA have indicated that they did not rely on any turbocharger system operating above 950°C (NHTSA/EPA 2014), although, as described earlier, many systems are currently in production with 1050°C capability.
- The charge air cooler and exhaust manifold could have significantly higher costs than shown in the cost teardown studies.
- The powertrain control module (PCM) costs could benefit from further refinement, as was noted in the peer review of the pilot EPA/FEV cost study.
- Indirect costs applied to the direct manufacturing costs from these studies warrant revision upward due to concerns with engineering research and development for the required pace of technology introductions, testing capacity constraints, sustainability of vehicle segments, cost of capital, and stranded investment.

Direct Manufacturing Costs

The direct manufacturing costs of technologies for SI engines were estimated using one of the following processes:

1. EPA and NHTSA developed direct manufacturing costs of several technologies based on FEV teardown cost studies. These studies were critically reviewed. Updates or adjustments were applied, as required, to the teardown cost studies based on the committee's expertise and judgment that incorporated a review of available information, including other cost data from published studies, input from vehicle manufacturers and suppliers, and a review of retail prices adjusted to reflect direct manufacturing costs.
2. For technology costs that were not supported by teardown cost studies, the committee identified subsystem and major components of the technology similar to the

TABLE 2.20 Results of Binning Costs

Engine	Incremental to	Total Incremental Costs ($)	SGDI ($)	Turbocharging ($)	Downsizing ($)
1.6L I4 Turbo SGDI	2.4L I4 MPFI DOHC	532	213	404	(85)
2.0L I4 DOHC SGDI	3.0L V6 MPFI V6 DOHC	69	213	404	(547)
3.5L V6 DOHC Turbo SGDI	5.4L V8 MPFI three-valve SOHC	846	321	681	(155)

SOURCE: Olechiw (2009).

TABLE 2.21 High-Cost Components in High-Cost Subsystems for Turbocharged and Downsized Engines

Subsystem	Net Cost Impact to OEM ($)	Net Cost Impact to OEM ($)	Net Cost Impact to OEM ($)
Induction Air Charging Subsystem	2.4L I4 NA to 1.6L I4 TC	3.0L V6 NA to 2.0L I4 TC	5.4L V8 SOHC to 3.5L V6 DOHC
Turbocharger assembly	151.85	169.89	329.82
Charge air cooler	18.65	20.92	35.61
Tube assembly	18.76	53.93	42.61
Engine and vehicle assembly of air induction components	25.70	25.67	27.35
Multiple components (<$15 each)			
Total Subsystem	258.89	280.70	448.79
Fuel Induction Subsystem	**Net Cost Impact to OEM ($)**	**Net Cost Impact to OEM ($)**	**Net Cost Impact to OEM ($)**
High pressure pump	69.61	64.90	81.12
Fuel injectors Solenoid 7 hole	17.42	5.10	15.15
Fuel Rails	14.93	10.77	15.15
Multiple components (<$15 each)			
Total Subsystem	107.30	84.76	124.57
Engine Management, Engine Electronic and Electrical Subsystem	**Net Cost Impact to OEM ($)**	**Net Cost Impact to OEM ($)**	**Net Cost Impact to OEM ($)**
Powertrain Control Module (PCM) - Hardware	40.00	40.00	60.00
Multiple components (<$15 each)			
Total Subsystem	56.61	21.57	70.15

SOURCE: FEV Inc. (2009).

process used in the teardown cost studies. Costs were estimated for these subsystems and components by applying the committee's expertise, which incorporated a review of available information, including NHTSA's estimates and associated references, other cost data from published studies (such as shown in Table 2.29), input from vehicle manufacturers and suppliers, and a review of retail prices adjusted to reflect direct manufacturing cost.

Examples of both of these processes are provided in this section.

An example of the process for estimating direct manufacturing costs for the intake cam phasing system is shown in Table 2.22. The subsystems and components required for this technology installed on an engine are listed, together with estimated costs and comments on the sources of these costs. Also shown for comparison is EPA's and NHTSA's cost estimate. Since the cost estimates were generated for the 2012 MY, learning factors, as specified by NHTSA, were applied to the direct manufacturing costs to provide 2017 MY estimates in 2010 dollars, as shown in the table and the 2025 MY costs shown in later tables (EPA/NHTSA 2012b). NHTSA's learning factors include the effects of accumulated production volume and innovations in design and manufac-

TABLE 2.22 Example of Direct Manufacturing Cost (DMC) Estimates for Intake Cam Phasing (ICP) System

	I4 Engine	
Technology	DMC ($)	Source of Costs
Intake Cam Phasing (ICP)		
Cam phaser (1 per intake camshaft)	21.90	Committee's expertise/judgement
Up-sized oil pump	1.80	FEV: Half of cost of turbo oil pump upgrade[a]
Oil control valve - spool for filling and emptying	12.00	Committee's expertise/judgement
PWM output from low side driver of ECM	4.00	Committee's expertise/judgement
Oil drillings - inlet and return	1.00	Committee's expertise/judgement
Position feedback sensor	4.00	Committee's expertise/judgement
Cam phase trigger wheel - 4 pulses per revolution	1.00	Half of FEV cost of camshaft sprocket[a]
Revised cam driver cover	2.00	Committee's expertise/judgement
Wiring and connectors	1.79	FEV: 40% of cost of wiring[a]
Total: 2012 MY cost in 2010$	49.49	
DMC Learning Type 12, 2012 to 2017 Learning Factor - 0.86		
Total: 2017 MY direct manufacturing cost (2010$)	42.56	15% increase
Reference: EPA/NHTSA direct manufacturing costs 2017 MY (2010$)	37.00	

[a] FEV teardown cost study for turbocharged downsized engines.
SOURCE: NRC Committee; FEV (2009); EPA/NHTSA (2012b).

turing. Further discussion of learning factors is contained in Chapters 7 and 8. In the case of intake cam phasing shown in the table, the direct manufacturing cost was estimated to be 15 percent higher than EPA's and NHTSA's estimate (Martec 2008; EPA/NHTSA 2010). The committee took this computed cost to be the high most likely value and retained the Agencies' cost as the low most likely value.

Another example of the process for estimating direct manufacturing costs for the continuously variable valve lift system is shown in Table 2.23. In this case, the direct manufacturing cost was also estimated to be 15 percent higher than EPA's and NHTSA's estimate (EPA/NHTSA 2010). The committee took this computed cost to be the high most likely value and retained the Agencies' cost as the low most likely value.

An example of the process for assessing direct manufacturing costs for turbocharging and downsizing is shown in Table 2.24. In this case, direct manufacturing costs are based on the FEV teardown cost study so only selected subsystems and components were examined for possible updates and adjustments to costs. Two key components, the turbocharger assembly and the intercooler, were examined, and the resulting modifications to the costs, together with the sources of the revised costs, are shown in the table. For the case of turbocharging and downsizing, this approach leads to an estimate of direct manufacturing cost that is 15 percent higher than NHTSA estimates. The committee took this computed cost to be the high most likely value and retained the Agencies' cost as the low most likely value.

Indirect Costs: Estimation of Components for ICP

EPA, RTI International, and the Transportation Institute of the University of Michigan developed the concept of Indirect Cost (IC) multipliers to support the evaluation of costs for regulatory actions (Rogozhin et al. 2009). Chapter 7 discusses indirect cost multipliers in greater detail. For medium-complexity technology, typical of many SI engine technologies, the research and development (product development) costs were shown in the foregoing reference to amount to 5 percent of the indirect costs. The committee examined the details of the product development costs to implement intake cam phasing (ICP) technology as an example and found that an upward adjustment of the indirect cost multipliers was appropriate.

NHTSA considered ICP as a low-complexity technology with an indirect cost multiplier of 1.24, which would be comparable in complexity to the application of low rolling resistance tires. The product development steps required for ICP would consist of the following: (1) installation of the new hardware in the engine, (2) software development, (3) engine mapping for initial calibration optimization, (4) calibration development in the vehicle at all environmental conditions, (5) durability development, and (6) certification from EPA and the California Air Resources Board (CARB).

TABLE 2.23 Example of Direct Manufacturing Cost (DMC) Assessment for Continuously Variable Valve Lift (CVVL) System

	I4 Engine	
Technology	DMC ($)	Source of Costs
Continuous Variable Valve Lift (CVVL Similar to Valvematic)		
Intermediate shaft with finger follower	35.85	Committee's expertise/judgement
Internal shaft for roller finger followers	35.85	Committee's expertise/judgement
Roller finger followers - 8 required	34.02	Committee's expertise/judgement
Cylinder head casing with boring and shaft bearing caps	82.75	FEV: Adjusted cost for SGDI cyl head modif[a]
Electric motor actuator	29.84	Committee's expertise/judgement
PWM output from low side driver of ECM	4.00	Committee's expertise/judgement
ECU input for angle sensor	1.00	Committee's expertise/judgement
Angle position feedback sensor	4.00	Committee's expertise/judgement
Revised valve cover	3.00	Committee's expertise/judgement
Wiring and connectors	1.79	FEV: 40% of cost of wiring[a]
Total: 2012 MY cost in 2010$	232.10	
DMC Learning Type 12, 2012 to 2017 Learning Factor = 0.86		
Total: 2017 MY direct manufacturing cost (2010$)	199.61	15% increase
Reference: EPA/NHTSA direct manufacturing costs 2017 MY (2010$)	174.00	

[a] FEV teardown cost study for turbocharged downsize engine.
SOURCE: NRC committee; FEV (2009); EPA/NHTSA (2012b).

TABLE 2.24 Example of Direct Manufacturing Cost (DMC) Assessment for Downsizing and Turbocharging (TRBDS1) Technology

	I4 Engine			
Technology	EPA/NHTSA DMC ($)	Adjustment ($)	Revised DMC ($)	Source of Costs
Turbocharging and Downsizing (TRBDS1) 18 bar BMEP 33% Downsizing				
Overall DMC	335.00			FEV teardown cost study[a]
Selected Subsystems and Components				
Induction Air Charging System				
Turbocharger Assembly	152.00	38.00	190.00	Committee's expertise/judgement
Charge air cooler	19.00	11.00	30.00	Committee's expertise/judgement
Total: 2012 MY cost in 2010$	335.00	49.00	384.00	
DMC Learning Type 11, 2012 to 2017 Learning Factor = 0.86				
Total: 2017 MY direct manufacturing cost (2010$)	288.10		330.24	15% increase
Reference: EPA/NHTSA direct manufacturing costs 2017 MY (2010$)	288.00			

[a] FEV teardown cost study for turbocharged, downsized engine.
SOURCE: NRC committee; FEV (2009); EPA/NHTSA (2012b).

This is considerably more complex and more labor- and time-intensive than the addition of low resistance tires and would require confirmation of performance and durability. Consequently, the assignment of medium complexity to the ICP was considered to be more appropriate as a starting point in the analysis of indirect costs for this technology. NHTSA assigned medium-complexity indirect cost multipliers (ICMs) to most other SI engine technologies requiring the product development process described above.

A detailed review of indirect costs for the development and application of powertrain technologies was undertaken by the committee, using ICP as an example. The product development costs for the ICP technology, based on the above steps, were estimated in Table 2.25 to be $8.60 by amortizing the total product development cost over a 5-year period for an annual production volume of 50,000 units, which was assumed to be typical of a specific vehicle/engine combination application. The costs shown in Table 2.25 relate only to the application of the ICP technology to a specific vehicle/engine configuration and do not include engine design and development costs, which would be included in direct manufacturing costs. However, with more rapid introduction of technologies, the 5-year period can often be reduced significantly, which would increase the indirect cost per vehicle, since the indirect cost would be be amortized over a shorter time period.

The following adjustments to the IC multiplier resulting from the product development costs for the ICP technology found in Table 2.25 are shown in Table 2.26 and described below.

1. NHTSA's indirect cost multiplier for the medium-complexity ICP technology is 1.39, which yields an indirect cost of $22.54 (0.39 × $57.79) for the case of ICP.
2. The indirect cost multiplier for medium-complexity technology would allocate an amount equal to 5 percent of the indirect cost to product development, which equals $1.13 (0.05 × $22.54).
3. An estimated product development cost of $8.60 is shown in Table 2.26. Therefore, an increased indirect cost of $7.47 ($8.60 – 1.13) would be incurred for the ICP technology.
4. Adding the incremental indirect cost of $7.47 to the indirect cost of $22.54 yields a revised indirect cost of $30.01, which is equal to 52 percent of the direct manufacturing cost ($30.01/$57.79).

An example of applying ICP to a V6 engine is shown in Table 2.27. ICP applied to a V6 engine has double the direct manufacturing cost of applying ICP to an I4 engine. However, the moderate increases in the product development costs for the V6 application in the areas of engine design (for interfacing with each specific vehicle) and prototype hardware costs result in an increase in the ICM to 1.46 from NHTSA's estimated value of 1.39.

There are several reasons the committee determined that these indirect costs were appropriately associated with technologies providing fuel consumption reductions. These indirect costs are associated with the addition of an individual technology providing fuel consumption reductions. These technologies would be applied not in the reference case defined by EPA and NHTSA but instead only in the control case. And these technologies have been applied individually only after a technology has been developed and proven in research and advanced development. For example, DCP might be rolled out in a new model year with VVL being rolled out separately in a subsequent model year after the technology has been developed. Under this deployment scenario, assigning these costs to an individual technology is generally

TABLE 2.25 Product Development Cost Estimates for Intake Cam Phasing Technology Example

Indirect Costs for ICP Example	
Hardware and Labor	Cost ($)
Engine design for ICP - 1 person-year	150,000
Software development - 1 person-year	150,000
Calibration development - 2 person-years	300,000
Dynamometer operation (engineer, technicians)	250,000
Chassis dynamometer operation (engineer, technicians)	250,000
Prototype engines and control hardware	150,000
Prototype vehicles (2 calibration, 2 durability)	400,000
Technician - 2 person-years	200,000
Durability testing - 1 person-year	150,000
Certification - 1 person-year (engineer, technicians)	150,000
Total	2,150,000
Total IC costs 2012MY in 2010$ (amortized over 50,000 units for 5 years)	8.60

TABLE 2.26 Calculation of Revised ICM for Intake Cam Phasing-I4 Engine Technology Example for an I4 Engine

	Reference EPA/NHTSA Indirect Cost Process	Revised Product Development Cost ($)	Incremental Cost ($)	Revised IC
Direct Manufacturing Cost (DMC)	$57.79			
Indirect Cost Multiplier (ICM) = DMC+IC)/DMC	1.39			1.52
Indirect Cost (IC)	$22.54		$7.47	$30.01
Product Development Cost Percent of IC	5%			
Product Development Cost	$1.13	$8.60	$7.47	

appropriate, although there are some opportunities for combining the application of technologies, such as combining SGDI with turbocharging and downsizing. In such a case, many of the unique elements of the indirect costs for SGDI, particularly prototype hardware, software, and calibration, would be included in a consolidated program. And the committee understands that, with the increasing stringency of the CAFE/GHG standards and with further development of new technologies to a production-feasible level, there may be more instances of multiple technologies being introduced with a consolidation of development software and calibration processes.

This analysis used ICP as an example, and there are other insights resulting from the empirical examination of other SI engine technologies and multiple technologies that require the powertrain product development process involving experimental hardware and vehicles, software, calibration, durability testing, and EPA and CARB certification. Due to the uncertainties surrounding the ICMs, the committee generally assessed only the direct manufacturing costs for each technology. Chapter 7 discusses the committee's concern that an empirical basis for EPA/NHTSA's indirect cost multipliers is lacking. EPA presented evidence to the committee that, on average, the ICM method resulted in the ratio of total costs to direct manufacturing costs of approximately 1.50, which is consistent with feedback that the committee obtained from several vehicle manufacturers, with the NRC Phase I study, and with NHTSA studies supporting rulemaking prior to the 2012 rulemaking (EPA 2014e).

The committee also is recommending several modifications to the complexity levels assigned by NHTSA to the SI engine technologies, as shown in Table 2.28. The committee recommends that both intake cam phasing and cylinder deactivation be modified to a medium complexity level, rather than the low complexity level assigned by NHTSA, since each of these technologies involves similar product development steps associated with the costs shown in Table 2.25.

TABLE 2.27 Calculation of Revised ICM for Intake Cam Phasing-V6 Engine Technology Example

Example for ICP for a V6 Engine		
Product Development Cost ($)		
ICP	$2,150,000	
Incremental Cost for ICP for V6		
Engine Design	$150,000	
Prototype engines and control hardware	$150,000	
Total =	$2,450,000	
Total IC cost per unit in 2010$		$9.80

	Reference: EPA/NHTSA Indirect Cost Process	Revised Product Development Cost	Incremental Cost (IC)	Revised IC
Direct Manufacturing Cost (DMC)	$115.58			
Indirect Cost Multiplier (ICM) = (DMC+IC)/DMC	1.39			1.46
Indirect Cost (IC)	$45.08		$7.55	$52.62
Product Development Cost				
Percent of IC	5%			
Product Development Cost	$2.25	$9.80	$7.55	

TABLE 2.28 Complexity Levels for SI Engine Technologies

Technology	NHTSA Complexity Level	Recommended Complexity Level
Low friction lubricants	Low	Low
Engine friction reduction	Low	Low
Intake cam phasing	Low	Medium
Dual cam phasing	Medium	Medium
Variable valve lift	Medium	Medium
Cylinder deactivation	Low	Medium
Gasoline direct injection	Medium	Medium
Turbocharging and downsizing	Medium	Medium
Cooled EGR	Medium	Medium

Other Available Cost Data

The committee based the cost estimates discussed in this chapter on the latest available information provided by the FEV studies conducted for the final CAFE rulemaking, as well as input received from vehicle manufacturers and suppliers. Since turbocharged, downsized engines account for a significant part of the overall incremental costs for SI engines, the committee reviewed past studies that estimated the costs of these engines. A summary from the Northeast States Center for a Clean Air Future study (NESCCAF 2004) through the Phase 1 NRC study in 2011 to the International Council on Clean Transportation (ICCT)/FEV studies for GHG reductions in Europe in 2012 and the 2013 NRC study *Transitions to Alternative Vehicles and Fuels* are listed in Table 2.29. With a few exceptions, these previous studies are within +/- 15 percent of the NHTSA estimates contained in the TSD (EPA/NHTSA 2012b). Many of the previous studies evaluated mild forms of turbocharging without downsizing and did not envision the extent of turbocharged and downsized engines evaluated in the TSD for the final CAFE rule.

Summary of Costs of Spark-Ignition Engine Technologies

The direct manufacturing costs of technologies for reducing fuel consumption were estimated using one of the processes described earlier in the Direct Manufacturing Costs section of this chapter. A summary of the committee's low and high most likely direct manufacturing cost estimates for spark-ignition engine technologies are compared to NHTSA's estimates for an I4 engine in Table 2A.4 (Annex at end of this chapter). The committee's estimated direct manufacturing costs for I4, V6, and V8 engines are provided in Table 2A.2a, b, and c (Annex) for 2017, 2020, and 2025, respectively. The committee's most likely low estimates agreed with NHTSA's estimates, as did many of the most likely high estimates. However, the most likely high estimates were approximately 15 percent higher than NHTSA's costs for cam phasing, VVL, and 18 bar BMEP turbocharged, downsized engines.

Overall Summary of Spark-Ignition Engine Effectiveness and Costs

An overall summary of the committee's most likely estimates for I4 SI engine fuel consumption reduction effectiveness and cost for 2017, 2020, and 2025 MYs is shown in Tables 2.30a and b. These estimates are shown as technology pathways using only SI engine technologies to illustrate the combined, overall effectiveness and cumulative cost resulting from applying the technologies discussed earlier in the chapter. These SI engine technologies are listed in the order that they are discussed in this chapter and presented by NHTSA/EPA in the TSD. These tables show the committee's low and high most likely estimates, respectively. Figure 2.8 shows 2025 MY cumulative cost (2010 dollars) estimates plotted as a function of percent fuel consumption reduction. Several of the significant technologies are labeled on the plot. The first level of turbocharging and downsizing to 18 bar BMEP (TRBDS1) provides the largest individual reduction in fuel consumption. The addition of cooled EGR (CEGR1), together with the second level of turbocharging and downsizing to 24 bar BMEP (TRBDS2), provides the second largest reduction in fuel consumption. The next largest reduction in fuel consumption is provided by discrete variable valve lift (DVVL). Moving to the final level of turbocharging and downsizing, to 27 bar BMEP (CEGR2), was significantly more expensive but less effective than the previous technologies.

As discussed in Chapter 8, an important factor in developing a pathway is the order of applying the technologies, which is primarily done on the basis of cost effectiveness (cost per percent fuel consumption reduction). Some of the SI engine technologies defined by NHTSA and analyzed by the committee are being applied prior to the beginning of the 2017 MY to 2025 MY time frame to meet the 2016 MY CAFE targets. The committee developed an example pathway for a midsize passenger car in Chapter 8 (Tables 8.4 a and b). Using this pathway, the SI engine technologies that

TABLE 2.29 Other Available Cost Data for Turbocharged, Downsized Engines

		Incremental Mfg Cost[a] (2016 Euros)	Euros to Dollars ($)[b]	Estimated Total Cost ($)[c]	NHTSA 2017 Total Costs ($)	Other Costs Relative to NHTSA
Other Available Cost Data						
Turbocharged Downsized Engines						
FEV Study for ICCT - GHG Reduction for Europe - Brussels 01.02.2012						
Engine 3	Turbocharging	473.00	639	888	525	
2.4L I4-1.6L I4	Downsizing				-43	
	GDI				277	
	Total =			888	759	1.17
Engine 5	Turbocharging	854.00	1,153	1,603	885	
5.4L V8 3V to	Downsizing				62	
3.5L 4V DOHC	GDI				501	
	Total =			1,603	1,448	1.11
FEV Light-Duty Vehicle Cost Analysis - European Vehicle Market (Phase 2) September 27, 2012						
Cooled EGR	LP Cooled EGR			127		
	HP Cooled EGR			127		
	Total =			254	249	1.02
NRC 2013, Transitions to Alternative Vehicles and Fuels (2050)						
ICE (Assume ICE only technologies)				1,652	1,830	0.9
NRC 2011, Phase 1 Report						
I4 Engine	Turbocharging and Downsizing			490	482	1.02
V8 Engine	Turbocharging and Downsizing			790	806	0.98
NESCCAF 2004						
	Variable Geometry Turbocharging (I4, V6, V8)			400	525	0.76
Expert Input						
I4 Engine	Turbocharging and Downsizing			750	525	1.43
	Approximately +/-15%, except for NESCCAF and OEM					

[a] European labor costs estimated to be 20% higher than in the U.S.
[b] Euros to Dollars = $1.35.
[c] ICM = 1.39 (same as applied by NHTSA.

might be applied to bring the null vehicle[9] up to the content of a typical 2008 MY vehicle consisted of intake cam phasing and dual cam phasing together with other non-SI engine technologies discussed in Chapter 8. Additional technologies that might be applied by the 2016 MY, based on selecting the technologies with the lowest cost per percent fuel consumption reduction, included low friction lubricants – level 1 and engine friction reduction – level 1 together with other non-SI engine technologies. The SI engine technologies that might be applied during the 2017 to 2025 MY time frame were subsequently identified, together with other non-SI engine technologies. The effectiveness and cost of these technologies are shown in Tables 2.30a and b and summarized in Table 2.31. Approximately an 8 percent reduction in fuel consumption may be achieved by SI engines from the null vehicle to the 2016 MY. For the 2017 to 2025 MYs, the SI engine may achieve approxi-

[9] The null vehicle concept was developed by EPA and NHTSA as a reference point against which effectiveness and cost can be consistently measured (Olechiw 2014). It is defined as a vehicle having the lowest level of technology in the 2008 MY. Technologies are first added to bring the null vehicle into compliance with the 2016 standards, followed by compliance with the 2021 and 2025 standards. The concept is particularly important because, even though NHTSA and EPA use different compliance models, the effectiveness values determined by both Agencies are relative to the same null package; each compliance model uses the same base data. This committee applied the null vehicle concept to illustrate effectiveness and cost in an example pathway.

TABLE 2.30a Low Most Likely Estimates of SI Engine Fuel Consumption Reduction Effectiveness and Costs for 2017, 2020 and 2025 (2010 dollars)

SI Engine Only Pathway - Low Most Likely Direct Manufacturing Costs

Low Most Likely Cost Estimates Paired with High Most Likely Effectiveness Estimates

Possible Technologies[a]	% FC Reduction	Reduction Multiplier	2017 Cost Estimates	2020 Cost Estimates	2025 Cost Estimates	2017 Cost/ Percent FC ($/%)
Low Friction Lubricants - 1 LUB1	0.7%	0.993	$3	$3	$3	$4.29
Engine Friction Reduction - 1 EFR1	2.6%	0.974	$48	$48	$48	$18.46
Low Friction Lub - 2 & Engine Friction Red - 2 LUB2_EFR2	1.3%	0.987	$51	$51	$51	$39.23
Intake Cam Phasing ICP	2.6%	0.974	$37	$35	$31	$14.23
Dual Cam Phasing DCP (vs. ICP)	2.5%	0.975	$31	$29	$27	$12.40
Discrete Variable Valve Lift DVVL	3.6%	0.964	$116	$109	$99	$32.22
Continuously Variable Valve Lift CVVL (vs. DVVL)	1.0%	0.990	$58	$55	$49	$58.00
Cylinder Deactivation - NA for I4 DEACD	0.0%	1.000				
Stoichiometric Gasoline Direct Injection SGDI (Required for TRBDS)	1.5%	0.985	$192	$181	$164	$128.00
Turbocharging & Downsizing - 1 TRBDS1 33% DS 18 bar BMEP	8.3%	0.917	$288	$271	$245	$34.70
Turbocharging & Downsizing - 2 TRBDS2 50% DS 24 bar BMEP	3.5%	0.965	-$92	-$89	-$82	-$26.29
Cooled EGR - 1 CEGR1 50% DS 24 bar BMEP	3.5%	0.965	$212	$199	$180	$60.57
Cooled EGR - 2 CEGR2 56% DS 27 bar BMEP	1.4%	0.986	$364	$343	$310	$260.00
SI Engine Only (incl LUB & EFR)	28.2%	0.718	$1,308	$1,235	$1,125	$46.31
Null Vehicle - 2016 MY	8.2%	0.918	$119	$115	$109	$14.60
SI Engine 2017 - 2025 MY	17.9%	0.821	$613	$578	$526	$34.25
SI Engine After 2025	4.9%	0.951	$576	$542	$490	$118.74

[a] Null vehicle: I4, DOHC, naturally aspirated, 4 valves/cylinder PFI fixed valve timing and 4 speed AT.

mately a 17 to 18 percent reduction in fuel consumption at an estimated direct manufacturing cost in the range of $526 to $705. The high most likely estimated cost is the result of increased costs for several of the technologies and the lower effectiveness of some of the technologies together with the replacement of continuously variable valve lift (CVVL) with the higher cost cooled EGR (CEGR1) to provide the additional reduction in fuel consumption to achieve the 2025 MY CAFE target, as can be seen by comparing the pathways in Tables 8.4a and b.

Fuel Economy and Performance Trade-offs

From 1980 to 2009, there were significant gains in automotive technology, but those gains have applied to improved performance and safety rather than fuel economy, as shown in Figure 2.9. Horsepower more than doubled and 0 to 60 mph times decreased by 35 percent from 14.3 seconds to 9.5 seconds. Average vehicle weight increased 27 percent during the same period, primarily due to increased vehicle size as well as reinforced structures and added equipment such as airbags for improved safety. Fuel economy remained relatively unchanged in the period, with only a 2.9 percent

TABLE 2.30b High Most Likely Estimates of SI Engine Fuel Consumption Reduction Effectiveness and Costs for 2017, 2020, and 2025 (2010 dollars)

SI Engine Only Pathway - High Most Likely Direct Manufacturing Costs

High Most Likely Cost Estimates Paired with Low Most Likely Effectiveness Estimates

Possible Technologies[a]	% FC Reduction	Reduction Multiplier	2017 Cost Estimates	2020 Cost Estimates	2025 Cost Estimates	2017 Cost/ Percent FC ($/%)
Low Friction Lubricants - 1 LUB1	0.7%	0.993	$3	$3	$3	$4.29
Engine Friction Reduction - 1 EFR1	2.6%	0.974	$48	$48	$48	$18.46
Low Friction Lub - 2 & Engine Friction Red - 2 LUB2_EFR2	1.3%	0.987	$51	$51	$51	$39.23
Intake Cam Phasing ICP	2.6%	0.974	$43	$41	$36	$16.54
Dual Cam Phasing DCP (vs. ICP)	2.5%	0.975	$35	$33	$31	$14.00
Discrete Variable Valve Lift DVVL	3.6%	0.964	$133	$125	$114	$36.94
Continuously Variable Valve Lift CVVL (vs. DVVL)	1.0%	0.990	$67	$63	$56	$67.00
Cylinder Deactivation - NA for I4 DEACD	0.0%	1.000				
Stoichiometric Gasoline Direct Injection SGDI (Required for TRBDS)	1.5%	0.985	$192	$181	$164	$128.00
Turbocharging & Downsizing - 1 TRBDS1 33% DS 18 bar BMEP	7.7%	0.923	$331	$312	$282	$42.99
Turbocharging & Downsizing - 2 TRBDS2 50% DS 24 bar BMEP	3.2%	0.968	-$96	-$92	-$86	-$30.00
Cooled EGR - 1 CEGR1 50% DS 24 bar BMEP	3.0%	0.970	$212	$199	$180	$70.67
Cooled EGR - 2 CEGR2 56% DS 27 bar BMEP	1.4%	0.986	$364	$343	$310	$260.00
SI Engine Only (incl LUB & EFR)	27.2%	0.728	$1,383	$1,307	$1,189	$50.89
Null Vehicle - 2016 MY	8.2%	0.918	$129	$125	$118	$15.83
SI Engine 2017 - 2025 MY	17.1%	0.829	$678	$640	$705	$39.64
SI Engine After 2025	4.4%	0.956	$576	$542	$366	$132.17

[a] Null vehicle: I4, DOHC, naturally aspirated, 4 valves/cylinder PFI fixed valve timing and 4 speed AT.

increase in average light-vehicle fuel economy between 1981 and 2009. The rise in fuel economy that began in 2005 was due to the increase in the standards for light trucks.

Performance and horsepower have marketing appeal, and this appeal may continue even as the CAFE standards are increased. In this environment, vehicle manufacturers will need to consider the trade-offs between continuing to increase performance and reducing fuel consumption. The magnitude of this trade-off was evaluated using available published information. In a recent study, the relationship between performance as measured by 0 to 60 mph acceleration time and power to weight ratio was determined from multiple data sources; the results are shown in Figure 2.10 (Berry 2010). From this graph, it can be seen that a 10 percent decrease in 0 to 60 mph time from a typical value of 8 seconds requires approximately a 10 percent increase in power/weight ratio, although this relationship is dependent on the initial value of the 0 to 60 mph time. This relationship was also derived from fundamental principles in Appendix L.

The effect of power-to-weight ratio on fuel consumption was determined from the empirical expression developed by NHTSA, which is shown earlier in this chapter in the section Effectiveness of Turbocharged, Downsized Engines. For a 3,500 lb vehicle, a 10 percent increase in power-to-weight

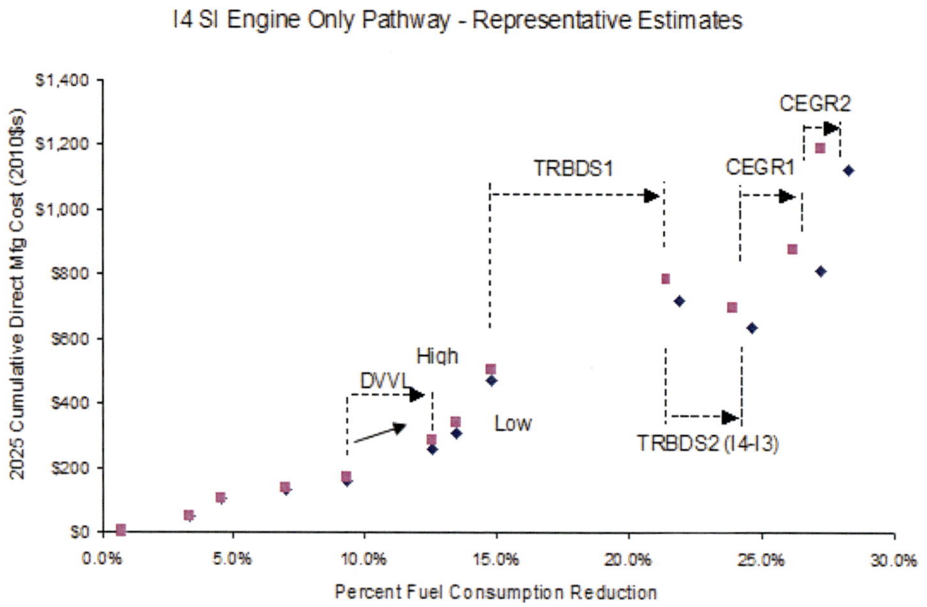

FIGURE 2.8 Cumulative 2025 direct manufacturing costs (2010 dollars) (with low estimates shown as diamonds and high estimates shown as squares) versus percent fuel consumption reduction for an example I4 SI engine pathway.

TABLE 2.31 Estimated Percent Fuel Consumption Reductions and Direct Manufacturing Costs for I4 SI Engine Technologies for Midsize Car in the Selected Time Frames

Time Frame	Fuel Consumption Reduction (%)	2025 MY Direct Manufacturing Cost (2010$)
Null vehicle to 2016 MY	8.2	109 - 118
2017 MY to 2025 MY	17.1 - 17.9	526 - 705

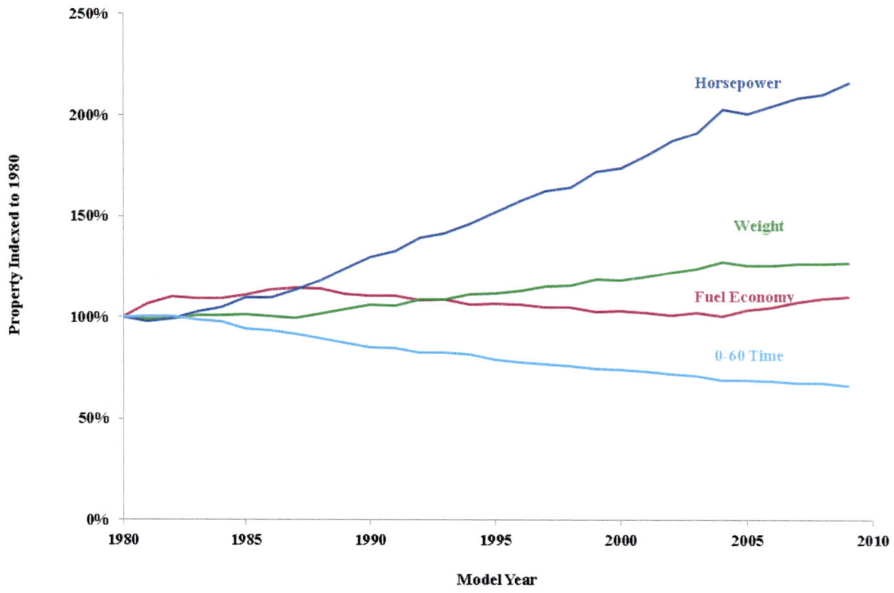

FIGURE 2.9 Changes in horsepower, 0 to 60 time, weight, and fuel economy, 1980 to 2009.
SOURCE: DOE (2010).

ratio was calculated to result in a 2.6 percent increase in fuel consumption. Consistent with this calculated result, Knittel (2011) estimated that reducing horsepower and torque by 1 percent increases fuel economy by roughly 0.3 percent, and a similar result was found by Michalek et al. (2004) in a study that used vehicle models to determine the effect of changes in engine power on fuel economy. Combining this result with the previous relationship indicates that a 10 percent decrease in 0 to 60 mph time will result in a 2.6 percent increase in fuel consumption. These results will vary depending on the initial value of 0 to 60 mph acceleration time and the weight of the vehicle, but the directional trend will remain.

In contrast to increasing performance, decreasing performance can provide significant reductions in fuel consumption. For a 3,500 lb vehicle, a 10 percent increase in 0 to 60 mph time from a typical average value of 8 seconds can result from approximately a 10 percent decrease in power/weight ratio. A 10 percent decrease in power-to-weight ratio was calculated from the NHTSA empirical expression to result in a 3.2 percent reduction in fuel consumption. In contrast to this method for reducing fuel consumption, a similar reduction in fuel consumption can be obtained by applying technologies that may have cost effectiveness values in the range of $25 to over $50 per percent reduction in fuel consumption.

The final CAFE rule states that the CAFE standards "should not . . . affect vehicles' performance attributes" and the "technology cost and effectiveness estimates . . . reflect this constraint" (EPA/NHTSA 2012a). Although constant performance attributes were assumed in the technology effectiveness estimates, several manufacturers told the committee that vehicle performance would continue to be increased, as it was over the time period shown in Figure 2.10, due to competitive pressures.

SI Technologies and Off-Cycle Fuel Economy

There exists a gap between the fuel economy experienced on-road and that evaluated in the mandated test cycles. Deviation of real-world fuel economy from EPA window sticker value, as well as from the CAFE compliance values, is expected to increase as some additional SI fuel economy technologies are applied to vehicles. When a vehicle is driven more aggressively, such as at higher speeds and higher acceleration rates than specified by the FTP75 and the Highway Fuel Economy Test (HWFET) drive cycles used for CAFE compliance, more fuel is consumed. If the vehicle has a conventional, naturally aspirated engine, the fuel consumption outside the CAFE drive cycles differ from on-cycle fuel consumption due to the gradual changes in BSFC values on the fuel consumption map of the engine and the increased power requirements at higher speeds or accelerations rates.

For a turbocharged, downsized engine, changes in the BSFC values on the fuel consumption map outside the CAFE drive cycles can be greater than with a naturally aspirated engine. For example, with a highly turbocharged and downsized engine, higher speeds may require enrichment to limit exhaust temperature to protect the turbocharger and catalyst. This enrichment would result in a greater increase in fuel consumption than would be experienced with a naturally aspired engine. A similar effect would occur during higher acceleration rates. This growing gap is not unique to SI technologies and may increase with use of technologies optimized for the test cycles. The possibly growing discrepancy between compliance and on-road fuel economy is discussed further in Chapter 10.

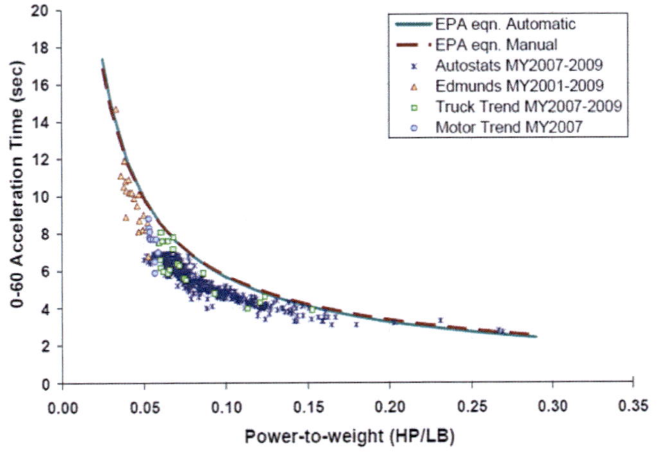

FIGURE 2.10 Performance as indicated by 0 to 60 mph acceleration time versus power-to-weight ratio.
SOURCE: Berry (2010). © 2010 Massachusetts Institute of Technology. Used with permission.

FUEL CONSUMPTION REDUCTION TECHNOLOGIES – NOT INCLUDED IN FINAL CAFE RULE ANALYSIS

This section discusses technologies reviewed but not included in the Agencies' quantitative analysis, such as ethanol flex-fuel vehicles, vehicles fueled with compressed natural gas, lean burn engines, and homogeneous charge, compression ignition (HCCI) engines. The fuel consumption reductions for technologies using alternative fuels are shown in gasoline gallons equivalent (gge) in Table 2A.1 (Annex tables at the end of the chapter). Shown in parentheses after the gasoline gallons equivalent are the actual reductions in CAFE fuel economy resulting from the application of the utility factor in the case of flexible or bi-fuel vehicles and the petroleum equivalency factor (PEF) where applicable.

Ethanol Flexible Fuel Vehicles

Flexible-fuel vehicles (FFVs) allow more than one fuel to be used in a single tank. In the United States, these are ethanol FFVs that, in the case of LDVs, can be fueled with any mix of gasoline and ethanol from 0 percent to 85 percent ethanol. The primary technologies are a corrosion-resistant fuel system, including the fuel injectors, fuel lines, and fuel pump together with the control system for sensing and maintaining stoichiometry in the engine with the prevailing fuel mixture in the tank. The incremental direct manufacturing cost for an ethanol FFV is estimated to be $100 for a large car with a V6 engine, while the cost is lower for an I4 engine and higher for a V8 engine, since the incremental cost is partially dependent on the number of corrosion-resistant injectors (Woodall 2010).

For several manufacturers, FFVs account for a large percentage of total production, in part because FFVs generate credits toward CAFE compliance. In MY 2012, 17 percent of sales overall were FFVs, with some manufacturers at much higher percentages: GM's sales were 44 percent FFVs, Chrysler's, 35 percent, and Ford's, 28 percent (EPA 2014c). An FFV achieves significantly lower miles per gallon operating on 85 percent ethanol (E85) than it does operating on gasoline, because the energy density (Btu/gal) of E85 is 27-37 percent lower than that of gasoline. On an energy-equivalent basis, however, an FFV's ethanol and gasoline fuel economies are very similar. E85 supply is limited in most of the U.S., and NREL data indicate that less than 1 percent of the fuel used by the nation's more than 11 million FFVs is ethanol (Graves 2014; Moriarty 2013).

Ethanol has an RON of 109, with an AKI of 99.5. Although this high octane rating has the potential to provide for an increase in fuel economy by increasing the compression ratio, such optimization for ethanol is not possible since the compression ratio of an FFV is limited by operation on gasoline, typically with an RON of 91.

For CAFE compliance purposes, the fuel economy of an FFV is measured on gasoline and on E85, the fuel economy on E85 is adjusted to reflect its petroleum content, and the gasoline and adjusted-E85 fuel economy values are weighed using a utility factor to reflect use of E85. As noted above, fuel economy on a per volume basis (mpg) is lower when operating on E85 than when operating on gasoline, typically about 68% of the gasoline mpg. The E85 fuel economy is adjusted by dividing by the Petroleum Equivalency Factor (PEF) of 0.15 to reflect that E85 is considered to consist of only 15 percent petroleum-derived fuel. An FFV is currently considered to use E85 for 50 percent of the time, resulting in a utilization factor of 0.5. This results in an overall certification fuel economy for an FFV calculated using the following equation:

$$FFV\ mpg = \frac{1}{\left(\dfrac{1-utility\ factor}{gasoline\ mpg}\right)+\left(\dfrac{utility\ factor}{E85\ mpg}\right)\times PEF}$$

For an FFV with measured fuel economy of 25 mpg on gasoline and 17 mpg on E85, for example, this results in a certification fuel economy of 41 mpg, or 1.64 times the fuel economy on gasoline. This 64 percent increase under CAFE standards in certification fuel economy for FFVs illustrates why many manufacturers currently make FFVs a large percentage of their total production fleet. The 64 percent increase is equivalent to a 40 percent reduction in fuel consumption. However, there is a cap (1.2 mpg for 2014 MY cars and 2014 MY trucks, separately) on the amount a manufacturer may increase its fleet average fuel economy under the CAFE program using this calculation of fuel economy for FFVs. The cap will be phased out beginning in MY 2016 and will reach zero by 2020.

For MY 2017 and 2018, manufacturers will continue to calculate the CAFE fuel economy using a 50/50 harmonic average of the fuel economy for the alternative fuel and the conventional fuel. The fuel economy for the alternative fuel continues to be increased by dividing it by 0.15, the PEF. After 2019, the CAFE fuel economy will weight the FFV fuel economy of the two values using the same real-world weighting factor that is used under the EPA program. In contrast to the CAFE program, EPA is implementing changes to the GHG program after 2015; these changes include establishment of the E85 weighting factor and elimination of the 0.15 multiplier for E85 GHG emissions. Recently, EPA finalized an E85 weighting factor of 0.14[10] for MY 2016-2018, which manufacturers may use for weighting CO_2 emissions for purposes of FFV certification (EPA 2013a; EPA 2014d). EPA will take action to establish weighting factors for MY 2019 and beyond after a full review of updated information.

[10] 0.14 is the weighting of CO_2 emissions for E85 fuel and 1 – 0.14 is the weighting of CO_2 emissions for conventional fuel. Through 2015 MY, the weighting factors are 0.5 for CO_2 emissions for E85 fuel multiplied by 0.15 and 1 – 0.5 for CO_2 emissions for conventional fuel.

Although this weighting factor is not applicable to the CAFE program in this time frame, a common factor will apply after the 2019 MY. Upon transitioning to the real-world weighting factors, the CAFE and GHG programs will no longer cap the amount by which FFVs can raise manufacturers' average fuel economy. The crediting of alternative fueled vehicles in the CAFE and GHG programs is discussed further in Chapter 10.

Compressed Natural Gas Vehicles

Natural gas has various properties that make it appealing as a vehicle fuel, including its abundance, relatively low cost at present in the United States, and high octane rating (120+ AKI compared to 87 AKI for regular gasoline) (AFDC n.d.). With regard to efficiency, however, commercialized compressed natural gas (CNG) vehicles have not demonstrated an advantage over gasoline vehicles. Table 2.32 compares key attributes of Honda's Civic Natural Gas vehicle with a similar gasoline model. For the past several years, the Civic CNG vehicle has been the only OEM-dedicated natural gas vehicle in the light-duty market.

Since natural gas has a substantially higher octane rating than gasoline, allowing a higher compression ratio, this capability is applied to the Civic Natural Gas vehicle to partially offset the lower peak power of the engine, which comes from the displacement of air for combustion in the cylinders by natural gas. Other consumer considerations are the natural gas vehicle's reduced range and trunk space, both due to the low energy density of CNG, and an incremental retail price of over $9,500 relative to the gasoline model. Annual sales have been less than 1,500 units, mostly to corporate and government fleets. The $9,500 incremental price of the Honda Civic CNG vehicle was used as the basis for estimating an incremental direct manufacturing cost of $6,000 for a CNG vehicle by using an ICM of 1.5. Since this cost is based on a very low production volume, reductions in this cost are expected if higher volumes were to develop.

The final CAFE rule provides an incentive for producing CNG vehicles. The CAFE fuel economy of a CNG vehicle is determined by dividing its fuel economy in equivalent miles per gallon of gasoline by the PEF of 0.15 (EPA/NHTSA 2012a). Therefore, a CNG vehicle with a combined CAFE fuel economy of 30 mpgge would have a 200 mpg CAFE fuel economy rating.

Although CNG has the potential to reduce operating costs and CO_2 emissions, consumer acceptance issues, including lack of refueling stations, slow and noisy refueling process, fewer model choices, and perceived danger of CNG, must be successfully addressed. Several European manufacturers are currently marketing CNG vehicles in Europe and Asia, which could be brought to the United States if adequate infrastructure existed. Some interest was expressed by the natural gas industry and some automakers in extending the CNG multipliers beyond 2021.

Bi-fuel CNG/Gasoline Vehicles

Bi-fuel CNG/gasoline vehicles are equipped with a natural gas tank and a gasoline tank and can switch operation from one fuel to the other. Bi-fuel vehicles may attract more customer interest than dedicated CNG vehicles. Low natural gas prices generated considerable interest in natural gas for heavy-duty applications that has extended to bi-fuel versions of the 2013 Chevrolet Silverado, GMC Sierra, Ford Super Duty, and Ram heavy-duty pickups trucks and the Ford E Series van (autonet.ca 2013).

Ford recently announced that the 2014 MY F-150 light-duty pickup with the 3.7L V6 engine can be ordered with a "prep" option from the factory for natural gas, which can then be sent to an outfitter for conversion to CNG operation. The "prep" option includes hardened valves, valve seats, and pistons and piston rings and has a retail cost of $315 (Ford News Center 2013). The F-150 will be able to operate on either natural gas or gasoline through separate fuel systems. The CNG conversion, which includes fuel tanks, fuel lines, and unique fuel injectors, will cost $6,000 to $9,500 depend-

TABLE 2.32 Comparison of 2012 MY Gasoline and Natural Gas Honda Civic

	Gasoline	Natural Gas
Displacement	1.8L	1.8L
Compression Ratio	10.6:1	12.7:1
Power	140 hp	110 hp
EPA Fuel Economy - MPG (City/Hwy/Combined)	28 / 39 / 32	27 / 38 / 31[a]
Fuel Capacity	13.2 gal	8.03 GGE[a] (3600 psi CNG tank)
Range (Using Combined MPG)	422 miles	249 miles
Cargo Volume	12.5 cu. ft.	6.1 cu. ft.
Retail Price	$17,545	$27,095

[a] GGE (Gasoline Gallons Equivalent).
SOURCE: Honda.com and Edmunds.com (2013).

ing on fuel tank capacity. General Motors announced that the 2015 MY Chevrolet Impala will include a powertrain that switches from compressed natural gas to gasoline, with a total driving range of up to 500 miles (Krasny 2013). Costs were not available at the time of the announcement.

These bi-fuel LDVs can switch operation from one fuel to the other, which minimizes the range anxiety resulting from the dearth of CNG refueling stations. This flexibility comes at the cost of having a sub-optimal engine design and combustion as well as two separate tanks and fuel lines for both the gasoline and the natural gas.

In the CAFE program for MYs 2017–2019, the fuel economy of dual fuel vehicles will be determined in the same manner as specified in the MY 2012–2016 rule. Beginning in MY 2020, in order to use the utility factor based on estimated usage of CNG, dual fuel CNG vehicles must have a minimum CNG range-to-gasoline range ratio of 2.0, and gasoline can only be used when the CNG tank is empty. Any dual fuel CNG vehicle that does not meet this requirement would use a utility factor of 0.50, the value that has been used in the past for dual fuel vehicles under the CAFE program. For a dual fuel CNG vehicle with a fuel economy of 25 mpg using gasoline and 95 percent of the gasoline fuel economy on a mpgge basis using natural gas, the overall CAFE fuel economy would be 43.2 mpg ($1/\text{mpg} = 0.5/25 + 0.5/(0.95 \times 25)/0.15$), or 1.73 times the fuel economy on gasoline, which is equivalent to a 42 percent reduction in fuel consumption. EPA provides multipliers for both dedicated and dual fuel CNG vehicles for MYs 2017-2021 that are equivalent to the multipliers for PHEVs.

Lean Burn Gasoline Direct Injection

Lean burn, spray-guided fuel injection systems operating at higher injection pressures than conventional direct injection engines have a potential for a 5 to 15 percent reduction in fuel consumption, resulting in a thermal efficiency approaching that of a diesel engine (NHTSA 2009). NHTSA has stated that when combined with advanced NO_x aftertreatment systems, lean-burn GDI engines may be a possibility in North America.

Tier 3 ultra low sulfur gasoline (less than 15 ppm S) may stimulate renewed interest in lean-burn GDI engines. NO_x aftertreatment systems for lean burn engines consist of lean NO_x traps (LNTs), which preferentially store sulfate compounds from the fuel, thereby reducing NO_x storage capacity over time. As a consequence, the system must undergo periodic desulfurization by operating at a net-fuel-rich condition at high temperatures in order to retain NO_x trapping efficiency (EPA/NHTSA 2012b). Previous experience in Europe with production lean burn engines indicated that they did not provide the expected reductions in fuel consumption, partially due to frequent desulfurization of the lean NO_x trap that required periodic rich operation of the engine. The reduction in sulfur content may permit the use of a lean NO_x trap (LNT) without requiring frequent desulfurization, though at high cost (MARTEC 2010).

EPA Tier 3 motor vehicle emission and fuel standards (EPA 2014b) will reduce gasoline sulfur content to 10 ppm on an annual average basis, starting on January 1, 2017. This level is similar to levels already being achieved in California, Europe and Japan. EPA will continue to cap sulfur levels at 80 ppm and 95 ppm at the refinery gate and at the pump, respectively. Concerns remain with this requirement for the elevated sulfur levels at the retail pump, with the cap remaining at 95 ppm. Whether an average fuel sulfur requirement, rather than a sulfur cap on all fuel, would be sufficient to accommodate a technology such as lean burn in the real world remains a subject of discussion. Several vehicle manufacturers provided comments to EPA that lean burn engines may require 20 ppm and 25 ppm limits at the refinery gate and downstream, respectively, which are significantly lower than the limits specified by EPA.

Only a few vehicle manufacturers mentioned consideration of lean burn engines in their future CAFE compliance plans. Mercedes claims that lean burn can provide a 7 percent reduction in fuel consumption. In 2014, Mercedes had lean burn in seven of their vehicles in Europe, although uncertainty remains regarding whether the Tier 3 fuel requirements for fuel sulfur content are low enough for lean burn technologies to be introduced into the United States. The committee estimated an $800 direct manufacturing cost for a lean burn system for an I4 engine by accounting for a lean NO_x trap, direct injection, and an ignition system upgrade for the dilute combustion.

Electric Assist Turbocharging

An electric motor can be added to assist a turbocharger at low engine speeds to mitigate unwanted performance characteristics such as turbocharger lag and low boost pressure. Connecting a motor to the turbocharger shaft can provide the extra boost needed to overcome the torque deficit at low engine speeds (Uchida 2006). Both electric turbochargers and superchargers are being developed. Honeywell, Valeo, and BorgWarner have been reported to be developing electrically assisted turbochargers.

BMW has been reported to be working on a hybrid turbocharger, in which the compressor and turbine can be coupled to an electric motor-generator with clutches. The turbine and compressor rotate on different shafts, and clutches can be used to couple them to the motor-generator. Between the turbine and compressor is an electric motor. In full throttle acceleration, the compressor is driven by the motor. In this process, the time that would have been required for exhaust gases to spin up a traditional turbine to its operating speed is almost eliminated. When the turbine has reached its operating speed, a clutch couples it to the motor-generator. At this condition, the electric motor-generator functions in the generator mode. The resultant current flows to the battery,

TABLE 2.33 Fuel Consumption Reduction Test Results from Eaton EAVS Supercharger System and Comparison to NHTSA Estimates

Test Results for Vehicle with Eaton EAVS Supercharger System	FTP-75 (mpg)	HWY (mpg)	Combined (mpg)	Fuel consumption (gal/100 mi)
2.8L Naturally Aspirated Engine	22.00	35.60	26.57	3.764
1.4L with Eaton EAVS Supercharger	30.44	52.04	37.43	2.672
Fuel Consumption Reduction =				29%

Comparison of Fuel Consumption Reductions Using NHTSA Estimates for Each Function/Technology

Functions	Eaton EAVS-SC (% FC reduction)	Turbocharged Downsized Engine with Added Technologies (% FC Reduction)
50% downsizing	20.1	20.1
Stop-start	2.1	2.1
Mild Hybrid (Reduced effectiveness for EAVS-SC)	2.2	6.55
Improved Accessories 1	1.22	1.22
Improved Accessories 2	2.36	2.36
Multiplicative Total =	26%	29%

and the surplus load from the generator is used to control the turbine's speed (Spinelli 2011).

A variation of the electrically assisted turbocharger is BorgWarner Turbo Systems' eBooster. This system employs an electric motor to drive a compressor and may be positioned either ahead of or behind the turbocharger. Unlike conventional electrically assisted turbochargers, the eBooster concept has two stages, similar to a series of two turbo-machines. As a result, the two units' pressure ratios are multiplied. This system is currently under development in close cooperation with various customers (BorgWarner n.d.). Audi has been reported to be working on a similar system called an "electric biturbo." This system employs an electric compressor to rapidly provide a low-rpm performance improvement. It is combined with a traditional turbocharger, which is used to achieve greater power at the top end. Volvo recently announced a triple-boost engine with two parallel turbochargers linked to an electrically powered compressor to eliminate the lag in boost pressure at the lower engine speeds (Birch 2014).

Another variation is Eaton's Electrically Assisted Variable Speed (EAVS) supercharger, which includes the following additional features: stop-start, mild hybrid, and improved accessories (Tsourapas et al. 2014). The system combines a variable speed supercharger with engine stop-start functions together with regenerative (mild hybrid) capabilities using a planetary gear set to couple the engine, supercharger and motor. The supercharger is combined with a small electric motor having approximately one-third the power of traditional mild hybrids and a battery to provide engine boost at any speed without lag. Eaton reported that a 50 percent downsized engine with the EAVS supercharger system provided a 29 percent reduction in fuel consumption, as shown in Table 2.33. Also shown in Table 2.33 is a comparison of the fuel consumption reduction that could be achieved by combining the features in the Eaton vehicle by using NHTSA's estimates for each technology. This analysis yielded a 26 percent reduction in fuel consumption, which was in the range of Eaton's measured data from its experimental test vehicle. In September 2014, DOE awarded a $1.75M cost-sharing project to Eaton to demonstrate an electrically assisted supercharger operating with the energy from a waste heat recovery system.

The committee's estimated incremental direct manufacturing costs for the EAVS supercharger system when applied to a midsize car with an I4 engine was approximately $1,300 and when applied to a large car with a V6 engine was approximately $1,000, as shown in Table 2.34. These estimated costs are within the range estimated by using NHTSA's costs for a turbocharged, downsized engine together with the costs for the additional functions provided by the EAVS supercharger system listed in Table 2.33. Applying the EAVS supercharger system to a V6 engine results in a lower cost than applying it to an I4 engine because of the larger savings in downsizing the V6 engine to an I4 engine as compared to downsizing an I4 engine to an I3 engine.

HCCI for Gasoline Fueled Engines

Concept of Operation and Expected Benefits

In homogeneous charge, compression ignition (HCCI) engines, also known as low-temperature combustion engines, a premixed charge of fuel and air is compressed until

TABLE 2.34 Estimated Direct Manufacturing Cost for the Eaton EAVS Supercharger System

	EAVS Supercharger with 50% SI Engine Downsizing				
	EAVS SC - I4 to I3 Midsize Car		EAVS SC - V6 to I4 Large Car		
	Components	2020 MY NRC Estimate	Components	2020 MY NRC Estimate	Comments
		I4 to I3		V6 to I4	
IC Engine Size		2.4L to 1.2L		2.8L to 1.4L	
	EAVS SC	$1,050	EAVS SC	$1,050	Expert estimate
	Battery	$505	Battery	$505	NHTSA RIA p. 330
System Cost to OEM	Total =	$1,555	Total =	$1,555	
	Downsizing I4 - I3	-$161	Downsizing V6 - I4	-$465	TSD Table 3-32
	Alternator[a]	-$52	Alternator[a]	-$52	
	12 V Battery[a]	-$15	12 V Battery[a]	-$15	
	Starter motor[a]	-$26	Starter motor[a]	-$26	
Cost Reductions	Total =	-$254	Total =	-$558	
Vehicle Net Impact		$1,302		$998	
	Service parts[a]	Retail Price ($)	Estimated Direct Mfg Cost ($)		
	Alternator	348	52		
	Battery	100	15		
	Starter motor	170	26		

[a] DMC~15% of retail price.

it auto-ignites, with heat release occurring throughout the cylinder volume rather than in a flame front (Najt and Foster 1983; Thring 1989). This combustion concept is particularly challenging at high loads since the high levels of dilution necessary to limit the pressure rise rate to ensure acceptable NVH levels may not be achievable. At lighter loads, the dilution, which increases thermodynamic efficiency and lowers pumping losses and peak combustion temperatures, results in reduced fuel consumption, NO_x and particulate emissions. As a result, the HCCI engine created high expectations for overcoming some of the disadvantages of the SI engine and the compression ignition (CI) engine and has, for many years, attracted significant interest and research investment (Epping et al. 2002; Kulzer et al. 2006).

The factors that contribute to the improved efficiency of an HCCI engine include the following:

- Higher compression ratio than conventional SI engines;
- Lean or dilute operation with air and/or residuals providing a higher specific heat ratio for improved thermodynamic efficiency;
- Low temperatures and rapid heat release that reduce heat losses; and
- Low pumping losses due to dilute operation relative to a throttled SI engine.

Progress to Date

Many challenges in applying the HCCI concept to light-duty vehicles continue to be addressed in a range of research programs. Control of auto-ignition phasing has been achieved with a variety of methods that affect the thermodynamic state and the chemical composition of the charge. VVT to control HCCI combustion phasing with high amounts of residuals, sometimes called controlled auto ignition (CAI), appears to be the most promising method. CAI relies on increasing cylinder residuals with exhaust valve re-breathing or a negative valve overlap (NVO) time period during which the intake and exhaust valves are closed for the first part of each piston's intake stroke, creating a high vacuum in the associated cylinder. As the intake valve opens, the high pressure differential elevates turbulent mixing intensity, which contributes to a leaner air/fuel ratio and reduced exhaust-gas emissions (Willand et al. 1998; Filipe and Stein 2002).

To achieve unthrottled operation with high residuals, HCCI engines generally rely on low valve lifts at low loads. Consequently, a VVL mechanism with at least two lift settings is expected to be necessary for high load operation and transition to SI operation. Internal dilution can be controlled with the NVO strategy, since it offers fast cycle-to-cycle control of the initial cylinder conditions necessary to achieve autoignition. Successful HCCI operation relies on combustion feedback, so in-cylinder pressure sensing in every cylinder is expected to be necessary. The speed range of HCCI is also limited due to the chemically driven combustion process, which includes a delay prior to initiation. At high loads the heat release becomes increasingly violent and eventually could lead to damaging knock or even thermal run away (Chiang et al. 2007a; Olsson et al. 2002; Thring 1989). At high loads, the combustion phasing must be retarded to avoid knocking. Late phasing, however, involves incomplete combustion, where unburned fuel from one cycle adds to the injected fuel of the next cycle, causing significant cycle-to-cycle variability (Thring 1989; Koopmans 2001; Hellstrom 2012). Although start of injection (SOI) control based on combustion feedback was shown to reduce this cyclic variability, the overall fuel consumption and NO_x emissions began to deteriorate (Gerdes 2012; Hellstrom 2012).

As a result of these issues, a gasoline engine must return to the SI mode at high loads (Kulzer et al. 2007). In 2009 General Motors demonstrated a vehicle with HCCI operation from idle to speeds up to 60 mph by using multiple injections and multiple ignitions (Green Car Congress 2009b; Yun et al. 2009). Combustion-induced noise was identified as an issue that needed to be resolved. No drive cycle emissions or fuel consumption results have been reported from this project.

Current Projects

Current efforts focus on (1) managing the HCCI-SI combustion mode switches for covering the entire speed-load range and (2) extending the gasoline HCCI range to avoid the need for mode switches. Due to the limited load range for successful HCCI operation, recently shown to be below 3 bar BMEP, mode switching from HCCI combustion to conventional SI combustion and back is required for covering the full speed and load range of an engine (Nuesch 2014).

The mode switch control problem is difficult due to the different nature of each combustion mode (Koopmans 2001). Switching into the HCCI mode is particularly demanding, with only indirect control of the combustion. The modes operate under significantly different conditions, with HCCI often operating unthrottled and lean while the SI mode is throttled and stoichiometric (Cairns and Blaxill 2005a, 2005b; Nier et al. 2012). Because of the dynamics in the air path, the transitions require several engine cycles, during which neither mode operates under normal conditions, and they incur fuel penalties. The combustion is generally less efficient during the transitions than in either stabilized mode.

Another issue for HCCI engines is the coordination of the engine modes with the exhaust aftertreatment system. Although reduction catalysts may not be necessary with the low NO_x emissions in the lean HCCI mode, the three-way catalyst (TWC) needs to be fully warmed up and ready to convert engine-out emissions when the engine switches to the SI mode. HCCI has low exhaust temperatures and might not maintain the catalyst light-off temperature after prolonged operation (Nier and Karrelmeyer 2011). Moreover, during an HCCI to SI mode switch, the engine might need to run rich to deplete the TWC oxygen storage to expedite the NO_x conversion in the SI mode. During these rich periods the fuel penalty of mode switching increases and can be detrimental to the overall fuel economy. The ACCESS project, discussed later in this section, found that maintaining tailpipe emissions levels equivalent to those of a super ultra-low emission vehicle during mode switching eliminated the fuel consumption benefits of HCCI.

Mode switching could be avoided if the load range for HCCI combustion mode could be extended to the full operating range of the engine. Investigations of the combination of ignition, multiple injections, positive valve overlap (PVO), and boosting or supercharging are under way for extending HCCI operation, and these investigations are discussed in Appendix M. Some of the projects that are investigating HCCI-SI mode switching and extending the gasoline HCCI load range are summarized below, with additional details provided in Appendix M.

Advanced Combustion Control Enabling Systems and Solutions (ACCESS) (AVL, Bosch, Emitech, Stanford, University of Michigan)

The Advanced Combustion, Controls, Enabling Systems and Solutions (ACCESS) project, partially funded by a DOE grant, is focused on coordinating multi-mode combustion events over the engine drive cycle operating conditions. The project goal is to improve fuel economy by 25 percent by implementing part-load HCCI operation in a turbocharged, downsized engine meeting the California tailpipe emission standards for super ultra-low emission vehicles (SULEVs). A turbocharged, downsized 2.0L I4 engine is being used to replace the naturally aspirated 3.6L V6 engine. A 5 percent fuel economy improvement relative to a turbocharged, downsized engine was projected for the HCCI combustion mode on the FTP75 cycle, based on multicylinder engine data. However, this potential fuel consumption benefit of lean HCCI was eliminated under the SULEV emission constraints due to the need to switch to a fuel-rich mode after lean operation to deplete the oxygen storage and restore the three-way catalyst (TWC) NO_x conversion efficiency. Stoichiometric HCCI and spark-assisted HCCI (SACI) are now the emphasis of this project, with a vehicle demonstration planned at the end of the DOE contract in December 2014.

The committee estimated that HCCI, applied to an I4 engine, would have an incremental direct manufacturing cost of $450. This estimate includes the costs for a cooled EGR system, four combustion pressure sensors, and an electronic control system with appropriate signal processing, and input/output capabilities, together with the necessary wiring and connectors.

Homogeneous Charge Compression Ignition (HCCI) (ORNL and Delphi HCCI)

Oak Ridge National Laboratory (ORNL) and Delphi are investigating ways to expand the load where gasoline HCCI can be achieved (Szybist et al. 2012, 2013) on a single-cylinder version of a 2.0L four-cylinder engine with the compression ratio increased to 11.85:1. Recent results have shown that the load for HCCI operation in a naturally aspirated engine can be increased from 3.5 bar IMEP to 6.5 bar IMEP, with high levels of boost pressure up to 1.9 bar to provide additional air dilution as well as with the use of external EGR. Under boosted conditions, NO_x emissions remained low (<0.01 g/kWh).[11]

Gasoline Direct Injection Compression Ignition (GDICI) (University of Wisconsin and General Motors)

The University of Wisconsin and General Motors have investigated the use of 87 AKI regular-grade gasoline in a high-speed, direct-injection, light-duty compression ignition engine to extend the low-temperature combustion (LTC) regime to high loads (Ra et al. 2011). This system is called gasoline direct injection compression ignition (GDICI). The investigation found that GDICI operation of a light-duty engine was feasible under full load conditions of 16 bar indicated mean effective pressure (IMEP), thereby significantly extending the low-emission combustion concept (Ra et al. 2012). The engine had a compression ratio of approximately 16.5:1 and operated with multiple injections, specifically double- and triple-pulse injections. Both particulate matter (PM) and NO_x emissions were reduced to levels of about 0.1 g/kg-f while achieving an indicated specific fuel consumption (ISFC) as low as 173 g/kW-hr with a triple pulse fuel injection strategy (Ra et al. 2012).

Gasoline Direct Injection Compression Ignition (GDCI) (Delphi, Hyundai, Wayne State University, University of Wisconsin, Wisconsin Engine Research Center)

Similar to the University of Wisconsin and General Motors GDICI engine, Delphi and Hyundai are developing, under a DOE contract, a Gasoline Direct Injection Compression Ignition (GDCI) engine with the goal of achieving full-time, low-temperature combustion using multiple late injections over the entire engine speed-load map from idle to full load (Sellnau et al. 2012). Complete mixing of all the fuel in a homogeneous charge is averted with late injections, as this would cause rapid burning of the whole mixture. Regular unleaded gasoline (90.6 RON) and unleaded gasoline with 10 percent ethanol (91.7 RON) fuels are being used in this engine. Compression ratios between 14:1 and 16.2:1 were evaluated. Low-temperature combustion was demonstrated from 2 to 18 bar IMEP by maintaining 40 percent EGR at high loads and using inlet air heating at low loads (Confer et al. 2012). A minimum ISFC of 181 g/kW-hr was obtained with NO_x emissions less than 0.2 g/kW-hr, although this NO_x level would exceed the 2025 Tier 3 standards for a midsized car (see footnote 11). A multi-cylinder engine with the GDCI combustion system has been built, and testing was under way as of May 2013 (Confer et al. 2013). Demonstration of this engine in a vehicle, including cold starting and transient operation, was scheduled to be completed by the end of the DOE contract in 2014. In September 2014, DOE awarded a $10 million cost-sharing project to Delphi to accelerate the development of the GDCI low-temperature combustion technology. The committee estimated that the GDCI engine could have an incremental direct manufacturing cost approximately equal to that of an advanced diesel engine, which is estimated in Chapter 3 at $2,572 (2010 dollars) in 2025 for a midsize car relative to a baseline engine.

HCCI Engines Using Other Fuels

The previous section focused on the application of HCCI combustion to gasoline-fueled engines. However, research is under way to apply HCCI combustion across the spectrum of fuels and fuel injection processes. This spectrum of HCCI combustion research is depicted graphically in Figure 2.11. In addition to gasoline at one end of the spectrum and diesel fuel at the other end, multiple fuels in combination with port-injection and direct-injection scenarios with a mixture of low- and high-reactivity fuels are also being investigated within this spectrum. Alternative and low-carbon fuels are also considered within this spectrum. The combustion technologies for improving the classical diesel engine shown on the right-hand side of Figure 2.11 are discussed in Chapter 3.

The committee has made the following assessments of HCCI for gasoline-fueled engines:

- HCCI has been projected to provide up to 5 percent reduction in fuel consumption relative to a turbocharged, downsized engine.
- Maintaining SULEV (equivalent to Tier 3 or Tier 2 Bin 2) emissions during mode switching between lean HCCI and stoichiometric SI combustion modes can eliminate the HCCI fuel consumption reduction benefit if a TWC is used.
- Lean HCCI with its currently limited operating range requiring mode switching to SI combustion with TWC aftertreatment alone is not a suitable low-emission,

[11] NO_x emissions reported in g/kg-f and g/kW-h can be approximately related to each other and to the standard in g/mi as follows (using on-road energy from Table 3-62, TSD, for a 3500 lb midsize car):

(NO_x g/kg-f) × (ISFC g-f/kW-hr)/(1,000 g/kg) = NO_x g/kW-hr

(0.1 g/kg-f) × (180 g-f/kW-hr)/1,000 =0.018 g/kW-hr

(NO_x g/kW-hr) × (.277 kW-hr/mi) = NO_x g/mi

(0.018 g/kW-hr × (.277 kW-hr/mi) = 0.005 g/mi

Reference: Tier 3 Standard: 0.030 g/mi NMOG + NO_x (EPA 2014b) (approximately 0.020 g/mi for NO_x).

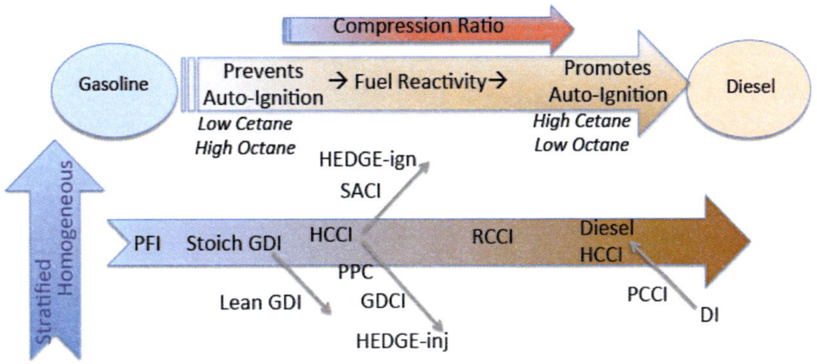

FIGURE 2.11 Advanced combustion concept spanning the range from gasoline SI to diesel CI engines. NOTE: SI, spark ignition; GCI, gasoline compression ignition; RCCI, reactivity controlled compression ignition (dual-fuel); and CDC, conventional diesel combustion. SOURCE: Daw (2013). Oak Ridge National Laboratory.

high-efficiency concept. Lean exhaust aftertreatment may be required to realize efficiency improvements.

- Further research and development is needed in advanced turbocharging and multiple-lift valvetrains for maintaining high dilution and low pumping losses to support widening the LTC combustion range.
- The benefits of HCCI relative to conventional engines might not be substantial as new technologies, including EGR dilution and unthrottled operation facilitated by VVL, are added to conventional combustion engines.

The committee forecasts that, although HCCI is not likely to have an impact on CAFE by the 2025 MY since the technology is still in the laboratory, it might have a role by the 2030 timeframe if the potential fuel consumption benefits can be demonstrated at Tier 3 emission standards.

FUEL CONSUMPTION REDUCTION TECHNOLOGIES – NOT CONSIDERED IN FINAL CAFE RULE ANALYSIS

In contrast to the previous section, which discusses technologies reviewed but not included in the Agencies' quantitative analysis, this section discusses fuel economy technologies not considered at all in the Agencies' analysis, including high compression ratio engines, ethanol-boosted engines, dedicated EGR technologies, variable compression ratio and displacement engines, camless valvetrains, and waste heat recovery.

High Compression Ratio with High Octane Gasoline

Fuel Consumption Reduction Potential

Increasing the octane rating of gasoline raises resistance to knock and therefore allows higher compression ratios in the engine. Higher compression ratios, in turn, can lead to higher brake thermal efficiency, although the incremental improvement in efficiency diminishes as the compression ratio increases. The effects of compression ratio on brake thermal efficiency, together with ITE and mechanical efficiency, are shown in Figures 2.12a for full load conditions and Figure 2.12b for part load conditions. The derivation of these figures is described in Appendix N. These figures illustrate the following effects of compression ratio on brake thermal efficiency:

- At full load, brake thermal efficiency increases, but at a decreasing rate, with increasing compression ratio, similar to ITE.
- At part load, up to 3 percent reduction in fuel consumption for naturally aspirated engines might be realized if compression ratio is increased from today's typical level of 10:1 to approximately 12:1, which is approximately a 1.5 percent reduction in fuel consumption per 1.0 compression ratio increase. Possibly greater reductions in fuel consumption might be realized for turbocharged engines capable of operating at higher boost pressures without knock so that further downsizing could be realized. Increasing gasoline octane from 91 RON of regular grade gasoline to 95 RON has been estimated to facilitate operation at a 12:1 compression ratio.
- At part load, nearly insignificant improvements in brake thermal efficiency on the CAFE test cycles are expected to be obtained by increasing compression ratio beyond approximately 12:1 due to the increasingly lower mechanical efficiency.

Current production vehicles use a range of compression ratios. The EPA Fuel Economy Guide (EPA 2012b) identifies vehicles that specify the use of premium gasoline. The Wards-MAHLE Engine Specification Chart was used to provide the compression ratio for the vehicles listed by EPA. Analyzing this information for 2012 MY intermediate

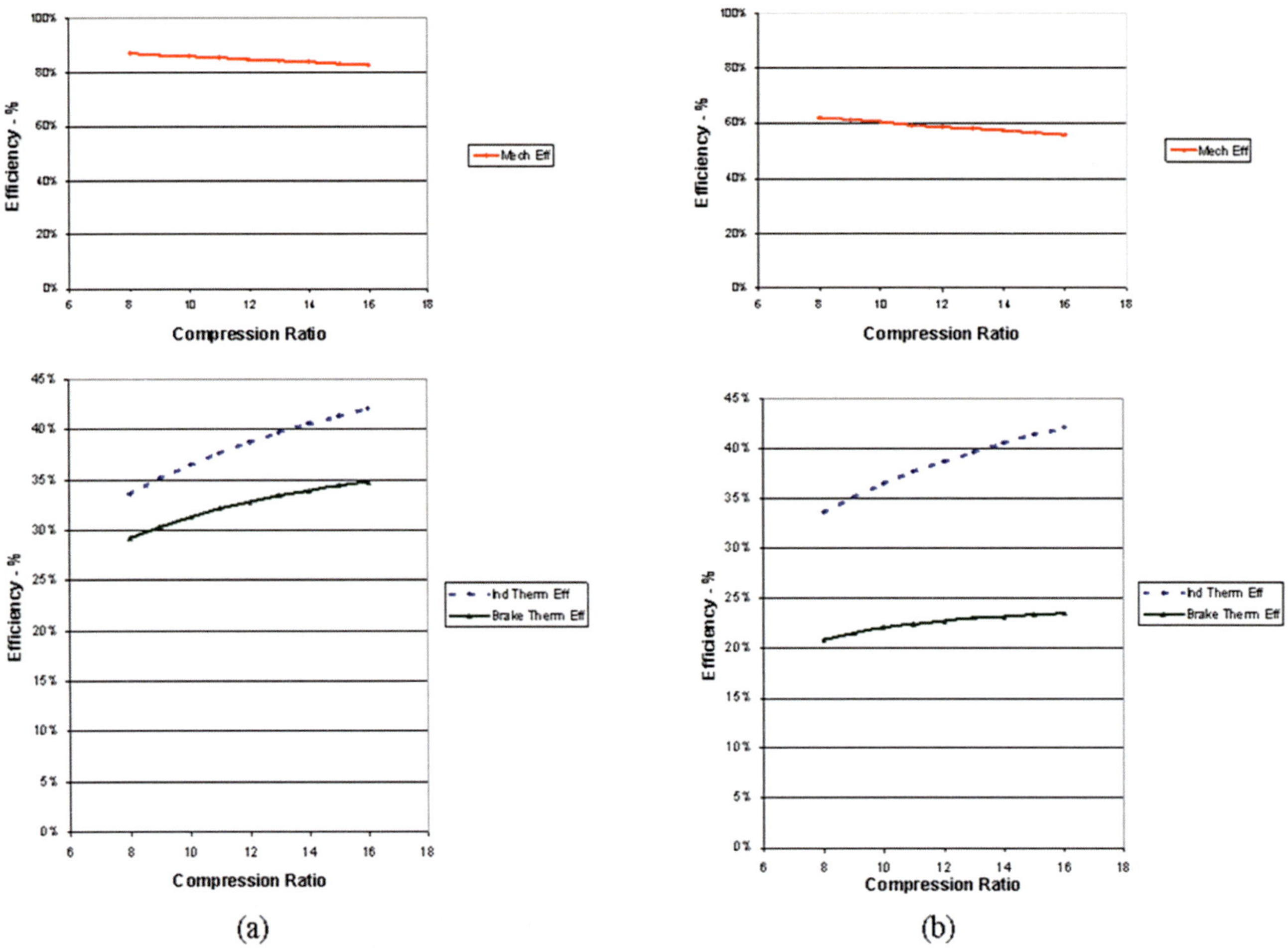

FIGURE 2.12 Effects of compression ratio on brake thermal efficiency, indicated thermal efficiency, and mechanical efficiency for (a) full load and (b) part load conditions representative of CAFE test cycles.
SOURCE: Developed from data in Heywood (1988).

and large cars (excluding turbocharged and direct injected engines) provided the following results:

Gasoline AKI Rating	Compression Ratio (Avg +/- Std Dev)
Regular gasoline (87 AKI)	10.3 +/- 0.2
Premium gasoline (91 AKI)	10.7 +/- 0.9

These results indicate that specifying premium gasoline facilitated an average 0.4 compression ratio increase, but the reasons why the full potential of a 1.0 to 2.0 compression ratio increase was not realized with the 4 AKI increase with premium gasoline in these comparisons are not known (Chow 2013).

Significant increases in compression ratio to approximately 11.5:1 have appeared in recent production vehicles while regular grade AKI 87 gasoline is still specified. This increase of approximately two compression ratios is projected to provide approximately a 3 percent reduction in fuel consumption. The control of knock at full load may be problematic; spark retard and cooling the mixture through enrichment are the usual methods for controlling knock at these conditions. These methods, however, will result in greater deterioration of customer fuel economy at high load conditions beyond the EPA test cycle. The committee estimated an incremental direct manufacturing cost for an increase in compression ratio would be approximately $50 for strengthened pistons and reduced tolerances to maintain the higher nominal compression ratio.

Fuel Octane Issues

Two octane numbers are associated with a fuel: a research octane number (RON), which is determined with a Cooperative Fuels Research (CFR) test engine running at a low speed of 600 rpm to represent engines at part throttle conditions, and a motor octane number (MON), which is determined with the CFR test engine running at a higher speed of 900 rpm to represent severe high-load and high-throttle conditions. The octane rating for gasoline posted on pumps in the United States is the "anti-knock index," which is the average of the RON and MON.

Gasolines commonly available in the United States have the following ratings:

Grade	MON	RON	AKI
Regular	83	91	87
Premium	87	95	91

The heads of powertrain engineering for Chrysler, Ford and GM recently expressed support for an increase in octane for regular fuel from 91 RON to 95 RON, citing fuel efficiency benefits of 2-5 percent (Winter 2014). A recent MIT analytical study found that adopting 98 RON fuel as the new standard grade fuel could provide a 3 to 4.5 percent reduction in fuel consumption for naturally aspirated engines and a 3 to 7.3 percent reduction in fuel consumption for turbocharged engines (Chow et al. 2014). An alternative to refining high-octane gasoline is to increase octane through blending with higher octane components such as ethanol, which has an octane rating of about 113 (RFA 2013). To increase the use of biofuels in the United States, the EPA has issued waivers that allow gasoline to be sold with ethanol levels up to 15 percent by volume (increased from the previous limit of 10 percent) for MY 2001 and later vehicles (EPA 2011). A 15 percent ethanol blend could increase octane rating by 2-4 RON relative to current E10 regular grade gasoline (API 2010). With a higher minimum octane level, fuel consumption could be reduced by up to 5 percent, and the incremental direct manufacturing cost would be approximately $75 for an I4 engine resulting from strengthened pistons and reduced tolerances to maintain the nominal compression ratio.

Gasoline used in the United States today contains 10 percent ethanol on average (EIA 2013). Despite ethanol's high octane content, the E10 currently sold in the U.S. has a 91 RON (AKI 87), which is the same as regular unleaded gasoline (E0). This has resulted from the petroleum industry's reduction of the octane level in the gasoline blend stock used for E10. A vehicle manufacturer has suggested that, if the octane of the current gasoline blend stock were to be retained at current levels by refiners, the increased ethanol content may provide the necessary increase in octane level to 95 RON to facilitate higher compression ratio engines. Regular grade gasoline with a higher minimum octane level would need to be widely available before manufacturers might broadly offer engines with significantly increased compression ratios. EPA's Tier 3 program, which changes the certification test fuel to E10 with octane representative of today's level of 91 RON (87 AKI), does not contemplate the above scenario.

If a manufacturer were to design vehicles requiring higher octane fuel, such as E30 (30 percent ethanol by volume blend with gasoline), EPA's Tier 3 program would allow manufacturers to petition the EPA Administrator for approval of the use of a higher octane fuel (EPA 2014d). If a petition is pursued, the manufacturer must demonstrate that the operator would use such a fuel. EPA stated that "this could help manufacturers that wish to raise compression ratio to improve vehicle efficiency, as a step toward complying with the 2017 and later light-duty CAFE standards" (EPA 2014d). Due to the relatively low energy content of ethanol, a vehicle tested on E30 would experience a loss in volumetric fuel economy. For CAFE compliance, EPA protocols call for an adjustment to the certification fuel economy based on the certification fuel energy content (see Tier 3 discussion in the section "Future Emission Standards for Criteria Pollutant Emissions").

It should be noted that raising octane levels in gasoline would have impacts on the amount of energy used to produce the fuel and thus on the well-to-wheels GHG emissions. Raising octane rating through the refining process may involve the use of petroleum products more refined than gasoline, which increases the energy requirements. This raises the concern that the fuel efficiency benefits of using high-octane gasoline could be offset by an increase in full-fuel cycle greenhouse gas emissions (Green Car Congress 2013c). Work of the Japan Clean Air Program showed that increasing RON from 90 to 95 increased well-to-wheels GHG by about 1.5 percent (Szybist 2013). Where ethanol is used to increase octane, its source will determine well-to-wheels GHG emissions relative to gasoline. This remains a controversial topic.

High Compression Ratio, Exhaust Scavenging and Direct Injection

Several manufacturers are developing or producing engines with exceptionally high compression ratios while operating on regular grade (87 AKI) gasoline. The technologies being applied to these engines are described in this section.

Mazda Skyactiv Engine

Mazda has developed the Skyactiv technology, which it first introduced in the 2012 Mazda3. The 2.0L 155-hp four-cylinder Skyactiv engine in the Mazda3 was reported to consume 15 percent less fuel than its predecessor of the same displacement. Other improvements include 15 percent more torque, especially in the low-to-mid rpm range, a 10 percent weight reduction, and 30 percent less internal friction. The 12:1 compression ratio for this engine, which was increased to 13:1 in subsequent applications, uses 87 AKI octane regular gasoline in the United States. The key technologies applied in the Skyactiv engine include a high compression ratio, exhaust scavenging, and direct injection. A 4 into 2 into 1 exhaust manifold, dual variable valve timing, and direct multi-hole gasoline injection are used to prevent preignition and knock at the high compression ratios. Additional technologies include a new design of pistons, shorter combustion duration, and delayed ignition during start-up. A fundamental characteristic of the engine is its higher compression ratio, which improves brake thermal efficiency.

Several enablers led to the use of high compression ratios. Dual VVT allows the use of more aggressive cam timing profiles to fully purge hot exhaust gases from the cylinders. The 4 into 2 into 1 exhaust manifold allows the exhaust gas to be expelled from each cylinder without interference from a pressure pulse from another cylinder. A disadvantage is that the long exhaust manifold moves the catalyst farther away, making rapid light-off to reduce emissions during start-up difficult. Direct injection with fine fuel atomization provides evaporative cooling in the combustion chamber to further facilitate operation at the high compression ratio.

The fuel consumption reductions achieved by the Mazda Skyactiv technology were examined based on the EPA fuel economy data shown in Table 2.35. The compact Mazda3 with Skyactiv technology provides a 9-10 percent reduction in fuel consumption relative to the Ford Focus with a 2.0L naturally aspirated engine and the Toyota Corolla with a 1.8L naturally aspirated engine. The midsize Mazda6 with Skyactiv technology provides a 12-14 percent reduction in fuel consumption relative to the Hyundai Sonata with a naturally aspirated engine or the Ford Fusion with a turbocharged, downsized engine. Figure 2.12 shows that an increase in compression ratio from 9:1 to 13:1 can increase brake thermal efficiency at moderate loads by 7 percent, which appears to account for a significant portion of the reduction in fuel consumption shown in Table 2.35. The committee obtained an independent estimate of 10 percent reduction in fuel consumption for Skyactiv technology (Duleep 2014). The cost of the Mazda Skyactive engine is not known, but is expected to be considerably lower than the cost for a downsized, turbocharged engine. An independent cost estimate in the range of $250 was obtained for the current Skyactiv engine (Duleep 2014).

Recently, Mazda announced plans for its next generation Skyactiv2 engine, which was claimed to be 30 percent more efficient than the current Skyactiv engine by using a compression ratio of 18:1 and lean HCCI combustion. However, these two features alone are estimated to provide less than a 10 percent reduction in fuel consumption. No test results are available to confirm the feasibility or benefits of Mazda's announced features of the Skyactiv2 engine (Griemel 2014). The future path for the Skyactive approach beyond the present status is not clear, since the compatibility of turbocharging and downsizing with the exhaust scavenging system of the Skyactiv approach is unknown.

Atkinson Cycle Engines (for Non-Hybrid Vehicles)

Atkinson cycle engines have been used by Toyota in its hybrid vehicles since 1997. The Atkinson cycle engine with a high compression ratio enhances thermal efficiency but reduces torque. The Atkinson thermodynamic cycle is shown on a P-V diagram in Appendix D. In hybrid applications, the motor torque compensates for this reduction in engine torque. Hybrid applications of Atkinson cycle engines are discussed further in Chapter 4. Recently, Toyota announced that the issue with low torque has been overcome, and this development is expected to facilitate the application of

TABLE 2.35 Comparisons of EPA Fuel Economy for Mazda Vehicles with Skyactiv Technology

Vehicle	Engine/ Transmission	2014 MY EPA Combined FE (mpg)	2014 MY EPA Uncorrected Combined FE (mpg)	Power (Hp)[a]	Curb Weight (lb)[b]	Power/ Weight	Adjusted to Comparable Power/Wt[c] (EPA Combined FC in gal/100 mi)
Compact Cars							
Mazda 3 Skyactiv	2.0L, A-S6	33	45.1	155	2781	0.056	2.22
Ford Focus SFE	2.0L, AM-6	33	43.6	160	2960	0.054	2.32
Ford Focus	2.0L, AM-6	30	41.3	160	2960	0.054	2.45
Toyota Corolla	1.8L, AV-S7	32	43.5	140	2875	0.049	2.42
Midsize Cars							
Mazda 6 Skyactiv (w/o e-iloop)	2.5L, A-S6	30	40.7	184	3183	0.058	2.46
Hyundai Sonata	2.4L, A-6	28	36.6	182	3245	0.056	2.76
Ford Fusion	1.5L TC, A-S6	28	36.4	178	3427	0.052	2.85

[a] 2014 Wards Mahle Light-Duty Engine Specifications.
[b] Cars.com.
[c] Using equation shown earlier in Chapter 2 from EPA/NHTSA Technical Support Document, 2012.
SOURCE: EPA Fuel Economy Guide (2014).

Atkinson cycle engines in conventional vehicles (Yamada 2014). Toyota has called this Atkinson cycle engine ESTEC (Economy with Superior Thermal Efficient Combustion).

The ESTEC engine features a geometric compression ratio of 13.5:1, water-cooled EGR, and electrically actuated VVT. To compensate for the increase in gas temperature resulting from the compression ratio increase, Toyota has applied a 4 into 2 into 1 exhaust manifold to enhance purging exhaust gas from the combustion chamber. An intake port that provides high tumble enables rapid combustion to reduce the tendency for knocking at the high compression ratio. In addition, the temperature of the cylinder surface is optimized with new water jackets. Internal EGR is used at low loads, while cooled external EGR is phased in as the load is increased. Even with these features, retarded ignition timing is required at high loads to avoid knocking. ESTEC was reported to reduce fuel consumption by 11 percent at low loads and by 5 percent at higher loads near the maximum thermal efficiency operating condition. The ESTEC engine has many similarities to Mazda's previously discussed Skyactiv engine.

Ethanol-Boosted Direct Injection

The direct injection, ethanol-boosted, turbocharged SI engine concept was originated at the Massachusetts Institute of Technology (MIT), with commercialization being pursued by Ethanol Boosting Systems, LLC (EBS) (Bromberg et al. 2012a, 2012b). The ethanol-boosted, direct injection engine uses conventional port-injection of gasoline. Ethanol (E85) is then directly injected into the combustion chamber to eliminate knock by cooling the air/fuel mixture. Ethanol is added only under high-load conditions; otherwise, the engine operates like a conventional gasoline engine. Injecting ethanol raises the fuel octane rating, allowing for high compression ratios approaching 14:1 in a turbocharged engine with a manifold pressure of 3 bar. EBS said that it expects the ethanol usage of this engine would range from 2.5 to 5 percent or less but would be dependent on the driving cycle.

In a DOE cost-share project, Ford, in collaboration with AVL, demonstrated that a turbocharged 5.0L engine with this system provided a 75 percent increase in BMEP relative to a direct-injection, naturally aspirated gasoline engine operated at a stoichiometric air/fuel ratio. These results were used to estimate 25 to 30 percent better fuel economy than a conventional gasoline engine, if 43 percent downsizing were applied. These results would be comparable to diesel brake thermal efficiency levels.[12] EBS estimated that for a light-duty truck, the incremental cost of an ethanol boosted engine plus exhaust aftertreatment system would be approximately $1,500, which the committee converted to an estimated direct manufacturing cost of $1,000 for a V8 engine using an ICM of 1.5. EBS has suggested that the ethanol-boosted engine would be significantly less costly than a diesel engine with the required exhaust emission control systems. The disadvantage of this system is that two fuel tanks are required, both of which would need to be filled separately. The requirement for fueling the vehicle with two different fuels is a drawback for adoption of this technology, especially since E85 is not widely available throughout the United States.

Dedicated Exhaust Gas Recirculation

Dedicated EGR (D-EGR) is a concept developed by Southwest Research Institute (SwRI) in which an individual cylinder is dedicated to EGR production to mitigate issues associated with EGR control and tolerance (Alger and Mangold 2009). SwRI has tested a four-cylinder, turbocharged engine with one cylinder exhausting directly to the intake manifold in order to provide a constant EGR level of 25 percent. A schematic of the D-EGR engine concept is shown in Figure 2.13. In addition to the D-EGR system, the engine includes a belt-driven supercharger with a bypass valve, a turbocharger with a wastegate, an intercooler, and an EGR cooler. To ensure reliable ignition of the dilute mixtures, SwRI developed an advanced ignition system called dual-coil offset (DCO). This system provides high-energy, continuous discharge by using two inductive coils connected by a diode to a standard spark plug. SwRI's cost objective for the D-EGR system was less than $1,000. Assuming SwRI's objective was for total cost, the committee estimated a direct manufacturing cost objective of $667 for the D-EGR technology.

The dedicated cylinder that provides the EGR is operated with up to 40 percent excess fuel, provided by an extra port fuel injector, and creates hydrogen and carbon monoxide. SwRI found that reintroducing the hydrogen into the engine enhances dilution tolerance and provides improved combustion stability. SwRI also found that the hydrogen and carbon monoxide can increase the knock resistance of the engine, which facilitated the increase in compression ratio from 9.2:1 to 11.7:1. SwRI estimated that the resulting 1 percent hydrogen (RON ~ 130) and carbon monoxide (RON ~ 106) in the intake increased the effective octane from 90 to 93 RON, and adding 20 percent EGR further increased the effective octane to 103 RON. The rich combustion in the dedicated cylinder used for EGR production is likely to increase carbon formation, which could affect cylinder sealing integrity (rings and valves). Peak pressure and burn rate variations from the rich cylinder to the other cylinders may result in undesirable engine vibration and torque fluctuations.

SwRI has demonstrated the D-EGR concept in several experimental engines and in a vehicle. The D-EGR engine tested on a dynamometer reduced BSFC by approximately 10 percent at conditions encountered in the CAFE test cycles (Chadwell et al. 2014). The vehicle demonstration likewise showed that D-EGR could improve fuel economy

[12] If both the ethanol-boosted engine and the diesel had the same thermal efficiencies, then the diesel engine would have 11.5 percent better volumetric mpg (the ratio of heating values for diesel fuel (129,488 Btu/gal) to gasoline (116,090 Btu/gal).

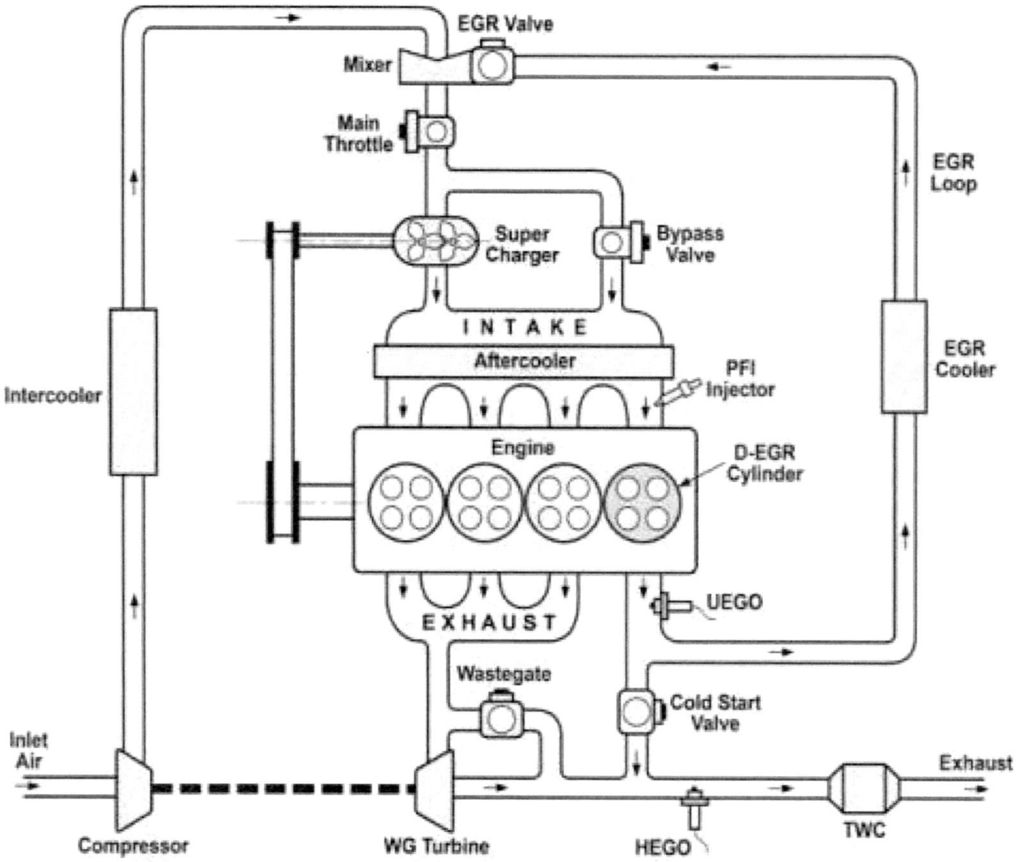

FIGURE 2.13 Schematic of SwRI D-EGR engine system.
SOURCE: Alger (2014). Reprinted with permission, Southwest Research Institute ©2013. *All Rights Reserved.*

by approximately 10 percent (Alger 2014), as shown in Table 2.36. These improvements do not involve engine downsizing; instead, significant portions of the improvements are attributable to the increase in compression ratio by 2.5 ratios and the addition of high rates of EGR as a diluent for combustion.

PSA Peugeot Citroën recently disclosed that it will commercialize high-efficiency gasoline engines with D-EGR, derived from the SwRI program (Green Car Congress 2013a). The new engines, expected to be available in Europe by 2018, will consume 10 percent less fuel than their predecessors, according to PSA. The estimated cost of $67 per percent fuel consumption reduction for the D-EGR system is higher than that of a turbocharged, downsized engine-level 2 with cooled EGR-level 1, but D-EGR has shown greater fuel consumption reduction effectiveness values in limited tests with experimental hardware. Some technical features of the D-EGR system may serve as enablers for increasing the effectiveness of other SI engine technologies, such as compression ratio increase and downsized, turbocharged engines.

Variable Compression Ratio

Variable compression ratio (VCR) is a technology that adjusts the compression ratio of an engine during operation to increase thermal efficiency while operating under varying loads. Lighter loads benefit from higher ratios to be more efficient while lower compression ratios are required at higher loads to prevent knock. Variable compression ratio engines change the volume above the piston at top dead center. The change is done dynamically in response to the load and driving demands. As turbochargers are used to increase the specific output of downsized engines, VCR becomes more desirable as an enabler for even higher boost pressures. Since some changes in effective compression ratio can be achieved with VVT, the improvements that might be obtained with VCR may be diminished. Nevertheless, several VCR concepts have recently been investigated. Some of these are shown in Figure 2.14 and are described in further detail in Appendix O. Several OEMs indicated that they are conducting research on VCR engines, but no technical details, future plans, or preferred approaches were provided.

TABLE 2.36 D-EGR Vehicle Demonstration

2012 Buick Regal	Engine Disp	Compression Ratio	Boost System	Trans	FTP (mpg)	HWFET (mpg)
Production EPA Certification Data[a]	2.0L	9.2:1	TC	MT	23.5	37.4
D-EGR System SwRI Data	2.0L	11.7:1	TC + SC	MT	25.7	40.2 (Est)[b]
Change (%)					9.5%	10.7%

[a] 2012 MY.
[b] Based on % change provided by SwRI.
SOURCE: Alger (2014).

The MCE5 Development (S.A.) Company was established in 2000 in Lyon, France, with the objective of developing VCR engine technology. MCE-5 has developed a multi-cylinder engine, which has been installed in demonstration vehicles. The compression ratio can be varied continuously from 6:1 to 15:1. A schematic of the MCE-5 technology is shown in Figure 2.14. The MCE-5 technology was also used in the Peugeot (PSA) system, and MCE-5 is reported to be working with many of the other European OEMs (MCE-5 2013). MCE-5 reported that the VCR engine provided a 5 percent reduction in fuel consumption on the NEDC test cycle. MCE-5 indicated that its technology would result in an additional cost of approximately $896 for a four-cylinder engine, which translates into a 2025 direct manufacturing cost of $597 (2010 dollars) by applying an ICM of 1.5.

Variable Displacement Engine

An ideal variable displacement engine would be able to adapt its displacement depending on the power demand, thereby minimizing pumping and friction losses incurred with light load operation of conventional engines. One approach for achieving a variable displacement engine is cylinder deactivation, previously discussed in this chapter. However, cylinder deactivation does not eliminate all of the pumping and friction losses associated with the larger displacement engine. Several concepts have been proposed for a fully variable displacement engine.

Numerous organizations have recently explored concepts for truly variable displacement engines. Evaluations of their potential are often based on theoretical analyses. Two recent examples of such concepts are discussed in this

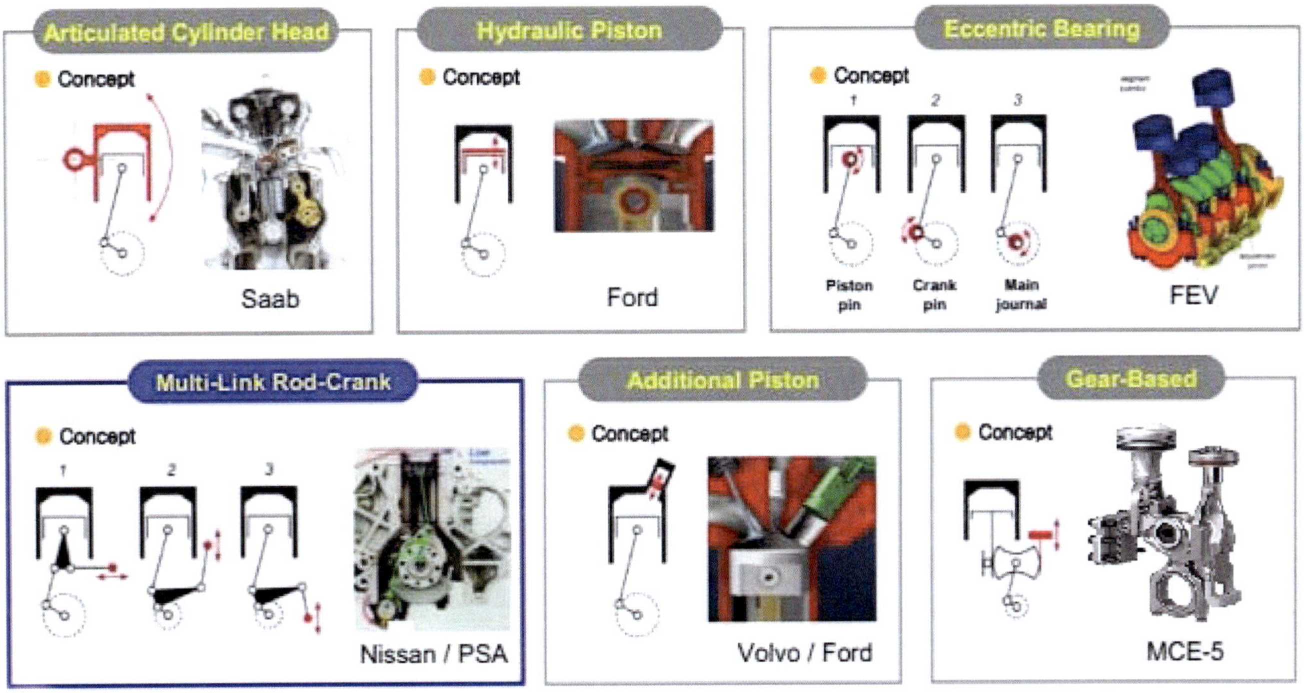

FIGURE 2.14 Variable compression ratio concepts.
SOURCE: Lee et al. (2011); MCE-5.

section, but the literature contains many other concepts for variable displacement engines, some of which have been evaluated by others and rejected as not feasible for a variety of technical reasons. Many challenges and technical hurdles exist that must be resolved before the fuel economy potential of an experimental variable displacement engine can be demonstrated.

Scalzo Automotive Research Pty Ltd is developing a concept called a Piston Deactivation Engine (PDE). This concept has been incorporated in an experimental 1.7L three-cylinder PDE that can operate with one, two, or three cylinders. This concept allows the piston of the deactivated cylinder to be stopped, or "parked," when not required, unlike the cylinder deactivation engines, where the pistons continue to operate through their full stroke. Figure 2.15.1 shows the engine in the active position, with a conventional crankshaft connected to the piston through an oscillator, which is a rocking, adjustable four-bar mechanism located on the opposite side of the cylinder relative to the crankshaft. Figure 2.15.2 shows the piston in the "parked" position. By rotating the adjustor relative to the oscillator, the lower pin of the piston connecting rod is positioned to be concentric with the rotational axis of the oscillator so that the piston motion is reduced to zero. Scalzo has estimated a reduction in fuel consumption in excess of 30 percent (Boretti and Scalzo 2011). Moreover, Scalzo has estimated an additional 5 to 10 percent fuel consumption reduction with the addition of VCR and a 10 to 15 percent fuel consumption reduction with the addition of both VCR and turbocharging. However, test data are not available to confirm these estimates by Scalzo. Future development plans for this engine concept have not been disclosed.

Another concept for a variable displacement engine has been proposed by Engine Systems Innovation, Inc. (ESI). This engine concept, which is configured in the form of a barrel or axial engine, is shown in Figure 2.16. Variable displacement is achieved by axially moving the carrier along the crankshaft. Moving the carrier axially to the left decreases the displacement of the engine and, at the same time, decreases the angle of oscillation of the nutator so that compression ratio can be maintained or appropriately modified at the reduced displacement. ESI has used the GT-Power engine simulation model to estimate that the VDE engine could provide a 23 percent reduction in fuel consumption relative to a baseline naturally aspirated engine. Hardware development plans for this engine concept have not been disclosed. However, since prototype engines demonstrating the estimated reductions in fuel consumption are not available, VDE technology is unlikely to have an impact on CAFE by 2025 or 2030.

Camless Valvetrain

Improvements in fuel consumption, torque, and emissions are projected to be achieved with flexible control of the valve timing, duration and lift. Camless valvetrains have long been investigated by numerous vehicle manufacturers, suppliers, and consulting engineering companies, and their studies have been extensively documented in technical publications. Some of these researchers have suggested that a camless valvetrain could provide 10 to 20 percent reduction in fuel consumption relative to a conventional cam and valve system with fixed timing. However, measured test results have not confirmed these early projections.

There are two concepts that have been considered for a camless valvetrain. One concept uses electrohydraulic valve actuation, under development by Sturman Industries, and the other uses electromagnetic valve actuation, under development by Valeo (Green Car Congress 2011a). None

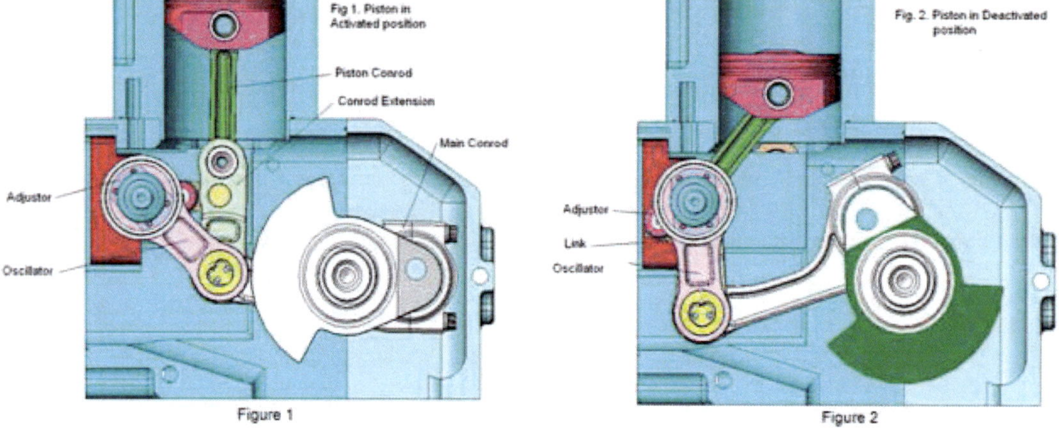

FIGURE 2.15 Scalzo variable displacement engine (VDE).
SOURCE: Scalzo (2014).

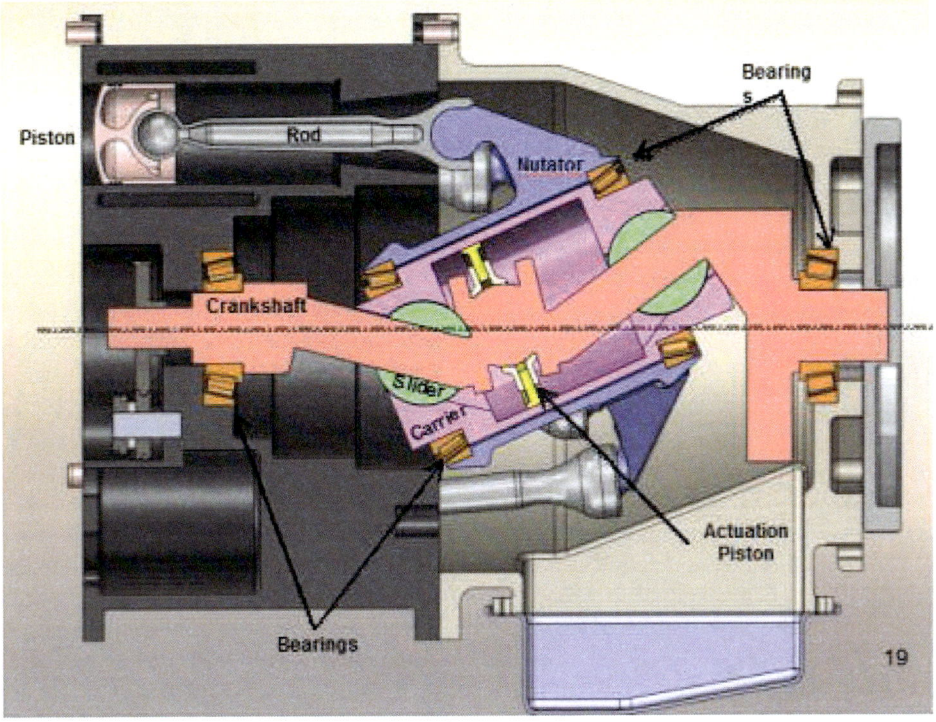

FIGURE 2.16 ESI variable displacement engine in a barrel or axial configuration.
SOURCE: Arnold (2014). Courtesy of Engine Systems Innovation, Inc. Continuously Variable Displacement Engine, US Patent 8,511,625.

of the investigated systems have progressed to a production status. This may be due to several factors. First, VVT and continuously variable valve lift systems can achieve many of the functional capabilities of a camless valvetrain. Second, many problems have been identified during the research and development of camless engines. These problems include high power consumption, accuracy at high speed, temperature sensitivity, weight and packaging issues, high noise, high cost, and undesirable engine failure modes in case of electrical problems.

Another approach to a camless valvetrain is under development by U.K.-based Camcon (Birch 2013). This system, called Intelligent Valve Actuation (IVA), is being designed to allow valve events (lift, timing, and period) to be optimized for every speed and load condition. This system uses a desmodromic valve gear for each valve. Since each valve is actuated by a cam, it is not a "camless" system, although conventional camshafts are eliminated. For full valve lift, an actuator rotates the cam to maximum lift and continues down the other side of the cam lobe and back to the valve closed position. For partial lift, the cam is partially rotated to achieve the target lift and then is returned to its base position. Camcon uses its proprietary Binary Actuation Technology (BAT) to provide the required multidirectional rotation and position control of the cam. Camcon has claimed a 15 percent improvement in fuel economy, which is considerably more than NHTSA's 3.6-4.9 percent estimate for VVL and 4.1-5.5 percent for dual (intake and exhaust) cam phasing. Fuel consumption results or estimated costs for the system applied to an engine are not available. Camcon is reported to be seeking a supplier to commercialize the system.

Waste Heat Recovery

Approximately one-third of the fuel energy supplied to an SI gasoline engine is lost as exhaust enthalpy, and another one-third is lost as heat rejected to coolant. Therefore, recovering a portion of this energy continues to receive attention in research efforts to improve the efficiency of passenger cars. Turbocharged engines already employ a form of exhaust energy recovery by extracting exhaust energy to drive the compressor to provide more airflow to the engine for increased power. Beyond turbocharging, current research on waste heat recovery is focusing on thermoelectric generators (TEG) and organic Rankine cycle (ORC) systems (Saidur et al. 2012). It should be noted that all waste heat recovery systems rely on a low-grade heat source, meaning that the temperature differentials are not very large and thus tend to have low efficiency potential.

A thermoelectric generator (TEG) converts thermal energy from different temperature gradients between the hot and cold ends of a semiconductor into electric energy. The primary challenge in using a TEG is its low thermal efficiency, which is typically less than 4 percent (Saidur et al.

2012), but future thermoelectric materials have the potential to reach higher efficiencies and power densities (Karri et al. 2011). Materials to improve the conversion efficiency of TEGS, like BiTe (bismuth telluride), CeFeSb (skutterudite), ZnBe (zinc–beryllium), SiGe (silicon–germanium), SnTe (tin telluride) and new nanocrystalline or nanowire thermoelectric materials, are currently in the development stage.

DOE has had continuing research programs on the use of TEGs for waste heat recovery from internal combustion engines. In 2011, DOE completed the installation of TEG systems in BMW (X6) and Ford (Lincoln MKT) vehicles (LaGrandeur 2011). At 65 mph, the Ford vehicle generated approximately 450 W of electrical power at an exhaust temperature of approximately 250°C. Over 700 W of electrical power were generated at an exhaust temperature of approximately 500°C. Assuming the Ford vehicle requires approximately 15 kW of engine power at 65 mph, and assuming that the 450 W of power output from the TEG could be converted to 360 W of mechanical power (assuming 80 percent conversion efficiency), then the fuel consumption of the Ford vehicle could potentially be reduced by approximately 2.5 percent. In October 2011, DOE initiated a follow-on to the TEG program for passenger vehicles with the objective of achieving a 5 percent improvement in fuel economy over the US06 drive cycle and determining the economic feasibility of manufacturing TEG systems in quantities of 100,000 per year. Honda has reported that a TEG containing the appropriate combination of elements with different temperature properties could provide about a 3 percent improvement in fuel economy (Mori et al. 2011). At an estimated cost of $1.00/W, a TEG capable of 700 W would have an estimated direct manufacturing cost of $700 (Green Car Congress 2014a).

Another waste heat recovery system is the organic Rankine bottoming cycle. Because of the low-grade heat sources, the efficiency of the cycle depends on the selected working fluids and operating conditions of the system. Over the last 10 years, interest in the Rankine bottoming cycle has prompted some automotive manufacturers to investigate its potential. Researchers have reported that Honda and BMW (Turbosteamer) have achieved a decrease in fuel consumption of approximately 10 percent for passenger cars (Endo et al. 2007; Freymann et al. 2008). However, little information on ORCs has been available in the past 5 years, suggesting that further development of ORCs has slowed, making it unlikely that they could reach production status for light-duty vehicles by 2025 (even though development continues for heavy-duty applications).

Exhaust heat recovery systems have been reported to provide reductions in fuel consumption ranging from marginal to 4 percent (assumed to be on the FTP75 drive cycle). Issues associated with TEG are that the TEG unit in the exhaust increases back pressure (which lowers output and reduces efficiency), the energy output during the test cycles will be much less than 450 W, and the materials costs are high. Another use of waste heat recovery is to use exhaust heat for rapid warm-up of the engine and transmission oil to reduce friction, as discussed earlier in this chapter in the section "Engine Friction Reduction."

CONTROL SYSTEMS, MODELS, AND SIMULATION TECHNIQUES

Control systems, models, and simulation techniques are enablers for many of the fuel consumption reduction technologies considered by NHTSA and EPA in their analysis. Current and future engines have a large number of control variables, which must be optimized to realize the maximum reduction in fuel consumption consistent with other vehicle requirements, including the control of emissions and acceptable driveability. Powertrain control systems are expected to be interfaced in the near future with traffic information, navigation systems, and vehicle-to-vehicle communications for more efficient and safe vehicle operation. Controls are important to consider since they are central to the implementation of so many of the technologies considered to improve fuel economy. Though these are discussed here with the SI technologies, they are also important when considering compression ignition diesel engine technologies, electrified powertrains, and improved transmission, which are discussed in Chapters 3-5. As discussed at the conclusion of this section, a shortage of human resources in this area might impact launches of fuel reduction technologies that require sophisticated controls.

Traditional SI engine control systems used classical single-input, single-output (SISO) controllers, which could be optimized independently using proportional-integral-derivative (PID) controllers and were tuned for fast and robust regulation to specified setpoints. Examples of setpoints include (1) idle speed controlled by the throttle actuator, (2) air-to-fuel ratio controlled by the pulsewidth of the fuel injectors, and (3) knock control managed by the spark timing. The control typically consists of constant values of the setpoints, the gains for the PID controllers, and the feedforward map of the actuator to provide fast response during changes in speed and load conditions. The addition of new technologies with new degrees of freedom for engine optimization has led to the significant growth of control variables. Some of the primary control variables in current engines may include the following:

- Air fuel ratio,
- Direct fuel injection timing and duration for multiple injections,
- Spark timing with multiple ignition strikes,
- Intake cam phasing and duration,
- Exhaust cam phasing and duration,
- Intake valve lift,
- Boost pressure with turbocharger wastegate or variable turbine nozzles (VTN),

- Cooled EGR rate,
- Cylinder deactivation, and
- Interface with transmission and vehicle controls.

Each of these variables influences fuel consumption and emissions and consequently must be precisely controlled over the engine's operating range. Developing the control strategy for these variables requires the use of automated optimization techniques. Applying optimization techniques requires the development of a multidimensional map of the engine, which provides the steady-state fuel consumption and emissions as functions of the above variables. From this map, an analytical model is developed in which fuel consumption and emissions are expressed as functions of the above variables as well as engine speed and load. Using this model, optimization techniques are applied to minimize fuel consumption while complying with other constraints for emissions and driveability. Commercially produced techniques, such as Matlab's Model-Based Calibration Toolbox, are available for the entire process from mapping the engine through final optimization. Extremum seeking (ES) techniques have also been developed to locate optimum operating points in minimum time without having detailed maps or response surfaces (Popovic et al. 2006). Final chassis dynamometer and on-road testing are required to calibrate and modify the models if necessary.

In addition to defining the setpoint maps for scheduling the control variables at steady state, the transition from one setpoint to the next must be designed using dynamic models and model-based control. Controlling the transition from one setpoint to another for many air path and combustion states is critical for achieving good drivability as well as meeting increasingly stringent emission levels with optimized fuel economy. The complexity associated with dynamic subsystem interaction needs to be accounted for early in the control design process. In some cases, hierarchical controllers impose master-slave sequences of actions depending on the control authority and the bandwidth of the actuators. Nonlinear phenomena in the air path, the combustion process, and the exhaust aftertreatment system are intensified by actuator and sensor saturation and require advanced gain-scheduling approaches.

Architectures, components, and sensors are being optimized to handle the increase in control system complexity. Virtual sensors, also known as estimators, are being developed to replace hardware sensors for cost reductions and improved reliability by eliminating sensors that are often located in hostile environments. A potential candidate for a virtual sensor, which is facilitated by advanced processors, is exhaust temperature, which can be calculated from real-time transient heat transfer analyses. Alternatively, real sensors may be augmented with model-based estimators for increasing the measurement speed of response, filtering sensor noise, or, possibly, providing redundancy for on-board diagnostic functions.

Simulation techniques permit the design of control systems before prototype hardware is available. Engine simulation models, such as GT-Power, have been modified to run in real time so that transient control systems can be developed in parallel with the development of prototype engine hardware (Gamma Technologies n.d.). Real-time models, applied to hardware in the loop systems, have facilitated the development of control systems where the "hardware" may progress from the microprocessor or engine control unit (ECU) to the actual engine on a dynamometer for final optimization.

System partitioning is being altered to reduce cost and packaging. Use of the Controller Area Network (CAN) to transfer control signals from non-powertrain modules to the engine control unit has resulted in the elimination of some sensors. Faster microprocessors have saved space with the consolidation of previously distributed microprocessors and the application of virtual sensors. As modern automotive control approached 20 million lines of software code in single microprocessors, formal verification methods are necessary to confirm the functionality of control systems.

As illustrated in this section, the complexity of powertrain control systems has been intensifying due to the added features described in this chapter and the need for overall systems optimization. Other vehicle systems are also experiencing similar growth. As a result, North American vehicle manufacturers and suppliers have a critical need to find and hire many types of engineers, including mechanical, software, electrical, and manufacturing (Sedgwick 2013). Software engineers are in particularly high demand. This high demand is putting strain on the engineering resources of vehicle manufacturers and suppliers and might impact new vehicle launches containing fuel consumption reduction technologies. The manufacturers and suppliers are competing with Silicon Valley for the software, electrical, and control engineering talent.

FUTURE EMISSION STANDARDS FOR CRITERIA POLLUTANT EMISSIONS

EPA finalized the Tier 3 emission and fuel rules in April 2014. These standards are designed to reduce air pollution from passenger cars and trucks (EPA 2014b) and are important to consider since they may make a possible fuel economy technology more difficult or more expensive to implement, or they may enable other technologies. Though these standards are discussed here with the SI engine technologies, they are also important when considering compression ignition (CI) diesel engine technologies discussed in Chapter 3.

Starting in 2017, the Tier 3 rule sets new vehicle emission standards and lowers the sulfur content of gasoline. EPA established new tailpipe emission standards for the sum of non-methane organic gases (NMOG) and nitrogen oxides (NO_x), presented as NMOG + NO_x, and for PM that would apply to all light-duty vehicles, as shown in Table 2.37. The proposed NMOG and NO_x tailpipe standards for light-duty

TABLE 2.37 EPA Tier 3 Emission Standards for LDVs, light-duty trucks (LDTs), and medium- duty passenger vehicles (MDPVs) and Schedule for Phasing-in Tier 3 PM Standards

	Table I-1 Tier 3 LDV, LDT, and MDPV Fleet Average FTP NMOG+NO*x* Standards (mg/mi)								
Model year	2017[a]	2018	2019	2020	2021	2022	2023	2024	2025 and later
LDV/LDT1[b]	86	79	72	65	58	51	44	37	30
LDT2,3,4 and MDPV	101	92	83	74	65	56	47	38	30

[a] For LDV and LDTs above 6,000 lbs GVWR and MDPVs, the fleet average standards apply beginning in MY 2018.
[b] These standards apply for a 150,000 mile useful life. Manufacturers can choose to certify some or all of their LDVs and LDT1s to a useful life of 120,000 miles. If a vehicle model is certified to the shorter useful life, a proportionally lower numerical fleet-average standard applies, calculated by multiplying the respective 150,000 mile standard by 0.85 and rounding to the nearest mg.

	Table I-3 Phase-in for Tier 3 FTP PM Standards (mg/mi)					
Model year	2017[a]	2018	2019	2020	2021	2022 and later
Phase-in (percent of U.S. sales)	20[b]	20	40	70	100	100
Certification Standard (mg/mi)	3	3	3	3	3	3
In-Use Standard (mg/mi)	6	6	6	6	6	3

[a] For LDV and LDTs above 6,000 lbs GVWR and MDPVs, the fleet average standards apply beginning in MY 2018.
[b] Manufacturers comply in MY 2017 with 20 percent of their LDV and LDT fleet under 6,000 lbs GVWR, or alternatively with 10 percent of their total LDV, LDT, and MDPV fleet.
SOURCE: EPA (2014b).

vehicles represent approximately an 80 percent reduction from today's fleet average and a 70 percent reduction in per-vehicle PM standards. EPA is extending the regulatory useful life period during which the standards apply from 120,000 miles to 150,000 miles. EPA is also implementing more stringent evaporative emission standards, which represent about a 50 percent reduction from current standards.

NMOG + NO_x Standards

Fleet-average NMOG + NO_x emissions are calculated by the manufacturer, using weighted average emissions of each model year's vehicles. This, in turn, is compared with the pertinent standard for the given model year. Proposed NMOG + NO_x standards for light-duty vehicles and trucks, defined as vehicles below 8,500 lb Gross Vehicle Weight Rating (GVWR), and medium-duty passenger vehicles, defined as vehicles between 8,500 and 10,000 lb GVWR, are as follows:

- 30 mg/mi, as determined using the US06 (high-speed, high-acceleration) component of the Federal Test Procedure (FTP), by 2025. Today's fleet average is 160 mg/mi.
- 50 mg/mi, as determined using the Supplemental Federal Test Procedure (SFTP), by 2025. The current fleet average is roughly 200 mg/mi.

PM Standards

The PM standards are applied to individual vehicles separately, on a per-vehicle basis, rather than applied as a fleet average. PM standards for LDVs, LDTs, and MDPVs are as follows:

- As determined on the FTP, the standard is 3 mg/mi for all vehicles and model years beginning with 20 percent of the fleet in 2017 and rising to 100 percent of the fleet in 2021. For reference, the current standard is 10 mg/mi.
- As determined using the US06 component of the SPTP, the PM standard through 2018 is 10 mg/mi and 6 mg/mi for 2019 and later model years.

Fuel Standards

Gasoline sulfur reductions finalized by EPA will make emission control systems more effective and assist manufacturers in complying across the fleet. In addition, the gasoline sulfur standards would bring substantial immediate benefits because they would result in reduced emissions for existing vehicles. In order to meet federal standards, EPA will require an annual average standard of 10 ppm of sulfur by January 1, 2017. Also, the present 80 ppm refinery gate and 95 ppm downstream cap are to be maintained by EPA. The Tier 3 gasoline sulfur standards are consistent with levels already reached in California, Europe, Japan, and South Korea, among others.

Changes to Emissions Test Fuel

The test fuel used in the federal emissions tests is being updated by EPA in order to more realistically match current in-use gasoline. This also allows for adjustments that may account for anticipated sulfur and ethanol content. The updated test fuel will be 10 percent ethanol by volume (the present test fuel is 0 percent ethanol) and will also have reduced octane and sulfur (to bring it in line with the Tier 3 specifications). Octane will be lowered to around 87 AKI (R+M)/2 to be representative of in-use fuel. In addition, EPA is specifying the first test fuel requirements for E85.

Because ethanol has a lower heating value than gasoline (76,330 Btu/gal vs. 116,090 Btu/gal) and the CAFE fuel economy is defined in terms of miles per gallon of fuel, the same vehicles tested with gasoline having 10 percent ethanol will yield approximately 3.4 percent lower fuel economy than those tested with gasoline having zero ethanol content. For CAFE purposes, an existing fuel economy equation for gasoline, which has been used since 1988, includes modification to the value of the test fuel energy content. This modification is applied in order to determine what the fuel economy would be if the 1975 baseline test fuel was used (Memorandum to Tier 3 Docket 2013). This equation contains an "R-factor," which is employed because the fuel economy difference is not linearly proportional to the difference in the test fuel's heating value. Since 1988, an R-factor of 0.6 has been employed. However, EPA is currently investigating the suitability of this value.

The proposed Tier 3 emission standards are similar to the California LEV III emission standards that were approved in January 2012. Therefore, the impact of the Tier 3 standards on vehicle technologies would apply equally to the California LEV III standards discussed in the next section. The proposed fuel standard, which would reduce sulfur to 10 ppm, was considered more of an enabler for higher mileage durability than an enabler for the introduction of lean burn. Only one manufacturer specifically cited the proposal as a possible enabler for lean burn, strongly dependent on the downstream fuel sulfur levels at the pump.

Technologies

Tier 3 emission control technologies identified by EPA for large light-duty truck applications are shown in Figure 2.17. Large LDTs will face greater difficulty than other LDVs in meeting the Tier 3 NMOG + NO_x standards at the 30 mg/mi level. For this vehicle segment, the technologies identified by EPA to provide the 77 percent reduction in emissions, taking into account compliance margins, are these: addition of a hydrocarbon adsorber, reduced thermal mass, increased catalyst active materials, secondary air injection, and calibration changes. Catalyst efficiencies for NO_x and NMOG when going from Tier 2 to Tier 3 levels reflect the beneficial effects of reducing gasoline sulfur levels to 10 ppm (from a level of 30 ppm). Compliance margins were reduced from 60 to 50 percent in going from Tier 2 to Tier 3 standards to account for anticipated improvement for in-use deterioration and variability.

The Tier 3 final rule states that EPA does not expect the Tier 3 emission standards to result in "any discernible changes in vehicle fuel economy" (EPA 2014b). However, applying the following three emission control technologies suggested in the final rule could result in less than a 0.5 percent increase in fuel consumption, as described in Appendix P. Secondary air injection for oxidation of unburned HC emissions may be implemented using an electrically driven air pump requiring approximately 100 W for the first 60 seconds of the FTP test cycle, resulting in approximately a 0.015 percent increase in fuel consumption. The addition of a hydrocarbon adsorber for controlling cold start HC emissions will increase back pressure, which could result in approximately a 0.06 percent increase in fuel consumption. Calibration changes consisting of spark retard and increased idle speed for 30 seconds for faster catalyst warm-up could result in approximately a 0.24 percent increase in fuel consumption. Although the combined magnitude of these increases in fuel consumption is expected to be small, they will nevertheless increase the task of complying with the CAFE standards. Tier 3 gasoline sulfur standards may be an enabler for lean burn technology to provide a reduction in fuel consumption, although few vehicle manufacturers revealed lean burn in their future CAFE plans.

California LEV III Standards

"Advanced Clean Car Rules," a set of car and light-duty truck emissions rules through 2025, was approved by the California Air Resources Board in January 2012. The rules will be gradually phased in, and they will include

- LEV III amendments to California Low Emission Vehicle regulations. Two regulations are bundled together under the LEV III cover:
  - —LEV III emission standards for criteria emissions for vehicle MY 2015-2025 and
  - —GHG emission standards for vehicle MY 2017-2025.
- ZEV (zero emission vehicle) regulation.
- Clean fuel outlets regulation.

California's LEV III emission standards, shown in Table 2.38, are very similar to the Tier 3 standards, with the exception of the PM standard, which is reduced to 1 mg/mi by 2028. The committee found that some manufacturers thought that the 3 mg/mi PM standard could be met without a gasoline particulate filter (GPF). However, there may be a possibility, particularly with gasoline direct injection engines, that the 1 mg/mi PM standard would require a GPF, with an associated increase in fuel consumption due to additional exhaust back pressure on the engine and the possible need for periodic regeneration of the collected particulate matter.

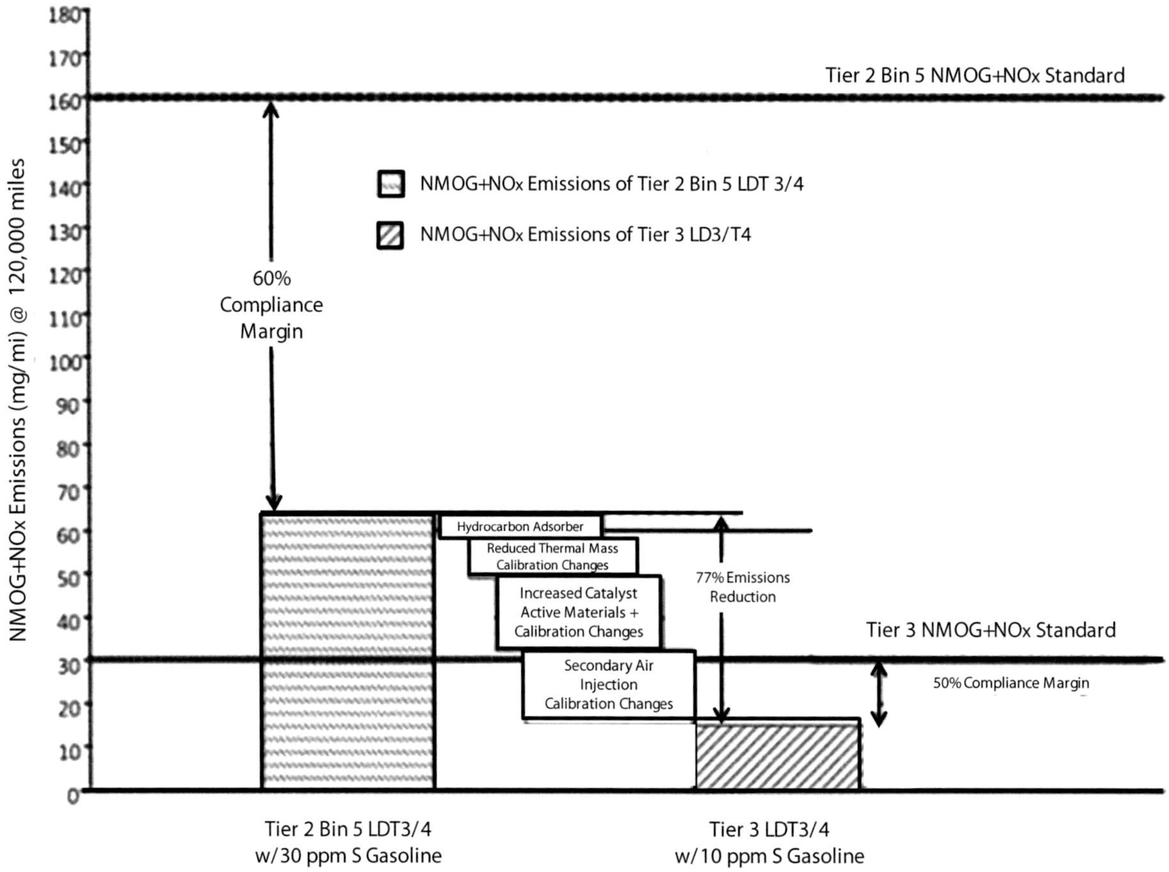

FIGURE 2.17 Tier 3 emission technologies for large, light-duty truck compliance.
SOURCE: EPA (2013).

TABLE 2.38 California LEV III Emission Standards

LEV III Emission Standards, Durability 150,000 miles, FTP-75

Vehicle Types	Emission Category	$NMOG+NO_x$ (g/mi)	CO (g/mi)	HCHO (mg/mi)	PM[a] (g/mi)
All PCs, LDTs ≤ 8500 lbs GVW, and All MDPVs	LEV160	0.160	4.2	4	0.01
	ULEV125	0.125	2.1	4	0.01
	ULEV70	0.070	1.7	4	0.01
	ULEV50	0.050	1.7	4	0.01
	SULEV30	0.030	1.0	4	0.01
	SULEV20	0.020	1.0	4	0.01

[a]Applicable only to vehicles not included in the phase-in of the final PM standards shown below.

LEV III Particulate Matter Emission Standards, FTP-75

Vehicle Type	PM Limit (mg/mi)	Phase-in
PCs, LDTs, MDPVs	3	2017-2021
	1	2025-2028

SOURCE: DieselNet (2013).

OTHER CONSIDERATIONS

Wide Range of Fuels Makes It Harder to Calibrate for Fuel Economy

Gasolines are defined by regulations in which properties and test methods are clearly specified. Several government and state bodies are responsible for defining U.S. gasoline standards. Standards have been developed by the American Society for Testing Materials (ASTM), the Society of Automotive Engineers (SAE), the U.S. Environmental Protection Agency (EPA), and the California Air Resources Board (CARB) (Hamilton 1996). These standards are used to directly reduce emissions or enable technologies that reduce emissions. Several examples of EPA's standards for gasoline are listed in Appendix Q.

The recommended gasoline for most cars is regular 87 AKI. A fuel's octane rating is representative of the unburned end gases' ability to resist spontaneous autoignition. The driver of a vehicle has the responsibility to fuel the vehicle with the AKI rating recommended in the owner's manual for the given engine. The AKI rating must be displayed on a yellow sticker on the gas pump. If fuel with suboptimal octane rating is used, combustion may be less efficient than it would be with the optimal octane level. If this occurs, power and fuel economy will be adversely affected.

Engines are usually calibrated using the fuel that will be recommended in the owners' manual. Calibrations generally use fixed values for control parameters, such as spark timing, based on engine mapping test results that provide optimum fuel economy. The exception to this is the use of a sensor to detect knock at high loads so that the spark timing can be retarded to protect the engine from catastrophic damage. This calibration process generally results in approaching the optimum fuel economy for the specific engine, transmission, and vehicle hardware.

Impact of Low Carbon Fuels on Achieving Reductions in GHG Emissions (California LCFS 2007—Alternative Fuels and Cleaner Fossil Fuels CNG, LPG)

The low-carbon fuel standard (LCFS) is a rule that was enacted by California in 2007 and is the first low-carbon fuel standard mandate in the world. The LCFS directive requires that, by 2020, California's transportation fuels will decrease in carbon intensity by 10 percent. Decreased emissions from the tailpipe, as well as all other production and distribution emissions and any other emissions associated with the use of transport fuels in California, will contribute to the 10 percent reduction. As such, the entire life cycle of the fuel is affected by the California LCFS, making the standard a "well-to-wheels" or "seed-to-wheels" emission standard. Details of the LCFS and the approaches being considered for achieving LCFS compliance are discussed in Appendix R.

Importance of Alternative Fuels Infrastructure

As noted in this and later chapters, there are several vehicle technologies that are dependent on alternative fuels. The success of these technologies will be dependent on the infrastructure for these fuels.

There has been noteworthy industry investment in infrastructure for refueling with natural gas in transportation applications. There were 1,242 CNG and 74 LNG fueling stations in the United States as of September 2013 (AFDC n.d.). Clean Energy Fuels and Pilot Flying J truck stops are building America's Natural Gas Highway, a national network of natural gas refueling stations. Seventy of the planned 150+ stations had been constructed by February of 2013. Clean Energy also constructed 127 stations in 2012. These stations have varied uses, including applications associated with transportation, waste, and aviation. Shell and Travel Centers for America have reached an agreement for Shell to construct natural gas filling stations at up to 100 travel centers. The California Energy Commission (CEC), through the Alternative and Renewable Fuel and Vehicle Technology Program, has contributed more than $16 million to infrastructure for natural gas refueling.

Advanced vehicle technologies may also consume the alternative fuels electricity and hydrogen. Plug-in electric vehicles (PEVs) and fuel-cell vehicles (FCVs) employ, respectively, electricity and hydrogen, which promise to be important components of LCFS compliance, especially as the program reaches maturity.

Importance of Treating Vehicle Technology and Fuels as a System

It has long been recognized that vehicle technology and fuels are a system. In the earliest days of the automotive industry, achieving compatibility of engines and available fuels resulted in compression ratios as low as 4.5:1 in the 1908 Ford Model T. Through the ongoing development of fuels and engines, compression ratios have increased to the range of 9:1 to 11:1 with 87 AKI fuels, resulting in significant improvements in engine thermal efficiencies.

The efforts to achieve compatibility of engines and fuels have been fostered by the OEMs as well as several organizations, including the Society of Automotive Engineers (SAE) and the Coordinating Research Council (CRC). SAE fosters the exchange of information through their Fuels and Lubricants, Engines, Alternative Powertrains and other vehicle-related organizations and associated journals. The CRC aims to encourage cooperative science research in order to advance the improved blends of fuels, lubricants, and associated equipment. Also, the CRC promotes cooperation with the government on issues that are relevant on a national or international scale. The Sustaining Members of CRC are the American Petroleum Institute (API) and a group of automobile manufacturers (Chrysler, Ford, General Motors, Honda,

Mitsubishi, Nissan, Toyota, and Volkswagen). The Alliance of Automotive Manufacturers (AAM) reiterated that the EPA Tier 3 emission standards must continue to treat vehicles and fuels as a system (AAM 2013).

The 2017-2025 CAFE standards will lead to further efforts to ensure compatibility of engines and fuels. Some examples where engines and fuels will need to continue to be treated as systems include the following:

- Any high BMEP engines will need to be developed as a system with available and future anticipated fuels.
- The further reduction of sulfur may enable the application of lean burn engines with suitable lean NO_x aftertreatment.
- The possibility of E30 as a commercial fuel, as suggested by the option to use E30 as a certification fuel in the Tier 3 standards, or the availability of higher octane gasolines, may facilitate the development of higher compression ratio engines.
- The continuing research into low-temperature combustion (LTC) processes, such as HCCI or SACI processes, will rely on the autoignition characteristics of available fuels.

The OEMs that the committee met with were focused on treating vehicles and fuels as a system.

FINDINGS AND RECOMMENDATIONS

Finding 2.1 (Overall Fuel Consumption Reduction Effectiveness and Costs) Spark ignition engines are dominant in light-duty vehicles today and are expected to remain dominant, with further reductions in fuel consumption beyond 2025. Spark ignition engine technologies combined (improved lubricants, lower engine friction, variable valve timing and lift, direct injection, cooled exhaust gas recirculation and downsizing/turbocharging) were estimated to provide an overall reduction in fuel consumption of 27 to 28 percent from the null vehicle, which is within the range of NHTSA's estimates. For an example midsize car, the spark ignition engine technologies that might be applied in the 2017 to 2025 time frame were estimated to provide approximately a 17 to 18 percent reduction in fuel consumption and these technologies had a cumulative direct manufacturing cost of approximately \$526 to \$705. These results will vary with vehicle type, engine type, and the vehicle manufacturer's overall CAFE compliance plan.

Finding 2.2 (Fuel Consumption Reduction Effectiveness Compared to EPA and NHTSA) EPA and NHTSA defined the fuel consumption reduction effectiveness of technologies relative to a spark ignition engine in a null vehicle. NHTSA also defined effectiveness relative to previous technologies that had already been applied according to decision tree paths, and these effectiveness values were generally lower due to negative synergies. The committee developed low and high most likely effectiveness estimates relative to the baseline engine of the null vehicle. The committee's estimates of effectiveness agreed with many of NHTSA's estimates. For several technologies, the committee's high estimates agreed with NHTSA's estimates, while the low estimate for the 18 bar BMEP engine was 1 percentage point lower than the high estimate of 14.9 percent, was 2 percentage points lower than the high estimate of 20.1 percent for the 24 bar BMEP engine, and was 1 percentage point lower than the high estimate of 3.5 percent for cooled exhaust gas recirculation. The low estimates resulted from the following factors: (1) reduced compression ratio for 87 AKI (91 RON) gasoline in the United States, (2) spark retard to preclude knock at elevated boost pressure levels, (3) wider ratio transmissions and/or modified torque converters to compensate for turbocharger lag, (4) reduced effectiveness of EGR, and (5) EPA certification test results.

Recommendation 2.1 (Fuel Consumption Reduction Effectiveness) Full system simulation is acknowledged to be the most reliable method for estimating fuel consumption reductions for technologies before prototype or production hardware becomes available for testing. The committee compliments EPA and NHTSA on initiating their full system simulation programs. For spark ignition engines, these simulations should be directed toward the most effective technologies that could be applied by the 2025 MY to support the midterm review of the CAFE standards. The simulations should use either engine maps based on measured test data or an engine-model-generated map derived from a validated baseline map in which all parameters except the new technology of interest are held constant. Full system simulations for spark ignition engine technologies should be confirmed whenever possible using vehicle results to ensure that fuel octane and drivability requirements have been included.

Finding 2.3 (Cost) The committee developed low and high most likely direct manufacturing cost estimates for the spark ignition engine technologies. The committee's low estimates agreed with NHTSA's estimates as did many of the high estimates, while the high estimates were approximately 15 percent higher than NHTSA's estimated direct manufacturing costs for cam phasing, variable valve lift, and 18 bar BMEP turbocharged, downsized engine. Additional costs of up to nearly \$200 may result from turbocharger upgrades for higher temperature operation, ignition upgrades for reliable ignition with cooled exhaust gas recirculation, the addition of torsional vibration dampers to transmissions for smaller displacement engines, and vehicle integration components for noise, vibration, and harshness control and thermal management.

Recommendation 2.2 (Updated Cost Teardown Studies) Since teardown cost studies are acknowledged to be the most reliable cost estimating methodology, and recognizing the uncertainties in some of the cost studies performed 5 years ago that supported the final CAFE rulemaking, NHTSA and EPA should consider updated teardown cost studies of the latest spark ignition engine technologies identified in the final CAFE rule, as well as cost studies for spark ignition engine technologies anticipated but not currently in production, to support the midterm review of the CAFE standards. Enhanced validation through market testing, in which quotes are obtained from suppliers, should be included in these studies. Vehicle integration costs, including noise, vibration, and harshness control measures, ignition systems upgrades, transmission upgrades with torsional damping, and installation components for engine mounts, heat protection, cooling, wiring and connectors, and other installation requirements should be included in these studies.

Finding 2.4 (Other Technologies by 2025 Not Considered by EPA/NHTSA) Several technologies beyond those considered by EPA and NHTSA might provide additional fuel consumption reductions for spark ignition engines or provide alternative approaches at possibly lower costs for achieving reductions in fuel consumption by 2025. These technologies include (1) higher compression ratio with current regular grade gasoline with an estimated effectiveness of up to 3 percent and a 2025 MY direct manufacturing cost of $50 to $100; (2) higher compression ratio with higher octane regular grade gasoline, if it were to become widely available, with estimated effectiveness of up to 5 percent and a direct manufacturing cost of $75 to $150; (3) high compression ratio with exhaust scavenging and direct injection (also known as Skyactiv, Atkinson cycle) with effectiveness ranging up to 10 percent and a direct manufacturing cost of approximately $250 to $500; (4) electrically assisted, variable-speed supercharger with an effectiveness of approximately 26 percent and a direct manufacturing cost of approximately $1,000 to $1,300; and (5) lean burn facilitated by low-sulfur fuel with an effectiveness of up to 5 percent and direct manufacturing costs of up to $800 to $1,000, although significantly less advantage is expected when compared to an engine with cooled exhaust gas recirculation. Although EPA and NHTSA estimated minimum effectiveness for cylinder deactivation in four-cylinder engines, recent introductions in Europe merit continued investigation of this technology.

Finding 2.5 (Alternative Fuel Technologies) Although alternative fuel technologies have little or even negative effects on miles per gallon gasoline equivalent (mpgge) used to compare energy consumption, several of these technologies may impact CAFE by 2025 by benefiting from application of the Petroleum Equivalency Factor (PEF). Compressed natural gas-gasoline bi-fuel vehicles are already being introduced with the potential to reduce petroleum consumption in these vehicles by up to 43 percent, but at an estimated direct manufacturing cost of up to $6,000 or $7,800, although potentially lower costs could be realized with learning and if higher volumes develop. After 2019, when real-world weighting factors must be applied, the effectiveness of flexible-fuel vehicles for achieving CAFE benefits will likely be significantly reduced. Ethanol-boosted, direct injection engines that are turbocharged and downsized have shown the potential for 20 percent fuel savings relative to baseline engines with a direct manufacturing cost of approximately $750 to $1,000, although the requirement for two fuels significantly diminishes the attractiveness of this concept.

Finding 2.6 (Other Technologies after 2025 Not Considered by EPA/NHTSA) After 2025, several other technologies beyond those considered by EPA and NHTSA might provide additional fuel consumption reductions or alternative approaches for spark ignition engines. These technologies include: (1) variable compression ratio with the potential for 5 percent reduction in fuel consumption and an estimated direct manufacturing cost of $600 to $900; (2) dedicated EGR (D-EGR) with the potential for 10 percent reduction in fuel consumption and an estimated cost of $667; (3) spark assisted, homogeneous charge compression ignition (SI-HCCI) combustion process with a potential for up to 5 percent reduction in fuel consumption relative to a turbocharged, downsized engine and an estimated direct manufacturing cost of approximately $450 to $550 assuming three way catalyst (TWC) aftertreatment; (4) gasoline direct injection compression ignition (GDCI) with up to 5 percent reduction in fuel consumption when applied to a turbocharged, downsized engine and an estimated direct manufacturing cost of approximately $2,500 to $3,750 relative to a baseline engine; and (5) waste heat recovery with the potential for up to 3 percent reduction in fuel consumption and an estimated direct manufacturing cost of approximately $700 to $1,050. Since approximately 60 to 70 percent of fuel energy is lost as heat to coolant or exhaust enthalpy, albeit low quality heat, additional research on waste heat recovery technologies may be called for to determine the estimated benefit of this technology.

Finding 2.7 (HCCI) Homogeneous charge compression ignition (HCCI) for gasoline engines, also known as low-temperature combustion, is estimated to provide up to 5 percent reduction in fuel consumption. HCCI issues include limited range of loads for successful operation and the difficulty of controlling switching between HCCI and SI combustion modes. Recent results from a DOE-sponsored project found that the potential fuel consumption benefit of lean HCCI was eliminated under SULEV (Tier 3 or Tier 2 Bin 2) emission constraints because of the need to switch to a fuel-rich mode after lean operation to deplete the stored oxygen in order to restore the three-way catalyst (TWC) NO_x conversion efficiency. Recent research with stoichiometric HCCI and other efforts with multiple direct injections of gasoline to provide partially premixed combustion have extended the range of HCCI to full load on a single-cylinder laboratory engine, and confirmation is under way on a multicylinder engine.

Although HCCI is not likely to have an impact on CAFE by 2025 MY since the technology is still in the laboratory, it might have a role by the 2030 time frame after the full benefits of turbocharged and downsized engines have been realized.

Finding 2.8 (Mazda Skyactiv Technology) Mazda introduced a gasoline engine in 2012 with Skyactiv technology, which consists of enablers that primarily facilitate operation at a high 13:1 compression ratio while using 87 AKI (91 RON) regular gasoline. The enablers of the high compression ratio are enhanced exhaust scavenging and direct injection. Fuel consumption reduction effectiveness is estimated to be up to 10 percent with a direct manufacturing cost of approximately $250. Mazda recently announced plans for its next-generation Skyactiv2 engine, which was claimed to be 30 percent more efficient than the current Skyactiv engine by using a compression ratio of 18:1 and lean HCCI combustion. However, these two features alone are estimated to provide up to 5 to 10 percent additional reduction in fuel consumption. No test results are available to confirm the feasibility or benefits of Mazda's announced features of the Skyactiv2 engine. Additionally, the compatibility of turbocharging and downsizing with the exhaust scavenging system of the Skyactiv approach is unknown.

Finding 2.9 (High Octane Gasoline) Increasing octane from 87 AKI (91 RON) of regular grade gasoline to 91 AKI (95 RON) has the potential to provide 3 to 5 percent reduction in fuel consumption for naturally aspirated engines if compression ratio is increased by 2 ratios from today's typical level, and possibly even greater reductions in fuel consumption for turbocharged engines by allowing operation at higher boost pressures for further downsizing. Future availability of ethanol in the United States may raise the ethanol content of gasoline from the current E10 to E15. If the octane of the current gasoline blend stock were to be retained at current levels by the refiners, the increased ethanol content might provide the increase in octane level needed to facilitate higher compression ratio engines. However, regular grade gasoline with a higher minimum octane level would need to be widely available before manufacturers could broadly offer engines with significantly increased compression ratios. EPA's Tier 3 program, which changes the certification test fuel to E10 with octane representative of today's level of 91 RON (87 AKI), does not contemplate the above scenario. However, EPA's Tier 3 program does allow manufacturers to use high-octane gasoline for testing vehicles that require premium if they can demonstrate that such a fuel would be used by the operator.

Recommendation 2.3 (High Octane Gasoline) EPA and NHTSA should investigate the overall well-to-wheels CAFE and GHG effectiveness of increasing the minimum octane level and, if it is effective, determine how to implement an increase in the minimum octane level so that manufacturers would broadly offer engines with significantly increased compression ratios for further reductions in fuel consumption.

Finding 2.10 (Tier 3 Emission Standards) EPA's Tier 3 emission standards will require approximately an 80 percent reduction in NMOG + NO_x emissions and a 70 percent reduction in particulate matter (PM). These standards may result in less than a 0.5 percent increase in fuel consumption due to (1) the possible use of an electrically driven secondary air injection pump; (2) the addition of a hydrocarbon adsorber which would increase back pressure; and (3) calibration changes. Tier 3 gasoline sulfur standards may be an enabler for lean burn technology although few vehicle manufacturers revealed lean burn in their future CAFE plans. Some manufacturers are concerned that California's LEV III requirement of 1 mg/mi PM by 2028 may require gasoline particulate filters on gasoline direct injection engines.

Finding 2.11 (Off-Cycle Fuel Economy) The relative deviation of real-world fuel economy from CAFE compliance values is expected to increase as more advanced fuel economy technologies defined by EPA and NHTSA are applied to achieve the 2017-2025 MY CAFE targets. Turbocharged, downsized engines may use fuel enrichment at high loads to manage temperatures and spark retard to control knock, both of which will deteriorate fuel economy more than in a larger naturally aspirated engine in on-road conditions.

Finding 2.12 (Critical Need for Engineering Skills) The complexity of powertrains, control systems, and vehicles is accelerating with the addition of multiple fuel consumption reduction technologies and the need for overall systems optimization. New vehicle launches nearly doubled in 2014 over the preceding year. North American vehicle manufacturers and suppliers have a critical need for many types of engineers, including mechanical, electrical, and manufacturing. This high demand is straining the engineering resources and might impact new vehicle launches containing fuel consumption reduction technologies. Vehicle manufacturers and suppliers are competing with other sectors, especially the IT sector, for software, electrical, and control engineering talent.

REFERENCES

AAM (Alliance of Automotive Manufacturers). 2013. U.S. EPA Tier 3 Rulemaking and Market Gasoline Standards. Auto Industry Perspective Presentation to OMB, March 22. http://www.whitehouse.gov/sites/default/files/omb/assets/oira_2060/2060_03222013b-1.pdf.

Abuelsamid, S. 2009. Hyundai's New 2.4l Gdi Four Cylinder Hits up to 200hp in Sonata, 10% More Mpg. Autobloggreen, November 23. http://www.autoblog.com/2009/11/23/hyundais-new-2-4l-gdi-four-cylinder-hits-up-to-200hp-in-sonata/.

AFDC (Alternative Fuels Data Center). n.d. http://www.afdc.energy.gov/data_download/. Accessed September 28, 2013.

Alger, T. 2010. SwRI's HEDGE Technology for High Efficiency, Low Emissions Gasoline Engines. DEER Conference, September 29.

Alger, T.F. 2014. High Efficiency Engines of the Future. Presentation to the National Research Council Committee on Assessment of Technologies for Improving Light-Duty Vehicle Fuel Economy, Phase 2. Washington, DC, February 13.

Alger, T.F., and B.W. Mangold. 2009. Dedicated EGR: A new concept in high efficiency engines. Presented at the Society of Automotive Engineers Congress, SAE International Technical Paper 2009-01-0694.

Alger, T., B. Mangold, C. Roberts, and J. Gingrich. 2012. The Interaction of fuel anti-knock index and cooled EGR on engine performance and efficiency. SAE Int. J. Engines 5(3):1229-1241. doi:10.4271/2012-01-1149.

Ameri, M. 2010. Energy and exergy analyses of a spark-ignition engine. Int. Journal of Exergy (IJEX) 7(5).

American Petroleum Institute (API). 2010. Determination of the Potential Property Ranges of Mid-Level Ethanol Blends. Washington, D.C.

Arnold, S. 2014. Continuously Variable Displacement Engine. Engine Systems Innovation, Inc. Provided to the National Research Council Committee on Assessment of Technologies for Improving Light-Duty Vehicle Fuel Economy, Phase 2. February 21.

Autoweek. 2013. Honda Goes Turbo with New Engine Family. November 18. http://autoweek.com/article/car-news/honda-goes-turbo-new-engine-family.

Berry, I. 2010. The Effects of Driving Style and Vehicles Performance on the Real-World Fuel Consumption of the U.S. Light–Duty Vehicles. MIT MS Thesis, February.

Bickerstaffe, S. 2012. Ford 1.0 EcoBoost. Automotive Engineer, February 1.

Birch, S. 2013. 'Valve-by-wire' gives a lift to combustion technology. SAE Automotive Engineering Magazine, June 17. http://articles.sae.org/12246/.

Birch, S. 2014. Volvo "triple boost" engine uses twin turbos plus e-compressor. Automotive Engineering Magazine, October 23. http://articles.sae.org/13626.

Blaxill, H. 2012. MAHLE Downsizing Demonstrator. Presentation to the National Research Council Committee on Assessment of Technologies for Improving Light-Duty Vehicle Fuel Economy, Phase 2. Dearborn, Michigan, September 27.

Boretti, A., and J. Scalzo. 2011. Piston and valve deactivation for improved part load performances of internal combustion engines. SAE International, Technical Paper 2011-01-0368.

BorgWarner. n.d. eBoost by BorgWarner. http://www.3k-warner.de/products/eBooster.aspx. Accessed August 20, 2013.

Bromberg, L., D.R. Cohn, and J.B. Heywood. 2012a. Alcohol boosted turbo gasoline engines. White Paper for National Petroleum Council Study, MIT and Ethanol Boosting Systems, LLC (EBS).

Bromberg, L., D.R. Cohn, and J.B. Heywood. 2012b. Impact of ethanol blends in SI engines: 'Increasing the impact of ethanol.' MIT Plasma Science and Fusion Center and Ethanol Boosting Systems, LLC (EBS). International Workshop on Ethanol Combustion Engines, Sao Paolo, Brazil, October 4.

Brooke, L. 2013. Chrysler Sees the ICE Future. SAE Automotive Engineering International, October 1.

Brown, S.F. 2000. Closing In On the Camless Engine: It Would Save Fuel, Run Cleaner, and Respond Better to the Driver's Right Foot. Fortune Magazine, May 29.

Cairns, A., and H. Blaxill. 2005a. The effects of combined internal and external exhaust gas recirculation on gasoline controlled authoignition. SAE Technical Paper 2005-01-0133.

Cairns, A., and H. Blaxill. 2005b. Lean boost and exhaust recirculation for high load controlled autoignition. SAE Technical Paper 2005-01-3744.

CARB (California Air Resources Board). 2009. Staff Report: Initial Statement of Reasons: Proposed Regulation to Implement the Low Carbon Fuel Standard. Public Hearing to Consider the Proposed Regulation to Implement the Low Carbon Fuel Standard. Sacramento, California, March 5.

CARB. 2012. California 2015 and Subsequent Model Criteria Pollutant Exhaust Emission Standards and Test Procedures and 2017 and Subsequent Model Greenhouse Gas Exhaust Emission Standards and Test Procedures for Passenger Cars, Light-Duty Trucks, and Medium-Duty Vehicles. Adopted March 22.

Carley, L. 2007. Are You Ready for GF-5 Oils? Underhood Service, January 1. http://www.underhoodservice.com/are-you-ready-for-gf-5-oils/.

Caswell, D.A. 1984. Controlled Variable Compression Ratio Piston for an Internal Combustion Engine. U.S. Patent No. 4,469,055 A, Sept 4.

Chadwell, C., T. Alger, J. Zuehl, and R. Gukelberger. 2014. A Demonstration of Dedicated-EGR on a 2.0 L GDI Engine. SAE Int. J. Engines 7(1): 434-447. doi: 10.4271/2014-01-1190.

Cheah, L., C. Evans, A. Bandivadekar, and J. Heywood. 2007. Factor of Two: Halving the Fuel Consumption of New U.S. Automobiles by 2035. Publication No. LFEE 2007-04 RP.

Chiang, C.J., A.G. Stefanopoulou, and M. Jankovic. 2007. Nonlinear Observer-based Control of Load Transitions in Homogeneous Charge Compression Ignition Engines. IEEE Trans. Control Syst. Technol. 15(3):438-448.

Chow, E. 2013. Exploring the Use of Higher Octane Gasoline for the U.S. Light-Duty Vehicle Fleet, MS Thesis, MIT, June.

Chow, E., J. Heywood, and R. Speth. 2014. Benefits of Higher Octane Standard Gasoline for the U.S. Light-Duty Vehicle Fleet. SAE Technical Paper 2014-01-1961. doi:10.4271/2014-01-1961.

Chrysler. 2013. Written comment regarding the EPA Proposed rule Entitled Control of Air Pollution From Motor Vehicles: Tier 3 Motor Vehicle Emissions and Fuel Standards, published in the Federal Register on May 21, 2013. Docket ID No. EPA-HQ-OAR-2011-0135-4326, July 1. http://www.regulations.gov/#!documentDetail;D=EPA-HQ-OAR-2011-0135-4326.

Coltman, D., J.W.G. Turner, R. Curtis, D. Blake, B. Holland, R.J. Pearson, A. Arden, and H. Nuglisch. 2008. Project Sabre: A Close-Spaced Direct Injection 3-Cylinder Engine with Synergistic Technologies to achieve Low CO2 Output. SAE Paper 2008-01-0138.

Colucci, J. 2012. Improving Auto Fuel Economy via Fuel Changes. Presentation to the National Research Council Committee on Assessment of Technologies for Improving Light-Duty Vehicle Fuel Economy. Dearborn, Michigan, December 3.

Colucci, J. 2013. Higher Octane: Opportunities, Challenges, and Pathways towards Market Introduction. Presentation to SAE International High Octane Fuels Symposium. Washington, D.C., January 29.

Confer, K. 2014. 2013 DOE Vehicle Technologies Review – Gasoline Ultra Fuel Efficient Vehicle. DOE Merit Review ACE064, May 17.

Confer, K., J. Kirwan, M. Sellnau, J. Juriga, and N. Engineer. 2012. Gasoline Ultra Fuel Efficient Vehicle Program Update. DOE DEER Conference, October 16.

Confer, K.A., J. Kirwan, and N. Engineer. 2013. Development and Vehicle Demonstration of a Systems-Level Approach to Fuel Economy Improvement Technologies. SAE 2013-01-0280. doi:10.4271/2013-01-0280.

Cruff, L., M. Kaiser, S. Krause, H. Roderick et al. 2010. EBDI® - Application of a Fully Flexible High BMEP Downsized Spark Ignited Engine. SAE Technical Paper 2010-01-0587. doi:10.4271/2010-01-0587.

Dana. n.d. Active Warm-Up. http://www.dana.com/wps/wcm/connect/dext2/dana/markets/light+vehicle/thermal+management/active+warmup/active+warm-up. Accessed July 22, 2013.

Daw, S., Z. Gao, V. Prikhodko, S. Curran, and R. Wagner. 2013. Modeling Emissions Controls for RCCI Engines. Engine Research Center Symposium, June.

Dick, A., J. Greiner, A. Locher, and F. Jauch. 2013. Optimization Potential for a State of the Art 8-Speed AT. SAE 2013-01-1272.

DieselNet. 2013. Cars and Light-Duty Trucks – California. Emission Standards. http://www.dieselnet.com/standards/us/ld_ca.php. Accessed July 12, 2013.

Diesel Fuel News. 2012. Toyota/Shell Study: 12% Fuel-Economy Boost with FT Diesel in Optimized Engine. Diesel Fuel News 6(33).

DOE (Department of Energy). n.d. Energy Efficient Technologies. www.fueleconomy.gov. http://www.fueleconomy.gov/feg/tech_adv.shtml. Accessed July 24, 2013.

DOE. 2010. Vehicle Technologies Fact Sheet #630, July 5.

DOE. 2012a. www.fueleconomy.gov.

DOE. 2012b. FY 2012 Progress Report for Advanced Combustion Engine Research and Development. Ford, DOE-ACE-2012AR, December.

DOE. 2012c. FY 2012 Progress Report for Advanced Combustion Engine Research and Development. General Motors, DOE-ACE-2012AR, December.

Drake, M., and D. Haworth. 1990. Advanced Gasoline Engine Development using Optical Diagnostics and Numerical Modeling. http://www.combustioninstitute.org/documents/Drake_Haworth_Talk_with_videos.pdf. Accessed September 29, 2013.

Duleep, G. 2014. New Technologies to Meet 2025 CAFE Standards. Presentation to the National Research Council Committee on Assessment of Technologies for Improving Light-Duty Vehicle Fuel Economy, Phase 2. Washington, D.C., June 24.

EIA (Energy Information Administration). 2013. Frequently Asked Questions. http://www.eia.gov/tools/faqs/faq.cfm?id=27&t=4. Accessed September 10, 2013.

Endo, T., S. Kawajiri, Y. Kojima, K. Takahashi, T. Baba, S. Ibaraki et al. 2007. Study on Maximizing Energy in Automotive Engines. SAE 2007-01-0257, Presented at SAE World Congress & Exhibition, April, Detroit, Michigan.

EPA (Environmental Protection Agency). n.d. Gasoline. Last updated March 3, 2014. http://www.epa.gov/otaq/fuels/gasolinefuels/index.htm. Accessed August 21, 2013.

EPA. 2012. Fuel Economy Guide. Available at http://www.fueleconomy.gov/feg/pdfs/guides/FEG2012.pdf.

EPA. 2012. Regulatory Impact Analysis: Final Rulemaking for 2017-2025 Light-Duty Vehicle Greenhouse Gas Emission Standards and Corporate Average Fuel Economy Standards. PA-420-R-12-016.

EPA. 2013a. Control of Air Pollution from Motor Vehicles: Tier 3 Motor Vehicle Emission and Fuel Standards. Proposed Rule, Federal Register, Vol. 78, No. 98.

EPA. 2013b. EPA Proposes Tier 3 Tailpipe and Evaporative Emission and Vehicle Fuel Standards. EPA-420-F-13-018a.

EPA. 2014a. Light-Duty Automotive Technology, Carbon Dioxide Emissions, and Fuel Economy Trends: 1975 Through 2014. EPA-420-R -14-023.

EPA. 2014b. Control of Air Pollution from Motor Vehicles: Tier 3 Motor Vehicle Emission and Fuel Standards; Final Rule. Federal Register, Vol. 79, No. 81.

EPA. 2014c. Greenhouse Gas Emission Standards for Light-Duty Vehicles, Manufacturer Performance Report for the 2012 Model Year. http://www.epa.gov/otaq/climate/documents/420r14011.pdf.

EPA. 2014d. E85 Flexible Fuel Vehicle Weighting Factor for Model Year 2016-2018 Vehicles. November 12.

EPA. 2014e. Mid-Term Evaluation Technologies and Costs – 2022-2025 GHG Emissions Standards. Briefing to the National Research Council Committee on Fuel Economy of Light-Duty Vehicles, Phase 2, Ann Arbor, Michigan, July 31.

EPA/NHTSA (National Highway Traffic Safety Administration). 2010. Joint Technical Support Document Rulemaking to Establish Light-Duty Vehicle Greenhouse Gas Emission Standards and Corporate Average Fuel Economy Standards. EPA-420-R-10-901.

EPA/NHTSA. 2012a. 2017 and Later Model Year Light-Duty Vehicle Greenhouse Gas Emissions and Corporate Average Fuel Economy Standards. EPA 40 CFR 85, 86, 600; NHTSA 49 CFR 523, 531, 533, 536, 537 (August 28, 2012): 74854.

EPA/NHTSA. 2012b. Joint Technical Support Document, Final Rulemaking 2017-2025 Light-Duty Greenhouse Gas Emission Standards and Corporate Average Fuel Economy Standards. EPA-420-R-12-901.

Epping, K., S. Aceves, R. Bechtold, and J. Dec. 2002. The Potential of HCCI Combustion for High Efficiency and Low Emissions. SAE Technical Paper 2002-01-1923. doi:10.4271/2002-01-1923.

Eriksson, L., S. Frei, C. Onder, and L. Guzzella. 2002. Control and Optimization of Turbocharged Spark Ignited Engines. IFAC 15th Triennial World Congress, Barcelona, Spain.

Ernst, K. 2012. Audi Developing Its Own Electrically-Assisted Turbocharging. Motor Authority, September 19. http://www.motorauthority.com/news/1079270_audi-developing-its-own-electrically-assisted-turbocharging.

Ethanol Turbo Boost for Gasoline Engines: Diesel and Hybrid Equivalent Efficiency at an Affordable Cost. Ethanol Boosting Systems, LLC. Excerpt from a presentation made on November 27, 2007 to the NRC Committee charged with the evaluation of Fuel Economy of Light-duty Vehicles. http://www.ethanolboost.com/EBS_Overview.pdf.

FEV, Inc. 2009. Light-Duty Technology Cost Analysis Pilot Study. U.S. Environmental Protection Agency, EPA-420-R-09-020.

FEV, Inc. 2013. Light-Duty Technology Cost Analysis, Report on Additional Case Studies. U.S. Environmental Protection Agency, EPA-420-R-13-008.

Filipe, D., and R.A. Stein. 2002. Spark ignition engine with negative valve-overlap. US Patent 6394051 B1. Ford Global Technologies, Inc. http://www.google.com/patents/US6394051.

Ford Media Center. 2014. All-New Ford F150 2.7-Liter EcoBoost V6 Engine Delivers V8 Capability and Performance, https://media.ford.com/content/fordmedia/fna/us/en/news/2014/07/22/all-new-ford-f-150-2-7liter-ecoboost-v6-engine-delivers.html.

Ford News Center. 2013. First CNG-Capable 2014 Ford F-150 Rolls Off the Line in Kansas City. November 21, 2013. http://corporate.ford.com/news-center/press-releases-detail/first-cng-capable-2014-ford-f-150-rolls-off-the-line-in-kansas.

Freymann R., W. Strobl, and A. Obieglo. 2008. The Turbosteamer: A system introducing the principle of cogeneration in automotive applications. MTZ 69:20-27.

Fuel Economy Race Brings Expensive Oil to Inexpensive Cars. Consumer Reports News, September 19, 2012. http://news.consumerreports.org/cars/2012/09/fuel-economy-race-brings-expensive-0w-20-synthetic-oil-to-inexpensive-cars.html.

Gorzelic, P., E. Hellström, A. Stefanopoulou, L. Jiang, and S. Gopinath. 2012. A Coordinated Approach for Throttle and Wastegate Control in Turbocharged Spark Ignition Engines. 24th Chinese Control and Decision Conference. Taiyuan, CN, 1524-1529.

Graves, R. 2014. Opening Comments, Presentation to SAE 2014 High Octane Fuels Symposium. Washington, D.C., January 21.

Green Car Congress. 2007. FEV Displays Turbocharged, Direct-Injected, E85 Variable Compression Ratio Engine. Green Car Congress, April 16. http://www.greencarcongress.com/2007/04/fev_displays_tu.html.

Green Car Congress. 2009a. Fiat Introduces Multiair Electro-Hydraulic Valve-Timing System. Green Car Congress, March 4. http://www.greencarcongress.com/2009/03/fiat-introduces.html.

Green Car Congress. 2009b. GM's HCCI Demonstrator Combines a Set of Enabling Technologies and Strategies for Extending Operating Range. Green Car Congress, May 28. http://www.greencarcongress.com/2009/05/gm-hcci-20090528.html.

Green Car Congress. 2009c. Fiat Launches MultiAir Combined with Start/Stop on the MiTo. Green Car Congress, June 15. http://www.greencarcongress.com/2009/06/mito-20090615.html.

Green Car Congress. 2011a. Sturman Industries Electro-hydraulic Valvetrain Enables Peak Efficiencies Exceeding 40 Percent in Natural Gas Engine with Fuel Consumption Reduction of up to 18 Percent. Green Car Congress, June 25.

Green Car Congress. 2011b. U. of Wisconsin RCCI Combustion Work Progressing; Modeled 53% Gross Indicated Efficiency in a Light-Duty Engine Could Result in 2x Fuel Savings Compared to SI Gasoline. Green Car Congress, May 31. http://www.greencarcongress.com/2011/05/u-of-wisconsin-rcci-combustion-work-progressing-modeled-53-gross-indicated-efficiency-in-a-light-dut.html.

Green Car Congress. 2012. New Volkswagen Polo BlueGT With Cylinder Deactivation Offers Gasoline Engine Fuel Economy Of Up To 52 Mpg US. Green Car Congress, March 6, 2012. http://www.greencarcongress.com/2012/03/new-volkswagen-polo-bluegt-with-cylinder-deactivation-offers-gasoline-engine-fuel-economy-of-up-to-5.html.

Green Car Congress. 2013a. PSA to Commercialize SwRI-developed Dedicated EGR Technology in High-efficiency Gasoline Engines by 2018. Green Car Congress, February 4.

Green Car Congress. 2013b. Schaeffler Bringing Thermal Management Module to North American Market; Up to 4% Improvement in Fuel Economy. Green Car Congress, July 12. http://www.greencarcongress.com/2012/07/schaeffler-20120712.html.

Green Car Congress. 2013c. 2013 SAE International High Octane Fuels Symposium: The potential for high octane fuels (Part 1). Green Car Congress, January 31. http://www.greencarcongress.com/2013/01/hofs-20130131.html.

Green Car Congress. 2014a. Mercedes Brings Lean-burn, Stratified BlueDirect to CLA-Class for 7% Boost in Fuel Economy. Green Car Congress, February 7. http://www.greencarcongress.com/2014/02/20140207-mb.html.

Green Car Congress. 2014b. GMZ Energy Announces New, High-Power Thermoelectric Module: TG16-1.0. Green Car Congress, October 1.

Greimel, H. 2014a. Mazda Pins Improved MPG on Skyactiv 2. Automotive News, January 6.

Greimel, H. 2014b. Toyota's Massive Engine Overhaul. Automotive News, July 14.

Grundy, J., L. Kiley, and E. Brevick. 1976. AVCR 1360-2 High Specific Output-Variable Compression Ratio Diesel Engine. SAE International Paper No. 760051.

Gamma Technologies. n.d. GT-POWER Engine Simulation Software: Engine Performance Analysis Modeling. http://www.gtisoft.com/upload/Power.pdf.

Hamilton, B. 1996. "Gasoline FAQ." http://www.faqs.org/faqs/autos/gasoline-faq/part1/.

Haraldsson, G., P. Tunestal, B. Johansson, and J. Hyvonen. 2004. HCCI Closed-Loop Combustion Control Using Fast Thermal Management. SAE 2004-01-0943.

Hellstrom, E., A.G. Stefanopoulou, J. Vavra, A. Babajimopoulos, D. Assanis, L. Jiang, and H. Yilmaz. 2012. Understanding the dynamic evolution of cyclic variability at the operating limits of HCCI engines with negative valve overlap. SAE Int. J. Engines 5(3): 995-1008.

Heywood, J.B. 1988. Internal Combustion Engine Fundamentals. New York: McGraw-Hill.

Hitomi, M. 2013. Mazda's Technology Pathway to Meet New Fuel Economy/GHG Standards. Presentation to the National Research Council Committee on Assessment of Technologies for Improving Light-Duty Vehicle Fuel Economy, Phase 2. Washington, D.C., September 27.

Impact of Low Carbon Fuels to Achieve Reductions inn GHG Emissions (California LCFS 2007 – Alternative Fuels and Cleaner Fossil Fuels CNG, LPG).

Ingram, A. 2013. Chevy, GMC Bi-Fuel Natural Gas Pickup Trucks Now in Production. Green Car Reports, November 9. http://www.greencarreports.com/news/1080377_chevy-gmc-bi-fuel-natural-gas-pickup-trucks-now-in-production.

Ishibashi, Y., and H. Morikawa. 2010. A Macroscopic Understanding of the Controlled Auto-Ignition for Vehicle Engines. SAE 2010-32-0086.

J.D. Power. 2012. Engine-Cylinder Deactivation Saves Fuel. McGraw Hill Financial, February 24. http://autos.jdpower.com/content/consumer-interest/gnHI9wQ/engine-cylinder-deactivation-saves-fuel.htm.

Karri, M.A., E.F. Thacher, and B.T. Helenbrook. 2011. Exhaust energy conversion by thermoelectric generator: Two case studies. Energy Conversion and Management 52(3): 1596-611.

Knittel, C. 2011. Automobiles on steroids: Product attribute trade-offs and technological progress in the automobile sector. American Economic Review 101(7):3368-99. http://www.aeaweb.org/articles.php?doi=10.1257/aer.101.7.3368.

Kobayashi, A., T. Satou, H. Isaji, S. Takahashi, and T. Miyamoto. 2012. Development of New I3 1.2L Supercharged Gasoline Engine. SAE Technical Paper 2012-01-0415. doi:10.4271/2012-01-0415.

Koopmans, L., and I. Denbratt. 2001. A Four Stroke Camless Engine, Operated in Homogeneous Charge Compression Ignition Mode with Commercial Gasoline. SAE 2001-01-3610.

Krasny, R. 2013. GM to Sell Car Next Year Powered by Gasoline or Natural Gas. Reuters, October 16. http://www.reuters.com/article/2013/10/16/autos-gm-naturalgas-idUSL1N0I510I20131016. Accessed January 25, 2014.

Kulzer, A., A. Christ, M. Rauscher, C. Sauer, G. Würfel, and T. Blank. 2006. Thermodynamic Analysis and Benchmark of Various Gasoline Combustion Concepts. SAE 2006-01-0231.

Kulzer, A., J-P. Hathout, C. Sauer, R. Karrelmeyer, W. Fischer, and A. Christ. 2007. Multi-Mode Combustion Strategies with CAI for a GDI Engine. SAE Technical Paper 2007-01-0214. doi:10.4271/2007-01-0214.

LaGrandeur, J.W. (Douglas T. Crane, PI). 2012. Thermoelectric Waste Heat Recovery Program for Passenger Vehicles. DOE Vehicle Technologies Program Annual Merit Review, May 18.

Land, K. 2013. Daimler's Technology Pathway to Meet New Fuel Economy/GHG Standards. Presentation to the National Research Council Committee on Assessment of Technologies for Improving Light-Duty Vehicle Fuel Economy, Phase 2. Washington, D.C., September 27.

Lavoie, G., E. Ortiz-Soto, A. Babajimopoulos, J. Martz, and D. Assanis. 2012. Thermodynamic sweet spot for high-efficiency, dilute, boosted gasoline engines. Int. J. Engine Res. 14(3): 260-278.

Lawal, B., and I. Garba. 2013. Evaluation of parameters affecting the performance of spark-efficient engine. Int. Journal of Eng. Research and Technology 2(4). www.ijert.org.

Lee, H-W., W.G. Kin, M-R. Cho, J-W. Cho, and S. H. Lee. 2011. Concept Development of a Variable Compression Ratio Engine Using TRIZ. Japan TRIZ Symposium 2010 Paper. Presented at the Sixth TRIZ Symposium in Japan, Atsugi, Kanagawa, Japan, September 9-11.

Leone, T. 2014. Overview of Efficiency Advantages of High-Octane Fuels. Presentation to SAE High Octane Fuels Symposium. Washington, D.C., January 21.

Lodi, F.S. 2008. Reducing Cold Start Fuel Consumption through Improved Thermal Management. Masters Thesis, University of Melbourne.

Longer, Lower, Wider, More Aero, and Stylish Corolla Unveiled. Automotive Engineering Magazine, SAE International Online, June 10, 2013. http://www.sae.org/mags/sve/desgn/12234.

Lubrizol. 2010. GF-5 Performance Requirements. http://www.gf-5.com/the_story/performance/.

Manofsky, L., J. Vavra, D. Assanis, and A. Babajimopoulos. 2011. Bridging the Gap between HCCI and SI: Spark Assisted Compression Ignition. SAE 2011-01-1179.

Martec Group. 2008. Variable Costs of Fuel Economy Technologies. Prepared for Alliance of Automobile Manufacturers.

MARTEC. 2010. Technology Cost and Adoption Analysis Impact of Ultra-Low Sulfur Gasoline Standards. Prepared for the American Petroleum Institute.

MCE-5. www.mce-5.com. Accessed July 9, 2013.

Mercedes-Benz. 2013. Written comment regarding the EPA Notice of Proposed Rulemaking published in the Federal Register on May 21, 2013. Docket ID No. EPA-HQ-OAR-2011-0135-4676, July 9. http://www.regulations.gov/#!documentDetail;D=EPA-HQ-OAR-2011-0135-4676.

Merkelbach, B. 2009. Continental Introduces First Turbocharger for Gasoline Engines. ATZ Online, March 4. http://www.atzonline.com/index.php;do=show/id=9303/alloc=1.

Michalek, J.J., P.Y. Panos, and S.J. Skerlos. 2004. A Study of Fuel Efficiency and Emissions Policy Impact on Optimal Vehicles Design Decisions. Transactions of the ASME, Vol. 126, November.

Miller, T., W. Van Dam, and G. Parsons. 2012. Recent Developments in GF-6, the New North American Gasoline Engine Oil Performance

Category: Part 1: The New J300 Viscosity Grade; Implications and Formulation Trade-offs. SAE Technical Paper 2012-01-1707. doi:10.4271/2012-01-1707.

Morey, B. 2014. Cooled EGR Shows Benefits for Gasoline Engines. Automotive Engineering Magazine, September 17. http://articles.sae.org/13530/.

Mori, M., T. Yamagami, M. Sorazawa, T. Miyabe et al. 2011. Simulation of fuel economy effectiveness of exhaust heat recovery system using thermoelectric generator in a series hybrid. SAE Int. J. Mater. Manuf. 4(1):1268-1276. doi:10.4271/2011-01-1335.

Moriarty, K. 2013. E85 Deployment. National Renewable Energy Laboratory (NREL), EIA Biofuels Workshop, March 20. http://www.eia.gov/biofuels/workshop/presentations/2013/pdf/presentation-05-032013.pdf.

Murphy, T. 2011. Mercedes Says Dirty Fuel Problematic in U.S. WardsAuto, March 16. http://wardsauto.com/ar/mercedes_dirty_fuel_110315.

Najt, P.M., and D.E. Foster. 1983. Compression-Ignited Homogeneous Charge Combustion. SAE Int. Congress and Exposition. SAE 830264.

NESCCAF (Northeast States Center for a Clean Air Future). 2004. Reducing Greenhouse Gas Emissions from Light-DutyMotor Vehicles. http://www.nesccaf.org/documents/rpt040923ghglightduty.pdf.

NHTSA (National Highway Traffic Safety Administration). 2009. Final Regulatory Impact Analysis: Corporate Average Fuel Economy Standards for MY 2011 Passenger Cars and Light Trucks. U.S. Department of Transportation Office of Regulatory Analysis and Evaluation, National Center for Statistics and Analysis.

NHTSA. 2012. Final Regulatory Impact Analysis: Corporate Average Fuel Economy for MY 2017-MY2025 Passenger Cars and Light Trucks. Office of Regulatory Analysis and Evaluation, National Center for Statistics and Analysis.

NHTSA/EPA. 2014. NHTSA and EPA Responses to NRC Committee on Technologies for Improving the Fuel Economy of Light Duty Vehicles, Phase 2 Questions Received March 10, 2014. May 2.

Nier, T., A. Kulzer, and R. Karrelmeyer. 2012. Analysis of the Combustion Mode Switch Between SI and Gasoline HCCI. SAE Technical Paper 2012-01-1105. doi:10.4271/2012-01-1105.

NRC (National Research Council). 2011. Assessment of Fuel Economy Technologies for Light-Duty Vehicles. Washington, D.C.: The National Academies Press.

Nüesch, S., E. Hellström, L. Jiang, and A.G. Stefanopoulou. 2014. Mode Switches among SI, SACI, and HCCI Combustion and their Influence on Drive Cycle Fuel Economy. 2014 American Control Conference (ACC).

Olechiw, M. 2009. Binning of FEV Cost to GDI, Turbo-charging, and Engine Downsizing. Docket EPA-HQ-OAR-2009-0472, March 25.

Olechiw, M. 2014. Baseline Vehicles. Email communication to the committee, June 27.

Olsson, J.O., P. Tunestal, B. Johansson, S. Fiveland, R. Agama, M. Willi, and D. Assanis. 2002. Compression Ratio Influence on Maximum Load of a Natural Gas Fueled HCCI Engine. SAE 2002-01-0111.

Owen, K., T.R. Coley and C.S. Weaver. 2005. Automotive Fuels Reference Book. 3rd ed., SAE International, Warrendale, Pa.

Polovina, D., D. Mckenna, J. Wheeler, J. Sterniak, O. Miersch-Wiemers, A. Mond, and H. Yilmaz. 2013. Steady State Combustion Development of a Downsized Multi-Cylinder Engine with Range Extended HCCI/SACI Capability. SAE 2013-01-1655.

Popovic, D., M. Jankovic, S. Magner, and A. R. Teel. 2006. Extremum seeking methods for optimization of variable cam timing engine operation. IEEE Transactions on Control Systems Technology 14(3). doi: 10.1109/TCST.2005.863660.

Quantum Devices, Inc. 2010. Brushless Motors vs. Brush Motors, What's the Difference? August 27. http://quantumdevices.wordpress.com/2010/08/27/brushless-motors-vs-brush-motors-whats-the-difference/.

Ra, Y., P. Loeper, M. Andrie, R. Krieger, D. Foster, R. Reitz, and R. Durrett. 2012. Gasoline DICI Engine Operation in the Ltc Regime Using Triple Pulse Injection. SAE Technical Paper 2012-01-1131, April.

Ra, Y., P. Loeper, R. Reitz, M. Andrie, R. Krieger, D. Foster, R. Durrett, V. Gopalakrishnan, A. Plazas, R. Peterson, and P. Szymkowicz. 2011. Study of High Speed Gasoline Direct Injection Compression Ignition (GDICI) Engine Operation in the LTC Regime. SAE Technical Paper 2011-01-1182, April.

Reese, R. 2013. A MultiAir/MultiFuel Approach to Enhancing Engine System Efficiency. DOE Vehicle Technologies Program Annual Merit Review and Peer Evaluation. Presented May 17, Arlington, Virginia.

RFA (Renewable Fuels Association). 2013. Ethanol Facts: Engine Performance. http://www.ethanolrfa.org/pages/ethanol-facts-engine-performance. Accessed August 26, 2013.

Ricardo, Inc. 2011. Computer Simulation of Light-Duty Vehicle Technologies for Greenhouse Gas Emission Reduction in the 2020-2025 Timeframe. U.S. Environmental Protection Agency, EPA-420-R-11-020.

Ricardo, Inc. 2012. Calculation of Friction in High Performance Engines. Presented at Ricardo Software European User Conference, April 20.

Rogozhin, A., M. Gallaher, and W. McManus. 2009. Automobile Industry Retail Price Equivalent and Indirect Cost Multipliers. Prepared for EPA by RTI International and University of Michigan Transportation Research Institute. RTI Project Number 0211577.002.004.

Russ, S. 1996. A Review of the Effect of Engine Operating Conditions on Borderline Knock. SAE Technical Paper 960497. doi:10.4271/960497.

SAE International. Automotive Engineering International, July 2, 2013, Vol. 121, No. 5.

Saidur, R., M. Rezaei, W.K. Muzammil, M.H. Hassana, S. Pariaa, and M. Hasanuzzaman. 2012. Technologies to recover exhaust heat from internal combustion engines. Renewable and Sustainable Energy Reviews 16 (2012): 5649-5659.

Saporito, N. 2014. The Technicalities behind GM's New Engine Family. GM Inside News, March 19.

Scalzo Automotive Research Pty, Ltd. 2012. The Piston Deactivation Engine. http://www.scalzoautomotiveresearch.com/technology.

Schechter, M., and M. Levin. 1996. Camless Engine. SAE International Technical Paper 960581. doi:10.4271/960581.

Schmuck-Soldan, S., A. Königstein, and F. Westin. 2011. Two-Stage Boosting of Spark Ignition Engines. Internationales Wiener Motorensymposium Europe S.r.l., Torino.

Scott, D. 2003. Camless Valvetrain Ready for Production. WardsAuto, January 31. http://wardsauto.com/news-amp-analysis/camless-valvetrain-ready-production.

Sedgwick, D. 2013. Suppliers Race to Hire Engineers. Automotive News, December 16.

Sellnau, M., J. Sinnamon, K. Hoyer, and H. Husted. 2012. Full time Gasoline Direct-Injection Compression Ignition (GDICI) for high efficiency and low NO_x and PM." SAE Paper 2012-01-0384, April.

Sierra Research, Inc. 2008. Basic Analysis of the Cost and Long-Term Impact of the Energy Independence and Security Act Fuel Economy Standards. Report No. SR2008-04-01, April 24.

Sjoberg, M. 2012. Advanced Lean-Burn DI Spark Ignition Fuels Research. DOE Annual Merit Review. Presented May 15, Washington, D.C. http://energy.gov/sites/prod/files/2014/03/f10/ft006_sjoberg_2012_o.pdf.

Sniderman, D. 2012a. Better Fuel Efficiency Through a Better Oil Pump. ASME.org, May. https://www.asme.org/engineering-topics/articles/automotive/better-fuel-efficiency-through-a-better-oil-pump.

Sniderman, D. 2012b. Using Waste Engine Heat in Automobile Engines. ASME.org, June. https://www.asme.org/engineering-topics/articles/automotive/using-waste-engine-heat-in-automobile-engines.

Spinelli, M. 2011. Will BMW's Electric Turbocharger Eliminate Turbo Lag? Jalopnik, November 2. http://jalopnik.com/5855317/will-bmws-electric-turbocharger-end-turbo-lag.

Swedberg, S. 2013. Thumbs Up for SAE 16 Vis Grade. Lube Report, January 3. http://www.imakenews.com/lng/e_article002579890.cfm. Accessed June 10, 2013.

SwRI to Launch Preignition Prevention Program (P3) Consortium. Southwest Research Institute News, San Antonio, Texas, October 6, 2010.

Szybist, J., K. Edwards, M. Foster, K. Confer, and W. Moore. 2013. Characterization of Engine Control Authority on HCCI Combustion as the High Load Limit Is Approached. SAE International 2013-01-1665, April.

Thring, R. H. 1989. Homogeneous charge compression ignition (HCCI) engines. SAE Int. Fall Fuels and Lubricants Meeting and Exhibition. SAE 892068.

Truett, R. 2013. GM's Cruze Diesel Ghostbuster. Automotive News, May 20.

Truett, R. 2014. Feel Free to Salute the Redesigned Ford F-150. Automotive News, January 27.

Tsourapas, V., et al. 2014. Eaton's electrically assisted TVSR supercharger with variable speed, mild-hybrid and engine start/stop functionalities. JSAE, May 2014.

Tula Technology. n.d. http://www.tulatech.com. Accessed July 26, 2013.

Turner, J.W.G., A. Popplewell, R. Patel, T. R. Johnson et al. 2014. Ultra boost for economy: Extending the limits of extreme engine downsizing. SAE Int. J. Engines 7(1): 387-417. doi:10.4271/2014-01-1185.

Uchida, H. 2006. Trend of turbocharger technologies. R&D Review of Toyota CRDL 41(3): 1-8.

Vis, P.J. 2013. Ford Fusion 2013 Uses NASA Space Tech: EcoBoost Turbo is Built to Last with Space Shuttle Superalloy. Vis Labs.

Volkswagen. 2013. Set up for EPA Tier-3 Control of Air Pollution from Motor Vehicles: Tier 3 Motor Vehicle Emission and Fuel Standards. Written comment regarding the EPA Notice of Proposed Rulemaking published in the Federal Register on May 21, 2013. Docket ID No. EPA-HQ-OAR-2011-0135, July 1. http://www.regulations.gov/#!documentDetail;D=EPA-HQ-OAR-2011-0135-4299.

Wade, W.R., J. White, C. Jones, C. Hunter, et al. 1984. Combustion, Friction and Fuel Tolerance Improvements for the IDI Diesel Engine. SAE Technical Paper 840515. doi:10.4271/840515.

Weall, A., K. Edwards, M. Foster, K. Confer, and W. Moore. 2012. HCCI Load Expansion Opportunities Using a Fully Variable HVA Research Engine to Guide Development of a Production Intent Cam-Based VVA Engine: The Low Load Limit. SAE Paper No. 2012-01-1134, April 2012.

Wheeler, J., D. Polovina, V. Frasinel, O. Miersch-Wiemers, A. Moond, J. Sterniak, and H. Yilmaz. 2012. Design of a 4-Cylinder GTDI Engine with Part-Load HCCI Capability. SAE Paper 2012-01-0287, April.

Willand, J., R-G. Nieberding, G. Vent, and C. Enderle. 1998. The Knocking Syndrome: Its Cure and Its Potential. SAE Int. Fall Fuels and Lubricants Meeting and Exhibition. SAE 982483.

Winter, D. 2014. SAE Powertrain Panel: Higher-Octane Gas Could Improve Fuel Economy. WardsAuto, April 9. http://wardsauto.com/vehicles-technology/sae-powertrain-panel-higher-octane-gas-could-improve-fuel-economy?NL=WAW-08&Issue=WAW-08_20140414.

Witzenburg, G. 2013. Chrysler V-6 Shines in Both Cars and Trucks. WardsAuto, August 20. http://wardsauto.com/vehicles-amp-technology/chrysler-v-6-shines-both-cars-and-trucks.

Woodall, B. 2010. GM seeks more U.S. ethanol refueling stations. Reuters, February 16.

Yang, J. 2005. Potential applications of thermoelectric waste heat recovery in the automotive industry. Pp. 155-159 in International Conference on Thermoelectrics.

Yun, H., N. Wermuth, and P. Najt. 2009. Development of Robust Gasoline HCCI Idle Operation Using Multiple Injection and Multiple Ignition (MIMI) Strategy. SAE Technical Paper 2009-01-0499. doi:10.4271/2009-01-0499.

Zareei, J., and A.H. Kakaee. 2013. Study of the effects of ignition timing on gasoline engine performance and emissions. Eur. Transp. Res. Rev. 5:109-116.

Zoia, D.E. 2014. New Small Engines, 48V Mild Hybrids on Way From Fiat Chrysler. WardsAuto, May 6. http://wardsauto.com/vehicles-technology/new-small-engines-48v-mild-hybrids-way-fiat-chrysler.

ANNEX TABLES

TABLE 2A.1 NRC Committee's Estimated Fuel Consumption Reduction Effectiveness of SI Engine Technologies

Percent Incremental Fuel Consumption Reductions: NRC Estimates					
Technologies		Midsize Car I4 DOHC	Large Car V6 DOHC	Large Light Truck V8 OHV	
Spark Ignition Engine Technologies	Abbreviation	Most Likely	Most Likely	Most Likely	Relative To
NHTSA Technologies					
Low Friction Lubricants - Level 1	LUB1	0.7	0.8	0.7	Baseline
Engine Friction Reduction - Level 1	EFR1	2.6	2.7	2.4	Baseline
Low Friction Lubricants and Engine Friction Reduction - Level 2	LUB2_EFR2	1.3	1.4	1.2	Previous Tech
VVT- Intake Cam Phasing (CCP - Coupled Cam Phasing - OHV)	ICP	2.6	2.7	2.5	Baseline for DOHC
VVT- Dual Cam Phasing	DCP	2.5	2.7	2.4	Previous Tech
Discrete Variable Valve Lift	DVVL	3.6	3.9	3.4	Previous Tech
Continuously Variable Valve Lift	CVVL	1.0	1.0	0.9	Previous Tech
Cylinder Deactivation	DEACD	N/A	0.7	5.5	Previous Tech
Variable Valve Actuation (CCP + DVVL)	VVA	N/A	N/A	3.2	Baseline for OHV
Stoichiometric Gasoline Direct Injection	SGDI	1.5	1.5	1.5	Previous Tech
Turbocharging and Downsizing Level 1 - 18 bar BMEP 33%DS	TRBDS1	7.7 - 8.3	7.3 - 7.8	6.8 - 7.3	Previous Tech
Turbocharging and Downsizing Level 2 - 24 bar BMEP 50%DS	TRBDS2	3.2 - 3.5	3.3 - 3.7	3.1 - 3.4	Previous Tech
Cooled EGR Level 1 - 24 bar BMEP, 50% DS	CEGR1	3.0 - 3.5	3.1 - 3.5	3.1 - 3.6	Previous Tech
Cooled EGR Level 2 - 27 bar BMEP, 56% DS	CEGR2	1.4	1.4	1.2	Previous Tech
Other Technologies					
By 2025:					
Compression Ratio Increase (with regular fuel)	CRI-REG	3.0	3.0	3.0	Baseline
Compression Ratio Increase (with higher octane regular fuel)	CRI-HO	5.0	5.0	5.0	Baseline
Compression Ratio Increase (CR~13:1, exh. scavenging, DI (aka Skyactiv, Atkinson Cycle))	CRI-EXS	10.0	10.0	10.0	Baseline
Electrically Assisted Variable Speed Supercharger[a]	EAVS-SC	26.0	26.0	26.0	Baseline
Lean Burn (with low sulfur fuel)	LBRN	5.0	5.0	5.0	Baseline
After 2025:					
Variable Compression Ratio	VCR	Up to 5.0	Up to 5.0	Up to 5.0	Baseline
D-EGR	DEGR	10.0	10.0	10.0	TRBDS1
Homogeneous Charge Compression Ignition (HCCI) + Spark Assisted CI[b]	SA-HCCI	Up to 5.0	Up to 5.0	Up to 5.0	TRBDS1
Gasoline Direct Injection Compression Ignition (GDCI)	GDCI	Up to 5.0	Up to 5.0	Up to 5.0	TRBDS1
Waste Heat Recovery	WHR	Up to 3.0	Up to 3.0	Up to 3.0	Baseline
Alternative Fuels[c]:					
CNG-Gasoline Bi-Fuel Vehicle (default UF = 0.5)	BCNG	Up to 5 Incr [42]	Up to 5 Incr [42]	Up to 5 Incr [42]	Baseline
Flexible Fuel Vehicle (UF dependent, UF = 0.5 thru 2019)	FFV	0 [40 thru 2019, then UF TBD]	0 [40 thru 2019, then UF TBD]	0 [40 thru 2019, then UF TBD]	Baseline
Ethanol Boosted Direct Injection (CR = 14:1, 43% downsizing) (UF~0.05)	EBDI	20 [24]	20 [24]	20 [24]	Baseline

[a] Comparable to TRBDS1, TRBDS2, SS, MHEV, IACC1, IACC2.
[b] With TWC aftertreatment. Costs will increase with lean NO_x aftertreatment.
[c] Fuel consumption reduction in gge (gasoline gallons equivalent) [CAFE fuel consumption reduction].

TABLE 2A.2a NRC Committee's Estimated 2017 MY Direct Manufacturing Costs of SI Engine Technologies

2017 MY Incremental Direct Manufacturing Costs (2010$): NRC Estimates					
Technologies		Midsize Car I4 DOHC	Large Car V6 DOHC	Large Light Truck V8 OHV	
Spark Ignition Engine Technologies	Abbreviation	Most Likely	Most Likely	Most Likely	Relative To
NHTSA Technologies					
Low Friction Lubricants - Level 1	LUB1	3	3	3	Baseline
Engine Friction Reduction - Level 1	EFR1	48	71	95	Baseline
Low Friction Lubricants and Engine Friction Reduction - Level 2	LUB2_EFR2	51	75	99	Previous Tech
VVT- Intake Cam Phasing (CCP - Coupled Cam Phasing - OHV)	ICP	37 - 43	74 - 86	37	Baseline for DOHC
VVT- Dual Cam Phasing	DCP	31 - 35	72 - 82	37 - 43	Previous Tech
Discrete Variable Valve Lift	DVVL	116 - 133	168 - 193	37 - 43	Previous Tech
Continuously Variable Valve Lift	CVVL	58 - 67	151 - 174	N/A	Previous Tech
Cylinder Deactivation	DEACD	N/A	139	N/A	Previous Tech
Variable Valve Actuation (CCP + DVVL)	VVA	N/A	N/A	157	Baseline for OHV
Stoichiometric Gasoline Direct Injection	SGDI	192	290	277 - 320	Previous Tech
Turbocharging and Downsizing Level 1 - 18 bar BMEP 33%DS	TRBDS1	288 - 331	-129 to -86	942 - 1,028	Previous Tech
V6 to I4 and V8 to V6			-455* to -369*	841* to 962*	
Turbocharging and Downsizing Level 2 - 24 bar BMEP 50%DS	TRBDS2	182	182	308	Previous Tech
I4 to I3		-92* to -96*			
Cooled EGR Level 1 - 24 bar BMEP, 50% DS	CEGR1	212	212	212	Previous Tech
Cooled EGR Level 2 - 27 bar BMEP, 56% DS	CEGR2	364	364	614	Previous Tech
V6 to I4				-524* to -545*	
Other Technologies					
By 2025:					
Compression Ratio Increase (with regular fuel)	CRI-REG	50	75	100	Baseline
Compression Ratio Increase (with higher octane regular fuel)	CRI-HO	75	113	150	Baseline
Compression Ratio Increase (CR~13:1, exh. scavenging, DI (aka Skyactiv, Atkinson Cycle))	CRI-EXS	250	375	500	Baseline
Electrically Assisted Variable Speed Supercharger	EAVS-SC	1,302	998	N/A	Baseline
Lean Burn (with low sulfur fuel)	LBRN	800	920	1,040	Baseline
After 2025:					
Variable Compression Ratio	VCR				Baseline
D-EGR	DEGR				TRBDS1
Homogeneous Charge Compression Ignition (HCCI) + Spark Assisted CI[a]	SA-HCCI				TRBDS1
Gasoline Direct Injection Compression Ignition	GDCI				Baseline
Waste Heat Recovery	WHR				Baseline
Alternative Fuels:					
CNG-Gasoline Bi-Fuel Vehicle	BCNG	6,000	6,900	7,800	Baseline
Flexible Fuel Vehicle	FFV	75	100	125	Baseline
Ethanol Boosted Direct Injection (incr CR to 14:1, 43% downsizing)	EBDI	740	870	1,000	Baseline

*Costs with reduced number of cylinders, adjusted for previously added technologies. See Appendix T for the derivation of turbocharged, downsized engine costs.
[a] With TWC aftertreatment. Costs will increase with lean NO_x aftertreatment.

TABLE 2A.2b NRC Committee's Estimated 2020 MY Direct Manufacturing Costs of SI Engine Technologies

2020 MY Incremental Direct Manufacturing Costs (2010$): NRC Estimates					
Technologies		Midsize Car I4 DOHC	Large Car V6 DOHC	Large Light Truck V8 OHV	
Spark Ignition Engine Technologies	Abbreviation	Most Likely	Most Likely	Most Likely	Relative To
NHTSA Technologies					
Low Friction Lubricants - Level 1	LUB1	3	3	3	Baseline
Engine Friction Reduction - Level 1	EFR1	48	71	95	Baseline
Low Friction Lubricants and Engine Friction Reduction - Level 2	LUB2_EFR2	51	75	99	Previous Tech
VVT- Intake Cam Phasing (CCP - Coupled Cam Phasing - OHV)	ICP	35 - 41	70 - 81	35- 41	Baseline for DOHC
VVT- Dual Cam Phasing	DCP	29 - 33	67 - 76	35 - 41	Previous Tech
Discrete Variable Valve Lift	DVVL	109 - 125	158 - 182	N/A	Previous Tech
Continuously Variable Valve Lift	CVVL	55 - 63	142 - 163	N/A	Previous Tech
Cylinder Deactivation	DEACD	N/A	131	147	Previous Tech
Variable Valve Actuation (CCP + DVVL)	VVA	N/A	N/A	261 - 301	Baseline for OHV
Stoichiometric Gasoline Direct Injection	SGDI	181	273	328	Previous Tech
Turbocharging and Downsizing Level 1 - 18 bar BMEP 33%DS	TRBDS1	271 - 312	-122 to -81	877 - 958	Previous Tech
V6 to I4 and V8 to V6			-432* to -349*	779* - 891*	
Turbocharging and Downsizing Level 2 - 24 bar BMEP 50%DS	TRBDS2	172	172	289	Previous Tech
I4 to I3		-89* to -92*			
Cooled EGR Level 1 - 24 bar BMEP, 50% DS	CEGR1	199	199	199	Previous Tech
Cooled EGR Level 2 - 27 bar BMEP, 56% DS	CEGR2	343	343	579	Previous Tech
V6 to I4				-522* to -514*	
Other Technologies					
By 2025:					
Compression Ratio Increase (with regular fuel)	CRI-REG	50	75	100	Baseline
Compression Ratio Increase (with higher octane regular fuel)	CRI-HO	75	113	150	Baseline
Compression Ratio Increase (CR~13:1, exh. scavenging, DI (aka Skyactiv, Atkinson cycle))	CRI-EXS	250	375	500	Baseline
Electrically Assisted Variable Speed Supercharger	EAVS-SC	1,302	998	N/A	Baseline
Lean Burn (with low sulfur fuel)	LBRN	800	920	1,040	Baseline
After 2025:					
Variable Compression Ratio	VCR				Baseline
D-EGR	DEGR				TRBDS1
Homogeneous Charge Compression Ignition (HCCI) + Spark Assisted CI[a]	SA-HCCI				TRBDS1
Gasoline Direct Injection Compression Ignition	GDCI				Baseline
Waste Heat Recovery	WHR				Baseline
Alternative Fuels:					
CNG-Gasoline Bi-Fuel Vehicle	BCNG	6,000	6,900	7,800	Baseline
Flexible Fuel Vehicle	FFV	75	100	125	Baseline
Ethanol Boosted Direct Injection (incr CR to 14:1, 43% downsizing)	EBDI	740	870	1,000	Baseline

* Costs with reduced number of cylinders, adjusted for previously added technologies. See Appendix T for the derivation of turbocharged, downsized engine costs.
[a] With TWC aftertreatment. Costs will increase with lean NO_x aftertreatment.

TABLE 2A.2c NRC Committee's Estimated 2025 MY Direct Manufacturing Costs of SI Engine Technologies

2025 MY Incremental Direct Manufacturing Costs (2010$): NRC Estimates					
Technologies		Midsize Car I4 DOHC	Large Car V6 DOHC	Large Light Truck V8 OHV	
Spark Ignition Engine Technologies	Abbreviation	Most Likely	Most Likely	Most Likely	Relative To
NHTSA Technologies					
Low Friction Lubricants - Level 1	LUB1	3	3	3	Baseline
Engine Friction Reduction - Level 1	EFR1	48	71	95	Baseline
Low Friction Lubricants and Engine Friction Reduction - Level 2	LUB2_EFR2	51	75	99	Previous Tech
VVT- Intake Cam Phasing (CCP - Coupled Cam Phasing - OHV)	ICP	31 - 36	63 - 73	31 - 36	Baseline for DOHC
VVT- Dual Cam Phasing	DCP	27 - 31	61 - 69	31 - 36	Previous Tech
Discrete Variable Valve Lift	DVVL	99 - 114	143 - 164	N/A	Previous Tech
Continuously Variable Valve Lift	CVVL	49 - 56	128 - 147	N/A	Previous Tech
Cylinder Deactivation	DEACD	N/A	118	133	Previous Tech
Variable Valve Actuation (CCP + DVVL)	VVA	N/A	N/A	235 - 271	Baseline for OHV
Stoichiometric Gasoline Direct Injection	SGDI	164	246	296	Previous Tech
Turbocharging and Downsizing Level 1 - 18 bar BMEP 33%DS	TRBDS1	245 - 282	-110 to -73	788 - 862	Previous Tech
V6 to I4 and V8 to V6			-396* to -316*	700* - 800*	
Turbocharging and Downsizing Level 2 - 24 bar BMEP 50%DS	TRBDS2	155	155	261	Previous Tech
I4 to I3		-82* to -86*			
Cooled EGR Level 1 - 24 bar BMEP, 50% DS	CEGR1	180	180	180	Previous Tech
Cooled EGR Level 2 - 27 bar BMEP, 56% DS	CEGR2	310	310	523	Previous Tech
V6 to I4				-453* to -469*	
Other Technologies					
By 2025:					
Compression Ratio Increase (with regular fuel)	CRI-REG	50	75	100	Baseline
Compression Ratio Increase (with higher octane regular fuel)	CRI-HO	75	113	150	Baseline
Compression Ratio Increase (CR~13:1, exh. scavenging, DI (aka Skyactiv, Atkinson Cycle))	CRI-EXS	250	375	500	Baseline
Electrically Assisted Variable Speed Supercharger	EAVS-SC	1,302	998	N/A	Baseline
Lean Burn (with low sulfur fuel)	LBRN	800	920	1,040	Baseline
After 2025:					
Variable Compression Ratio	VCR	597	687	896	Baseline
D-EGR	DEGR	667	667	667	TRBDS1
Homogeneous Charge Compression Ignition (HCCI) + Spark Assisted CI[a]	SA-HCCI	450	500	550	TRBDS1
Gasoline Direct Injection Compression Ignition	GDCI	2,500	2,875	3,750	Baseline
Waste Heat Recovery	WHR	700	805	1,050	Baseline
Alternative Fuels:					
CNG-Gasoline Bi-Fuel Vehicle	BCNG	6,000	6,900	7,800	Baseline
Flexible Fuel Vehicle	FFV	75	100	125	Baseline
Ethanol Boosted Direct Injection (incr CR to 14:1, 43% downsizing)	EBDI	740	870	1,000	Baseline

* Costs with reduced number of cylinders, adjusted for previously added technologies. See Appendix T for the derivation of turbocharged, downsized engine costs.

[a] With TWC aftertreatment. Costs will increase with lean NO_x aftertreatment.

TABLE 2A.3 NRC Estimates of Low and High Most Likely Effectiveness Values (As a Percent Reduction in Fuel Consumption) Relative to NHTSA Estimates for SI Engine Technologies for I4 Engines

Technology	Abbrev.	Low and High Effectiveness Estimates Compared to NHTSA Estimates	Rationale and Analysis	Concerns
Low Friction Lubricants – Level 1	LUB1	0.5% - 0.8% Agree with NHTSA	Change in oil viscosity estimated to provide 0.5 % FC reduction.	All engines, such as high output, turbocharged engines may not be able to use low friction lubricants.
Engine Friction Reduction – Level 1	EFR1	2.0% - 2.7% Agree with NHTSA	Need to include dual cooling circuit for early warm-up for full benefit.	No manufacturer divulged comprehensive plans for incorporating low friction technologies.
Low Friction Lubricants and Engine Friction Reduction Level 2	LUB2_ EFR2	1.04% - 1.37% Agree with NHTSA	Change in viscosity estimated to provide the FC reduction, but engine design changes are required.	Requires 0W-16, currently being introduced and 0W-12, which has not yet been defined by SAE. Engine design changes not defined.
Variable Valve Timing – Intake Cam Phasing	ICP	2.1% - 2.7% Agree with NHTSA	Feasible based on lower pumping losses and higher thermodynamic efficiency.	
Variable Valve Timing – Dual Cam Phasing (over ICP)	DCP	2.0% - 2.7% Agree with NHTSA	Feasible based on lower pumping losses and higher thermodynamic efficiency.	Up to 4.5% FC reduction reported by vehicle manufacturer for combined DCP benefit.
Continuous Variable Valve Lift	CVVL	3.6% - 4.9% Agree with NHTSA	Feasible based on lower pumping losses.	A manufacturer reported up to 5% FC reduction with CVVL.
Cylinder Deactivation on DOHC	DEACD	0.44% - 0.66% Agree with NHTSA	Additional reductions in pumping and friction are small.	
Stoichiometric Gasoline Direct Injection	SGDI	1.5% Agree with NHTSA	Evaporative cooling allows higher CR and increased power for downsizing.	An enabler for high BMEP engines, but manufacturers are concerned about cost and particulate emissions.
Turbocharging and Downsizing – Level 1 (18 bar BMEP)	TRBDS1	Low: 11.1% - 13.9% High: 12.1% - 14.9% Agree with NHTSA	FC reductions in production vehicles generally lower. CR reduction and spark retard for knock control. Drivability enhancements.	NHTSA estimates reflect a range of applications. Assume 87 AKI (91 RON) fuel and good drivability with higher driveline ratios or modified torque converters.
Turbocharging and Downsizing – Level 2 (24 bar BMEP)	TRBDS2	Low: 14.4% - 18.1% High: 16.4 - 20.1% Agree with NHTSA	FC reductions in production vehicles generally lower. CR reduction and spark retard for knock control. Drivability enhancements.	NHTSA estimates reflect a range of applications. Assume 87 AKI (91 RON) fuel and good drivability with higher driveline ratios or modified torque converters.
Cooler Exhaust Gas Recirculation (EGR) – Level 1 (24 bar BMEP) (Incremental)	CEGR1	Low: 2.5% - 2.6% High: 3.5% - 3.6% Agree with NHTSA	Reduction of additional pumping losses limited after applying previous technologies	Effectiveness of EGR for knock control needs to be demonstrated. Reduced combustion speeds remain an issue at full load.
Cooled Exhaust Gas Recirculation (EGR) – Level 2 (27 bar BMEP) (Incremental)	CEGR2	1.0% - 1.4% Incremental Agree with NHTSA	FC reductions are theoretically possible, but increased friction and pumping losses may dominate.	Many manufacturers indicated 25 bar BMEP is the limit for turbocharging.

TABLE 2A.4 NRC Low and High Most Likely Direct Manufacturing Cost Estimates Relative to NHTSA Estimates for SI Engine Technologies for I4 Engine

		2025 Direct Manufacturing Cost			
		NRC Estimates ($)		NHTSA Estimates ($)	
Technology	Abbrev.	Low	High		Rationale
Low Friction Lubricants – Level 1	LUB1	3	3	3	Relative difference in cost of 0W-20 oil compared to 5W-30 oil.
Engine Friction Reduction – Level 1	EFR1	48	48	48	Greater risk for cost to increase for partially defined friction reductions actions.
Low Friction Lubricants and Engine Friction Reduction Level 2	LUB2-EFR2	51	51	51	Costs of 0W-12 oil without final specifications and associated engine changes are unknown
Variable Valve Timing – Intake Cam Phasing	ICP	31	42	31	Cost increase with inclusion of all components for VVT system.
Variable Valve Timing – Dual Cam Phasing (over ICP)	DCP	27	36	27	Cost increase with inclusion of all components for VVT system.
Discrete Variable Valve Lift	DVVL	99	114	99	Cost increase with inclusion of all components for CVVL system.
Continuous Variable Valve Lift	CVVL	49	56	49	Cost increase with inclusion of all components for CVVL system.
Cylinder Deactivation on DOHC	DEACD	118 (for V6)	118 (for V6)	118 (for V6)	Greater risk for higher cost. (Needs further study)
Stoichiometric Gasoline Direct Injection	SGDI	164	164	164	Based on cost teardown study, which is generally accepted.
Turbocharging and Downsizing – Level 1 (18 bar BMEP)	TRBDS1	248	282	248	Cost increase due to increased turbocharger and intercooler cost.
Turbocharging and Downsizing – Level 2 (24 bar BMEP)	TRBDS2	155 (incremental)	155 (incremental)	155 (incremental)	Turbocharging cost estimated to be 1.5 times cost for 18 bar BMEP engine.
Cooler Exhaust Gas Recirculation (EGR) – Level 1 (24 bar BMEP)	CEGR1	180 (incremental)	180 (incremental)	180 (incremental)	In range estimated by manufacturers and suppliers
Cooled Exhaust Gas Recirculation (EGR) – Level 2 (27 bar BMEP)	CEGR2	310 (incremental)	310 (incremental)	310 (incremental)	Turbocharging cost estimated to be 2.5 times cost for 18 bar BMEP engine.

TABLE 2A.5 EPA Fuel Economy Data Examples of Downsizing and Turbocharging

Table 1. EPA Fuel Economy Data Examples of Downsizing and Turbocharging
Automatic Transmission Unless Noted

				EPA Label			CAFE Unadjusted			Comparable Power to Weight Ratio CAFE Unadjusted			Adjusted % FC Reduction using LPM		
Model Year	Vehicle	Engine	Percent Downsizing	Comb FE MPG	% FE Improvement	% FC Reduction	Comb FE MPG	% FE Improvement	% FC Reduction	Adjusted MPG	% FE Improvement	% FC Reduction	Engine Features	Adjusted EPA Data for TRBDS	LPM Estimate for TRBDS
2014	Cadillac CTS	3.6L Nat. Asp.		22			28.4			30.2			ICP, GDI		
		2.0L Turbo	44%	23	4.5%	-4.3%	30.5	7.4%	-6.9%		1.1%	-1.0%	ICP, GDI, 21 bar BMEP	-1.0%	-14.1%
2015	Chev. Cruze	1.8L Nat. Asp.		27			35.1			35.2			DCP, MFI		
		1.4L Turbo	22%	30	11.1%	-10.0%	40.1	14.2%	-12.5%		14.0%	-12.3%	DCP, MFI, 18 bar BMEP	-12.3%	-10.4%
2015	Chev. Sonic	1.8L Nat. Asp.		28			37.8			38.4			DCP, MFI		
		1.4L Turbo	22%	31	10.7%	-9.7%	41.4	9.5%	-8.7%		7.7%	-7.2%	DCP, MFI, 18 bar BMEP	-7.2%	-9.0%
2014	Dodge Dart (Prem. Fuel)	2.4L Nat. Asp		27			36.2			37.6			DCP, DVVL, MFI		
		1.4L Turbo	42%	31	14.8%	-12.9%	41.1	13.5%	-11.9%		9.3%	-8.5%	DCP, DVVL, MFI, 21 bar BMEP	-8.5%	-10.2%
2014	Ford Edge	3.5L Nat. Asp.		22			28.6			30.5			LUB1, DCP, MFI,		
		2.0L Turbo	43%	24	9.1%	-8.3%	31.8	11.2%	-10.1%		4.4%	-4.2%	DCP, GDI, 21 bar BMEP	-3.3%	-10.8%
2014	Ford Escape	2.5L Nat. Asp.		25			32.9			32.5			LUB1, ICP, MFI		
		1.6L Turbo	36%	26	4.0%	-3.85%	34.6	5.2%	-4.9%		6.6%	-6.1%	LUB1, DCP, GDI, 18 bar BMEP	-2.7%	-8.7%
2015	Ford Explorer	3.5L Nat. Asp.		20			25.6			27.1			LUB1, DCP, MFI,		
		2.0L Turbo	43%	23	15.0%	-13.0%	29.8	16.4%	-14.1%		10.0%	-9.1%	DCP, GDI, 21 bar BMEP	-8.3%	-13.2%
2014	Ford F150	5.0L Nat. Asp		17			22.1			22.0			LUB1, DCP, MFI		
		3.5L Turbo	30%	18	5.9%	-5.6%	23.8	7.7%	-7.1%		8.1%	-7.5%	DCP, GDI, 18 bar BMEP	-6.7%	-10.0%
2014	Ford Fiesta MT SFE	1.6L Nat. Asp.		31			41.4			41.0			LUB1, DCP, MFI		
		1.0L Turbo	38%	36	16.1%	-13.9%	48.4	16.9%	-14.5%		18.0%	-15.3%	LUB1, EFR1,DCP, GDI, 18 bar BMEP, AERO, LLR	-10.4%	-8.8%
2015	Ford Fusion	2.5L Nat. Asp.		26			34.6			34.4			LUB1, ICP, MFI		
		1.5L Turbo	36%	28	7.7%	-7.1%	36.4	5.2%	-4.9%		5.8%	-5.5%	LUB1, DCP, GDI, 18 bar BMEP	-1.4%	-10.4%
2015	Ford Taurus	3.5L Nat. Asp.		23			29.5			31.5			LUB1, DCP, MFI,		
		2.0L Turbo	43%	26	13.0%	-11.5%	33.8	14.6%	-12.7%		7.1%	-6.7%	DCP, GDI, 21 bar BMEP	-6.0%	-13.3%
2014	Hyundai Sonata	2.4L Nat. Asp.		28			36.6			32.6			LUB1, DCP, GDI		
		2.0L Turbo	17%	25	-10.7%	12.0%	32.2	-12.0%	13.7%		-1.2%	1.2%	DCP, GDI, 18 bar BMEP	0.4%	-11.0%
2015	Kia Forte 5	2.0L Nat. Asp.		28			37.4			35.1			ICP, GDI		
		1.6L Turbo	20%	24	-14.3%	16.7%	31.9	-14.7%	17.2%		-9.0%	9.9%	DCP, GDI, 18 bar BMEP	13.2%	-11.0%
2014	VW Passat	2.5L Nat. Asp.		25			31.9			31.9			ICP, MFI		
		1.8L Turbo	28%	28	12.0%	-10.7%	36.2	13.5%	-11.9%		13.5%	-11.9%	ICP, GDI, 18 bar BMEP	-10.5%	-11.6%

3

Technologies for Reducing Fuel Consumption in Compression-Ignition Diesel Engines

INTRODUCTION

The compression-ignition (CI) diesel engine has long been used in the over-the-road, heavy-duty-vehicle sector of trucks and buses in the United States and is recognized as the most fuel-efficient internal combustion engine. However, the diesel engine has not penetrated the U.S. light-duty vehicle market, consisting of Class 1 and 2a[1] passenger cars and light trucks, with less than 1 percent of new LDVs sales in 2014. In contrast, diesel engines have significantly penetrated the Class 2b pickup and van market, with approximately 50 percent market share in 2014, and the light-duty passenger car and light truck vehicle market in Europe, with approximately 56 percent market share in 2013.

This chapter begins with a review of the fundamentals of CI engines and their role in the 2017-2025 MY final CAFE rulemaking. The next section discusses the available technologies for reducing fuel consumption, carbon dioxide, and criteria emissions in advanced diesel engines. This is followed by a discussion of estimated incremental costs of diesel engines relative to baseline gasoline engines. The incremental retail prices of diesel-powered vehicles relative to gasoline-powered vehicles are reviewed together with the current and projected future diesel vehicle offerings. The low market share penetration of diesel-engine-powered vehicles in the United States currently and the role of these vehicles in achieving the 2017-2025 CAFE standards will be discussed. New and emerging technologies, including the use of alternative fuels in diesel engines and advanced combustion systems, are discussed. The chapter concludes with the committee's findings and recommendations regarding diesel engines. Estimates of effectiveness in fuel consumption reductions and costs are developed throughout the chapter, and a complete set of estimates for the diesel technologies applied to a midsize car, a large car, and a large light truck are provided in Table 3A.1 for effectiveness of reduction in fuel consumption and Tables 3A.2a, b, and c for direct manufacturing costs (see annex at end of this chapter).

COMPRESSION IGNITION ENGINE EFFICIENCY FUNDAMENTALS

Light-duty CI engines operating on diesel fuels have the highest thermodynamic cycle efficiency of all light-duty engine types. The diesel thermodynamic cycle efficiency advantage over the more common spark-ignition (SI) gasoline engine stems from three major factors: the diesel engine's use of lean mixtures, its lack of need for throttling the intake charge, and its higher compression ratios. The diesel thermodynamic cycle is shown on a P-V diagram and compared with the Otto cycle representation of the SI engine in Appendix D.

Lean fuel mixtures are thermodynamically more efficient than rich mixtures because of the higher ratio of specific heats. Such mixtures are enabled by the diesel combustion process. In this process, diesel fuel, which has chemical and physical properties to ensure that it self-ignites readily, is injected into the cylinder late in the compression stroke. This ability to operate on overall lean mixtures allows diesel engine power output to be controlled by limiting the amount of fuel injected without resorting to throttling the amount of air inducted. This attribute leads to the second major factor enabling the higher efficiency of the diesel engines, namely the absence of throttling during the intake process, which otherwise leads to negative pumping work. Finally, the diesel combustion process needs higher compression ratios to ensure ignition of the heterogeneous mixture without spark ignition. The higher diesel compression ratios (e.g., 16-18 versus 9-11 for SI gasoline engines) improve thermodynamic expansion efficiency, although some of the theoretical gain is lost due to increased ring-to-bore wall friction from higher

[1] The Federal Highway Administration (FHWA) organizes vehicles into classifications. Light-duty Class 1 and 2a vehicles are under 8,500 lb. gross vehicle weight (GVW), although medium-duty passenger vehicles (MDPVs) from 8,500 lb to 10,000 lb. GVW are included. Class 2b pickups are classified as heavy-duty trucks with a gross vehicle weight rating (GVRW) between 8,500 lb and 10,000 lb as classified by the FHWA.

cylinder pressures and friction from larger bearings required to withstand the resulting higher loads in the engine.

The diesel-engine-powered vehicles also achieve more miles per gallon than SI gasoline powered vehicles due to the higher heating value of diesel fuel (128,450 Btu/gal) vs. gasoline (116,090 Btu/gal). The approximately eleven percent higher heating value results in an eleven percent better fuel consumption on a volumetric basis. At the pump, diesel fuel also costs more than regular gasoline in most areas of the United States. For the first 6 months of 2014, the average price of diesel fuel was $3.87/gal, while regular gasoline was $3.59/gal, or approximately 8 percent higher costs for diesel fuel. However, early projections of diesel fuel prices in 2015 expect a national average of $2.84/gal and a rise to $3.24 in 2016 (EIA 2015). The effect of this price decrease on diesel vehicle penetration in the U.S. market is currently unknown.

The exhaust emissions from diesel engines have been regulated since the 1960s for light-duty diesels and 1973 for heavy-duty diesels. The current regulations of the Environmental Protection Agency (EPA) and the California Air Resources Board (CARB) require control of criteria emissions of hydrocarbons (HC), carbon monoxide (CO), oxides of nitrogen (NO_x), and particulate matter (PM). The CO emissions of diesels are inherently low due to lean combustion, and HC emissions are low compared to gasoline engines. The NO_x and PM emissions have been controlled through engine technology, but recently more stringent standards have resulted in aftertreatment being used for PM (which is reduced with diesel particulate filters [DPF]) and NO_x (which is controlled with selective catalytic reduction [SCR]). This is discussed in greater detail later in the chapter.

Although diesel fuel has higher energy content per gallon than gasoline, it also has a higher carbon density that results in approximately 15 percent more carbon dioxide emitted per gallon of diesel fuel relative to a gallon of gasoline. Diesel produces 10,180 g of carbon dioxide per gallon when burned, while gasoline produces 8,887 g of carbon dioxide per gallon (EPA/NHTSA 2012a). The EPA/National Highway Traffic Administration (NHTSA) Joint Technical Support Document refers to the additional carbon dioxide released from the burning of a gallon of diesel, relative to the burning of a gallon of gasoline, as the "carbon penalty" (EPA/NHTSA 2012b). Due to the cited "carbon penalty,"a diesel vehicle yields greater fuel economy improvements compared to its CO_2 emissions reduction improvements. Another consideration is that diesel fuel is generally slightly more efficient to refine than gasoline, and there is a potential CO_2 and energy benefit when refining crude oil to diesel as compared with refining crude to gasoline. This possible offset is not accounted for in the Agencies' regulations.

This issue is amplified in the final CAFE rule, which states that the "163 g/mi [carbon dioxide standard] would be equivalent to 54.5 mpg, . . . [assuming] gasoline fueled vehicles (significant diesel fuel penetration would have a different mpg equivalent)" (EPA/NHTSA 2012a). EPA and NHTSA cite the additional carbon dioxide released from burning diesel fuel, compared to burning gasoline, as one of the reasons why manufacturers might not invest significantly in diesel engine technologies as a way to comply with the CAFE and GHG standards for MY 2017-2025 (EPA/NHTSA 2012b).

EPA/NHTSA 2017-2025 CAFE Rulemaking

The Agencies' CAFE rulemaking for the 2017-2025 time frame relies heavily on the analysis done in the previous 2012-2016 rulemaking. The 2017-2025 rulemaking acknowledges the benefits of diesel engines regarding reduced pumping losses, improved torque, diesel fuel's higher energy content compared to gasoline, and lean combustion. In spite of these benefits, EPA and NHTSA's Joint Technical Support Document (TSD) recognizes the challenges that manufacturers will face regarding tailpipe emissions of diesel vehicles due to the "carbon penalty" and the NO_x reductions required in the U.S. Tier 2 Bin 2 standards. In addition, it is recognized that diesels will also need to meet the EPA Tier 3 rules (EPA/NHTSA 2012b; EPA 2014a) introduced in March 2014 (see Chapter 2).

In order to meet the stricter CAFE/GHG standards, the Agencies acknowledged the potential need for vehicle manufacturers to include diesels in their product strategies. According to the TSD, several vehicle manufacturers have indicated to the Agencies that diesels will be part of their strategy to meet the midterm goals. Manufacturers that produce more diesel-engine-powered vehicles have also informed the Agencies that they expect diesel technologies to be part of a feasible strategy for reducing fuel consumption, carbon dioxide, and NO_x emissions in the future.

In analyzing the technology, the TSD discusses the challenges of reducing diesel emissions to meet future requirements while acknowledging the fuel consumption reduction benefits of diesel engines. The approach to reducing emissions will include a combination of improvements to the combustion system to reduce emissions leaving the engine and improvements to the aftertreatment system. Technologies to improve the combustion system include fuel systems with higher injection pressures and multiple injection capabilities, advanced control systems, higher levels of cooled exhaust gas recirculation (EGR) to reduce NO_x emissions, and advanced turbocharger systems. The aftertreatment system will continue to consist of a diesel oxidation catalyst followed by a DPF, or a catalyzed DPF and an SCR system for NO_x reductions.

During the analysis, the Agencies used performance as the equalizing metric to compare diesel and gasoline vehicles. For smaller vehicles, the Agencies applied an I4 diesel engine with a displacement of 2.0L to replace a larger displacement I4 gasoline engine. For large cars and mid-sized trucks, a large I4 diesel engine with a displacement of 2.8L was used instead of a larger displacement V6 gasoline engine. For large light trucks and performance cars, a V6 diesel engine with a 4.0 L displacement was used instead of the larger displacement V8 gasoline engine. It was also assumed that all new diesels would include SCR aftertreatment.

In the cost analysis, there have been four major changes from the 2012-2016 rulemaking to the recent analysis. The first is that only SCR-based systems would be used for NO_x control for all diesel engines, whereas lean NO_x traps had been previously used for I4 engine applications. Second, the Agencies assumed that the vehicle manufacturers would meet Tier 2 Bin 2 emission standards rather than the previously assumed Tier 2 Bin 5 standards. Tier 2 Bin 2 emissions standards were assumed to require catalyst volumes 20 percent larger. Third, the Agencies updated the platinum group metals cost[2] from the March 2009 values used for the 2012 to 2016 CAFE standards to the February 2011 values, which resulted in a 69 percent increase in cost per troy ounce. As of November 2014, the value of the platinum group metals is lower by 32 percent from the February 2011 value. EPA acknowledged that there is not a good option for handling platinum group metal costs but elected to be transparent by using the most recent price and reporting its basis. Finally, a \$50 direct manufacturing cost was added for improvements associated with fuel and urea controls.

FUEL CONSUMPTION REDUCTION EFFECTIVENESS

NRC Phase 1 Study

The NRC Phase 1 study used the EPA full system simulation carried out by Ricardo, Inc. in 2008 to assess the fuel consumption and CO_2 reductions of diesel engines relative to gasoline engines in three vehicle classes (NRC 2011). The results from this full system simulation were used to develop the committee's estimate of fuel consumption reduction effectiveness, as described in this section. The engine system layout studied by Ricardo is shown in Figure 3.1 for a six-cylinder diesel engine in a passenger car. The following system components are included in this diesel engine layout.

Gas Handling System

A single-stage, variable nozzle turbocharger with air-to-air charge air cooling was used for boosting for I4 and V6 diesel engines for cars. For the large light-truck applications with a V6 diesel engine, boosting was accomplished with a two-stage, series sequential turbocharging system. The low-pressure turbine was fixed geometry with a wastegate. The high-pressure turbocharger included a variable nozzle turbine. High levels of cooled EGR were provided with a single-stage, high-pressure EGR system and an EGR cooler, although additional cooling was expected to be required in the 2017 to 2025 time frame. The EGR system included a cooler bypass to aid in cold start, light load emissions, and transient operation. Ricardo indicated that this configuration would likely require an EGR diesel oxidation catalyst (DOC) to mitigate fouling issues in the EGR cooler and intake system.

Combustion System

The geometric compression ratio of the engine was 17.5:1. The fuel system was a high-pressure common rail (HPCR) system with 1,800 bar solenoid-operated injectors. Glow plugs were used to aid in cold start, with one or more having cylinder pressure sensing capability for adaptation to fuel cetane variations.

Aftertreatment

Aftertreatment included a DOC, a DPF, and a Lean NO_x Trap (LNT) (Figure 3.1). However, SCR was used for the V6 diesel engine for a large, light truck.

The results of the Ricardo full system simulations are summarized in the upper portion of Table 3.1. As will be discussed later in this chapter, the table shows that the reductions in CO_2 emissions for the diesel relative to the gasoline engine are approximately 10 percentage points less than the reductions in fuel consumption. In this study, Ricardo evaluated the diesel powertrains in combination with other non-diesel fuel consumption reduction technologies, which were only applied to the diesel vehicles and not to the baseline gasoline vehicles. To normalize the results of these full system simulations, the committee applied estimates of the fuel consumption reduction effectiveness of these technologies from the NHTSA regulatory impact analysis (RIA), as shown in the lower portion of Table 3.1 (NHTSA 2012). After normalizing the full system simulation results for comparable technologies applied to both the diesel and gasoline vehicles, the diesel vehicles were shown to provide fuel consumption reductions of 26 to 29 percent relative to the gasoline vehicles.

For the 2017-2025 CAFE final rule, NHTSA combined "conversion to diesel" and "conversion to advanced diesel" into one technology labeled "advanced diesel." Conversion to advanced diesel was defined in the NRC Phase 1 study as consisting of downsizing, downspeeding, friction reduction, and combustion improvements. The normalized fuel consumption reductions for diesel engines listed in Table 3.1 already included some of the features considered as part of NHTSA's description of "conversion to advanced diesel". Specifically, the Ricardo full systems simulations included downsizing with a 20 percent reduction in displacement. Another feature of the conversion to advanced diesel engine was downspeeding, but this technology was defined as increasing transmission ratios beyond 6, which would be accounted for as a transmission technology and is not included in the

[2] In March 2009, platinum cost \$1,085 per troy ounce; in February 2011 it was \$1,829; and in November 2014 it was \$1,240.

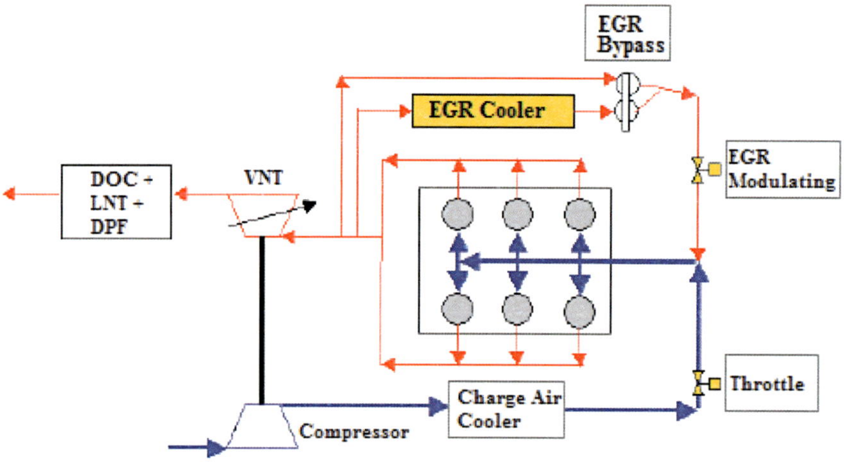

FIGURE 3.1 Schematic of the V6 diesel engine system for a passenger car used in the Ricardo full system simulation.
SOURCE: EPA/Ricardo (2008).

TABLE 3.1 Estimated Fuel Consumption and Carbon Dioxide Reductions for Diesel Engines Relative to Gasoline Engines

Ricardo Full System Simulation Results Based on EPA (2008)	Small MPV	Full-Size Car	Truck
Gasoline engine	2.4L I4	3.5L V6	5.4L V8
Transmission	4-sp AT	5-sp AT	4-sp AT
Diesel engine	1.9L I4	2.8L I4	4.8L V8
Transmission	6-sp DCT	6-sp DCT	6-sp DCT
Gasoline CO_2 (g/mi)	316	356	517
Diesel CO_2 (g/mi)	247	273	391
Reduction (%)	21.8	23.3	24.2
Gasoline fuel consumption (gal/100 mi)	3.596	4.051	5.883
Diesel fuel consumption (gal/100 mi)	2.449	2.707	3.877
Reduction (%)	31.9	33.2	34.1
Normalization of diesel vs. gasoline engines without non-diesel features	Small MPV	Full-Size Car	Truck
Gasoline engine	2.4L I4	3.5L V6	5.4L V8
Transmission	4-sp AT	5-sp AT	4-sp AT
Diesel engine	1.9L I4	2.8L I4	4.8L V8
Transmission	6-sp DCT	6-sp DCT	6-sp DCT
Deletion of non-diesel features added only to diesel vehicle simulation (reduction in fuel consumption [%][a])			
6-sp AT replacing 4-sp AT	2.0	2.0	2.1
6-sp DCT replacing 6-sp AT	4.1	3.8	3.8
Efficient accessories and high efficiency alternator	1.2	1.0	1.6
Electric power steering	1.3	1.1	0.8
Multiplicative total	8.4	7.7	8.1
Normalized diesel fuel consumption (gal/100 mi)	2.655	2.915	4.191
Percent reduction for diesel w/o added features vs. gasoline (%)	26.2	28.0	28.7

[a] Percent reductions in fuel consumption taken from NHTSA RIA (2012).
NOTE: AT, automatic transmission; DCT, dual clutch transmission; MPV, midsize passenger vehicle
SOURCE: NRC (2011) Table 5.1.

accounting for diesel engine fuel consumption reduction effectiveness improvements.

Applying the foregoing considerations, the estimated fuel consumption reductions for NHTSA's advanced diesel were derived from the estimates for "conversion to diesel" shown in Table 3.1 by including the applicable technologies for "conversion to advanced diesel," and the results are shown in Table 3.2. The fuel consumption reductions estimated by the committee agreed closely with NHTSA's estimates shown in the last line of Table 3.2.

2014 MY Diesel Vehicle Fuel Consumption Reductions-EPA Certification Data

The EPA certification fuel economy (two-cycle combined CAFE) data of comparable current vehicles with gasoline and diesel engines were used to determine the reduction in fuel consumption provided by the diesel vehicles. The 36 diesel-powered 2014 MY vehicles were identified from the Diesel Technology Forum website. The fuel economy comparison for all of these vehicles is provided in Annex Table 3A.3 at the end of this chapter. The fuel consumption reduction data for diesel vehicles shows a range of 14 percent to 36 percent. A summary of the results for several of the comparisons of vehicles with automatic transmissions is shown in Table 3.3. The comparisons in this table were normalized to approximately equal performance by using the methodology described in Chapter 2. The diesel power was adjusted first to account for the characteristic higher peak torque relative to a naturally aspirated gasoline engine with comparable power. The performance of the gasoline engines were subsequently normalized to the adjusted power-to-weight ratio of the diesel vehicles using NHTSA's formula (discussed in Chapter 2). The normalization resulted in minor changes to the results, since there were only small differences in power-to-weight ratio for the diesel and gasoline vehicles being compared in most cases. The reductions in fuel consumption provided by the diesel vehicles ranged from a low of 6.7 percent to a high of 28.9 percent compared to gasoline engine vehicles from the same manufacturer.

Table 3.3 shows the effectiveness of today's diesel vehicles compared to their gasoline versions as having an average of 20 percent reduction in fuel consumption and a range of 6.7-28.9 percent reduction in fuel consumption using EPA's normalized power-to-weight ratio conversions. Several reasons may be responsible for this lower average result and the wide range of results. First, many of the 2014 MY diesel vehicles had not fully implemented all of the features defined for the advanced diesel engine, which the committee estimates to result in a 29.0-30.5 percent reduction in fuel consumption. Second, there may be differences in the features of the diesel and baseline gasoline vehicles

TABLE 3.2 Summary of Diesel Vehicle Fuel Consumption Based on NRC Phase One Study (percent reduction in fuel consumption)

Technology Improvements	I4	V6	V8
Conversion to diesel[a] (%)	26.2	28.0	28.7
Conversion to advanced diesel			
Downsizing[b] (%) 20% reduction in displacement Two-stage turbocharger	4.0[b]	4.0[b]	4.0[b]
Downspeeding[c] (%) — Increasing transmission ratios beyond 6	1.5[c]	1.5[c]	1.5[c]
Friction reduction (%) Dual pressure oil pump Nonrecirculating low-pressure fuel pump	1.5	1.5	1.5
Combustion improvement (%) Greater than 2,000 bar injection pressure Piezo Injectors	3	3	3
Multiplicative total (%)	4.5	4.5	4.5
Total diesel fuel consumption reduction (%)	29. 5	31.3	31.9
NHTSA estimated fuel consumption reduction (%)	29.4	30.5	29.0

[a] From Table 3.2 of the NRC (2011) report.
[b] Downsizing by 20 percent was already included in the Ricardo full system simulation and is not included in the accounting for the diesel engine fuel consumption reduction effectiveness.
[c] The downspeeding benefit is attributed to an increased number of transmission ratios beyond 6. This reduction in fuel consumption is accounted for as a transmission technology and is not included in the accounting for the diesel engine fuel consumption reduction effectiveness.
SOURCE: NRC (2011), Tables 5.8 and G.1.

TABLE 3.3 Comparison of Diesel and Gasoline Vehicle Fuel Consumption Using EPA Certification Data Normalized for Power-to-Weight Ratio

Vehicle	Diesel Vehicle Power (hp)	Diesel Vehicle Weight wt (lb)	Diesel Power – to – Weight Ratio (hp/lb)	Adjusted Power (hp) for Torque[a]	Gasoline Vehicle Power (hp)	Gasoline Vehicle Weight wt (lb)	Gasoline Power – to – Weight Ratio (hp/lb)	Gasoline gpm Mod/ Gasoline gpm	Diesel gpm % Reduction Relative to Gas gpm Mod
Audi A6	240	4,178	0.057	0.061	220	3,726	0.059	1.01	26.3
Audi A7	240	4,167	0.058	0.061	310	4,167	0.074	0.93	22.6
Audi A8L	240	4,564	0.053	0.056	333	4,365	0.076	0.90	18.1
Audi Q5	240	4,475	0.054	0.057	220	4,079	0.054	1.02	22.0
Audi Q7	240	5,412	0.044	0.047	280	5,192	0.054	0.97	15.0
BMW 328i	181	3,460	0.052	0.055	180	3,410	0.053	1.02	28.9
BMW 535i	255	4,255	0.060	0.064	241	3,814	0.063	1.00	23.7
Chevy Cruze	151	3,471	0.044	0.046	138	3,206	0.043	1.02	22.5
Porsche Cayenne	240	4,797	0.050	0.053	300	4,398	0.068	0.92	6.7
VW Beetle	140	3,157	0.044	0.047	170	2,948	0.058	0.93	15.0
VW Golf	140	3,120	0.045	0.048	170	3,102	0.055	0.95	21.6
VW Jetta	140	3,457	0.040	0.047	115	2,804	0.041	1.05	28.0
VW Passat	140	3,494	0.040	0.042	170	3,221	0.053	0.93	23.1
VW Touareg	240	4,974	0.048	0.051	280	4,711	0.059	0.96	11.9

[a]Adjusted hp/wt to account for approximately 6 percent more torque of diesel during 0-60 mph acceleration (power equal to gasoline engine.
NOTE: gpm, gal/mile; Mod, modified for equal power/weight.

being compared that were not identified in the EPA database. Third, there may be a difference in the maturity of the diesel and gasoline engines that are being compared.

These comparisons of diesel and gasoline vehicles provided an opportunity to determine if a relationship existed between downsizing and reduction in fuel consumption for the diesel vehicles. However, no significant correlation was found due in part to the wide range of results. However, engines with the larger percentage downsizing consistently provided reductions in fuel consumption with generally higher values.

COMBUSTION IGNITION ENGINE CRITERIA EMISSION REDUCTION

Criteria Emission Standards

Various studies used throughout this chapter reference a variety of criteria emission standards. The referenced federal and California emission standards are listed in Table 3.4. The NO_x standard is the same for the federal Tier 2 Bin 5 and California ULEV II standards, although the non-methane organic gases (NMOG) standard is significantly lower for the ULEV II standard. Vehicles can be certified to the ULEV II standard and be compliant under the Tier 2 Bin 5 standard. The NO_x standard for Tier 2 Bin 2 is significantly lower than the Tier 2 Bin 5 standard. A similar level of NO_x and NMOG control is provided with the recently published federal Tier 3 Bin 30 and California LEV III standards (since NMOG + NO_x = 0.03 g/mi is equivalent to NMOG = 0.01 g/mi plus NO_x = 0.02 g/mi). Relative to the Tier 2 Bin 2 standard, the PM standard has been lowered from 0.010 g/mi to 0.003 g/mi for the federal Tier 3 Bin 30 and the California LEV III standards. By 2028, the California LEV III PM standard is reduced further to 0.001 g/mi.

HC/CO Control

In spite of relatively low exhaust temperatures, the control of HC/CO has traditionally been relatively easy for diesel engines due to the relatively low levels of these constituents emitted from conventional diesel combustion. However, that situation has changed as the diesel combustion process has been modified to reduce combustion-gas temperatures, which reduces exhaust temperatures even further. As the combustion temperatures have been reduced, HC/CO emissions have risen. The DOC was introduced around 1996 to reduce hydrocarbon emissions and to reduce the soluble organic fraction of the particulate matter. As a result of the reduced exhaust temperatures noted above, the DOC is being moved closer to the turbocharger outlet to increase the temperature of the catalyst to increase its conversion effi-

TABLE 3.4 Criteria Emission Standards (g/mile) Relevant in This Study to Diesel Engines in Light-Duty Vehicles

Emission Standard	Federal or California	NMOG[a] + NO_X	NMOG[a]	CO	NO_x	PM	Mileage
ULEV II	California		0.055	2.1	0.07	0.010	120,000
Tier 2 Bin 5	Federal EPA		0.09	4.2	0.07	0.010	120,000
Tier 2 Bin 2	Federal EPA		0.01	2.1	0.02	0.010	120,000
Tier 3 Bin 30	Federal EPA	0.03		1.0	NA	0.003	150,000
LEV III	California	0.03		1.0	NA	0.003	150,000
LEV III	California	0.03		1.0	NA	0.001	By 2028

[a]For diesel-fueled vehicles, NMOG means non-methane hydrocarbons (NMHC).

ciency. Oxidation catalyst coatings are being added to diesel particulate filters (with DPFs thus becoming catalyzed DPFs) for additional HC/CO control and to provide the temperature rise required for regeneration, as discussed the next section.

Particulate Control

Particulate matter is being controlled using DPFs. These PM filters are quite effective, filtering out 90 to 99 percent of the PM from the exhaust stream, making diesel engines more attractive from an environmental impact point of view. However, PM accumulates in the filters and imposes additional back pressure on the engine's exhaust system, thus increasing pumping work done by the engine. This increase in pumping work increases fuel consumption. In addition, there is a second fuel consumption penalty caused by the additional fuel required to regenerate the filter by oxidizing retained PM. The low exhaust temperatures encountered in light-duty automotive applications of these filters are insufficient to passively oxidize the accumulated PM. As a result, exhaust temperatures must be increased by injecting fuel (most frequently in the engine cylinder after combustion has been completed) which will be oxidized as the fuel and exhaust gas mixture passes over the DOC or catalyzed DPF. These hot gases flow directly to the DPF to oxidize the PM retained in the filter. Engine control algorithms for filter regeneration not only must sense when the filters need to be regenerated and control the regeneration without overheating the filter, but these algorithms must also contend with other events like the driver turning off the engine while regeneration is underway, thus leaving an incompletely regenerated filter. When the vehicle is then restarted, the control algorithms must appropriately manage either completion of the regeneration or start of a new filling and regeneration cycle. These algorithms have become quite sophisticated, with the result that PM filter systems are quite reliable and durable.

NO_x Control

There are two approaches to aftertreatment of NO_x emissions: NO_x storage and reduction catalysts, which are also called lean NO_x traps, and selective catalytic reduction devices using ammonia as a reducing agent for NO_x.

NO_x Storage Catalysts

NO_x storage catalysts (NSC) utilize a typical monolith substrate that has barium and/or potassium as well as a precious metal (e.g., platinum) in its coatings. These coatings adsorb NO_x from the exhaust gas to form nitrates, thereby storing the NO_x in the catalyst. As NO_x is adsorbed from the exhaust, adsorption sites on the surface of the coating fill up. Once all the coating sites are filled with adsorbed NO_x, the NSC can no longer adsorb additional NO_x so it passes through the NSC without being adsorbed. Since this pass-through would not be acceptable, before the catalyst is completely filled the NSC must be regenerated to purge the adsorbed NO_x and free the sites to adsorb additional NO_x. By supplying the NSC with a rich exhaust stream containing CO and hydrogen, the CO and H_2 molecules desorb the NO_x from the catalyst surface and reduce it to N_2, H_2O, and carbon dioxide. Therefore, like the PM filter, the NSC operates in a cyclic fashion, first filling with NO_x and then purging the NO_x.

Selective Catalytic Reduction

Selective catalytic reduction (SCR) has been applied to heavy-duty diesel engines in the United States since 2007 and also in Europe. SCR was introduced in the U.S. in light-duty vehicle applications in 2009 on some Mercedes, BMW, and VW diesel vehicles. This system, called BlueTEC, was jointly developed by three manufacturers. SCR functions by injecting ammonia in the form of an aqueous solution of urea into the exhaust stream, which then passes over a copper-zeolite or iron-zeolite SCR catalyst. The aqueous solution of ammonia is called diesel exhaust fluid (DEF) in the U.S. and AdBlue in Europe, and must be carried on board the vehicle in sufficient quantities. The ammonia reacts with the NO_x over the SCR catalyst, reducing the NO_x to N_2 and water. The amount of urea that needs to be supplied to the SCR catalyst depends on the level of NO_x in the exhaust and therefore depends on driving conditions, but for light-duty vehicles it is a small fraction of the fuel flow.

Combined NSC and SCR Systems

Another strategy that has been proposed is to use a system in which the NSC is followed by SCR without external urea addition. Under some operating conditions with the appropriate washcoat formulation, NSCs can convert NO_x to ammonia, which is undesirable for an NSC-only system and hence must be cleaned up before exiting the exhaust system. However, by following the NSC with a SCR catalyst without urea injection, which is generally called passive SCR, the SCR will capture and store the ammonia generated by the NSC and use it to reduce NO_x. Since the amount of ammonia generated by the NSC is not large, the passive SCR unit will have low conversion efficiencies but can be a useful supplement to the NSC system. This approach has been used by Mercedes in its BlueTEC I system in Europe.

SCRF, SCR on a DPF

The SCR-on-filter (SCRF) technology consists of applying an SCR coating on the DPF in the current DOC + DPF + SCR system. This system results in positioning the SCRF closer to the engine for improved light-off, improved low-temperature/cold-start performance, and lower back pressure. Several SCRF configurations are currently being developed. One configuration consists of two layers of catalyst, with the first coat using a silica and Ce/Zr SCR formulation and with the second coat using Fe/zeolite or vanadium (Michelin et al. 2014). Another configuration consists of a Cu-zeolite formulation for better high-temperature durability (Johansen et al. 2014). An alternative concept proposes a DOC + SCRF + underfloor SCR with the SCRF optimized for high NH_3 storage and high NO_x conversion and the underfloor SCR catalyst used for NH_3 slip management and extra NO_x conversion during high load conditions (Wang et al. 2007).

In summary, diesel emission control continues to include basic improvements in diesel combustion and extensive use of aftertreatment technology. Improvements are continuing to be researched. Aftertreatment technology is expensive and results in small increases in fuel consumption due to the extra fuel required to regenerate the DPF and the increased back pressure resulting from the DPF. In addition, the operating costs will increase as the cost of the DEF must be added to the cost of the fuel.

DIESEL ENGINE AND DIESEL VEHICLE COST DATA

Costs from NRC Phase 1 2011 Report

In the absence of teardown cost studies for current diesel engines, costs were developed using input from previous studies together with costs contained in the EPA/NHTSA Technical Support Document. Developing updated costs included several steps. First, the costs for diesel engines developed in the NRC Phase 1 report *Assessment of Fuel Economy Technologies for Light-Duty Vehicles* were reviewed and used for diesel engines complying with the Tier 2 Bin 5 emissions standards. In the next step, costs of the appropriate features of NHTSA's "conversion to advanced diesel" were added to the Tier 2 Bin 5 diesel engines. Finally, the incremental costs to achieve Tier 3 emission standards were developed and added to the Tier 2 Bin 5 advanced diesel. These steps are described in the following sections.

Diesel Costs at Tier 2 Bin 5 Emission Levels—From NRC Phase 1 Report

Incremental costs of diesel engines at Tier 2 Bin 5 emissions levels were developed in the NRC Phase 1 report. These costs were developed from a Martec study in 2007-2008 using a methodology referred to as a bill of materials process (Martec 2008). The Martec study is described in Annex 3B. Martec sought input from vehicle manufacturers and suppliers with the goal of reaching consensus and agreement on the cost estimates. The costs developed in the Phase 1 study are summarized in Table 3.5. Cost are included for the common rail fuel injection system; the variable geometry turbocharger with air-to-air charge air cooler; electrical upgrades; engine upgrades; high-pressure and low-pressure EGR system; Tier 2 Bin 5 emissions control system; and onboard diagnostics and associated sensors. Credits were provided for SI-engine-related components that were deleted.

CONVERSION TO ADVANCED DIESEL—FROM NRC PHASE 1 REPORT

The "conversion to advanced diesel" required adding the costs for engine downsizing, friction reduction, a low-pressure EGR system, and high-pressure piezo injections. The costs of these features are listed in Table 3.6 and added to the previously developed diesel costs to provide the advanced diesel incremental costs. Engine downsizing requires the application of two-stage turbocharging for all engine configurations. Additionally, to maintain equal power and torque, the downsized engines will have to operate at higher brake mean effective pressure (BMEP) levels, which require higher-pressure-compatible bearings, stronger materials, higher-temperature-capable valve seats, more expensive head gasket materials, and noise, vibration and harshness (NVH)-control technologies, all of which would increase the cost. The incremental cost for the EGR system results from the addition of two-stage turbocharging to the advanced diesel engines. Friction reductions were achieved with a dual-pressure oil pump and a non-recirculating low-pressure fuel pump. The common rail fuel injection pressure for the advanced diesel was increased from 1,800 bar to over 2,000 bar, and piezo injectors replaced solenoid injectors.

TABLE 3.5 Diesel Engine and Vehicle Incremental Direct Manufacturing Costs at ULEV II (Tier 2 Bin 5) Emissions from NRC Phase One Study (2008 dollars)

SI Gasoline Engine CI Diesel Engine	SI: 2.4L I4 CI: 2.0L I4	SI: 4.0L-4.2L V6 CI: 3.5L V6	SI: 5.3L-6.2L V8 CI: 3.5L V6
Common rail 1,800 bar fuel system[1]	675	911	911
Variable geometry turbocharger (VGT) with air-air charge air cooler	375	485	830[a]
Electrical upgrades: starter motor, alternator, battery can cabin heater	125	167	167
Engine upgrades: cam, crank, con rods, bearings, pistons, oil lines, countermeasures to engine/vehicle NVH	161	194	194
HP/LP EGR system[2,b]	215	226	226
DOC+DPF+SCR (NSC for I4)[c] (Included PGM cost in 2009$) (Included urea dosing system)	688 (PGM 597) (NSC)	964 (PGM 296) (urea sys. 363)	1,040 (PGM 372) (urea sys. 363)
Onboard diagnostics and sensors[3]	154	227	227
Total diesel incremental cost	2,393	3,174	3,595

Credits included in costs shown above	I4 to I4	V6 to V6	V8 to V6
[1] SI engine content deleted	(32)	(48)	(48)
[2] PFI emissions and evaporative system deleted	(245)	(343)	(200)
[3] Switching oxygen sensors deleted	(18) for 2	(36) for 4	(36) for 4

NOTES: HP/LP, high-pressure/low-pressure; NVH, noise, vibration, and harshness; PFI, port fuel injection; PGM, platinum group metals; NSC, NO_x storage and reduction catalysts.
[a] One variable geometry turbocharger and one fixed geometry turbocharger.
[b] HP EGR system is shown on Ricardo diesel system schematic.
[c] Tier 2, Bin 5.
SOURCE: NRC (2011).

TABLE 3.6 Summary of Tier 2 Bin 5 Diesel and Conversion to Advanced Diesel Incremental Costs (dollars)

	Average Direct Manufacturing Cost (2008 dollars)		
	I4	V6	V8
Conversion to diesel	2,393	3,174	3,595
Conversion to advanced diesel[a]			
Downsizing	50	75	75
Two-stage turbocharger	375	375	0
Dual-pressure oil pump	5	6	6
Non-recirculating low-pressure fuel pump	10	12	12
EGR system enhancement		95	95
High-pressure (>2,000 bar) piezo injectors	80	120	120
Total	520	683	308
Advanced diesel incremental cost	2,913	3,857	3,903

[a] From Table 5.8 of NRC Phase 1 report (NRC 2011).

TIER 3 FROM TIER 2 BIN 5 INCREMENTAL COSTS – FROM TSD

The incremental costs to achieve Tier 2 Bin 2 emission standards (similar to Tier 3 for NMOG + NO_x, as shown in Table 3.4) relative to Tier 2 Bin 5 emission standards were developed by comparing NHTSA's direct manufacturing costs for the 2017-2025 CAFE standards at Tier 2 Bin 2 emission standards with the direct manufacturing costs for the 2012-2016 CAFE standards at Tier 2 Bin 5 emissions standards. This comparison is shown in Table 3.7, in which the incremental costs for Tier 2 Bin 2 emissions controls were developed. This comparison included the following steps. First, the GDP modifier was applied to NHTSA's estimated cost for diesel at Tier 2 Bin 5 emission standards to convert them from 2007 dollars to 2010 dollars. Next, reverse learning was applied to NHTSA's 2017 costs (2010 dollars) for the diesel at Tier 2 Bin 2 emission standards to convert them to 2012 costs. Finally, the difference between these two costs yields the incremental direct manufacturing costs of the Tier 2 Bin 2 emissions control systems over the Tier 2 Bin 5 emission control systems.

Cost Estimates for Advanced Diesel at Tier 2 Bin 2 Emissions

The final step in determining the diesel engine costs at Tier 3 emission levels required adding the incremental Tier 2 Bin 2 emission control costs to the advanced diesel engine costs, as shown in Table 3.8. The GDP multiplier was applied to the advanced diesel engine cost for 2009 (2008 dollars) shown in Table 3.6 to bring these cost up to the 2010 dollar level. Then, appropriate learning factors were applied to bring these costs to the 2012 base year and then to the 2017 MY. The Tier 2 Bin 2 emission costs were added to these advanced diesel engine costs to yield the advanced diesel costs shown in the table.

The committee's estimated direct manufacturing costs, derived using the process shown in Table 3.8, are listed in Table 3.9 for 2017, 2020, and 2025 and compared with NHTSA's estimated costs. As shown in the table, the committee's estimated direct manufacturing costs are approximately 31 to 47 percent higher than NHTSA's estimated costs. The NRC Phase 1 report was peer reviewed and published with the basis for the advanced diesel costs. In addition, the incremental costs used to estimate the cost of the Tier 2 Bin 2 emission control system relative to the Tier 2 Bin 5 emission control system were derived directly from NHTSA's

TABLE 3.7 Derivation of the Committee's Estimated 2017 Direct Manufacturing Costs at Tier 2 Bin 2 Emission Standards Relative to Tier 2 Bin 5 Emission Standards

Step	Source	Year	Year$	Standard Car (I4 2.0L)	Large Car (V6 3.0L)	Large Truck (V6 4.0L)
1	2012-2016 rulemaking Tier 2 Bin 5	2012	2007$	$1,697	$2,399	$2,676
2	Apply GDP multiplier (1.04)	2012	2010$	$1,765	$2,495	$2,766
3	2017-2025 rulemaking Tier 2 Bin 2	2017	2010$	$2,059	$2,522	$2,886
4	Apply reverse learning 2017 to 2012 (0.89)	2012	2010$	$2,367	$2,899	$3,317
5	Cost for Tier 2 Bin 2 over Tier 2 Bin 5 (Subtract 2 from 4)	2012	2010$	$602	$404	$551

TABLE 3.8 Derivation of the Committee's Estimated 2017 Diesel Engine Direct Manufacturing Costs at Tier 2 Bin 2 Emissions

	Year	Year$	Standard Car (I4 1.6L)	Large Car (V6 2.8L)	Large Truck (V8 3.5L)
NRC 2011 Estimate (Tier 2 Bin 5)	2009	2008$	$2,913	$3,857	$3,903
Apply GDP Multiplier (1.02)[a]	2009	2010$	$2,971	$3,934	$3,981
Apply Learning (1.18 and 1.04)[b]	2017	2010$	$2,421	$3,206	$3,244
Tier 2 Bin 2 Emission Costs	2017	2010$	$ 602	$ 404	$ 551
Diesel Engine Costs	2017	2010$	$3,023	$3,565	$3,795

[a] GDP multiplier of 1.02 applied for 2008$s.
[b] Yearly costs due to learning are reduced by 3 percent per year from 2009 to 2015 and 2 percent for the years 2016 and 2017.

TABLE 3.9 Comparison of NRC Direct Manufacturing Costs (All Costs in 2010 dollars) for Advanced Diesel Engine at Tier 2 Bin 2 Emissions with EPA/NHTSA Estimates

	MY 2017		MY 2020		MY 2025	
Advanced Diesel Engine at Tier 2 Bin 2 Emissions	NRC Estimated Most Likely DMC[a]	NHTSA Estimated DMC[a]	NRC Estimated Most Likely DMC[a]	NHTSA Estimated DMC[a]	NRC Estimated Most Likely DMC[a]	NHTSA Estimated DMC[a]
Midsized car I4 diesel engine	3,023	2,059	2,845	1,938	2,572	1,752
Large car V6 diesel engine	3,565	2,522	3,356	2,374	3,034	2,146
Large light truck V6 diesel engine	3,795	2,886	3,571	2,716	3,228	2,455

[a]Relative to baseline gasoline engine.

estimated costs. Additionally, the committee's estimated costs are directionally consistent with input received from the vehicle manufacturers.

These advanced diesel costs were derived from NHTSA's estimates for meeting Tier 2 Bin 2 emission standards. As shown in Table 3.4, the Tier 3 emissions standards, which were enacted after the final CAFE rule was issued, have essentially the same NMOG + NO_x emission requirements as the Tier 2 Bin 2 standards. However, the Tier 3 standards have the additional requirement of 0.003 g/mi PM, which is a significant reduction from the Tier 2 Bin 2 PM requirement of 0.010 g/mi. A review of CARB certification test results shows that some current diesel cars are achieving 0.001 g/mi PM in certification testing, even though the current standard is 0.010 g/mi (CARB 2014). Therefore, current technology DPFs may be adequate for meeting the Tier 3 PM requirement. More test results are needed to confirm that the 0.003 g/mi PM standard would not require additional costs, as assumed in the cost analysis of the advanced diesel engine.

The committee examined NHTSA's cost estimates by deducing the aftertreatment costs of $688 for Tier 2 Bin 5 emission standards and the $602 for the Tier 2 Bin 2 emission standards from NHTSA's cost estimate of $2,059 for an I4 advanced diesel and found that the incremental diesel engine cost without emissions control systems would be $768. This estimate appears to be low when the common rail fuel injection system alone is estimated to cost $675 (from Table 3.6) without considering the cost of the turbocharger, the base engine upgrade, the EGR system, and the electrical system upgrades. This illustration suggests that an updated teardown cost study of a modern diesel engine is needed.

OTHER COSTS ESTIMATES

The committee reviewed other sources of cost estimates for diesel emission reduction technologies and compared them with the cost estimates developed in this report. The report of the International Council on Clean Transportation (ICCT), *Estimated Cost of Emission Reduction Technologies for Light-Duty Vehicles* (2012), provided costs for diesel engine aftertreatment systems meeting Euro 6 standards (Sanchez et al. 2012, Tables 4-11 through 4-14). The estimated costs for Euro 6 diesel engine aftertreatment system components consisting of DOC, DPF, and LNT or SCR are shown in Table 3.10 for several different engine displacements.

The 2012 ICCT report also estimated costs of emission control technologies that included engine modifications and aftertreatment systems for European diesel Euro 6 and U.S. Tier 2 Bin 5 standards as shown in Table 3.11 (Sanchez et al. 2012, Tables 4-16 through 4-18).

A wide range of cost estimates were found in reviewing other sources. The large variations were due to inconsistent emission standards ranging from Euro 6 to Tier 2 Bin 5 and a lack of detail in defining the engine and aftertreatment systems. However, the ICCT estimates for aftertreatment system costs ranging from $648 to $1,011 for Euro 6 emission standards are relatively consistent, with the NRC Phase 1 report

TABLE 3.10 Direct Manufacturing Costs[a] of Diesel Aftertreatment System Components for EURO 6 Standards (U.S. dollars)

Diesel Engine Displacement	1.5L	2.0L	2.5L	3.0L
DOC	62	78	99	116
DPF	266	332	402	468
LNT	320	413	509	602
SCR	418	453	494	526/633[b]

[a] Long-term costs as defined by ICCT = 0.8x (DMC + overhead + warranty cost); dollar year not defined.
[b] For meeting Tier 2 Bin 5 standard.
SOURCE: ICCT (2012).

TABLE 3.11 Direct Manufacturing Costs[a] of Emission Reduction Technologies for Euro 6 Standards and for Tier 2 Bin 5 Standards (U.S. dollars)

	Euro 6[b]		Tier 2 Bin 5[c]	
	<2.0L	>2.0L	2.0L I4	3.0L V6
Engine[d]	$699	$800	$736	$817
Aftertreatment	$648[e]	$1,011[e]	$823[e]	$1,217[f]
Total Cost	$1,347	$1,811	$1,559	$2,035

[a] Long term costs as defined by ICCT = 0.8x (DMC + overhead + warranty cost); dollar year not defined.
[b] Euro 6 standards (g/km) = 0.09 HC/0.5 CO/0.08 NO_x/0.0045 PM.
[c] Tier 2 Bin 5 standards (g/km) = 0.056 NMHC/2.5 CO/0.04 NO_x/0.006 PM.
[d] A/F control and engine-out emissions controls, consists of: 50% of fuel system, 50% of turbocharger, 50% of intercooler, 50% of VGT, EGR valves, EGR cooling system.
[e] Includes DOC, DPF, LNT.
[f] Includes DOC, DPF, SCR.
SOURCE: ICCT (2012).

showing aftertreatment costs ranging from $688 to $1,040 for Tier 2 Bin 5 emission standards.

Summary of Diesel Engine Technologies

Many of the technologies applied in the conversion to a diesel engine and the additional conversion to an advanced diesel engine were previously reviewed in the NRC Phase 1 report or the EPA/NHTSA Technical Support Document. EPA and NHTSA relied primarily on the study conducted by their contractor Ricardo for the technologies in the TSD (Ricardo Inc. 2011). These technologies are listed in the upper part of Table 3.12 and include not only technologies for the diesel engine, engine emission control, and aftertreatment systems, but also for the vehicle integration of the engine, which includes the cooling system, battery and electrical systems, torsional vibration damper, noise, vibration and harshness controls, and accommodation of the increased weight of the engine. Some of these technologies are directed toward new and emerging technologies, which will be discussed in a later section of this chapter. The lower part of Table 3.12 lists technologies that were not considered in the NRC Phase 1 study or the EPA/NHTSA TSD and that consist almost exclusively of new and emerging technologies.

Diesel Vehicle Incremental Cost

Diesel-engine-powered vehicles cost more than similar gasoline-engine-powered vehicles for a variety of reasons. Incremental costs are incurred not only due to the engine, engine emission control system, and aftertreatment system, but also due to vehicle integration costs related to the cooling system; electrical system components; NVH control; and handling the increased weight of the diesel engine. Some of these incremental costs for diesel-engine-powered vehicles were considered by EPA/NHTSA, but a detailed accounting was not provided in the TSD or other supporting documentation to determine if all of these costs were included. In contrast, the costs developed in the NRC Phase 1 study, which the committee used as the beginning point for the cost estimates in this report, included costs for nearly all of the following systems except for the cooling system, torsional vibration damper, and handling of the increased engine weight. This complete accounting of costs is partly responsible for the committee's higher cost estimates for the diesel engines relative to NHTSA's estimates.

Engine

The core diesel engine (cylinder block, cylinder head, crankshaft, camshaft, connecting rod and piston) needs to carry the loads from the increased compression ratio and the higher cylinder firing pressures. This results in the application of materials with higher strength such as cast iron or steel instead of aluminum, and castings and forgings instead of other types of lower-cost forming. The diesel engine also includes other components that cost more than a gasoline direct-injected, turbocharged engine. These components include the fuel injection system, which includes the fuel injection pump (with 2,000 bar pressure capability assumed for the advanced diesel, but rising in some applications to 2,500 bar), the fuel rail, high pressure fuel lines, and injectors. The diesel engine also includes a more expensive turbocharger that is likely to include variable geometry turbine nozzles to assist in driving the EGR to the intake system and an air-to-air intercooler or water-to-air aftercooler. A VVT system may also be employed in the future.

Engine Emission Control System

The diesel engine emission control system consists of the common rail fuel injection system, the electronic control module (ECM), the EGR system (which may be a low-pressure loop, a high-pressure loop, or a dual loop with both systems), EGR coolers, and various sensors and wiring

TABLE 3.12 Diesel Engine Technologies Considered by the Agencies and the NRC Phase One Study

Technologies Considered by the NRC Phase One Study	Technologies Considered by both the EPA/NHTSA TSD and the NRC Phase One Study	Technologies Considered by the EPA/NHTSA Technical Support Document
• Downspeeding • Air-air intercooler and ducts • Pressure-sensing glow plugs • Combustion improvements • Reduced friction • Reduced accessory loads, water pump, fuel pump, etc. • NVH countermeasures • Two-state turbocharging with air-water intercooler • Dual-pressure oil pump • Cylinder-pressure-sensing glow plugs • 2,000 bar piezo-actuated injector • Improved low-pressure fuel pump • Engine (cam, crank, connecting rods, pistons) • High-pressure EGR with cooling • Low-pressure EGR with cooling	• Downsizing • High-pressure common rail fuel system • VGT • Larger battery, alternator, and starter motor • Glow plugs	• Powertrain mounting • Supplemental heater • Transmission modifications (damper) • Sound insulation enhancement • Smaller radiator
Technologies Not Considered in NRC Phase One Study or EPA/NHTSA TSD		
• Narrow-speed-range operation • Cylinder-pressure-sensing fuel injectors • Direct-acting piezo injectors • Variable valve timing (VVT) • Variable valve lift (VVL) • Onboard diagnostics (OBD) • Reduced compression ratio	• Preturbine DOC • Mass reduction • Mechanical turbocompounding • Electrical turbocompounding • Electrification of front engine accessory drive (FEAD) • Closed-loop combustion control with in-cylinder pressure sensors • Electric ancillaries (coolant pump, oil pump, etc.)	• Increased compression ratio • Advanced boosting technologies, e-turbines, e-compressors, superchargers • Calibration system optimization (engine + transmission + aftertreatment)

harnesses. The onboard diagnostic (OBD) system must also be included to detect deterioration in these systems.

Aftertreatment System

The diesel aftertreatment system consisting of DOC, DPF, and SCR and the associated costs were discussed earlier in this chapter. However, the new technology of applying SCR catalyst coating on the DPF to provide a single-substrate SCR filter (SCRF) may lead to lower costs and improved emission reduction. Vehicle manufacturers see little prospect of significantly reducing aftertreatment costs, since U.S. Tier 3 standards and the California LEV III standards will tend to increase aftertreatment volumes and platinum group metal usage. The prospects of replacing urea injection systems with a lower cost ammonia system also appear to be minimal.

Cooling System

The diesel cooling system is more complex than for a gasoline engine and consists of the radiator, the air-to-air cooler, the EGR cooler, the thermostat and housing, and the water pump. The cooling system must dissipate heat from the engine cooling system, from the intercoolers or aftercoolers, from the EGR cooler, from the oil cooler, and from any other water-cooled component (such as a fuel cooler).

Battery, Starter (Cranking) Motor, and Alternator

Due to the higher torque needed to crank a diesel engine and the cold starting requirements, the manufacturers use a larger capacity battery with diesels than with gasoline engines. The starter motor and alternator also need to be larger in capacity.

Vibration Damper and Clutch

Diesel engines have higher firing torque pulse amplitude than gasoline engines, so they need more sophisticated torsional vibration damping. Torsional vibration dampers are generally included in automatic transmissions with lockup torque converters. However, the cost of a two-torsional damper system could increase the cost of the torque converter by 20 percent, and the centrifugal pendulum absorber could increase the cost of the torque converter by 50 percent. These incremental costs need to be included in the transmission costs for diesel engines.

Noise, Vibration, Harshness Issues

Significant reductions have been made in diesel engine noise as a result of emission control, common rail fuel systems, and diesel pilot injection. Manufacturers are installing other noise control systems such as engine covers, hood covers, firewall insulation, under-engine panels, tuned muffler systems, special glass on windows, and triple door seals (some of these features are also used on gasoline engines). The vibration and harshness issues are controlled with the torsional vibration damper (discussed previously) and with engine mounts. In some cases, active mounts may be required.

Increased Weight

If the diesel engine installation results in increased weight compared to the gasoline engine installation (although this may not be the case with downsized diesel engines) the increased weight could affect the design of the brakes and tires and possibly structural and suspension components.

Total Cost of Ownership

As discussed above, diesel-powered vehicles cost more than gasoline-powered vehicles. The University of Michigan Transportation Research Institute (UMTRI) did a study for the Robert Bosch Corporation in March 2013, *Total Cost of Ownership: Gas Versus Diesel Comparison.* The results show that diesel vehicles had a higher resale value than their gasoline counterparts after 3 years/45,000 miles and 5 years/75,000 miles. With the diesel vehicle's better fuel economy, fuel costs for these periods were lower in spite of higher pump prices for diesel fuel. The total cost of ownership (TCO), considering depreciation, fuel, repairs, fees, taxes, insurance, and maintenance, was lower for all light-duty vehicles with diesels. For instance, for the VW Jetta and Golf, the TCOs were $3,128 and $5,013, respectively, less for diesels after 3 years/45,000 miles and $5,475 and $1,506 less for diesels after 5 years/75.000 miles. The incremental costs for these diesel vehicles are shown in Table 3.13. Although the price of the diesel engine option for a vehicle is higher than the gasoline engine vehicle, the resale value of the diesel option is also higher and the fuel costs are lower, so the total cost of ownership is lower.

Incremental Retail Price of Current Diesel Vehicles

Retail incremental prices for the diesel engine option in several vehicles are reviewed in this section. The following incremental diesel engine prices were recently published and the sources are noted:

- Chrysler RAM 1500. The diesel engine costs $2,860 more than the same truck with the V8 Hemi engine, which is a $1,900 premium over the standard V6 Pentastar engine. Therefore, the diesel is a $4,760 premium over the standard gasoline engine. (USA Today 2013).
- Chrysler Grand Cherokee. The diesel engine bears a $4,500 premium: "The diesel engine adds a $4,500 premium to the price, putting the sticker of a top end diesel-powered Grand Cherokee at $53,490, including shipping" (Automotive News February 25, 2013).
- Porsche Cayenne. The base price for the vehicle with the diesel engine of $56,725 is $6,900 more than the base price for a gasoline Cayenne.
- GM/Chevrolet Cruze. The Cruze diesel with 2LT trim vs. Cruze Eco with 1LT trim is $25,695 vs. $20,490. The difference is $5,205, except when the same trim is specified, the incremental cost is $3,224.

TABLE 3.13 Prices of Gasoline and Diesel Equivalent Vehicles (2013 dollars)

Vehicle	With Gasoline Engine	With Diesel Engine	Diesel Incremental Price
VW Beetle	19,995	24,195	4,200
VW Golf 2-Door	18,095	24,495	6,400
VW Passat SE	23,945	26,295	2,350
VW Jetta SE	20,420	23,196	2,776
VW Touareg	47,535	51,035	3,500
Chrysler Ram vs V6			4,760
Chrysler Ram vs V8			2,860
Chrysler Cherokee			4,500
Porsche Cayenne			6,900
Chevrolet Cruze			3,224

SOURCE: Volkswagen Web site (www.vw.com) and references given previously.

- Daimler/Mercedes GLK 250. The all-wheel drive 2.5L I4 diesel GLK 250 has a sticker price of $39,495. However, the I4 diesel GLK is priced $500 less than the 3.5L V6 gasoline engine GLK 350 with all-wheel drive, possibly reflecting the cost savings of downsizing the diesel engine with fewer cylinders.
- Volkswagen. Using Volkswagen's website and the feature "build and price," the diesel and gasoline equivalent vehicles were compared and are shown in Table 3.13 together with the foregoing comparisons.

In summary, the latest diesel engine incremental prices, based on prices in the current market for the limited number of vehicles evaluated, show that diesel-powered vehicles are priced higher than gasoline vehicles by an average of $4,147. These incremental prices are within the range of the committee's estimated direct manufacturing costs, which suggests that these vehicle manufacturers may not be adequately recovering their overhead for their diesel vehicle offerings.

Diesel Product Offerings

Vehicle manufacturers introduced 22 additional diesel-powered light-duty vehicles to the U.S. market in 2013, which brings the total to 42 diesel offerings. By 2017, 54 diesel-powered light-duty vehicles are expected to be available in the U.S. market (Shuldiner 2013). A listing of anticipated new diesel offerings for 2014 through 2017 is provided in Table 3.14. Sales of 2014 MY diesel light-duty vehicles for the first 6 months of 2014 (see Table 3.14) totaled 75,509 and made up 0.93 percent of the new vehicle market (Cobb 2014).

Diesel Market Penetration

EPA and NHTSA developed a possible cost-effective compliance path in the final 2017-2025 CAFE rule that showed 0 percent penetration of diesel engines by 2025. This low market penetration for diesel engines is in contrast to the announced introductions of new diesel engine models by European and Asian vehicle manufacturers listed in Table 3.14. The diesel engine market penetration in the U.S. also increased to 0.93 percent in 2013 and EPA projected a 1.5 percent penetration in 2014 (EPA 2014b). However, sales of diesels in 2014 only reached 0.84 percent of all light- duty vehicle sales (Cobb 2015).

Vehicle manufacturers must also consider the costs for CAFE compliance. The cost effectiveness values of various fuel consumption reduction technologies, defined as the incremental cost per percent reduction in fuel consumption ($/% FC), are discussed in Chapter 8. An I4 diesel engine in a midsize car that can provide a 29.4 percent reduction in fuel consumption and with an estimated most likely direct manufacturing cost of $3,023 has a cost effectiveness of $103 per percent reduction in fuel consumption. Although this cost effectiveness is competitive with that of hybrid vehicles, it is about twice the cost per percent fuel consumption reduc-

TABLE 3.14 Light-Duty Diesel Vehicle Models

Year	Manufacturer	Vehicle Model
2014	Volkswagen	Jetta, Passat, Golf, Tuareg, Beetle
	Porsche	Cayenne
	Audi	Q7, Q5, A6, A7, A3, A8, A6 3.0 TDI Premium Plus
	Mercedes	GLK, ML, GL, E, S, R
	BMW	X5, 3 Series, 5 Series, 328d, 535d
	GM/Chevrolet	Cruze
	Chrysler/Jeep	Grand Cherokee
	Chrysler	RAM 1500
2015 (Announced)	Audi	A3 TDI, A4 TDI, A3 Sportback TDI
	Volkswagen	Golf Sportwagen TDI
	Porsche	Macan Diesel
2016 (Announced)	Volkswagen	Cross Blue Plug-in Hybrid TDI
	Nissan	Titan with Cummins 5.0L V8
	GM/Chevrolet	Colorado, GMC Canyon
	Toyota	Tundra with Cummins 5.0L V8
	Mercedes Benz	G-Class SUV BlueTEC
2017 (Announced)	Audi	R8 TDI
	Volkswagen	Golf GTD

tion for an SI engine with all of NHTSA's fuel consumption reduction technologies applied. In addition, the SI engine with all of NHTSA's technologies has been estimated to provide fuel consumption reductions within 2 percentage points of the diesel engine. These cost effectiveness considerations may be a limiting factor in the market availability and subsequent penetration of diesel engine vehicles in the United States.

NEW AND EMERGING TECHNOLOGIES

This section will review new and emerging technologies that can reduce fuel consumption of diesel engines, including the use of alternative fuels. Progress in the ongoing research into alternative combustion systems, generally classified as low-temperature combustion, will also be reviewed. A complete set of estimates for these diesel technologies applied to a midsize car, a large car, and a large light truck are provided in Tables 3A.1 for effectiveness of reduction in fuel consumption and Tables 3A.2a, b, and c for direct manufacturing costs (annex at end of chapter).

Alternative Fuel Technologies

Utilization of alternative fuels in diesel engines has been receiving the attention of researchers at EPA, DOE and the national laboratories, vehicle manufacturers, universities, and the fuel industry. Some of the research programs/projects are related to reducing criteria emissions, but others are focused on reducing fuel consumption. A primary consideration for conventional diesel engines is the specification of fuel with a lower cetane number in the United States than in Europe and other parts of the world. Another consideration is the specification of biofuels designated B5 to B100 (where B5 indicates 5 percent biodiesel and 95 percent petroleum diesel) to ensure that these fuels can be a "drop-in fuel" for current diesel engines. Most engine manufacturers currently allow up to B20, but some manufacturers only allow up to B5.

The use of some alternative fuels that do not have cetane numbers within the current specification range must address the issue that the diesel CI engine does not tolerate low cetane or high octane fuels very well. Gasoline, with a very low cetane number, can be compression ignited in a diesel engine by using a higher compression ratio, about 23:1, but this concept has been applied only in military engines. The use of natural gas, also with a very low cetane number, has been receiving increased attention for heavy-duty diesel engines, but the principles should be applicable to light-duty diesel engines as well. Using natural gas in a diesel engine is accomplished by fumigating the natural gas (mixing the natural gas with the combustion air before it enters the combustion chamber) or injecting it directly into the cylinder, and igniting the gas with a pilot injection of diesel fuel. The direct injection of diesel fuel results in autoignition of a small kernel of diesel fuel, which propagates and ignites the natural gas fuel mixture without spark ignition. Natural gas combustion benefits from the high compression ratios. This process is categorized by EPA as a dual-fuel engine because it blends natural gas with diesel fuel.

Research and development is under way to develop dual-fuel compressed natural gas (CNG)/diesel engines with higher efficiency. Laboratory experiments and simulations show that a dual-fuel CNG/diesel modification of a 2.0L VW TDI engine achieved brake thermal efficiencies up to 39.5 percent using natural gas/diesel ratios of 98 percent/2 percent, at high loads (Ott et al. 2013). The high engine-out NO_x emissions at high loads were reduced using a cost-effective three-way catalyst with stoichiometric operation instead of a NO_x aftertreatment system. Accurate and robust dual-fuel combustion requires in-cylinder pressure measurements and advanced feedback control of the start and duration of the diesel injection. Control of transient operation, cold-start issues, and engine-out methane leakage are still under investigation. In contrast to a dual-fuel engine, a bi-fuel engine, which is also called a "switchable" system, can switch between diesel fuel and CNG (CNG Solutions 2012).

Advanced Combustion Systems

Low-temperature combustion technologies are methods for lowering criteria emissions while also having the potential to improve brake thermal efficiency and reduce fuel consumption in diesel engines. The direction for diesel combustion system technology development has been toward more premixed combustion and away from traditional diesel engine diffusion-type combustion. Higher levels of dilution, provided by large amounts of EGR, together with earlier injection and longer ignition delays, reduce both average and local temperatures and allow more mixing time, so that the local fuel-air ratios will be significantly leaner. This combination of lower temperatures and locally leaner mixtures minimizes the occurrence of diffusion flame combustion, which results in reduced NO_x and PM emissions, as shown in Figure 3.2.

The combustion strategies that utilize this approach have been given many different names in the literature, including LTC (low-temperature combustion), PCI (premixed CI), and PCCI (premixed-charge CI). All of these partially homogeneous charge strategies drive the combustion process in the direction of HCCI (homogeneous-charge compression ignition), as shown in Figure 3.3. The term HCCI in its purest form refers to virtually homogeneous rather than partially homogeneous charge. To utilize these premixed forms of combustion, a number of measures are used to reduce temperatures and improve mixing of the charge. The simplest and most effective measure is increased EGR. In addition to increased EGR, lowering the compression ratio also reduces mixture temperatures and, as a bonus, allows increasing engine power without exceeding cylinder-pressure design

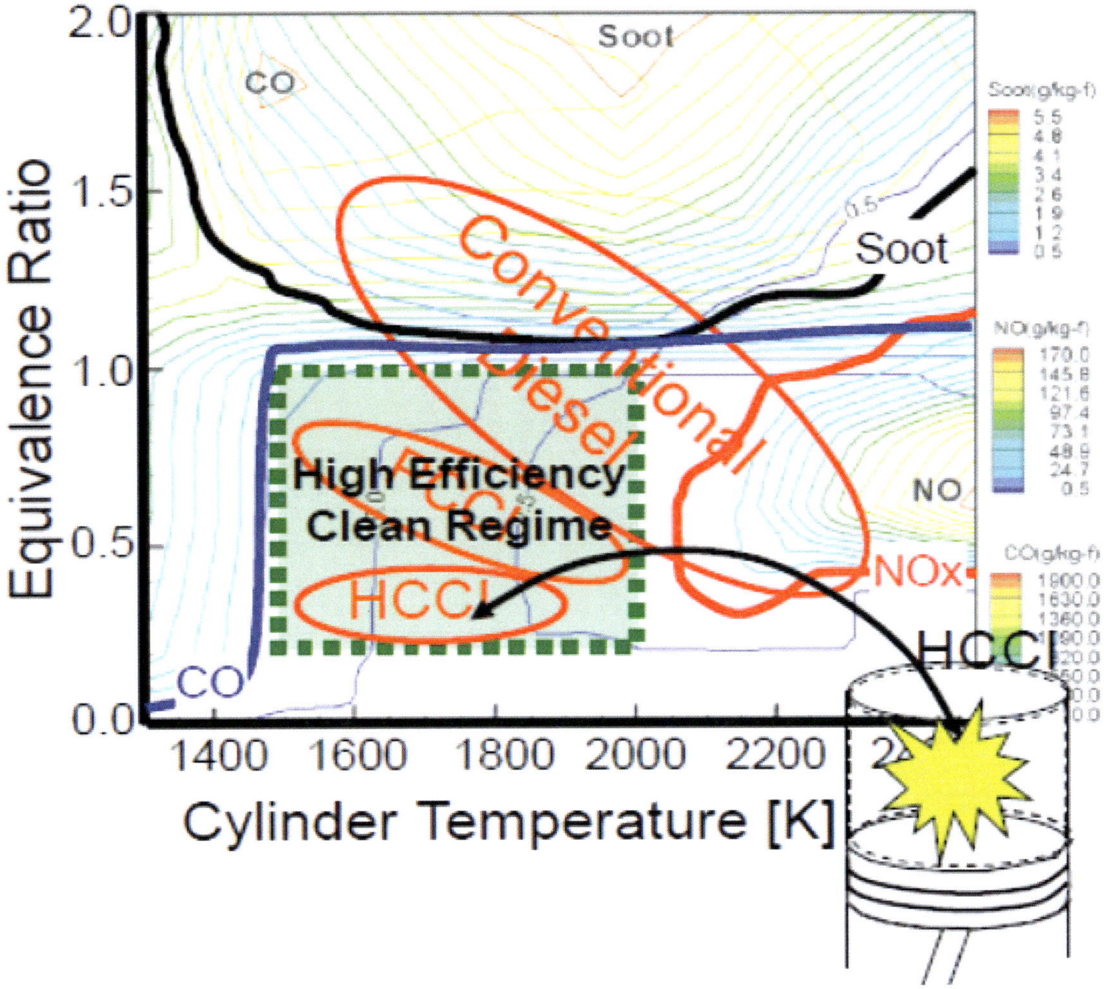

FIGURE 3.2 HCCI combustion regime with lean equivalence ratios and low temperatures for low NO_x and PM emissions.
SOURCE: Figure courtesy of Prof. R.D. Reitz, Engine Research Center, University of Wisconsin-Madison (2010).

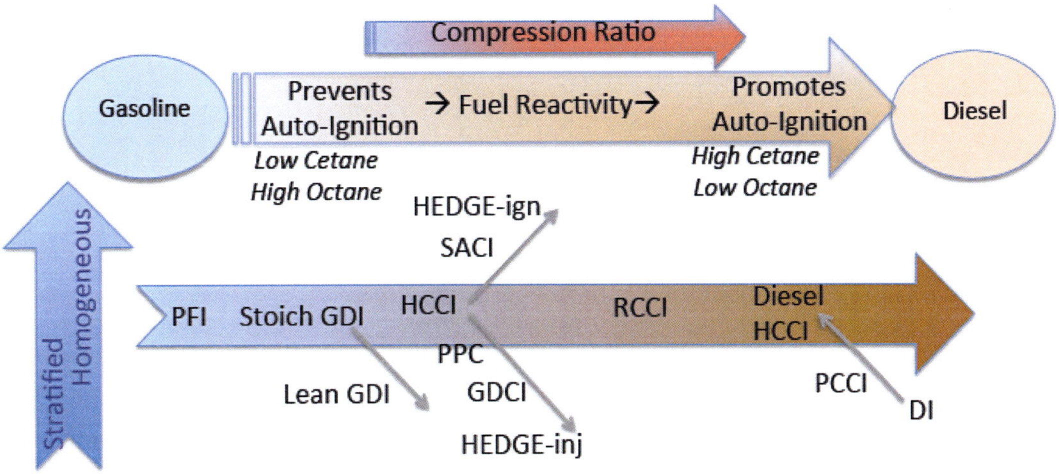

FIGURE 3.3 Advanced combustion concepts spanning the range from gasoline SI to diesel CI engines showing various fuels and stratification strategies.
SOURCE: Daw et al. (2013), Oak Ridge National Laboratory (ORNL).

limits. However, lower compression ratios present challenges in developing acceptable cold-start performance in spite of improved glow plugs and glow plug controls. Technologies being developed to support the move in combustion technology toward premixed low-temperature combustion include piezo-actuated fuel injectors with higher injection pressures, cylinder-pressure-based closed-loop control, higher-pressure, two-stage turbocharger systems, and dual high- and low-pressure loop EGR systems.

In a conventional diesel engine, with the high reactivity of diesel fuel, autoignition follows fuel injection after a small delay. The injection is typically timed close to top dead-center (TDC) to accomplish high efficiency. Without premixed fuel-air mixtures in the end-gas, knock does not occur, so that high compression ratios can be used to achieve high fuel efficiency. The relative absence of premixed charge mixtures in conventional diesel engines is predominantly responsible for the higher emissions.

A modification being applied to diesel engines is the use of multiple injections, instead of a single injection, to allow fuel and air-mixing to create more homogenous conditions (Dec 2009). The partially mixed conditions reduce rich-lean spatiotemporal distributions to significantly reduce smoke. Experimental investigation of the benefits and the limits of this approach, called premixed charge compression ignition (PCCI), are under way for both light- and heavy-duty diesel engines (Hanson et al. 2010; Ra et al. 2011; Simescu et al. 2003; Song-Charng et al. 2004). To lengthen the autoignition delay to allow better mixing, a high EGR rate is used, which also reduces NO_x emissions. Even with very high levels of EGR, diesel HCCI can be achieved in only a narrow operating range due to excessive rates of pressure rise at high loads. Although considerable effort has been pursued in controlling and optimizing diesel HCCI engines in Europe due to the significant cost benefit of simplifying the diesel exhaust aftertreatment, significantly wider operating ranges have not yet been realized.

In an effort to extend the range of PCCI, the University of Wisconsin has been pursuing the combustion concept of a dual fuel PCCI, called reactivity-controlled compression ignition (RCCI), where gasoline is injected by PFI and diesel fuel is directly injected, as shown in Figure 3.4. Both fuels mix in the cylinder, and the cetane number of the mixed fuel controls the ignition timing (Kokjohn et al. 2009). Despite the flexibility offered by the dual fuel, mode switches between RCCI and conventional direct injection (DI) are necessary at high load (Daw et al. 2013). Mode switches between RCCI and DI should be smoother than in pure HCCI-DI (or gasoline HCCI-SI mode, as described in Chapter 2) and easier to execute since dual-fuel percentages can be controlled more

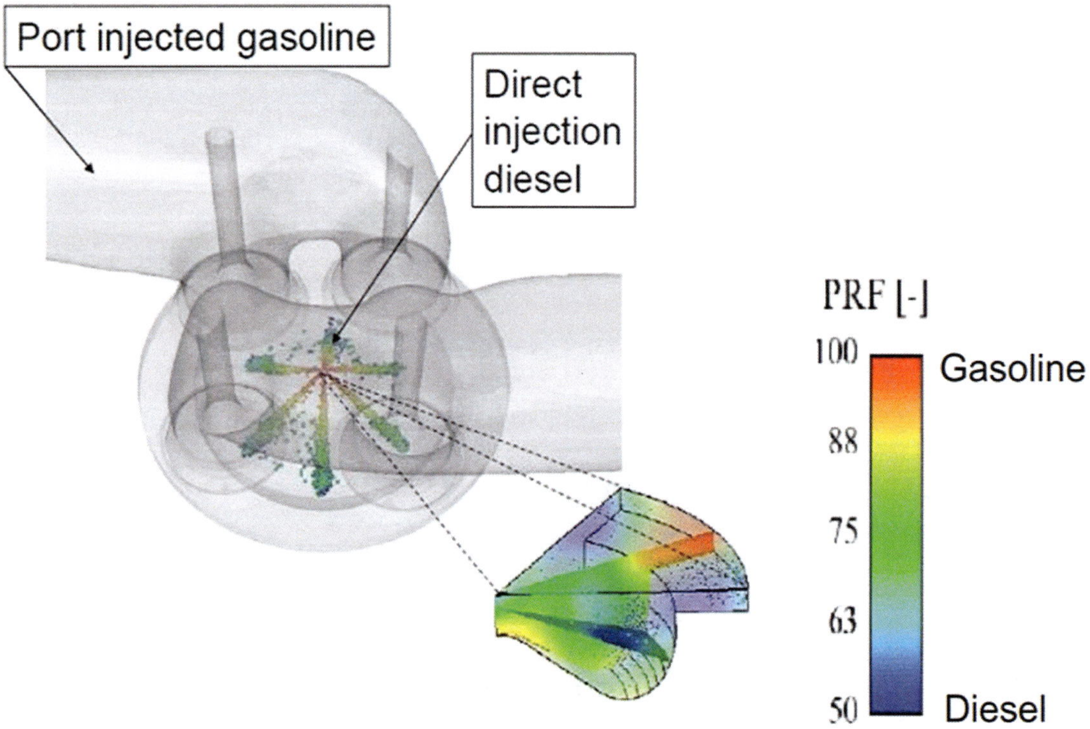

FIGURE 3.4 Schematic of RCCI combustion system with port-injected gasoline and DI diesel fuel showing the mixing of the two fuels in the combustion chamber.
SOURCE: Ra and Reitz (2011). Figure courtesy of Prof. R.D. Reitz, Engine Research Center, University of Wisconsin-Madison.

accurately and faster than the air path. Nevertheless, with mode switching, the full conventional diesel aftertreatment system and associated cost and complexity will be necessary.

Various alternative fuels such as a gasoline blend of up to 30% ethanol (E30) and a diesel blend of up to 20% biodiesel (B20) are being explored for extending RCCI to higher loads. However, a more important consideration is the inability to extend RCCI to low loads. Researchers at ORNL have found that the conventional diesel mode at idle provides similar efficiencies and NO_x levels, but significantly lower HC and CO emissions than RCCI (Curran 2014). Exhaust temperatures below 200°C with high HC and CO emissions present challenges for the application of RCCI with current oxidation catalysts.

The status of RCCI was reported at the 2014 DOE Annual Merit Review. A light duty diesel with RCCI was projected to have up to 30 percent improvement in fuel economy, or 23 percent reduction in fuel consumption, compared to a PFI naturally aspirated gasoline engine. However, this improvement is significantly less that the 29.0 to 30.5 percent estimated by NHTSA for a conventional advanced diesel engine in a light-duty vehicle. RCCI was shown to have 52 percent coverage of the non-idling portion of the Federal Test Procedure (FTP) cycle and 74 percent coverage of the highway fuel economy test (HWFET) cycle (Curran 2014). In summary, alternative combustion systems are still in the research phase, and although progress is being reported, significant utilization of HCCI or derivative concepts is unlikely until after 2025.

Summary of New and Emerging Technologies

New and emerging technologies have the potential to provide additional reductions in fuel consumption of diesel engines. Both short-term technologies, which are already appearing in production vehicles, and longer term technologies expected in the future are discussed in this section. The committee's estimates of fuel consumption reduction effectiveness and costs for the technologies likely to be applied in the future are shown in Tables 3A.1 and 3A.2 in the annex (at the end of the chapter). The committee recommends that EPA and NHTSA evaluate the benefit of these technologies applied to the latest technology diesel engines using full system simulation supported by engine mapping. Cost analyses of these technologies are also recommended.

Downsizing

The two categories of downsizing need to be distinguished. As discussed earlier in the chapter, conversion from a naturally aspirated gasoline engine to a diesel engine included approximately a 20 percent downsizing of the diesel engine combined with turbocharging. The fuel consumption reduction effectiveness of the downsizing in the conversion to diesel has already been taken into account in the diesel engine effectiveness estimates.

Further downsizing of the diesel engine can provide additional reductions in fuel consumption and is already taking place. Opel recently introduced a 1.6L diesel engine in Europe that will replace its existing 1.7L engine and lower power versions of their 2.0L engine (Green Car Congress 2013a). Downsizing a diesel engine will be less effective in reducing fuel consumption than downsizing a gasoline engine since pumping losses are already low in the diesel engine. Downsizing is considered a short-term opportunity for reducing the fuel consumption of diesel engine.

Downspeeding

Downspeeding an engine by designing it to operate in a narrower speed range is being applied as a means of reducing engine friction as well as a method of keeping the engine in the most efficient areas of the engine map while reducing transmission losses. Ricardo estimated that operation in a narrow speed range could reduce fuel consumption by 2 percent to 4 percent after optimizing combustion for the narrower speed range (Ricardo Inc. 2011). To maintain vehicle performance, increased boost pressures are required. Downspeeding requires proper selection of final drive ratio and the narrow speed range is facilitated by an increase in the number of ratios in the transmission. Downspeeding with increased boost pressure is estimated to provide 2.5 percent reduction in fuel consumption. The estimated cost of downspeeding was based on estimates for light heavy-duty diesel engines in pickup trucks developed by EPA/NHTSA (EPA/NHTSA 2011). The higher peak combustion pressure with downsizing was estimated to require $9 additional expenditure for strengthening the cylinder head, and the higher pressure boost system was estimated to require $16 more for a turbocharger with improved efficiency and $3 for an improved EGR cooler, for a cumulative direct manufacturing cost of downspeeding of $28 (2010 dollars) for the 2017 MY and $24 for the 2025 MY.

Common Rail Fuel System

The baseline fuel injection pressure for the advanced diesel is over 2,000 bar, but many companies are investigating increased pressures up to 3,000 bar, and some are already offering 2,500 bar systems. The fuel injectors are tending to evolve from solenoid-controlled to piezo-controlled and to a direct-acting-piezo ceramic actuator for opening and closing the injector needle three times faster than conventional servo-hydraulic-actuated systems (Delphi n.d.). Denso has introduced a 2,500 bar system that reduced fuel consumption by 3 percent, reduced particulate matter by 50 percent, and reduced NO_x emissions by 8 percent (Denso 2013). Mazda is using piezo injectors with 12 holes to facilitate cold starting in its low compression ratio Skyactiv diesel (Coleman 2014). Injectors are available with built-in pressure sensors to measure fuel injection pressure in real time and control the fuel injection quantity and timing of each injector

(Denso 2013). Delphi announced in 2014 that its 2,700 bar F2E distributed pump common rail system is in volume production for heavy-duty vehicles (Meek et al. 2014). The committee estimated that increasing injection pressures to 2,500 to 3,000 bar will provide 2.5 percent reduction in fuel consumption with direct manufacturing cost of $26 (2010 dollars) for the 2017 MY and $22 for the 2025 MY. This estimated cost was based on estimates for light heavy-duty diesel engines in pickup trucks developed by EPA/NHTSA (2011). The increased injection pressure was estimated to require $3 additional cost for the enhanced fuel pump, $10 additional cost for the enhanced fuel rail, and $13 additional cost for the enhanced fuel injectors. This estimated cost may be revised upward as the design details and associated costs of recently introduced production systems become known.

Turbocharging

The baseline I4 advanced diesel has one variable geometry turbocharger with air-air charge cooling or air-water charge cooling, while the V6 advanced diesel in a large light truck is envisioned to have two-stage turbocharging with one variable geometry turbocharger and one fixed geometry turbocharger. The next generation technology will include broader applications of two-stage turbocharging as engines are further downsized or downspeeded. Some two-stage turbocharger systems are already in production (with or without two-stage cooling and with or without variable nozzle turbines). The next-generation Nissan Titan with a 5.0L Cummins diesel will have a two-stage turbocharger system (Tingwell 2012). Increased boost pressures are included in the committee's estimated effectiveness and costs for the downspeeding technology discussed earlier.

Other boosting systems, including electrically assisted turbochargers, turbocharger plus supercharger, and turbochargers with electrically driven compressors, have been proposed for diesel engine applications. Another technology with potential for reducing fuel consumption is an asymmetric-twin scroll turbine housing in which cylinders supplying EGR and operating with high back pressures operate with a different scroll than the cylinders that are not supplying EGR and operating with low back pressure (Markus 2008). Turbocompounding, in which a second turbine following the turbocharger is geared to the crankshaft, is being used in heavy-duty applications. The Ricardo analysis showed that turbocompounding in which the second turbine is coupled to an electrical generator could provide 4 percent to 5 percent reduction in fuel consumption on the highway cycle. Turbocharger friction reduction may provide less than 1 percent additional reduction in fuel consumption.

Exhaust Gas Recirculation

Most current diesel applications of EGR use the high-pressure loop or the "short loop system." These systems have fast response times but can result in EGR cooler fouling. Low-pressure EGR, or "long loop system," has been proposed, possibly alone but more likely as a combination of high-pressure and low-pressure EGR, or "dual loop system." Low-pressure EGR has slow response but provides clean EGR without fouling, can be cooler, and provides more stable cylinder-to-cylinder EGR distribution. A dual-loop EGR system was found in a research program to provide a 2.1 percent reduction in fuel consumption together with a 50 percent reduction in NO_x and a 22 percent reduction in soot emissions (Zamboni and Capobianco 2013). Another study showed that low-pressure EGR could provide up to a 3 percent reduction in fuel consumption (Czarnowski et al. 2008). Ricardo estimated that a low-pressure EGR circuit for increased EGR flow rate in conjunction with a separate low-temperature cooling circuiting will reduce fuel consumption by 2 to 4 percent (Ricardo Inc. 2011). Dual-loop EGR has been implemented in VW's latest 2.0L diesel engine (EA288). The committee estimated that a low-pressure EGR loop will provide 3.5 percent reduction in fuel consumption with 2025 direct manufacturing costs of $113 to $141.

Friction

Reducing engine friction is an effective path for reducing fuel consumption. The baseline advanced diesel engine defined by NHTSA, which provides 29.4 to 30.5 percent reduction in fuel consumption, as discussed previously in this chapter, already includes the friction reduction technologies that were described for the SI engine in Chapter 2. NHTSA's decision trees illustrate that these friction reduction technologies were applicable to both the SI engine path and the advanced diesel engine path. These friction reduction technologies included low friction lubricants level-1, engine friction reduction level-1, and low friction lubricants together with engine friction reduction level-2, which, taken together, were estimated to provide 3.5 to 4.9 percent reduction in fuel consumption. As for SI engine friction reduction, which was discussed in Chapter 2, achieving these friction reductions requires attention to the details of many engine components and design details, including bearings, piston rings, two-stage oil pumps, deactivated water pumps, thermal management, dual-chamber oil pans, reduced back pressure, and lower compression ratio. The applicability of the low-friction lubricants to turbocharged, downsized SI engines presents significant technical issues and similar issues are expected with advanced diesel engines.

Since diesel engines have higher friction levels than spark-ignition engines, additional reductions in friction may be possible. The trend towards lower compression ratios may provide additional opportunities for friction reduction. Reduction of parasitic losses in driving the high pressure fuel pump may provide additional reductions in friction. The committee estimated that additional friction reductions of approximately 2.5 percent might be possible at an incremental cost of $96 for the 2017 MY and $82 for the 2025 MY based on estimates for light heavy-duty diesel engines in pickup trucks developed by EPA/NHTSA (EPA/NHTSA 2011). The estimated cost could be significantly higher if technologies such as low-friction diamond-like coatings, two-stage oil pumps, thermal management systems involving split cooling, and dual-chamber oil pans need to be applied.

Combustion

The diesel engine combustion is largely controlled due to criteria emissions of hydrocarbons, particulate matter and oxides of nitrogen. Within these constraints, there are some opportunities for further reductions in fuel consumption. Ricardo estimated that closed loop combustion control (CLCC) using cylinder pressure sensors in glow plugs could reduce fuel consumption by less than 1 percent, while also providing compensation for changes in the fuel cetane quality, which becomes increasingly important with the increased use of biodiesel fuel. General Motors developed a V6 diesel engine in Europe that included CLCC (Green Car Congress 2007). Since CLCC also improves NO_x and PM emissions, these improvements can be traded for further reductions in fuel consumption. CLCC has been implemented in VW's latest 2.0L diesel engine (EA288). With this combination of benefits, the committee estimated that CLCC could provide up to 2.5 percent reduction in fuel consumption at a 2025 cost of $58 to $87. Research on HCCI and RCCI, discussed in the previous section, is expected to require CLCC for feedback control of combustion timing.

Aftertreatment

The typical aftertreatment system in the production of light-duty diesel vehicles currently is a combination of DOC, DPF, and SCR. An improvement in the SCR from 96 percent to 98 percent efficiency will allow the engine-out NO_x to be increased and provides a corresponding improvement in fuel economy. Currently the highest SCR efficiency is 96 percent and occurs in a relatively narrow range of temperatures, as shown in Figure 3.5.

The primary proposed aftertreatment improvement is the application of the SCR catalyst on the DPF, or SCR filter (SCRF), so that NO_x and PM can be controlled on one substrate. This development is expected to reduce the cost and size of the system while maintaining emission control capabilities. The enabler for the SCRF is advanced zeolite SCR technology that can withstand the active filter regenerations.

Other promising concepts have been investigated in the past. One concept is the use of solid ammonia in place of diesel exhaust fluid (DEF). Several companies have tested solid ammonia systems (amines system), but issues remain in achieving an effective and customer friendly system. Another system that received attention in the past was a NO_x catalyst that generated ammonia within the catalyst that was being developed by Honda. In the two-layer structure of the catalyst, one layer adsorbed NO_x from the exhaust gas. The engine air-fuel ratio was periodically set to run rich, so that the NO_x in the adsorption layer would react with hydrogen obtained from the exhaust gas to produce ammonia. The adsorbent material in the upper layer temporarily adsorbed

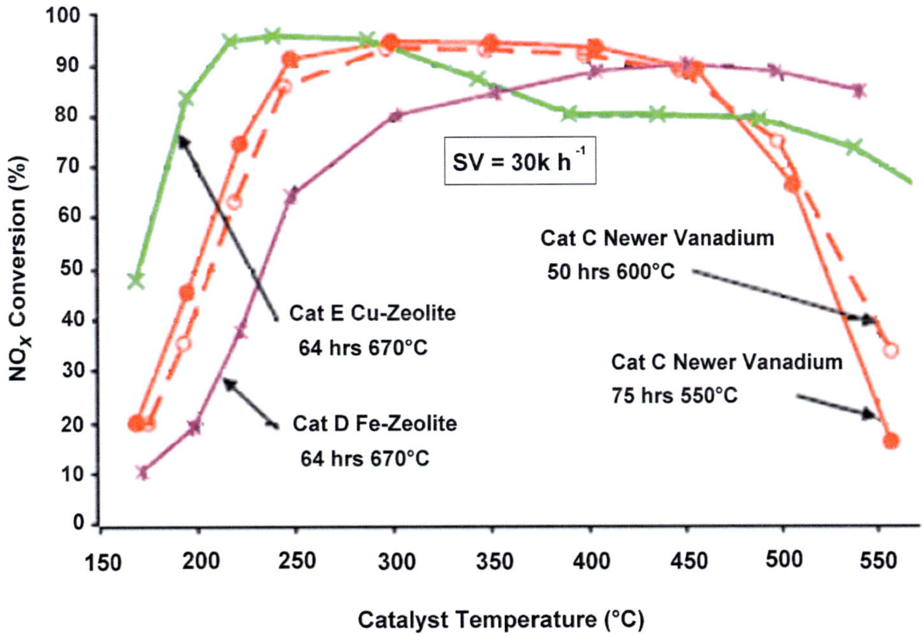

FIGURE 3.5 Relative NO_x conversion efficiency for vanadium and zeolite SCR catalysts for NO as the NO_x species. Cu-zeolites have the best low-temperature performance, and Fe-zeolites are best at higher temperatures.
SOURCE: Girard (2009). Reprinted with permission from SAE paper 2009-01-0121 Copyright © 2009 SAE International.

the NH_3. The resulting ammonia was used later in a reaction that converted the NO_x in the exhaust into nitrogen. Although a promising concept for eliminating the SCR system with urea, the development of this system appears to have been terminated, possibly due to the delicate balance between the ammonia generation and use together with the fuel consumption required for the rich engine operation.

Variable Valve Actuation and Intake Port Control

Variable valve actuation includes variable valve lift (VVL), variable valve timing (VVT), and variable intake ports. The baseline-advanced diesel engine does not have any of these features, but some are in production today. Mitsubishi introduced VVT into production in 2010 on a 1.8L DOHC I4 diesel engine in Japan (BBC Top Gear 2010). It lowered the compression ratio in this engine to reduce fuel consumption and advanced the closing of the intake valve to increase the effective compression ratio for cold-starting.[3] Advanced valve timing also provided for a reduced idle speed of 600 rpm. The lift of one intake valve was reduced to enhance swirl (Mitsubishi 2010). In 2013, VW introduced VVT, actuated by cam phasing, for reducing internal emissions. This system enables the engine to achieve a higher effective compression during cold-start and low NO_x and particulate emissions (Green Car Congress 2013b). The VW diesel engine also has swirl control valves in the intake port. Combinations of VVT and intake port control are candidates for further analysis.

Mass Reduction

Diesel engines are heavier than gasoline engines because they are designed to withstand significantly higher peak combustion pressures than gasoline engines. In the past, diesel engines were designed with gray iron cylinder heads and cylinder block (crankcase). However, many modern light-duty diesel engines are now being designed for reduced mass with aluminum cylinder heads and either aluminum or compacted graphite iron (CGI) cylinder blocks. Other features that are providing mass reductions include integration of the charge air cooler into the intake manifold and exhaust aftertreatment close to the engine (Schmidt 2012).

[3] Part load brake thermal efficiency reaches a broad maximum value around a compression ratio of 16:1, as shown in Figure 2.13. As compression ratio is increased above 16:1, mechanical efficiency decreases faster (due to higher engine friction) than the increase in indicated thermal efficiency, so that brake thermal efficiency begins to decrease. In contrast to diesel engines, which may benefit from reducing compression ratio to approach 16:1, gasoline engines, with compression ratios in the range of 9:1 to 10:1 can benefit from increasing compression ratio if adequate knock-resistant fuel is available, although the rate of increase of indicated thermal efficiency diminishes at higher compression ratios.

Waste Heat Recovery

Nearly two-thirds of the input energy to the diesel engine is lost as exhaust enthalpy or heat rejected to coolant. Although this is low-quality heat (since the temperature difference relative to ambient is small compared to, for example, exhaust gas entering the turbocharger), recovering a portion of this energy could provide reductions in fuel consumption. Several approaches have been considered, including thermoelectric generators (TEGs) and an organic Rankine cycle system (Briggs 2010). Additionally, turbocompounding could also be considered, as previously discussed in this chapter. The committee estimated that waste heat recovery might provide up to 2.5 percent reduction in fuel consumption, with a 2025 cost of $700 for midsized cars, $805 for large cars, and $1,050 for large light trucks. The effectiveness of waste heat recovery for a diesel may be less than for a gasoline engine due to the lower exhaust temperatures of the diesel. As discussed in Chapter 2, an approach using TEGs could incur costs of approximately $700.

Other Technologies

In addition to the foregoing technologies, many of which are already being incorporated in production engines, other technologies are being developed to further reduce fuel consumption of diesel engines. These technologies are discussed in this section.

Higher Expansion Ratios

Variable intake valve timing has already been introduced for controlling the effective compression ratio. Similarly, variable exhaust valve timing may be applied to provide higher expansion ratios, particularly for the lighter load conditions of the dominant CAFE test cycles.

Exhaust Manifold Integrated in the Cylinder Head

Integrated exhaust manifolds in cylinder heads have been introduced in gasoline engines and are now being considered for diesel engines. The shortened path to the turbocharger in a diesel engine can provide reduced heat loss for faster warm-up of the aftertreatment system and provide more energy to the turbocharger for faster transient response times and increased boost pressures. In addition, the integration of the exhaust manifold and cylinder head can reduce weight and cost (D'Ambrosio et al. 2013).

Exhaust Manifold with Air Gap Insulation

Internal heat transfer in an exhaust manifold is the dominant mode of heat loss in an exhaust manifold in the first 30 seconds of operation on the FTP test cycle. Air gap insulation in the exhaust manifold will reduce heat loss from the exhaust gas to provide faster warm-up of the aftertreatment system.

Improved EGR Coolers

Additional reductions in NO_x emissions and improved performance can be realized with lower EGR temperatures achieved through more effective EGR coolers. BorgWarner has developed a cooler design that combines the best attributes of plate-and-shell coolers with plate-and-insert heat exchangers. These EGR coolers are claimed to provide improved heat rejection from the EGR with less soot and hydrocarbon buildup (BorgWarner 2014).

Reduced Piston Cooling

Modifications to the piston design and to the combustion process have the potential to reduce the piston temperature so that less oil will be needed to cool the pistons and a smaller oil pump can be used, thereby reducing engine friction. Alternatively, higher-temperature piston materials can also reduce the amount of oil needed to cool the pistons. In 2015, Toyota introduced a 2.8L diesel engine with a maximum brake thermal efficiency of 44 percent that was achieved by using a thermal barrier coating to reduce heat loss during combustion together with numerous other design features (Mihalascu 2015). The thermal insulation consisted of Thermo Swing Wall Insulation Technology and silica-reinforced porous anodized aluminum on the pistons.

Advanced Actuators (EGR, Wastegates, Exhaust Throttles)

Electric actuators as replacements for pneumatic actuators can provide faster transient response for reduced emissions and improved performance.

Low-Temperature Combustion

Research in low temperature combustion is continuing. The pursuit of HCCI in its purest form can be achieved only in a narrow operating range and will require mode switching to conventional diesel combustion, together with complete conventional diesel aftertreatment systems. PCCI, which uses multiple injections, may extend the operating range but with higher emissions. With these limitations, research is focusing on the dual-fuel RCCI process using port fuel injection of gasoline and direct injection of diesel fuel. Mode switches between RCCI and conventional diesel combustion may still be needed so that the complete conventional diesel aftertreatment system would be required. Since HCCI for diesel engines is still in the research phase, it should be considered a technology that may be applied after 2025.

Gasoline Direct Injection Compression Ignition

This combustion system is discussed in Chapter 2 since it uses only gasoline fuel. However, the combustion system is also being investigated in diesel engines as a means to achieve lower fuel consumption with lower emissions.

Opposed Piston Engine

Achates Power, Inc. has been investigating a two-stroke opposed piston diesel engine as an alternative to the conventional four-stroke diesel engine. In a recent publication, Achates reported on its study, which showed that an opposed piston diesel engine could provide over 30 percent improvement in fuel economy when compared to an equivalent four-stroke diesel engine in a light-duty full-size 5,500 lb pickup truck application (Redon et al. 2014).

Alternative Fuels for Reduced Carbon Dioxide

The Department of Energy is working on further increasing the availability of advanced biofuels. Today, virtually all the renewable diesel fuel being produced in the United States is biodiesel, although next-generation renewable diesel fuels are quickly being developed. Today, most diesel engines can run on high-quality blends of biodiesel with few modifications. Next-generation, drop-in renewable diesel fuels are expected to provide further benefits. Some other alternative fuels, including those with higher cetane number, such as dimethyl ether and Fisher-Tropsch fuels, and fuels with lower cetane values, such as natural gas, may require modifications to adapt to these fuels, particularly in the areas of the fuel injection system and the addition of closed-loop combustion control system (Diesel Technology Forum 2014).

Improved Accessories

The improved accessories at level-1 and level-2 discussed in Chapter 2 and 6 will equally benefit diesel engine vehicles. Included in these improved accessories are an electric water pump, electric cooling fan, and high-efficiency alternator.

Transmissions

The increase in the number of ratios and the reduction in parasitic losses in future transmissions are expected to provide similar benefits to those described in Chapter 5 for the SI engine.

FINDINGS AND RECOMMENDATIONS

Finding 3.1 The committee estimated that the fuel consumption reduction effectiveness of an advanced diesel as defined by NHTSA and meeting Tier 3 emission standards ranges from 29.0 to 30.5 percent relative to a baseline naturally aspirated gasoline engine, which agrees with NHTSA's estimates. The advance diesel engine includes a 1,800 bar common rail fuel injection system, variable geometry turbocharger, high-pressure cooled EGR, and low-friction lubricants and engine friction reductions defined by NHTSA. The incremental direct manufacturing cost for advanced diesel engines is estimated to be 31 to 47 percent higher than NHTSA's estimates when all of the costs associated with a diesel engine, emission control system, and vehicle integration are included. However, these costs are not based on teardown cost studies.

Recommendation 3.1 EPA and NHTSA should expand their full system simulations supported by mapping the latest diesel engines that incorporates as many of the latest technologies as possible, as discussed in this chapter. EPA and NHTSA should conduct a teardown cost study of a modern diesel engine with the latest technologies to provide an up-to-date estimate of diesel engine costs. The teardown study should evaluate all costs, including vehicle integration, which includes the cooling system, torsional vibration damper, electrical systems, including starter motors, batteries and alternator, noise, vibration, and harshness (NVH) control technologies, and vehicle costs resulting from the increased weight of the diesel engine. The study should include an analysis of the increased residual value of a diesel-powered vehicle.

Finding 3.2 The EPA certification fuel economy (uncorrected CAFE) data for comparable current vehicles with gasoline and diesel engines showed that diesel engines provided an average of 20 percent reduction in fuel consumption, which is lower than the committee's estimate of 29.0 to 30.5 percent. Several factors may be responsible for this lower average result. First, many of the 2014 MY diesel vehicles had not fully implemented all of the features discussed in this chapter for the advanced diesel engine. Second, there may be inconsistencies in features between the diesel and baseline gasoline vehicles being compared that were not identified in the EPA database. Third, there may be a difference in the maturity of the diesel and gasoline engines that are compared.

Finding 3.3 Diesel fuel has a higher energy content per gallon than gasoline, but it also has a higher carbon density, which causes approximately 15 percent more carbon dioxide to be emitted per gallon (often referred to as the "carbon penalty"). Therefore, a manufacturer that invests in diesel technology and that can meet the CAFE standards will have greater difficulty meeting the GHG standards than if the manufacturer had invested in gasoline engine technologies. The "carbon penalty" is one reason manufacturers might not choose to invest heavily in diesel engine technologies as a way to comply with the CAFE and GHG standards for MYs 2017-2025.

Finding 3.4 In the current market, vehicles with diesel engine are priced an average of more than $4,000 more than comparably equipped gasoline vehicles. Moreover, based on the committee's estimated direct manufacturing costs, these prices may be lower than the total cost of the diesel engine to the manufacturer when all overhead costs are included. However, a recent study found that the total cost of ownership considering depreciation, fuel, repairs, fees, taxes, insurance, and maintenance was lower for light-duty vehicles with diesel engines compared to those with gasoline engines.

Finding 3.5 The most significant short-term opportunity for further reduction in fuel consumption beyond the level of the advanced diesel engine is through downsizing and downspeeding with a narrow operating speed range. To maintain vehicle performance, increased boost pressures are required. Downspeeding is accomplished with proper selection of final drive ratio, and the narrow speed range can be accomplished with an increase in the number of ratios in the transmission.

Finding 3.6 Although diesel engines provide significant reductions in fuel consumption, additional improvements will be possible in the long-term by the application of new technologies. Some of the new technologies that are expected to be implemented in the next ten years are closed-loop combustion control and fuel quality sensing using fuel injectors or glow plugs with built-in pressure sensors; fuel injection systems capable of 2,500-3,000 bar injection pressures; improved aftertreatment systems such as SCR catalyst applied to the DPF; low- and high-pressure exhaust gas recirculation (EGR); friction reduction; and waste heat recovery.

Finding 3.7 Homogeneous charge compression ignition (HCCI) can only be achieved in a narrow operating range and will require mode switching to conventional diesel combustion together with complete conventional diesel aftertreatment systems. With these limitations, research is focusing on the dual-fuel reaction-controlled compression ignition (RCCI) process. Mode switches between RCCI and conventional diesel combustion may still be needed. A light-duty diesel with RCCI was projected to have up to 23 percent reduction in fuel consumption, compared to a port fuel injection (PFI) naturally aspirated gasoline engine. However, this improvement is significantly less that the 29.0 to 30.5 percent estimated by NHTSA for a conventional advanced diesel engine. RCCI was shown to have 52 percent coverage of the nonidling portion of the Federal Test Procedure (FTP) cycle and 74 percent coverage of the Highway Fuel Economy Test (HWFET) cycle. Alternative combustion strategies are still in the research phase, and, although progress is being reported, significant utilization of HCCI or derivative concepts is unlikely until after 2025.

Finding 3.8 EPA and NHTSA developed a possible cost-effective compliance path in the final 2017-2025 CAFE rule, which showed zero percent penetration of diesel engines in 2025, although they recognized that each manufacturer would determine its own compliance path. This prognosis for low diesel engine market penetration comes even as European and Asian vehicle manufacturers announce new diesel engine introductions. The committee found that most manufacturers and suppliers did not agree with a near-zero penetration for diesel engines, even though some manufacturers do not have any diesel-powered vehicles in their product plans for the United States. However, the diesel's 2025 MY direct manufacturing cost effectiveness is approximately $100 per percent fuel consumption reduction, which significantly exceeds the cost effectiveness of approximately $50 per percent fuel consumption reduction for a sparkignition engine having nearly comparable fuel consumption reduction when all of NHTSA's defined technologies are applied.

REFERENCES

BBC Top Gear. 2010. Mitsubishi ASX. http://www.topgear.com/uk/mitsubishi/asx/road-test/1.8-did-driven.

Belzowski, B., and P. Green. 2013. Total Cost of Ownership: A Gas Versus Diesel Comparison. University of Michigan Transportation Research Institute, March.

BorgWarner. 2014. Passenger Vehicle EGR Coolers. http://www.borgwarner.com/en/Emissions/products/Pages/PC-EGR-Coolers.aspx. Accessed December 6, 2014.

Briggs, T.E., K.D. Edwards, S.J. Curran, E.J. Nafziger, and R.M. Wagner. 2010. Performance of an Organic Rankine Cycle Waste Heat Recovery System for Light Duty Diesel Engines. U.S. Department of Energy DEER Conference. Detroit, Michigan, September.

California Energy Commission. 2007. "Full Fuel Cycle Assessment Well to Tank Energy Inputs, Emissions, and Water Impacts," February. http://www.energy.ca.gov/2007publications/CEC-600-2007-002/CEC-600-2007-002-D.PDF.

CARB (California Air Resources Board). 2014. Home page. http://www.arb.ca.gov/homepage.htm.

Cobb, J. 2014. Dashboard, June 24. http://www.hybridcars.com/june-2014-dashboard/.

Cobb, J. 2015. December 2014 Dashboard. Hybridcars.com, January 6. http://www.hybridcars.com/december-2014-dashboard/.

Coleman, D. 2014. What's all this Skyactiv anyway. Mazda. Accessed December 5, 2014. http://www.cargroup.org/assets/speakers/presentations/36/coleman_dave.pdf.

Curran, S., A. Dempsey, Z. Gao, V. Prikhodko, J. Parks, D. Smith. and R. Wagner. 2014. High Efficiency Clean Combustion in Multi-Cylinder Light-Duty Engines. 2014 DOE Hydrogen Program and Vehicle Technologies Annual Merit Review ACE016, May 14. http://energy.gov/sites/prod/files/2014/07/f17/ace016_curran_2014_o.pdf.

Curran, S., R. Hanson, R. Wagner, and R. Reitz. 2013. Efficiency and Emissions Mapping of RCCI in a Light-Duty Diesel Engine. SAE Technical Paper 2013-01-0289. doi:10.4271/2013-01-0289.

Curran, S., Z. Gao, J. Szybist, and R. Wagner. 2014. Fuel Effects on RCCI Combustion: Performance and Drive Cycle Considerations. 2014 CRC Workshop on Advanced Fuels and Engine Efficiency, February.

Czarnowski, R., V. Joergl, O. Weber, J. Shutty, and P. Keller. 2008. Can Future Emissions Limits be Met with a Hybrid EGR System Alone? U.S. Department of Energy DEER Conference. Detroit, Michigan, August.

D'Ambrosio, S., A. Ferrari, E. Spessa, L. Magro, and A. Vassallo. 2013. Impact on performance, emissions and thermal behavior of a new integrated exhaust manifold cylinder head Euro 6 diesel engine. SAE Int. J. Engines 6(3):1814-1833. doi:10.4271/2013-24-0128.

Daw, S., Z. Gao, V. Prikhodko, S. Curran, and R. Wagner. 2013. Modeling Emissions Controls for RCCI Engines. Engine Research Center Symposium, Madison, Wisconsin, June 5.

Dec, J.E. 2009. Advanced compression-ignition engines—Understanding the in-cylinder processes. Proceedings of the Combustion Institute 32(2): 2727–2742. doi:10.1016/j.proi.2008.08.008.

Delphi. n.d. Delphi Direct Acting Light Duty Diesel Common Rail System. https://delphi.com/shared/pdf/ppd/pwrtrn/direct-acting-light-duty-diesel-crs.pdf. Accessed December 5, 2014.

DENSO. 2013. DENSO Develops a New Diesel Common Rail System with the World's Highest Injection Pressure. June 26. http://www.globaldenso.com/en/newsreleases/130626-01.html.

DieselNet. 2014. ECOpoint Inc. www.dieselnet.com.

Diesel Technology Forum. n.d. An Alternative Fuel Option. http://www.dieselforum.org/energy-and-environment/an-alternative-fuel-option. Accessed December 6, 2014.

DOE (U.S. Department of Energy). 2014. Download Fuel Economy Data. Office of Energy Efficiency and Renewable Energy. www.fueleconomy.gov/feg/download.shtml.

DOE VTO (U.S. Department of Energy, Vehicle Technologies Office). 2014. Annual Merit Review and Peer Evaluation Meeting, Washington, D.C., June 16-20. http://www4.eere.energy.gov/vehiclesandfuels/resources/merit-review/search.

EIA (U.S. Energy Information Administration). 2015. Prices. Short-Term Energy Outlook, March 10. http://www.eia.gov/forecasts/steo/report/prices.cfm.

EPA (Environmental Protection Agency). 2014a. Control of Air Pollution from Motor Vehicles: Tier 3 Motor Vehicle Emission and Fuel Standards; Final Rule. Federal Register 79(81).

EPA. 2014b. Light-Duty Automotive Technology, Carbon Dioxide Emissions, and Fuel Economy Trends: 1975 through 2014. EPA-420-S-14-001.

EPA/NHTSA (National Highway Traffic Safety Administration). 2010. Joint Technical Support Document, Final Rulemaking to Establish Light-Duty Vehicle Greenhouse Gas Emission Standards and Corporate Average Fuel Economy Standards. EPA-420-R-10-901.

EPA/NHTSA. 2011. Final Rulemaking to Establish Greenhouse Gas Emissions Standards and Fuel Efficiency Standards for Medium- and Heavy-Duty Engines and Vehicles Regulatory Impact Analysis. EPA-420-R-11-901, April.

EPA/NHTSA. 2012a. 2017 and Later Model Year Light-Duty Vehicle Greenhouse Gas Emissions and Corporate Average Fuel Economy Standards. EPA 40 CFR 85, 86, 600; NHTSA 49 CFR 523, 531, 533, 536, 537.

EPA/NHTSA. 2012b. Joint Technical Support Document, Final Rulemaking 2017-2025 Light-Duty Greenhouse Gas Emission Standards and Corporate Average Fuel Economy Standards. EPA-420-R-12-901.

EPA/Ricardo Inc. 2008. A Study of Potential Effectiveness of Carbon Dioxide Reducing Vehicle Technologies. Prepared for EPA by Ricardo, Inc. EPA420-R-08-004a, June.

Freitag, A. 2013. Robert Bosch LLC, Presentation to the National Research Council Committee on Assessment of Technologies for Improving Fuel Economy of Light-Duty Vehicles, Phase 2, Washington D.C., June 13.

Girard, J., C. Montreuil, J. Kim, G. Cavataio, and C. Lambert, Technical advantages of vanadium SCR systems for diesel NOx control in emerging markets, SAE Int. J. Fuels Lubr. 1(1):488-494, 2009, doi:10.4271/2008-01-1029.

Green Car Congress. 2007. GM Introduces New V-6 Clean Diesel: Diesel will Debut in Cadillac CTS in 2009 in Europe. Green Car Congress, March 6.

Green Car Congress. 2013a. Opel introducing new 1.6L diesel engine family; fuel consumption reduced by up to 10%, Euro 6 compliant, Green Car Congress, January 6.

Green Car Congress. 2013b. New diesel Volkswagen Golf GTD heading for the US; first driving impressions. Green Car Congress, July 16.

Hanson, R., S. Kokjohn, D. Splitter, and R. Reitz. 2010. An experimental investigation of fuel reactivity controlled PCCI combustion in a heavy-duty engine. SAE Int. J. Engines 3 (1):700-716. doi:10.4271/2010-01-0864.

Herold, R., M. Wahl, G. Regner, J. Lemke, and D. Foster. 2011. Thermodynamic Benefits of Opposed-Piston Two-Stroke Engine. SAE Technical Paper 2011-01-2216.

Honda. 2006. Honda Develops Next-Generation Clean Diesel Engine Capable of Meeting Stringent Tier II Bin 5 Emissions Requirements in the U.S. Honda Worldwide, September 25. http://world.honda.com/news/2006/c060925DieselEngine/.

Hussain, J., K. Palaniradja, N. Alagumurthi, and R. Manimaran. 2012. Effect of exhaust gas recirculation (EGR) on performance and emission characteristics of a three cylinder direct injection compression ignition engine. Alexandria Engineering Journal 51(4): 241-247.

Johannessen, T. 2012. 3rd Generations SCR System Using Solid Ammonia Storage and Direct Gas Dosing—Expanding the SCR Window for RDE (Real Driving Emissions). U.S. Department of Energy DEER Conference. Dearborn, Michigan, October 17.

Johansen, K., H. Bentzer, A. Kustov, K. Larsen, T.V.W. Janssens, and R. Barfod. 2014. Integration of Vanadium and Zeolite Type SCR Functionality into DPF in Exhaust Aftertreatment Systems—Advantages and Challenges. SAE Technical Paper 2014-01-1523. doi:10.4271/2014-01-1523.

Johnson, S. 2014. Diesel Efficiency and Associated Fuel Effects. CRC Workshop, February 25.

Kanda, T., T. Hakozaki, J. Uchimoto, N. Hatano, H. Kitayama and H. Sono. 2006. PCCI Operation with Fuel Injection Timing Set Close to TDC. SAE Technical Paper 2006-01-0920. doi 10.4271/2006-01-0920.

Kokjohn, S., R. Hanson, D. Splitter, and R. Reitz. 2009. Experiments and modeling of dual-fuel HCCI and PCCI combustion using in-cylinder fuel blending. SAE Int. J. Engines 2(2):24-39. doi 10.4271/2009-01-2647.

Majewski, W.A. 2011. Solid reductant storage for SCR systems. DieselNet Technology Guide. https://www.dieselnet.com/tech/cat_scr_solid.php.

Meek, G., R. Williams, D. Thornton, P. Knapp, and S. Cosser. 2014. F2E—Ultra High Pressure Distributed Pump Common Rail System. SAE Technical Paper 2014-01-1440. doi 10.4271/2014-01-1440.

Michelin, J., F. Guilbaud, A. Guil, I. Newbigging, E. Jean, M. Reichert, M. Balenovic, and Z. Shaikh. 2014. Advanced Compact SCR Mixer: BlueBox. SAE Technical Paper 2014-01-1531. doi 10.4271/2014-01-1531.

Mihalascu, D. 2015. Toyota Details Its New GD Family of Turbo Diesels. Carscoops.com, June 21. http://www.carscoops.com/2015/06/toyota-details-its-new-gd-family-of.html.

Mitsubishi. 2010. Clean Diesel Engine. http://www.mitsubishi-motors.com/en/spirit/technology/library/diesel.html.

Müller, M., T. Streule, S. Sumser, G. Hertweck, A. Nolte, and W. Schmid. 2008. The Asymmetric Twin Scroll Turbine for Exhaust Gas Turbochargers. ASME Paper No. GT2008-50614.

Nat G. Dual Fuel, Bi-Fuel, Dedicated? CNG Solutions. http://www.nat-g.com/why-cng/bi-fuel-dual-fuel-dedicated/. Accessed December 7, 2014.

NHTSA (National Highway Traffic Safety Administration). 2012. Final Regulatory Impact Analysis: Corporate Average Fuel Economy for MY2017-MY2025 Passenger Cars and Light Trucks. Office of Regulatory Analysis and Evaluation, National Center for Statistics and Analysis.

NRC (National Research Council). 2011. Assessment of Fuel Economy Technologies for Light-Duty Vehicles. Washington, D.C.: The National Academies Press.

Ott, T., C. Onder, and L. Guzzella. 2013. Hybrid-electric vehicle with natural gas-diesel engine. Energies 6(7): 3571-3592. doi 10.3390/en6073571.

Prikhodko, V., S. Curran, J. Parks, and R. Wagner. 2013. Effectiveness of diesel oxidation catalyst in reducing HC and CO emissions from reactivity controlled compression ignition. SAE Int. J. Fuels Lubr. 6(2):329-335. doi 10.4271/2013-01-0515.

Ra, Y., and R.D. Reitz. 2011. A combustion model for IC engine combustion simulations with multi-component fuels. Combustion and Flame 158:69-90.

Redon, F., C. Kalebjian, J. Kessler, N. Rakovec, et al. 2014a. Meeting Stringent 2025 Emissions and Fuel Efficiency Regulations with an Opposed-Piston, Light-Duty Diesel Engine. SAE Technical Paper 2014-01-1187. doi 10.4271/2014-01-1187.

Redon, F., J. Koszewnik, G. Regner, C. Kalebjian, J. Kessler, N. Rakovec, and J. Headley. 2014b. Meeting Stringent 2025 Emissions and Fuel Efficiency Regulations with an Opposed-Piston, Light-Duty Diesel Engine. SAE 2014 High Efficiency IC Engine Symposium.

Reese, R. 2012. A MultiAir/MultiFuel Approach to Enhancing Engine System Efficiency. DOE Vehicle Technologies Program Annual Merit Review ACE062.

Reitz, R.D. 2010. High Efficiency Fuel Reactivity Controlled Compression Ignition (RCCI) Combustion. U.S. Department of Energy DEER Conference. Detroit, Michigan, September 28.

Ricardo Inc. 2011. Computer Simulation of LDV Technologies for GHG Emission Reduction in the 2020-2025 Timeframe. EPA-420-R-11-020, December.

Sanchez, F.P., A. Bandivadekar, and J. German. 2012. Estimated Cost of Emission Reduction Technologies for Light-Duty Vehicles. ICCT (International Council on Clean Transportation). http://www.theicct.org/sites/default/files/publications/ICCT_LDVcostsreport_2012.pdf.

Schmidt, O. 2012. Volkswagen Group Powertrain and Fuel Strategy. http://www.cargroup.org/assets/speakers/presentations/40/schmidt_oliver.pdf.

Shuldiner, H. 2013. U.S. Light-Vehicle Diesel Offerings to Double This Year, Bosch Exec Says. WardsAuto, January 31. http://wardsauto.com/suppliers/us-light-vehicle-diesel-offerings-double-year-bosch-exec-says.

Simescu, S., S.B. Fiveland, and L.G. Dodge. 2003. An Experimental Investigation of PCCI-DI Combustion and Emissions in a Heavy-Duty Diesel Engine. SAE Technical Paper 2003-01-0345. doi 10.4271/2003-01-0345.

Song-Charng, K., A. Patel, and R. Reitz. 2004. Development and Application of Detailed Chemistry Based CFD Models for Diesel PCCI Engine Simulations. In Proc. of SAE Conference 2004-30-0030.

Tingwell, E. 2012. Next-Gen Nissan Titan Destined for Detroit, we detail its clever engine. Car and Driver, October 28.

Wang, D.Y., S. Yao, D. Cabush, and D. Racine. 2007. Ammonia Sensor for SCR NO_x Reduction. Delphi. U.S. Department of Energy DEER Conference. Dearborn, Michigan, October 16-19.

Woodyard, C. 2013. Ram brings diesel power back to light-duty pickups. USA Today, September 24. http://www.usatoday.com/story/money/cars/2013/09/21/ram-diesel/2844813/.

Zamboni, G., and M. Capobianco. 2013. Influence of high and low pressure EGR and VGT control on in-cylinder pressure diagrams and rate of heat release in an automotive turbocharged diesel engine. Applied Thermal Engineering 51(1-2):586-596.

ANNEX

TABLE 3A.1 NRC Committee's Estimated Fuel Consumption Reduction Effectiveness of Diesel Engine Technologies

		Midsize Car I4 DOHC	Large Car V6 DOHC	Large Light Truck V8 OHV	
Diesel Engine Technologies	Abbreviation	Most Likely	Most Likely	Most Likely	Relative To
NHTSA Technologies					
Advanced Diesel	ADSL	29.4	30.5	29.0	Baseline
Other Technologies					
Low Pressure EGR	LPEGR	3.5	3.5	3.5	ADSL
Closed Loop Combustion Control	CLCC	2.5	2.5	2.5	ADSL
Injection Pressures Increased to 2,500 to 3,000 bar	INJ	2.5	2.5	2.5	ADSL
Downspeeding with Increased Boost Pressure	DS	2.5	2.5	2.5	ADSL
Friction Reduction	FR	2.5	2.5	2.5	ADSL
Waste Heat Recovery	WHR	2.5	2.5	2.5	ADSL

TABLE 3A.2a NRC Committee's Estimated 2017 MY Direct Manufacturing Costs of Diesel Engine Technologies

		Midsize Car I4 DOHC	Large Car V6 DOHC	Large Light Truck V8 OHV	
Diesel Engine Technologies	Abbreviation	Most Likely	Most Likely	Most Likely	Relative To
NHTSA Technologies					
Advanced Diesel	ADSL	3,023	3,565	3,795	Baseline
Other Technologies					
Low Pressure EGR	LPEGR	133	166	166	ADSL
Closed Loop Combustion Control	CLCC	68	102	102	ADSL
Injection Pressures Increased to 2,500 to 3,000 bar	INJ	24	26	26	ADSL
Downspeeding with Increased Boost Pressure	DS	28	28	28	ADSL
Friction Reduction	FR	64	96	96	ADSL
Waste Heat Recovery	WHR	N/A	N/A	N/A	

TABLE 3A.2b NRC Committee's Estimated 2020 MY Direct Manufacturing Costs of Diesel Engine Technologies

		Midsize Car I4 DOHC	Large Car V6 DOHC	Large Light Truck V8 OHV	
Diesel Engine Technologies	Abbreviation	Most Likely	Most Likely	Most Likely	Relative To
NHTSA Technologies					
Advanced Diesel	ADSL	2,845	3,356	3,571	Baseline
Other Technologies					
Low Pressure EGR	LPEGR	125	157	157	ADSL
Closed Loop Combustion Control	CLCC	64	96	96	ADSL
Injection Pressures Increased to 2,500 to 3,000 bar	INJ	23	25	25	ADSL
Downspeeding with Increased Boost Pressure	DS	26	26	26	ADSL
Friction Reduction	FR	60	91	91	ADSL
Waste Heat Recovery	WHR	N/A	N/A	N/A	

TABLE 3A.2c NRC Committee's Estimated 2025 MY Direct Manufacturing Costs of Diesel Engine Technologies

		Midsize Car I4 DOHC	Large Car V6 DOHC	Large Light Truck V8 OHV	
Diesel Engine Technologies	Abbreviation	Most Likely	Most Likely	Most Likely	Relative To
NHTSA Technologies					
Advanced Diesel	ADSL	2,572	3,034	3,228	Baseline
Other Technologies					
Low Pressure EGR	LPEGR	113	141	141	ADSL
Closed Loop Combustion Control	CLCC	58	87	87	ADSL
Injection Pressures Increased to 2,500 to 3,000 bar	INJ	20	22	22	ADSL
Downspeeding with Increased Boost Pressure	DS	24	24	24	ADSL
Friction Reduction	FR	54	82	82	ADSL
Waste Heat Recovery	WHR	700	805	1,050	ADSL

TABLE 3A.3 Fuel Economy of Current Vehicles, with Gasoline and Diesel Engines, Using the EPA Database and the Diesel Technology Forum Website

					EPA Label MPG					CAFE MPG					
Year	Manufacturer	Model	Engine	Tran	City	Hwy	Comb	%FE	%FC	City	Hwy	Com	%FE	%FC	Line
2014	Audi	A6	3.0L 6 cyl gas	A	18	27	22			23.1	38.1	28.1			415
			3.0L 6 cyl diesel	A	24	28	29	32	24	30.7	52.5	37.7	34	26	414
2014	Audi	A7	3.0L 6 cyl gas	A	18	28	21			23.3	37.3	27.2			418
			3.0L 6 cyl diesel	A	24	38	29	38	28	30.7	52.4	37.7	39	28	417
2014	Audi	A8L	3.0L 6 cyl gas	A	18	28	21			22.3	37.3	27.2			613
			3.0L 6 cyl diesel	A	24	36	28	33	25	29.9	51.4	36.9	36	26	612
2014	Audi	Q5	3.0L 6 cyl gas	A	18	26	21			22.8	35.5	27.2			943
			3.0L 6 cyl diesel	A	24	31	27	29	22	29.2	43.5	34.3	26	21	944
2014	Audi	Q7	3.0L 6 cyl gas	A	16	22	18			19.4	30	23.1			1072
			3.0L 6 cyl diesel	A	19	28	22	22	18	22.8	39.1	28.1	22	18	1071
2014	BMW	328i	2.0L 4 cyl gas	A	23	35	27			29.2	50.1	35.9			238
		328d	2.0L 4 cyl diesel	A	32	45	37	28	27	41.6	64.8	49.6	38	27	236
2014	BMW	328i xDrive	2.0L 4 cyl gas	A	22	33	36			28.4	46.9	34.5			240
		328d xDrive	2.0L 4 cyl diesel	A	31	43	35	36	26	40.6	60.9	47.7	38	28	237
2014	BMW	328i xD S-Wagen	2.0L 4 cyl gas	A	22	33	26			28.2	45.9	34.1			713
		328d xD S-Wagen	2.0L 4 cyl diesel	A	31	43	35	38	26	40.6	60.9	47.7	40	28	712
2014	BMW	535i	3.0L 6 cyl gas	A	20	30	24			25.1	42.6	30.8			430
		535d	3.0L 6 cyl diesel	A	26	38	30	25	20	32.3	54.4	40.3	31	24	428
2014	BMW	535i xDrive	3.0L 6 cyl gas	A	20	29	23			25	40.9	30.3			432
		536d xDrive	3.0L 6 cyl diesel	A	26	37	30	30	23	33.4	52.5	39.9	32	24	429
2014	Chrysler Jeep	Gd. Cherokee 4x2	3.6L V6 gas	A	17	24	20			21.8	34.5	26.1			1060
			3.0L V6	A	22	30	25	25	20	28	42.5	33.1	27	21	1059
2014	Chrysler Jeep	Gd. Cherokee 4x4	3.6L V6 gas	A	17	24	19			21.1	33.3	25.3			1102
			3.0L V6 diesel	A	21	28	24	26	21	26.6	39.3	31.2	23	19	1101
2014	GM Chevrolet	Cruze	1.8L 4 cyl gas	A	22	35	27			28.5	49.1	35.1			480
			2.0L 4 cyl diesel	A	27	46	33	22	18	34.8	66.3	44.3	26	21	482
2014	Mercedes	E350	3.5L V6 gas	A	21	30	24			27.7	43.8	33			566
		E250	2.1L 4 cyl diesel	A	28	45	34	42	29	36.9	64.8	45.8	39	28	564
2014	Mercedes	E350 4 matic	3.5L V6 gas	A	21	29	24			26	40.7	31			569
		E250 BluTec 4 mat	2.1L 4 cyl diesel	A	27	42	32	33	25	35.6	59.6	43.5	40	29	565

continued

TABLE 3A.3 Continued

					EPA Label MPG					CAFE MPG					
Year	Manufacturer	Model	Engine	Tran	City	Hwy	Comb	%FE	%FC	City	Hwy	Com	%FE	%FC	Line
2014	Mercedes	ML350 4 matic	3.5L V6 gas	A	17	22	19			21.9	31.1	25.3			1126
		ML350 BluTec 4ma	3.0L V6 diesel	A	20	28	23	21	17	25.2	38.8	29.9	18	15	1127
2014	Mercedes	GL350 4 matic	4.7L V8 gas	A	14	19	16			17.3	26.2	20.4			1122
		GL350 BluTec 4ma	3.0L V6 diesel	A	19	26	22	38	27	24	36.3	28.3	39	28	1121
2014	Mercedes	GLK350 4 matic	3.5L V6 gas	A	19	25	21			23.3	34.7	27.3			1006
		GLK250 BluTec 4m	2.1L I4 diesel	A	24	33	28	33	25	31.1	46.9	36.8	35	26	1005
2014	Porche	Cayenne	3.6L 6 cyl gas	A	17	23	20			21.6	32.9	25.5			1133
		Cayenne Diesel	3.0L 6 cyl diesel	A	20	29	23	15	13	24.1	41	29.6	16	14	1135
2014	VW	Beetle Convertible	2.0L 4 cyl gas	M	23	31	26			27.8	43.9	33.3			217
			2.0L 4 cyl diesel	M	28	41	32	23	19	36.1	58	43.5	31	23	216
			2.0L 4 cyl gas	A	23	29	25			28.9	40.8	33.2			215
			2.0L 4 cyl diesel	A	28	37	31	24	19	36.4	51.9	42.1	27	24	214
2014	VW	Beetle	2.0L 4 cyl gas	A	24	30	26			29.2	42.3	33.9			382
			2.0L 4 cyl diesel	A	29	39	32	23	19	37.3	58	43.5	29	23	381
2014	VW	Beetle	2.0L 4 cyl gas	M	23	31	26			27.8	43.8	33.3			384
			2.0L 4 cyl diesel	M	28	41	32	23	19	36.1	58	43.5	29	23	383
2014	VW	Golf	2.0L 4 cyl gas	A	24	32	27			29.6	42.9	34.4			393
			2.0L 4 cyl diesel	A	30	42	34	26	21	39.1	59.3	46.2	34	26	390
			2.0L 4 cyl gas	M	21	31	25			25.7	40.9	30.9			394
			2.0L 4 cyl diesel	M	30	42	34	36	27	38.7	59.8	46	49	33	391
2014	VW	Jetta	2.0L 4 cyl gas	A	24	32	27			29.6	44.5	34.8			398
			2.0L 4 cyl diesel	A	30	42	34	26	21	39.1	59.3	46.2	33	24	397
			2.0L 4 cyl gas	M	23	33	26			28.3	45.2	34			402
			2.0L 4 cyl diesel	M	30	42	34	31	24	38.7	59.8	46	35	26	401
2014	VW	Passat	2.5L 5 cyl gas	A	22	31	25			27.1	40.8	31.9			607
			2.0L 4 cyl diesel	A	30	40	34	36	27	37.9	56.8	44.6	40	28	605
			1.8L 4 cyl gas	A	24	34	28			30.2	48	36.2			603
			2.0L 4 cyl diesel	A	30	40	34	21	18	37.9	56.8	44.6	23	19	605
			2.5L 5 cyl gas	M	22	32	26			26.1	42.8	31.7			608
			2.0L 4 cyl diesel	M	31	43	35	35	26	38.2	62.3	46.4	46	31	606
			1.8L 4 cyl gas	M	24	35	28			30.3	48.2	36.3			604
			2.0L 4 cyl diesel	M	31	43	35	25	20	38.3	62.3	46.4	28	21	606

continued

TABLE 3A.3 Continued

					EPA Label MPG					CAFE MPG					
Year	Manufacturer	Model	Engine	Tran	City	Hwy	Comb	%FE	%FC	City	Hwy	Com	%FE	%FC	Line
2014	VW	Jetta Sport Wagen	2.5L 5 cyl gas	A	23	30	26			27.1	41.9	32.2			743
			2.0L 4 cyl diesel	A	29	39	33	27	21	37.6	56.2	44.2	37	27	741
			2.5L 5 cyl gas	M	23	33	26			26.2	44.1	32.1			744
			2.0L 4 cyl diesel	M	30	42	34	31	24	38.7	59.8	46	43	30	742
2014	VW	Touareg	3.6L 6 cyl gas	A	17	23	19			21.3	31.6	25			1148
			3.0L 6 cyl diesel	A	20	29	23	21	17	24.1	41	29.6	18	16	1147

Accessed January 1, 2014. From: www.fueleconomy.gov/feg/download.shtml.

.	Chrysler	RAM 1500 4x2	3.6L 6 cyl gas	A	17	25	20			21.4	34.5	25.6			804*
2014	Chrysler	RAM 1500	3.0L V6 diesel	A	20	28	23	15	13	25.8	39	30.4	19	16	805*
2014	Chrysler	RAM 1500	3.6L 6 cyl gas 3.0L V6 diesel	A	15	21	17			18.1	28.4	21.6			832*
2014		RAM		A	19	27	22	29	23	24.5	37.2	30	39	28	833*
		1500													
		4X4													

Revised March 23, 2014, * accessed March 23, 2014 at www.fueleconomy.gov/feg/download.shtml.

Summary of Martec Study Methodology

Conducted in 2007 and 2008, the Martec Group study relied on information and meetings with vehicle manufacturers and suppliers to estimate costs of converting engines from SI to a performance-equivalent compression ignition (CI) system. The methodology is referred to as a bill of materials approach, which takes into account the engineering necessities as well as a likely marketing and sales perspective for assigning costs. Once the bill of materials was generated, Martec then sought input from vehicle manufacturers and suppliers, with the goal being to reach a consensus on the cost estimates that were developed. Assuming that a vehicle manufacturer would sell to both North American and European markets, the study estimated a high production volume of 500,000 units for all components, with the caveat that, depending on the region, some of the production volumes might not be reached before 2020.

Unlike other studies in the 2000-2008 time frame, the Martec study included an evaluation of the incremental costs increases from the aftertreatment systems. In the evaluation of aftertreatment, the costs of commodities and the regulatory standards in 2007 and 2008 were considered in combination with information from vehicle manufacturers and suppliers regarding consumer acceptance. Based on the costs and availability of the commodities at that time, Martec chose NSC aftertreatment systems for deployment on cars and an SCR-urea aftertreatment system for SUVs and trucks. The decision to use SCR-urea aftertreatment systems on the SUVs and trucks was based on the need to comply with emission compliance for ULEVII. As for the diesel particulate filters, their effectiveness with respect to given vehicle sizes and consumer acceptance of them was considered. The use of advanced cordierite was assumed for DPF systems based on its effectiveness and consumer acceptance.

The NRC Phase 1 study used many of the costs from the Martec study with some key deviations. The biggest difference between the NRC Phase 1 and Martec studies was the decision to remove cylinders when converting from an SI to a CI engine. For large cars, Martec downsized the engine from a V6 to an I4 when converting an SI engine to a performance equivalent CI engine. For trucks and SUVs, Martec downsized the number of cylinders to go from a V8 to a V6. The Phase 1 study made the decision not to remove cylinders and instead used a V6 as a baseline for trucks and SUVs. Therefore, the costs in the Phase 1 study do not include a credit for the removal of cylinders from either the large car or from SUV/truck classes of vehicles. Another difference in the cost analysis performed in the Phase 1 study and the Martec study was the decrease in the price of commodities from 2007 to 2009 (NRC 2011).

References

Martec Group. 2008. Variable Costs of Fuel Economy Technologies. Prepared for the Alliance of Automobile Manufacturers.

NRC (National Research Council). 2011. Assessment of Fuel Economy Technologies for Light-Duty Vehicles. Washington, D.C.: The National Academies Press.

4

Electrified Powertrains

FUEL EFFICIENCY FUNDAMENTALS OF ELECTRIFIED POWERTRAINS

Electrification of the powertrain is a potentially powerful method to reduce fuel consumption (FC) and hence greenhouse gases (GHGs). Electrification comes in a variety of forms, from the simplest stop-start systems with only an augmented alternator, to more complex hybrid systems that supplement the engine with an electric drive, to purely battery electric vehicles and finally, to fuel cell systems. This chapter starts with a brief review of the history of electrified powertrains in vehicles. Next, the various electrification architectures and technologies are discussed, including those in use today and those likely to be implemented to 2030. Finally, the implications of these technologies for fuel economy and cost are evaluated.

Electrically-propelled vehicles (EVs)[1] with lead acid batteries had a large share of the market in the early twentieth century, representing about equal numbers with both steam and internal combustion engines (ICEs). EVs lost out to ICEs because gasoline has a much higher energy density than batteries, enabling longer distance travel. Continuous development of the ICE resulted in low-cost, high-performance engines, while EV development, by contrast, stalled. Several experimental vehicles were developed from the mid-1960s to the early 2000s with little or no success in the market.

With increasing concern for reduced fuel consumption, investment in EV research was spurred by the Partnership for a Next Generation of Vehicles (PNGV) program starting in 1993, funded by the U.S. government and the domestic automakers (GM, Ford, and Chrysler). This public-private partnership resulted in prototype, midsize passenger vehicles getting about 80 mpg that were too expensive to make and sell. The market competitiveness of these vehicles was further hindered by a particularly low incentive for fuel economy in the late 1990s, when oil prices dropped to as low as $12/barrel in 1998. The PNGV program was replaced by the FreedomCar program under President Bush, and the focus shifted to fuel cell vehicles using hydrogen as the fuel (NRC 2010).

Greater market success for vehicle electrification came with the combination of ICEs and electric motors in hybrid vehicles (HEVs). The first commercially successful hybrid vehicle—the Prius—was available for sale in Japan in 1997 and was introduced worldwide in 2000. In addition to representing a significant engineering advance in fuel economy compared to conventional vehicles, the Prius's success was helped by two external factors. From 2000 to 2008, oil prices increased, hitting a high of $145/barrel. Energy security concerns in the United States increased as domestic production decreased and oil imports grew to account for over 50 percent of consumption in the mid-1990s (EIA 2014). In addition, concern increased globally about the effect of GHGs on climate change.

Recently, a new type of electrified powertrain called a plug-in hybrid electric vehicle (PHEV) has been introduced. PHEVs have both a large battery that can be charged from the grid and an internal combustion engine, making it possible to drive a larger percentage of miles, if not all miles, fueled by electricity rather than petroleum. More recently, stop-start technology is being incorporated into ICE vehicles. Currently, electrification technologies have achieved only low penetration volumes: 2.75 percent of the market for HEVs,

[1] The following terminology and abbreviations for electrified vehicles will be used in this chapter:

EVs (or xEVs): all vehicles where an electric motor provides all or part of the propulsion.

HEVs: hybrid vehicles, which have two sources of power: an internal combustion engine and an electric drive.

BEVs: battery electric vehicles, where the battery and an electric drivetrain is the source of motive power.

PHEVs: "Plug-in hybrid electric vehicles" are hybrid vehicles with a larger battery that can sustain drive for several miles with the ICE off and can be charged from the grid.

PEVs: "Plug-in electric vehicles" are EVs that derive at least some of their energy by plugging in to the electric grid (BEVs and PHEVs).

FCEVs: fuel cell electric vehicles, which are also known as fuel cell hybrid vehicles.

0.39 percent of the market for PHEVs, and 0.34 percent of the market for BEVs in 2014 (Cobb 2015). Automaker incentive to produce BEVs and FCEVs in the future will be greatly driven by the California zero emission vehicle (ZEV) mandate. The ZEV mandate will be discussed further in this chapter as well as in Chapter 10. The variety of vehicles with some degree of electrification is described in the next section.

TYPES OF ELECTRIFIED POWERTRAINS

Several different electrified vehicle powertrains have been developed and produced with varying commercial success. Typical electrified architectures are defined below and described in greater depth in terms of the relevant engineering principles, implementation of electrified components, and control system requirements. Each architecture is illustrated with example vehicles that are currently in production. In general, hybrid and plug-in electric technologies have been applied to smaller vehicles due to the torque requirements for larger vehicles, which are more difficult to satisfy with these technologies. In Annex Table 4A.1, the committee lists all EVs on sale in the United States in 2014.

Electric Vehicle Categories Defined

Hybrid (HEV) Architectures

- Stop-Start (SS). The engine is turned off when the vehicle pauses in traffic, restarting quickly when the vehicle needs to move again.
- Mild Hybrid (MHEV). A small electric motor and battery combined with an internal combustion engine (ICE) allows for assisted acceleration and regenerative braking.
- Strong Hybrid. A larger electric motor and battery combined with a downsized ICE allow for better regenerative braking as well as periods of electric motor drive.
  - —P2 Strong Hybrid. A parallel hybrid with a clutch connecting the single electrical motor and the engine crankshaft. The vehicle uses a conventional transmission.
  - —PS Strong Hybrid. A power-split hybrid with a planetary gear set that connects the engine, battery, and two electric motor/generators.

Plug-in Electric Vehicle (PEV) Architectures

- Plug-in Hybrid (PHEV). Strong hybrid with a downsized ICE, often a larger generator and battery for extended electric range, and the necessary electronics to charge the battery and therefore power the vehicle from the electric grid.
  - —Series PHEV. The ICE, generator, battery, motor, and transmission are all in series, so all drive to the wheels is ultimately provided by the electric motor, powered by either the battery or the ICE.
  - —PS PHEV. Power split similar to the PS strong hybrid but with a larger battery and the ability to drive with the engine off.
- Battery Electric Vehicle (BEV) Architecture. The only source of power is a large battery, charged from the grid, which drives the wheels via a motor connected in series.

Fuel Cell Electric Vehicle (FCEV) Architecture

- The source of power is a fuel cell that generates electricity from a fuel such as hydrogen, either to charge a battery or to drive a motor to power the wheels.

Hybrid Vehicle Fuel Efficiency Fundamentals

HEVs derive all of their energy from petroleum fuel, but compared to conventional vehicles, they use that fuel more efficiently to power the vehicle. Most hybrid vehicles use an internal combustion engine, a battery, and one or more electrical machines. External combustion engines are excluded, although one of the earliest recent hybrids had a Stirling engine (Agarwal et al. 1969). Fuel cell hybrid electric vehicles do not have a combustion engine and will be discussed later in the chapter.

HEVs reduce fuel consumption relative to conventional vehicles in three ways: by implementing regenerative braking, reducing idle, and enabling engine downsizing.

Regenerative Braking

During braking, the kinetic energy of a conventional vehicle is converted into heat in the brakes and is thus lost. An electric motor/generator connected to the drivetrain can act as a generator and return a portion of the braking energy to the battery for reuse. This is called regenerative braking. Regenerative braking is most effective in urban driving and in the urban dynamometer driving schedule (UDDS) cycle, in which about 50 percent of the propulsion energy ends up in the brakes (NRC 2011, 18). Different architectures have different options for regenerative braking, but in the ideal case and with 100 percent efficiency, fuel consumption can be reduced by half for urban driving.

Enabling Engine Downsizing and Efficient Operation

Hybrid powertrains can enable engine downsizing, more efficient use of engine power for motoring, and electrification of accessories. Downsizing the engine and operating closer to its maximum power improves its efficiency. The maximum engine power is many times that required for moving the vehicle along a level road at constant speed. As is discussed in Chapters 1, 2 and 5, the engine is relatively inefficient at light loads. The electrical machine in hybrids augments the engine power to maintain performance, allowing the use of a smaller engine that operates closer to its best efficiency. Additionally, when motoring, if not all the engine power is needed for propulsion, some of it can be used to recharge the battery.

Hybridization also allows for lower power, more efficient operation, and the use of more efficient engines such as the Atkinson cycle engine. Hybridization and the associated move from mechanical, belt-driven accessories to electricity-driven accessories can increase or decrease efficiency. Powering accessories when needed, rather than whenever the engine is on, reduces energy consumption, as illustrated in Table 2.19 of Chapter 2. However, converting mechanical energy to electrical and back to mechanical incurs losses, for example in the case of the air conditioning compressor.

Engine downsizing can lead to a situation on long steep grades where the vehicle does not have the acceleration the driver expects. If the battery is used to provide additional power to pass several slow moving vehicles, it is conceivable that the battery state of charge (SOC) will drop below the minimum and drivers may not have the passing power they expect. A similar situation exists for PHEVs and BEVs toward the end of their electric range. Insufficient acceleration on steep grades is common in some conventional vehicles with a small engine, for instance vehicles with small three-cylinder engines. Vehicle manufacturers make judgments on how much power to provide to balance performance and cost for both hybrid and conventional vehicles.

Hybrid and Electric Vehicle Architectures

Hybridization of the drivetrain has been implemented to varying degrees, increasing from stop-start and mild hybrids, through strong hybrids to plug-in hybrid electric vehicles, each with increasing degrees of electrification. The varying hybrid architectures, the technology used to implement them, and considerations of cost and consumer acceptance related to that technology are described below.

Stop-Start and Mild Hybrids

Stop-start and mild hybrid systems have minimal electrification and therefore exhibit both the smallest costs and the least fuel consumption reductions. Stop-start systems in vehicles have an augmented starter motor for a quick start and a standard alternator that can accept some of the braking energy. In its simplest form, stopping the engine stops fuel consumption. The reduction in fuel consumption is minimal for these simple stop-start systems, estimated to be 2.1 percent. Mild hybrids also incorporate a motor/generator, either bolted to the crankshaft or connected via a belt. It is used as a generator when the driver applies the brake, and it acts as a motor assisting the engine during acceleration. The SOC is monitored so that the electric motor can start the engine reliably. As the size of the motor/generator is increased, progressively more regenerative braking can be used and then the motor can provide assist during acceleration, thus permitting the engine to be downsized.

The control system for stop-start and mild hybrids is constrained by the battery state and typically involves anticipation of the driver intent. For stop-start systems, the engine should stop when the vehicle stops and quickly restart using the more powerful starter motor. In vehicles with manual transmissions with a neutral gear, the driver's intent to idle or to launch is clear. In the case of powertrains with automatic transmissions, both the driver intent and the power flow path through the converter complicate the control of the start-stop system where the motor is connected to the crank-shaft. Control strategies for electrified powertrains are further described later in the chapter.

A potential problem with stop-start systems is customer acceptance relative to fuel consumption savings that are on the order of 5 percent for on-road fuel economy (2.1 percent in compliance fuel economy). Restarting the engine creates noise and vibration, which may not be acceptable in the U.S. market, accustomed to automatic transmissions and smooth accelerations from a stop. Another problem of stop-start is that in order to provide cooling during stops, an electrically driven air-conditioning compressor may have to be added, increasing cost. Stop-start systems are finding wide acceptance in Europe, but, with the possible exception of GM in some models, not in the U.S. Stop-start systems are credited with off-cycle benefits that are discussed in Chapters 6 and 8.

One of the more interesting mild hybrid systems is the GM eAssist used in a Buick LaCrosse (Hall 2012; Hawkins et al. 2012). It has a 15 kW induction motor/generator that has three functions: (a) a starter with enough power for an instant start; (b) a generator to keep the battery charged; and (c) a motor to augment the engine during accelerations. It is connected to the engine through an augmented belt. The lithium ion battery, which has 500 Wh capacity, provides 18 kW during acceleration. For a mild hybrid it has remarkable performance: an improvement in fuel economy from 27.8 to 38.1 mpg (37 percent) or a 27 percent reduction in fuel consumption compared to the standard LaCrosse. This dramatic performance improvement is not only due to hybridization, since the hybrid has a smaller engine, downsized from a 3.6L V6 to a 2.4L I4, and other features to reduce fuel consumption. In addition to the regenerative braking and smaller engine, this improvement is accomplished by aggressive fuel cutoff when the driver's foot is lifted off the accelerator, underbody panels to reduce aerodynamic drag, and a smaller fuel tank to reduce mass. GM claims that the hybrid LaCrosse has performance similar to the conventional LaCrosse, since the electric motor augments the engine in providing acceleration; however, the 0-60 time for the hybrid is slower than that of the conventional, indicating they do not have the same performance. Results were quite different when eAssist was applied to the Chevrolet Malibu. For the 2013 model year (MY), combined fuel economy for the eAssist Malibu was 38.7 mpg vs 34.8 mpg for the conventional, an 11 percent improvement in mpg and a 10 percent reduction in fuel consumption. For the 2014 MY, the fuel economy of the eAssist Malibu with a 2.4L I4 engine was the same as that of the Malibu with a more efficient 2.5L I4 engine with stop-start.

Battery life is a key element in the design of electrified powertrains and is affected by the conditions of battery use, including the SOC swing. Among other considerations, the allowed swing in the SOC depends on the mode of operation. In hybrids, battery current reverses many times during driving since the electric drive augments the engine during accelerations and recovers energy during braking. DOE specifications call for the battery to survive 300,000 cycles (U.S. DRIVE 2013, 5). To achieve that lifetime, automakers use a narrow swing. GM used a big enough battery so that normally the swing in the SOC is 20 percent of that of the full state of charge. The Prius has had an excellent record of battery life using a NiMH battery with a SOC swing of 20 percent (The Clean Green Car Co. n.d.; Ransom 2011).[2] (See also Ingram 2013 for evidence of real-world battery life validation.) In contrast, the Agencies concluded that in the 2017-2025 time frame, mild hybrids would be able to use 40 percent of the SOC and thus in their calculations they assume a battery that has half the cost and size of the LaCrosse's (EPA/NHTSA 2012a). As described in the section on battery technology, there is no high-power vehicle battery technology targeted for use before 2025 such as could accommodate a 40 percent SOC swing. Vehicle manufacturers acknowledged that, as extended in-use experience is obtained, the battery SOC swing may be increased for all electrified powertrains.

Strong Hybrids

Strong hybrid electric vehicles use a larger motor/generator and battery than mild hybrids, allowing for recapture of energy through regenerative braking as well as greater engine downsizing. The systems are much more costly to implement than stop-start or mild hybrid systems, but they generally exhibit much better fuel economy. The two primary types of strong hybrids on the market, the parallel (P2) and the power split (PS), are described below.

P2 Hybrids

The P2 architecture (see Figure 4.1) has a clutch between the engine and the motor/generator, which provides two advantages: (1) the engine friction does not reduce regenerative braking, and (2) using the transmission, the motor/generator can spin at higher speed to recover more energy. This architecture provides the option of a PHEV if a bigger battery and motor/generator are installed. Installing a larger motor/generator may be a problem in transversely mounted engines. This architecture has the advantage over the power split in that there is no double energy conversion during certain operating conditions. However, the motor/generator operates at a low speed since it is normally coupled to the engine and is therefore larger for the same power. A challenge for P2 configurations is being able to maintain a good drive quality because the clutch connects and disconnects the engine during operation. Some current implementations of the P2 include an augmented cranking motor for smoother engine start.

The P2 control system is more complex than for MHEV since the motor/generator provides regenerative braking as well as acceleration, and so the clutch needs to be disengaged to maximize regeneration at certain speeds and depending on the force applied to the brake pedal. Also, there are more variables to be controlled, including engine speed and torque and transmission gear as well as power flows to and from the battery. The SOC must be monitored so that the battery is not overcharged going down a long hill. Coordination of the motor/generator and the service brakes is necessary since using the service brakes extensively minimizes the energy recovered, but the motor/generator cannot provide emergency braking alone. Coordination becomes particularly important on icy surfaces so that the wheels do not lock and cause skidding.

One example of a P2 hybrid is the Hyundai Sonata (Hyundai Motor Company 2014) that gets 51.5 mpg combined vs. 36.6 mpg for the conventional version (DOE 2014a), a 28.9 percent reduction in fuel consumption (Table 4.5). The standard Sonata has a 2.4L engine rated at 177 hp, with a compression ratio of 11.3:1. The Sonata hybrid has a 2.4L engine with a lower rating of 159 hp, with a compression ratio of 13:1, indicating that this is an Atkinson cycle version of the base engine, with lower power output but improved efficiency. The Sonata hybrid also has a 47 hp permanent magnet electric motor to provide more total power than the standard Sonata (Hyundai 2014).

Power Split Hybrids

Power split hybrids rely on a planetary gear set whose three inputs are the engine, the motor, and the generator (see Figure 4.2). The generator is used to charge the battery and the motor is connected to the wheels to provide additional torque during acceleration and to recover energy from the wheels to recharge the battery. This architecture has several advantages:

- Some of the power goes to the wheels in the most efficient manner: through gears.

[2] In exceptional circumstances, such as going down a long hill, the swing may exceed 20 percent, but the battery size is determined for normal operation.

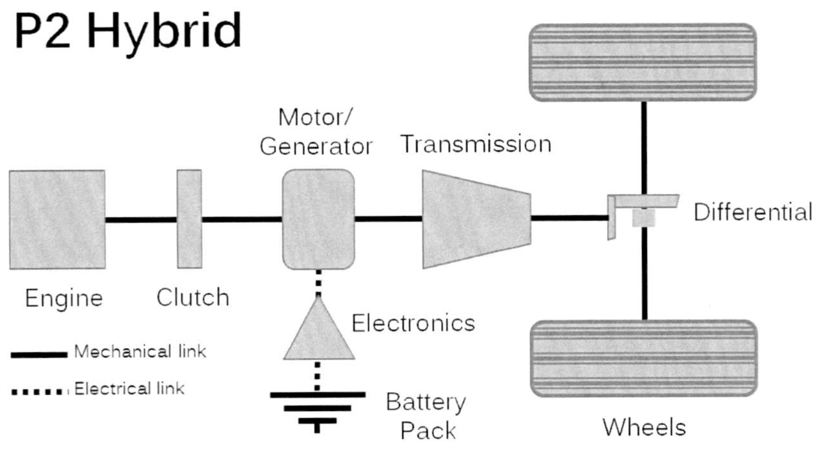

FIGURE 4.1 P2 hybrid architecture showing the motor/generator coupled to the engine through a clutch.

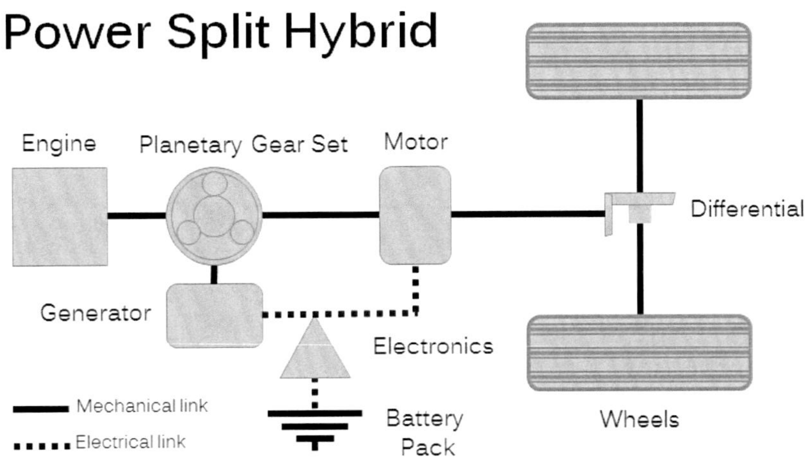

FIGURE 4.2 Power split hybrid architecture showing the separate generator and motor electrically connected via the battery and also via a planetary gear set.

- When engine power is not needed to move the vehicle, some of the power goes to the generator to charge the battery. This raises the engine output and thus improves engine efficiency.
- The motor, battery, and generator can be sized to handle only a fraction of the peak engine power and thus minimize costs.

To some extent, the energy flow is controlled by controlling the engine and generator speeds and the power electronics that control the generator output. However, the kinematics of a planetary three-input gear set require that separately the sum of the power and the torque of the three inputs add up to zero, ignoring the losses in the gears (Meisel 2009, 2011). As a result, some of the engine power has to go through the generator-battery-motor loop, which is less efficient than the direct mechanical connection. For a detailed discussion of this, see Meisel (2009, 2011).

The control system must meet requirements similar to those for the P2 hybrid's control system for motoring and especially antilock (ABS) braking. In this case there are even more variables since engine speed and torque as well as speed and power flows to the generator and motor are all subject to the requirement that torque and power at the planetary gear need to sum to zero.

Both Toyota and Ford have had great success with the PS architecture. In 2014, Toyota had 66 percent of hybrid sales and Ford 12 percent (annex Table 4A.1). The Toyota Camry LE hybrid has 57.4 mpg as the EPA certification fuel economy, a 50.3 percent increase in fuel economy, equivalent to a 33.5 percent decrease in fuel consumption compared to the regular Toyota Camry, which registers 38.2 mpg. Similarly, the Ford Fusion Hybrid SE gets 66.1 mpg, a 91.2 percent higher mpg (47.7 percent lower fuel consumption) compared to the 2.5 L Fusion conventional vehicle. It could be that the Ford hybrid has additional fuel saving features: By carefully choosing the gear ratios and the generator control, the system can be tuned to give best results for a specified test cycle, leading to larger gaps between test-cycle and real-world fuel economy. Ford's and Toyota's test-cycle fuel economies for urban and highway driving, as a result, do not comport well with the fuel economy experienced by a typical driver. This can be seen by comparing the certification fuel economy values with those on the vehicle's label, which is adjusted to better reflect real-world fuel economy. For the PS hybrid Ford Fusion, for example, the ratio of label to compliance value is about 0.71 while for the conventional Fusion, the ratio is about 0.75 (EPA 2014a). Separately, Ford recently reduced the amount of fuel economy claimed for the C-Max hybrid, due to aerodynamic differences between it and the Fusion as well as problems with the coastdown procedure and resulting dynamometer settings (Woodyard 2013; EPA 2014b).

A disadvantage of the power split architecture is that when towing or driving under other real-world conditions, performance is not optimum. GM, in collaboration with Chrysler and BMW, tried to correct this by adding another planetary gear set to produce what they called a two-mode drive. Meisel (2009, 2011) goes into great depth comparing the two drives. A more recent paper also discusses this two-mode drive (Arata et al. 2011). The EPA website shows a reduction in fuel consumption of 26.3 percent for the 2013 two-mode model, which is less than other hybrids. The lower efficiency gain as recorded in the test cycle may reflect the fact that the drive was optimized for off-cycle driving or towing and not just for the certification cycles. In any case, this drive was poorly received in the market and was abandoned by Chrysler, BMW, and GM.

Plug-in Hybrids

PHEVs are the next step from hybrids to full BEVs. They differ from hybrids because they can obtain electricity by charging the battery from the electrical grid, allowing for some portion of their drive to be powered without petroleum. Similar to conventional hybrids, they still have an engine, but they generally have a larger battery and an electric drive capable of propelling the vehicle with the engine turned off. There are various architectures, some of which have been carried over from HEVs. The most popular have batteries that can provide all-electric ranges of about 12, 20, or 40 miles. The advantages of PHEVs over HEVs include these:

- Reduced petroleum consumption because some energy comes from the grid.
- Significantly lower cost per mile driven electrically than with gasoline and an ICE (NRC 2015).
- Reduced GHG emissions depending on the fraction of miles driven electrically and on where and when the PHEV is charged, as the emissions of electric power generators vary by place and time (Anair and Mahmassani 2012). The DOE offers a calculation of these emissions. For example, the Chevrolet Volt emits 300 g CO_2/mi, including upstream emissions, for the grid region encompassing much of Michigan. The emissions for much of California, in the CAMX region, result in 200 g CO_2/mi (DOE 2014b).
- Reduced tailpipe emissions (NO_x, CO, and HC).
- Convenience of operating without liquid fuel much of the time, depending on the driving distance between charges (for instance when commuting).
- In contrast to BEVs, PHEVs are similar to HEVs in their ability to fuel with petroleum upon exhaustion of the battery, to provide range similar to conventional vehicles with easy refueling.

The main disadvantage of PHEVs is cost. To obtain full performance in an all-electric mode they require not only a larger battery but in most cases larger motors and power electronics. The exception is the Toyota Prius PHEV, which appears to use an otherwise identical electrical system as the hybrid, with the addition of a bigger battery. The Prius PHEV has limited capability in the all-electric mode with limited acceleration (0 to 60 in 11.3 s), an all-electric range of 6 miles (MotorTrend n.d.) and a top speed of 62 mph (Siler 2010). PHEV performance is monitored and vehicle controls switch between electric-only, mixed and engine-only drive depending on driving demands and battery charge status. PHEVs with smaller batteries, motors or power electronics will switch more often into mixed or engine-only mode than PHEVs with more robust electric powertrains, though in all cases, controls will switch on the engine to provide power to meet driving demands.

Simple Series PHEV

As shown in Figure 4.3, the engine in a series PHEV drives a generator that charges the battery by means of relatively simple electronics. The battery powers the motor through the main electronics and during regenerative braking returns power to the battery. This system has several advantages:

- The engine can run at constant speed and at full load when charging the battery. Thus it can operate at its maximum efficiency and with simpler emission controls.
- The motor is designed to provide full power and full vehicle performance in all-electric mode.
- The large motor can act as a generator to recover the maximum amount of regenerative braking.

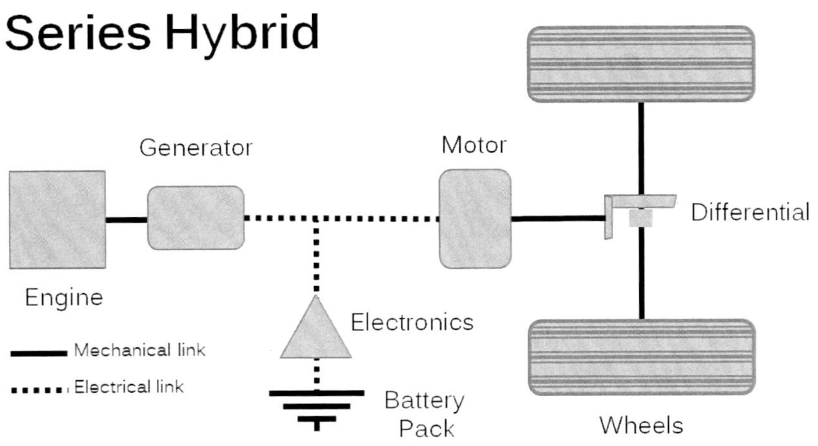

FIGURE 4.3 Series hybrid architecture, as used in PHEV applications.

The control system is relatively simple for series PHEVs. The battery provides all the power until the SOC reaches a low level and the engine starts, providing power to charge the battery. Care must be taken that the battery, even in its depleted state, can provide the necessary acceleration. Also, charging the battery during a long downhill drive is monitored to avoid overcharge. As in other hybrids, antilock braking needs to coordinate service brakes and motor torque.

The prime example of a vehicle with series architecture is the GM Volt. It has a battery with a nominal 16 kWh energy rating, but GM chose to operate in most conditions between 20 percent and 80 percent SOC to ensure performance after 100,000 miles of service and an 8-year warranty (GM-Volt.com 2010). The battery cells are made by LG Chem in Korea. EPA and DOE certify the all-electric range as 38 miles. The vehicle has a relatively small engine, 63 kW, since GM required it to maintain performance with the battery depleted. When the engine is turned on to provide charging, it can be connected to the wheels mechanically at high speeds. This avoids the double energy conversion (mechanical to electrical to mechanical) to improve efficiency. Acceleration is provided by the motor, rated at 111 kW. The generator is matched to the 63 kW engine and is rated at 54 kW. Additionally, GM had Goodyear design special low-rolling-resistance tires for improved fuel economy on the Volt.

The main disadvantage of the Volt is high cost. Since it has the largest battery of PHEVs, a large motor and a generator, as well as full power electronics, the Volt costs more than other PHEVs. EPA rates the Volt running on engine-only at 48.4 mpg combined compared to a similarly sized Cruze at 35.1 mpg, showing 27.5 percent lower fuel consumption (DOE n.d.a). However its great advantage is that it is able to carry out a large fraction of all trips in an all-electric mode using no petroleum-based fuel (Gordon-Bloomfield 2012a).

Honda Two-Motor Series Hybrid (2M)

Figure 4.4 (Higuchi et al. 2013) illustrates the new 2M architecture which Honda calls Intelligent Multi-Mode Drive (iMMD). In the EV mode, a clutch connects the drive motor to the wheels. If a charge sustaining mode is used, the engine can provide energy to the battery. In hybrid mode, electric power can come from the engine/generator and the battery. An interesting feature of this architecture is that the size of the various components, engine, motor, and generator can be optimized separately. Honda chose to eliminate a multispeed transmission, or CVT, and instead drive the wheels through a fixed gear ratio in the engine mode. As Figure 4.5 shows, at highway speeds the engine alone drives the vehicle.

By varying the size of the battery, Honda offers both a conventional HEV and a PHEV version of the Accord (Honda n.d.a, n.d.b). The hybrid achieves fuel economy of 69.9 mpg with a 2.0L I4 engine, while a regular Accord has 40.2 mpg with a 2.4L I4 engine. If we assume that the additional performance in the hybrid is provided by the electric motor, then the iMMD provides a 57.6 percent reduction in fuel consumption, or a 73.6 percent improvement in mpg vs. the regular Accord. The PHEV contains a 6.7 kWh battery and has an electric range of 13 miles. Presumably the limited electric range is a result of the limited size of the battery, therefore reducing the cost of the battery. Unlike the PS architecture, however, the motor is larger since it has to provide full power at low speeds. Both the hybrid and the PHEV have the same engine (141 hp at 6,200 rpm) and electric motor (124 kW).

Both the hybrid and the PHEV Accords are new in 2014. Early reviews indicate a concern that the fuel economy in a test drive "failed to achieve numbers even close to its EPA ratings" (Thomas 2014) and that while the vehicle delivers good speed and styling, there are some NVH issues (Csere 2013; Edmunds.com). As Figure 4.4 shows, engaging and

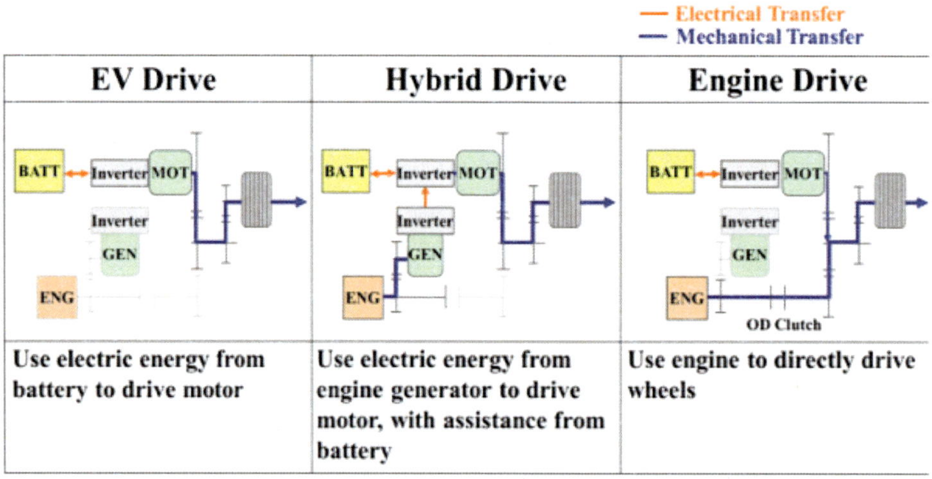

FIGURE 4.4 Honda two-motor series architecture, showing the clutch-modulated connection between the battery, engine, motor and generator, and the wheels.
SOURCE: Higuchi et al. (2013) Reprinted with permission from SAE paper 2013-01-1476. Copyright © 2013 SAE International.

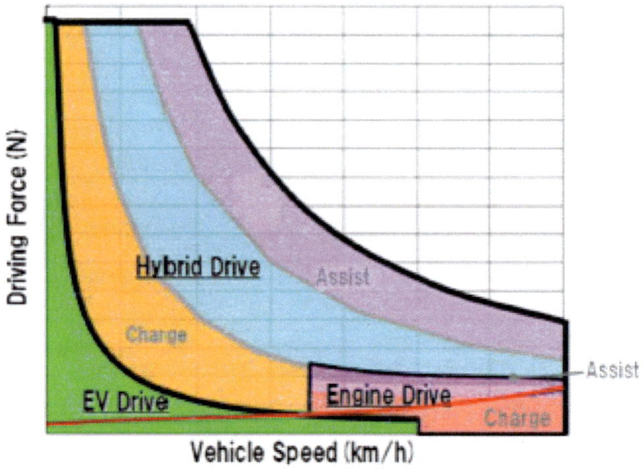

FIGURE 4.5 Honda two-motor series showing modes of operation at various vehicle speeds and driving forces.
SOURCE: Higuchi et al. (2013) Reprinted with permission from SAE paper 2013-01-1476. Copyright © 2013 SAE International.

disengaging several clutches is necessary, which may lead to such NVH issues (see Honda n.d.a, n.d.b).

Power Split PHEVs

These are essentially the same as the power split HEVs with a bigger battery and the ability to drive with the engine turned off. The vehicles in the market are made by Toyota and Ford. One problem is that the HEV drive motor is usually too small to provide full acceleration. Toyota's solution is to have the engine come on when more power is needed, while Ford has installed a larger motor in its PHEV. The net result is that the Toyota Prius PHEV can operate in the UDDS cycle in all-electric mode for only 6 miles but must use "blended operation" in the more demanding cycles. In contrast, the GM Volt can meet both UDDS and the Highway Fuel Economy Test (HWFET) cycles in all-electric mode. Since the greatest cost in PHEVs is the battery, it will be interesting to see whether the Toyota succeeds better in the market than the GM Volt, which has higher performance but much higher cost.

Battery Electric Vehicles

Unlike hybrids, BEVs derive all their propulsion from energy stored in the battery, so their range is limited by battery size. Battery energy density and cost improvements are therefore central to improved performance of BEVs. Originally, lead acid batteries were used in BEVs. The invention of the nickel metal hydride battery (NiMH) offered a roughly twofold improvement in energy density over the lead acid battery at increased cost and the battery continues to be used for HEVs. The NiMH battery range was still not sufficient for a BEV, as demonstrated by the EV1 vehicle produced by GM. A few hundred of these were produced but were recalled and scrapped by GM. Although the lithium ion battery was invented at Exxon in the 1970s (Whittingham 1973, 1976) and used in small electronic devices and computers in the 1990s, it was not used for vehicle propulsion until 2006 in a limited production Tesla Roadster. Tesla implemented lithium ion batteries more seriously in the Model S in 2012. Other early users of lithium ion batteries were the GM PHEV Volt, described above, and the Nissan Leaf BEV.

The structure of a BEV drivetrain is quite simple in comparison to that of the HEV drivetrain. A schematic of a BEV is shown in Figure 4.6. It requires a large, full performance electric motor and power electronics similar to those described for the Volt. It also requires a much larger battery. The 2014 Leaf has an 80 kW motor and a 24 kWh battery (Nissan n.d.) and according to EPA has a range of 84 miles (DOE n.d.b). The Tesla model S has a 265 mile range with an 85 kWh battery.

The Agencies calculate battery cost from the battery size and the cost/kWh, both of which vary by the degree of electrification of the vehicle. Battery size is scaled based on the assumed vehicle range, vehicle weight, SOC swing, and power/distance requirements of the vehicle (Tesla n.d.). The Agencies' allowed SOC swing is 80 percent (10-90 percent) for BEVs and 70 percent (15-85 percent) for PHEVs (EPA/NHTSA 2012a, 3-147), which is appropriate for those architectures. Once the size of the battery was determined, the cost was evaluated via the BatPaC model. The model included varying cost/kWh for BEV, PHEV, and HEV batteries, primarily due to their focus on energy in the former and power in the latter. The BatPaC model assumes that energy-optimized batteries will require fewer, thicker components, while power-optimized batteries require more, thinner components, increasing the relative cost of the power-optimized battery. The resulting difference in cost/kWh between a PHEV battery and EV battery is exaggerated, however. For example, the Leaf uses 21.5 kWh and the Volt, 10 kWh. This should lead to a greater than 2 to 1 ratio of cost for the battery direct manufacturing cost (DMC). The Agencies reported battery DMCs for standard-size passenger vehicles with 15 percent applied weight reduction for the EV75 similar to the Leaf: $11,174, and for the PHEV40 similar to the Volt, $8,642. This leads to a ratio of 1.29 to 1, making the Leaf appear much less expensive relative to the Volt.

Fuel Cell Hybrid Electric Vehicles

Fuel cell hybrids have an architecture similar to series hybrids, as shown in Figure 4.7, with the engine and generator replaced by a fuel cell. For automotive applications the

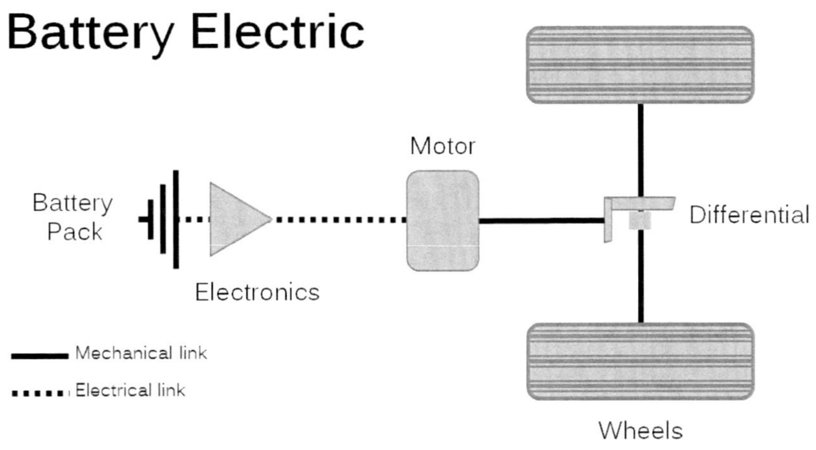

FIGURE 4.6 Battery electric vehicle architecture.

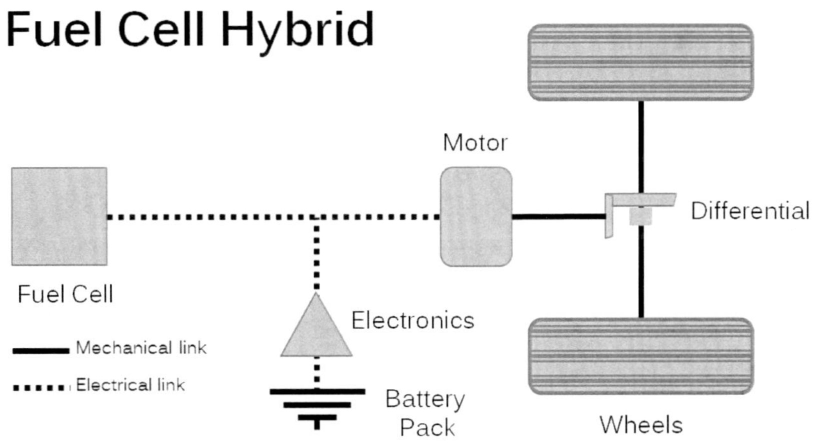

FIGURE 4.7 Fuel cell hybrid vehicle architecture.

fuel cell is powered by hydrogen stored on board. Although they are currently not in mass production, FCEVs offer the possibility of high-efficiency, petroleum-free transportation just like BEVs but without the range limitations of the battery. A significant hurdle to fuel cell vehicle deployment is the extensive infrastructure required. A more detailed discussion of the technology prospects for FCEVs is given later in this chapter.

Enabling Technologies for Vehicle Electrification

Internal Combustion Engines for Hybrid Vehicles

The majority of hybrid vehicles use an SI engine with what is known as the Atkinson cycle. In its modern version, this uses a conventional SI engine with the intake valves held open beyond bottom dead center. This allows flow from the combustion chamber to the inlet manifold and effectively reduces the compression ratio while providing a higher expansion ratio after the mechanical compression ratio is increased to approximately 13:1. The net effect is better engine efficiency (brake specific fuel consumption, BSFC) but lower power output. In hybrids this is acceptable since acceleration is helped by the motor, and efficiency is of paramount importance.

With PHEVs that use the power split architecture, the Atkinson cycle is maintained. In the GM Volt with a series configuration, the engine essentially operates at wide open throttle and there is no need to change the valve timing. Another new idea for hybridization is the use of a very small engine as an emergency backup. The BMW i3, for example, is basically a BEV but has the option of having a small motorcycle engine to extend the range.

Diesel engines can be used in hybrid vehicles; however, none are currently sold in the United States. As stated by some automakers, because diesel hybrids do offer the ultimate technology to reduce petroleum fuel consumption in an ICE, they will perhaps one day be offered for sale in the U.S.

Supervisory Control Strategies in Hybrid Electric Vehicles

HEVs balance multiple power flows in the powertrain. A supervisory controller in HEVs manages power flows among powertrain components such as the engine (or the fuel cell), a motor, and a battery (or capacitors) to minimize cost functions such as fuel consumption, emissions, battery life, and drivability. Many supervisory control strategies for hybrid vehicles have been proposed to fully exploit hardware potential and optimize or negotiate various objectives. Most supervisory control strategies can be classified into rule-based and optimal control approaches.

The rule-based strategies are simple heuristics based on regenerative braking and load leveling. Load leveling is attempting to maintain the engine operation within predetermined regions where fuel efficiency is relatively high. A rule-based control is advantageous for ease of implementation and for the effectiveness of SOC regulation, and it can influence the fuel economy indirectly by tuning the engine and battery operating regions (Hochgraf et al. 1996; Jalil et al. 1997). However, the rule-based strategies do not directly minimize cost functions such as fuel consumption, emissions, and battery life.

On the other hand, optimal strategies are based on optimal control theory and can produce implementable power management by directly manipulating a cost function that weighs the various high-level objectives to be accomplished. Optimal power management relies on models of the torque and efficiency characteristics of all the components that participate in the power flow. These high-level models of the engine (speed, torque, efficiency) and the battery (SOC, voltage, efficiency) make it possible to cascade the cost function objectives to an implementable optimal power split sequence. Additional models are used to cascade the high-level power split commands to the low-level actuator settings such as engine fueling and battery current. For an effective strategy, the supervisory power management needs to be informed of and account for the engine and battery constraints for a reasonable time horizon. Thus the supervisory controller receives signals from the engine control unit (ECU), as discussed in Chapter 2, and the battery management system (BMS), discussed later in this chapter.

To minimize fuel consumption, a truly optimal decision depends on the future decisions where storing some power from the battery now to increase its SOC will pay off later, at a time when an acceleration might be achieved with electric power instead of forcing the engine to operate in a less-efficient region. In real-time implementation on-board a vehicle, information about the future acceleration commands from the driver (driver intent over the entire drive cycle) is obviously very limited. Automated and connected vehicles could provide this necessary future information to the supervisory controllers from vehicle-to-grid (V2G) and vehicle-to-vehicle (V2V) communication along with traffic and terrain information. Currently, however, optimal supervisory methodologies address the inherent dependency on the future information in various ways. Important optimal supervisory methods that have been implemented are discussed below.

The Equivalent Consumption Minimization Strategy (ECMS) and Dynamic Programming (DP) are the most widely studied methods for guaranteeing optimality (Pisu and Rizzoni 2007; Serrao et al. 2011; Lin et al. 2003; Liu and Peng 2008). In an ECMS approach, the total energy to be minimized is considered to be the sum of fuel consumption and battery energy over a driving cycle (Pisu and Rizzoni 2007; Serrao et al. 2011). Typically, an equivalent factor is introduced to convert the battery energy to equivalent fuel energy. In Serrao et al. (2011), adaptive ECMS (A-ECMS) has been proposed to adjust this equivalent factor in real time.

While deterministic DP is used to solve an optimal control problem when the entire driving cycle is given, stochastic DP solves the optimal control when a driving cycle is undetermined (Lin et al. 2003; Liu and Peng 2008). Apart from the causality issue, the use of DP on-board the vehicle is not simple since the computational effort in calculating the optimal decision increases exponentially with the number of state and control variables, also known as "curse of dimensionality." This strategy requires vast computing resources that are not easily accessible, although secure cloud computing might provide these resources in the future. Nevertheless, thorough off-line analysis of DP results for various drive cycles provides good insight into the nature of optimal solutions (Serrao et al. 2011; Liu and Peng 2008). It should be noted that solutions from ECMS and DP are almost identical when fuel consumption is considered as a primary objective to be minimized. Recently, Model Predictive Control (MPC)-based algorithms have been developed for a supervisory controller in HEVs (Di Cairano et al. 2013; Kim et al. 2015) utilizing a prediction of acceleration demands for a short horizon in the future using driver learning methods (Sun et al. 2014).

The performance of the aforementioned strategies in terms of fuel consumption is reported in Pisu and Rizzoni (2007), Liu and Peng (2008), and Di Cairano (2013). Careful assessments are required because the hybrid architectures considered in the studies differ one from the other (i.e., parallel in Pisu and Rizzoni 2007, series in Di Cairano et al. 2013, and power split in Liu and Peng 2008). Nonetheless, optimal model-based strategies always outperform rule-based strategies over various driving cycles such as UDDS, the HWFET cycle, and US06. Table 4.1 shows that the effectiveness of optimal control-based strategy is demonstrated not only by simulations but also by experiments. It is noted that the reported improvements in fuel economy are significantly affected by driving patterns. As expected, the benefits of vehicle electrification are greater when driving in the city, with frequent stop-and-go operation, than when driving on highways.

Although the field of developing supervisory controllers for HEVs seems to be mature, it is still evolving as other objectives such as drivability and emissions are introduced. For instance, in Opila et al. (2012, 2013) authors investigated the influence of the supervisory controller on the frequency of engine on/off and gear shifting. In Kum et al. (2013), authors studied optimal clutch and motor control strategies that resolve drivability concerns during engine starts. The committee found that well-tuned controls were critical in consumer acceptance of new fuel economy technologies such as stop-start. Furthermore, much effort has been devoted to develop supervisory controllers for diesel-powered hybrid vehicles that can substantially decrease emissions such as generated smoke during transients. In Kim et al. (2015) and Nüesch et al. (2014), authors have formulated optimal control problems by applying MPC and DP, respectively, and shown significant reduction in emissions, pointing to possible simplification and cost reduction of the diesel exhaust after-treatment.

Motors and Power Electronics

Electric motors have been used in vehicle accessories for over a century, but their use expanded rapidly when high-energy magnets were invented in the 1980s. With the emphasis on reduced weight, they are now in use in applications such as electric power steering and engine cooling fans, among others. Increasingly, motors are also used to power vehicle motion. Electric vehicles in the 1970s and 1980s used brush-type traction dc motors that were replaced by induction motors starting with the EV1 in the mid-1990s. With the availability of new high-energy magnets, induction motors were replaced by higher efficiency permanent magnet motors starting with the Toyota Prius. Practically all xEVs use permanent magnet motors. Neodymium, one of the rare earth materials essential in high energy magnets, is mostly mined in China and in 2011 experienced a temporary almost tenfold spike in price (Piggott 2011). This forced a reconsideration of induction motors, which Tesla, GM, and Toyota have used in vehicles. Companies restarted mines for rare earth materials in countries other than China, and permanent magnet motors will presumably regain the market.

Motor cost is one component of the cost of vehicle electrification. The Agencies' analysis of hybridization cost was based on an FEV teardown study of the Ford Fusion and the

TABLE 4.1 Relative Fuel Economy Improvements Obtained Between Optimal Control Strategies and Rule-Based Strategies in Simulations and Experiments of Various Fuel Economy Drive Cycles (percent). The Optimal Control Strategies MPC and DP Consistently Outperform the Rule-Based Strategies

Controller	UDDS Simulation	US06 Simulation	UDDS Test
Rule-based 1	+0.00	+0.00	+0.00
Rule-based 2	+1.85	+1.9	+1.34
MPC	+3.75	+4.01	+5.73
DP	<+7.00%	<+6.00%	-

SOURCE: Di Cairano (2013).

Fusion HEV, a PS hybrid (EPA 2011, 12). In their analysis, the Agencies used the power of the machine to scale cost data obtained from the Fusion to other vehicle classes as well as to the P2 hybrid system (EPA 2011, 125). Use of power to scale motor cost is incorrect. Materials costs scale with motor volume, and volume scales with torque, not power. Fundamentally, the rotor diameter squared multiplied by the rotor length is proportional to torque (Alger 1970; Pyrhönen et al. 2009). Scaling by power rather than torque is important when comparing motors designed to operate at different speeds. The P2 motor is inline with the engine and transmission and has the same revolutions per minute (rpm) as the engine. This constraint is not present in the PS hybrid, thereby allowing the use of a higher speed and smaller motor. Power is equal to torque times speed, so a slow motor, such as is used in a P2 architecture, will have a higher torque and be much heavier than a PS motor that is designed to operate at much higher speed and lower torque. For example, a PS motor that operates at 6,000 rpm will weigh half as much as a P2 motor operating at 3,000 rpm. Also, the PS architecture appears to be more effective in reducing fuel consumption, as illustrated in Table 4.6.

Power electronics are needed to perform two functions in electrified powertrains: (1) they convert the direct current (dc) provided by the battery into an alternating current (ac) of controlled amplitude and frequency to power the electric motors, and (2) they convert the grid power (120/240 V ac) to dc to charge the battery. The technology for powering the motor from the battery has been developed for industrial use over the last 60 years; the main problems are improving efficiency and reducing size while providing adequate cooling. Research is ongoing in the use of wide band gap (WBG) materials in place of silicon and in the development of high-temperature, high-frequency capacitors. Presumably to meet the needs of xEVs, the development of power electronics devices using WBG materials has increased, and power electronics may very well find limited application in vehicles by 2020 (Nikkei 2014). Since devices using WBG materials operate at higher temperatures, their advantage will be reduced package size, easier cooling, and, possibly, higher efficiency. Cost will continue to be an issue.

Choice of voltage for the vehicle electrical system has been based on safety, electric loads, efficiency, and available technologies. The electric system powering accessories and the starter operates at 12 V, powered by a lead acid battery. Attempts made in the past to replace the 12 V system with a higher voltage, specifically with 48 V, did not succeed because of safety concerns. At voltages higher than about 24 V, a break in the wire could create a sustainable arc that could ignite the insulation, causing a fire. xEVs have, in addition, a high-voltage system for the battery and electric drive. Typically this starts at 48 V for stop-start systems and goes up to 500 V. The 12 V system is still used to power all low-power accessories, lights, and the like. To prevent fires, all voltages higher than 12 V use special color-coded wires, heavier insulation, and special connectors. Many xEVs use an electrically driven air-conditioning compressor to provide cooling when the engine is stopped. This is normally driven from the high-voltage battery (Green Car Congress 2014, 2015). Note that these vehicles have a 12 V system for accessories, and often they have a dc to dc converter to make sure the 12 V battery is charged since it provides essential functions.

Power electronics for charging the battery are in development. The simplest way is to use a controlled rectifier, perhaps with a dc to dc voltage booster. In this way the devices used for driving the motor can be reused for charging since the two functions are not performed at the same time. This does not provide galvanic isolation between the battery and the plug, however, and most automakers have used a more complex circuit with high-frequency conversion that allows transformer coupling. For PEVs, an interesting development is to charge the battery without plugging in to an outlet. This can be done by inductive coupling between two coils: one in the ground and the second on the vehicle. The two coils can be separated by as much as 12 inches and the radiated field can be controlled so that it does not exceed harmful levels (Miller and Onar 2013). It does not seem likely that such a feature will have much effect on adoption of PEVs and thus will have minimal impact on fuel economy.

Cooling

Cooling is critical for batteries, power electronics, and motors. Eventually, all heat generated has to be rejected to the ambient, but how it is done affects both effectiveness and cost. The heat can be rejected to a liquid or directly to the air. Some automakers, like GM, have used the vehicle's refrigeration systems for cooling. For electrified powertrains, battery cooling is most critical since battery life depends essentially on two factors: deterioration due to temperature during shelf life and the number of charge and discharge cycles (Steffke et al. 2013; Lohse-Busch et al. 2013). Time will tell how critical cooling is for battery life, although there are indications that air cooling for the Leaf battery may be inadequate (Gordon-Bloomfield 2012b). Ideally one would like to have one liquid cooling circuit for the whole powertrain, but the lower temperature required for batteries than for either power electronics or motors makes a single system difficult to optimize. WBG materials offer the possibility of using engine coolant since they can handle higher temperatures.

Other Nonbattery Components

Vehicle electrification requires more than the addition of motors and batteries and, for BEVs, removal of the ICE and transmission. The preceding sections of this chapter describe the potential for engine technology changes, as well as required motors, power electronics, and cooling systems to enable electrification. The chapter describes more sophisticated software and algorithms for the controls hybrids may require, though these technologies are becoming ever more important in SI engine vehicles as well. Less obvious changes may also be required, especially for PEVs, where the battery is larger and the ICE is either absent or likely to be turned off for large portions of the duty cycle. For example, the ICE provides much of the climate control within a vehicle, so in its absence, systems must be added to heat, cool, and defrost the vehicle.

To evaluate the technologies required for hybridization as well as their costs, the Agencies used an FEV teardown of a Ford Fusion Hybrid, which has a PS architecture. The results were used to estimate the costs for a P2 architecture. Apart from the size of the P2 motor discussed above, the committee agrees with the teardown costs of the PS hybrid as applied to that architecture and the P2 architecture.

The Agencies proceeded to use the hybrid Fusion teardown results to estimate the costs of PHEVs and BEVs. In the opinion of the committee, the Agencies did not fully identify and evaluate the costs of changes inherent to PHEVs and BEVs; these include the body system, brake system, climate controls, a complex dc/dc converter to maintain headlight intensity for the 12 V battery during engine stops, power distribution and control, on-vehicle charger, supplemental heating, high-voltage wiring, battery discharge systems, purchase and installation of a home charger in the case of PEVs and removal of the ICE and transmission in the case of BEVs (EPA/NHTSA 2012a). The scaling of the nonbattery costs from the PS teardown does not adequately estimate the costs for these and other components required for PHEVs and BEVs. Discussions with several automakers and one supplier knowledgeable in the complexities of wiring and other subsystems identified added costs for more extensive wiring, power cables with RFI suppression, and sealed fuel tanks, among other unestimated or underestimated costs. Overall, the committee finds that the range of nonbattery costs includes a low estimate equal to the Agencies' estimates and a high estimate up to approximately \$1,300 and \$500 above the Agencies' estimates for a midsize PHEV 40 and an EV75 in 2025, respectively. These cost increments were developed by multiplying the Agencies' nonbattery costs, exclusive of the charger, by a factor of 1.5 for both the PHEV 40 and the EV75. The committee's estimate of the charger cost was the same as that of the Agencies.

Batteries

Batteries are required in conventional vehicles for electric starting of the vehicle. For the improved fuel economy provided by regenerative braking and electrically powered drive, larger batteries are required. Current vehicles on the market typically use nickel metal hydride (NiMH) or lithium ion (Li-ion) batteries for hybrid vehicles and Li-ion batteries for battery electric vehicles. These high-power, high-energy, large-volume, heavy batteries are the most significant incremental cost for vehicle electrification (Whittingham 2004). Improvements in battery chemistry and engineering are thus critical to reducing the cost of HEVs and BEVs. In this section, current and near-term (to 2025) battery technologies will be discussed. The focus will be on Li-ion batteries since NiMH batteries are not expected to see significant development. Longer-term battery technologies will be discussed in the next section.

The rechargeable, lithium-ion battery was first introduced as a commercially viable product by the Sony Corporation in the early 1990s following more than two decades of research in the field (Whittingham 2004). Since that time, Li-ion technology has matured to the point of dominating the consumer electronics market. State-of-the-art Li-ion batteries now enable portable electronic devices, which have changed the way we live and communicate. The success of Li-ion batteries has prompted a surge in research and development aimed at harnessing the energy-storage capabilities of Li-ion chemistries for more advanced applications such as PEV transportation. However, despite the success of Li-ion with respect to consumer electronics, transportation applications are considerably more demanding, particularly in terms of battery life, safety, and cost (USABC 2013a, 2013b). As such, several major challenges must be addressed and overcome if Li-ion is to power the next fleet of light-duty vehicles. While there are currently many novel Li-ion-related technologies under investigation (Yang et al. 2011), this section will present an overview of current research on some of the most promising near-term technologies related to rechargeable Li-ion systems, including next-generation cathodes, anodes, and electrolytes.

Lithium Ion

Basic Operating Principles

Figure 4.8 shows a schematic of a typical Li-ion battery consisting of two electrodes (cathode and anode), a separator, and a liquid electrolyte that permeates the system. The cathode, or positive electrode, is lithiated on discharge, while the anode, or negative electrode, is lithiated on charge. As in Figure 4.8, when electrical current is applied to charge the cell, lithium ions move out of the cathode ($Li_{1-x}CoO_2$) and become trapped inside the anode storage medium, which is usually graphitized carbon (Li_xC_6). When the battery is discharged, the lithium ions travel back to the cathode and produce an external electrical current.

Cathodes

Table 4.2 includes a list of current commercial cathodes and the relevant battery metrics for comparison. Layered $LiCoO_2$ (shown in Figure 4.8) has been the standard Li-ion chemistry for almost 30 years, largely because of its high volumetric energy density. During cell operation at 3.0-4.2 V, the practical capacity of the $LiCoO_2$ electrodes is approximately 150 mAh/g, ~50 percent of its theoretical value

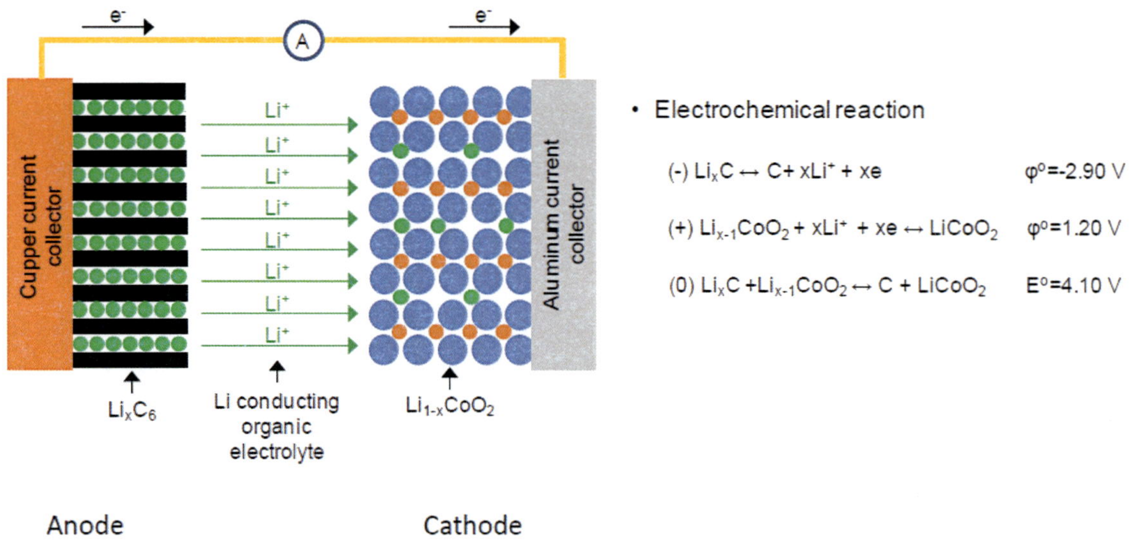

FIGURE 4.8 Working Li-ion battery utilizing a $LiCoO_2$ cathode and a graphite anode having aluminum and copper current collectors, respectively. The electrolyte permeates the entire system and, together with the separator, allows for the diffusion of positively charged Li^+ ions but not negatively charged electrons. Electrons must travel through the external circuit, constituting an electric current that powers the attached device (load) on discharge.
SOURCE: Amine et al. (2014). Reproduced with permission.

TABLE 4.2 Best-Case (low rate), Practically Achievable Working Voltages and Capacities vs. Li/Li^+ for Various Cathode Materials

Material	Voltage (Ave. vs Li/Li^+)	Capacity (mAh/g)	Crystal Density (g/cm^3)	Tap Density (g/cm^3)	Specific Energy (Wh/kg)	Volumetric Energy (Wh/L)
$LiCoO_2$	3.8	150	5.10	2.9	570	2,907
$LiNi_{1/3}Mn_{1/3}Co_{1/3}O_2$ (NMC)	3.7	170	4.75	2.5	629	2,988
$LiNi_{0.8}Co_{0.15}Al_{0.05}O_2$ (NCA)	3.7	185	4.85	2.5	685	3,322
$LiMn_2O_4$	4.0	110	4.31	2.5	440	1,896
$LiFePO_4$	3.4	160	3.60	1.5	544	1,958
$0.5Li_2MnO_3•0.5LiMO_2$[a]	3.6	250	4.30	1.8	828	3,870
$LiMn_{1.5}Ni_{0.5}O_4$[a]	4.7	135	4.40	2.0	635	2,794

[a] Not yet commercialized.
NOTE: Crystal densities are theoretical values, while tap densities represent typical, practical values, determined experimentally as the actual weight per unit volume occupied by a given material. Volumetric energy densities are calculated using the crystal densities for comparison because the optimum, final electrode densities will vary among materials. M = Mn, Ni, or Co in $LiMO_2$. Capacity and voltage targets for $0.5Li_2MnO_3•0.5LiMO_2$ are based on DOE's end-of-life goals for composite materials; crystal density is calculated as an average of Li_2MnO_3 and $LiMn_{0.5}Ni_{0.5}O_2$.

(273 mAh/g), due to the surface reactivity and instability of the delithiated $Li_{1-x}CoO_2$ structure (Jeong et al. 2011; Lee et al. 2012). This instability, the high possibility of thermal runaway in inadequately controlled batteries, and the relatively high cost of cobalt have led to efforts to find alternative cathode materials to $LiCoO_2$ that provide Li-ion cells with superior energy density, rate capability, safety, and cycle life.

Several alternative cathode materials to $LiCoO_2$ have been exploited by the Li-ion battery industry over the past decade. They include compositional variations of the layered $LiCoO_2$ structure, such as $LiNi_{0.8}Co_{0.15}Al_{0.05}O_2$ (NCA) (Bang et al. 2006; Lee, S.H. et al. 2013); spinel electrodes derived from $LiMn_2O_4$, such as lithium-rich compounds in the $Li_{1+x}Mn_{2-x}O_4$ system (Park et al. 2008; Gu et al. 2013); and $LiFePO_4$,which has an olivine-type structure (Padhi et al. 1997a, 1997b; Yuan et al. 2011).

Although NCA provides a slightly higher practical capacity (160-185 mAh/g) than $LiCoO_2$, its thermal instability on delithiation, which is due to the presence of the high valance Ni, compromises the safety of Li-ion cells. On the other hand, spinel $LiMn_2O_4$ and olivine $LiFePO_4$ electrodes are significantly more stable to lithium extraction than the layered Co- and Ni-based electrodes (both structurally and thermally), but they deliver relatively low practical capacities in a lithium cell above 3 V, typically 110-160 mAh/g at moderate current rates. It became clear by the end of the 1990s that alternative structurally stable, high-potential cathode materials (>3 V) with rate capabilities and capacities superior to those achievable with standard $LiCoO_2$-, $LiMn_2O_4$-, and $LiFePO_4$-type electrodes were required. As an alternative to increase energy, $LiNi_xMn_yCo_zO_2$ ($x < 1$, $y < 1$, and $z < 1$) (NMC) was developed for potential use in automotive applications and offers capacities similar to NCA (Zonghai et al. 2013; Amine et al. 2011). As a practical reference, Chevy's PHEV Volt, with a ~40 mile all-electric range, uses a physically blended NMC-based/$LiMn_2O_4$ cathode material. To drive the cost down and/or the driving range up, significant advancements in cathode technology are needed beyond NMC. Recently, Argonne National Laboratory has developed a family of high-energy-density lithium- and manganese-rich xLi_2MnO_3•$(1 - x)LiMO_2$ (M = Mn, Ni, Co) composite cathodes by structurally integrating a Li_2MnO_3 stabilizing component into an electrochemically active $LiMO_2$ (M = Mn, Ni, Co) electrode (Thackeray et al. 2007; Sun et al. 2005). The relatively high Mn content in these high-energy cathode materials lowers material costs, while the excess lithium boosts specific capacity to 250 mAh/g between 4.6 and 2.5 V and therefore significantly improves the energy density of the battery cell to 900 Wh/kg. However, in practical cells, when these high-energy NMC oxides are cycled against graphite, deliverable capacity decreases dramatically with cycle number along with a significant decay of cell discharge voltage (Li, Y. et al. 2013; Bettge et al. 2013) and a severe loss of energy density, which hinders its practical application in electric vehicles. Table 4.2 summarizes the battery specification of various lithium ion cells, including a high-energy NMC cathode, and Table 4.3 summarizes performance of all battery systems used in electric vehicles currently on the market.

Anode Materials for Rechargeable Li-Ion Batteries

As will be discussed below, lithium batteries with metallic lithium anodes offer the highest theoretical capacity of almost all conventional battery types and in principle should provide the highest energy density of all lithium batteries, primary or secondary, since lithium metal has an extremely high specific capacity (3,860 mAh/g) and lower negative redox potential (-3.04 V vs. standard hydrogen electrode (SHE)) (Aurbach and Cohen 1996).

To avoid the technical hurdles posed by lithium metal as the anode material, lower specific capacity carbon-based materials, such as graphite (372 mAh/g), are most commonly used. In order to overcome the capacity limit of current technology, materials such as Sn and Si (Hou et al. 2013; Deng et al. 2013; Menkin et al. 2014), which form alloys with lithium, are potentially more attractive anode candidates since they can incorporate larger amounts of lithium (Figure 4.9). Among these metals, silicon-based anodes are particularly attractive because of their higher theoretical specific capacity of approximately 4,200 mAh/g (ca. $Li_{4.4}Si$), which is far larger than that of graphite and oxide materials (Ge et al. 2013). However, the application of bulk silicon anode faces one major problem: During the reaction that forms the silicon–lithium alloy (corresponding to the insertion of lithium in the negative electrode during the charging process), the volume expansion from the delithiated phase to the lithiated phase may reach 380 percent (Figure 4.10). This high expansion, followed by a contraction of the same amplitude upon discharging rapidly leads to irreversible mechanical damage to the electrode and eventually leads to a loss of contact between the negative electrode and the underlying current collector, which causes a rapid capacity fade during cycling. Furthermore, silicon usually possesses low electrical conductivity, which has the effect of kinetically limiting the use of the battery. A significant effort is under way to enable this system by designing conductive binders that can minimize any particle isolation or by incorporating Si in graphene sheet to keep good conductivity at the electrode level during cycling (Wang et al. 2013; Wu et al. 2013; Ye et al. 2014).

TABLE 4.3 Li-ion Battery Systems in Electric Vehicles

EV	Cathode	Anode	Battery Supplier	Type of Cell[a]	Number of Cells	Electric Energy (kWh)	Power (kW)	Specific Energy Density (Wh/kg)	Electric Range (mi)
Tesla Model S	NCA (layered)	Carbon (layered)	Panasonic	C	>7,000	85	270	116 (pack)	265
Chevy Volt	LMO (spinel)	Carbon (layered)	LG Chem	P	>200	16	111	88 (pack)	40
Nissan Leaf	LMO (spinel)	Carbon (layered)	Nissan NEC JV	P	192	24	90	140 (pack)	84
Honda Fit	NCM (layered)	LTO (layered)	Toshiba Corp.	P	432	20	92	100 (cell)	82
BYD E6	LFP (olivine)	Carbon (layered)	BYD	P	96	48	75	—	186

[a] C, cylindrical; P, prismatic.

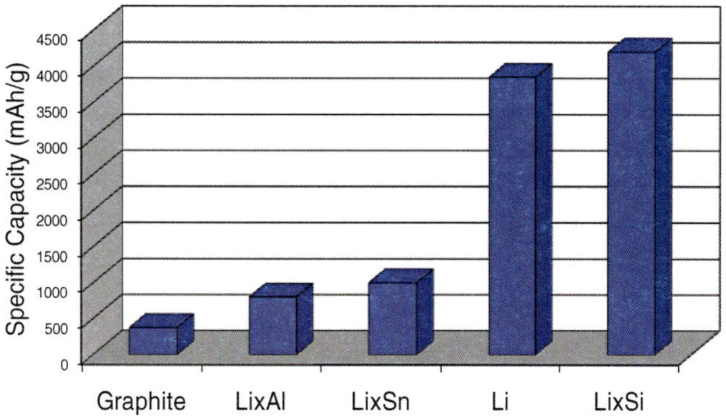

FIGURE 4.9 Specific capacities of graphite, Li_xAl, Li_xSn, Li and Li_xSi anodes (mAh/g).
SOURCE: Amine et al. (2014). Reproduced with permission.

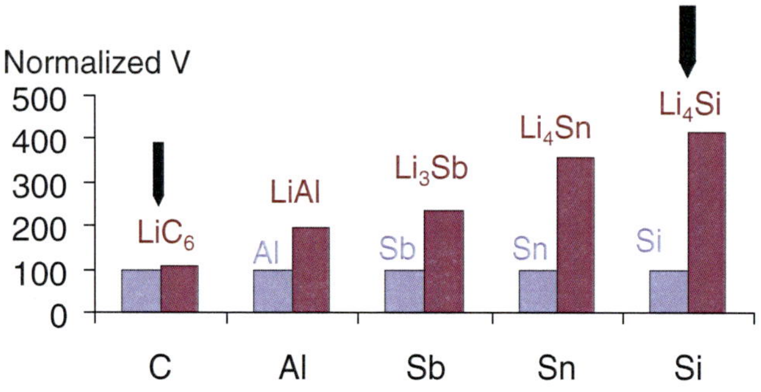

FIGURE 4.10 Volume expansion of different Li-metal alloys, including Li_4Si.
SOURCE: Committee-generated from data reported in patent WO 2005076389 A2.

Battery Management Systems

A battery management system (BMS) for Li-ion battery packs is responsible for ensuring that all the battery cells operate within prescribed intervals of voltage, temperature, and Li-ion concentration, as shown in Figure 4.11. To this end, the predominant role of the BMS is real-time estimation and prediction of the battery states and their proximity to the limits—SOC, state of power (SOP), and state of health (SOH).

Battery SOC describes the remaining energy of a battery, which is equivalent to the ubiquitous fuel gauge of a conventional vehicle. Information on the battery SOC is very important for supervisory controllers in electrified vehicles to determine power flows to maximize system efficiency. Many studies have been conducted to develop methods for accurate SOC estimation. These methods can be divided into three categories: coulomb counting, voltage-inversion, and model-based estimation.

Coulomb counting relies on the integration of the current drawn from and supplied to a battery over operation time (Ng et al. 2009). This method is advantageous owing to its simple structure and ease of implementation. However, sensor accuracy, temperature-dependent capacity, and calibration of the initial battery SOC make it difficult to accurately estimate subsequent battery SOC. On the other hand, the voltage-inversion method utilizes the one-to-one relationship between voltage and battery SOC (Pop et al. 2005; Dubarry et al. 2013). That is, the available capacity is determined by measuring terminal voltage during battery discharge operations. However, it is not easy to provide constant discharge current during battery operation. Corrections due to current

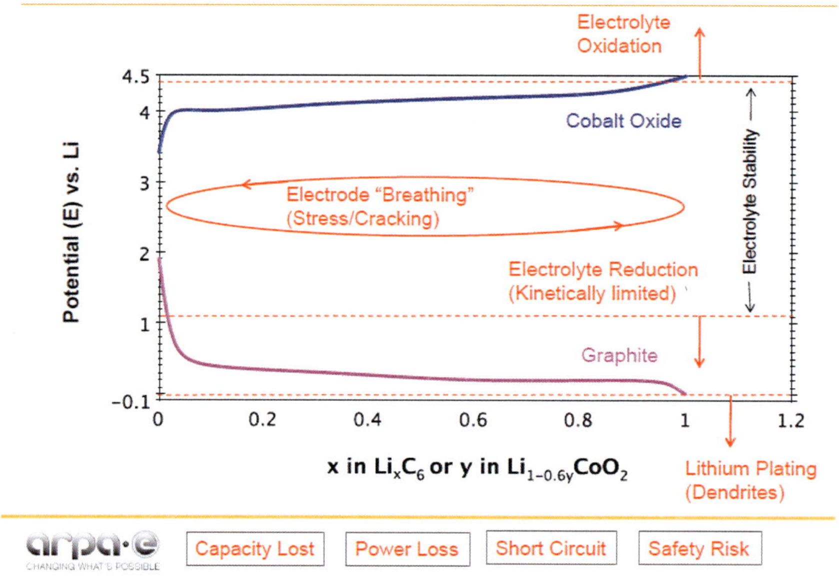

FIGURE 4.11 The battery management system protects each cell from a variety of detrimental conditions as described by Ilan Gur, ARPA-E program manager of the AMPED program, in his opening remarks at the 2014 AMPED review meeting.

and temperature-dependent SOC make this method more complicated than traditional coulomb counting.

Various model-based methods with current and voltage measurement (closed-loop estimators) have been developed for battery SOC estimation in an effort to overcome the drawback and merge the benefits of the coulomb counting and voltage-inversion methods (Plett 2004a; Lee et al. 2008; Di Domenico et al. 2010; Smith et al. 2010; Kim 2010; Rahimian et al. 2012; Xiong et al. 2013). The closed-loop SOC estimator relies on coulomb counting that is modified by an error between the estimated voltage and measured cell voltages. Clearly, a voltage prediction is necessary for forming the voltage error, and many recent efforts have targeted computationally efficient and physics-based models that emulate the electrochemical cell behavior. The estimation gain can be computed using various techniques such as pole placement, sliding mode observer, and Kalman filter, including extended Kalman filter and unscented Kalman filter.

Battery SOP refers to the constant power that can be safely drawn from or provided to a battery over a certain period of time. Estimating the battery SOP is of vital importance in protecting Li-ion batteries from overheating as well as overcharging/discharging. Much effort has been devoted to developing model-based methods to estimate battery SOP in real time (Plett 2004a; Smith et al. 2010; Xiong et al. 2013; Moura et al. 2013; Kim et al. 2013). Battery SOP estimation is also important for battery thermal management for applications with limited cooling for Li-ion batteries.

Battery SOH as an indicator of battery degradation defines the present performance of a battery relative to its fresh condition. The performance degradation of a battery may be the result not of a single mechanism but of several complicated mechanisms. Nonetheless, degradation mechanisms can lead to either a decrease in total available capacity or an increase in internal resistance. Thus, various model-based estimation techniques have been developed to identify those parameters with voltage and temperature measurement (Kim 2010; Verbrugge and Tate 2004; Goebel et al. 2008; Plett 2004b; Kim and Cho 2011; Lin et al. 2013).

For electrified vehicles, which require high voltage levels, large banks of series-connected cells are used to satisfy the power demand. Generally, a battery pack consists of hundreds of individual cells. Since aging, use, and calendar life lead to cell-to-cell variability, BMS should be able to equalize cells, referring to cell balancing or cell equalizing, in order to prevent individual cell overcharge or overdischarge. Cell balancing methods can be divided into two categories: dissipative and nondissipative. Dissipating methods equalize the cells by extracting energy from the higher charged ones and dissipating it on shunts or resistors (Asumadu et al. 2005) or selectively disconnecting imbalanced cells from the battery pack (Shibata et al. 2001). Nondissipating methods, on the other hand, can be divided into discharge equalizing systems, like multioutput transformers (Kutkut et al. 1999), charge-equalizing systems, like the distributed Cuk converter (Chen et al. 2009; Park et al. 2009), and bidirectional equalizing systems, like a switched capacitor or an inductor circuit (Moo et al. 2003; Speltino et al. 2010). It should be noted that each approach, regardless of its advantages and drawbacks, relies on an estimated SOC to perform cell balancing.

In summary, the BMS is critical for the performance (SOP estimation), utilization (SOC estimation), degradation (SOH estimation), and, finally, the safety of the battery pack. The processor, the voltage and current sensors, wiring harness, and switching network for the cell-balancing add to the cost of a battery. The accuracy of the BMS and confidence in its functionality are also responsible for defining the battery SOC range, hence influencing the battery size and thus the vehicle cost. Finally, the BMS influences the vehicle fuel consumption indirectly, by informing the hybrid electric vehicle supervisory controller about the battery status and availability (SOC, SOP) and hence defining the operating window for the internal combustion engine.

High-Power vs. High-Energy Batteries

Batteries are designed to store energy and deliver it at needed rates, producing the power required to move the vehicle. There are trade-offs in choice of battery chemistry and battery component design to maximize energy or power. For example, batteries that are designed to contain as much charge as possible, as used in BEVs, are designed for higher energy. In contrast, batteries that must survive more charge and discharge cycles, such as those used in HEVs, are designed with higher power in mind. As discussed previously in reference to mild hybrids in particular, HEV batteries are currently designed to be oversized in terms of energy in order to limit the SOC swing to 20 percent for a battery's lifetime. Designing battery materials that can provide more power would enable a larger SOC swing for HEVs, enabling the necessary acceleration and regenerative braking with use of a smaller battery, which could lower the battery cost. There are attempts to design such batteries, and one technology that can allow a 40 percent swing in SOC is a spinel LMO coupled with LTO. That system is low voltage (2.5 V), however, which means that cells must be added to get the voltage required for the pack. This will lead to higher costs. In addition, the spinel has a dissolution issue that requires overdesigning the battery pack, leading to significant cost increases. As of now, there is no system under development that is targeting high power that can be used in 2025. Most of the materials under development are seeking to provide high energy to enable lowering the cost of BEVs and PHEVs.

Modeling Estimates of Future Battery Costs

The recent penetration of lithium-ion (Li-ion) batteries into the vehicle market has prompted interest in projecting and understanding the costs of this family of chemistries. The Battery Performance and Cost (BatPaC) model is a calculation method that was developed at Argonne National Laboratory for estimating the manufacturing cost and performance of Li-ion batteries for electric-drive vehicles including HEVs, PHEVs, and BEVs. The BatPaC model is a publicly available bottom-up design and cost model developed for the Li-ion chemistries with support from the U.S. Department of Energy Vehicle Technologies Office. BatPaC has gone through multiple public and private peer reviews sponsored by EPA (ICF 2011) and has been used in the analysis of the 2017-2025 CAFE/GHG rule (EPA/NHTSA 2012a). A detailed description of the BatPaC model is available in Nelson et al. (2011 and 2012).

To date, a number of cost models for various levels of detail have been published in different forms (Anderman et al. 2000; Barnett et al. 2009, 2010; Dinger et al. 2010). The cost of a battery will change depending on the materials chemistry, battery design, and manufacturing process (Gallagher et al. 2011; Nelson et al. 2009; Santini et al. 2010). Therefore, it is necessary to account for all three areas with a bottom-up cost model for Li-ion battery packs used in automotive transportation. The cost of the designed battery is calculated by accounting for every step in the Li-ion battery manufacturing process. The assumed annual production level directly affects each process step. The total cost to the original equipment manufacturer (OEM) calculated by the model includes the materials, manufacturing, and warranty costs for a battery produced in the year 2020 (in 2010 dollars). A user of the model will be able to recreate the calculations and, perhaps more important, understand the driving forces for the results. Almost every variable in the calculation may be changed by the user to represent a system different from the default values pre-entered into the program.

The distinct advantage of using a bottom-up cost and design model is that the entire power-to-energy space may be traversed to examine the correlation between performance and cost. The BatPaC model accounts for the physical limitations of the electrochemical processes within the battery. Thus, unrealistic designs are penalized in energy density and cost, unlike cost models based on linear extrapolations. Additionally, the consequences for cost and energy density from changes in cell capacity, parallel cell groups, and manufacturing capabilities are easily assessed with the model. New proposed materials may also be examined to translate bench-scale values to the design of full-scale battery packs providing realistic energy densities and prices to the OEMs.

Enabling Technologies for Vehicle Electrification to 2030

The committee's statement of task seeks advice on the fuel economy technologies expected to be available between 2020 and 2030. In order for electric vehicles to become truly mainstream, significant breakthroughs are required in their energy storage systems. As compared to today's extant Li-ion technologies, vehicular batteries must achieve lower cost, improved safety, longer driving ranges, less refueling time, and less environmental impact. Fuel cell vehicles face challenges in increasing durability and decreasing cost as well as in the hydrogen supply infrastructure. Both systems, and hydrogen fuel cell vehicles in particular, require deployment of an infrastructure for refilling with electricity or hydrogen. The following section describes the battery and fuel cell technologies likely to be in use in some portion of the fleet by 2030.

Batteries

From 2020 to 2025, the existing cathode chemistries, including NMC cathodes rich in nickel, will likely be refined and the trade-offs between safety, cost, energy density, and power will be better understood. On the anode side, the industry is predicting a blend of mostly graphite with 5 to 10 percent Si-based anode. As a result, the cell energy density will likely increase by 30-50 percent compared to today's energy performance. By ~2025-2030, the industry is predicting that the use of stabilized high-energy-density lithium-and manganese-rich $x\text{Li}_2\text{MnO}_3\bullet(1-x)\text{LiMO}_2$ (M = Mn, Ni, Co) composite cathodes and a blend of graphite and up to 20 percent silicon anode will double the energy density of today's lithium ion. As a result, the battery will cost significantly less.

Beyond Li-Ion Systems

As discussed above, the rechargeable Li-ion batteries have transformed portable electronic devices and likely will play a key role in the electrification of transport in the near- to midterm. However, the inherent energy density of the current Li-ion technology is not sufficient for the long-term needs of extended-range BEVs. In this section, the committee provides a brief overview of three systems beyond Li-ion—rechargeable Li-S, Li-air, and magnesium batteries— and addresses some of the key challenges for each individual system.

Li-Metal Anodes

Two major technical bottlenecks prevent the realization of a successful rechargeable Li metal battery (Wu et al. 2014). One is the growth of lithium dendrites during the repeated charge/discharge cycles, which severely compromises the rechargeability of each lithium cell and could also pose a serious safety hazard because of the potential internal short circuit if these dendrites penetrate the separators and contact the cathode directly. The other is the low coulombic efficiency during the repeated cycles, although this can be partially compensated for by an excess amount of lithium. Overcoming these hurdles presents an enormous challenge to the lithium battery industry. Recently, researchers demonstrated that the growth of the lithium dendrites can be partially prevented through either a physical blocking mechanism using solid-state poly(ethylene oxide) copolymer electrolytes (Balsara et al. 2009) or a self-healing mechanism that uses novel electrolyte additives (Ding et al. 2013). However, these mechanisms are effective only under very limited conditions—that is, at high temperature or low current density. Therefore, work is needed to look for a more reliable way to prevent dendrite growth in order to push the lithium anode for broad applications. Despite these obstacles, significant efforts are still being made to capitalize on and exploit the advantages of the metallic lithium systems such as Li-S and Li-air batteries, with a big assumption that these obstacles can be overcome eventually.

Li-S Batteries

The rechargeable Li-S cell operates by reduction of S at the cathode upon discharge to form a series of soluble polysulfide species (Li_2S_8, Li_2S_6, Li_2S_4) that combine with Li to ultimately produce solid Li_2S_2 and Li_2S at the end of discharge, as illustrated in Figure 4.12. Upon charging, Li_2S_2/Li_2S are converted back to S via similar soluble polysulfide intermediates formed in the discharge process and lithium plates to the nominal anode, making the cell reversible. This contrasts with conventional Li-ion cells, where the lithium ions are intercalated in the anode and cathodes, and consequently the Li-S system allows for a much higher lithium storage density (Barghamadi et al. 2013; Xiulei and Nazar 2010; Yang et al. 2013a).

The Li-S battery, if based on the reaction $S_8 + 16\ Li = 8\ Li_2S$, operates at an average voltage of 2.15 V and has a theoretical specific capacity of 1,675 mAh/g-S. This leads to an energy density of 2,600 Wh/kg (2,800 Wh/L) that is five times higher than that of the conventional Li-ion intercalation battery. Sulfur is an abundant material available on a large scale and at low cost as a side product of petroleum and mineral refining, which makes it attractive for low-cost and high-energy rechargeable lithium batteries. Furthermore, the unique feature of the Li-S chemistry provides inherent chemical overcharge protection, enhancing safety, particularly for high-capacity multi-cell battery packs (Yang et al. 2013a).

Although sulfur-based electrochemical cells had already been reported in 1962, the electronically insulating nature of sulfur, the solubility of intermediately formed polysulfides in common liquid organic electrolytes, and the use of metallic lithium as a negative electrode have still not been solved satisfactorily. In addition, the formed polysulfides in the electrolyte migrate to a lithium metal anode and are electrochemically reduced (Yan et al. 2013), resulting in low coulombic efficiency and rapid capacity fade in Li-S batteries.

Recently, the interest in Li-S-based secondary batteries has been steadily increasing thanks to the design of new nanostructure materials that may be able to overcome issues related to the conductivity of bulk materials (Xiulei et al. 2009; Yang et al. 2013b; Zheng et al. 2011). Moreover, the development of new electrolytes, binder materials, and cell design concepts in general has led to significant advances in the field of Li-S-based secondary batteries within the last few years (Barghamadi et al. 2013; Xiulei and Nazar 2010). There is no doubt that Li-S batteries remain attractive for the longer term because of their inherently high energy content, high power capability, and potential for low cost, although they are still in the development stage.

Li-Air Batteries

Li-air batteries could theoretically provide the needed order of magnitude improvement in energy density because

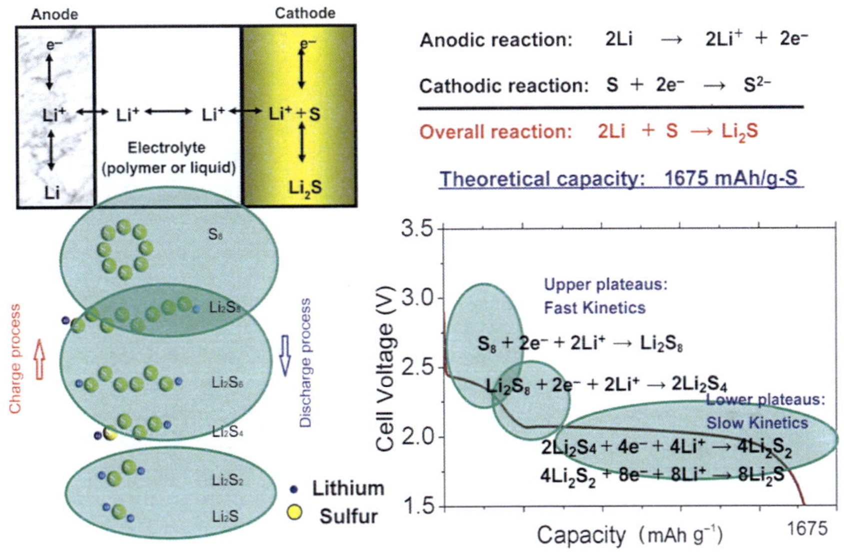

FIGURE 4.12 Scheme of a Li-S cell and its electrochemical reactions.
SOURCE: Amine et al. (2014). Reproduced with permission.

they do not need to store their oxidant (Bruce et al. 2011; Bruce et al. 2012). Whereas state-of-the-art Li-ion batteries have achieved 150-200 Wh/kg (of the 900 Wh/kg theoretically possible) at the cell level, Li-air batteries have the potential to achieve 3,620 Wh/kg (when discharged to Li_2O_2 at 3.1 V) or 5,200 Wh/kg (when discharged to Li_2O at 3.1 V). When the "free" oxygen supplied during discharge and released during charge is not included in the calculation, Li-air cells offer ~11,000 Wh/kg. This is basically identical to the lower heating value for gasoline which is ~13,000 Wh/kg when oxygen is supplied externally. Unlike any other battery technology, Li-air energy density is competitive with that of liquid fuels.

During discharge of the Li-air cell, Li is oxidized to Li^+ as a metallic Li anode, conducts through an electrolyte made up of a non-aqueous solvent and a Li salt, and reacts with O_2 from the air on a cathode made of carbon, a catalyst, and a binder deposited on a carbon paper substrate, as shown in Figure 4.13. The Li-air technology has the potential to significantly reduce the cost well below that of Li-ion technology due to the higher specific energy densities and the lower cost of the proposed cell components, in particular of the carbon-based cathode materials versus the nickel, manganese, and cobalt oxides used in Li-ion battery cathodes. The non-aqueous electrolyte is preferred, as it has been shown to have higher theoretical energy densities than aqueous electrolyte designs (Zheng et al. 2008).

Current Li-air batteries are still in the experimental stage, and the realization of the high theoretical energy densities and practical application of this technology have been limited by the low power output (i.e., low current density), poor cyclability, and low energy efficiency of the cell. These limitations are caused by the materials and system design:

(1) Unstable electrolytes. The current non-aqueous carbonate electrolytes are volatile, unstable at high potentials, easily oxidized, and reduced at the lithium anode in the presence of crossover oxygen. This seriously limits cycle life (Freunberger et al. 2011; McCloskey et al. 2011a).

(2) Lithium electrode poisoning due to oxygen crossover and reaction with the electrolyte destroys the integrity and functioning of the cell and shortens its cycle life (Assary et al. 2013).

(3) Li_2O_2 and/or Li_2O deposition on the carbon cathode surface or within the pores. This creates clogging and restricts the oxygen flow, lowering capacity (Lu et al. 2011; Lu and Shao-Horn 2013).

(4) Inefficient cathode structure and catalysis. Commonly used carbons and cathode catalysts do not access the full capacity of the oxygen electrode and cause significant charge overpotentials. This lowers rates (Li, F. et al. 2013; Shao et al. 2012; Shao et al. 2013).

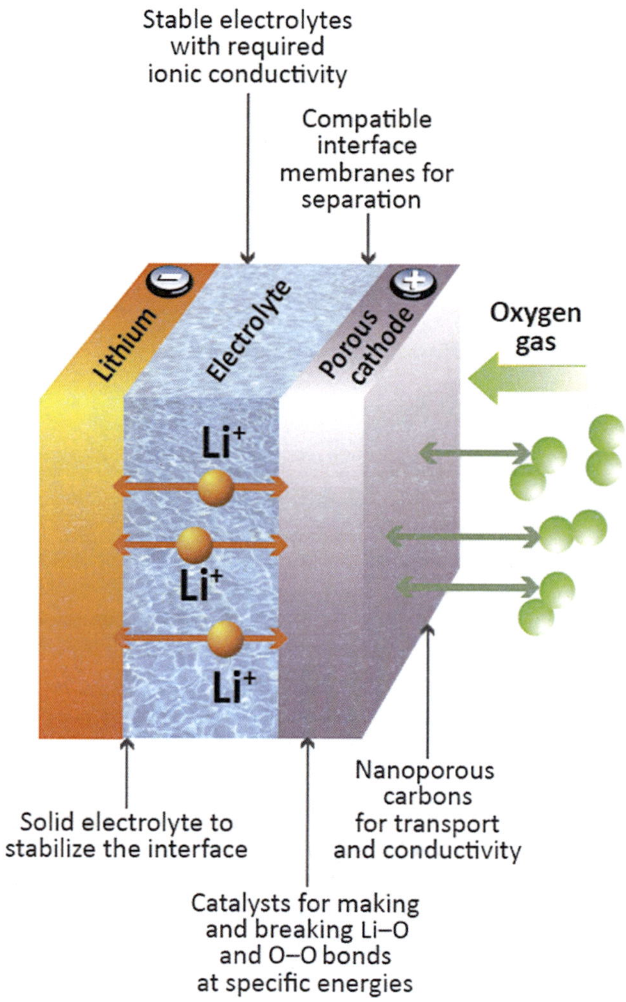

FIGURE 4.13 Diagram of a non-aqueous Li-air battery.
SOURCE: Amine et al. (2014). Reproduced with permission.

It has recently become apparent that the electrolyte plays a key role in Li-air cell performance (McCloskey et al. 2011b; Black et al. 2012; Jung et al. 2012). The oxygen anion radical O_2^- intermediate or other reduction species that may be formed during the discharge process can be highly reactive and may cause the electrochemical response to be dominated by electrolyte decomposition rather than the expected lithium peroxide formation. The overall result is the consumption of the alkyl carbonate electrolyte.

Although electrolyte stability is of paramount importance, cathode materials also represent a major technology challenge in Li-air cell development (Li, F. et al. 2013; Lu and Amine 2013; Shao et al. 2012, 2013). The ultimate goal is to determine how to effectively increase the specific capacity and power capability of Li-air cells yet still achieve long cycle life. Attaining that goal strongly depends on the materials and their microstructures in the O_2-breathing cathode (Lu et al. 2010; Oh et al. 2012).

Though it offers a high theoretical energy density, in practice a Li-air battery may reach an energy density only twice that of a Li-ion battery. Decreasing the lithium metal content may be limited by the difficulty of manufacturing thin lithium metal electrodes, resulting in about a four times excess lithium used. More significantly, if Li-air batteries must use pure O_2 rather than ambient air, then the size and weight of an oxygen tank must be taken into account in the energy density calculation. Li-air technology is likely to take more than 30 years before a real practical prototype can be developed and used to power an electric vehicle.

Rechargeable Magnesium Batteries

Magnesium-based batteries are, in principle, a very attractive alternative to other batteries, including Li batteries. Mg is much less expensive than Li because Mg is abundant in the Earth's crust. Mg and its compounds are usually less toxic and safer than Li compounds because Mg is stable when exposed to the atmosphere. Mg is also lightweight, which, in theory, could enhance the volumetric energy density of the cell. A rechargeable magnesium battery has been regarded as highly promising technology for energy storage and conversion since its first working prototype was ready for demonstration about a decade ago, and it could compete with lead-acid or Ni-Cd batteries in terms of energy density and self-discharge rate (Yoo et al. 2013). Since Mg provides two electrons per atom with electrochemical characteristics similar to Li, Mg batteries can offer a theoretical specific capacity of 2,205 mAh/g. Proper design and architecture should lead to Mg-based batteries with energy densities of 400-1,100 Wh/kg for an open circuit voltage in the range of 0.8-2.1 V, which would make it an attractive candidate for electric vehicles, electrical grid energy storage, and stationary back-up energy.

Possible future directions to achieve the goal of the high-energy-density Mg batteries include (1) developing high-capacity/low-voltage Mg-S (or other equivalent high-capacity redox couples) cathodes and (2) employing moderate-capacity/high-voltage Mg ion intercalation cathodes. To become practical, Mg batteries are still required to attain a specific energy comparable to that of state-of-the-art Li-ion batteries. Additionally, because of the low rate of Mg^{2+} diffusion, this system very likely will not ultimately provide enough power capability to power an electric vehicle but would remain an attractive candidate for electrical grid energy storage and stationary back-up energy.

All Solid-State Batteries

Solid-state lithium battery designs have the potential to deliver at least two times the volumetric energy density of conventional Li-ion batteries at less than half the cost per kilowatt-hour. This approach eliminates binders, separators, and liquid electrolytes. By eliminating these components, one can get around 95 percent of the theoretical energy density of the active materials. Solid-state batteries could herald a breakthrough in electrified driving because they are more compact and offer higher energy density than state-of-the-art Li-ion batteries (see Figure 4.14). In the absence of a thermally sensitive solid-electrolyte interphase, solid-state batteries intrinsically have a higher tolerance to thermal abuse and are much safer than Li-ion batteries using a flammable electrolyte. In addition, the solid-state electrolyte is mechanically strong enough to efficiently suppress the growth of lithium dendrites, which might cause an internal short inside a lithium battery using a liquid electrolyte, so it can enable the use of lithium metal as the anode for high energy-density batteries.

Solid-state batteries generally have a low power density, primarily because of two physical limitations associated with solid-state electrolytes: (1) low Li-ion conductivity

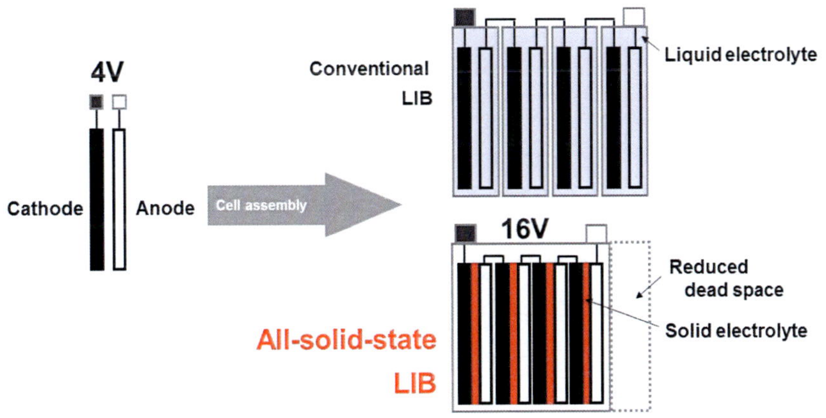

FIGURE 4.14 Cell design comparison between a conventional Li-ion battery and an all-solid-state battery.
SOURCE: Kotani (2013).

inside the electrolyte and (2) low ionic conductivity across the solid-solid interface. In principle, solid-state electrolytes are a class of single-ion conductor, in which the Li-ion can diffuse inside the solid while anions are immobilized. The disadvantage of a single-ion conductor is that the anion cannot establish a concentration gradient to assist the transfer of Li-ions in the electrolyte and the electrolyte-electrode material interface. Moreover, the engineering design of a solid-state battery has to be balanced between energy density and power density. A thicker cathode film is ideal to maximize the loading of active components for a high energy density. On the other hand, a thick electrode extends the diffusion path of both Li-ions and electrons during the normal charge/discharge operation, leading to a decrease in the power density. Currently, the design of an efficient electrolyte/cathode interface holds the greatest promise to boost the power density of solid-state batteries without sacrificing their energy density (Ohta et al. 2012).

Toyota is leading all-solid-state battery development and is planning to use solid-state lithium batteries as early as the 2020s. Since 2012, Toyota has managed to achieve fivefold increases in the power output of its experimental solid-state batteries (Kotani 2013). Although the current technology is still in the laboratory stage, Toyota expects it to be ready for cars in the early 2020s. If technology development is successful, the batteries could give BEVs a range of more than 300 miles on a single charge. Their current solid-state battery's energy density is around 400 Wh/L, compared with a maximum of around 300 Wh/L for Li-ion batteries. Toyota aims to increase the density to between 600 and 700 Wh/L by 2025.

Fuel Cells

The committee believes that fuel cell technology will be part of the vehicle mix in 2030. From the Final CAFE Rule in the Federal Register it is noted that

- Fuel cell electric vehicles were considered, but deemed not ready in the 2017-2025 timeframe (EPA/NHTSA 2012b, 62706)
- EPA is providing incentive multipliers for fuel cell electric vehicles for CO_2 compliance purposes, similar to the multipliers for EVs and PHEVs, in the 2017-2021 MY time frame to promote the increased application of these technologies in the program's (i.e., the CAFE Rule's) early model years (EPA/NHTSA 2012b, 62628). Incentives for AFVs within the CAFE and GHG programs are discussed further in Chapter 10.

The proton exchange membrane (PEM) fuel cell is the selected fuel cell technology for the automotive sector as it can be applied to all vehicle classes and platforms. The major automakers (Daimler, Toyota, Honda, Hyundai, GM, Ford, BMW, and Nissan) are working on solutions and vehicle applications. The following teams have emerged as these automakers move toward commercialization: Daimler/Ford/Nissan, Toyota/BMW, and GM/Honda. Hyundai is pursuing FCEVs independently.

This section presents an evaluation of today's status of PEM fuel cell technology and the plans communicated by the major automakers for future deployment worldwide. The hydrogen infrastructure plans under development to support FCEVs will be discussed in the next section. While the committee does not expect a significant impact on CAFE in the 2025 time period, it will be valuable to understand the development of both the technology and the hydrogen infrastructure to achieve the future prospects of this technology. Additionally, the increasingly stringent ZEV mandate may drive deployment of FCEVs.

The FCEV consists of a fuel cell system, hydrogen storage, power electronics, an electric drive motor/generator, and, typically, a small battery pack to collect regenerative braking and provide additional energy during cold start. The fuel cell system consists of an anode supply system for hydrogen, a cathode supply system for air (oxygen), a thermal management system, other supporting hardware known as balance of plant (BOP), as well as the controls to integrate the electrical power generation into electric vehicle type architectures. The power electronics, electric drive motor/generator, and battery pack have been strongly influenced by much of the work done to date on both HEVs and BEVs. These PEM fuel cell systems produce DC electricity electrochemically (as do battery-type vehicles) and have operating temperatures of 60 to100°C. The basic PEM technology concept and corresponding system architectures lend themselves very well to the transportation sector.

It is very difficult to give cost numbers for a technology still under development. According to an October 16, 2013, DOE Report (Record #13012), the cost of an 80 kW_{net} automotive polymer electrolyte membrane (PEM) fuel cell system based on 2013 technology and operating on direct hydrogen is projected to be $67/kW when manufactured at a volume of 100,000 units/year and $55/kW at 500,000 units/year (Spendelow and Marcinkoski 2014). Automakers that are part of the U.S. DRIVE partnership participated in the vetting of the report. Key assumptions from this cost analysis report are shown in Table 4.4.

Current costs at low volume (fewer than 10,000 units) appear to be closer to $300-$500/kW leading to systems costs for 100 kW systems of $30,000 to $50,000. These costs clearly need to be driven down through improved materials, better system integration, and greater volumes. An indication of the progress that has been made is that in 2014 Toyota announced that the price of its vehicle to be produced in 2015 would be around $69,000 (Reuters 2014), but price, particularly for a newly introduced technology, may not be indicative of costs to the manufacturer trying to develop a market.

The automakers take different approaches to the fuel cell system. An automotive system shown in schematic form in

TABLE 4.4 2013 DOE Report Key Assumptions of Cost Analyses for Fuel Cell System

Characteristic	Unit	2007	2008	2009	2010	2011	2012	2013
Stack power	kW_{gross}	90	90	88	88	89	88	89
System power	kW_{net}	80	80	80	80	80	80	80
Cell power density	mW_{gross}/cm^2	583	715	833	833	1110	984	692
Peak stack temperature	°C	70-90	80	80	90	95	87	97
PGM loading	mg/cm^2	0.35	0.25	0.15	0.15	0.19	0.20	0.15
PGM total content (gross)	g/kW_{gross}	0.6	0.35	0.18	0.18	0.17	0.20	0.23
PGM total content (net)	g/kW_{net}	0.68	0.39	0.20	0.20	0.19	0.22	0.25
Pt cost	$/tr oz.	1100	1100	1100	1100	1100	1100	1500
Stack cost	$\$/kW_{net}$	50	34	27	25	22	20	27
Balance of plant cost	$\$/kW_{net}$	42	37	33	25	26	26	27
System assembly and testing	$\$/kW_{net}$	2	2	1	1	1	1	1
System cost	$\$/kW_{net}$	94	73	61	51	49	47	55

SOURCE: Spendelow and Marcinkoski (2014).

Figure 4.15 includes the hydrogen storage and vehicle cooling systems (Mathias 2014).

Some systems use compressor/expanders and others use compressors only to improve overall efficiencies, but at a cost impact. Some use humidifiers to ensure good proton conductivity of the membrane, while others are driving material developments for self-humidifying membranes. In an automotive application, as described, gaseous hydrogen is supplied to the anode side via onboard storage tank(s), and air (oxygen) is supplied to the cathode side under pressure through an electrically driven compressor to improve the operating performance of the system. In order to achieve the high power densities required for packaging in a vehicle, the stacks are liquid cooled and are configured in a series fashion of 200-400 cells depending on the application and system requirements, as shown in Figure 4.16.

Stack power densities today are on the order of 2-3 kW/L. More important from a vehicle perspective are the whole sys-

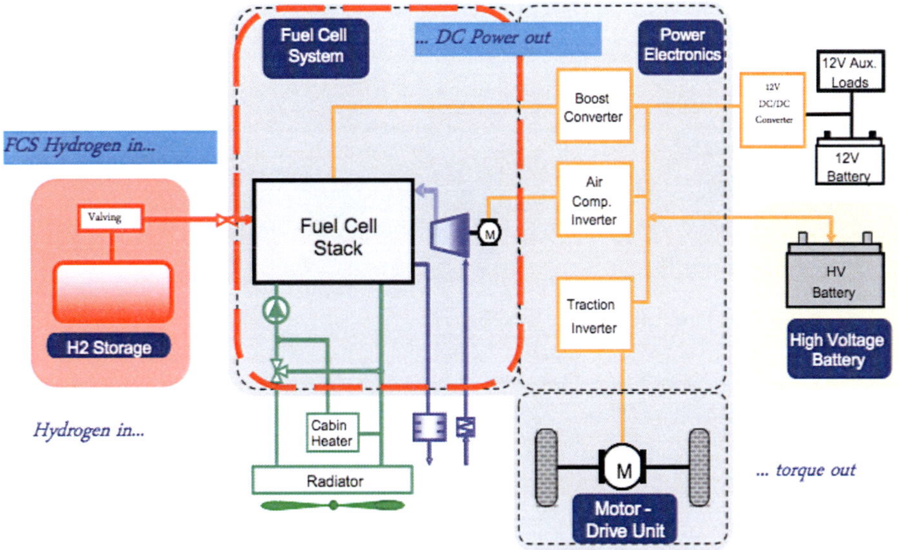

FIGURE 4.15 Diagram of a fuel cell vehicle, including hydrogen storage, the fuel cell stack, power electronics, and batteries.
SOURCE: M. Mathias, Honda/GM Fuel Cell Partnership – Moving from Technical to Commercial Viability, SAE 2014 Hybrid and Electric Vehicle Technologies Symposium.

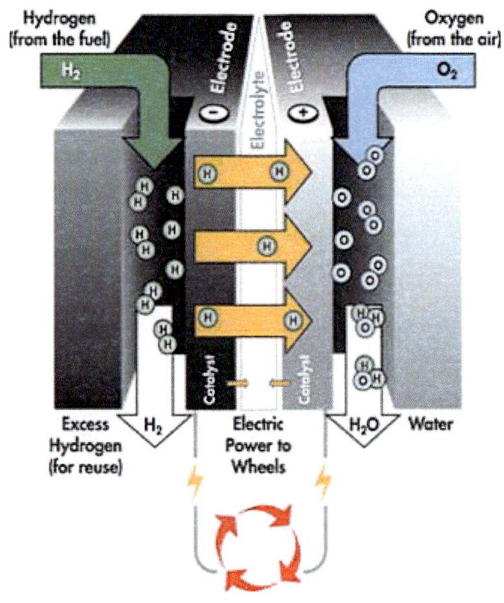

FIGURE 4.16 Schematic of a hydrogen fuel cell. Hydrogen (H_2) is oxidized at the anode, separating electrons and protons. Electrons pass through the external circuit, doing electrical work before reaching the cathode. In parallel, protons move through the electrolyte to the cathode, where protons and electrons reduce oxygen to water to complete the electrochemical circuit.
SOURCE: Battery University (2003). Sponsor: Cadex Electronics Inc.

tem power and gravimetric power densities, which directly impact the ability to package these systems in the various vehicle platforms.

System efficiencies can approach the DOE target of 60 percent at the 25 percent load points in the typical driving ranges. Figure 4.17 shows actual system efficiencies from four learning demonstration teams participating in the National Renewable Energy Laboratory (NREL) vehicle study program (Wipke 2012). These efficiency numbers include hydrogen as the fuel, electrical power required to run the air compressor and other ancillary power requiring devices, termed the balance of plant (BOP), and the delivery of DC electric power to the inverter. Fuel cell systems can be very efficient from off idle to approximately 25 percent load, where the majority of normal passenger car operation occurs.

Automotive companies have been developing the PEM technology for more than 20 years. Major accomplishments in the areas of durability, cost, and packaging have allowed the construction of prototype fleets of 10s and 100s of vehicles. Much work has been done with the materials supply base for the membranes, catalysts, metal plate materials, and gas diffusion layers, and suppliers are working on the aforementioned BOP components.

An example of cost reduction in materials is in the catalyst area. The last several years have seen a focus on reduction of platinum group metal (PGM) content, Pt alloys, novel support structures, and non-PGM catalysts. Catalyst cost is projected to be the largest contributor to overall system level costs at high volume. Demonstrated small-scale performance at overall catalyst levels of 0.16 g_{PGM}/kW has been demonstrated with the 3M NSTF membrane electrode assembly, which would yield approximately 14.4 gr of PGM catalyst for an 89 kW stack as referenced in the 2013 DOE Annual Merit Review (Satyapal 2013). Current prototype vehicle fleets generally use 50-70 grams of Pt for a 100 kW stack, so much work needs to be done in the scale-up and engineering of full automotive size systems to get to levels currently achievable in the lab.

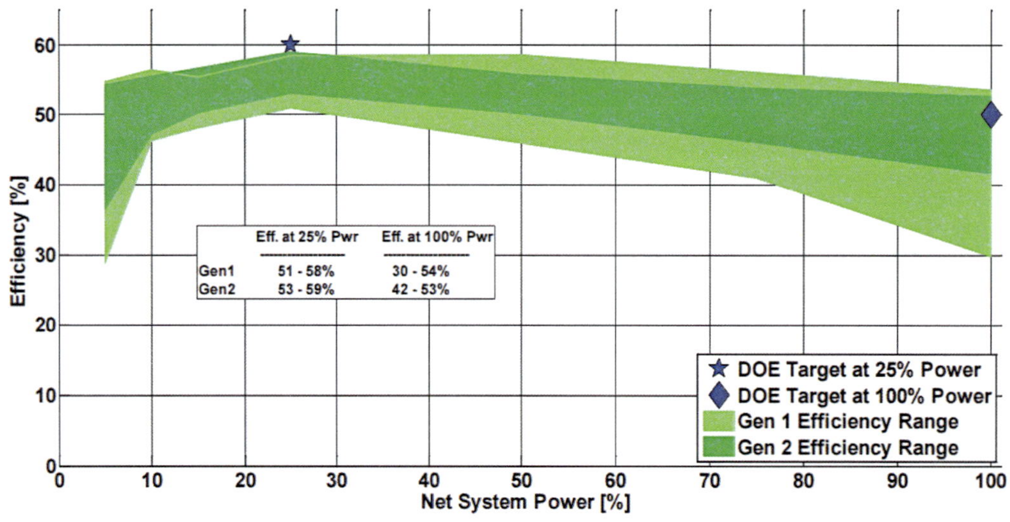

FIGURE 4.17 Fuel cell system efficiency at various vehicle power loads.
SOURCE: Wipke (2012).

The automakers have announced commercial vehicle sales in the 2015-2017 timeframe. To support their expectations, in 2013 they announced the following partnerships, mentioned earlier: (1) Toyota and BMW, (2) Daimler, Ford, and Nissan, and (3) GM and Honda, all of which were to assist in the product commercialization phase. Additionally, Hyundai has a significant internal program and publicly announced on November 20, 2013, plans to offer its next-generation Tucson fuel cell vehicle for the U.S. market for just $499 per month, including unlimited free hydrogen refueling and At Your Service Valet Maintenance at no extra cost. The first vehicles were delivered to lessees in June 2014 at several Southern California Hyundai dealers (Voelcker 2014). This first commercial implementation of FCEVs shows EPA compliance fuel consumption of 71.1 miles per gallon gas equivalent and a range of 265 miles. These programs and partnerships between the major players are being put in place to help reduce the engineering costs associated with new technology developments and more closely focus suppliers of both the fuel-cell-specific materials and the BOP components and subsystems such as compressors, sensors, and ancillary equipment. These partnerships also give credibility to the technology and its status. Further proof that these partnerships are accelerating commercialization plans is Toyota's July 2014 announcement of a 2015 vehicle priced at $69,000 in Japan (Gremeil 2014). Additionally, Honda is expected to release a retail FCEV in 2016, supplanting its lease program of the small scale production Honda FCX Clarity.

The automakers are being forthright regarding their perspectives on fuel cells and the challenges yet to be overcome. Toyota has gone on record (Ohnsman 2013) that the automotive fuel cell propulsion system is the system of the future from its perspective and has significant advantages over battery vehicles in the matters of range and refill (or recharge time in the case of batteries). Other automakers have made similar public announcements. Honda, for example, recently announced improvements to the hydrogen vehicle filling process to shorten times and make it more customer-friendly (Honda 2014).

To meet commercial high-volume product targets there are still several hurdles to overcome. Validated durability is a key at both the catalyst and the membrane level. The automakers, national labs, and the supplier base have done a tremendous amount of work to understand fundamental failure mechanisms and then address them through materials development, design improvements, and system-level controls refinements. Material developments in the catalyst area include improvements in the actual catalyst support, alloys of various materials, core/shell type technologies to improve effectiveness and reduce total Pt loading, and several alternatives with non-precious metal catalyst materials and concepts. Thinner supported-type membranes improve voltage performance and efficiencies as well as water management issues. Materials that can conduct and operate at lower levels of relative humidity are the ultimate goal that the supply community continues to pursue. These developments are continuing and occurring globally, with key suppliers and development programs in the United States, Japan, and Europe as the OEM engineering and commercialization programs move forward.

Enabling Infrastructure for PEVs and FCEVs

Gasoline- and diesel-powered vehicles, including conventional HEVs, use the extensive existing petroleum fueling infrastructure. Similarly, PEVs and FCEVs require an infrastructure for fueling on electricity or hydrogen, but this infrastructure may or may not resemble the existing petroleum infrastructure. Much electric infrastructure exists in private and public buildings and may be co-opted for electric vehicle charging; however, public electric fueling infrastructure is still in development. Hydrogen fueling infrastructure is in an even earlier stage of development and cannot rely on existing infrastructure; this represents a higher barrier to FCEV deployment than does the electricity infrastructure for PEV deployment. The infrastructure needs for PEVs and FCEVs are discussed further below.

PEV Charging Infrastructure

Electric fueling infrastructure is as important for PEVs as petroleum fueling infrastructure is for ICEs. The infrastructure that develops to fuel PEVs may not resemble the gas station model that has developed for ICEs, however. Because PEVs currently refuel more slowly than gasoline vehicles, and because an existing infrastructure of power lines and outlets reaches nearly every building, many PEV drivers have found it convenient to refuel at home or other locations where their vehicle remains parked for long periods. Some workplaces have chosen to implement charging at their parking facilities (NRC 2015). In some areas of high PEV deployment, public charging infrastructure is developing. While all PEVs can use 120 and 240 V charging infrastructure, some BEVs such as the Nissan Leaf and Tesla Model S vehicles can also use DC fast charging stations. Both Nissan and Tesla are building national networks of charging stations. Among current commercial models, only the Model S is practical for cross-country travel, as its charging time is much shorter than its range (NRC 2015).

FCEV Hydrogen Fueling Infrastructure

Several parts of the world are preparing for fuel cell vehicles by developing a hydrogen infrastructure for refueling. The hydrogen fuel for FCEVs is required to be very high purity to ensure optimum performance. Work needs to be done to determine trade-offs with performance and life with lower grades of industrially-produced hydrogen. Most notable infrastructure developments are occurring in California, Germany, Japan,

South Korea, and the U.K. Global deployments of hydrogen refueling are shown in Figure 4.18.

The implementation of a hydrogen infrastructure is required for FCEVs to reach a high volume of adoption. There has been much discussion and debate over when and if such an infrastructure should be realized. In the United States, this has become politicized, but in other areas of the world FCEVs and hydrogen are being considered in a more long-term context to reduce CO_2 footprints and enable other applications of the technology. As of late 2013, there were 10 public hydrogen fueling stations in the U.S., nine of them in California (DOE 2014c). The California Fuel Cell Partnership reports that if all currently planned and funded stations are built as expected, there will be 37 in the state by 2015 (Elrick 2013). In estimating fueling stations needed to support upcoming FCEVs, the Partnership has identified "68 strategically placed stations required to be operational by the beginning of 2016," as shown in Figure 4.18.

While South Carolina is the only other state with a public fueling station at present, a group of entities were recently awarded $500,000 from the Texas Emission Reduction Program (TERP) to partially fund the building of the first public hydrogen fueling station in that state at the Port of Houston (Curtin and Gangi 2013). Several additional states have expressed a commitment to provide infrastructure to support forthcoming fuel cell vehicles. On October 24, 2013, eight U.S. governors signed an agreement to support ZEVs and the necessary infrastructure developments (Carroll 2013), setting "a collective target of having at least 3.3 million zero emission vehicles on the road in our states by 2025 and to work together to establish a fueling infrastructure that will adequately support this number of vehicles." A federal tax credit of up to 30 percent of the cost, not to exceed $30,000, is available for the installation of hydrogen fueling equipment. The credit expired December 31, 2014 (DOE 2005). There is also a tax credit in place of $0.50 per gallon of liquefied hydrogen sold for the purpose of fueling vehicles.

FUEL CONSUMPTION BENEFITS

Hybrid Fuel Consumption Estimates

The Agencies estimate the effectiveness of various electrification technologies as described above. For hybrids, a large amount of certification data exists for comparison to the Agencies' estimates; however, it is impossible to directly validate the EPA/NHTSA estimates of fuel consumption by comparison to vehicles in the market for two reasons:

(1) Performance of the conventional and hybrid vehicles is not always the same. For example, hybrids generally have faster acceleration 0 to 30 mph than conventional vehicles and less for 0 to 60. Detailed data are not listed by the EPA website and are not reliably available to compare hybrid and conventional vehicle performance.
(2) A single vehicle comprises a package of technologies, so isolating the effect of hybridization alone can be difficult as manufacturers may modify hybrids with fuel-saving features beyond the electrified powertrain. For example, in the Buick LaCrosse, GM covered the underbody to reduce drag and implemented aggressive regenerative braking (Hawkins et al. 2012).

Additionally, while certification data exist for current and past model years, the standards are binding to 2021 for fuel economy and 2025 for GHGs. Technologies will

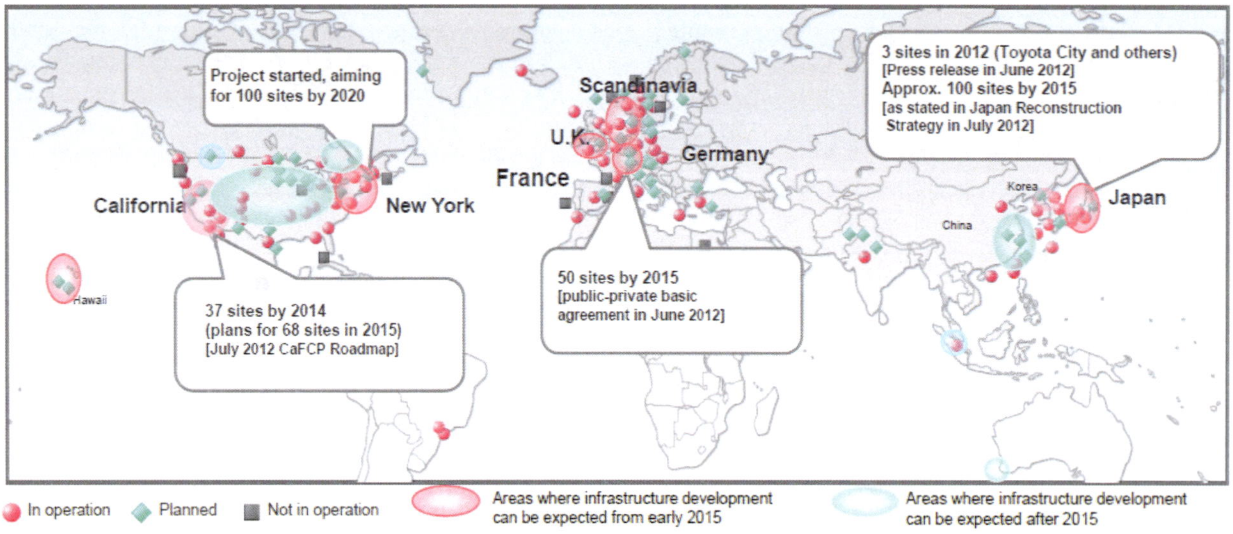

FIGURE 4.18 Projected worldwide locations of hydrogen stations.
SOURCE: Toyota (2014).

be improved, developed, and even abandoned in this time frame. Table 4.5 compares the versions of 2014 MY hybrids with their conventional equivalents. The table lists separately examples of MHEV, P2, and PS architectures. The left-hand column shows the reduction in fuel consumption assumed by the Agencies (EPA/NHTSA 2012a, 3-112).

MHEV

The data for the GM eAssist used as an example in the technical support document (TSD) for an MHEV illustrate the difficulty in using conventional comparator vehicles to derive the benefit of hybridization alone. The Malibu and the LaCrosse are similar vehicles and yet they show different fuel consumption improvements upon hybridization. The 2014 LaCrosse shows an increase of 37 percent in fuel economy between the conventional and hybrid versions, while the 2013 Malibu shows an 11 percent increase in fuel economy. The Malibu comparison was used to determine the fuel consumption reduction of MHEVs because the conventional and hybrid models have the same size engine and are thus more comparable than the LaCrosse models, which do not have the same engine size. In the case of the LaCrosse, the hybrid's smaller engine as well as greater aerodynamic improvements introduces new fuel saving technologies beyond the MHEV alone. GM claims that because of the boost that the 15 kW motor provides, the LaCrosse eAssist has acceleration similar to that of the conventional LaCrosse despite a smaller engine (Hall 2012; Hawkins et al. 2012). In 0-60 mph performance, however, the 8.6 sec performance of the Buick LaCrosse eAssist with the 2.4L I4 engine differs significantly from the 6.4 sec 0-60 mph performance of the Buick LaCrosse with the 3.6L V6 engine, so a comparison of fuel economy at equal performance cannot be obtained from these two vehicles (Gale 2012). For similar reasons, the Jetta and the Fusion models shown in Table 4.5 were not used directly to determine the fuel consumption reduction benefits of the P2 and PS architectures, respectively.

Using these estimates, the committee concludes that the effect of hybridization is a 10 percent reduction in fuel consumption for the mild hybrid. Note that the Agencies assumed an incremental fuel consumption reduction effectiveness of 6.5 percent for mild hybridization of midsize passenger vehicles. The addition of stop-start incremental effectiveness of 2.1 percent leads to a comparative total effectiveness for a mild hybrid vehicle of 8.6 percent. The TSD reports vehicle simulation resulting in a total fuel consumption effectiveness of 11.6 percent for mild hybrid midsize passenger vehicles relative to the baseline vehicle (EPA/NHTSA 2012a, 3-75, table 3-19; NHTSA 2012).

P2

The Agencies base their high reduction in fuel consumption for the P2 on a Ricardo study (2012) and claim the effectiveness of the P2 hybrid used in this final rulemaking is 48.6 percent for a midsize passenger car (EPA/NHTSA 2012a, 3-124). This total effectiveness estimate includes the transmission effectiveness of 18.7 percent (for the decision tree pathway, including improved controls/externals, six-speed automatic transmission with improved internals, eight-speed dual clutch transmission, high-efficiency gearbox, and shift optimizer) . Removing the transmission effectiveness, leaving out stop-start effectiveness of 2.1 percent and the effectiveness of ISG of 6.5 percent, the total incremental effectiveness of P2 alone is 33.6 percent, as can be seen in the NHTSA decision tree. This is larger than indicated by examples in Table 4.5. Part of their reasoning seems to be that by 2017, automakers will find ways to improve the P2 system. Although this may be possible and is used for the committee's high estimate of fuel consumption reduction, it does not seem to be the case in 2014. The Sonata hybrid shows an increase in mpg of only 40.7 to 40.0 percent, resulting in the low estimated fuel consumption reduction of 28 percent for P2 hybrids. The Jetta, like the LaCrosse, has a downsized, turbocharged engine, so it combines two technologies and is not indicative of what can be achieved with hybridization alone. Further comparisons of P2 vehicles with their conventional analogs are available in Annex Table 4A.2. For the next rulemaking, EPA is developing a full system simulation tool, ALPHA, that will be able to simulate SI and hybrid vehicles and will be publicly available, while NHTSA will be relying upon full system simulation from Argonne National Laboratory using the Autonomie simulation model. In early studies, the ALPHA model had been used to estimate the effectiveness of P2 and PS hybrids and was successfully validated against current examples of these architectures to within 5 percent of the test fuel economy (Lee, S.D. et al. 2013; Lee, B. et al. 2013). ALPHA may improve estimation of fuel consumption reduction for strong hybrids.

PS

In developing the effectiveness of the PS architecture, the TSD states that "In MYs 2012-2016 final rule, EPA and NHTSA used a combination of manufacturer-supplied information and a comparison of vehicles available with and without a hybrid system from EPA's fuel economy test data to estimate that the effectiveness is 19 to 36 percent for the classes to which it is applied. The estimate would depend on whether engine downsizing is also assumed. In the CAFE incremental model, the range of effectiveness used was 23 to 33 percent as engine downsizing is not assumed (and accounted for elsewhere)" (EPA/NHTSA 2012a). As the table shows for the hybrid Ford Fusion and Toyota Camry, both lie at the upper end of the Agency estimates without any downsizing. The NRC estimate of PS effectiveness is based off the Agency estimate at 33 percent on the low end, and the Camry hybrid-conventional comparison of a fuel consumption reduction of 33.5 percent at the high end.

TABLE 4.5 Comparison of Effectiveness Estimated by the Agencies with EPA Certification Fuel Consumption and Fuel Economy Data of Actual Vehicles. Further Examples of P2 Hybrid Vehicles Are Available in Annex Table 4A.2.

Agencies' FC reduction projection for midsize vehicle	Example Models	Engine Size (L) and Type	Transmission	Certification City FE (mpg)	Certification Hwy FE (mpg)	Certification Combined FE (mpg)	Certification Combined FC (g/100 mi)	Fuel Consumption Change (%)	Fuel Economy Improvement (% mpg)
MHEV 6.6%	2013 Malibu regular[a]	2.5	S6	28.34	48.18	34.79	2.87	Baseline	Baseline
	2013 Malibu eAssist[a]	2.4	S6	31.98	51.88	38.65	2.59	−10.0	11.1
	2014 LaCrosse regular	3.6 V6	S6	22.46	39.25	27.81	3.60	Baseline	Baseline
	2014 LaCrosse eAssist	2.4 I4	S6	31.50	51.20	38.10	2.62	–27.0	37.0
P2 33.6%	2014 Sonata regular	2.4 I4	AM6	30.42	48.79	36.63	2.73	Baseline	Baseline
	2014 Sonata hybrid	2.4 I4	AM6	48.00	56.60	51.52	1.94	–28.9	40.7
	2014 Sonata hyb ltd	2.4 I4	AM6	47.70	56.40	51.26	1.95	–28.5	40.0
	2014 Jetta regular	2.0	S6	28.10	41.50	32.88	3.04	Baseline	Baseline
	2014 Jetta hybrid	1.4 T	AM 7	57.50	65.30	60.77	1.65	–45.9	84.8
PS 33%	2014 Camry	2.5 I4	S6	32.00	50.00	38.19	2.62	Baseline	Baseline
	2014 Camry hyb LE	2.5 I4	Var	58.50	56.10	57.40	1.74	–33.5	50.3
	2014 Camry hyb XLE/SE	2.5 I4	Var	55.07	54.56	54.84	1.82	–30.4	43.6
	2014 Fusion	2.0 I4 T	S6	28.50	46.60	34.54	2.90	Baseline	Baseline
		1.5 I4 T	S6	29.63	50.38	36.37	2.75	Alternative baseline	Alternative baseline
		2.5 I4	S6	28.30	47.38	34.56	2.89	Alternative baseline	Alternative baseline
	2014 Fusion hybrid	2.0 I4	Var	65.07	67.34	66.07	1.51	–47.7	91.3

[a] The 2013 Malibu was used because the 2014 model added stop-start to the conventional vehicle.

Fuel Consumption Measurement

Hybridization of the ICE drivetrain with the addition of an electrical system improves fuel economy and reduces GHG emissions. For compliance purposes, this can be recorded in a straightforward way by measuring the fuel consumed in the test cycles, as for a typical ICE vehicle, and described in the J1711 SAE standard (Hybrid-EV Committee 2010).[3] Calculating GHG emissions and fuel economy is more complex for vehicles that receive some of their energy from the electric grid (PHEVs and BEVs), that do not use a liquid fuel (BEVs and FCEVs), or that utilize a fuel that does not produce CO_2 at the tailpipe (hydrogen FCEVs). Including the energy used and carbon emitted to generate electricity or hydrogen adds to the complication, as discussed in Chapter 10 (DOE 2014d). The recently completed NRC report *Overcoming Barriers to Deployment of Plug-in Electric Vehicles* (NRC 2015) discusses some of the complexities of estimating carbon emissions from electricity generation used to fuel BEVs and PHEVs.

COSTS

The preceding sections describe the technologies used for vehicle electrification, their likely penetrations and estimates of their effectiveness when implemented to 2025, as well as technologies likely to be used to 2030. Additionally, the committee described its estimates of costs, especially noting where they differed from the Agencies' cost estimates. Particular areas that the Agencies should reexamine in the midterm evaluation include the cost of technologies required for consumer acceptance of stop-start, the cost of motors for strong hybrids, in particular the P2 system, and the non-battery component costs for PEVs. The committee's range of most likely costs and effectiveness values are collected in Table 4.6 and used in Chapter 8.

In addition to the costs of individual technologies, the CAFE/GHG standards make assumptions about production volume in order to estimate costs. For the years 2020 and 2025 the Agencies assumed a North American volume of 450,000 and a corresponding degree of learning to estimate costs (EPA/NHTSA 2012a, 3-111). Although this figure may be relevant for some conventional powertrain technologies, in the opinion of the committee this is optimistic for electrified powertrains. Using Ford as an example, 450,000 units sold in 2017 would constitute approximately 15 percent of its 2013 sales. Even if all xEV sales are combined, it is highly unlikely that the xEV share of the market will be that high for Ford. Since unit costs are higher for low volumes, this assumption leads to an underestimate of the cost of hybridization.

The Agencies' analysis is contradictory in assuming a high volume for the purpose of calculating costs and a low volume for technology penetration to 2025. Only 2 percent of the fleet is projected to be hybrids in the Agencies' compliance demonstration path. Despite the low projected production volume and the high volume used when calculating the component costs, the battery costs for 2012 seem reasonable. This could be due to one of two things:

(1) Factories are not utilized to capacity and suppliers are selling at low prices or
(2) As noted in the TIAX study presented on August 13, 2013 to the NRC Committee on Overcoming Barriers to Electric Vehicle Deployment (Sriramulu and Barnett 2013), the economics of scale kick in at 60,000 units per supplier rather than at 450,000 units for the market as a whole, as assumed by the Agencies.

Table 4.6 collects the committee's range of most likely fuel consumption reduction measurements and direct manufacturing costs for a midsize car with an I4 engine in 2014. The fuel consumption effectiveness values were generally equal to those estimated by NHTSA. For the P2 and PS, different lower bounds were estimated for the fuel consumption reduction based on comparisons between 2014 hybrids and their conventional counterparts.

The committee's cost estimates include the Agencies' costs, which the committee judged to be valid lower estimates of costs for electrification technologies in 2025, reflecting an optimistic scenario of technology development and implementation. The committee's range of most likely costs also included higher values, reflecting committee expert judgment of the costs of technologies required to implement electrified powertrains. For MY2025, these higher costs were due to +\$50 additional nonbattery technologies needed for integration of stop-start (+100 in MY 2017 and +75 in MY 2020), 1.5 × nonbattery technologies costs for BEV and PHEV powertrains, 1.3 × battery costs for the MHEV and P2 that reflect a more conservative SOC swing, and 1.4 × costs for properly sizing the P2 motor by torque rather than power. Justification for these cost increases is described earlier in the chapter.

FINDINGS AND RECOMMENDATIONS

Finding 4.1 In hybrids, electric current reverses direction many times during driving. To ensure long battery life, DOE specifications call for 300,000 "shallow" cycles, and mild hybrids such as the Buick eAssist use a state of charge swing of 20 percent. In projecting mild hybrid costs, the Agencies sized the battery based on an assumed 40 percent SOC swing, thus making the Agencies estimate of the battery of the mild hybrid half the size and half the cost of current implementations.

[3] See 40 CFR § 1066.501I U 1066 F – Electric Vehicles and Hybrid Electric Vehicles at http://www.ecfr.gov/cgi-bin/retrieveECFR?gp=1&SID=99734ec227a2cf053c111fd96e3b22c2&ty=HTML&h=L&mc=true&n=pt40.33.1066&r=PART#sp40.33.1066.f.

TABLE 4.6 Summary of NRC Estimates of Direct Manufacturing Costs and Fuel Consumption Effectiveness for Electrification Technologies for a Midsize Car Replacing an I4 Engine

Electrification Technology	NRC Most Likely Fuel Consumption Reduction[a] (%)	NHTSA Estimated Fuel Consumption Reduction[a] (%)	NRC Most Likely 2025 MY DMC Costs (2010$)	NHTSA Estimated 2025 MY DMC Costs (2010$)
SS	2.1[b]	2.1[b]	225 - 275	225
MHEV	6.5[b]	6.5[b]	888 - 1018	888
P2	28.9 - 33.6	33.6	2,041 - 2,588	2,041
PS	33 - 33.5	33	2,671	2,671
PHEV40	N/A	65.1	8,325 - 9,672	8,325
EV75	N/A	87.2	8,451 - 8,963	8,451

[a] Relative to baseline unless otherwise noted.
[b] Relative to previous technology.

Recommendation 4.1 For the midterm review, the Agencies should consult with battery manufacturers and automakers to determine the appropriate size of the battery for hybrids. Battery life is a key element in electrified powertrains, and premature failure should be avoided.

Finding 4.2 Battery cost is the dominant cost for PHEVs and BEVs. It is a function of energy and power requirements, battery chemistry, and required battery life. Battery life depends on the number of cycles required, the stability of the chemistry to cycling at the required state of charge swing, the thermal and stress evolution it undergoes, and its shelf life. Due to rapid development of battery technology, there are no real-world data to validate battery life beyond those from simulations and accelerated aging tests, so the appropriate state of charge swing to meet the conventional powertrain warranty of 8 years and 100,000 miles is unknown. GM is sizing the battery more conservatively than Nissan by using a smaller swing in SOC and, accordingly, a larger battery. The Agencies accepted the state of charge swings of the two automakers and assumed a higher cost per kilowatt hour for the power-optimized battery of the PHEV.

Recommendation 4.2 Proper sizing of the battery is essential for appropriate assessment of both cost and lifetime, parameters particularly critical for the extensive battery requirements of PHEVs and BEVs. For their midterm review, the Agencies should examine auto manufacturers' experiences of battery life to determine the appropriate state of charge swing for PHEVs and BEVs so that they can assign costs appropriately.

Finding 4.3 The Agencies determine that the P2 architecture is likely to be the dominant strong hybrid technology, based on projected cost and effectiveness of P2 vs. PS hybrids. The cost estimate is partly based on the assumption that electric motors scale as power. In fact, the rotor volume and cost depend entirely on torque and hence the cost of electric motors scale with torque. The P2 motor is inline with the engine and transmission and has the same rotational speed as the engine. This constraint is not present with the PS hybrid, thereby allowing the use of a higher speed, smaller motor. Also, to minimize NVH, it appears that some automakers are not using the crankshaft-mounted electric motor for starting but are augmenting the conventional cranking motor and 12 V battery. The effectiveness of the PS hybrid models now available in the market as compared to the effectiveness of their conventional analogs with the same engine show that the PS architecture provides hybrids with significantly greater reduction in fuel consumption than similar P2 hybrids and their conventional analogs.

Recommendation 4.3 The cost of a P2 hybrid may possibly be higher than predicted by the Agencies and comparable to that of the PS hybrid for comparable performance. For the midterm review, the Agencies should undertake a teardown of the next generation PS and P2 architectures to update cost. Full system simulation of P2 and PS architectures should be undertaken to estimate effectiveness for the midterm review.

Finding 4.4 The committee is not in a position to precisely determine the cost increases for electrified powertrains. Based on inputs from automakers, battery suppliers, and independent consultants, it is the opinion of the committee that the battery cost estimates used by the Agencies are broadly accurate, while the cost of the nonbattery elements is too low, perhaps by a factor of as much as 2. To conservatively estimate these unaccounted-for costs, the committee used a high cost estimate of 1.5 times the Agencies' estimates for the nonbattery components for BEVs and PHEVs.

Recommendation 4.4 At the time of the midterm review there will be several vehicles with electrified powertrains in the market. The Agencies should commission teardown studies of the most successful examples of (1) stop-start, (2) strong hybrids (PS, P2, and two motor architectures),

(3) PHEV20 and PHEV40, and (4) BEV100. At that time there will be better estimates of volumes for each type in the 2020 to 2025 time frame so that a better estimate of cost can be calculated.

Finding 4.5 Lithium-sulfur and lithium-air batteries will not be used in vehicles in sufficient numbers by 2030 to affect fuel consumption, and it may take more than 20 years before they are in the mass market. These technologies are still in the development stage and have many challenges related to poor efficiency, poor cycle life, and serious safety concerns due to the use of very reactive lithium metal.

Finding 4.6 Limited volumes of FCEVs were introduced in California in 2014 by Hyundai and are being introduced by Toyota in 2015. FCEVs will have minimal impact, if any, on 2025 CAFE compliance based on current automaker plans for market introduction but may become more important by 2030. A coordinated national plan for H_2 infrastructure deployment will be required if successful, high-volume FCEV deployment is to be realized.

REFERENCES

Agarwal, P., R. Mooney, and R. Toepel. 1969. Sir-Lec I, A Stirling Electric Hybrid Car. SAE Technical Paper 690074. doi:10.4271/690074.

Alger, P. 1970. Induction Machines. Gordon and Breach Science Publishers, Inc. Basel, Switzerland. Equation 3.9, p. 75.

Amine, K., R. Kanno, and Y. Tzeng. 2014. Rechargeable lithium batteries and beyond: Progress, challenges, and future directions. MRS Bulletin 39(5): 395-401.

Amine, K., Z. Chen, Z. Zhang, J. Liu, W. Lu, Y. Qin, J. Lu, L. Curtis, and Y.-K.J. Sun. 2011. Mechanism of capacity fade of MCMB/ $Li_{1.1}[Ni_{1/3}Mn_{1/3}Co_{1/3}]_{0.9}O_2$cell at elevated temperature and additives to improve its cycle life. Mater. Chem. 21: 17754.

Anair, D., and A. Mahmassani. 2012. State of Charge: Electric Vehicles' Global Warming Emissions and Fuel-Cost Savings across the United States. Union of Concerned Scientists, June. http://www.ucsusa.org/assets/documents/clean_vehicles/electric-car-global-warming-emissions-report.pdf.

Anderman, M., F. Kalhammer, and D. MacArthur. 2000. Advanced Batteries for Electric Vehicles: An Assessment of Performance, Cost, and Availability. California Air Resources Board, June. http://www.arb.ca.gov/msprog/zevprog/2000review/btapreport.doc. Accessed December 17, 2010.

Arata, J., M. Leamy, J. Meisel, K. Cunefare, and D. Taylor. 2011. Backward-looking simulation of the Toyota Prius and General Motors two-mode power-split HEV powertrains. SAE Int. J. Engines 4(1):1281-1297. doi:10.4271/2011-01-0948.

Assary, R.S., J. Lu, P. Du, X. Luo, X. Zhang, Y. Ren, L.A. Curtiss, and K. Amine. 2013. The effect of oxygen crossover on the anode of a Li-O(2) battery using an ether-based solvent: insights from experimental and computational studies. ChemSusChem 6(1): 51.

Asumadu, J.A., M. Haque, H. Vogel, and C. Willards. 2005. Precision Battery Management System. Proceedings of the IEEE Instrumentation and Measurement Technology Conference 2: 1317-1320. doi: 10.1109/IMTC.2005.1604361.

Aurbach, D., and Y. Cohen. 1996. The application of atomic force microscopy for the study of Li deposition processes. Journal of the Electrochemical Society 143(11): 3525.

Balsara, N., M. Singh, H.B. Eitouni, and E.D. Gomez. 2009. U.S. pat. app. no. 0263725 A1.

Bang, H.J., H. Joachin, et al. 2006. Contribution of the structural changes of LiNi0.8Co0.15Al0.05O2 cathodes on the exothermic reactions in Li-ion cells. Journal of the Electrochemical Society 153(4): A731-A737.

Barghamadi, M., A. Kapoor, and C. Wen. 2013. A review on Li-S batteries as a high efficiency rechargeable lithium battery. Journal of the Electrochemical Society 160(8): A1256.

Barnett, B., D. Ofer, C. McCoy, Y. Yang, T. Rhodes, B. Oh, M. Hastbacka, J. Rempel, and S. Sririramulu. 2009. PHEV Battery Cost Assessment. 2009 DOE Merit Review, May. http://www1.eere.energy.gov/vehiclesandfuels/pdfs/merit_review_2009/energy_storage/es_02_barnett.pdf.

Barnett, B., J. Rempel, D. Ofer, B. Oh, S. Sriramulu, J. Sinha, M. Hastbacka, and C. McCoy. 2010. PHEV Battery Cost Assessment. 2010 DOE Merit Review, June. http://www1.eere.energy.gov/vehiclesandfuels/pdfs/merit_review_2010/electrochemical_storage/es001_barnett_2010_o.pdf.

Battery University. 2013. BU-210: Fuel Cell Technology. http://batteryuniversity.com/learn/article/fuel_cell_technology.

Bettge, M., Y. Li, K. Gallagher, Y. Zhu, Q. Wu, W. Lu, I. Bloom, and D.P. Abraham. 2013. Voltage fade of layered oxides: Its measurement and impact on energy density. Journal of the Electrochemical Society 160(11): A2046.

Black, R., B. Adams, et al. 2012. Non-aqueous and hybrid Li-O_2 batteries. Adv. Energy Mater. 2(7): 801-815.

Bruce, P.G., L.J. Hardwick, and K.M. Abraham. 2011. Lithium-air and lithium-sulfur batteries. MRS Bull. 36: 506.

Bruce, P.G., S.A. Freunberger, L.J. Hardwick, and J-M. Tarascon. 2012. Li-O_2 and Li-S batteries with high energy storage. Nat. Mater. 11(2): 19.

Carroll, R. 2013. 8 U.S. States Band Together to put Millions of Zero-Emissions Cars on the Road. Reuters, October 24.

Chen, M., Z. Zhang, Z. Feng, J. Chen, and Z. Qian. 2009. An improved control strategy for the charge equalization of lithium ion battery. Twenty-Fourth Annual IEEE Applied Power Electronics Conference and Exposition: 186–189. doi: 10.1109/APEC.2009.4802653.

Cobb, J. 2015. December 2014 Dashboard. Hybridcars.com, January 6. http://www.hybridcars.com/december-2014-dashboard/.

Csere, C. 2013. 2014 Honda Accord Plug-In Hybrid. Car and Driver, May. http://www.caranddriver.com/reviews/2014-honda-accord-plug-in-test-review.

Curtin, S., and J. Gangi. 2013. State of the States: Fuel Cells in America 2013. U.S. Department of Energy Energy Efficiency and Renewable Energy Fuel Cell Technologies Office. Fuel Cells 2000, Breakthrough Technologies Institute, Washington D.C. http://www1.eere.energy.gov/hydrogenandfuelcells/pdfs/state_of_the_states_2013.pdf.

Deng, M.-J., D.-C. Tsai, W.-H. Ho, C.-F. Li, and F.-S. Shieu, 2013. Electrolytic deposition of Sn-coated mesocarbon microbeads as anode material for lithium ion battery. Applied Surface Science 285: 180.

Di Cairano, S., W. Liang, I.V. Kolmanovsky, M.L. Kuang, and A.M. Phillips. 2013. Power smoothing energy management and its application to a series hybrid powertrain. Control Systems Technology, IEEE Transactions on 21(6): 2091-2103.

Di Domenico, D., A.G. Stefanopoulou, and G. Fiengo. 2010. Lithium-ion battery state of charge and critical surface charge estimation using an electrochemical model-based extended Kalman filter. Journal of Dynamic Systems, Measurement, and Control 132(6): 0613021–11. doi: 10.1115/1.4002475.

Ding, F.W., Xu, G.L. Graff, J. Zhang, M.L. Sushko, X. Chen, Y. Shao, M.H. Engelhard, Z. Nie, J. Xiao, S. Liu, P.V. Sushko, J. Liu, and J.-G. Zhang. 2013. Dendrite-free lithium deposition via self-healing electrostatic shield mechanism. Journal of the American Chemical Society 135(11): 4450.

Dinger, A., R. Martin, X. Mosquet, M. Rabl, D. Rizoulis, M. Russo, and G. Sticher. 2010. Batteries for Electric Vehicles: Challenges, Opportunities, and the Outlook to 2020. The Boston Consulting Group. http://www.bcg.com/documents/file36615.pdf. Accessed December 11, 2010.

DOE (Department of Energy). n.d.a. Compare Side-by-Side Fuel Economy: 2013 Chevrolet Volt. http://www.fueleconomy.gov/feg/Find.do?action=sbs&id=32655. Accessed September 3, 2013.

DOE. n.d.b. Compare Side-by-Side Fuel Economy: 2014 Chevrolet Spark EV, 2012 Nissan Leaf, 2013 Nissan Leaf, 2014 Chevrolet Volt. http://www.fueleconomy.gov/feg/Find.do?action=sbs&id=33640&id=32154&id=33558&id=33900. Accessed September 4, 2013.

DOE. 2005. Hydrogen Fuel Infrastructure Tax Credit. Alternative Fuels Data Center. http://www.afdc.energy.gov/laws/law/US/351.

DOE. 2014a. www.fueleconomy.gov.

DOE. 2014b. Beyond Tailpipe Emissions: Greenhouse Gas Emissions for Electric and Plug-In Hybrid Electric Vehicles. http://www.fueleconomy.gov/feg/Find.do?action=bt2.

DOE. 2014c. Alternative Fueling Station Locator. Alternative Fuels Data Center. http://www.afdc.energy.gov/locator/stations/.

DOE. 2014d. Sources and Assumptions for the Electric and Plug-in Hybrid Vehicle Greenhouse Gas Emissions Calculator. http://www.fueleconomy.gov/feg/label/calculations-information.shtml.

DOE/OBES (Office of Basic Energy Sciences). 2007. Basic Research Needs for Electrical Energy Storage. Report of the Basic Energy Sciences Workshop on Electrical Energy Storage, April 2-4. http://web.anl.gov/energy-storage-science/publications/EES_rpt.pdf.

Dubarry, M., C. Truchot, B.Y. Liaw, K. Gering, S. Sazhin, D. Jamison, and C. Michelbacher. 2013. Evaluation of commercial lithium-ion cells based on composite positive electrode for plug-in hybrid electric vehicle applications: III. Effect of thermal excursions without prolonged thermal aging. Journal of the Electrochemical Society 160(1): A191-A199. doi: 10.1149/2.063301jes.

Edmunds.com. Features and Specs for the Honda Accord Plug-In Hybrid Base. http://www.edmunds.com/honda/accord-plug-in-hybrid/2014/.

EIA (Energy Information Agency). 2014. Petroleum Weekly Archive. http://www.eia.gov/oog/info/twip/twiparch/110525/twipprint.html.

Elrick, B. 2013. Hydrogen fuel cell electric vehicles: The California Example. Presentation to the National Research Council Committee on Assessment of Technologies for Improving Light-duty Vehicle Fuel Economy, Irvine, CA, October 14.

EPA (Environmental Protection Agency). 2011. Light Duty Technology Cost Analysis, Power-Split and P2 HEV Case Studies. Prepared for EPA by FEV, Inc. EPA-420-R-11-015. http://www.epa.gov/otaq/climate/documents/420r11015.pdf.

EPA. 2014a. Enforcement and Compliance History Online. http://echo.epa.gov/.

EPA. 2014b. Fuel Economy Testing and Labeling. OEPA-420-F-14-015. http://www.epa.gov/fueleconomy/documents/420f14015.pdf.

EPA/NHTSA (National Highway Traffic Safety Administration). 2012a. Joint Technical Support Document, Final Rulemaking 2017-2025 Light-Duty Greenhouse Gas Emission Standards and Corporate Average Fuel Economy Standards. EPA-420-R-12-901.

EPA and NHTSA. 2012b. 2017 and Later Model Year Light-Duty Vehicle Greenhouse Gas Emissions and Corporate Average Fuel Economy Standards. EPA 40 CFR 85, 86, 600; NHTSA 49 CFR 523, 531, 533, 536, 537. October 15.

Fitzgerald, A.E., and C. Kingsley Jr. 1952. Electric Machinery. McGraw Hill.

Freunberger, S.A., Y. Chen, Z. Peng, J.M. Griffin, L.J. Hardwick, F. Barde, P. Novak, and P.G.J. Bruce. 2011. Reactions in the rechargeable lithium-O_2 battery with alkyl carbonate electrolytes. Am. Chem. Soc. 133: 8040.

Gale, Z. 2012. First Test: 2012 Buick LaCrosse eAssist. Motor Trend, February 27. http://www.motortrend.com/roadtests/sedans/1202_2012_buick_lacrosse_eassist_first_test/.

Gallagher, K.G., P.A. Nelson, and D.W. Dees. 2011. PHEV Battery Cost Assessment. 2011 DOE Merit Review Presentation, Washington, D.C., May 9-12. http://www1.eere.energy.gov/vehiclesandfuels/pdfs/merit_review_2011/electrochemical_storage/es111_gallagher_2011_o.pdf.

Ge, M., X. Fang, J. Rong and C. Zhou. 2013. Review of porous silicon preparation and its application for lithium-ion battery anodes. Nanotechnology 24(42): 422001.

GM-Volt.com. 2010. Chevrolet Volt Battery Warranty Details and Clarifications. GM-Volt.com, July 19. http://gm-volt.com/2010/07/19/chevrolet-volt-battery-warranty-details-and-clarifications.

GM-Volt.com. n.d. Chevy Volt Specs. GM-Volt.com. http://gm-volt.com/full-specifications/.

Goebel, K., B. Saha, A. Saxena, J. Celaya, and J. Christophersen. 2008. Prognostics in battery health management. Instrumentation Measurement Magazine, IEEE 2008(11): 33-40. doi: 10.1109/MIM.2008.4579269.

Gordon-Bloomfield, N. 2012a. 95% Of All Trips Could Be Made In Electric Cars, Says Study. Green Car Reports, January 13. http://www.greencarreports.com/news/1071688_95-of-all-trips-could-be-made-in-electric-cars-says-study.

Gordon-Bloomfield, N. 2012b. Nissan Buys Back Leaf Electric Cars Under Arizona Lemon Law. Green Car Reports, September 28. http://www.greencarreports.com/news/1079475_nissan-buys-back-leaf-electric-cars-under-arizona-lemon-law.

Green Car Congress. 2014. Audi moving ahead with 48V system in vehicles; mild hybrids. Green Car Congress, August 25. http://www.greencarcongress.com/2014/08/audi-moving-ahead-with-48v-system-in-vehicles-mild-hybrids.html.

Green Car Congress. 2015. Hyundai showcasing new Tucson 48V Hybrid Concept and diesel Plug-in-Hybrid Concept at Geneva. Green Car Congress, March 8. http://www.greencarcongress.com/2015/03/20150308-hyundai.html.

Gremeil, H. 2014. Toyota fuel cell car to cost $69,000 in Japan, debut in U.S., Europe in 2015. Autonews, June 25. http://www.autonews.com/article/20140625/oem05/140629935/toyota-fuel-cell-car-to-cost-$69000-in-japan-debut-in-u.s.-europe-in.

Gu, M., I. Belharouak, et al. 2013. Formation of the spinel phase in the layered composite cathode used in Li-ion batteries. ACS Nano 7(1): 760-767.

Hall, L. 2012. 2012 Buick LaCrosse eAssist. Hybridcars.com, February 7. http://www.hybridcars.com/2012-buick-lacrosse-eassist/.

Hawkins, S., F. Billotto, D. Cottrell, A. Houtman, S. Poulos, R. Rademacher, K. Van Maanen, and D. Wilson. 2012. Development of General Motors' eAssist powertrain. SAI Int. J. Alt. Power. 1(1):308-323. doi:10.4271/2012-01-1039.

Higuchi, N., O. Sunaga, M. Tanaka, and H. Shimada. 2013. Development of a new two-motor plug-in hybrid system. SAE Int. J. Alt. Power 2(1): 135-145.

Honda. n.d.a. 2015 Accord Hybrid Specifications. http://automobiles.honda.com/accord-hybrid/specifications.aspx. Accessed January 18, 2014.

Honda. n.d.b. 2014 Accord Plug-In Specifications. http://automobiles.honda.com/accord-plug-in/specifications.aspx. Accessed January 18, 2014.

Honda. 2014. Honda R&D Installs Advanced Fast-Fill Hydrogen Refueling Station. Honda News Release, March 3.

Hou, X., H. Jiang, Y. Hu, Y. Li, J. Huo, and C. Li. 2013. In situ deposition of hierarchical architecture assembly from Sn-filled CNTs for lithium-ion batteries. ACS Applied Materials & Interfaces 5(14): 6672.

Hybrid – Ev Committee. 2010. Recommended Practice for Measuring the Exhaust Emissions and Fuel Economy of Hybrid-Electric Vehicles, Including Plug-in Hybrid Vehicles. J1711 Standard. SAE, June 8. http://standards.sae.org/j1711_201006/.

Hyundai Motor Company. 2014. 2015 Sonata Hybrid: Specifications. Accessed January 18, 2013. https://www.hyundaiusa.com/sonata-hybrid/specifications.aspx.

ICF International. 2011. Peer Review of the Draft Report "Modeling the Cost and Performance of Lithium-Ion Batteries for Electric-Drive Vehicles." Prepared for the U.S. Environmental Protection Agency, March.

Ingram, A. 2013. Owner of 100,000-Mile Nissan Leaf Electric Car to be Honored Monday. Green Car Reports. http://www.greencarreports.com/news/1089091_owner-of-100000-mile-nissan-leaf-electric-car-to-be-honored-monday.

Jalil, N., N.A. Kheir, and M. Salman. 1997. A rule-based energy management strategy for a series hybrid vehicle. Proceedings of the American Control Conference: 689-693.

Jeong, S., S. Park, et al. 2011. High-performance, layered, 3D-$LiCoO_2$ cathodes with a nanoscale Co_3O_4 coating via chemical etching. Advanced Energy Materials 1(3): 368-372.

Jung, H.-G., J. Hassoun, et al. 2012. An improved high-performance lithium-air battery. Nat. Chem. 4(7): 579-585.

Kim, I.-S. 2010. A technique for estimating the state of health of lithium batteries through a dual-sliding-mode observer. IEEE Transactions on Power Electronics 25(4): 1013–1022. doi: 10.1109/TPEL.2009.2034966.

Kim, J., and B.H. Cho. 2011. State-of-charge estimation and state-of-health prediction of a Li-ion degraded battery based on an EKF combined with a per-unit system. IEEE Transactions on Vehicular Technology 60(9): 4249-4260. doi: 10.1109/TVT.2011.2168987.

Kim, Y., A. Salvi, A. Stefanopoulou, and T. Ersal. 2015. Reducing soot emissions in a diesel series hybrid electric vehicle using a power rate constraint map. IEEE Transactions on Vehicular Technology 64(1): 2-12.

Kim, Y., S. Mohan, J.B. Siegel, and A.G. Stefanopoulou. 2013. Maximum Power Estimation of Lithium-ion Batteries Accounting for Thermal and Electrical Constraints. ASME Dynamic Systems Control Conference, Palo Alto, California. doi: 10.1115/DSCC2013-3935.

Kotani, Y. 2013. Toyota Research Activity Toward Next Generation Battery. Presented at the 6th International Conference on Advanced Lithium Batteries for Automobile Applications, Chicago, IL, September 9-11.

Kum, D., H. Peng, and N. K. Bucknor. 2013. Control of engine-starts for optimal drivability of parallel hybrid electric vehicles. Journal of Dynamic Systems, Measurement, and Control 135(2): 021020-021020-10.

Kutkut, N., H.L.N. Wiegman, D. Divan, and D. Novotny. 1999. Design considerations for charge equalization of an electric vehicle battery system. IEEE Transactions on Industry Applications 35(1): 28-35. doi: 10.1109/28.740842.

Lee, B., S. Lee, J. Cherry, A. Neam, et al. 2013. Development of Advanced Light-Duty Powertrain and Hybrid Analysis Tool. SAE Technical Paper 2013-01-0808. doi:10.4271/2013-01-0808.

Lee, S., J. Kim, J. Lee, and B. Cho. 2008. State-of-charge and capacity estimation of lithium-ion battery using a new open-circuit voltage versus state-of-charge. Journal of Power Sources 185(2): 1367-1373. doi: 10.1016/j.jpowsour.2008.08.103.

Lee, S.D., B. Lee, J. McDonald, L. Sanchez, et al. 2013. Modeling and Validation of Power-Split and P2 Parallel Hybrid Electric Vehicles. SAE Technical Paper 2013-01-1470. doi:10.4271/2013-01-1470.

Lee, S.H., C. S. Yoon, et al. 2013. Improvement of long-term cycling performance of Li $Ni_{0.8}Co_{0.15}Al_{0.05}O_2$ by AlF_3 coating. Journal of Power Sources 234: 201-207.

Lee, S.W., C. Carlton, et al. 2012. The nature of lithium battery materials under oxygen evolution reaction conditions. J. Am. Chem. Soc. 134(41): 16959-16962.

Li, F., T. Zhang, and H. Zhou. 2013. Challenges of non-aqueous Li-O_2 batteries: Electrolytes, catalysts, and anodes. Energy Environ. Sci. 6(4): 1125.

Li, Y., M. Bettge, B. Polzin, Y. Zhu, M. Balasubramanian, and D.P. Abraham. 2013. Understanding long-term cycling performance of $Li_{1.2}Ni_{0.15}Mn_{0.55}Co_{0.1}O_2$–graphite lithium-ion cells. Journal of the Electrochemical Society 160(5): A3006.

Lin, C.C., H. Peng, J. W. Grizzle, and J. Kang. 2003. Power management strategy for a parallel hybrid electric truck. IEEE Transactions on Control Systems Technology 11: 839-849.

Lin, X., H.E. Perez, J.B. Siegel, A.G. Stefanopoulou, Y. Li, R.D. Anderson, Y. Ding, and M.P. Castanier. 2013. Online parameterization of lumped thermal dynamics in cylindrical lithium ion batteries for core temperature estimation and health monitoring. IEEE Transactions on Control System Technology 21(5): 1745-1755. doi: 10.1109/TCST.2012.2217143.

Liu, J., and H. Peng. 2008. Modeling and control of a power-split hybrid vehicle. Control Systems Technology, IEEE Transactions on (16)6: 1242-1251.

Lohse-Busch, H., M. Duoba, E. Rask, K. Stutenberg, V. Gowri, L. Slezak, and D. Anderson. 2013. Ambient Temperature (20°F, 72°F and 95°F) Impact on Fuel and Energy Consumption for Several Conventional Vehicles, Hybrid and Plug-In Hybrid Electric Vehicles and Battery Electric Vehicle. SAE Technical Paper 2013-01-1462. doi:10.4271/2013-01-1462.

Lu, J., and K. Amine. 2013. Recent research progress on non-aqueous lithium-air batteries from Argonne National Laboratory. Energies 6(11): 6016.

Lu, J., L. Li, J.–B. Park, Y.–K. Sun, F.Wu, and K. Amine. 2014. Aqueous and non-aqueous Li-air Battery. Chem. Rev. 114: 5611–5640.

Lu, Y.-C., D. G. Kwabi, et al. 2011. The discharge rate capability of rechargeable Li-O(2) batteries. Energy Environ. Sci. 4(8): 2999-3007.

Lu, Y.-C., and Y. Shao-Horn. 2013. Probing the reaction kinetics of the charge reactions of nonaqueous Li-O_2 batteries. J. Phys. Chem. Lett. 4(1): 93-99.

Lu, Y.-C., Z. Xu, H.A. Gasteiger, S. Chen, K. Hamad-Schifferli, and Y.J. Shao-Horn. 2010. Platinum–gold nanoparticles: A highly active bifunctional electrocatalyst for rechargeable lithium–air batteries. Am. Chem. Soc. 132(35): 12170-12171.

Mathias, M. 2014. Honda/GM Fuel Cell Partnership – Moving from Technical to Commercial Viability. SAE 2014 Hybrid and Electric Vehicle Technologies Symposium Presentation, La Jolla, California, February 13.

McCloskey, B.D., et al. 2011a. On the efficacy of electrocatalysis in Li-O_2 batteries. J. Am. Chem. Soc. 133(45): 18038.

McCloskey, B.D., et al. 2011b. Solvents' critical role in nonaqueous lithium-oxygen battery electrochemistry. J. Phys. Chem. Lett. 2(10): 1161-1166.

McDowell, M., I. Ryu, S.-W. Lee, C. Wang, W. Nix, and Y. Cui. 2012. Studying the kinetics of crystalline silicon nanoparticle lithiation with in situ transmission electron microscopy. Advanced Materials 24(45): 6034-6041.

Meisel, J. 2009. An Analytic Foundation for the Two-Mode Hybrid-Electric Powertrain with a Comparison to the Single-Mode Toyota Prius THS-II Powertrain. SAE Technical Paper 2009-01-1321. doi:10.4271/2009-01-1321.

Meisel, J. 2011. Kinematic study of the GM front-wheel drive two-mode transmission and the Toyota Hybrid System THS-II transmission. SAE Int. J. Engines 4(1):1020-1034. doi:10.4271/2011-01-0876.

Menkin, S., Z. Barkay, D. Golodnitsky, and E. Peled. 2014. Nanotin alloys supported by multiwall carbon nanotubes as high-capacity and safer anode materials for EV lithium batteries. Journal of Power Sources 245: 345.

Miller, J., and O. Onar. 2013. Wireless Power Transfer Systems. IEEE Transportation Electrification Conference (ITEC13), Dearborn, Michigan, June 17.

Moo, C. S., Y.-C. Hsieh, and I.S. Tsai. 2003. Charge equalization for series-connected batteries. IEEE Transactions on Aerospace and Electronic Systems 39(2): 704-710. doi: 10.1109/TAES.2003.1207276.

MotorTrend. n.d. New Toyota Prius Plug-In Reviews, Specs and Pricing. http://www.motortrend.com/new_cars/04/toyota/prius_plug_in/.

Moura, S.J., N.A. Chaturvedi, and M. Krstic. 2013. Constraint management in Li-ion batteries: A modified reference governor approach. American Control Conference (ACC): 5332, 5337. June 17-19. doi: 10.1109/ACC.2013.6580670.

Nelson, P.A., D.J. Santini, and J. Barnes. 2009. Factors Determining the Manufacturing Costs of Lithium-Ion Batteries for PHEVs. International Electric Vehicles Symposium EVS-24, Stavanger, Norway.

Nelson, P., K. Gallagher, I. Bloom, and D. Dees. 2011. Modeling the Performance and Cost of Lithium-Ion Batteries for Electric Vehicles. ANL-11/32. Chemical Sciences and Engineering Division, Argonne National Laboratory, Argonne, IL.

Nelson, P.A. K.G. Gallagher, and I. Bloom. 2012. BatPaC (Battery Performance and Cost) Software.

Ng, K. S., C.-S. Moo, Y.-P Chen, and Y.-C. Hsieh. 2009. Enhanced coulomb counting method for estimating state-of-charge and state-of-health of lithium-ion batteries. Applied Energy 86(9): 1506-1511. doi: 10.1016/j.apenergy.2008.11.021.

NHTSA (National Highway Traffic Safety Administration). 2012. Final Regulatory Impact Analysis, Corporate Average Fuel Economy for MY 2017-MY2025 Passenger Cars and Light Trucks. August.

Nikkei. 2014. New Power Chip Will Let Toyota Hybrids Burn Even Less Fuel. Nikkei Asian Review, May 21. http://asia.nikkei.com/Tech-Science/Tech/New-power-chip-will-let-Toyota-hybrids-burn-even-less-fuel.

Nissan. n.d. Compare Leaf S and SV Specs. NissanUSA.com. http://www.nissanusa.com/electric-cars/leaf/versions-specs/. Accessed September 4, 2013.

NRC (National Research Council). 2010. Review of the Research Program of the FreedomCAR and Fuel Partnership, Third Report. Washington, D.C.: The National Academies Press.

NRC. 2011. Assessment of Fuel Economy Technologies for Light-Duty Vehicles. Washington, D.C.: The National Academies Press, pp. 12-23.

NRC. 2015. Overcoming Barriers to Electric Vehicle Deployment. Washington, D.C.: The National Academies Press.

Nüesch, T., M. Wang, P. Isenegger, C. H. Onder, R. Steiner, P. Macri-Lassus, and L. Guzzella. 2014. Optimal energy management for a diesel hybrid electric vehicle considering transient PM and quasi-static NO_x emissions. Control Engineering Practice 29: 266-276.

Oh, S. H., R. Black, E. Pomerantseva, J.-H. Lee, and L.F. Nazar. 2012. Synthesis of a metallic mesoporous pyrochlore as a catalyst for lithium–O_2 batteries. Nat. Chem. 4(12): 1004.

Ohnsman, A. 2013. Toyota Seek Prius-Like Success with 2015 Fuel-Cell Model. Bloomberg, June 26. http://mobile.bloomberg.com/news/2013-06-26/toyota-seeks-prius-like-success-with-2015-fuel-cell-model.html?cmpid=yhoo.

Ohta, S., T. Kobayashi, J. Seki, and T. Asaoka. 2012. Electrochemical performance of an all-solid-state lithium ion battery with garnet-type electrolyte. Toyota Central R&D Labs, Journal of Power Sources 202: 332– 335.

Opila, D.F., X. Wang, R. McGee, R. B. Gillespie, J. A. Cook, and J. W. Grizzle. 2012. An energy management controller to optimally trade off fuel economy and drivability for hybrid vehicles. Control Systems Technology, IEEE Transactions on 20(6): 1490: 1505.

Opila, D.F., X. Wang, R. McGee, R.B. Gillespie, J.A. Cook, and J.W. Grizzle. 2014. Real-world robustness for hybrid vehicle optimal energy management strategies incorporating drivability metrics. Journal of Dynamic Systems, Measurement, and Control 136(6): 061011-061011-10.

Padhi, A.K., K.S. Nanjundaswamy, et al. 1997a. Phospho-olivines as positive-electrode materials for rechargeable lithium batteries. Journal of the Electrochemical Society 144(4): 1188-1194.

Padhi, A.K., K.S. Nanjundaswamy, et al. 1997b. Effect of structure on the Fe^{3+}/Fe^{2+} redox couple in iron phosphates. Journal of the Electrochemical Society 144(5): 1609-1613.

Park, H.-S., C.-E. Kim, C.-H. Kim, G.-W. Moon, and J.-H. Lee. 2009. A modularized charge equalizer for an HEV lithium-ion battery string. IEEE Transactions on Industrial Electronics 56(5): 1464-1476. doi: 10.1109/TIE.2009.2012456.

Park, S. H., S. H. Kang, et al. 2008. Physical and electrochemical properties of spherical $Li_{1+x}(Ni_{1/3}Co_{1/3}Mn_{1/3})_{(1-x)}O_2$ cathode materials. Journal of Power Sources 177(1): 177-183.

Piggott, H. 2011. Price of Neodymium Finally Levels Off. Scoraig Wind Electric News, September 16. http://scoraigwind.co.uk/2011/09/price-of-neodymium-finally-levels-off/.

Pisu, P., and G. Rizzoni. 2007. A comparative study of supervisory control strategies for hybrid electric vehicles. IEEE Transactions on Control Systems Technology 15(3): 506-518.

Plett, G.L. 2004a. High-performance battery-pack power estimation using a dynamic cell model. IEEE Transactions on Vehicular Technology 53(5):1586-1593. doi: 10.1109/TVT.2004.832408.

Plett, G.L. 2004b. Extended Kalman filtering for battery management systems of LiPB-based-HEV battery packs: Part 3. State and parameter estimation. Journal of Power Sources 134: 277-292. doi: 10.1016/j.jpowsour.2004.02.033.

Pop, V., H.J. Bergveld, P.H.L. Notten, and P.P.L. Regtien. 2005. State-of-the-art of battery state-of-charge determination. Measurement Science and Technology 16(12): R93–R110. doi:10.1088/0957-0233/16/12/R01.

Pyrhönen, J., T. Jokinen, and V. Hrabovcová. 2009. Design of rotating electrical machines. Chichester, West Sussex, United Kingdom: Wiley, p. 284 egn 6.2.

Rahimian, S.K., S. Rayman, and R.E. White. 2012. State of charge and loss of active material estimation of a lithium ion cell under low Earth orbit condition using Kalman filtering approaches. Journal of the Electrochemical Society 159(6): A860-A872. doi: 10.1149/2.098206jes.

Ransom, K. 2011. Ten-Year Old Toyota Hybrid Priuses Defy Early Critics. Aol Autos, March 29. http://autos.aol.com/article/toyota-prius-reliability/.

Reuters. 2014. Japan Moves to Fast-Track Cars Powered by Hydrogen Fuel Cells. The New York Times, June 25. http://www.nytimes.com/2014/06/26/business/international/japan-bets-big-on-cars-powered-by-hydrogen-fuel-cells.html?_r=0.

Ricardo. 2012. Calculation of Friction in High Performance Engines. Presented at Ricardo Software European User Conference, April 20.

Santini, D.J., K.G. Gallagher and P.A. Nelson. 2010. Modeling the Manufacturing Costs of Lithium-Ion Batteries for HEVs, PHEVs, and EVs. International Electric Vehicles Symposium EVS-25, Shenzhen, China.

Satyapal, S. 2013. Fuel Cell Technologies Update. DOE Annual Merit Review, Hydrogen and Fuel Cell Technical Advisory Committee Presentation, Washington, D.C., April 23. http://www.hydrogen.energy.gov/pdfs/htac_apr13_1_satyapal.pdf.

Serrao, L., S. Onori, and G. Rizzoni. 2011. A comparative analysis of energy management strategies for hybrid electric vehicles. ASME Journal of Dynamic Systems, Measurement and Control 133: 1-9.

Shao, Y., F. Ding, J. Xiao, J. Zhang, W. Xu, S. Park, J.-G. Zhang, Y. Wang, and J. Liu. 2013. Making Li-air batteries rechargeable: Material challenges. Adv. Funct. Mater. 23(8):987.

Shao, Y., S. Park, J. Xiao, J.-G. Zhang, Y. Wang, and J. Liu. 2012. Electrocatalysts for nonaqueous lithium-air batteries: Status, challenges, and perspective. ACS Catal. 2(5):844.

Shibata, H., S. Taniguchi, K. Adachi, K. Yamasaki, G. Ariyoshi, K. Kawata, K. Nishijima, and K. Harada. 2001. Management of serially-connected battery system using multiple switches. Proceedings of International Conference on Power Electronics and Drive Systems 2: 508-511. doi: 10.1109/PEDS.2001.975369.

Siler, S. 2010. 2012 Toyota Prius Plug-In Hybrid. Car and Driver. http://www.caranddriver.com/reviews/toyota-prius-2012-toyota-prius-plug-in-hybrid-review.

Smith, K.A., C.D. Rahn, and C.-Y. Wang. 2010. Model-based electrochemical estimation and constraint management for pulse operation of lithium ion batteries. IEEE Transactions on Control Systems Technology 18(3): 654-663. doi: 10.1109/TCST.2009.2027023.

Speltino, C., A. Stefanopoulou, and G. Fiengo. 2010. Cell equalization in battery stacks through state of charge estimation polling. Proceedings of American Control Conference: 5050-5055.

Spendelow, J., and J. Marcinkoski. 2014. Fuel Cell System Cost. DOE Fuel Cell Technologies Office Record, June 13. http://www.hydrogen.energy.gov/pdfs/14012_fuel_cell_system_cost_2013.pdf.

Sriramulu, S., and B. Barnett. 2013. Technical, Manufacturing, and Market Issues Associated with xEV Batteries. Presentation to the NRC Committee on Overcoming Barriers to EV Deployment. Washington, D.C., August 13.

Steffke, K.W., S. Inguva, D. Van Cleve, and J. Knockeart. 2013. Accelerated Life Test Methodology for Li-Ion Batteries in Automotive Applications. SAE Technical Paper 2013-01-1548. doi:10.4271/2013-01-1548.

Sun, C., S. Moura, X. Hu, J.K. Hedrick, and F. Sun. 2014. Dynamic traffic feedback data enabled energy management in plug-in hybrid electric vehicles. Control Systems Technology, IEEE Transactions on PP(99): 1,1. doi: 10.1109/TCST.2014.2361294.

Sun, Y.K., S.T. Myung, M.H. Kim, J. Prakash, and K. Amine. 2005. Synthesis and characterization of $Li[(Ni_{0.8}Co_{0.1}Mn_{0.1})_{0.8}(Ni_{0.5}Mn_{0.5})_{0.2}]O_2$ with the microscale core-shell structure as the positive electrode material for lithium batteries. Journal of the American Chemical Society 127:13411.

Tesla. n.d. Specs. http://www.teslamotors.com/models/specs.

Thackeray, M. S.-H. Kang, C.S. Johnson, J.T. Vaughey, R. Benedek, and S.A. Hackney. 2007. Li_2MnO_3-stabilized $LiMO_2$ (M=Mn, Ni, Co) electrodes for high energy lithium-ion batteries. Journal of Materials Chemistry 17: 3112.

The Clean Green Car Company. n.d. Toyota Prius II Battery Pack. VibrantPlanet.com. http://web.archive.org/web/20080225234612/ http://www.cleangreencar.co.nz/page/prius-battery-pack. Accessed January 18, 2014.

Thomas, D. 2014. 2014 Honda Accord Hybrid Expert Reviews. Cars.com National. http://www.cars.com/honda/accord-hybrid/2014/expert-reviews?reviewId=60097. Accessed January 18, 2014.

Toyota. 2014. Special Feature 01: Always Better Cars Making always better cars in order to exceed customer expectations. http://www.toyota-global.com/sustainability/features/car/.

USABC (United States Advanced Battery Consortium). 2013a. USABC Goals for Advanced Batteries for EV Vehicles. USCAR (United States Council for Automotive Research) Energy Storage System Goals.

USABC 2013b. USABC Goals for Advanced Batteries for PHEVs for FY 2018 to 2020 Commercialization. USCAR (United States Council for Automotive Research) Energy Storage System Goals.

U.S. DRIVE (United States Driving Research and Innovation for Vehicle Efficiency and Energy Sustainability). 2013. Electrochemical Energy Storage Technical Team Roadmap. U.S. Department of Energy Office of Energy Efficiency and Renewable Energy.

Verbrugge, M., and E. Tate. 2004. Adaptive state of charge algorithm for nickel metal hydride batteries including hysteresis phenomena. Journal of Power Sources 126: 236-249. doi: 10.1016/j.jpowsour.2003.08.042.

Voelcker, J. 2014. First 2015 Hyundai Tucson Fuel Cell Delivered in California. Green Car Reports, June 11. http://www.greencarreports.com/news/1092640_first-2015-hyundai-tucson-fuel-cell-delivered-in-california.

Wang, B., X. Li, T. Qiu, B. Luo, J. Ning, J. Li, X. Zhang, M. Liang, and L. Zhi. 2013. High volumetric capacity silicon-based lithium battery anodes by nanoscale system engineering. Nano Letters 13(11): 5578.

Whittingham, M. S. 1973. U.S. Patent 4009052 (Chalcogenide battery) and U.K. Patent, 1468416.

Whittingham, M. S. 1976. Electrical energy storage and intercalation chemistry. Science 192 (4244): 1126-1127. doi:10.1126/science.192.4244.1126. http://adsabs.harvard.edu/abs/1976Sci...192.1126W.

Whittingham, M.S. 2004. Lithium batteries and cathode materials. Chem. Rev. 104(10): 4271.

Wipke, K. 2012. National Fuel Cell Electric Vehicle Learning Demonstration Final Report. NREL/TP-5600-54860.

Woodyard, C. 2013. Ford Lowers C-max Hybrid's mpg Rating, Offers Rebates. USA Today, August 15. http://www.usatoday.com/story/money/cars/2013/08/15/ford-cmax-mp/2660371/. Accessed September 3, 2013.

Wu, M., J.E.C. Sabisch, X. Song, A.M. Minor, V.S. Battaglia, and G. Liu. 2013. In situ formed Si nanoparticle network with micronsized Si particles for lithium-ion battery anodes. Nano Letters 13(11): 5397.

Xiong, R., H. He, F. Sun, X. Liu, and A. Liu. 2013. Model-based state of charge and peak power capability joint estimation of lithium-ion battery in plug-in hybrid electric vehicles. Journal of Power Sources 229(0): 159 –169. doi: 10.1016/j.jpowsour.2012.12.003.

Xiulei, J., and L.F. Nazar. 2010. Advances in Li-S batteries. Journal of Materials Chemistry 20: 9821.

Xiulei, J., T. Kyu, and L.F. Nazar. 2009. A highly ordered nanostructured carbon-sulphur cathode for lithium-sulphur batteries. Nature Materials 8: 500.

Yan, D., X. Kai, X. Shizhao, and H. Xiaobin. 2013. Shuttle phenomenon – The irreversible oxidation mechanism of sulfur active material in Li–S battery. Journal of Power Sources 235: 181.

Yang, Y., G. Zheng, and Y. Cui. 2013a. Nanostructured Sulfur Cathodes. Chemical Society Reviews 42: 3018.

Yang, Y., G. Zheng, and Y. Cui. 2013b. A membrane-free lithium/polysulfide semi-liquid battery for large-scale energy storage. Energy & Environmental Science 6: 1552.

Yang, Z., J. Zhang, M.C. Kintner-Meyer, X. Lu, D. Choi, J.P. Lemmon, and J. Liu. 2011. Electrochemical energy storage for green grid. Chem. Rev. 111(5): 3577. doi: 10.1021/cr100290v.

Ye, Y.-S., X.-L. Xie, J. Rick, F.-C. Chang, and B.-J. Hwang. 2014. Improved anode materials for lithium-ion batteries comprise non-covalently bonded graphene and silicon nanoparticles. Journal of Power Sources 247(0): 991.

Yoo, H.D., I. Shterenberg, Y. Gofer, Y.G. Gershinsky, N. Pour, and D. Aurbach. 2013. Mg rechargeable batteries: An on-going challenge. Energy & Environ. Science 6: 2265.

Yuan, L.-X., Z.-H. Wang, et al. 2011. Development and challenges of LiFePO4 cathode material for lithium-ion batteries. Energy & Environmental Science 4(2): 269-284.

Zheng, G., Y. Yang, J.J. Cha, S.S. Hong, and Y. Cui. 2011. Hollow carbon nanofiber-encapsulated sulfur cathodes for high specific capacity rechargeable lithium batteries. Nano Letters 11: 4462.

Zheng, J. P., R.Y. Liang, M. Hendrickson, and E.J. Plichta. 2008. Theoretical energy density of Li–air batteries. J. Electrochem. Soc. 155(6): A432.

Zonghai, C., R. Yang, L. Eungje, C. Johnson, Q. Yan, and K. Amine. 2013. Study of thermal decomposition of $Li_{1-x}(Ni_{1/3}Mn_{1/3}Co_{1/3})_{0.9}O_2$ using in-situ high-energy x-ray diffraction. Adv. Energy Mater 3: 729.

ANNEX TABLES

TABLE 4A.1 List of xEV Models, Sales Volumes, Total Hybrid Market Share, and Architecture for CY2014

Hybrids				
Manufacturer	Model	Sales CY 2014	U.S. Hybrid Share 2014	Architecture
Toyota	Prius Liftback	122,776	0.2715	PS
Toyota	Prius C	40,570	0.0897	PS
Toyota	Camry Hybrid	39,515	0.0874	PS
Ford	Fusion Hybrid	35,405	0.0783	PS
Toyota	Prius V	30,762	0.0680	PS
Hyundai	Sonata	21,052	0.0466	P2
Ford	C-Max Hybrid	19,162	0.0424	PS
Lexus	CT200h	17,673	0.0391	PS
Toyota	Avalon Hybrid	17,048	0.0377	PS
Lexus	ES Hybrid	14,837	0.0328	PS
Honda	Accord Hybrid	13,997	0.0310	i-MMD
Kia	Optima Hybrid	13,776	0.0305	P2
Lincoln	MKZ	10,033	0.0222	PS
Lexus	RX 400 / 450 h	9,351	0.0207	PS
Subaru	XV Crosstrek Hybrid	7,926	0.0175	Other
Buick	Lacrosse Hybrid	7,353	0.0163	MH w IMA
Honda	Civic Hybrid	5,070	0.0112	MH w IMA
Honda	Insight	3,965	< 0.01	MH w IMA and CVT
Toyota	Highlander Hybrid	3,621	< 0.01	PS
Honda	CR-Z	3,562	< 0.01	MH w IMA and CVT or M6
Infiniti	Q50 Hybrid	3,456	< 0.01	PS
Nissan	Pathfinder Hybrid*	2,480	< 0.01	P2
Volkswagen	Jetta Hybrid	1,939	< 0.01	P2
Infiniti	QX 60 Hybrid*	1,678	< 0.01	P2 with CVT
Chevrolet	Malibu Hybrid	1,018	< 0.01	SS
Buick	Regal Hybrid	662	< 0.01	MH
Porsche	Cayenne Hybrid	650	< 0.01	P2
Chevrolet	Impala Hybrid	565	< 0.01	SS
Acura	ILX Hybrid	379	< 0.01	P2
Audi	Q5 Hybrid	283	< 0.01	P2
Lexus	GS 450h	183	< 0.01	PS
Infiniti	Q70 Hybrid	180	< 0.01	
Mercedes	E400H	158	< 0.01	P2
BMW	Active(335ih)	151	< 0.01	Other
Acura	RLX Hybrid	133	< 0.01	
BMW	Active (535ih)	112	< 0.01	Other
Chevrolet	Tahoe Hybrid	65	< 0.01	2-Mode
Lexus	LS 600h	65	< 0.01	PS
BMW	7-Series Hybrid	45	< 0.01	Other
Cadillac	Escalade Hybrid	41	< 0.01	2-Mode
GMC	Yukon Hybrid	31	< 0.01	2-Mode

Hybrids (continued)				
Manufacturer	Model	Sales CY 2014	U.S. Hybrid Share 2014	Architecture
Volkswagen	Touareg Hybrid	30	< 0.01	P2
Chevrolet	Silverado Hybrid	24	< 0.01	2-Mode
Mercedes	ML450H	20	< 0.01	
Mercedes	S400HV Hybrid	10	< 0.01	P2
GMC	Sierra Hybrid	6	< 0.01	2-Mode
Overall Hybrid Share of LDV Market			**2.75%**	

*Estimated.

EVs and PHEVs				
Manufacturer	Model	Sales CY 2014	U.S. PEV Share 2014	Architecture
Nissan	Leaf	30,200	0.2545	BEV
Chevrolet	Volt	18,805	0.1584	Series
Tesla	Model S	16,550	0.1394	BEV
Toyota	Prius Plug In	13,264	0.1118	P2
Ford	Fusion Energi	11,550	0.0973	P2
Ford	C-Max Energi	8,433	0.0711	P2
BMW	i3	6,092	0.0513	PHEV & BEV*
Smart	forTwo EV	2,594	0.0219	BEV
Ford	Focus EV	1,964	0.0165	BEV
Fiat	500E	1,503	0.0127	BEV
Cadillac	ELR	1,310	0.0110	PHEV
Toyota	RAV4 EV	1,184	<0.01	BEV
Chevrolet	Spark	1,145	<0.01	BEV
Porsche	Panamera S E-Hybrid	879	<0.01	PHEV
Mercedes	B-Class Electric	774	<0.01	BEV
BMW	I8	555	<0.01	PHEV
Honda	Accord Plug In	449	<0.01	PHEV
Honda	Fit EV	407	<0.01	BEV
Kia	Soul EV	359	<0.01	BEV
Volkswagen	e-Golf	357	<0.01	BEV
Mitsubishi	i	196	<0.01	BEV
Porshe	Cayenne S E-Hybrid	112	<0.01	PHEV
Overall PEV Share of the LDV Market			**0.72%**	

*PHEV/BEV breakdown unknown.
SOURCE: Cobb (2015).

TABLE 4A.2 Further Examples of P2 Hybridization Effectiveness in Vehicles in MY 2014

2014 Models	Engine Size (L) and Type	Transmission	Certification City FE (mpg)	Certification Hwy FE (mpg)	Certification Combined FE (mpg)	Certification Combined FC (gal/100 mi)	Fuel Consumption Change (%)	Fuel Economy Improvement (% mpg)
Audi Q5 Regular	3.0L V6 T	S8	29.18	43.5	34.26	2.92	Baseline	
Q5 Hybrid	2.0L I4 T	S8	30.40	39.9	34.05	2.94	0.6	-0.6
BMW 335i SS	3.0L V6 T	S8	27.45	45.8	33.5	2.99	Baseline	
Active Hybrid 3	3.0L V6 T	S8	32.70	46.41	37.71	2.65	-11.2	12.6
BMW 750LI SS	3.0L V6 T	S8	24.07	40.14	29.36	3.41	Baseline	
Active Hybrid 7L	3.0L V6 T	S8	28.21	43.13	33.41	2.99	-12.1	13.8
Infiniti Q60 FWD	3.5L V6	AV-S7	24.70	36.65	28.95	3.45	Baseline	
QX60 Hybrid FWD	2.5L I4 SC	AV-S7	35.20	41.10	37.63	2.66	-23.1	30.0
Infiniti Q70 Regular	3.7L V6	A-S7	22.48	36.10	27.08	3.69	Baseline	
Q70 Hybrid	3.5L V6	A-S7	38.25	47.90	42.06	2.38	-35.6	55.3
Nissan Pathfinder 2WD	3.5L V6	AV	24.64	36.98	28.99	3.45	Baseline	
Pathfinder Hybrid 2WD	2.5L I4 SC	AV	34.87	40.82	37.31	2.68	-22.3	28.7
Porsche Cayenne Regular	4.8L V8	A8	19.50	31.60	23.56	4.24	Baseline	
Cayenne Hybrid	3.0L V6 T SC	A8	25.10	33.10	28.16	3.55	-16.3	19.5

5

Transmissions

INTRODUCTION

The basic function of a transmission, together with the differential, is to reduce the relatively high engine output speeds to the vehicle's slower wheel speeds and to increase the torque applied to the wheels. Figure 5.1 illustrates a six-speed transmission enabling engine torque variation across changing engine speeds. Integral to many automatic transmissions are hydraulic torque converters, which also provide significant torque multiplication under starting and low speed conditions. Transmission design affects vehicle fuel consumption in two ways: First, increasing the number of gear ratios and providing a larger ratio spread allows the internal combustion engine to operate more often in regions of high efficiency. These design features concurrently provide smaller change steps to maintain nearer-to-optimum engine speeds. Second, reducing parasitic losses within the transmission improves transmission efficiency and reduces vehicle fuel consumption. In addition to these considerations, adapting transmissions to new fuel-efficient, turbocharged, downsized engines with fewer cylinders and to diesel engines with higher torque fluctuations generally necessitates additional torsional vibration damping.

In addition to pursuing improvements in conventional automatic transmissions, which dominate light-duty vehicles in the United States, a variety of alternative transmission designs are being developed and introduced into production. Dual-clutch transmissions (DCTs), with significantly lower parasitic losses, have already been introduced in some production vehicles and are providing reductions in fuel consumption. However, drivability and consumer acceptance issues remain for the most efficient dry-clutch DCTs. The continuously variable transmission (CVT) has seen a recent resurgence, and its penetration in the new vehicle fleet has increased significantly. Although the CVT provides the ideal ratio for any operating condition, the full potential of this attribute is not realized because the parasitic losses can exceed those of a conventional automatic transmission.

These approaches are being pursued by vehicle manufacturers and suppliers and are discussed in this chapter. The first section discusses the fundamentals of transmissions and covers design architecture, number of ratios and ratio spreads, and parasitic losses within various transmissions. The second section discusses specific transmission-related technologies, some considered and others not considered by the Corporate Average Fuel Economy (CAFE) rule analysis. In the second section, the committee's estimates of the effectiveness of each technology are presented, along with the committee's approach to estimating costs. The chapter concludes with the committee's findings and recommendations on transmissions.

TRANSMISSION FUNDAMENTALS FOR ACHIEVING FUEL CONSUMPTION REDUCTIONS

The energy distribution in a typical gasoline vehicle is shown in Figure 5.2. Driveline losses, which include the transmission, differential, and final drive gear, consume 5 to 6 percent of the energy input to the vehicle. Transmission losses alone consume approximately 4 percent of the energy. Since engine losses are approximately 70 percent, engine gross output energy is 30 percent of the input energy of the gasoline fuel. Thus, transmission losses of 4 percent of the energy input equal approximately 13 percent of the engine gross output. Consequently, if transmission losses could be reduced by 15 percent, fuel consumption would be reduced by 2 percent ($0.15 \times 4\% = 0.6\%$ reduction in losses; $0.6\%/30\% = 2\%$ FC reduction).

Transmission Architectures

Transmissions have been categorized by the Environmental Protection Agency (EPA) as automatic, CVT, and manual. EPA's estimated market share for each type of transmission in 2014 is shown in Figure 5.3. Automatic transmissions include conventional planetary automatic transmissions and DCTs. The architectures of these transmissions as well as CVTs are discussed in this section.

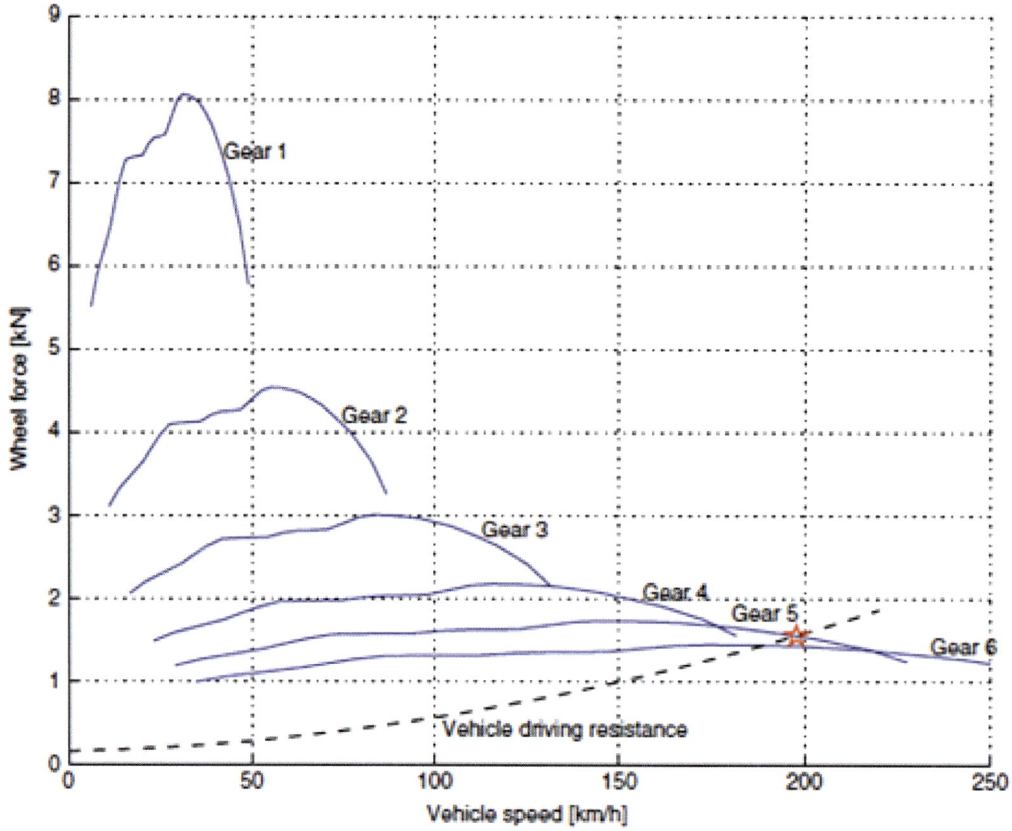

FIGURE 5.1 Wheel force, which is proportional to wheel torque, versus vehicle speed, which illustrates multiplication of engine torque and reduction in engine speed provided by a six-speed transmission. The vehicle resistance line determines maximum vehicle speed.
SOURCE: Reprinted from Eriksson and Nielsen (2014) with permission.

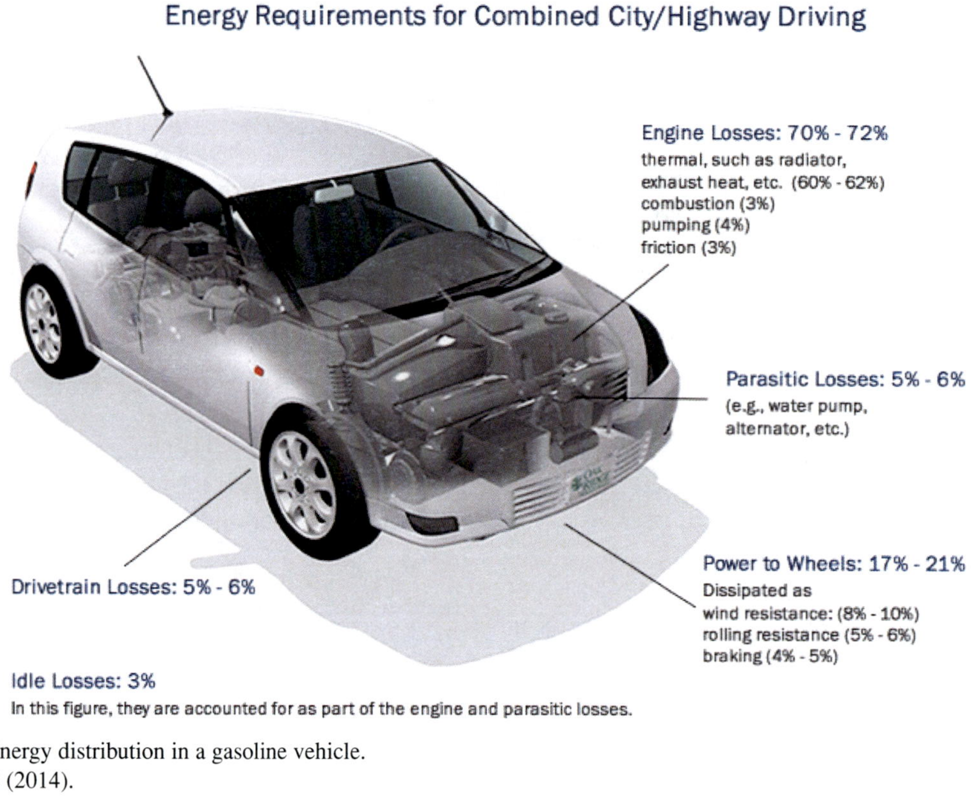

FIGURE 5.2 Energy distribution in a gasoline vehicle.
SOURCE: DOE (2014).

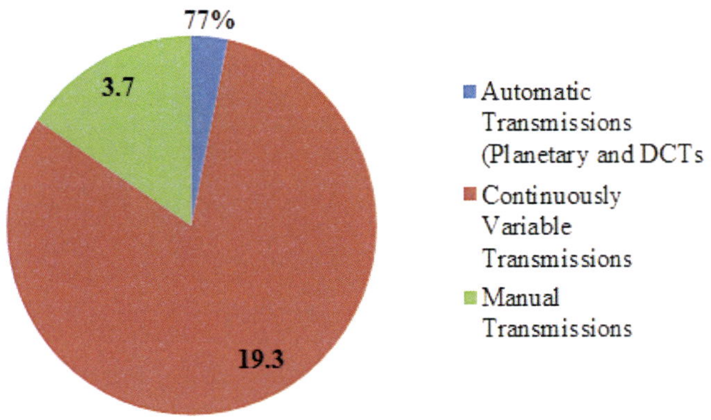

FIGURE 5.3 Market share of different types of transmissions in 2014.
SOURCE: Data from EPA (2014).

Planetary Automatic Transmission

Automatic transmissions (ATs) are currently the dominant transmission in the United States and are likely to remain the leading choice through 2025 (EPA 2014). Most automatic transmissions use planetary gear sets and are popular in large part due to their ease of operation and smooth launch feel off of idle. They have also benefited from over 70 years of continuing development and improvements in cost effectiveness. The smooth launch feel is made possible by the torque converter that not only provides a fluid coupling between the engine and the driveline but also provides significant torque multiplication at launch.

Conventional automatic transmissions generally use planetary gear sets to transfer power and multiply engine torque to the drive axle. A simple planetary gear set, shown in Figure 5.4, consists of three parts: a sun gear, a planet carrier, and a ring gear. A limited number of gear ratios are available from a single planetary gear set. Gear sets can be combined to increase the number of available gear ratios. A modern transmission will have various configurations of planetary gear sets to provide the various gear ratios required for the vehicle. The following three types of planetary gear sets are generally used in automatic transmissions:

- Simpson gear set has two planet carriers and two ring gears with a common sun gear. This provides three forward gears plus neutral and reverse.
- Ravigneaux gear set has two sun gears and two planet carriers with a common ring gear. This provides four forward gears plus neutral and reverse.
- Lapelletier gear set connects a simple planetary gear set to a Ravigneaux gear set. This provides six to eight forward gears.

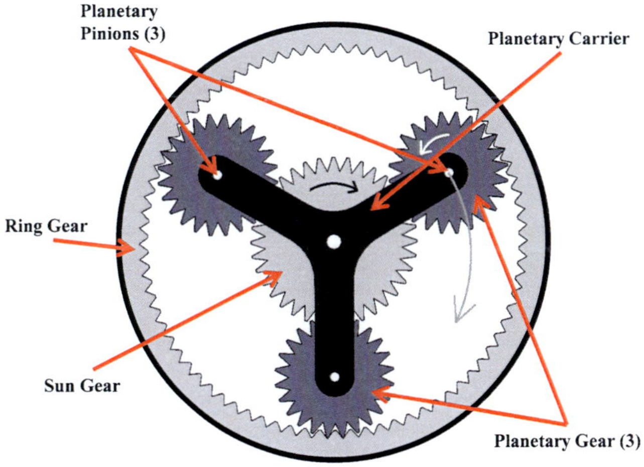

FIGURE 5.4 Planetary gear set configuration.

Six-Speed Automatic Transmissions

Six-speed automatic transmissions are currently widely used in the market, while seven-, eight-, and nine-speed transmissions are also in production, although with smaller market shares. A typical six-speed planetary transmission is shown in Figure 5.5 and includes the torque converter, the planetary gear set, clutches, the gerotor oil pump, and the valve body. A transmission control unit (TCU) is used to activate multiple solenoid valves that apply hydraulic pressure through the valve body to actuate or release multiple clutches and brakes that are applied or released to control the output speed ratio. The efficiency of conventional automatic transmissions ranges from 86 to 94 percent, where transmission efficiency is defined as the power output divided by the input power, multiplied by 100.

Eight-Speed Automatic Transmissions

Recent introductions of eight-speed transmissions have included the ZF 8HP45 transmission in several Chrysler products, including Ram pickup trucks and rear wheel drive (RWD) cars and numerous vehicles from several European manufacturers. The General Motors 8L90 transmission was introduced in the 2015 MY large pickup trucks and large SUVs and the Chevrolet Corvette sports car. The ZF eight-speed transmission, shown in Figure 5.6, consists of a torque converter, four planetary gear sets, and five shifting elements (brakes A and B; clutches C, D, and E). The use of five shifting elements is notable since the outgoing six-speed automatic transmission also used five shifting elements.

Nine-Speed Automatic Transmissions

Several nine-speed automatic transmissions have also been recently introduced. These include the front wheel drive (FWD) ZF 9HP, recently introduced in the Jeep Cherokee, and the Mercedes 9G-Tronic, recently introduced in a Mercedes E350 with a 3.0L diesel engine (Daimler 2013).

Ten-Speed Automatic Transmissions

Ford Motor Company and General Motors announced in 2013 that they are jointly developing nine- and ten-speed automatic transmissions for reduced fuel consumption and improved performance, particularly with smaller engines (Healey and Woodyard 2013). In 2014, VW announced that it had plans for a ten-speed dual clutch transmission for better fuel economy.

Due to advancements in architecture optimization, the increased ratios in these new transmissions are being implemented with minimal increase to package size, component count, and cost. Software optimization tools are being used to develop architectures requiring fewer elements by utilizing some of the elements (e.g., planetary gear elements) for multiple speeds. However, these new transmissions have high development costs and long design and validation phases. After defining a suitable layout, a new transmission generally takes 5 years for design, development, and implementation in production.

Dual-Clutch Transmission

DCTs are architecturally similar to manual transmissions but add automated shifting and typically utilize two coaxial input shafts and two clutches to shift between the two input shafts, as shown in Figure 5.7. This enables DCTs to perform a clutch "hand-off," where the clutch of the currently utilized gear opens as the clutch of the next gear to be engaged closes. With precise control of the releasing clutch and the engaging

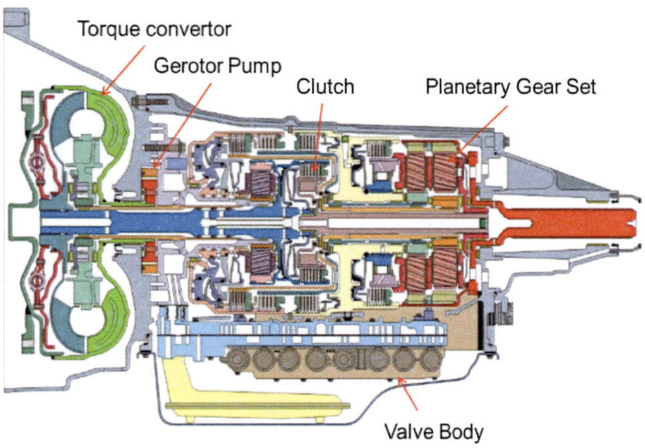

FIGURE 5.5 Typical six-speed planetary automatic transmission.
SOURCE: Copyright © 2006 ATSG (Automatic Transmission Service Group).

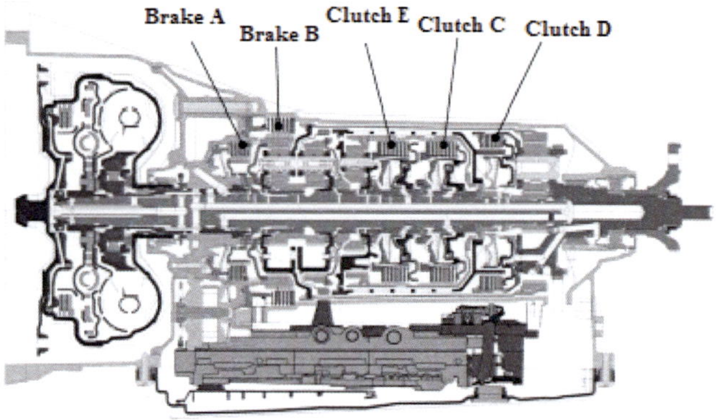

FIGURE 5.6 Cross section of the ZF eight-speed automatic transmission – 8HP45.
SOURCE: Dick et al. (2013). Reprinted with permission from SAE paper 2013-01-1272 Copyright © 2013 SAE International.

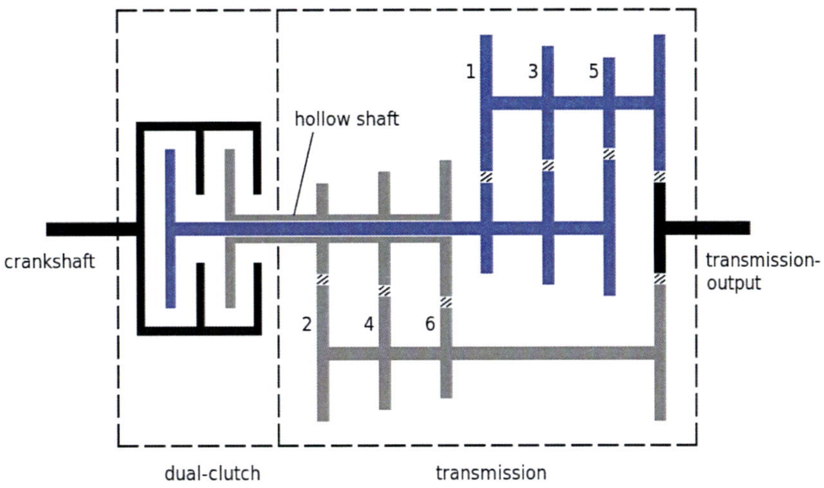

FIGURE 5.7 Schematic of a dual clutch transmission.
SOURCE: C-Lover, Wikimedia Commons, http://commons.wikimedia.org/wiki/File:Dual-clutch_transmission.svg, public domain, accessed February 13, 2015.

clutch for the next gear, smooth shift quality approaching that of a conventional automatic transmission can be achieved. DCTs have parasitic losses similar to a manual transmission, which are significantly lower than a conventional planetary automatic transmission because of the benefit of on-demand pumps, use of splash lubrication, and minimized clutch drag losses since only two clutches are used.

Although DCTs offer significant potential for minimizing transmission losses, they have faced customer acceptance obstacles in the United States, primarily since the U.S. market is accustomed to the smooth feel of a torque converter during launch, which is difficult to replicate with DCTs. The launch feel becomes more of an issue with downsized, turbocharged engines that provide less transient torque at launch. These engines, when paired with a DCT, often require a "launch gear" to provide suitable acceleration from a stop that equals the feel provided by a torque converter.

The lack of smooth launch performance with DCTs has prompted Honda to announce the development of an eight-speed DCT for the 2015 MY, featuring the first use of a torque converter in a DCT (Carney 2014). This should provide the smooth low-speed driving dynamics of a traditional automatic transmission with a gearbox that is more efficient. As an additional benefit, torque multiplication of the torque converter will improve acceleration from idle.

Another method OEMs are using to increase the market penetration of DCTs is to implement the wet-clutch version instead of the dry-clutch version. While the current dry-clutch DCT can offer 0.5 percent to 1 percent lower fuel consumption, current designs sacrifice drivability and suffer from poor customer acceptance. This difference in drivability and consumer acceptance can be seen in the comparison of two of Volkwagen's MY 2015 vehicles, the VW Golf and the VW Polo. The Golf, with a wet-clutch DCT, has received many positive reviews and awards, while the Polo, with a dry-clutch DCT, has received poor reviews for transmission-related drivability.

Continuously Variable Transmission

The continuously variable transmission (CVT) is becoming more popular due to its simple mechanical design and potential fuel economy benefits. EPA estimated that the U.S. market share of CVTs would increase to 19.3 percent in 2014. In contrast, the compliance demonstration scenario for the 2017-2025 standards within the Technical Support Document (TSD) has a zero percent penetration for CVTs. Several 2014 MY vehicles with CVTs are among the vehicles with high fuel economy ratings, as shown in Figure 8.11 in Chapter 8. A typical CVT consists of two cone-shaped pulleys and a connecting belt, as shown in Figure 5.8. By moving the cone-shaped pulley halves axially, the pulleys are able to produce a continuous variation in the ratio of the engine input speed to the driveline output speed. This continuous variation in ratios allows the engine to operate at its most efficient condition for the power level required. The CVT also benefits from the lack of discrete shift events of conventional automatic transmissions, preventing customer issues with possible shift harshness. In the near term, the belt-driven CVT is likely to remain the only type that will have any market penetration in the light-duty market before 2020. There is the possibility that the toroidal CVT may be offered in light-duty applications in the 2020 to 2030 time frame. The toroidal architecture has been tested by multiple OEMs, but these OEMs have currently chosen to go with conventional belt designs for production.

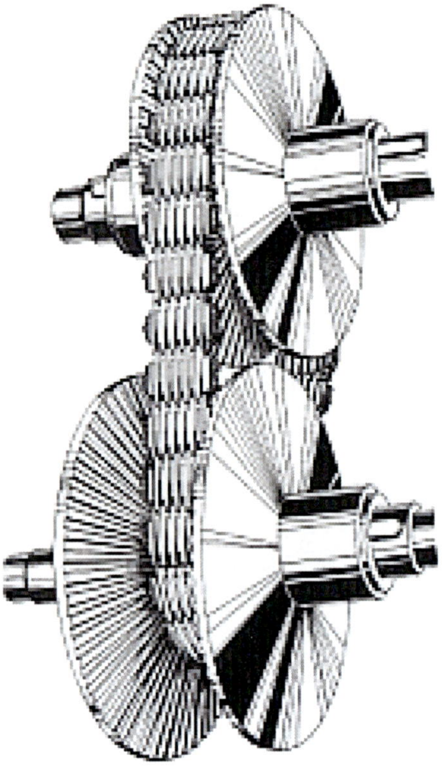

FIGURE 5.8 CVT with details of the steel belt.
SOURCE: Büdeler Naumann, Wikimedia Commons, http://commons.wikimedia.org/wiki/File:Pivgetriebe.png, accessed March 2, 2015.

The Dana Variglide continuously variable planetary (CVP) technology is another promising CVT technology (Dana Holding Corp. 2014). This technology was derived from the Fallbrook NuVinci CVP. While currently not in any known near-term production plans, this technology may be offered in the 2020 to 2030 time frame.

In contrast to this advantage, drawbacks to the CVT include slow consumer acceptance due to nontraditional engine sounds and vibrations and concerns about the materials used in the manufacturing of the belt. A term commonly used to describe the feel of most CVTs is "rubber band" because the connection between the driver's throttle input and the vehicle's acceleration response is often not as direct as with a conventional planetary automatic transmission. Some manufacturers have instituted a calibration from the control of a CVT that mimics an automatic transmission, but this can result in less-than-optimal fuel economy. However, in spite of the previously poor acceptance of CVTs in the U.S. market, OEMs have begun targeting specific vehicle types for CVTs as a method of reintroducing CVTs into the U.S. marketplace. An electronically controlled CVT (eCVT), used in powersplit hybrids, consists of a planetary gear set shown in Figure 4.2.

Manual Transmission

Manual transmissions (MTs) are estimated by EPA to have a market share in the U.S. of only 3.7 percent, in part because manual transmissions require the driver to manually actuate the clutch and change the gear. The manual transmissions in current vehicles are generally cheaper to manufacture, lighter in weight, better performing, and more fuel efficient than all but the newest automatic transmissions. Additionally, while more driver effort is required to operate a manual transmission than an automatic transmission, manual transmissions have a simpler mechanical design, as shown in Figure 5.9. In a manual transmission, the gears on the output shaft and the parallel layshaft are always engaged with each other. A selected gear is subsequently engaged to the output shaft with the use of a synchronizer. The collar of the synchronizer makes frictional contact with the gear before the dog teeth make contact to engage the output shaft. The synchronizers are engaged with forks that are controlled by rods engaged by the shift lever.

The lower cost of a manual transmission results not only from the simplicity of design but also from the absence of a transmission control unit (TCU), which generates costs for control software and calibration. Manual transmissions have the highest efficiency due to their inherently low parasitic losses. Because they are usually splash-lubricated from gears spinning in the oil sump, manual transmissions usually do not require the oil pump or forced cooling that most automatic transmissions require. These factors contribute to a manual transmission's ability to transfer torque with only about 4 percent energy loss of the engine's gross output, compared to 13 percent loss for conventional automatic transmissions.

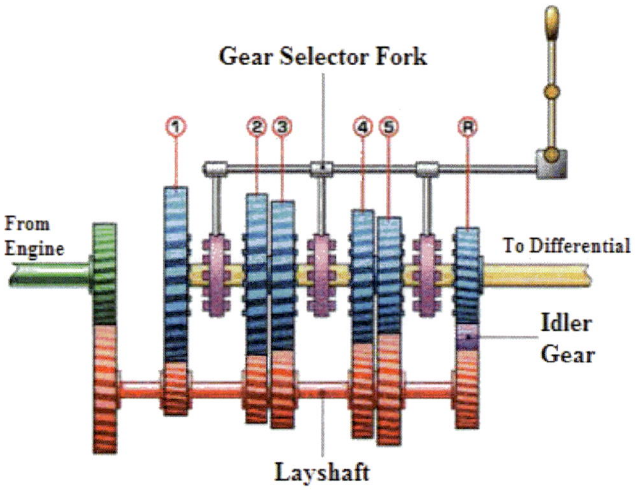

FIGURE 5.9 Manual transmission.
SOURCE: Brain (2000). Reprinted courtesy of HowStuffWorks.com. http://auto.howstuffworks.com/transmission3.htm.

Automated Manual Transmission

Automated manual transmissions (AMTs) are essentially manual transmissions with either electromechanical or electrohydraulic actuators added to automate both the clutch and gear selection. This transmission architecture promised parasitic losses nearly equivalent to manual transmissions, but with the same ease of operation as a conventional automatic transmission. While the AMT was successful in delivering low parasitic losses, it had significant deterioration in driving comfort when the fully automatic mode was not engaged. Figure 5.10, which is a plot of vibration dose value (VDV), a common metric used to measure shift comfort, shows that AMTs are significantly worse than conventional automatic transmissions and dual clutch transmissions. In short, the single clutch on an AMT introduced a lag in acceleration that drivers found uncomfortable. Due to this issue, AMTs are not considered acceptable for the U.S. market.

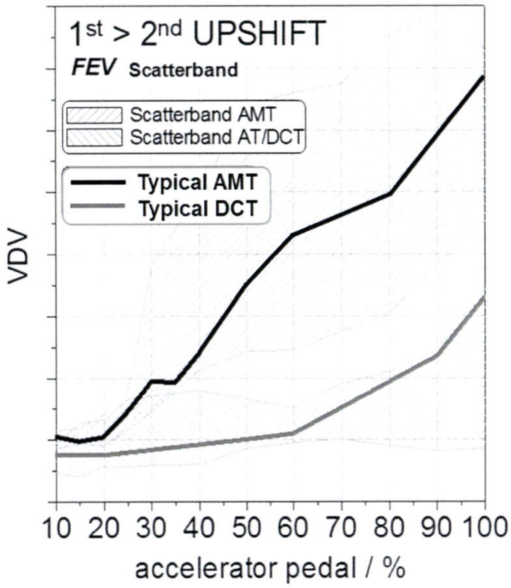

FIGURE 5.10 Comparison of vibration dose value (VDV) for automated manual transmission (AMT) with conventional automatic transmission (AT) and dual clutch transmission (DCT). SOURCE: Govindswamy, Baillie, and D'Anna (2013).

Number of Ratios and Ratio Spread

There are two primary methods employed in transmissions to reduce vehicle fuel consumption. The transmission ratios and the ratio spread are chosen so that the engine can operate at the lowest available BSFC[1] condition for the power level required.[2] The upper portion of Figure 5.11 shows the engine operating conditions on the CAFE cycle for a vehicle with a six- speed automatic transmission, and the lower portion of the figure shows the BSFC island map overlaid with several lines of constant power. The green dots show the lowest BSFC value for each constant power line. The upper plot shows that many of the operating conditions are between 1,000 rpm and 1,500 rpm and are close to the minimum BSFC condition for the power required. Some of the other conditions are at higher loads and speeds, but the lines of constant power are generally contained within the same region of BSFC values. There is, however, some opportunity to move some of the operating conditions at higher engine speeds toward the best BSFC values for reduced fuel consumption. The dominant trend toward reduced engine speeds for reduced fuel consumption is limited by NVH and drivability concerns.

An Argonne National Laboratories' (ANL) study of the impact of transmission technologies on fuel efficiency evaluated the fuel economy improvements of six-speed and eight-speed automatic transmissions. The ratios used for this study

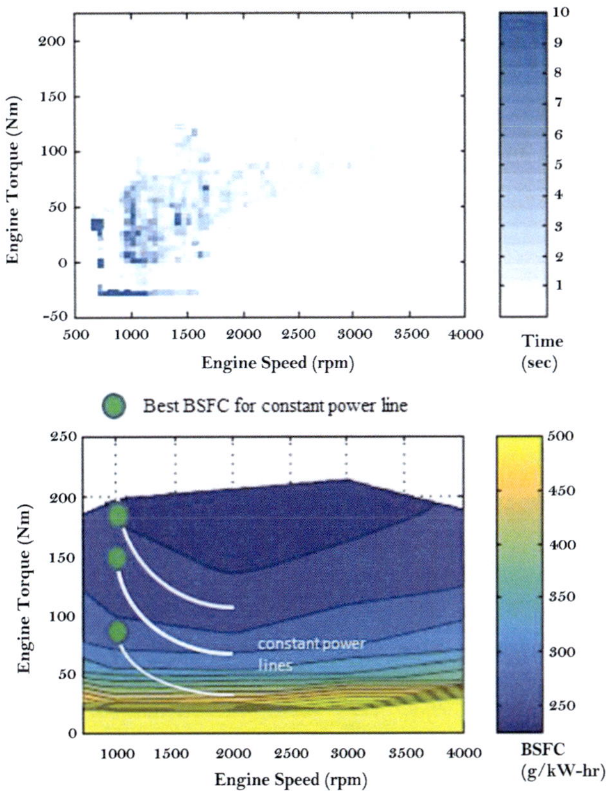

FIGURE 5.11 Engine operating conditions on the CAFE cycle for a vehicle with a six-speed automatic transmission and a BSFC island map overlaid with several lines of constant power. SOURCE: Developed from Middleton et al. (2015).

[1] The brake-specific fuel consumption (BSFC) is equal to an engine's fuel consumption (g/hr) divided by brake power (kW), which is the usable output power of an engine. BSFC is proportional to the inverse of engine efficiency (usable work/fuel energy input). A BSFC map displays islands of constant BSFC as a function of engine torque, or BMEP, versus engine speed. *Brake*-specific fuel consumption is typically measured with the engine operated on a dynamometer.

[2] The gear, or transmission, ratio is the ratio of the rotational speed of the driving gear to the rotational speed of the final drive gear.

are listed in Table 5.1; they represent ratios of transmissions currently on the market.

In addition to providing increased torque multiplication from idle and significantly lower engine speeds at high vehicle speeds, reducing the steps between each gear is another enabler for reduced fuel consumption. The reduced steps between each gear allow the engine to operate closer to the minimum fuel consumption condition for the power required for every operating condition. The ability of these reduced steps to facilitate operation closer to the minimum fuel consumption condition is illustrated in Figure 5.12. A line of constant power is overlaid on a BSFC island map. Operating conditions on this constant power line are shown for the six-speed and eight-speed transmissions. As illustrated, the eight -speed transmission is capable of operating closer to the lowest available BSFC condition shown at an engine speed of 1,000 rpm. In contrast, the engine with the six-speed transmission operates at a speed of more than 200 rpm over the engine speed provided by the eight-speed transmission. As a result, for this example of a constant power condition, the eight-speed transmission would operate with 5 percent lower fuel consumption. This example also illustrates the benefit of a CVT with continuously variable ratios to reach the lowest available BSFC condition, without having the constraints of a stepped ratio transmission.

Ratio spread is defined as the first gear ratio divided by the top gear ratio. As additional gears have been added to transmissions, with lower first gears and higher top gears, the ratio spread has also increased significantly over the years, shown in Figure 5.13. The highest ratio spread of 9.8:1 is shown for the Jeep Cherokee with the ZF nine-speed automatic transmission. The larger ratio spread provides increased torque multiplication off idle and significantly lower engine speeds at high vehicle speeds, which reduces fuel consumption.

TABLE 5.1 Gear Ratios for Five-, Six- and Eight-Speed Transmissions Representative of Current Transmissions on the Market

Gear Number	1	2	3	4	5	6	7	8
Reference 5-speed automatic	2.56	1.55	1.02	0.72	0.52			
6-speed transmissions	4.15	2.37	1.56	1.16	0.86	0.52		
8-speed transmissions	4.6	2.72	1.86	1.46	1.23	1	0.82	0.52

SOURCE: Moawad and Rousseau (2012).

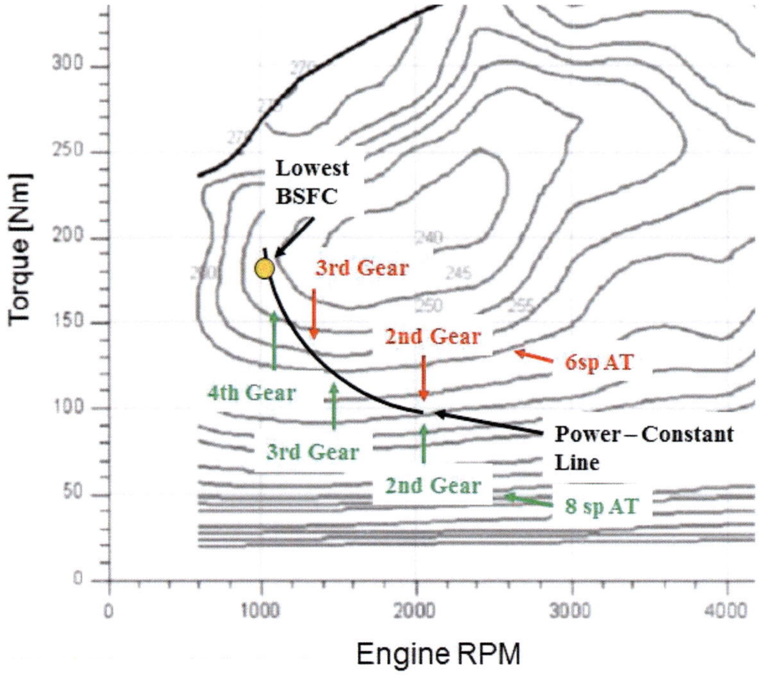

FIGURE 5.12 Fuel consumption benefits of an eight-speed compared to a six-speed automatic transmission, shown as an overlay on a BSFC island map.
SOURCE: Adapted from Dick et al. (2013). Reprinted with permission from SAE paper 2013-01-1272. Copyright © 2013 SAE International.

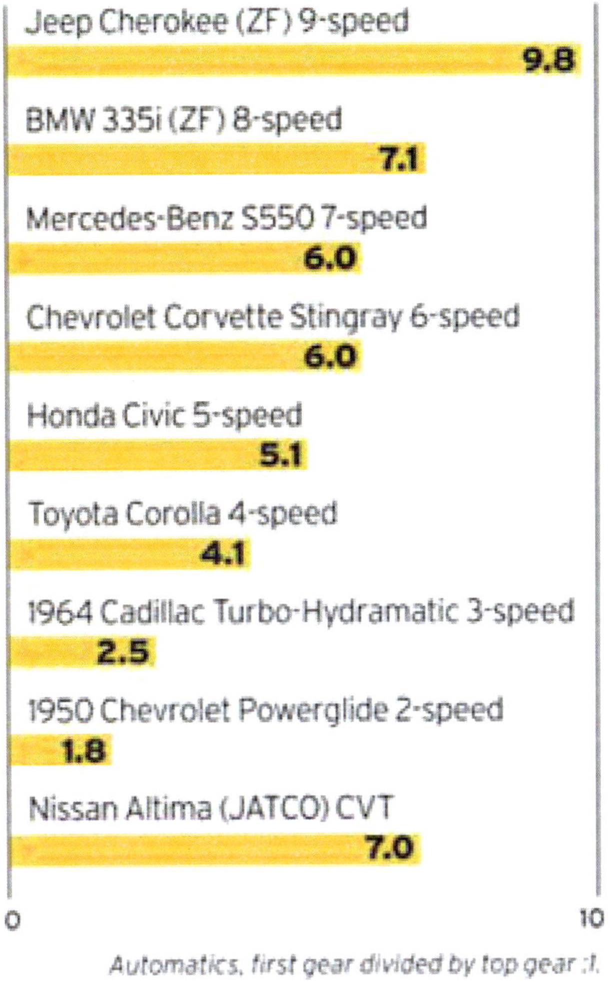

FIGURE 5.13 Increase in ratio spread over the past 65 years.
SOURCE: Sherman (2013). Published in *Car and Driver* magazine, December 2013.

A comparison of the engine operating envelopes for a vehicle with a six-speed automatic transmission and an eight-speed automatic transmission is shown in Figure 5.14. Compared to the six-speed transmission, the eight-speed transmission results in the engine operating at higher BMEP[3] levels over a narrower speed range and at lower average speeds, all of which tend to reduce fuel consumption.

As automatic transmissions and dual clutch transmissions trend toward higher numbers of ratios and ratio spreads, diminishing benefits for fuel consumption reduction are anticipated. Recent studies, such as the one by Getrag (Eckl and Lexa), shown in Figure 5.15, have indicated that only minimal reductions in fuel consumption are achieved beyond 7 ratios and a ratio spread of 8.5. This result depends on the engine and vehicle specifications, so the results may not apply to all vehicles. Additionally, the results from Figure 5.15 are based on the New European Drive Cycle (NEDC) and have the potential for different results on U.S. drive cycles. There is evidence later in this chapter indicating that the benefits of reducing transmission parasitic losses can be greater than the benefits of moving from seven speeds to nine. Moving from an 8.5 ratio spread to 10 also does not provide significant fuel economy benefits but may provide improved performance.

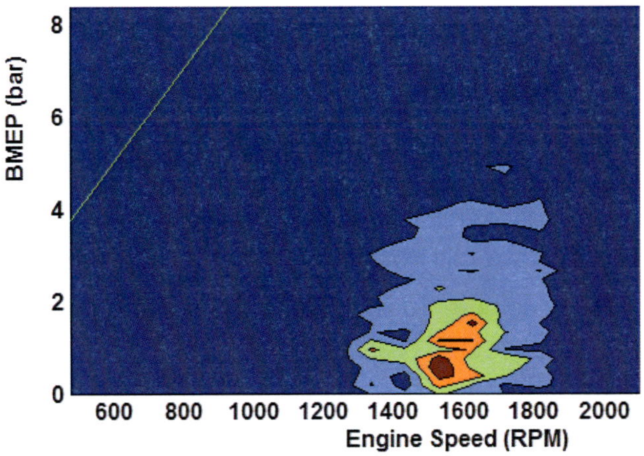

FIGURE 5.14 Engine operating conditions for six-speed (left) and eight-speed (right) automatic transmissions on the FTP-75 drive cycle. Note that the color scale shows the density of the operating points, where dark blue represents no operating points and dark red represents the highest density of operating points. The eight-speed transmission (right) compared to the six-speed transmission (left) results in a narrower range of engine speeds and higher level of BMEPs, which result in improved fuel economy.
SOURCE: Shidore et al. (2014).

The committee has also found, through full system simulations conducted by the University of Michigan, that as the engines incorporate more new technologies to improve fuel consumption, including variable valve timing and lift, direct injection, and turbocharging and downsizing, the benefits of increasing transmission ratios or switching to a CVT diminishes. Similar results have been reported from other modeling studies. As the engine efficiency map improves, the penalty of having a larger ratio step between gears is significantly reduced compared to the example shown in Figure 5.12 for a naturally aspirated engine. As discussed in Chapter 8, the benefit of an eight-speed transmission over a six-speed transmission is reduced by approximately 15 percent when added to a modestly turbocharged, downsized engine instead of a naturally aspirated engine.

There are, however, other reasons for going to higher ratio spreads and speeds. DCTs, for example, do not have the

[3] Brake mean effective pressure, or BMEP, is the torque per cubic inch of engine displacement. It is used to evaluate an engine's efficiency when producing torque at a given engine displacement. The higher the BMEP of an engine, the more work it produces for a given engine displacement. This is a theoretical value and does not represent actual in-cylinder pressures of the engine.

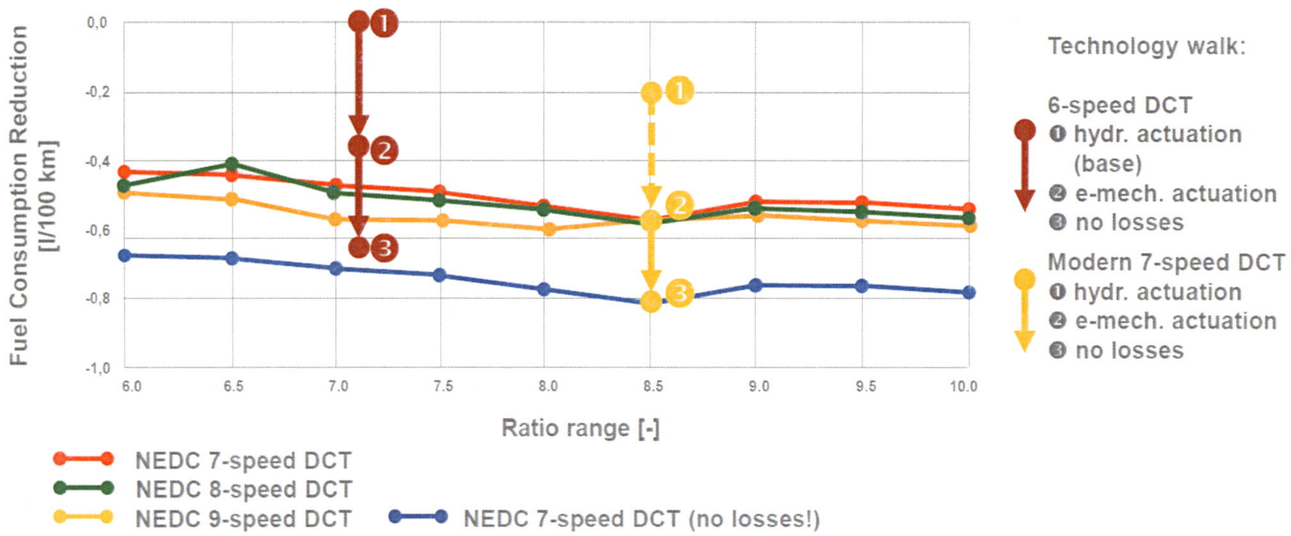

FIGURE 5.15 Fuel consumption reduction as a function of ratio range and number of speeds (ratios).
SOURCE: Eckl and Lexa (2012).

torque multiplication provided by a torque converter. In order to achieve acceptable vehicle launch feel, an ultrahigh first gear, significantly exceeding the 4.6:1 shown in the previous example, may be required to compensate for the lower transient torque available at launch from turbocharged, downsized engines. A very low final ratio can also be implemented to enable lower engine speed at highway speed. Some OEMs have already implemented such low ratio gears, which can maintain highway speed but require a downshift for any acceleration or grade. The two requirements for a high first gear and low final gear can result in an overall ratio spread of up to 10.5 for a dual clutch transmission. To maintain the ratio steps in the same range as smaller ratio spread transmissions, thus preventing unpleasant shift feel, a larger number of gears will be required, particularly for DCTs without torque converters. The first example of this is Volkswagen's announcement of the ten-speed DCT discussed previously. Conventional automatic transmissions are not expected to require the large ratio spreads of DCTs to achieve acceptable launch feel because torque converters provide torque multiplication during launch. The Honda DCT that utilizes a torque converter will also not require a higher number of speeds for customer satisfaction.

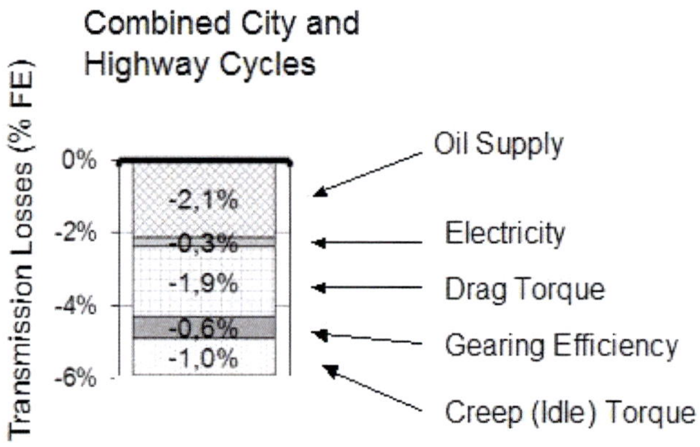

FIGURE 5.16 Transmission losses in a modern eight-speed automatic transmission.
SOURCE: Dick et al. (2013). Reprinted with permission from SAE paper 2013-01-1272. Copyright © 2013 SAE International.

Parasitic Losses

The second method for reducing fuel consumption involves improving efficiency by reducing parasitic losses of the transmission. Losses in a modern eight-speed automatic transmission operated over the combined CAFE city and highway cycles are shown in Figure 5.16. The total parasitic losses result in a 6 percent loss in fuel economy for this example transmission, and the categories of loss are identified in the figure, with oil supply, drag torque, and creep or idle torque being the three largest categories. Since the EPA/Ricardo study (2011) showed that baseline, four-speed automatic transmissions in the 2010 MY had losses of approximately 10 percent; the estimates were scaled to this level to represent the 2010 MY baseline that NHTSA used in its analysis of transmission technologies. The 2010 MY baseline transmission losses are listed in Table 5.2. Technologies for further reductions in transmission losses for each category shown in Table 5.2 are reviewed in this section.

TABLE 5.2 Transmission Losses Estimated for a 2010 Baseline Automatic Transmission

Losses	Losses - % FE
Oil Supply	3.6
Electricity	0.5
Drag Torque	3.2
Gearing Efficiency	1.0
Creep (Idle) Torque	1.7
Total Losses	10.0

NOTE: Transmission is four-speed automatic with torque converter.

Oil Supply

Automatic transmissions require an oil pump for lubrication and for hydraulic pressure for clutch clamping. The pressure range for automatic transmissions and dual clutch transmissions is generally dictated by the clamping force required on the clutch plates to transmit torque without slipping. Typically, automatic transmissions operate with oil pressures between 5 and 20 bar. An example plot of clutch torque capacity versus pressure is shown in Figure 5.17. The relationship between torque and pressure varies with the clutch diameter, number of plates, friction material, and oil properties.

The dominant loss in the oil supply category results from the oil pump. Automatic transmissions have typically used gerotor-type pumps driven off the torque converter. Alternative pump systems for oil supply are shown in Figure 5.18. These systems range from reducing the main supply pressure, using a more efficient vane pump, applying a dual-stage vane pump, and using a variable displacement pump. The variable displacement vane pump has the greatest potential to reduce losses and improve fuel economy by 1 percent, relative to a conventional fixed gerotortype pump. The high efficiency gearbox—level 1, discussed later in the chapter—includes the variable displacement vane pump.

Recently introduced transmissions have incorporated off-axis dual displacement or variable displacement vane pumps, as shown in Figure 5.19. These pumps are often driven by a chain off the main input axis and sometimes with a speed ratio change in order to operate the pump in a more efficient speed range. The variable displacement allows for quick response to provide high pressures required for high torque clutch engagement, but since the majority of operating points are at lower steady-state torque, the pump can operate at a significantly lower displacement for much of the drive cycle. Application of off-axis pump designs can also allow the pump to be downsized to reduce parasitic losses. This ability to operate the pump at a speed different from the torque converter's speed enables a smaller diameter pump since the speeds can be higher.

Further reductions in oil pump energy losses can be achieved with a dual pump system. The energy losses for a dual pump system are compared to a variable displacement vane pump over the NEDC driving cycle in Figure 5.20 for both conventional automatic transmissions and CVTs. Studies have shown that utilizing a small vane pump that is sized to cover the majority of driving conditions and a second electric auxiliary pump to cover high pressure demands can provide approximately a 2 percent reduction in fuel

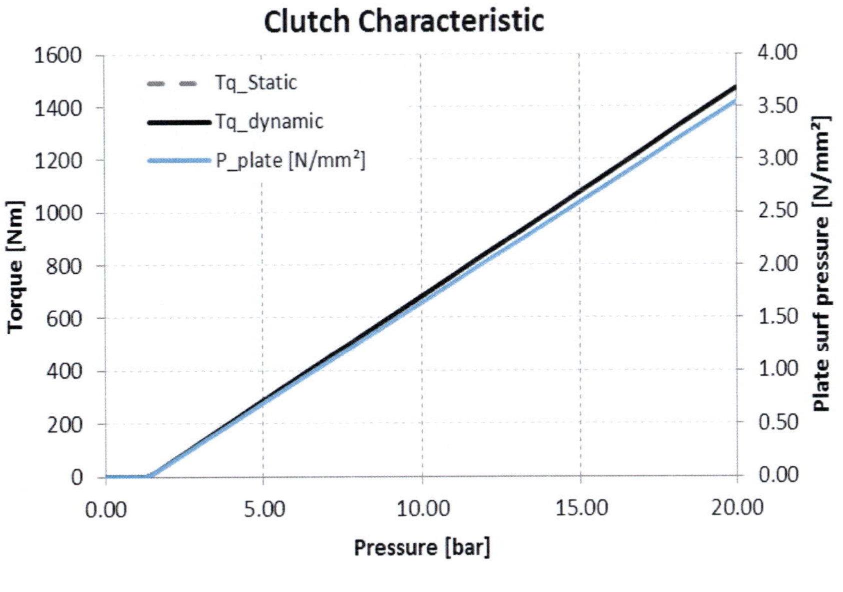

FIGURE 5.17 Typical clutch torque capacity as a function of hydraulic pressure.

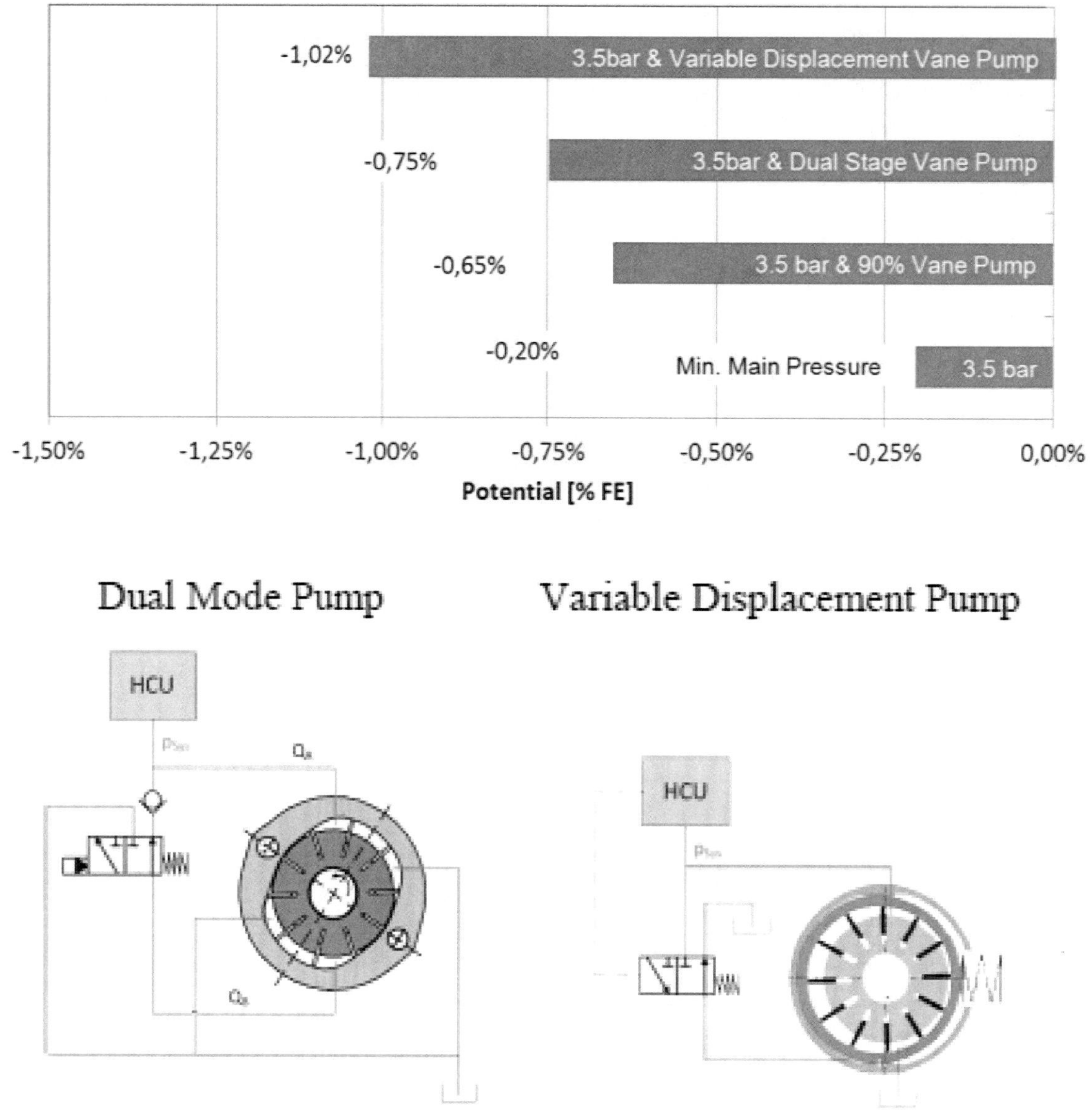

FIGURE 5.18 Alternative pump systems for oil supply.
SOURCE: Dick et al. (2013). Reprinted with permission from SAE paper 2013-01-1272 Copyright © 2013 SAE International.

consumption. The Mercedes 9G-Tronic transmission uses this type of dual pump system. However, the improvement in fuel consumption comes at the cost of a second pump. Alternatively, a single, variable displacement vane pump could be used to reduce costs, but this would sacrifice a small percentage of the efficiency improvement of the dual pump system for conventional automatic transmissions, as shown in Figure 5.20. The smaller, more frequently operating pump of the dual pump system can be better optimized for its operating conditions than the larger, single variable displacement pump. During periods of relatively low pressure demand, such as steady speed highway driving, the smaller, optimized pump would operate the majority of the time. The high efficiency gearbox—level 2, discussed later in the chapter—includes a dual pump system. As shown in Figure 5.20, the oil pump for the CVT has higher energy losses than a conventional automatic transmission. Consequently, improvements in the oil pump for the CVT can provide larger reductions in energy losses in this transmission.

High efficiency gearbox—level 3—involves removing a full-time oil pump from the automatic transmission and using an on-demand electric pump for lubrication/cooling and electromechanical shifting. Similar to some DCTs that use electric motors for shifting, an electric motor, ball ramp, and axial bearing can be used to shift a wet clutch in a conventional automatic transmission. A fully functional

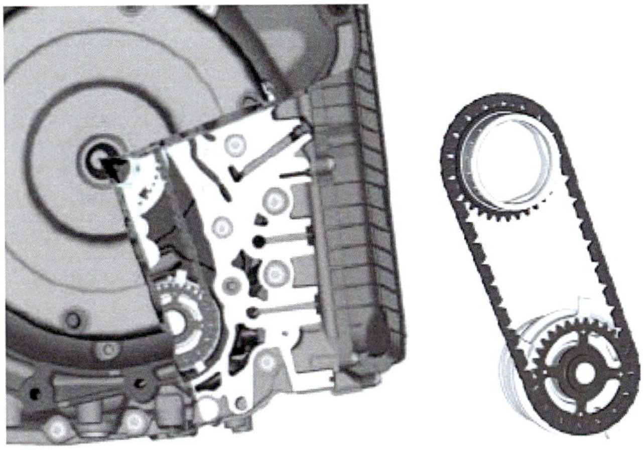

FIGURE 5.19 Off-axis double stroke vane pump.
SOURCE: Gartner and Ebenhock (2013). Reprinted with permission from SAE paper 2013-01-1276. Copyright © 2013 SAE International.

automatic transmission with this technology was developed by FEV; the shifting element is shown in Figure 5.21. The system has the additional benefit of locked end positions so that no power is required to maintain the clamp load on a clutch. Eliminating the pump can result in up to a 3.6 percent reduction in fuel consumption for conventional automatic transmissions, as indicated in Table 5.2. Production-proven technology would require a redesign of an automatic transmission architecture. As a result, this technology is considered applicable after 2025.

Drag Torque

Clutches

The second largest loss is due to drag torque from the clutches, brakes, bearings, and seals. Transmissions that use wet clutches, which include conventional automatic transmissions and DCTs, will incur drag losses from open (disengaged) clutch packs rotating in oil. The loss is caused by the shearing of oil between the rotating, open clutch plates. The smaller gaps between the plates and higher viscosity oil will result in a higher shearing loss. While DCTs will only have one open clutch at any given time, an automatic transmission may have from two to four. Additionally, a DCT will deselect the synchronizer for cruising conditions to reduce the loss to synchronizers rather than open the clutches. The clutch packs are used in automatic transmissions to change ratio by locking elements of planetary gear sets. A typical six-speed automatic transmission may have five clutch packs, while an eight- or nine-speed may have five or six. New transmission designs will attempt to keep as few clutches open as

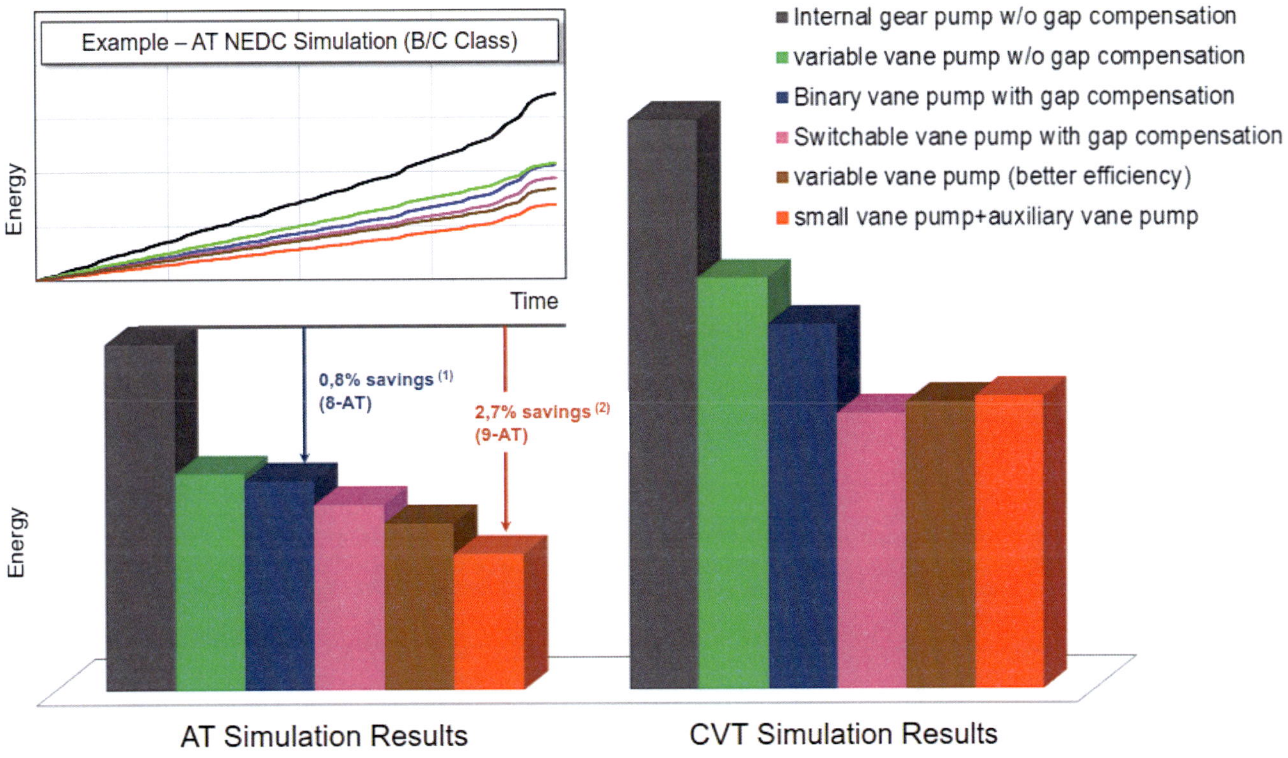

FIGURE 5.20 Transmission oil pump designs for reduced energy consumption in conventional automatic transmissions and CVTs.
SOURCE: Shulver (2013).

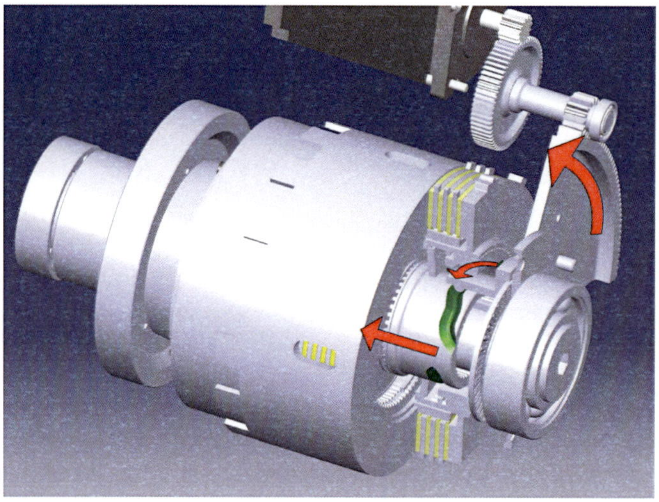

FIGURE 5.21 Electric motor with ball ramp and axial bearing for shifting a wet clutch in a conventional planetary automatic transmission. SOURCE: Govindswamy, Baillie and D'Anna (2013).

possible in each gear due to the drag losses. In several new transmission architectures, two clutch packs are open at any given time, leaving the other three or four closed. Clutch drag losses vary significantly due to differences in clutch sizes and overall transmission architecture.

New advancements in clutch plate technology have resulted in significant reductions in drag losses. The friction material on the clutch plates today generally includes grooves to optimize oil flow through the plates, which enhances cooling and reduces losses. Placing wave springs between the plates, as shown in Figure 5.22, creates wider gaps between the plates to allow oil to flow more freely between the plates. The larger the gap, the less oil shearing, but the larger gaps can have negative effects on clutch response time. The springs also add parts and cost. Figure 5.23 shows results from a Borg Warner study (Martin 2012) indicating that a 90 percent reduction in clutch drag loss can be achieved from a baseline of flat friction plates. While not shown in Figure 5.23, other components that should be considered are dual-rate clutch pistons since they are able to increase running clearance and still offer acceptable performance. Since most current automatic transmissions currently use clutch plates with some grooving in them, a modern transmission may achieve a clutch drag loss reduction of approximately 50 percent. Computational fluid dynamics can be used to create groove patterns that remove the oil from the shearing interface.

Clutch Slip

While it is most efficient to close a clutch as quickly as possible, the abrupt closure can cause unpleasant noise, vibration, and/or harshness (NVH). When shifting between gears in an automatic transmission or a DCT, some amount of clutch slip is required for smoothness, but slip has the negative effect of generating considerable heat, which requires cooling flow, and results in an increase in fuel consumption. As engine torsional vibrations continue to increase with turbocharged, downsized engines, clutch slip is increasingly being used for the lockup torque converter clutch of conventional automatic transmissions and in DCTs to reduce the transmission of these torsionals to the wheels. This slip across a clutch results in a power loss. Conventional automatics sometime utilize micro-slip as a means to improve NVH. This process involves slipping the clutch very slightly

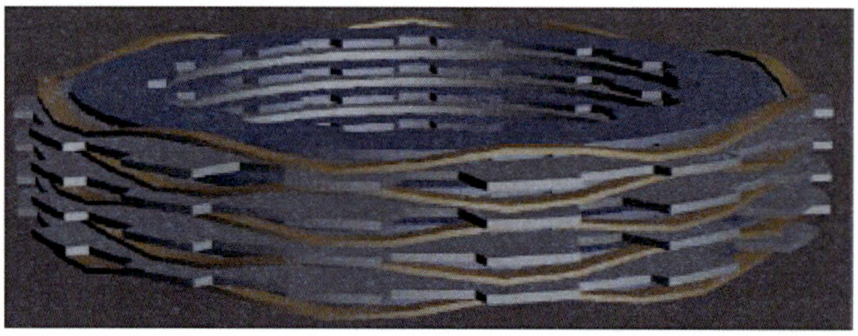

FIGURE 5.22 Wave springs for separating clutch plates.
SOURCE: Martin (2012).

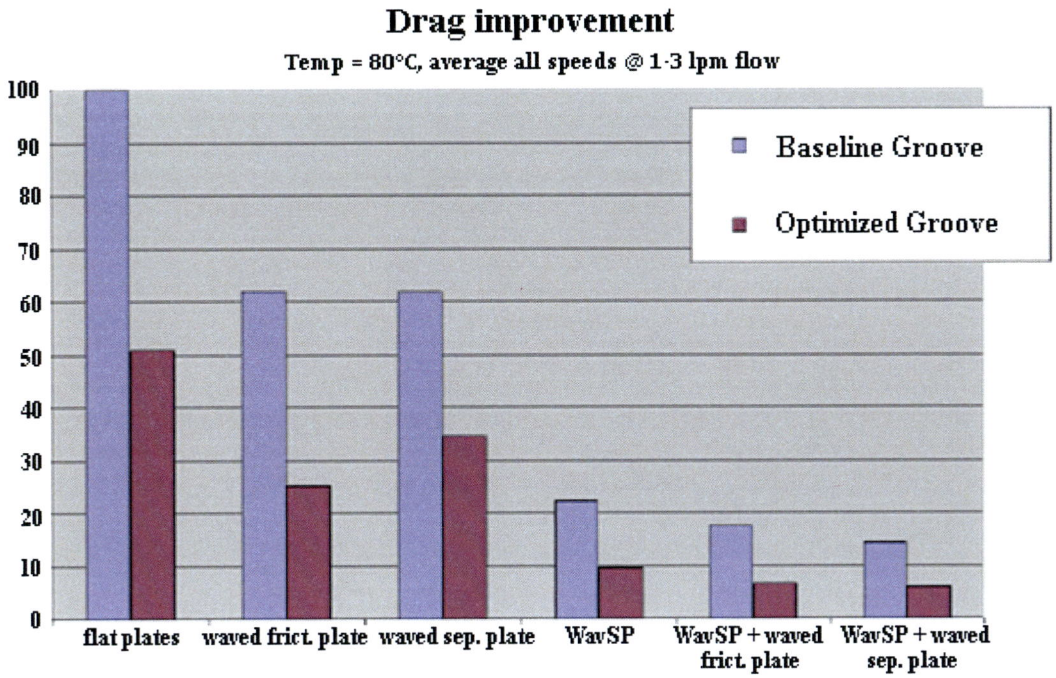

FIGURE 5.23 Improvements in clutch drag.
SOURCE: Martin (2012).

to smooth out some of the vibrations in the driveline that would have been transmitted by a locked, direct coupling. However, this practice is quite inefficient when compared to a damper and should generally be avoided if possible. This loss can be approximately 100 W at cruise speeds (resulting from 20 rpm micro-slip and 50 Nm torque). Usually, micro-slip is only used when passing through a small rpm range known to cause unacceptable NVH. In DCTs using a clutch for launch, a considerable amount of power is lost during the initial vehicle launch through clutch slip that is used to prevent engine stall and ensure smooth acceleration.

Churning

Churning losses in a transmission occur when a component, such as a gear, synchronizer, or clutch pack, rotates through an oil bath. In manual transmissions and DCTs, some amount of churning is required, as those transmissions typically depend on splash lubrication for their internal components. Automatic transmissions and CVTs typically employ forced lubrication, which results in less churning loss but requires a pump to pressurize oil. Oil levels in a transmission that result in rotating components churning the oil and in energy loss need to be kept as low as possible while still providing adequate lubrication.

The viscosity of transmission oil changes significantly with temperature. Transmissions operating with fully warmed-up oil have significantly lower spin losses, as shown in Figure 5.24. It is therefore desirable to warm up the transmission oil as quickly as possible. A rapid warm-up system for transmission oil was estimated to provide a fuel consumption reduction of 0.8 percent at a 2017 direct manufacturing cost of $45-$63.

Bearings and Seals

Bearings and seals used in transmissions also contribute to drag torque losses. Replacing the widely used tapered roller bearings with angular contact ball bearings has been shown to provide a 50 to 75 percent reduction in bearing friction in dual clutch transmissions, manual transmissions, and differentials. However, not all applications will allow for replacing tapered roller bearings due to the duty-cycle and architecture requirements. This is especially true in some highly loaded differentials.

Newly developed low-friction seals for transmission bearings can reduce seal friction by 50 percent to provide an overall reduction in bearing friction loss of approximately 10 percent (NSK 2014). Seal friction was reduced by narrowing the seal lip and modifying its shape. The seal lip shape was improved to stabilize the contact pressure around its edge.

Drag Losses

Low-Friction Lubricants

Similar to the use of low-viscosity engine oil for reduced fuel consumption, low-viscosity automatic transmission fluids (ATFs) can be used for additional reductions in fuel

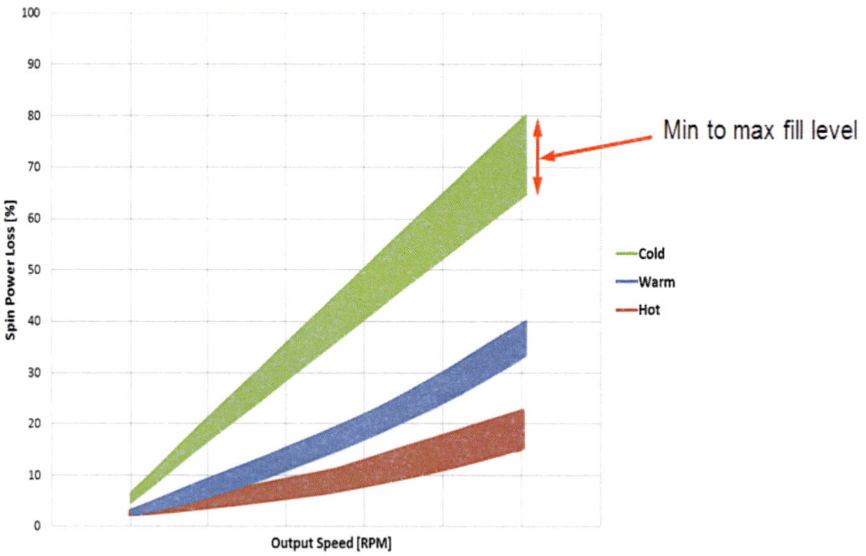

FIGURE 5.24 Spin loss vs. temperature and oil level.
SOURCE: Baillie et al. (2014).

consumption. While low-viscosity oil can result in reduced spin loss, it can also result in increased wear on gears and bearings. In a gear mesh, oil is used to create a thin film between the gear teeth to prevent metal to metal contact. For this purpose, higher viscosity oil is usually preferred, and using lower viscosity oil can necessitate making gears and bearings larger to reduce the unit loads on the gear teeth and bearings for acceptable life with lower viscosity oil, in turn raising the cost. The results presented by Noles (2013), shown in Figure 5.25, indicate that about a 2 percent fuel consumption reduction was obtained on the FTP 75 cycle by switching to the lowest viscosity oil. It is unclear if this transmission could still pass a minimum 150,000 mile durability requirement with this low viscosity ATF. Further investigation of the fuel consumption reductions and associated durability are required. The committee has assumed a conservative 0.5 percent benefit from lower viscosity oil at a 2017 direct manufacturing cost of $50, which is included in the high efficiency gearbox (level 2) estimate. The cost is primarily due to the need to review and update gear and bearing sizes. Validation testing will be required to ensure durability and performance, as well as the need to potentially update calibrations based on different hydraulic circuit performance.

Gearing Efficiency

Gears are one of the most efficient and cost effective means of transferring torque in a transmission. However, when torque is transmitted through gears, some amount of this torque is lost to heat as the gear teeth slide together, squeezing oil between them. These losses can be reduced by improving the surface finish on the gears. Superfinishing the gears and using various coatings on the gear teeth can provide a reduction in fuel consumption but at an increased cost. There is currently limited information available that correlates gear finish to fuel economy. The committee assumes that approximately 0.3 percent is a representative estimate for the reduction in fuel consumption with superfinishing the gear teeth in an automatic transmission.

Parasitic Loss Differences in Transmission Architectures

The differences in parasitic losses among the main transmission architectures are summarized below:

Automatic Transmissions
- Constant pump operation for clutch pressure and cooling,
- Forced lubrication,
- Open clutch drag,
- Torque converter inefficiencies while unlocked (idle and low speed), and
- Micro-slip.

Dual-Clutch Transmissions
- On-demand pump operation (or electromechanical actuation),
- Similar spin loss to a MT (for dry DCTs and next generation wet DCTs),
- Splash or on-demand lubrication,
- Only one open clutch in a wet DCT during certain operating conditions, and
- No torque converter inefficiencies.

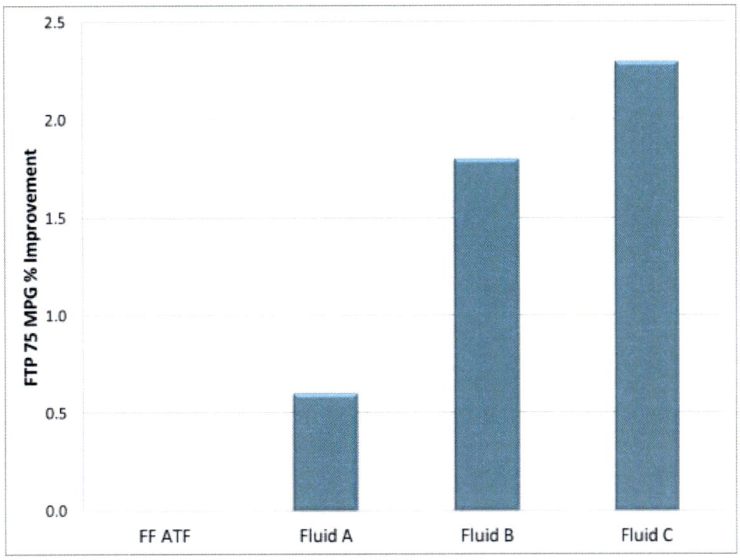

FIGURE 5.25 Fuel economy improvements with low friction lubricants in a 2.0L, four-cylinder SUV equipped with a FWD six-speed automatic transmission.
SOURCE: Noles (2013). Infineum, presented at 2013 SAE Transmission Symposium.

Continuously Variable Transmissions

- Pressure required by the sheaves at high torque,
- Low number of open clutches (usually zero or one),
- Torque converter inefficiencies while unlocked (idle and low speed), and
- Micro-slip.

The relative power losses for conventional automatic transmissions, DCTs, and CVTs are shown as a function of input speed in Figure 5.26 and as a function of input torque in Figure 5.27. Conventional automatic transmissions have parasitic losses from clutches, the torque converter, and constant pump operation. DCTs typically have the lowest losses, as shown in Figure 5.26, due largely to benefits from an on-demand pump, splash lubrication, and fewer open clutches. CVTs are traditionally penalized due to high pressures required for the sheaves to maintain clamp-load on the belt. However, the latest generation of CVTs is approaching conventional automatic transmission loss levels due to advanced controls, improved piston/belt designs, and reduced clamping pressures. CVTs benefit from having only a single clutch that is closed for drive and a brake that is closed for reverse.

Figure 5.27 shows that the DCT has the lowest losses while the CVT and conventional automatic transmission have similar losses at low input torques. However, the losses of the CVT increase significantly as input torque increases due to the high pressures required to maintain the clamping force. These results depend on specific designs, but significant "over clamping" may be used to ensure no belt slip during low-speed transient maneuvers. Such over clamping significantly increases the losses.

The vast majority of DCT losses over a typical fuel economy drive cycle can be attributed to load-independent drag and splash losses, as illustrated in Figure 5.28. For over 90 percent of the NEDC test cycle, the engine operates at less than 50 percent of rated torque. At 50 percent of rated torque, the average CVT losses are 85 percent higher than for a DCT, as shown in Figure 5.27. These losses continue to increase for high torque loads, so that a degradation in customer fuel economy would be anticipated for driving conditions beyond the CAFE test cycles.

DCTs were introduced into the U.S. market in production vehicles with six speeds and wet clutches. Dry clutch DCTs were also introduced in production and are capable of providing an additional 0.5 to 1 percent fuel consumption reduction at a direct manufacturing cost savings of approximately \$50 to \$60 relative to a wet clutch DCT. However, a dry clutch DCT has a limited maximum torque capability so that its use is limited to applications with smaller engines, despite having more efficiency than a wet clutch. The dry clutches do not require oil cooling flow and therefore do not contribute to oil churning losses that are incurred with wet clutches. The difference in losses between wet and dry clutch DCTs is shown in Figure 5.29. However, advances in wet clutch design, such as using on-demand cooling to create a "damp" clutch, have reduced the parasitic losses typically associated with a wet clutch. New clutch plate designs and leakage-free actuation have also contributed to reducing the losses of a wet clutch. The next generation wet clutch DCTs are expected to have

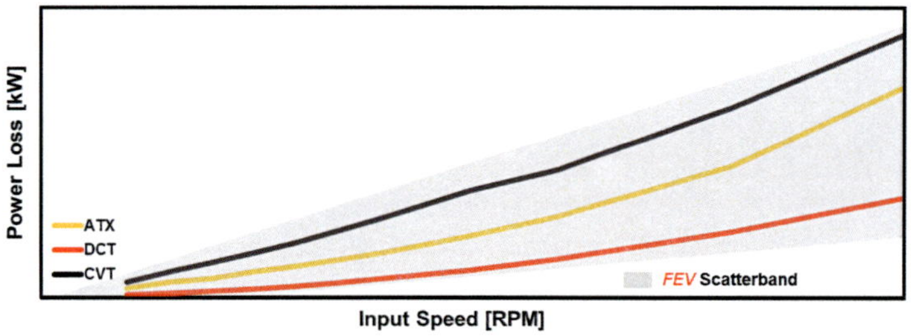

FIGURE 5.26 Transmission power losses as a function of input speed.
SOURCE: Govindswamy, Baillie, and D'Anna (2013).

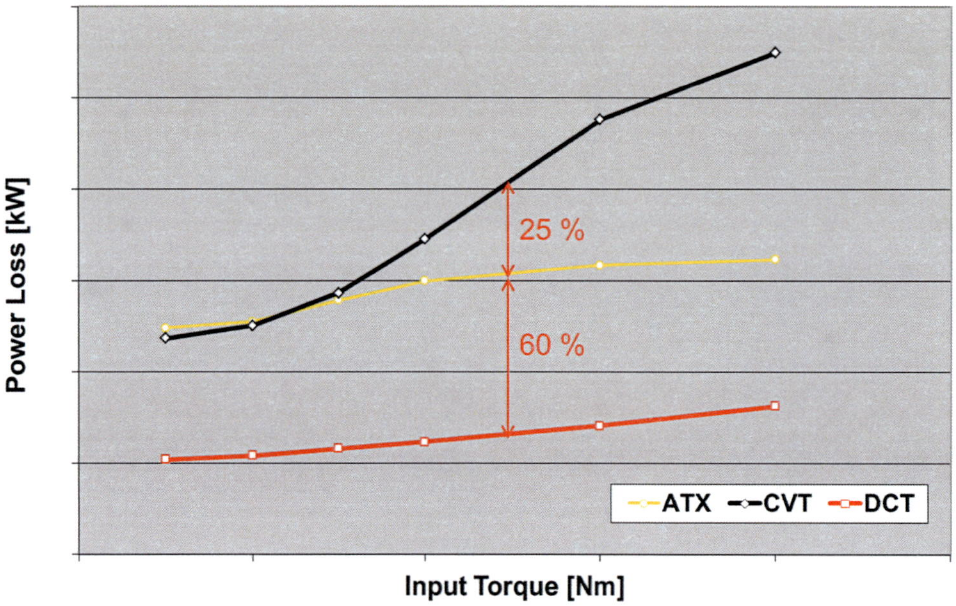

FIGURE 5.27 Transmission power losses as a function of input torque.
SOURCE: Baillie et al. (2014).

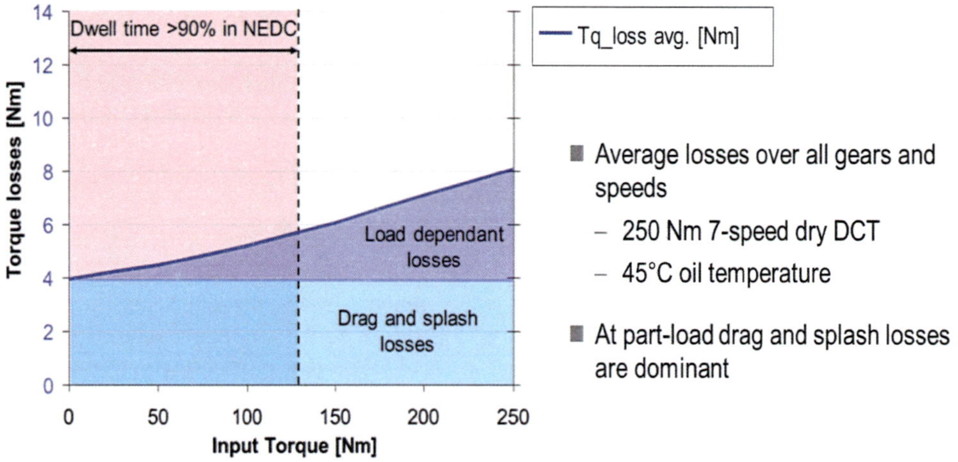

FIGURE 5.28 Torque losses as a function of input torque for a dual clutch transmission.
SOURCE: Baillie et al. (2014).

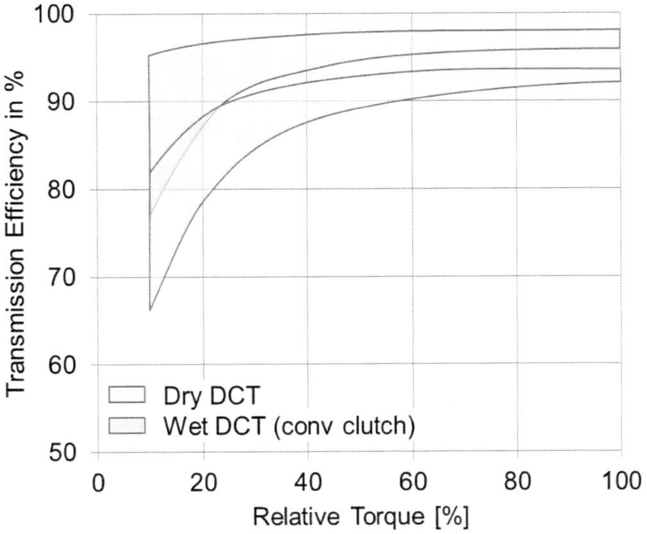

FIGURE 5.29 Comparison of wet and dry DCT efficiencies.
SOURCE: Baillie et al. (2014).

efficiencies approaching the lower end of the range shown for dry clutch DCTs.

FUEL CONSUMPTION REDUCTION TECHNOLOGIES CONSIDERED IN THE FINAL CAFE RULE ANALYSIS

This section discusses the committee's fuel consumption reduction and direct manufacturing cost estimates and compares them to NHTSA's estimates. Fuel consumption reduction effectiveness and direct manufacturing costs are generally presented for a midsize car with an I4 engine for simplicity. However, a complete set of estimates for transmission technologies for a midsize car with an I4 dual-overhead camshaft (DOHC) engine, a large car with a V6 DOHC engine, and a large light truck with a V8 overhead valve (OHV) engine are provided in Table 5A.1 for effectiveness and Table 2A.2a, b, and c (Annex) for 2017, 2020, and 2025 for direct manufacturing costs, respectively.

Learning Factors

In the following discussion of costs, several of the estimated costs shown in the TSD are derived from EPA/FEV teardown cost analyses. These costs are generally valid for the 2012 MY and are shown in 2010 dollars. The TSD shows costs for the 2017 through the 2025 MY, which are generally derived from the estimates for the 2012 MY. The estimates for the later years are derived by applying learning factors to the 2012 MY estimates. NHTSA applied learning curve Type 11 to most of the transmission technologies and Type 12 to continuously variable transmissions; these learning factors are shown in Table 5.3. The committee has continued to apply the same learning factors to the cost estimates in this chapter. Learning factors are discussed in greater detail in Chapter 8.

Improved Automatic Transmission Controls/Externals

Improved automatic transmission controls (IATCs) are defined in the NHTSA RIA (NHTSA 2012) as consisting of Aggressive Shift Logic—Level 1 (ASL1) and Early or Aggressive Torque Converter Lockup. ASL1 reduces fuel consumption by operating the engine at lower speeds and higher loads by shifting into a higher gear when possible. The degree to which the engine can operate at lower speed is limited by factors such as NVH and engine "lugging."

Early torque converter lockup is used to reduce the high parasitic losses that occur in a torque converter when it is open and to provide reductions in fuel consumption. Modern transmissions tend to lock up the torque converter just off idle and keep it locked for most driving conditions above 1,000 to 1,300 rpm. In situations where total lockup is not possible due to NVH concerns, the option of partial lockup, or micro-slip, which results in some clutch slip, can be employed. The torque converter, when open, is a fluid coupling that serves as a torsional vibration damper for torsional vibrations

TABLE 5.3 Learning Factors for Most Transmission Technologies

Learning Factor	2012	2017	2020	2025
Type 11	1.00	0.87	0.81	0.74
Type 12	1.1	0.95	0.89	0.81

Base year with LF - 1.00 is 2015 for Type 12

at low engine speeds. The ability to lock up the converter at low engine speeds has been extended by enhancing the ability of the torsional vibration dampers. Improved dampers and micro-slip control of the lockup clutch have enabled earlier lockup speeds.

NHTSA assumes that IATC will be applied to the baseline transmission before upgrading the transmission, but generally IATC is applied at the same time as an upgrade to a six-speed transmission is applied. The committee's estimates of effectiveness and costs for IATC are shown in Table 5.4.

Shift Optimization

Shift optimization, which is also called Aggressive Shift Logic—Level 2 (ASL2), is defined by NHTSA as a strategy that selects the appropriate transmission ratio to keep the engine operating near its most efficient point for a given power demand. During development of this strategy, Ricardo estimated that fuel economy benefits of up to 5 percent can be obtained when compared to typical MY 2010 shift maps. In this strategy, the transmission controller continuously evaluates all possible gear options that would provide the necessary tractive power (while limiting the adverse effects on driveline NVH) and selects the gear that lets the engine run in the most efficient operating zone. Ricardo acknowledged in its report that the ASL2 (shift optimization) strategy currently adversely affects drivability and, hence, consumer acceptability. The optimum shift strategy for fuel economy can often result in NVH issues or driver discomfort as the transmission experiences shift "busyness" due to frequent changes in gear ratios. NHTSA recognized that deteriorating NVH is a limiting factor in the degree of shift optimization possible and made attempts to prevent the transmission from shifting more often than in a baseline test case. Many vehicle manufacturers provided the same feedback and indicated that they had implemented similar aggressive shift strategies, only to provide updated calibrations to limit the aggressive shift strategy to reduce customer complaints about frequent shifting.

Because of this consistent feedback, the committee recommends that shift optimization is not available to provide NHTSA's estimated 3.9 to 4.1 percent reduction in fuel consumption. This is particularly significant since shift optimization was shown by NHTSA to be nearly a no-cost technology, which gave it the best cost effectiveness (dollars per percent fuel consumption reduction) of any of the technologies defined by NHTSA for the 2017 to 2025 CAFE rulemaking. New shift optimization algorithms continue to be applied to new transmission and vehicle combinations, but this is considered an integral part of the calibration process for any new transmission/vehicle and not a separate technology. In order to recognize that there are always continuous improvements being made in the areas of controls and NVH, the committee has applied a 0.5 to1.0 percent fuel consumption reduction to shift optimization (ASL2). A 2025 direct manufacturing cost of $22 equal to the cost for ASL1 was estimated for shift optimization and is primarily the result of NVH-related hardware necessary to allow the more aggressive shift strategy.

Six-Speed Automatic Transmissions

Effectiveness

Argonne National Laboratories used its Autonomie vehicle simulation tool to provide NHTSA with fuel consumption reduction results for several transmission technologies in support of the final CAFE rulemaking (Moawad and Rousseau 2012). ANL's results from simulating a vehicle with six- and eight-speed automatic transmissions are shown in Table 5.5. Two different levels of efficiency were evaluated for the eight-speed transmission. These results were used by the committee for evaluating NHTSA's fuel consumption reduction effectiveness estimates.

NHTSA estimated in the TSD (EPA/NHTSA 2012) that a six-speed automatic transmission can provide a 2.0 percent reduction in fuel consumption relative to a four-speed automatic transmission. The Autonomie vehicle simulation results shown in Table 5.5 indicate a 0.77 percent reduction for a six-speed relative to a five-speed transmission. Extrapolating the Autonomie results to a six-speed from a four-speed transmission is estimated to provide more than

TABLE 5.4 Estimated Fuel Consumption Reductions and 2025 MY Direct Manufacturing Costs (2010 dollars) for Improved Automatic Transmission Controls/Externals

Improved Automatic Transmission Controls/Externals (IATC)	NRC Estimated Most Likely Fuel Consumption Reduction (%)[a]	NHTSA Estimated Fuel Consumption Reduction (%)[a]	NRC Estimated Most Likely MY2025 DMC (2010$)[a]	NHTSA Estimated MY 2025 DMC (2010$)[a]
Aggressive Shift Logic - Level 1 (ASL1)			22	22
Early Torque Converter Lockup			20	20
Overall IATC	2.5 - 3.0	3.00	42	42

[a]Relative to baseline.

TABLE 5.5 Autonomie Vehicle Simulation Fuel Consumption Results and Percentage Improvements for Automatic Transmissions with a 2.2L Naturally Aspirated Engine

Fuel Consumption (L/100 km)			
Conventional - Automatic Transmission	FTP	HFET	Combined
5-speed - 92% efficiency	7.64	5.50	6.50
6-speed - 92% efficiency	7.62	5.44	6.45
Improvement (%)	0.38	1.11	0.77
6-speed - 92% efficiency	7.62	5.44	6.45
8-speed - 92% efficiency	7.53	5.29	6.33
Improvement (%)	1.08	2.61	1.90
8-speed - 92% efficiency	7.53	5.29	6.33
8-speed - 96% efficiency	7.31	5.13	6.14
Improvement (%)	2.92	3.05	2.99

SOURCE: Moawad and Rousseau (2012).

twice the fuel consumption reduction shown in the table, or approximately 2 percent reduction. Based on these results and other feedback received from vehicle manufacturers and suppliers, the committee's low most likely estimate of 2 percent agrees with NHTSA's estimate. A high most likely estimate of 2.5 percent was based on feedback received by the committee.

Cost

The TSD for the 2017-2025 MY CAFE standards indicates that a six-speed automatic transmission has a 2017 direct manufacturing cost of –\$13 (savings) relative to a four-speed automatic transmission. In contrast to this estimate, the TSD for the earlier 2012-2016 MY CAFE standards (EPA/NHTSA 2010) developed a cost of \$101 for a six-speed automatic transmission relative to a four-speed automatic transmission. The derivation of the \$101 direct manufacturing cost is summarized in Table 5.6, based on the following steps:

- The FEV teardown cost analysis determined that the six-speed transmission was \$106 less costly than the five-speed transmission (EPA/FEV 2010). The TSD indicated that this "counterintuitive" result was attributed to the six-speed transmission having a Lepelletier-type gear set instead of a conventional planetary gear set, which requires an additional one-way clutch. The issues with the FEV cost teardown study are described in Box 5.1.
- To relate the six-speed transmission cost to a four-speed transmission, the TSD applied a \$91 cost estimate from the NPRM for a five-speed transmission relative to a four-speed transmission.
- The TSD proceeded to average the non-Lepelletier gear set cost with the Lepelletier gear set cost (the FEV cost teardown estimate of –\$106). NHTSA estimated the cost of a six-speed transmission without the non-Lepelletier gear set at \$215 (2007 dollars) relative to four-speed transmission. Subtracting the \$91 (four-speed to five-speed transmission) gave a cost of \$124 for the non-Lepelletier cost of a six-speed transmission relative to a five-speed transmission.
- Averaging the \$124 for the non-Lepelletier gear set transmission with the FEV cost of –\$106 for the Lepelletier gear set transmission yielded an average cost of \$9 for a six-speed transmission relative to a five-speed transmission.
- The \$9 cost (five-speed to six-speed) was added to the \$91 cost (four-speed to five-speed) to provide a cost of \$101 (2007 dollars) for a six-speed transmission relative to a four-speed transmission.

Subsequent to the 2012-2016 MY TSD, the EPA/NHTSA 2017-2025 MY Technical Support Document estimated a direct manufacturing cost of –\$13 (savings) for a six-speed automatic transmission relative to a four-speed automatic

TABLE 5.6 Derivation of Direct Manufacturing Costs for Automatic Transmission (2007 dollars)

Technology	Non-Lepelletier Cost Used in 2011 CAFE Analysis	NPRM	FEV Teardown (Lepelletier-like)	Final Rule	Comments
5-Speed Relative to 4-Speed Automatic		\$91	N/A	\$91	Final Rules uses NPRM Value
6-Speed Relative to 5-Speed Automatic			–\$106	\$9	\$215 – \$91 = \$124 [\$124 + (–\$106)]/2 =\$9
6-Speed Relative to 4-Speed Automatic	\$215	\$153	N/A	\$101	\$91 + \$9 = \$101 (values are rounded)

NOTE: Blank cells represent values not considered in this analysis; N/A means that FEV did not conduct a teardown study of the technology. Refer to text for more detail on the Comments column.
SOURCE: EPA/NHTSA (2010).

BOX 5.1
FEV Cost Teardown Study Issues:
Six-Speed versus Five-Speed Automatic Transmission

The FEV cost teardown study (EPA/FEV 2010) had the following issues: (1) the baseline was not a four-speed automatic transmission but was a five-speed transmission, and (2) the six-speed transmission had a Ravigneaux gear set, whereas the five-speed transmission has three planetary gear sets, which is more complex than the Ravigneaux gear set. The three planetary gear sets require nine control elements (four disc clutches, three disc brakes, and two sprag clutches), whereas the six-speed transmission with the Ravigneaux gear set only requires six control elements (two disc clutches, three disc brakes and one sprag clutch). FEV recognized these issues by providing the following statement in their study:

> In regard to the 5-speed automatic transmission, many of innovative ideas implemented into the 6-speed automatic could have been incorporated into a new 5-speed if it were to be redesigned. The most obvious NTA (New Technology Advances) would be adopting a similar Ravigneaux geartrain design, which could conceivably have the same financial benefit recognized by the 6-speed automatic.

This was an unusual teardown cost analysis since the five-speed transmission contained more hardware (i.e., approximately 150 more parts), and was generally more complex, than the six-speed transmission. As a result, the six-speed transmission established a zero-cost baseline from which an incremental cost for the five-speed was established. The majority of the incremental cost increase of the five-speed over the six-speed was associated with the two additional clutch packs, the need for a counter shaft assembly, and some additional gearing.

transmission, which appears to have resulted from using only the case with the Lepelletier gear set. Applying the approach used by NHTSA in Table 5.6, which considers both Lepelletier and non-Lepelletier gear sets rather than the Lepelletier gear set alone, the committee developed incremental direct manufacturing costs for both gear sets, as shown in Table 5.7. The costs derived in Table 5.7 were adjusted to 2010 dollars from 2007 dollars by the GDP factor of 1.04. Then the appropriate learning factors were applied to derive the 2017, 2020, and 2025 direct manufacturing costs. The committee's estimated effectiveness and costs for six-speed automatic transmissions relative to four-speed automatic transmissions are summarized in Table 5.8. Accounting for the two different gear sets separately is preferable, although this information is not readily available to EPA in the manufacturers' certification applications. Since most light-duty vehicles will have transmissions with at least six speeds by the 2016 MY, this is not a critical cost issue for the 2017 to 2025 MY time frame but does affect total costs from the null or 2008 MY baseline vehicle.

Previous estimates for transmissions with increased ratios are shown in Table 5.9. These previous estimates are higher than the current estimate, but the committee believes that this reflects maturity, especially for the six-speed automatic transmission. Many current six-speed automatic transmissions have been on the market long enough to have undergone refinements directed toward improving efficiency and reducing costs. For example, the GM 6T40 FWD transmission is currently in its third generation.

Eight-Speed Automatic Transmissions

Effectiveness

NHTSA estimated that an eight-speed automatic transmission can provide a 4.6 percent reduction in fuel consumption relative to a six-speed automatic transmission, which is significantly larger than the 1.9 percent estimated by the Autonomie vehicle simulation results shown in Table 5.5. Vehicle manufacturers indicated that the fuel consumption reductions shown by NHTSA in the TSD and RIA were overstated. The Autonomie simulations appear to confirm this finding. Based on these findings, together with consistent input received from vehicle manufacturers and suppliers, the committee's most likely effectiveness estimate ranges from 1.5 to 2.0 percent. NHTSA's estimate of 4.6 percent is assumed to include not only the benefits of upgrading from a six-speed transmission to an eight-speed transmission, but also the benefits of the efficiency improvements that are discussed in the following section.

Cost

EPA contracted FEV to conduct a cost teardown study of an eight-speed automatic transmission relative to a six-speed automatic transmission and concluded that the eight-speed transmission would have an incremental cost of $61.84 (EPA/FEV 2011). This analysis compared the ZF eight-speed 8HP70 RWD transmission with the ZF six-speed 6HP26 RWD transmission. The six-speed transmission incorporates a Lepelletier automatic transmission gearing

TABLE 5.7 Derivation of Direct Manufacturing Costs for Six-Speed Automatic Transmissions from Four-Speed Automatic Transmissions

Technology	Non-Lepelletier Gearset Cost Used in 2011 CAFE Analysis	NPRM	Lepelletier-Like Gearset Cost from FEV Teardown Analysis	Comments
5-Speed Automatic from 4-Speed Automatic[a]		\$91	\$91	Apply NPRM Cost
6-Speed Automatic from 5-Speed Automatic[a]			–\$106	FEV Teardown Cost
6-Speed Automatic from 4-Speed Automatic[a]	\$215		–\$15	
Adjusted to 2010\$	\$224		–\$16	1.04 GDP Factor
Adjusted for Learning				
2017	\$195		–\$13	Learning Factor = 0.87
2020	\$181		–\$12	Learning Factor = 0.81
2025	\$165		–\$11	Learning Factor = 0.74

[a]2007 dollars.

TABLE 5.8 Estimated Fuel Consumption Reductions and 2025 MY Direct Manufacturing Costs (2010 dollars) for Six- and Eight-Speed Automatic Transmissions

6- and 8-Speed Automatic Transmissions	NRC Estimated Most Likely Fuel Consumption Reduction (%)[a]	NHTSA Estimated Fuel Consumption Reduction (%)[a]	NRC Estimated Most Likely MY2025 DMC (\$)[a]	NHTSA Estimated MY2025 DMC (\$)[a]
6-Speed Automatic from 4-Speed Automatic (Lepelletier Type)	2.0 - 2.5	2.0	–11	–11
6-Speed Automatic from 4-Speed Automatic (Non-Lepelletier Type)	2.0 - 2.5	N/A	165	N/A
8-Speed Automatic from 6-Speed Automatic (Lepelletier Type)	1.5 - 2.0	4.6[b]	47 - 115	47

[a]Relative to baseline, except when noted.
[b]NHTSA estimate is consistent with NRC estimate when HEG1 effectiveness of 2.7% is included.

TABLE 5.9 Other Available Direct Manufacturing Cost Estimates for Transmission Technologies Relative to 2007 Four-Speed Automatic Transmissions (dollars)

Transmission Type	EEA 2007(\$)	Martec 2008(\$)	EPA/FEV Teardowns 2010 and 2011(\$)	OEM Input 2013-2014(\$)
5-Speed AT	133			
6-Speed AT	133	215	–105.53[a]	40 to 530
7-Speed AT	170			
8-Speed AT		425	61.84[a]	50 to 150[b]

[a]The 6-speed FEV teardown is relative to a 5-speed, and the 8-speed is relative to a 6-speed.
[b]The 8-speed OEM costs are relative to a 6-speed.

configuration, which uses a single planetary gear set along with a Ravigneaux gear set. The eight-speed transmission implemented a new gearing system, consisting of four planetary gear sets controlled by an equivalent number of shift elements (consisting of three disc clutches and two brakes), as compared to the ZF six-speed transmission. FEV provided the following caveat regarding this cost analysis, indicating that the analysis was only for the addition of gears and that other efficiency technologies were not included in the analysis:

> Note that when the 8-speed transmission was redesigned, several other functional and performance updates not driven

by the added gear ratios were incorporated (e.g., modified hydraulic control strategy, spool valve material, friction discs, as well as a newly-developed torque converter). These modifications were not estimated in the analysis since they are independent of the gear ratio addition and modifications.

The committee reviewed the FEV cost teardown study of the eight-speed automatic transmission, especially in light of FEV's caveat, and found that numerous enhancements had probably been included to ensure suitable functionality of a transmission with increased number of ratios and increased frequency of shift events. These enhancements were not related to reduced losses for efficiency improvements, which are discussed separately in the next section. Such differences can also result when a specific teardown study that involves one example of the new technology and one example of the outgoing technology may not be representative of the new technology when it is implemented across the entire fleet of new vehicles (see Box 5.2). In addition to the EPA/FEV teardown cost estimate of $61.84, the committee's analysis of possible incremental costs for an eight-speed transmission is shown in Table 5.10. The analysis shows that the costs could range from $61.84 up to $156 (2010 dollars). Adjusting these costs for learning yielded 2025 MY direct manufacturing costs ranging from $47 to $115, as shown in Table 5.8.

The cost of the components for the efficiency improvements were not included in the cost to upgrade from a six-speed transmission to an eight-speed transmission, which the committee agrees is the correct method of cost analysis, since the cost of the efficiency improvements can be applied to most transmissions as a separate set of technologies. However, those improvements contribute to a significant portion of the ZF eight-speed transmission's improved efficiency. If typical transmissions are used as references for costs and benefits, such as the upgrade of a six-speed automatic transmission to a production eight-speed automatic transmission like the ZF 8H, then the costs and benefits of the eight-speed transmission upgrade need to be combined with those of the high-efficiency gearbox—level 1 (HEG1), discussed later in this section, to be representative of all of the technologies in the eight-speed transmission. Therefore, the costs and benefits of the eight-speed transmission, as defined above and shown in Table 5.8, must be combined with the technologies included in the HEG1 for 2017 MY and beyond.

BOX 5.2
Teardown Cost Study Issues: Eight-Speed Automatic Transmission and Dual-Clutch Transmission

Teardown cost studies have been used by the Agencies to significantly improve cost estimates of new technologies that may be applied to meet future CAFE targets. The teardown process involves selecting an example of the new technology that has been implemented in production and an example of the outgoing technology. The selection of these two examples is critically important in arriving at a representative estimate of the incremental cost of the new technology. However, there is a risk that this process may not provide estimates that are fully representative of the technology when it is implemented across the entire fleet of new vehicles. These concerns are particularly relevant to transmissions, which can often have significant variations in architectures, as discussed below.

Additionally, Tier 2 suppliers' profits may not be accurately reflected in the cost estimates. Patents and royalties may also influence estimated costs. Therefore, in addition to using the FEV teardown cost study, the committee also relied on its expertise, together with input from manufacturers and suppliers, to extend the range of cost estimates.

Eight-Speed Automatic Transmission versus Six-Speed Automatic Transmission

The FEV teardown cost study compared the ZF eight-speed 8HP70 automatic transmission with the ZF six-speed 6HP26 automatic transmission. The study noted that the eight-speed transmission implemented "a revolutionary gearing system." It further noted that "many of the innovative ideas implemented into the eight-speed automatic could have been incorporated into a new six-speed if it were to be redesigned." These differences will strongly influence the outcome of the teardown cost study.

Dual-Clutch Automatic Transmission versus Automatic Transmission

The foregoing reasons for variability of cost estimates apply equally to the DCT transmission. DCTs used by different manufacturers may also have different mechatronics for clutch and shift fork actuation. The actuation units can be electromechanical, electrohydraulic, or a mixture of both. The clutch modules also vary significantly among manufacturers, with some including a torsional damper and others relying on a damper in a separate dual mass flywheel. Therefore, one teardown cost estimate may not be representative of the entire fleet.

TABLE 5.10 Direct Manufacturing Cost Estimate for an Eight-Speed Automatic Transmission Relative to a Six-Speed Automatic Transmission

8-Speed versus 6-Speed Automatic Transmission
Possible Incremental Cost Compared to EPA/FEV Teardown Cost Estimate of $61.84 for 2012 MY (2010 dollars)

Technology	DMC ($)	Source of Costs
Modified hydraulic control strategy, spool valve, material, friction discs (Defined in EPA/FEV Teardown Study – Components that may have been included in 8-speed AT, but not related to reduced losses for efficiency improvements).	28	EPA/FEV estimate of $138.19 for cost of clutches in 6 sp AT[a] 20% of 6 sp AT cost of clutches estimated for modifications
Solenoids with enhanced response times and flow control (For improved shift quality with more frequen shifts of 8 speed AT)	23	EPA/FEV estimate of $45.99 for electrical controls of DCT[b] 1/2 of electrical controls estimated for enhanced solenoids
Enhanced speed and pressure sensors (For improved shift quality with more frequent shifts of 8 sp AT)	11	EPA/FEV estimate of $45.99 for electrical controls of DCT[b] 1/4 of electrical controls estimated for enhanced sensors
Case modifications (magnesium replacing aluminum) (to prevent weight increase)	32	DOE: $4 per pound with magnesium replacing aluminum[c] Applied to 8 pounds saved.
Total incremental costs	94	
EPA/FEV estimated cost – 8-sp AT versus 6-sp AT[d]	62	
Totat direct manufacturing cost: 2012 MY cost in 2010 dollars	156	
DMC Learning Type 11, 2012 to 2025 Learning Factor = 0.74		
Total 2025 MY Direct Manufacturing Costs (2010 dollars)	115	

[a] EPA/FEV (2010).
[b] EPA/FEV (2013).
[c] Powell et al. (2010).
[d] EPA/FEV (2011).

High-Efficiency Gearbox—Level 1 (HEG1), Level 2 (HEG2), and Level 3 (HEG3)

Effectiveness and Cost

Losses in automatic transmissions vary significantly, but the Ricardo (2011) analysis showed that a 2010 MY automatic transmission was approximately 90 percent efficient, so that the energy losses through the transmission were approximately 10 percent of the input energy. Since reducing these losses will be reflected directly in reduced fuel consumption, significant attention is being devoted to doing just that. Moreover, since technologies to reduce losses and improve transmission efficiency are being phased in primarily as new transmissions are being designed, two different levels of high-efficiency gearboxes were established to illustrate how these technologies might be phased in over time. In addition, a third level of a high efficiency gearbox was established as a technology that might be applicable beyond the 2025 MY. The reduction in losses, together with the accompanying reduction in fuel consumption and estimated 2017 costs (2010 dollars), are shown in Table 5.11. Appropriate learning factors were applied to the 2017 costs to derive the 2020 and 2025 costs. All of the technologies listed were discussed earlier in this chapter. The combination of HEG1 and HEG2 is shown to provide a slightly larger reduction in fuel consumption than NHTSA's estimate for HETRANS, but at a higher cost. The combination of the TSD estimates together with the committee's estimates resulted in the low and high most likely estimates for effectiveness and costs shown in Table 5.12.

Six- and Eight-Speed Dual-Clutch Transmissions

Effectiveness

The TSD indicates that a six-speed wet-clutch DCT transmission can provide a 3.4 to 3.8 percent reduction in fuel consumption relative to a six-speed conventional automatic transmission, based on the lumped parameter model. The NHTSA RIA indicates that a six-speed wet-clutch DCT can provide up to a 4.1 percent reduction in fuel consumption relative to a six-speed automatic transmission, and this value is used in NHTSA's decision trees. The range of 3.0 to 4.0 percent reduction in fuel consumption was considered to be the appropriate range for the six-speed wet-clutch DCT relative to a six-speed automatic transmission. The effectiveness of the dry DCT was estimated to provide 0.5 percent further reduction in fuel consumption relative to a wet-clutch DCT due to the reduced parasitic losses incurred by not needing to provide oil flow to the clutches.

The NHTSA RIA implies that the eight-speed DCT transmission provides the same 4.6 percent reduction in fuel consumption relative to a six-speed DCT as an eight-speed

TABLE 5.11 Reduction in Losses, Reduction in Fuel Consumption, and Estimated Costs for High-Efficiency Gearbox—Level 1 (HEG1), Level 2 (HEG2), and Level 3 (HEG3)

	Estimated Reduction in Losses (%)	Estimated Reduction in FC (%)	2017 Incremental Direct Manufacturing Costs ($)
High Efficiency Gearbox - Level 1 (HEG1)			
Oil Supply			
Low leakage valves and improved variable force solenoids		0.3	35
Off-axis, chain driven, dual mode/variable displacement pump		1.0	45
Drag Losses		1.0	40
Grooved friction material			
Wave spring separation of plates			
More efficient seals and bearings			
Total	17	2.3	120
High Efficiency Gearbox - Level 2 (HEG2)			
Oil Supply			
Dual pumps		1.0	45
Drag Losses			
Rapid warm-up of oil		0.8	54
Low friction synthetic AT fluid (improved cold temp. viscosity)		0.5	50
Gearing Efficiency			
Chemically superfinished gear teeth		0.3	45
Total	20	2.6	194
Overall (to 2025 MY)	37	4.9	314
High-Efficiency Gearbox—Level 3 (HEG3) (after 2025)			
Oil Supply			
On-demand electric oil pump and electromechanical shifting		1.6	150
Total	12	1.6	150
Overall (to 2025 MY)	49	6.5	

TABLE 5.12 Estimated Fuel Consumption Reductions and 2025 MY Direct Manufacturing Costs for High-Efficiency Gearboxes (2010 dollars)

High-Efficiency Gearbox (HEG)	NRC Estimated Most Likely Fuel Consumption Reduction (%)[a]	NHTSA Estimated Fuel Consumption Reduction (%)[a]	NRC Estimated Most Likely 2025 MY DMC Costs (2010$)[a]	NHTSA Estimated 2025 MY DMC Costs (2010$)[a]
HEG1	2.3 - 2.7	2.7[b]	102	N/A
HEG2 (HETRANS)	2.6 - 2.7	2.7	165	163
HEG1 + HEG2	4.9 - 5.4		267	
HEG3	1.6	N/A	128	N/A

[a] Relative to baseline.
[b] Derivation: Subtract revised 1.9% effectiveness of 8AT from NHTSA's 8AT effectiveness of 4.6%.

automatic transmission provides relative to a six-speed automatic transmission. The Autonomie vehicle simulation results shown in Table 5.13 indicate that the eight-speed DCT provides a 2.1 percent reduction in fuel consumption relative to a six-speed DCT with the same efficiency. A comparison at the same efficiency is appropriate, since the significant improvement in efficiency occurs when making the transition from a conventional automatic transmission to DCT. The committee's low and high most likely effectiveness estimates are shown in Table 5.14.

TABLE 5.13 Autonomie Vehicle Simulation Fuel Consumption Results and Percentage Improvements for Dual-Clutch Transmissions (DCTs) with a 2.2L Naturally Aspirated Engine

Fuel Consumption (L/100 km)			
Conventional - DCT	FTP	HWFET	Combined
6-speed - 92% efficiency	7.35	5.22	6.21
8-speed - 92% efficiency	7.23	5.09	6.08
Improvement (%)	1.63	2.51	2.1
8-speed - 92% efficiency	7.23	5.09	6.08
8-speed - 96% efficiency	7.05	4.94	5.91
Improvement (%)	2.49	3.04	2.78

TABLE 5.14 Estimated Fuel Consumption Reductions and 2025 MY Direct Manufacturing Costs for Six- and Eight-Speed Dual-Clutch Transmissions (DCTs)

Dual Clutch Transmissions (DCT)	NRC Estimated Most Likely Fuel Consumption Reduction (%)[a]	NHTSA Estimated Fuel Consumption Reduction (%)[a]	NRC Estimated Most Likely 2025 MY DMC Costs (2010 dollars)[a]	NHTSA Estimated 2025 MY DMC Costs (2010 dollars)[a]
6-sp DCT vs. 6-sp AT (dry clutch)	3.5 - 4.5	4.1	–127 to 26	–127[b]
6-sp DCT vs. 6-sp AT (wet clutch)	3.0 - 4.0	3.6	–75 to 75	–75[c]
8-sp DCT vs. 6-sp DCT (dry or wet)	1.5 - 2.0	4.6[d]	152	152

[a] Relative to baseline. Baseline 6-sp AT is Lepelletier type.
[b] NHTSA estimated cost of dry DCT is $61 less than wet DCT in 2017. Adjusting this difference by the learning factor (0.74/0.87) yields –$52. Adding -$52 to wet DCT cost of –$75 yields –$127.
[c] FEV teardown cost for 6-sp DCT was –$147. Adding correction of $50 yields –$97. Multiplying –$97 by 1.04 GDP factor and 0.74 learning factor yields –$75.
[d] Beneficial effects of HEG1, which appear to be included in the 4.6%, are not applicable to an 8-sp DCT from a 6-sp DCT.

Cost

The 2017-2025 TSD shows that a six-speed wet dual clutch transmission has an estimated 2007 direct manufacturing cost of –$147 (savings) relative to a six-speed conventional automatic transmission, based on the FEV teardown cost study (EPA/FEV 2011). Subsequent to this study, EPA and FEV issued a correction and adjusted the cost to –$97 (savings) (EPA/FEV 2013). The additional $50 cost was due to gear selecting solenoids and sensors, wiring harness, and communication drivers.

For the reasons cited in Box 5.2, an analysis of the EPA/FEV teardown cost study revealed that higher costs than those derived in the teardown study were possible. The committee's analysis of possible incremental costs for a six-speed DCT transmission over the EPA/FEV teardown cost estimate of –$97 is shown in Table 5.15. The analysis shows that the costs could range from –$97 up to $101 (2007 dollars). Adjusting these costs to 2010 dollatrs (with the GDP factor of 1.04) and for learning yielded 2025 MY direct manufacturing costs for a six-speed wet clutch DCT ranging from –$75 to $75 (2010 dollars), as shown in Table 5.14.

Cost estimates for a six-speed dry DCT relative to a six-speed automatic transmission are also shown in Table 5.14. The 2007 cost difference of –$61 (savings) for the dry clutch DCT relative to the wet clutch DCT estimated by NHTSA was used to develop the committee's cost estimates shown in Table 5.14.

The FEV teardown study of an eight-speed DCT from a six-speed DCT provided the cost shown in Table 5.16, which was used as the basis for the committee's direct manufacturing cost estimate for the eight-speed DCT (EPA/FEV 2011). This cost was adjusted to 2010 dollars from 2007 dollars by applying the GDP multiplier. The appropriate learning factors were applied to derive the 2017, 2020, and 2025 costs, with the 2025 MY costs summarized in Table 5.14. These cost estimates are within the range of estimates contained in feedback the committee received from vehicle manufacturers and suppliers.

The committee found that the currently high costs of DCTs stem from the relatively low sales volumes, compounded by the fact that DCTs used by different vehicle manufacturers have different mechatronics for clutch and shift fork actuation. The actuation units can be electrome-

TABLE 5.15 Direct Manufacturing Cost (DMC) Estimate for a Six-Speed Wet Dual-Clutch Transmission (DCT) Relative to a Six-Speed Conventional Automatic Transmission

6-Speed DCT versus 6-Speed Automatic Transmission Possible Incremental Cost over EPA/FEV Teardown Cost Estimate of –$97 for 2012 MY (2007 dollars)		
Technology	DMC ($)	Source of Costs
Clutch modules - enhancements	39	Expert estimate - 25% increase over EPA/FEV estimate of $155.11 for clutch modules in 6-sp DCT
Synchronizers - cost increase due to complexity	14	Expert estimate - increase over EPA/FEV estimate of $14.35 for synchronizer cost in Geartrain Subsystem
Mechatronics for clutch and shift actuation - enhancements	26	Expert estimate - 25% increase over EPA/FEV estimate of $103.23 for Mechanical Controls Subsystem
Differential gears (added for same function as 6-sp AT)	42	EPA/FEV estimate of $42.01 for cost of differential gears in 6-sp AT added to 6-sp DCT
Torsional vibration damper	45	Expert estimate - Up to 50% of EPA/FEV estimate of $90.74 for cost of torque converter
Case	32	Increased to equal EPA/FEV estimate for 6-sp AT case
Total incremental costs	198	
EPA/FEV estimated cost – 6-sp DT versus 6-sp AT	–97	
Total direct manufacturing cost: 2012 MY cost in 2007$	101	
Adjusted to 2010$	105	
DMC Learning Type 11, 2012 to 2025 Learning Factor = 0.74		
Total: 2025 MY Direct Manufacturing Cost (2010$)	75	

TABLE 5.16 Derivation of Eight-Speed Dual-Clutch Transmission (DCT) Costs from Six-Speed DCT Direct Manufacturing Costs

Technology	Direct Manufacturing Cost ($)	Comments
8-sp DCT from 6-sp DCT[a]	198	FEV teardown cost study
Adjusted to 2010 dollars	206	1.04 GDP Factor
Adjusted for Learning		
2017	179	LF = 0.87
2020	167	LF = 0.81
2025	152	LF = 0.74

[a] 2007 dollars.

chanical, electrohydraulic, or a mixture of both. The clutch modules vary significantly. Although the main difference is between wet and dry clutch configurations, other differences include the use of torsional dampers, while others rely on a damper in the separate dual mass flywheel. Since the hardware components from one DCT to another can vary significantly, a large variation in costs can be expected. This large variation in hardware components is partly responsible for DCTs not achieving significant cost reductions at current production volumes.

A seven-speed DCT would have significantly better cost effectiveness than an eight-speed DCT. In order to upgrade from a six-speed to an eight-speed DCT, in addition to the additional gear pairs, the eight-speed DCT requires an additional synchronizer, shift rail and fork, actuator, and position sensor. However, to upgrade to a seven-speed DCT, the unused side of one of the synchronizers in the six-speed DCT can be used, thereby eliminating the cost of a new synchronizer, shift rail and fork, actuator, and position sensors. The additional costs would consist primarily of the added gear pair, needle bearing, and potentially larger transmission housing. By eliminating a portion of the costs for mechanical controls and reducing the gear system incremental costs by half, a seven-speed DCT may have an incremental cost of approximately $60 over a six-speed DCT.

Dual-Clutch Transmission – High-Efficiency Gearbox

DCTs generally use hydraulic power for actuating the clutches and the transmission actuators, resulting in significant hydraulic losses. These losses can be reduced by replacing the hydraulic systems with electric motors for driving the clutches and the transmission shift fork actuators (Wagner et al. 2006). Although losses in a DCT are already significantly lower than a conventional automatic transmission, the electrically actuated DCT is expected to provide approximately 2 percent additional reduction in fuel consumption. The 2017 incremental direct manufacturing cost was estimated to be approximately \$150, which becomes a 2025 cost of \$127 (2010 dollars). The committee's estimates of effectiveness and costs for the DCT with a high-efficiency gearbox are shown in Table 5.17.

Dual-Clutch Transmission with Torque Converter

The smooth launch performance provided by a torque converter-equipped planetary automatic transmission is very difficult to duplicate with DCTs. Parking lot maneuvers can be especially difficult without the smooth performance of a torque converter. This lack of smooth launch performance has prompted Honda to announce the development of an eight-speed DCT for the 2015 MY, which features the first use of a torque converter in a DCT in addition to retaining the dual clutches. This should provide the smooth low-speed driving dynamics of a traditional automatic transmission but with a gearbox that is more efficient. As an additional benefit, torque multiplication through the torque converter will improve acceleration from idle. Since DCT clutches have required the use of a more costly dual-mass flywheel, adding the torque converter is believed by Honda to be no more expensive overall since it eliminates the need for the dual mass flywheel. The committee's low and high most likely estimates of effectiveness and costs for the DCT with a torque converter are shown in Table 5.17.

Torsional Vibration Damping

The torque generated by an internal combustion engine is not smooth over a single revolution of the engine; rather, it peaks at each cylinder firing. As the number of cylinders is reduced, the variation in torque over a revolution of the engine increases, reaching a worst case for three-cylinder engines. The high peak firing pressures of a diesel engine can further increase the variations in torque over a revolution of the engine. Torsional vibrations can lead to seat vibrations or noise at certain speeds, both of which reduce the comfort of the vehicle. Although an open torque converter diminishes the transfer of torque variations to the driveshaft, this capability is eliminated when the torque converter is locked up, which generally occurs during most operating conditions except for the initial launch of the vehicle from idle.

Torsional vibration damping mechanisms are generally applied to the torque converter and can consist of the following technologies:

- Turbine torsional damper: Applied to gasoline engine applications.
- Two-torsional damper system: Applied to four-cylinder gasoline and four- and six-cylinder diesel engines.
- Centrifugal pendulum absorber: Applied to three- and four-cylinder gasoline and three-, four-, and six-cylinder diesel engines.

Turbine torsional dampers are generally included in automatic transmissions with lockup torque converters. However, the cost of a two-torsional damper system could increase the cost of the torque converter by 20 percent, and the centrifugal pendulum absorber could increase the cost of the torque converter by 50 percent. These incremental costs need to be included in the transmission costs as turbocharged, downsized gasoline engines and diesel engines are applied to future vehicles. The unique costs for torsional vibration damping with dual clutch transmissions also need to be included in cost estimates.

Secondary Axle Disconnect

All-wheel-drive (AWD) vehicles continue to be popular, as shown by EPA's forecast that 31.2 percent of all light-duty vehicles in the United States would have four-wheel-drive in 2014 (EPA 2014). These vehicles incur a fuel consumption penalty due to the losses associated with the additional rotating components. By disconnecting the secondary driven axle and driveshaft, the rotating losses in the bearings and seals can be eliminated. AWD vehicles generally consist of permanently connected front- and rear-drive axles. During normal driving conditions, the rear axle is spinning but contributing no power to the vehicle

TABLE 5.17 Estimated Fuel Consumption Reductions and 2025 MY Direct Manufacturing Costs for Dual-Clutch Transmission Variants (2010 dollars)

Dual-Clutch Transmission Variants (DCT)	NRC Estimated Most Likely Fuel Consumption Reduction (%)[a]	NHTSA Estimated Fuel Consumption Reduction (%)[a]	NRC Estimated Most Likely 2025 MY DMC Costs (2010\$)[a]	NHTSA Estimated 2025 MY DMC Costs (2010\$)[a]
DCT- HEG	2.0	NA	127	NA
DCT - Torque Converter	Same as DCT	NA	Same as DCT	NA

[a] Relative to baseline.

propulsion. Significant energy is lost due to the friction in these drivelines.

The AWD disconnect clutch comprises a hydraulically-operated synchronizer clutch integrated in the input shaft on the power transfer unit (PTU) and an electrically operated dog clutch on the rear axle. An AWD disconnect clutch system allows the secondary axle and driveline branches to be disconnected from the primary axle. When used in connection with a rear axle disconnect, the secondary axle rotation can be stopped to eliminate the parasitic power losses. Disconnect systems consist of high-torque synchronizers that are energized with an electromagnetic actuator. The AWD disconnect clutch system was estimated by Schaeffler to offer up to 5 percent reduction in fuel consumption, which would provide an AWD vehicle with fuel consumption similar to that of a front wheel drive (FWD) vehicle (Lee 2010).

The NHTSA RIA estimates that a secondary axle disconnect system can provide a 1.4 percent reduction in fuel consumption. MY 2014 EPA certification fuel economy test data indicate that AWD vehicles have 3.8 to 7.2 percent higher fuel consumption than comparable non-AWD vehicles. If secondary axle disconnect systems could reduce the losses of the AWD system by 50 percent, the fuel consumption would be reduced by approximately 3 percent. The committee's low and high most likely effectiveness estimates are shown in Table 5.18.

The TSD shows the 2017 direct manufacturing cost of a secondary axle disconnect system to be $78. Although details of this cost estimate were not provided, the description in the TSD suggests that the cost did not include the electrically operated dog clutch on the rear axle. Applying the $22 cost estimate for a one-way clutch from the FEV cost teardown study of a six- speed automatic transmission, the 2017 direct manufacturing cost of the secondary disconnect system is estimated to be $100 ($78 from the TSD plus $22 for the one-way clutch). The committee's most likely direct manufacturing cost estimate is shown in Table 5.18.

FUEL CONSUMPTION REDUCTION TECHNOLOGIES NOT INCLUDED IN THE FINAL CAFE RULE ANALYSIS

Continuously Variable Transmission

NHTSA did not consider continuously variable transmissions (CVTs) in the TSD (2012) for the 2017-2025 rulemaking. However, EPA and NHTSA considered them in the TSD (2010) for the 2012-2016 rulemaking, which has been used together with other input to develop effectiveness and cost estimates. NHTSA estimated that a CVT could reduce fuel consumption by 2.2 to 4.5 percent relative to a four-speed automatic transmission. The committee contracted the University of Michigan to conduct a full system simulation that included replacing a six-speed automatic transmission with a CVT and found that the CVT provided a 1.2 percent reduction in fuel consumption. Combining this with a 2 percent effectiveness for a six-speed automatic transmission over a four-speed transmission indicates that the CVT could provide approximately 3.5 percent reduction in fuel consumption, which is within the range estimated by NHTSA. Some manufacturers' estimates significantly exceeded NHTSA's maximum range. This wide range of estimates is believed to reflect wide variations in losses in the CVT. The committee's low and high most likely estimates of effectiveness for the CVT are shown in Table 5.19.

In the TSD for the 2012-2016 rulemaking, NHTSA also provided an estimated 2012 MY direct manufacturing cost of $200 (2007 dollars) for the CVT relative to a four-speed automatic transmission. This cost was within the ranges provided by manufacturers. By applying the gross domestic product (GDP) multiplier of 1.04 and type 12 learning factor, a 2025 MY direct manufacturing cost estimate of $154 was developed, as shown in Table 5.19.

High-Efficiency Gearbox

As described earlier, CVTs have had higher losses than automatic transmissions. Some estimates have indicated that CVT efficiencies may be approximately 89 percent, whereas an automatic transmission could range from 90 to 96 percent, depending on the extent of the high efficiency gearbox technologies incorporated. Similar to automatic transmissions, CVTs can benefit from a reduction in losses. Major power losses occur with the hydraulic pump and the belt, in approximately equal proportions. Pump losses have already been reduced in CVTs with marginal control of the hydraulic pressure that provides adequate clamping pressure for the pulleys while still having adequate reserve pressure required for rapid downshifts. Further reductions in pumping losses can be achieved with a variable displacement hydraulic pump or a dual cavity pump. Additional reductions can be achieved with an on-demand electric pump, by increasing the coefficient of friction between the pulley and belt, and

TABLE 5.18 Estimated Fuel Consumption Reductions and 2025 MY Direct Manufacturing Costs for Secondary Axle Disconnect System

Secondary Axle Disconnect	NRC Estimated Most Likely Fuel Consumption Reduction (%)[a]	NHTSA Estimated Fuel Consumption Reduction (%)[a]	NRC Estimated Most Likely 2025 MY DMC Costs (2010$)[a]	NHTSA Estimated 2025 MY (2010$s) DMC Costs (2010$)[a]
SAX	1.4 - 3.0	1.4	86	66

[a] Relative to baseline. Baseline 6-sp AT is Lepelletier type.

by increasing the traction characteristics of the oil to reduce belt slip. Low friction coatings with high wear resistance may assist in reducing belt losses. Many of these areas are under development so effectiveness and costs are not well defined. However, these features have been estimated to reduce losses by approximately 20 to 25 percent, which would provide approximately a 3 percent reduction in fuel consumption. The higher efficiency CVT is estimated to include $50 for a variable displacement pump, an additional $35 for an on-demand electrically driven variable displacement pump and $40 for enhanced lubricants and low friction coatings with high wear resistance for an estimated 2017 direct manufacturing costs of $125, which becomes a 2025 cost of $107. The committee's low and high most likely estimates of effectiveness and costs for the CVT with a high-efficiency gearbox are shown in Table 5.19.

Nine- and Ten-Speed Automatic Transmissions

As discussed earlier in this chapter, nine-speed transmissions are currently in production and several announcements have been made regarding future production of nine- and ten-speed automatic transmissions. However, increasing the number of ratios in a transmission will have a diminishing beneficial effect on fuel economy and may increase the losses within the transmission. Based on the increase in the number of ratios from eight to ten alone, the committee estimated that a ten-speed automatic transmission may provide approximately a 0.3 percent reduction in fuel consumption with an estimated incremental direct manufacturing cost in 2025 of $75 (2010 dollars).

ZF Friedrichshafen AG recently announced that it would not follow others who have announced ten-speed automatic transmissions (Greimel 2014). ZF announced that its offerings would be limited to nine speeds. The CEO of ZF, Stefan Sommer, said, "We came to a limit where we couldn't gain any higher ratios. So the increase in fuel efficiency is very limited and almost eaten up by adding some weight and friction and even size of the transmission."

Effect of Engine Technology on Effectiveness of Increasing Transmission Ratios

In this chapter, the estimates for effectiveness of transmissions with increased ratios are relative to a baseline, naturally aspirated engine with four valves per cylinder, fixed valve timing and lift, port fuel injection, and a four-speed automatic transmission, unless otherwise noted. As technologies are added to spark ignition engines, fuel consumption is not only reduced, but the lower fuel consumption levels cover a broader range of engine speeds and loads. As a result, the effectiveness of increasing the number of ratios in a transmission is diminished. For example, Chapter 8 shows that, using NHTSA's methodology for synergies, the 5.0 percent effectiveness of a six-speed automatic transmission relative to a four-speed transmission applied a baseline, naturally aspirated engine is diminished to 1.6 percent when applied to an engine with additional fuel consumption technologies that include intake cam phasing, continuously variable valve lift, and turbocharging and downsizing to level 1. Likewise, the effectiveness of an eight-speed automatic transmission was reduced by 15 percent when added to a similar engine with additional fuel reduction technologies. As described in Chapter 8, similar results were found in full system simulations conducted by the University of Michigan. As engines incorporate more new technologies to reduce fuel consumption, the benefits of increasing transmission ratios or switching to a CVT diminishes.

TRANSMISSION CONTROLS

Control systems, models, and simulation techniques are enablers for many of the transmission technologies to reduce fuel consumption discussed in this chapter and previously highlighted in the discussion of spark ignition engine controls in Chapter 2. The function of early electronic controls for transmissions was gear shift scheduling as a function of accelerator pedal or throttle angle and vehicle speed (Kiencke and Nielsen 2000). The added flexibility of electronic scheduling over hydraulic controls provided opportunities to optimize the shift scheduling for multiple requirements, including optimized performance and fuel economy. Shift scheduling controls are also influenced by engine and vehicle conditions

TABLE 5.19 Estimated Fuel Consumption Reductions and 2025 MY Direct Manufacturing Costs for Continuously Variable Transmission and High Efficiency Gearbox

Continuously Variable Transmission (CVT)	NRC Estimated Most Likely Fuel Consumption Reduction (%)[a]	NHTSA Estimated Fuel Consumption Reduction (%)[a]	NRC Estimated Most Likely 2025 MY DMC Costs (2010$)[a]	NHTSA Estimated 2025 MY DMC Costs (2010$)[a]
CVT from 4-sp AT	3.5 - 4.5	2.2 - 4.5[b]	154	154[b]
HEG - CVT	3.0	NA	107	NA

[a] Relative to baseline.
[b] From 2012-2016 TSD.

as well as engine limitations, as shown in Figure 5.30, so that communication links are required between transmission and engine controllers unless the transmission control is integrated within the engine controller, as is the case for some manufacturers (Bai et al. 2013; Guzzella and Sciarretta 2007). The next advancement provided control of some hydraulic pressures to provide smoother engagements of clutches for improved customer satisfaction.

As the number of gears or ratios increased in transmissions, the complexity of the electronic controller, the number of sensor inputs and controller outputs, the software algorithms, the calibration parameters, and memory requirements increased at a faster rate. This nonlinear increase in control requirements as a result of the increase in number of gears or ratios over time is illustrated in Figure 5.31. This graph shows that an eight-speed automatic transmission requires 2 megabytes of read-only memory (ROM) consisting of algorithms, lookup tables, and calibration parameters. Progressing to a nine-speed transmission incurs a nonlinear increase to 3.5 megabytes of ROM.

Migrating from conventional automatic transmissions to dual clutch transmissions has been shown in this chapter

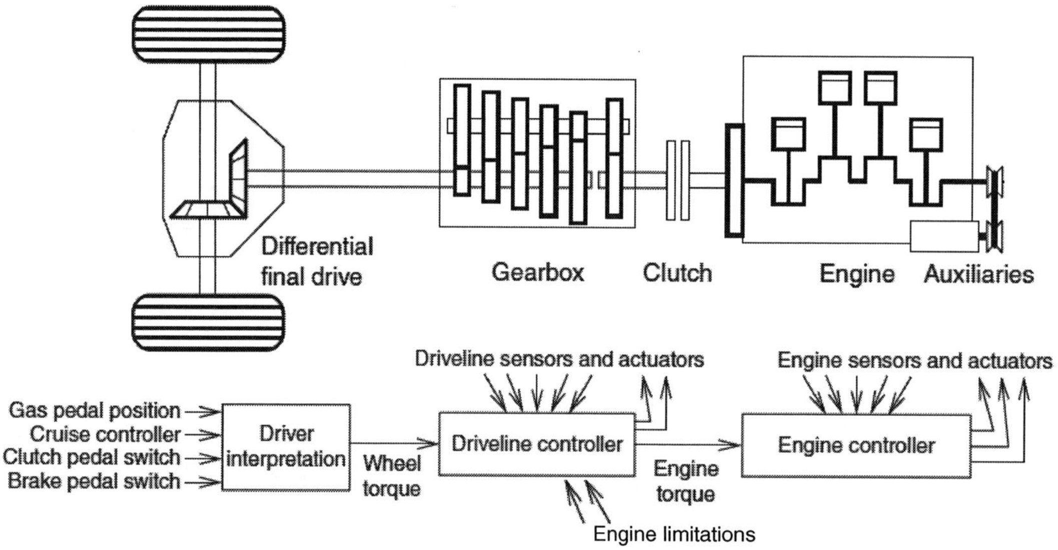

FIGURE 5.30 Schematic of signal flow and coordination between engine controller and driveline controller for gear ratio selection and clutch control.
SOURCE: Modified Figure 3.8 from Eriksson and Nielsen (2014). Reprinted with permission.

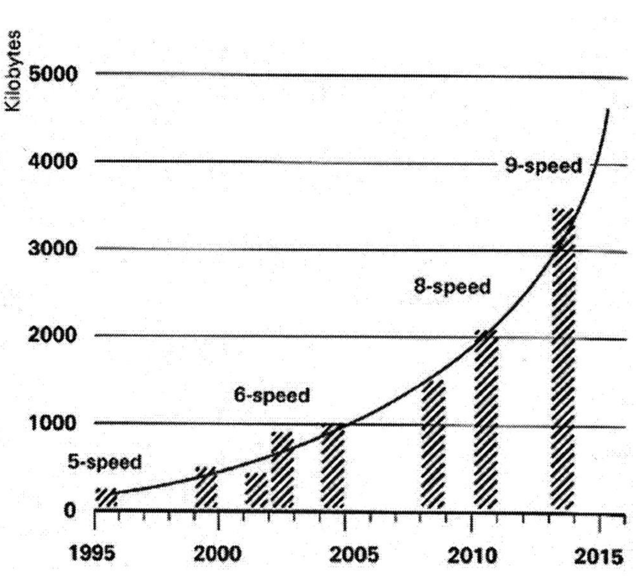

FIGURE 5.31 Increase in ROM requirements in transmission control units as a result of the increase in number of gears or ratios over time.
SOURCE: ZF Friedrichshafen AG (2013).

to provide significant reductions in fuel consumption. One of the enablers for DCT transmissions was the increased capability of the electronic controllers with high-data-rate signal processing. High-data-rate signal processing is used during the engagement and release of the clutches in a DCT transmission. To ensure smooth shift events, clutch pressures are modulated to ensure that the oncoming and offgoing target shaft speeds are achieved through utilization of either an adaptive learning system or a closed-loop control with high-data-rate feedback of shaft speeds. Direct feedback control of clutch pressures has also been found to provide improvements in shift smoothness in conventional automatic transmissions together with possible hardware simplifications; as a result, it is being applied to some conventional transmissions.

As discussed earlier, deficiencies in providing effective, smooth, and reliable clutch control in some early introductions of dual dry clutch transmissions were the underlying reason for drivability and warranty complaints (Vasca et al. 2011). These deficiencies result from the difficulty in developing an accurate and computationally efficient model of the relationship between the pressure applied by the clutch actuator or the actuator position and the torque transmitted through the clutch during the engagement phase (Zoppi et al. 2013; Oh and Choi 2014). This model-based predictive control methodology must accurately account for the slip-speed-dependent friction to ensure precise regulation of the slip acceleration during the lock-up phase in order to satisfy multiple objectives of low friction losses, minimum time for engagement, and driver comfort (Garofalo et al. 2001).

Although transmission control methodologies have been successfully developed and implemented to address quantifiable metrics such as reduction in fuel consumption or undesirable driveline oscillations, drivability metrics that relate to driver comfort and satisfaction are continuing to be developed (Kim et al. 2007). Applying limits to metrics such as vibration dose value (VDV)[4] or metrics related to the first derivative (acceleration) and second derivative (jerk) of vehicle velocity have led to significantly improved shift quality (Dorey and Holmes 1999). Limits on the number of shift events per unit time and the time between shift events may be used for real-time fuel economy optimization, but when the driver's perception does not correlate with the chosen limits, recalibration may be necessary to provide for the driver's comfort (Bai et al. 2013; Ngo et al. 2013). Other metrics in addition to those related to transmission shift quality may be needed to ensure that discontinuous powertrain functions, which occur with stop-start systems, multimode combustion switches with HCCI, and hybrid electric powertrains, can be provided while continuing to provide for the driver's comfort. Research by engineers, psychologists, and market analysts is continuing to be directed toward defining relationships that will ensure driver comfort during all modes of driving (Dorey and Holmes 1999; Skippon 2014). As transmission design and controls mature, the drivability and consumer acceptance issues associated with transmission designs having low market penetration are likely to be improved with advancements in computer simulations, modeling, control systems, and hardware designs.

As new transmission and engine technologies are incorporated in future powertrains, control of these powertrains will continue to focus on maximizing the reduction in fuel consumption. The increasing number of variables in the engine control, as discussed in Chapter 2, combined with the increasing number of variables in the transmission control, will require the continued application of optimization techniques to minimize fuel consumption and emissions while providing the performance and comfort expected by the driver.

FINDINGS AND RECOMMENDATIONS

Finding 5.1 New eight-, nine-, and ten-speed automatic transmissions are being introduced to replace six-speed automatic transmissions, which are currently dominant in light-duty vehicles. As transmissions trend towards higher numbers of gear ratios and ratio spreads, diminishing benefits for fuel consumption reductions are anticipated, which may be in the range of an additional 2 percent reduction. Studies have found that only minimal reductions in fuel consumption can be achieved beyond seven gear ratios. The benefits from parasitic loss reduction technologies, generally applied as new transmissions are introduced, can exceed the benefits from increasing the number of gear ratios and are reflected in the fuel consumption reductions achieved with these new transmissions.

Finding 5.2 Parasitic losses in typical transmissions are approximately 10 percent of the input energy, so that for a 15 percent reduction in these losses, a 2 percent reduction in fuel consumption could be expected. Reducing parasitic losses is focused on the oil supply system, drag torque resulting from clutches, bearings and seals, and gear losses. Fuel consumption reductions of approximately 5 percent, resulting from a 35 to 40 percent reduction in losses, may be possible by 2025. Opportunities may be available for further reductions in losses beyond 2025.

---

[4] Vibration dose value (VDV), used to assess intermittent vibration, is a cumulative measurement of the vibration level received over a specific period of time, such as a transmission shift event. *It is preferred* for cases where vibration may vary and be intermittent. The VDV formula uses the RMS acceleration raised to the fourth power and is known as the Root-mean quad method. This technique ensures that VDV is more sensitive to the peaks in the acceleration levels.

$$\mathrm{VDV} = \left(\int_0^T a^4(t)\,dt\right)^{1/4}$$

where VDV is the vibration dose value in $m/s^{1.75}$,
a(t) is the frequency-weighted acceleration in m/s^2, and
T is the total measurement period in seconds.
SOURCE: Gracey & Associates (n.d.).

Finding 5.3 As engines incorporate new technologies to improve fuel consumption, including variable valve timing and lift, direct injection, and turbocharging and downsizing, the benefits of increasing transmission ratios or switching to a CVT diminish. As the engine efficiency map improves, the penalty of having larger ratio steps between gears is significantly reduced compared to a naturally aspirated engine. The benefit of an eight-speed transmission over a six-speed transmission is reduced by approximately 15 percent when added to a modestly turbocharged, downsized engine compared to a naturally aspirated engine. However, the full benefits of parasitic loss reduction technologies (HEG) are still available.

Finding 5.4 Although dual-clutch transmissions can provide a 3.5 to 4.5 percent reduction in fuel consumption relative to conventional automatic transmissions, they are not likely to reach the high penetration rates predicted by EPA/NHTSA in the U.S. market. This is primarily due to customer acceptance issues stemming from dry-clutch DCTs with drivability that is different from that of conventional automatic transmissions. With fewer than NHTSA's anticipated applications of DCTs, other possibly less cost effective technologies will need to be introduced to compensate for the loss of the fuel consumption reduction benefits of widely applied DCTs. DCTs currently incur a modest cost increase over conventional, automatic transmissions, but costs could theoretically be as low as a conventional automatic transmission if high volumes were to be realized. Improved customer acceptance may be possible in the future with either wet clutch DCTs or dry clutch DCTs with a torque converter.

Recommendation 5.1 NHTSA and EPA should update the analyses of technology penetration rates for the midterm review to reflect the anticipated low DCT penetration rate in the U.S. market.

Finding 5.5 As the number of cylinders in the engine is reduced with downsizing and turbocharging, the variation in torque over a revolution of the engine increases, reaching a worst case for three-cylinder engines. These torsional vibrations can lead to seat vibrations or noise at certain speeds, both of which reduce the comfort of the vehicle. A variety of torsional vibration damping mechanisms can be added to the torque converter, but the cost of the torque converter could increase by as much as 50 percent. Additional mechanisms are expected to be required for torsional vibration damping with dual clutch transmissions. The cost of torsional damping systems needs to be included in transmission cost analyses.

Finding 5.6 The penetration of CVTs is increasing in the United States, and EPA estimated that CVT market share would reach 19.3 percent in 2014. The continuous variation in ratios provided by a CVT allows the engine to operate at its most efficient condition for the power level required. Several 2014 MY vehicles with CVTs are among the vehicles with high fuel economy ratings. The disadvantage of CVTs is that they have higher losses than conventional automatic transmissions; however, like conventional automatic transmissions, CVTs can benefit from a reduction in losses with increased costs.

Recommendation 5.2 NHTSA and EPA should add the CVT to the list of technologies applicable for the 2017-2025 CAFE standards.

Finding 5.7 Secondary axle disconnect systems are important since all-wheel-drive vehicles continue to be popular, with an estimated 31.2 percent of all light-duty vehicles in the United States having four wheel drive in 2014. If secondary axle disconnect systems could reduce the losses of the AWD system by 50 percent, fuel consumption reductions of up to 3 percent are expected.

REFERENCES

Bai, S., J. Maguire, and H. Peng. 2013. Dynamic Analysis and Control System Design of Automatic Transmissions. SAE International.

Baillie, C., B. Campbell, K. Govindswamy, and D. Tomazic. 2014. Advances in Parasitic Loss Reduction for Various Transmission Architectures. FEV North America, Inc. Presentation prepared for 8th International CTI Symposium, Rochester, MI.

Brain, M. 2000. How Manual Transmissions Work. http://auto.howstuffworks.com/transmission4.htm.

Carney, D. 2014. Honda's new 8-speed DCT uses a Torque Converter. SAE Automotive Engineering Magazine, August 6.

Daimler. 2013. New Nine-Speed Automatic Transmission debuts in the Mercedes-Benz E350 Blue Tec: Premier of the new 9G-Tronic. Daimler, July 24. http://media.daimler.com/dcmedia/0-921-1553299-1-1618134-1-0-1-0-0-0-0-1549054-0-1-0-0-0-0-0.html.

Dana Holding Corp. 2014. Dana Advances Development of VariGlide™ Continously Variable Planetary Technology. PR Newswire, May 19. http://www.prnewswire.com/news-releases/dana-advances-development-of-variglide-continuously-variable-planetary-technology-259791981.html.

Dick, A., J. Greiner, A. Locher, and F. Jauch. 2013. Optimization Potential for a State of the Art 8-Speed AT. SAE 2013-01-1272.

DOE (Department of Energy). 2014. Where the Energy Goes: Gasoline Vehicles. http://www.fueleconomy.gov/feg/atv.shtml. Accessed November 2, 2014.

Dorey, R., and C.B. Holmes. 1999. Vehicle Driveability – Its Characterisation and Measurement. SAE Technical Paper 1999-01-0949. doi:10.4271/1999-01-0949.

Eckl, B., and D. Lexa. 2012. How Many Gears do the Markets Need? GETRAG. International CTI Symposium, Berlin, Germany, December.

EPA (Environmental Protection Agency). 2014. Light-Duty Automotive Technology, Carbon Dioxide Emissions and Fuel Economy Trends, 1975 through 2014, October. EPA-420-R-14-023.

EPA/FEV. 2010. Light-Duty Technology Cost Analysis, Report on Additional Case Studies. U.S. Environmental Protection Agency, EPA-420-R-10-010, April.

EPA/FEV. 2011. Light-Duty Vehicle Technology Cost Analysis, Advanced 8- Speed Transmissions. U.S. Environmental Protection Agency, EPA-420-R-11-022, October.

EPA/FEV. 2013. Light-Duty Technology Cost Analysis, Report on Additional Case Studies – revised Final Report. U.S. Environmental Protection Agency, EPA-420-R-13-008, April.

EPA/NHTSA. 2010. Joint Technical Support Document: Rulemaking to Establish Light –Duty Vehicles Greenhouse Gas Emission Standards and Corporate Average Fuel Economy Standards, April.

EPA/NHTSA. 2012. Joint Technical Support Document, Final Rulemaking 2017-2025 Light-Duty Greenhouse Gas Emission Standards and Corporate Average Fuel Economy Standards. EPA-420-R-12-901.

Eriksson, L., and L. Nielsen. 2014. Modeling and Control of Engines and Drivelines (Automotive Series). John Wiley & Sons, SAE International, April.

Garofalo, F., L. Glielmo, L. Iannelli, and F. Vasca. 2001. Smooth Engagement for Automotive Dry Clutch. Proceedings of the 40th IEEE Conference on Decision and Control, Orlando, Florida, December: 529-534.

Gartner, L., and M. Ebenhock. 2013. The ZF automatic transmission 9HP48 transmission system, design and mechanical parts. SAE Int. J. Passeng. Cars - Mech. Syst. 6(2): 908-917. doi:10.4271/2013-01-1276.

Govindswamy, K., C. Baillie, and T. D'Anna. 2013. Choosing the Right Transmission Architecture Considering Customer Acceptance. SAE Int. Webinar, September 18.

Gracey & Associates. n.d. Vibration Dose: Definitions, Terms, Units and Parameters. Acoustic Glossary. http://www.acoustic-glossary.co.uk/vibration-dose.htm.

Greimel, H. 2014. ZF CEO: We're not chasing 10-speeds. Automotive News, November 23.

Guzzella, L., and A. Sciarretta A. 2007. Vehicle Propulsion Systems: Introduction to Modeling and Optimization, Third Edition. Springer.

Healey, J., and C. Woodyard. 2013. GM, Ford to jointly develop 10-speed transmissions. USA Today, April 15.

Kiencke, U., and L. Nielsen. 2000. Automotive Control Systems. Springer, SAE International.

Kim, D., H. Peng, S. Bai, and J.M. Maguire. 2007. Control of integrated powertrain with electronic throttle and automatic transmission. IEEE Transactions on Control Systems Technology 15(3), May.

Lee, B. 2010. All-Wheel Drive Disconnect Clutch System. Schaeffler SYMPOSIUM 2010: 360-64.http://www.schaeffler.com/remotemedien/media/_shared_media/08_media_library/01_publications/schaeffler_2/symposia_1/downloads_11/Schaeffler_Kolloquium_2010_27_en.pdf.

Martin, K. 2012. Transmission Efficiency Developments. SAE Transmission and Driveline Symposium: Competition for the Future, October 17-18. Detroit, Michigan.

Moawad, A., and A. Rousseau. 2012. Impact of Transmission Technologies on Fuel Efficiency – Final Report. DOE HS 811 667, August.

Ngo, V.-D., A. Jose, C. Navarrete, T. Hofman, M. Steinbuch, and A. Serrarens. 2013. Optimal gear shift strategies for fuel economy and driveability. Proc. IMechE Part D, Journal of Automobile Engineering 227(10): 1398-1413, October.

Noles, J. 2013. Development of Transmission Fluids Delivering Improved Fuel Efficiency by Mapping Transmission Response to Viscosity and Additive Changes. Presentation at the SAE Transmission & Driveline Symposium, Troy, Michigan, October 16-17. http://www.sae.org/events/ctf/2013/2013_ctf_guide.pdf.

NSK Europe. 2014. New Low-Friction TM-Seal for Automotive Transmissions. http://www.nskeurope.com/cps/rde/dtr/eu_en/nsk_innovativeproduct_IP-E-2066.pdf.

Oh, J., and S. Choi. 2014. Real-time Estimation of Transmitted Torque on Each Clutch for Ground Vehicles with Dual Clutch Transmissions. IEEE/ASME Transactions on Mechatronics, February.

Powell, B., J. Quinn, W. Miller, J. Allison, J. Hines and R. Beals. Magnesium Replacement of Aluminum Cast Components in a Production V6 Engine to Effect Cost-Effective Mass Reduction. http://energy.gov/sites/prod/files/2014/03/f8/deer10_powell.pdf. Accessed April 13, 2015.

Ricardo, Inc. 2011. Computer Simulation of Light-Duty Vehicle Technologies for Greenhouse Gas Emission Reduction in the 2020-2025 Timeframe. U.S. Environmental Protection Agency, EPA-420-R-11-020.

Sherman, D. 2013. CVT Transmissions. Car and Driver, December. http://www.caranddriver.com/features/how-cvt-transmissions-are-getting-their-groove-back-feature.

Shidore, N. et. al. 2014. Impact of Advanced Technologies on Engine Targets. Project VSS128, DOE Merit Review, June.

Shulver, D. 2013. Reduced Fuel Consumption Enabled by Optimized Transmission Pump Technology. Presentation at the SAE Transmission & Driveline Symposium, Troy, Michigan, October 16-17. http://www.sae.org/events/ctf/2013/2013_ctf_guide.pdf.

Skippon, S.M. 2014. How consumer drivers construe vehicle performance: Implications for electric vehicles. Transportation Research Part F: Traffic Psychology and Behaviour 23: 15-31.

Vasca, F., L. Iannelli, A. Senatore, and G. Reale. 2011. Torque transmissibility assessment for automotive dry clutch engagement. IEEE/ASME Transactions on Mechatronics 16(3): 564-573, June.

Wagner, U., R. Berger, M. Ehrlich, and M. Homm. 2006. Electromotoric Actuators for Double Clutch Transmissions. Proceedings of the 8th LuK Symposium.

ZF. 2013. Motion and Mobility. ZF Corporate Report. Friedrichshafen, Germany.

Zoppi, M., C. Cervone, G. Tiso, and F. Vasca. 2013. Software in the Loop Model and Decoupling Control for Dual Clutch Automotive Transmissions. 3d International Conference on Systems and Control, Algiers, Algeria, October.

ANNEX TABLES

TABLE 5A.1 NRC Committee's Estimated Fuel Consumption Reduction Effectiveness of Transmission Technologies

Technologies		Midsize Car I4 DOHC	Large Car V6 DOHC	Large Light Truck V8 OHV	
Transmission Technologies	Abbreviation	Most Likely	Most Likely	Most Likely	Relative To
NHTSA Technologies					
Improved Auto. Trans. Controls/Externals (ASL-1 & Early TC Lockup)	IATC	2.5 - 3.0	2.5 - 3.0	2.5 - 3.0	4 sp AT
6-speed AT with Improved Internals - Lepelletier (Rel to 4 sp AT)	NUATO-L	2.0 - 2.5	2.0 - 2.5	2.0 - 2.5	IATC
6-speed AT with Improved Internals - Non-Lepelletier (Rel to 4 sp AT)	NUATO-NL	2.0 - 2.5	2.0 - 2.5	2.0 - 2.5	IATC
6-speed Dry DCT (Rel to 6 sp AT - Lepelletier)	6DCT-D	3.5 - 4.5	3.5 - 4.5	N/A	6 sp AT
6-speed Wet DCT (Rel to 6 sp AT - Lepelletier) (0.5% less than Dry Clutch)	6DCT-W	3.0 - 4.0	3.0 - 4.0	3.0 - 4.0	6 sp AT
8-speed AT (Rel to 6 sp AT - Lepelletier)	8AT	1.5 - 2.0	1.5 - 2.0	1.5 - 2.0	Previous Tech
8-speed DCT (Rel to 6 sp DCT)	8DCT	1.5 - 2.0	1.5 - 2.0	1.5 - 2.0	Previous Tech
High-Efficiency Gearbox Level 1 (Auto) (HETRANS)	HEG1	2.3 - 2.7	2.3 - 2.7	2.3 - 2.7	Previous Tech
High-Efficiency Gearbox Level 2 (Auto, 2017 and Beyond)	HEG2	2.6 - 2.7	2.6 - 2.7	2.6 - 2.7	Previous Tech
Shift Optimizer (ASL-2)	SHFTOPT	0.5 - 1.0	0.5 - 1.0	0.5 - 1.0	Previous Tech
Secondary Axle Disconnect	SAX	1.4 - 3.0	1.4 - 3.0	1.4 - 3.0	Baseline
Other Technologies					
Continuously Variable Transmission with Improved internals (Rel to 6 sp AT)	CVT	3.5 - 4.5	3.5 - 4.5	N/A	Previous Tech
High-Efficiency Gearbox (CVT)	CVT-HEG	3.0	3.0	N/A	Previous Tech
High-Efficiency Gearbox (DCT)	DCT-HEG	2.0	2.0	2.0	Previous Tech
High-Efficiency Gearbox Level 3 (Auto, 2020 and beyond)	HEG3	1.6	1.6	1.6	Previous Tech
9-10 speed Transmission (Auto, Rel to 8 sp AT)	10SPD	0.3	0.3	0.3	Previous Tech

TABLE 5A.2a NRC Committee's 2017 MY Estimated Direct Manufacturing Costs of Transmission Technologies

Technologies		Midsize Car I4 DOHC	Large Car V6 DOHC	Large Light Truck V8 OHV	
Transmission Technologies	Abbreviation	Most Likely	Most Likely	Most Likely	Relative To
NHTSA Technologies					
Improved Auto. Trans. Controls/Externals (ASL-1 & Early TC Lockup)	IATC	50	50	50	Baseline 4 sp AT
6-speed AT with Improved Internals - Lepelletier (Rel to 4 sp AT)	NUATO-L	-13	-13	-13	IATC
6-speed AT with Improved Internals - Non-Lepelletier (Rel to 4 sp AT)	NUATO-NL	195	195	195	IATC
6-speed Dry DCT (Rel to 6 sp AT - Lepelletier)	6DCT-D	-149 to 31	-149 to 31	N/A	6 sp AT
6-speed Wet DCT (Rel to 6 sp AT - Lepelletier)	6DCT-W	-88 to 88	-88 to 88	-88 to 88	6 sp AT
8-speed AT (Rel to 6 sp AT - Lepelletier)	8AT	56 - 151	56 - 151	56 - 151	Previous Tech
8-speed DCT (Rel to 6 sp DCT)	8DCT	179	179	179	Previous Tech
High-Efficiency Gearbox Level 1 (Auto) (HETRANS)	HEG1	120	120	120	Previous Tech
High-Efficiency Gearbox Level 2 (Auto, 2017 and Beyond)	HEG2	194	194	194	Previous Tech
Shift Optimizer (ASL-2)	SHFTOPT	26	26	26	Previous Tech
Secondary Axle Disconnect	SAX	100	100	100	Baseline
Other Technologies					
Continuously Variable Transmission with Improved internals (Rel to 6 sp AT)	CVT	179	179	N/A	Baseline
High-Efficiency Gearbox (CVT)	CVT-HEG	125	125	N/A	Baseline
High-Efficiency Gearbox (DCT)	DCT-HEG	150	150	150	Baseline
High-Efficiency Gearbox Level 3 (Auto, 2020 and Beyond)	HEG3	150	150	150	Baseline
9-10 speed Transmission (Auto, Rel to 8 sp AT)	10SPD	75	75	75	Baseline

TABLE 5A.2b NRC Committee's 2020 MY Estimated Direct Manufacturing Costs of Transmission Technologies

Technologies		Midsize Car I4 DOHC	Large Car V6 DOHC	Large Light Truck V8 OHV	
Transmission Technologies	Abbreviation	Most Likely	Most Likely	Most Likely	Relative To
NHTSA Technologies					
Improved Auto. Trans. Controls/Externals (ASL-1 & Early TC Lockup)	IATC	46	46	46	Baseline 4 sp AT
6-speed AT with Improved Internals - Lepelletier (Rel to 4 sp AT)	NUATO-L	-12	-12	-12	IATC
6-speed AT with Improved Internals - Non-Lepelletier (Rel to 4 sp AT)	NUATO-NL	181	181	181	IATC
6-speed Dry DCT (Rel to 6 sp AT - Lepelletier)	6DCT-D	-138 to 28	-138 to 28	N/A	6 sp AT
6-speed Wet DCT (Rel to 6 sp AT - Lepelletier)	6DCT-W	-82 to 82	-82 to 82	-82 to 82	6 sp AT
8-speed AT (Rel to 6 sp AT - Lepelletier)	8AT	52 - 126	52 - 126	52 - 126	Previous Tech
8-speed DCT (Rel to 6 sp DCT)	8DCT	167	167	167	Previous Tech
High-Efficiency Gearbox Level 1 (Auto) (HETRANS)	HEG1	113	113	113	Previous Tech
High-Efficiency Gearbox Level 2 (Auto, 2017 and Beyond)	HEG2	183	183	183	Previous Tech
Shift Optimizer (ASL-2)	SHFTOPT	24	24	24	Previous Tech
Secondary Axle Disconnect	SAX	94	94	94	Baseline
Other Technologies					
Continuously Variable Transmission with Improved internals (Rel to 6 sp AT)	CVT	168	168	NA	Baseline
High-Efficiency Gearbox (CVT)	CVT-HEG	117	117	NA	Baseline
High-Efficiency Gearbox (DCT)	DCT-HEG	141	141	141	Baseline
High-Efficiency Gearbox Level 3 (Auto, 2020 and beyond)	HEG3	141	141	141	Baseline
9-10 speed Transmission (Auto, Rel to 8 sp AT)	10SPD	71	71	71	Baseline

TABLE 5A.2c NRC Committee's 2025 MY Estimated Direct Manufacturing Costs of Transmission Technologies

Technologies		Midsize Car I4 DOHC	Large Car V6 DOHC	Large Light Truck V8 OHV	
Transmission Technologies	Abbreviation	Most Likely	Most Likely	Most Likely	Relative To
NHTSA Technologies					
Improved Auto. Trans. Controls/Externals (ASL-1 & Early TC Lockup)	IATC	42	42	42	Baseline 4 sp AT
6-speed AT with Improved Internals - Lepelletier (Rel to 4 sp AT)	NUATO-L	-11	-11	-11	IATC
6-speed AT with Improved Internals - Non-Lepelletier (Rel to 4 sp AT)	NUATO-NL	165	165	165	IATC
6-speed Dry DCT (Rel to 6 sp AT - Lepelletier)	6DCT-D	-127 to 26	-127 to 26	N/A	6 sp AT
6-speed Wet DCT (Rel to 6 sp AT - Lepelletier)	6DCT-W	-75 to 75	-75 to 75	-75 to 75	6 sp AT
8-speed AT (Rel to 6 sp AT - Lepelletier)	8AT	47 - 115	47 - 115	47 - 115	Previous Tech
8-speed DCT (Rel to 6 sp DCT)	8DCT	152	152	152	Previous Tech
High-Efficiency Gearbox Level 1 (Auto) (HETRANS)	HEG1	102	102	102	Previous Tech
High-Efficiency Gearbox Level 2 (Auto, 2017 and Beyond)	HEG2	165	165	165	Previous Tech
Shift Optimizer (ASL-2)	SHFTOPT	22	22	22	Previous Tech
Secondary Axle Disconnect	SAX	86	86	86	Baseline
Other Technologies					
Continuously Variable Transmission with Improved internals (Rel to 6 sp AT)	CVT	154	154	NA	Baseline
High-Efficiency Gearbox (CVT)	CVT-HEG	107	107	NA	Baseline
High-Efficiency Gearbox (DCT)	DCT-HEG	127	127	127	Baseline
High-Efficiency Gearbox Level 3 (Auto, 2020 and beyond)	HEG3	128	128	128	Baseline
9-10 speed Transmission (Auto, Rel to 8 sp AT)	10SPD	65	65	65	Baseline

6

Non-Powertrain Technologies

INTRODUCTION

This chapter focuses on reducing fuel consumption with non-powertrain technologies. These technologies affect engine performance either directly or indirectly to reduce fuel consumption. The committee considers car body design (aerodynamics and mass), vehicle interior materials (mass), tires, and vehicle accessories (power steering and heating, ventilation, and air conditioning [HVAC] systems) as areas of significant opportunities for achieving near-term, cost-effective reductions in fuel consumption. These will be considered in some detail below.

The forces impeding vehicle motion on a level grade can be written as follows:

$$F = ma + R_a + R_{rr}$$

where ma is the inertial force, R_a is the aerodynamic resistance, and R_{rr} is the rolling resistance.

The total energy required for propulsion over the cycle is equal to the time integral of the positive product of force and velocity. The energy used to overcome inertial forces dominates in the FTP cycle, while the energy used to overcome aerodynamic resistance dominates in the highway cycle.

Collections of relatively low-cost vehicle technologies can have a positive impact on reducing fuel consumption. Low-rolling-resistance tires, improved vehicle aerodynamics, and electric power steering can reduce fuel consumption by about 10 percent with only moderate cost additions. Higher-efficiency air conditioning systems are available that better match cooling with occupant comfort while improving fuel economy. Electric and electric/hydraulic power steering also reduce the load on the engine by demanding power only when the operator turns the wheel, whereas older systems relied on hydraulic power supplied by the engine all the time.

This chapter is organized to discuss the major non-powertrain systems and their impact on fuel consumption and costs. It describes some of the issues that must be addressed prior to 2025 for the following technologies:

- Improvements in vehicle aerodynamics,
- Vehicle mass reduction,
- Improvements in tire rolling resistance,
- Improved vehicle accessories and HVAC, and
- Autonomous components and implementation.

AERODYNAMICS

Energy required to overcome drag does not depend on vehicle mass. It does depend on the size of the vehicle as represented by the frontal area.[1] For low-speed driving, about one-fourth of the energy delivered by the drivetrain goes to overcoming drag; for high-speed driving, one-half of the energy goes to overcoming drag. Vehicle drag coefficients (C_d) vary considerably, from 0.195 for the General Motors EV1 to 0.57 for the Hummer 2, with more typical values in the range of 0.25 to 0.38 for production vehicles. Vehicle drag can be reduced through both passive and active design changes. The drag coefficient can be lowered by more aerodynamic vehicle shapes with smaller influences from other factors, such as external mirrors, rear spoilers, frontal inlet areas, wheel well covers, and the vehicle underside.

Vehicles with higher C_d values (greater than 0.30) may be able to reduce the C_d by up to 10 percent at low cost without affecting the vehicle's interior volume. In the NRC Phase 1 report, *Assessment of Fuel Economy Technologies for Light Duty Vehicles* (NRC 2011), the committee's judgment was that a C_d of less than 0.25 would require significant changes that could include the elimination of outside rearview mirrors, total enclosure of the car underbody, and other costly modifications.

[1] The force required to overcome drag is represented by the product of the drag coefficient, C_d, the frontal area, A, and the square of speed, V. The formula is $F = \frac{1}{2} C_d AV^2$.

Fuel Consumption

Argonne National Laboratory estimated that, without engine modifications, a 10 percent reduction in aerodynamic drag would result in about a 0.25 percent reduction in fuel consumption for the urban cycle and a 2.15 percent reduction for the highway cycle. Under average driving conditions, a 10 percent reduction in drag resistance would reduce total fuel consumption about 2 percent (NRC 2011). If lower acceleration can be tolerated and the engine operates at the same efficiency, the improvement with a 10 percent reduction of aerodynamic drag could result in fuel consumption reduction as high as 3 percent.

The recent NRC report *Transitions to Alternative Vehicles and Fuels,* referred to as the 2050 Transitions Report in the remainder of this report, estimated aerodynamic improvements possible for the 2030 time frame (NRC 2013). That study's scenarios estimated a reduction in new-vehicle-fleet aerodynamic drag resistance for the midrange (high probability of attainment) case to average about 21 percent (4 percent reduction in fuel consumption) in 2030. For the optimistic case, the aerodynamic drag reductions are estimated to average about 28 percent in 2030.

The recent Technical Support Document (TSD) for the National Highway Traffic Safety Administration/Environmental Protection Agency (NHTSA/EPA) final rulemaking (EPA/NHTSA 2012b) considered an aerodynamic reduction in the 10-20 percent range. For the final rule, the Agencies considered two levels of improvements. The first level is that discussed in the 2017-2025 final rule and the 2012 TSD; it includes such body features as air dams, tire spats,[2] and perhaps one underbody panel. The second level of aero improvements includes such body features as active grille shutters, larger underbody panels, or low-profile roof racks. The 2012-2016 final rule estimated that a fleet average of 10-20 percent total aerodynamic drag reduction is attainable, which equates to incremental reductions in fuel consumption and CO_2 emissions of 2-3 percent (average 2.5 percent) for both cars and trucks. Several original equipment manufacturers (OEMs) have already aimed to achieve low drag coefficients of between 0.2 and 0.3 in their product lines, although these tend to be vehicles that have higher costs or are performance based. There are at least a half dozen mid-priced 2013 passenger cars advertising active grille shutters. In general, the additional data on improving vehicle aerodynamics provided to this committee by OEMs and Tier 1 suppliers have not challenged or contradicted the methodology and conclusions described in the NHTSA/EPA TSD.

Timing

Reductions of drag coefficient C_d by approximately 5 percent (up to 10 percent) have been taking place and will continue. Several OEMs expressed concern that reducing the drag coefficient too aggressively could have a negative impact on consumer acceptance. Additionally, several OEMs have already achieved C_d in the range 0.20 to 0.24. In the 2020-2025 time frame, 10-20 percent reductions in aerodynamic drag are plausible.

Costs

The Phase One report (NRC 2011) estimated that a 5 percent reduction in drag could be achieved with minimal cost through vehicle design. Slightly more aggressive reductions could be achieved by sealing the undercarriage and installing covers/shields (e.g., in the wheel well areas and on the underbody) costing between $10 and $100. A 10 percent reduction in aerodynamic drag would be an aggressive strategy calling for wind deflectors (spoilers) and possibly the elimination of sideview mirrors. The 2050 Transitions Report (NRC 2013) did not provide cost estimates related to aerodynamic improvements, although there would be significant cost increases associated with these technologies.

The Agencies' current estimates for direct manufacturing costs (DMC) for improvements in aerodynamic drag to a baseline vehicle in 2017 are $39 for Level 1 and $117 for Level 2 (Table 6.1). These estimates follow the same trend from the 2012-2016 rule, when NHTSA and EPA estimated the aero-level 1 (10%) total cost at $41 (2010 dollars) applicable in MY 2015. The second level of aero (20%) included such body features as active grille shutters, rear visors, larger underbody panels, or low-profile roof racks, with a DMC cost of $123 (2010 dollars). The committee concurred with the Agencies' cost estimates. Additionally, the committee assessed many of the other current studies on aerodynamic drag reductions in Table 6.2.

Barriers

Vehicles that exist today with C_d below or equal to 0.25 are usually specialty vehicles (e.g., sports cars and high-mileage vehicles like the Prius). While higher C_d vehicles (e.g., trucks, vans, and boxlike vehicles such as the Scion and the Flex) can reduce C_d, vehicle functionality could be diminished. If vehicle functionality (including "curbside appeal") is compromised, then the vehicle's appeal to the consumer would be reduced.

Elimination of sideview mirrors will require changes in safety regulations and improvement in vision systems, but studies believe that by removing sideview mirrors, drag reductions of 2-7 percent are possible (Hucho 2005). Tesla and the Alliance for Automobile Manufacturers have petitioned NHTSA on the topic of side mirror removal.

[2] Tire spats, or wheel fairings, are devices that cover the wheel well of a vehicle for the purpose of reducing aerodynamic drag.

TABLE 6.1 Agency-Estimated Direct Manufacturing Costs for Aerodynamic Drag Reductions—Levels 1 and 2 (2010 dollars)

Cost type	Aero	Incremental to	2017	2018	2019	2020	2021	2022	2023	2024	2025
DMC	Level 1	Baseline	39	38	37	37	36	35	35	34	33
DMC	Level 2	Aero-level 1	117	115	112	110	108	106	104	102	100

SOURCE: EPA/NHTSA (2012b).

TABLE 6.2 Summary of Other Studies and the Committee's Findings on the Direct Manufacturing Costs and Effectiveness of Aerodynamic Drag Reductions

Study	Year Published	Direct Manufacturing Cost ($)	C_d Reduction	FC Reduction	Comments
NRC Phase 1 Study	2011	$40 to $50	5-10%	1% to 2%	Wheel wells, underbody covers, body shape, mirrors
NRC 2050 Transitions Report	2011	N/A	21-28%	4-6%	Passive and active
EPA/NHTSA TSD	2012	$49 to $164	10-20%	2-3%	Passive and active
Current Study	2015	$49 to $165	10-20%	2-3%	Passive and active

MASS REDUCTION OPPORTUNITIES FROM VEHICLE BODY AND INTERIORS

The material trends in the automobile have been well established for more than 20 years, with an incremental, material substitution used to slowly introduce new materials that, in most cases, reduce mass. Figure 6.1 shows the decline in iron casting and an increase in high-strength steel, plastics/composites, and aluminum. Some vehicle subsystems have already made the lightweighting transition, such as the use of aluminum for powertrains and wheels. The change in vehicle composition is almost entirely due to the lightweighting impact of the new materials and, in some cases, to their potential to improve safety and crashworthiness.

Many expect these material trends to continue and even accelerate due to current fuel economy regulations. Mass reduction will be realized primarily through the use of more advanced high-strength steel for body structures, aluminum closure panels, and, in some cases, aluminum bodies.

At the edge of development, some structures may utilize advanced composite structures for the body (e.g., BMW i3), where carbon fiber systems allow for extremely lightweight and strong structures. But advanced composite structures will not be used in high sales volume vehicles for at least 10 years. Lightweighting technology deployment will vary depending on vehicle size and sales volume.

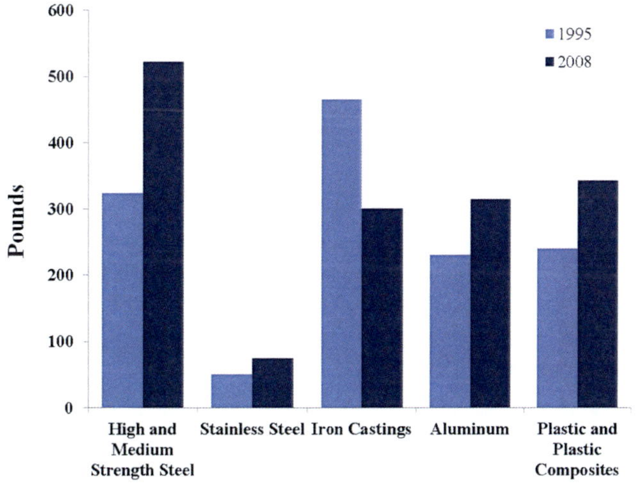

FIGURE 6.1 Selected material content per light-duty vehicle, 1995 and 2008.
SOURCE: DOE (2010).

Steel

High-strength steel has the advantage that it can be downgauged (made thinner) in many applications while still performing the same function as thicker, mild steel.[3] High-strength steels are traditionally viewed as those steels that have a tensile strength greater than 270 megapascals (MPa). A number of steel alloys in the 480-980 MPa range are routinely used for various structural components in the car, such as the front engine rails and some of the door beams and side pillars. Even stronger steels (1,000-1,500 MPa and higher) have been introduced for critical crash zones.

Plastics, Rubbers, and Composites

Plastics and composites offer significant long-term potential for reducing mass, but many challenges currently exist in broadening their application. The use of plastics, rubbers, and composites in automobiles is increasing with advances in chemistries and fabricating technologies. Many components inside and outside the vehicle now have fascias, lids, air foils, knobs, and other components made from composites because of advancements in colors, feel (soft skin feel), resistance to ultraviolet rays, and proper management of thermal expansion properties. Although composite materials are used throughout the car, not many applications are currently designed for structural crash management.

While many advances in these materials will occur to improve their performance, the use of fiber-reinforced materials (glass fiber or carbon fiber) for structural components is not expected to have significant penetration in the next 10 years (Figure 6.2). The growth in composites is largely constrained by cost and technical requirements (ability to join, thermal expansion differences, and a less-developed supply and recyclability chain).

Adhesives

Coatings, adhesives, and sealants are provided by the chemical industry. While all three product classes are important, adhesives pose the greatest challenge to OEMs and repair shops. Adhesives are a preferred joining method due to their superior joint bonding capabilities and ability to improve stiffness of the joining components. They can be sophisticated one-part or two-part and with or without a mechanical fastener or spot weld through the joint. Adhesives provide a mass reduction enabling technology because they can bond dissimilar materials, provide insulation if necessary, and tend to make a stronger bond than a localized joint (such as resistance spot welds or rivets), which can lead to further downgauging of material.

[3] Even as steel strength continues to increase, downgauging has limitations because other properties, such as stiffness, are important to the structure.

Aluminum

Aluminum is already a dominant material in powertrain components, heat exchangers, and road wheels and is an emerging material for all vehicle closures (e.g., 34 percent of MY 2012 vehicle hoods were aluminum and 48 percent are expected to be by MY 2015) (Ducker Worldwide 2014). In Europe and in the United States, the average car contains about 8 percent aluminum. This number is expected to double to 16 percent by 2025, driven primarily by the continued conversion from steel to aluminum.

One of the greatest challenges in manufacturing aluminum is joining. Traditional aluminum joining methods require a combination of welding, adhesive, and rivets, which limit joint configurations, present challenges at end-of-life recycling, and add cost to the process. In addition, aluminum is susceptible to galvanic corrosion when joined with dissimilar metals. Isolation of aluminum from other material through the use of adhesive or a material coating is typical to prevent galvanic corrosion. It is also pretreated to minimize corrosion and ensure paintability. In contrast to steel, aluminum does have the advantage in that it does not rust.

Aluminum generally takes three forms in the car: sheet, extrusion, and casting (a fourth form, forgings, can also be used). With the use of aluminum in cars forecast to grow (343 lb in 2012, growing to 550 lb in 2025), all three forms will increase in the average vehicle every year (Ducker Worldwide 2011). Extruded and welded aluminum bars are effective for front and rear rails (designed for crash management). The largest growth by far, however, is expected to be aluminum panels (see Table 6.3) for parts throughout the body. Aluminum sheet is cheaper than the other forms, and it is expected that the learning curve to develop sheet applications for high-volume vehicles will plateau as the industry learns by adding more applications every year.

Magnesium

Magnesium has the capability of providing 40 to 65 percent mass reduction in comparison to steel. Magnesium can be formed from sheet (like stamped steel), but is better used as thin-wall castings to maximize its weight reduction capabilities. Like aluminum, magnesium has galvanic corrosion concerns and must be isolated from other materials. Magnesium has a very limited infrastructure and knowledge base compared to aluminum and steel, but applications are appearing in production vehicles today (e.g., the liftgate inner on the Ford MKT). Other applications include steering column attachment, HVAC openings, pedal attachments, instrument panel structure attachments, door hinge attachments, spare tire modules, and A-pillar mounting attachments.

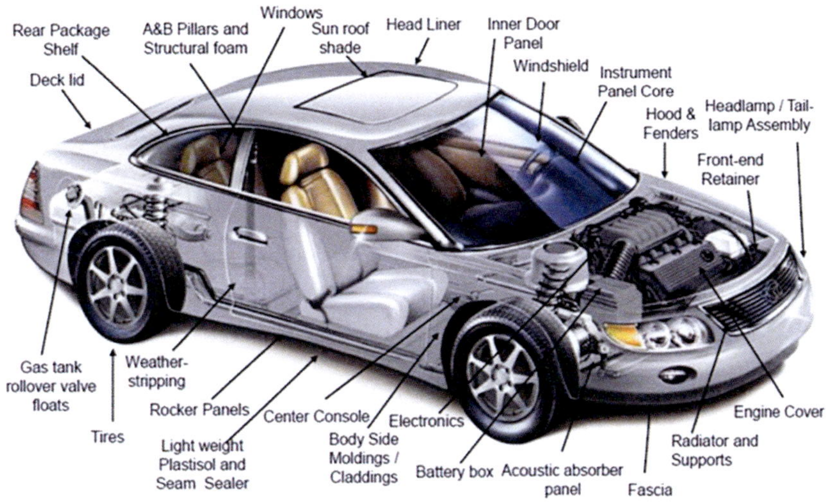

FIGURE 6.2 Auto part targets for lightweight plastics and rubber. Illustration provided by 3M Company.

TABLE 6.3 Distribution of Automotive Aluminum Utilization by Type

Automotive Aluminum Form	2012 (343 lb/vehicle) (%)	2025 (550 lb/vehicle) (%)
Casting	81	61
Sheet (rolled)	10	28
Extrusion	8	9
Forgings	1	2

SOURCE: Data adapted from Ducker Worldwide (2011).

Mixed-Material Car

The high-volume, mixed-material car is recognized as the longer-term, optimal approach to mass reduction, and most auto companies are headed in this direction. The mixed-material vehicle can be thought of as a vehicle that uses the most suitable material in each specific location to provide optimal performance and minimize weight at an appropriate cost. There are some cars on the market today that exemplify this goal; the McLaren P1 and the Audi TT are examples of vehicles that embody the mixed-material approach. The appeal of the mixed-material car is that it can enable weight reductions beyond what aluminum or high-strength steel alone can provide, and at lower cost.

Tooling and material costs for fabrication and joining are major decision factors in material selection. While the "optimized" vehicle will perhaps have a different mix of materials used for parts to optimize structure and mass, the corresponding costs for tooling and fabrication have to be considered. Parts made of different materials might mean more tooling costs.

Vehicle Design Challenges with Advanced Materials

The use of lightweight composites is an emerging trend in vehicles, and many of the barriers have been identified and are being studied. Some of the known barriers to use of lightweight composites include

- No ability to model time-, temperature-, and environment-dependent polymer composite properties,
- No integration of accurate composite models into engineering design tools,
- High-cost processing infrastructure,
- Long production times for structural composite parts, and
- Difficulties in identifying and repairing damaged composites.

These dynamic characteristics are one of the complications along with others such as the lack of long-term durability predictions and resultant overreliance on the build-and-test method for testing composite property behavior.

Modeling vehicle performance in the design stages is an intrinsic part of current vehicle development. More sophisticated models will need to be developed to support mixed-material vehicles.

Mass Reduction and Repairability

The use of new materials for automotive applications presents a challenge to the repair industry. For the greater part of a century, the automotive body structure and closures have been dominated by the use of mild and medium-strength steel. As a result, the repair industry is very good at assessing damage and repairing or replacing traditional steel parts when required. However, new materials do not follow the same assessment and repair rules as mild steel. In fact, the safe repair of damaged parts made of aluminum, composites, and even advanced high-strength steels requires specific methods and equipment. Ford has addressed some of these concerns with the introduction of the aluminum-intensive F-150 in MY 2015 by offering training courses on the properties of aluminum versus steel and offering advice and comments on retooling shops and equipment to both dealers and independent repair shops (Wernle 2014). With composites, parts that appear undamaged under visual inspection can still fail. The aerospace industry makes extensive use of ultrasonic testing to examine parts for flaws. Automakers are aware of these issues and are therefore reluctant to implement composites without first developing proper inspection techniques.

The automotive, insurance, and certification communities are responding to these safety issues. Companies such as Audi, Mercedes, Chrysler, and Honda either require specific certification or have created an approved network of collision repair shops. Such certification or network branding ensures some level of training to keep up with new repair standards for advanced lightweight materials. Insurers also require standards for the repair process and certification levels. The certification group, Inter-Industry Conference on Auto Collision Repair (I-CAR), is highly involved in the processes for repair of new materials and certification of repair shops.

Estimates of Mass Reduction Potential

The impetus behind lightweighting (mass reduction) of passenger cars and light-duty trucks is better performance and improved fuel economy. Lighter vehicles should handle better (e.g., responsiveness) and have improved stopping performance. The government's fuel economy standards are footprint-based and, by themselves, provide no incentive for downsizing vehicles.

Potential effects on safety, fuel economy, and vehicle costs have been analyzed where mass reduction is accomplished entirely through material substitution and smart design, which can reduce mass without changing a vehicle's functionality or safety performance and maintain structural strength. Three important aspects of lightweighting are these:

1. The fuel savings benefit of mass reduction is consistent among many mainstream vehicles. An industry estimate is that a 10 percent reduction in vehicle mass will produce approximately 6 to 7 percent reduction in fuel consumption for passenger cars and 4 to 5 percent reduction for light-duty trucks. A literature review of various studies relating mass reduction to fuel consumption reduction showed a range of 1.9 to 8.2 percent, with an average among the studies of 4.9 percent for every 10 percent in mass reduction (Cheah 2010).
2. The cost for mass reduction alternatives varies from negative (a cost savings) to several dollars per pound. It is generally acknowledged that the cost to reduce mass increases for each additional unit of mass eliminated on a vehicle.
3. The concept of mass decompounding[4] recognizes that, as vehicle mass is reduced, there are new opportunities to reduce additional mass and that these often minimize the overall cost increase. The most current studies cite opportunities for mass decompounding that range from 15 to 56 percent of the primary mass saved. Combining the information from these studies with the committee's expertise, the committee finds it likely that a reduced-mass vehicle would allow an additional 40 percent of the primary mass removed for cars and an additional 25 percent of the primary mass removed for trucks if decompounding strategies are implemented, assuming that the whole vehicle can be reoptimized for the new mass level. According to a recent study, primary mass reduction in the body provides the greatest potential for mass decompounding among subsystems, with engine and transmission subsystems providing the largest secondary mass reduction (Alonso et al. 2012). Subsystems that may offer potential mass decompounding will vary by vehicle design, but the most common opportunities for decompounding are those listed below (Bjelkengren, 2006):

   - Tires,
   - Wheels,
   - Powertrain,
   - Suspension system,
   - Braking system,
   - Bumpers,
   - Fuel and exhaust systems,
   - Steering system, and
   - Electrical systems and wiring.

   In this committee's analysis, the decompounding can be defined as:
   Decompounding = secondary mass reduction / primary mass reduction.

For a 10 percent mass reduction in midsized and large cars, 7.14 percent of the mass reduction is considered to be primary mass reduction and 2.86 percent of the mass removed from decompounding. For a 10 percent removal from light-duty trucks, 8 percent of the total mass removed would come from primary and 2 percent would come from mass decompounding, or secondary mass reduction (Table 6.4).

The committee reviewed the targets in the TSD for the 2017-2025 rule, shown in Table 6.5. It concluded that these are conservative targets because OEMs are likely to imple-

[4] Mass decompounding is the opportunity for additional, or secondary, mass reduction in a vehicle's design based on the new specifications of the newly designed vehicle following the initial, or primary, implemented mass reductions.

TABLE 6.4 Illustration of the Difference in the Distribution Between Primary Mass Reduction and Secondary Mass Reduction for a Total of 10 Percent Mass Reduction

Mass Reduction (%)	Cars	Trucks
Primary	7.14	8
Secondary	2.86	2
Total	10	10

TABLE 6.5 NHTSA-Estimated Maximum Mass Reduction for a Safety-Neutral Environment

Vehicle Type	Maximum Mass Reduction (%)
Subcompact	0
Midsize car	3.5
Large car	10
Minivan	20
Light truck	20

SOURCE: EPA/NHTSA (2012a).

ment more aggressive levels of mass reduction. Although OEMs tend to implement fuel consumption reduction technologies ranked in the order from highest to lowest cost effectiveness, there are technologies where other design considerations might dictate a different strategy. The committee feels that lightweighting is an example of a technology that might be implemented before technologies with a better cost effectiveness in terms of fuel consumption reduction because it offers other benefits.

Implementation of mass reduction techniques can provide several benefits that might be attractive to an OEM. Reducing vehicle mass can be even more attractive to consumers, and OEMs may perceive mass reduction techniques to be less risky than advanced engine or propulsion technologies. For light trucks, mass reductions can also increase towing and load capacities without any modifications to the powertrain. For hybrid and electric vehicles, a reduction in mass can allow the OEM to either increase the range or reduce the battery capacity to reduce cost while maintaining range. From a design perspective, lightweighting techniques can offer a proven method for reducing fuel consumption that, while complex, is not limited by the same functional requirements to the extent that powertrain or transmission technologies are.

The Department of Energy (DOE), with input from 135 participants, including representatives of 36 domestic and international automobile manufacturers, has identified five major vehicle component groups for lightweighting that can lead to an overall 20 percent mass reduction, a common industry target for 2020 (Schutte and Joost 2012) (see Table 6.6). The industry and a number of studies (two of which are summarized in Box 6.1) concur that the vehicle body offers the greatest opportunity for lightweighting relative to other parts of the car. The powertrain system, although already significantly lightened (for example, by using aluminum heads and blocks), receives additional benefit from the downsizing enabled by lightweighting other areas of the vehicle and by boosting the engine through turbocharging or supercharging. The chassis and suspension, like the body, have many parts and therefore significant lightweighting potential. The vehicle interior is already plastic intensive and is expected to stay so, though some opportunity exists to reduce the weight of plastic panels by further reducing the density of the plastics. Other places to reduce interior weight include seating and components behind the instrument panel. Overall, the greatest change in design and materials can be expected in the body and chassis/suspension due to aggressive mass reduction in those subsystems.

The long-term goal of the US Drive program sponsored by DOE is a 50 percent reduction in weight. The Partnership for a New Generation of Vehicles research effort from 1994 to 2002 was an early effort to conceptualize and build highly fuel-efficient vehicles. The mass reduction goal was 40 percent. Actual vehicles achieved a mass reduction of 20 to 30 percent.

From an aluminum/magnesium-intensive design, Lotus Engineering projected a 2020 potential for about a 20 percent weight reduction at zero cost and a 40 percent weight reduction at a cost of about 3 percent of total vehicle cost (Lotus Engineering 2010).

The Aluminum Association and Ducker Worldwide conducted a study that found all auto manufacturers are working on mass reduction as a critical technology to reduce fuel consumption (Ducker Worldwide 2011). Ducker found that "no single vehicle technology strategy can effectively achieve a 50+ mpg fuel economy target without *significant* weight reduction." Based on Ducker's estimation, the average weight

TABLE 6.6 DOE Assessment of Five Major Light-Duty Vehicle Component Groups Leading to Overall 20 Percent Mass Reduction by 2020

Light-Duty Vehicle Subsystem	Distribution of Vehicle Weight by Vehicle Group (Current Vehicles) (%)	Targets for Weight Reduction for Light-Duty Vehicles Through 2020 (%)
Body	23-28	35
Powertrain	25-28	10
Chassis and suspension	22-27	25
Interior	10-15	5
Closures and other	15-16	—
Complete vehicle total	100	20

SOURCE: DOE (2013).

BOX 6.1
Committee Summary of Two Studies on Reducing Vehicle Mass

"Mass Reduction for Light-Duty Vehicles for Model Years 2017-2025"
Principal Investigators: Electricore EDAG and GWU

In 2012, the DOT contracted with Electricore, EDAG, and George Washington University to design a midsized vehicle using lightweighted materials. The goal was to achieve maximum mass reduction within several performance and technological boundaries. Parameters for the design of the vehicle included maintaining vehicle footprint, retail price (with a 10 percent margin), performance, and safety. Production parameters stated that the material technology and engineering processes must be realistically achievable during the 2017-2025 time frame and should consider a vehicle volume of 200,000 vehicles. Additionally, only standard gasoline powertrains were to be considered – excluding hybrids, plug-in hybrids, and other electrified powertrains.

The resulting vehicle design claimed a 22.4 percent reduction in the overall mass of the vehicle. The estimated incremental cost of this design was $319 ($.96 per kg). In addition to the use of lightweighted materials, the powertrain was reduced from a displacement of 2.4 L (177 hp) to 1.8 L, with an accompanying reduction of 37 hp.

An Analysis of Impact Performance with Cost Considerations for a Low Mass Multi-Material Body Structure
Principal Investigators: Lotus Engineering

In 2009, the Energy Foundation contracted with Lotus Engineering to perform a study on mass reduction using a 2009 Toyota Venza as the baseline vehicle. Two scenarios were developed for this study with one vehicle being a high-production-volume vehicle with a standard spark ignition engine and one low-production vehicle with a hybrid powertrain developed by EPA. Unlike the study performed by EDAG and GWU, this study was aimed at removing 40 percent of the total mass from the vehicle while maintaining vehicle safety and footprint. Lotus approached this task with a full vehicle design approach and heavily utilized computer aided design. All materials available were considered and as much recycled material as possible was incorporated.

The resulting Lotus vehicle design was able to remove 241kg (or 37 percent) from the body-in-white Toyota Venza. The redesigned Venza met all current safety and performance standards while reaching a cost of only 3% more than the baseline vehicle (Lotus Engineering 2010).

of vehicles in 2025 will be reduced by 408 lb compared to the average 2008 vehicle. More advanced powertrains (e.g., battery electric vehicles and fuel cell vehicles) place greater value on vehicle mass reduction because of the cost premium associated with the powertrain.

The 2050 Transitions Report estimated a mid-range mass reduction potential in 2030 of 20 percent for passenger cars and 15 percent for light trucks and an optimistic reduction potential of 25 percent for passenger cars and 20 percent for light trucks (NRC 2013). The difference between mid-range and optimistic was primarily due to the ongoing trend toward comfort and convenience features, which add weight. The difference between passenger cars and light trucks was primarily that light trucks had an allowance for functionalities such as towing capacity, which might be constrained by fuel economy designs.

Factors That Constrain Future Mass Reduction and Fuel Consumption Improvements

Vehicle weight decreased rapidly in the late 1970s and early 1980s because of high fuel prices and implementation of the initial CAFE standards. Weight then increased significantly from the mid-1980s to the mid-2000s, when fuel prices fell and fuel economy standards were kept constant. Thus, based on history, projecting weight trends into the future is very uncertain. Technologies optimizing safety, comfort/convenience, and low emissions have contributed to an overall increase in vehicle mass over the past 30 years.

- *Safety.* Weight associated with increased safety measures is likely to be lower than in the past. The preliminary regulatory impact analysis for the proposed 2025 CAFE standards looked at weight increases for a variety of safety regulations, including proposed rules that would affect vehicles through 2025, and estimated a potential weight increase of 100-120 pounds (NHTSA 2012). That is about a 3 percent mass increase.
- *Comfort and convenience.* There has been an increase in the weight of vehicles due to increased luxury and comfort accessories. Continued weight increases are inconsistent with a future accompanied by strong CAFE standards. Manufacturers will have a strong incentive to reduce weight.

- *Towing capacity*. A performance constraint that might affect mass reduction for some light trucks relates to towing capacity. Towing limits are dominated by factors such as engine power, frame stiffness, axle and tire load ratings, and transmission load capacity. The overall weight of a vehicle is not the primary design restraint, but the mass associated with a stiff platform and axle/tire/powertrain design strongly influences the overall weight of a vehicle.

Timing

Model Years 2015-2020

Steel is the dominant materials strategy today and will be slow to phase out because of the extensive infrastructure developed over several decades. The infrastructure includes metallurgical knowledge, modeling software, forming processes (especially stamping presses and die making), assembly, welding, and painting. The repair and recycling industries are also steel-focused. Since the late 1980s, high-strength steels have been used to help with safety and mass reduction. Every year, the industry advances the steel strength and forming technology to compete with other materials. Today's high-volume, steel-intensive vehicles have aluminum in key locations, including hoods (about 30 percent of today's U.S. cars have aluminum hoods) and deck lids but not generally in structural areas. Future growth in aluminum parts will continue (closures, body structure, and bumpers) using a material substitution approach (i.e., the designs may not be optimized for aluminum but can still realize a positive benefit from the conversion). Based on input from the tool and die industry, there has been a significant upswing in the demand for aluminum parts. The expectation is that several aluminum closures will be introduced by MY 2015 and more structural applications for aluminum are also expected soon. This evolutionary step will be toward a high-volume, mixed-material vehicle made principally of the two materials, with a manageable level of complexity that continues to use much of the same steel infrastructure. Occasional use of magnesium is commonplace for small parts (brackets, instrument panel crossbars, seating brackets, etc.), and the use of plastics and composites will continue to increase in nonstructural areas.

Although aluminum bodies have been around for many years (semimonocoque or unibody), they have been directed at niche, high-end vehicles; Europe has been a leader (Audi and Jaguar). The trend toward the aluminum *unibody* is a more recent development for use in mainstream vehicles (over 50,000 per year), and the U.S. auto companies are evaluating this approach. Unibody is important because it is the dominant architecture used for mainstream vehicles today. Whether or not aluminum unibody vehicles migrate to higher volume vehicles will depend on how aggressive OEMs need to be to reduce mass (i.e., depending on fuel economy legislation and the availability of other fuel-saving technologies) and if aluminum processing costs come down.

Model Years 2020 to 2025

The production of optimized mixed-material vehicles using greater quantities of aluminum, magnesium, and composites is expected to become more widespread. Incremental steps will continue to be made each year with these materials on a case-by-case basis, using a material substitution approach (one part at a time) and leading eventually to the more complex optimized vehicle design beyond the next 10 years. There will still be significant opportunities to improve the vehicle structure beyond this time frame with additional mixed-material optimization.

Costs for Mass Reduction

Auto manufacturers recognize the need to reduce vehicle mass to improve performance and efficiency. Technologies that reduce mass without compromising crashworthiness are available. Thus, cost becomes the main constraint, although there remain other barriers, including supply chain challenges, integration into existing vehicle architectures, technology risk, and so on. It is generally recognized that mass increases in automobiles in recent years have resulted from improving personal comfort features, crashworthiness, performance attributes such as ride quality (noise, vibration, and harshness) and acceleration (bigger and heavier powertrains), and meeting regulatory requirements for safety (crashworthiness) and emissions. The use of advanced materials and design techniques has mitigated additional increases in mass from these consumer-oriented trends. An expected outcome of today's regulations for fuel economy and emissions is greater focus on net mass reduction. The shift in priority from merely mitigating mass increases to achieving net mass decreases across the fleet is expected to realize 15 percent less weight by 2025. There will be a cost to achieve this result, but evolving industry transformations will help to contain it. Automakers will also have to respond to future regulations that will necessitate additional mass (e.g., NHTSA estimates an additional 100 to 120 lb. to the vehicle through 2025), but the net weight reduction is still anticipated to be 15 percent.

Mass Reduction Pathways and Challenges

The pathways to lightweight vehicles are not substantially different across manufacturers for similar, competing car models. With exceptions for performance-oriented vehicle designs, the costs and complexity generally progress as follows:

- Mild steel to high-strength steel (for structural parts and components such as seats) and composites/plastics

for nonstructural or semistructural parts (trim, oil pan, wheel well, brackets);
- Steel to aluminum hang-on panels (hoods and deck lids) and limited use of small amounts of magnesium for brackets;
- Steel doors to aluminum doors, and additional aluminum in chassis components; and
- More aggressive use of high-strength steel, aluminum, magnesium, and composites for other structural components and, potentially, an aluminum-intensive body and chassis.

The Electricore/EDAG/GWU study of the 2011 Honda Accord developed design and cost analysis for four scenarios that reflect this progressive lightweighting strategy (Singh 2012). There are a number of reasons automotive manufacturers usually prefer smaller, incremental implementations of mass reduction techniques in vehicle designs as opposed to approaches that might require a complete vehicle redesign and an aggressive substitution of lightweight materials. A few are mentioned here:

- Limited or Constrained Resources: to launch new technologies, a company's access to resources such as staffing and materials can be a limiting factor.
- Risk Aversion: implementation of a new technology always carries new risks, and the tolerance for risk is limited. Lightweighting risks are related to crashworthiness, corrosion, noise, and vibration;
- Engineering Constraints and Design Considerations: sharing of components across multiple car platforms constrains flexibility in re-designs, including powertrain components and body and chassis parts. Standardized product design and processing methods that have been globally implemented require revision, with cascading effects on other products and processes; and
- New Material Supply Chain: the development of a reliable and robust supply chain can be obstacle to including lightweight materials in a vehicle design. A design requiring the use of a new material versus the development of a supply chain for a new material has always been a "chicken-and-egg" challenge that can impede innovation. For example, during the writing of this report, the aluminum supply chain is at capacity for at least 30 months due to the volume that will be consumed by the new aluminum-intensive 2015 Ford F-150 truck design.

The launch of the 2015 Ford F-150 is clearly seen as transformational and not incremental. The decision to produce a truck with an all-aluminum body is seen as a bold move. Though aluminum bodies have been produced before, they have not been used in a high-volume truck. The success of this product will be of interest to many people in both the aluminum supply chain and the automobile industry. If significant problems arise, they will hinder future aggressive lightweighting efforts; if successful, the trend toward high-volume, aluminum-intensive vehicles will accelerate.

The progression of lightweighting materials includes a progression to more diverse and, in some cases, complex processes. Automakers are in general agreement that a closer-to-optimal vehicle design is coming, and it will include a more diverse mix of materials (especially mild steel, high-strength steel, aluminum, magnesium, and composites). This is referred to as the mixed-material car, and the trend today is along this pathway. The mixed-material car will not be less crashworthy, and it will be better engineered for mass and performance. This diversity offers more potential to eliminate mass even while reducing costs. However, the transformation to a more complicated vehicle will take time. The modeling software (CAE) must be developed, and the supply chain steps for materials, tooling, fabricating, and joining will all become more diverse, perhaps in some cases reducing economies of scale (for example, it may prohibit the sharing of parts for a single set of tools across vehicles). An example of the complexity that comes with the mixed-material car is joining. In addition to spot welding (today's dominant joining technology for sheet metal), there will be continued growth in laser welding, friction stir welding, multiple grades of weld-bond adhesive, crimping, fastening, etc. Modeling software will be needed for the joining methods required for different materials, increasing the engineering investment. The industry is on this pathway, but it will take decades before coming close to realizing its full potential.

Mass Reduction Cost Considerations

Projecting the future cost to reduce mass is very difficult. Modest lightweighting opportunities arise regularly that may be very low cost (or even negative cost) because of technological advances in materials or related technologies, and these can be implemented on new vehicle models on a material substitution basis. While several idealized studies expect total vehicle lightweighting costs to be low, auto manufacturers generally see many factors that result in higher costs. The Honda Accord and Toyota Venza studies on mass reduction have yet to be proven feasible from the manufacturing, consumer acceptance, or engineering perspectives. When auto manufacturers develop physical prototypes of vehicles, they invariably add mass to achieve a variety of performance requirements. As mentioned earlier, when lightweighting, automotive manufacturers have many variables and performance constraints or objectives to consider that affect cost—for example, crashworthiness, stiffness, noise transmission, commonly shared parts, different product life cycles and system integration.

There are continual improvements in modeling software that have reduced the lead time and development costs for introducing new designs. These modeling tools are being developed in academia, industry, and government and nongovernment organizations. As the new materials and joining have evolved to improve the structure, the software has also evolved to simulate crashworthiness. With better modeling analysis, the development speed improves, and the need to add inefficiently designed reinforcements late in the program is significantly reduced. However, to remain useful, the modeling software must stay current with new materials and new joining techniques, which can be a challenge.

It is broadly accepted that the cost of reducing mass increases with the percent of mass reduction. The four scenarios from the NHTSA/Honda Accord study below demonstrate this. Honda has issued its report on the results under Scenario 2 and suggested that the actual weight savings under this scenario is only 53 percent of the anticipated study results: 175 kg instead of 332 kg. Honda did not directly address cost, but much of the weight difference would result in additional cost as material is added back to resolve the design problems. The committee recognizes that customer acceptance and vehicle safety are major concerns when developing any vehicle design that aims to implement significant mass-reducing techniques. It is also reasonable that these concerns would apply constraints to the vehicle design that limit the extent to which lightweighting techniques can be applied. However, the committee feels it would be an ineffective approach for an OEM to design and produce a lightweighted vehicle design that does not factor in these constraints early in the design phase and then revisit the design in order to meet safety requirements and customer acceptance issues. Thus the committee is not able to judge what the net effect would be of addressing Honda's concerns through clean sheet design.

The automotive industry today is generally operating under Scenario 1 from the EDAG Study (AHSS dominant), with movement toward Scenario 2 expected over the next few years (Table 6.7). Scenario 3, with aluminum body-in-white, could generally occur (across multiple models) in the 2020 to 2025 timeframe, but likely only for a few models of vehicles, and that may be held back if supply chain problems occur or costs are significantly higher than shown below. Again, this emphasizes the importance of the F-150 launch.

A brief compilation of several sources for estimating the cost of weight reduction are summarized in Table 6.8 along with comments regarding the studies.

Derivation of Cost per Pound of Mass Reduction from EDAG Study

The results of the NHTSA-sponsored study to evaluate mass reduction opportunities on the 2011 Honda Accord provide insight into opportunities for reducing vehicle weight. The chart below, taken from the study, illustrates the exponentially increasing cost as more weight is removed. There is general acceptance of the exponentially increasing cost curve for reducing mass, with the initial cost for lower levels of mass reduction starting at or below zero (i.e., cost savings). Progressing up the curve to reduce more weight incurs higher costs as different mass reduction strategies are employed (Figure 6.3).

A similar analysis has also been performed on the 2014 Silverado pickup truck, demonstrating an exponentially increasing cost curve as more mass is removed. The Silverado study is currently under peer review; however, as expected, the cost estimates to remove mass are greater than for the Honda vehicle. This is due, at least in part, to truck performance requirements for towing and cargo capacity that limit weight reduction, especially secondary mass decompounding with engine and transmission downsizing.

Automaker responses to independent mass reduction studies have been mixed. The studies offer creative insight into new design concepts, often using near-term-future technologies. However, they are also developed without many of the business constraints a manufacturer has to manage. For this reason, the mass reduction and cost estimates from independent studies are recognized as obtainable under ideal

TABLE 6.7 Summary of Results from Electricore/EDAG/GWU Study Sponsored by NHTSA

	Scenario (increasing aggressiveness):			
	1	2	3	4
BIW	AHSS	AHSS	Aluminum	Composite
Closures	AHSS	Aluminum	Aluminum	Aluminum/magnesium
Chassis	AHSS	Aluminum	Aluminum	Aluminum
Seats	AHSS	Magnesium	Magnesium	Magnesium
Mass savings (kg)	284	332	372	421
Total ($)	111	319	927	2,719
$/lb	0.18	0.44	1.13	2.94
Mass reduction(with powertrain) (%)	19.2	22.4	25.1	28.5

NOTE: BIW, body in white; AHSS, advanced high-strength steel.
SOURCE: Summary results from Electricore/EDAG/GWU study sponsored by NHTSA (2012).

TABLE 6.8 Analysis of Mass Reduction Studies and Results

Description of Study/Source	General Results	Comments
Toyota Venza Phase 2 Funded by EPA (+ International Council on Clean Transportation & Environment Canada) FEV, EDAG, and Munro Consultants Expand initial Lotus mass-reduction study and propose alternatives Target: 20% vehicle weight loss at minimum cost Use 2010 MY (2007 launch/3,772 lb.) Use current manufacturing technologies; cost effective for 2017-2020 production	• Strong emphasis on CAE optimization methodology. • Requires a comprehensive product development process. Consultants believe the optimization approach can be implemented. • High-strength steel, aluminum, component downsizing, thin glass, magnesium parts, lighter shocks, smaller wheels, downsize engine. • Vehicle: 689 lb reduced, 18% of vehicle. • Cost *saved is* $134/vehicle, $0.20 per lb saved (includes cost of tooling).	• Analysis is based on 2007 vehicle 10 years into future; doesn't consider added mass for crash, emissions, or driver comfort. • Some gauges and grades not commercially available. • Proposed magnesium, "too expensive" (except premium cars). • Thinner glazing and wheels transmit noise and vibrations.
2011 Honda Accord Funded by NHTSA Baseline vehicle: 27 mpg combined Electricore, EDAG, GWU Consultants Not to exceed 10% cost premium Technology/cost estimates for 2017-2025	• Simulated crashworthiness and overall vehicle performance. • Body mostly HSS with all-aluminum closure panels, some magnesium. • Recognized that magnesium doors were not practical. • 22.4% total vehicle weight savings (intermediate scenario) = 730 lb, resulting in $0.44/lb cost premium (slightly different results for different scenarios).	• Good study and identification of technologies are consistent with industry direction. • Overall performance of lightweighted vehicle is compromised. • Performance critique: handling, ride/comfort, noise, and safety (crashworthiness). • Business constraints: platform commonality. • OEM accepts 53% of downsizing/LW opportunity.
NRC, 2011, *Assessment of Fuel Economy Technologies for Light-Duty Vehicles,* "Non-Engine Technologies," Table 7.8	• 1%, $1.41/lb • 5%, $1.65/lb • 10%, $1.98/lb	• Estimates for other reductions: (%) / Low ($) / High ($) 1.5 / 1.28 / 1.53 5.0 / 1.50 / 1.80 7.5 / 1.65 / 1.98 10.0 / 1.80 / 2.16 15.0 / 2.01 / 2.19 20.0 / 2.22 / —
EPA/NHTSA, TSD, 2012	• Based on weighted average of various lightweighting studies. • Total cost = $4.36 × percent of mass reduction level (e.g., 10% mass reduction = $0.436/lb). • Maximum feasible mass reduction varies by vehicle size to meet safety neutrality requirement (0% for sub-compact and compact, 3.5% for midsize passenger car, 10% for large passenger car, 20% for minivan and light-duty truck). • NHTSA and EPA weighted scores independently. • Average of the two weighted scores used to reach a consensus value. • EPA estimate was $2.17 (e.g., 10% mass reduction = $0.217/lb). • NHTSA estimate was $6.49 (e.g., 10% mass reduction = $0.649/lb).	• Estimates are significantly less than industry estimates. • Data based on an incomplete set of studies.

continued

TABLE 6.8 Continued

Description of Study/Source	General Results	Comments
Auto manufacturer sentiment	• Pathways to 2025 will focus primarily on more high-strength steel and aluminum. • Magnesium and composites will have minimal impact. • 10% to 15% achievable by 2025. • DMC net costs for 3 companies: (1) \$1.80, (2) \$2.50 for about 12%-15% mass reduction, (3) \$2.22 for 7%).	• Auto manufacturers consistently much higher than mass reduction studies by independent consultants. • Mostly conversion to aluminum-intensive body components. • Higher number (\$2.50) not "optimized" vehicle with de-compounding (may be \$1.92/lb) assuming 30% compounding. • Estimate range: \$1.80 to \$1.92/lb.
EPA, NHTSA, CARB	• The relationship in the U.S. EPA/NHTSA 2012-2016 rulemaking assumed a constant \$1.32/lb for vehicle mass reduction up to 10%. • The 2010 joint TAR (EPA, NHTSA, & CARB) modified the cost using a curve resulting in \$0.43/lb for 10% mass reduction. • ARB weighted studies according to a formula that has multiple subcategories for each factor: $W_{study} = W_{design} \times W_{cost} \times W_{peer\ review}$ (LEV III GHG TSD, December 7, 2011).	• Change in cost estimation from \$1.32/lb to \$0.43/lb from the rulemaking to the joint TAR. • The heuristic weighting scheme and regression method for studies not well documented or validated. • Final scoring minimizing auto manufacturer input. Highest weight for 25 studies assigned to debated Lotus, 2010/Low Development Study.

NOTE: GHG, greenhouse gas; CARB, California Air Resources Board; TAR, technical assessment report; LEV, low emission vehicle.

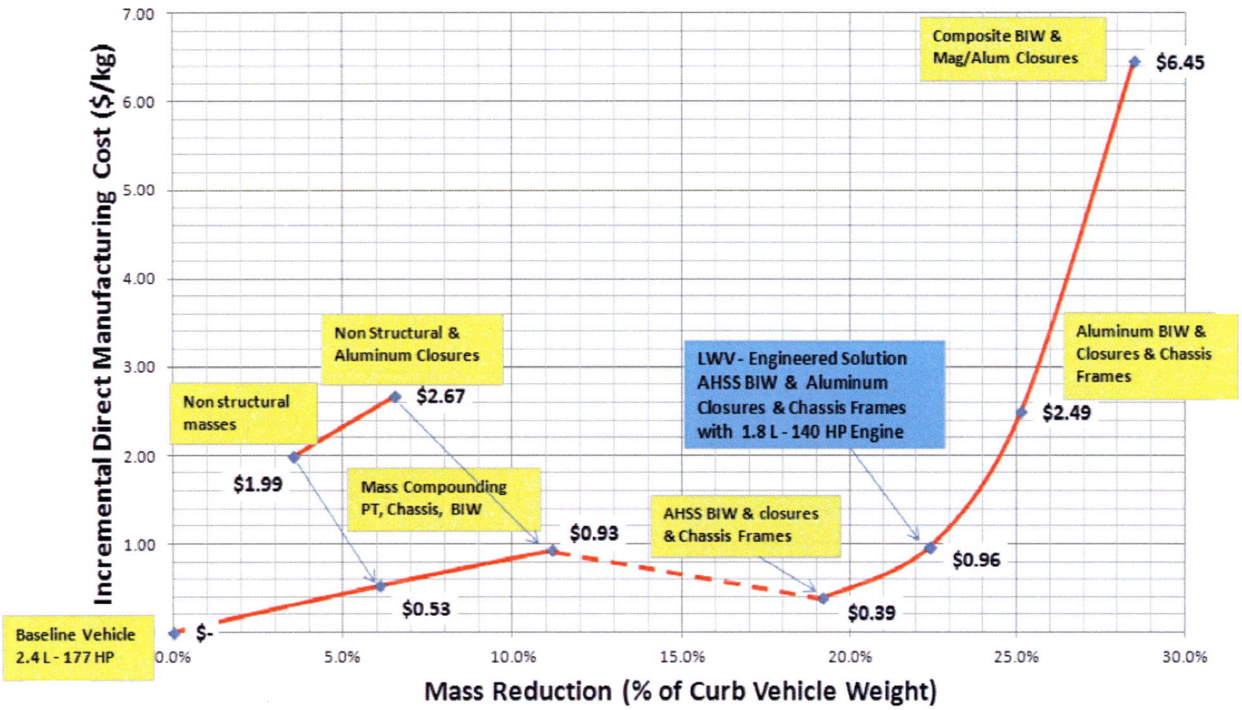

FIGURE 6.3 Cost per percent mass reduction from EDAG study of 2011 Honda Accord.
SOURCE: Singh (2012).

conditions and represent maximum mass reduction potential at the lowest potential cost. Several manufacturers have been consulted about lightweighting, and all have indicated that the cost to remove weight is much higher than the idealized studies indicate, generally starting at around $2.00/lb and increasing up to $4.00/lb or more (at levels of mass reduction from a few percent to 5-10 percent). In some cases, manufacturers support modest opportunities for "free" lightweighting (e.g., 1-2 percent). While there are opportunities to remove weight at low cost, concerns arise with the complexity of introducing new materials (e.g., magnesium and composite parts), their reliability over the life of the vehicle, and vehicle performance (vibration, structural performance such as stiffness, paint-ability, etc.).

Factors affecting mass reduction and cost that were raised by manufacturers include the following:

- Independent consultants are unaware of or unable to analyze complex interactions between vehicle subsubsystems affecting crashworthiness and other performance issues such as noise and vibration. The independent studies may provide generally good results, but they are incomplete. Many issues are only found when prototype vehicles are made and evaluated, generally resulting in countermeasures that add cost and weight.
- Given the competitive importance of ride and handling performance, automakers are very sensitive to technologies that affect this metric. Substituting advanced materials may be structurally sufficient but may adversely affect ride and handling, thus requiring various countermeasures to mitigate this unintended impact.
- Auto companies use many parts across multiple models or vehicle platforms and cannot, for practical reasons, optimize every part on every model of vehicle to maximize mass reduction. The sharing of parts is done for many reasons, including cost, quality, risk mitigation, and resource management. Similarly, some new materials/parts cannot be integrated easily into existing manufacturing facilities. Engines and transmissions are examples of systems used for multiple vehicles. In the Honda study, over 60 percent of the secondary mass savings was from downsizing the engine and transmission (see the section "Growing Impact of Global Platforms on Vehicle Design Optimization" in Chapter 7).

Committee's Mass Reduction Approach

The committee follows the approach taken in many of the studies described earlier, by estimating costs for various materials-based approaches to reducing vehicle mass. In the following section, increasingly aggressive percentages of removing mass from a vehicle model design are described in Scenarios 1 through 6. These scenarios are the committee's effort to generalize the selection of materials, engineering approaches, and common practices that OEMs will consider to achieve different percentages of mass reductions. It follows a progression where the lowest reductions are based on optimization and materials substitution; higher levels are achieved with replacing mild with high-strength steel and aluminum; and the highest levels are achieved through greater use of composites, including carbon fiber and other lighter metals such as magnesium. The scenarios do not include any weight additions that may be needed to meet future safety requirements. The two sets of values reported for these costs are based on two perspectives of how much mass reduction could be obtained for zero cost, a critical element for estimating the costs of mass reduction. The justification for applying these two different cost assumptions is based on two fundamental ideas. The committee considered both 0 percent and 6.25 percent to be plausible assumptions regarding the availability of zero-cost mass reduction. At the Society of Automotive Engineers (SAE) 2015 World Congress, a presentation from EPA highlighted possible subsystems that may offer zero net cost opportunities for mass reduction, with strategies such as implementing new component designs, material substitution and consolidation, and new material processing techniques (EPA/SAE 2015). These strategies entailed using new materials and designs in connecting rods and roller bearing and new materials, weather seals, and consolidating components in airbags. This approach is consistent with cost estimates for other technologies, where the committee's most likely estimates include two values that are meant to represent not the full uncertainty range but rather the different possible most likely values based on expert views represented by the committee.

In order to be consistent with other estimates of cost and fuel consumption benefits in this report, the committee considered these improvements relative to a 2008/2010-era null vehicle. This is a challenge as there is less certainty in terms of materials and design for such a vehicle than there is regarding other technologies. Based on the committee's expertise, such a vehicle is mainly steel, less than 10 percent aluminum, and a mix of other materials, with the steel being a mix of mild and high-strength steel, but with a higher fraction of mild steel. This is relevant for the use of vehicle-specific lightweighting studies by the Agencies. As described earlier, there have been several teardown/CAE studies to help assess the opportunity and cost for reducing mass in vehicles. These studies are difficult to generalize and apply to other vehicles because there is such wide variation across all vehicle models.

The committee's cost estimates also consider mass reductions due to decompounding. The mass reduction studies have shown that powertrain downsizing can have the greatest secondary mass reduction benefit. However, because of the long life cycles of powertrains (vis-à-vis car models) and the fact that individual powertrains are shared across multiple vehicle platforms, the cost analysis (below) assumes that powertrain downsizing occurs only when mass reduction is 10 percent or greater. Mass decompounding potential in trucks is less than in cars because of truck performance requirements, which significantly reduce the potential to downsize systems such as engine, transmission, wheels, tires, shocks, and brakes. For the purpose of this cost analysis, the committee assumes the mass decompounding potential in cars is 40 percent and in trucks, 25 percent. The effect of this difference on the cost estimates for trucks results in a 12 percent greater primary mass reduction cost per pound than in cars due to the 12 percent increase in primary mass removal required for the same total mass reduction (see Table 6-4). This assumption is applied by the committee throughout this analysis for all levels of mass reduction.

Scenario 1 – 2.5 Percent Mass Reduction

New materials and components are regularly developed over time that can reduce mass at negative to little or no cost, and there are often opportunities to introduce advancements to an existing vehicle. This occurs, for example, with advances in materials and design optimization. This scenario is one of the most debated because no vehicle is fully "optimized," and introducing many small incremental lightweighting changes may not be cost effective. Additionally, what may be considered an "optimized" vehicle design today will continue to evolve as design techniques and industry's increased experience with material substitution continue to improve over time. Manufacturers are also cautious about implementing some of these technologies because of concerns over customer satisfaction and possible compromises to vehicle performance. In the committee's cost analysis, no decompounding is applied for this level of mass reduction. As described in the committee's approach to mass reduction, the committee's estimates of costs for a 2.5 percent mass reduction are based on two perspectives on how much mass reduction could be obtained for zero cost. One value is based on the perspective that an OEM will be able to achieve a 2.5 percent mass reduction in a vehicle design at no additional cost. The second value is based on the perspective that any mass removed from a vehicle design would come at a cost, and the committee estimates that a 2.5 percent mass reduction will likely cost $0.25/lb.

Scenario 2 – 5 Percent Mass Reduction

Nonstructural mass reduction was achieved without secondary mass reduction at a cost that ranged from $1.99/kg to $2.67/kg for approximately 5 percent mass reduction (EDAG 2012). There are material substitution opportunities with some items, such as wiring harness (aluminum), plastic trim, instrument panel parts, battery, tires, and lighting. Many of the opportunities and concerns outlined in Scenario 1 will continue to hold true for Scenario 2. As with Scenario 1, no decompounding is applied in the committee's cost analysis for Scenario 2. Again, the committee recognizes that there will be circumstances where an OEM will be able to achieve a 5 percent mass reduction in a vehicle design at no additional cost. For circumstances that do not allow for any free mass reductions, the committee estimates that the cost of a 5 percent mass reduction to a vehicle design will likely be $0.50/lb.

Scenario 3 – 10 Percent Mass Reduction

The EDAG study resulted in a cost of $0.96/kg ($0.44/lb) for a 22.4 percent reduction in mass primarily using high-strength steel and aluminum closure panels. Necessary "countermeasures" (identified by Honda) to accommodate the mass reduction technologies and their additional mass requirements are listed below:

- Subframe safety, 0 lb (0 kg)
- Dashboard crashworthiness, 55.11 lb (25 kg)
- Side impact safety, 22.04 lb (10 kg)
- Rear crash safety, 33.07 lb (15 kg)
- Ride comfort; NVH; handling, 39.68 lb (18 kg)
- Other (miscellaneous), 15.43 lb (7 kg)
- Business conditions (platform parts), 88.18 lb (40 kg)
- Add-back for decompounding, 92.59 lb (42 kg)
  Total, 346.13 lb (157 kg)
  (346.13 lbs. reinstated by Honda to the EDAG study's initial 730 lbs.)

Decreasing the initial mass reduction by 157 kg and adding cost for the material used by the countermeasures (157 kg × $1.20/kg = $188) results in $2.90/kg[5] ($1.32/lb) for a net mass reduction of 11.8 percent. A 10 percent discount (estimated) to adjust for the added countermeasures and the higher cost for 11.8 percent mass reduction (versus 10 percent) reduces the cost to about $1.18/lb. In the committee's cost analysis, a 40 percent decompounding is assumed for cars and a 25 percent decompounding is assumed for trucks. Allowing for 6.25 percent of the weight reduction at zero cost, the committee estimates a cost of $0.44/lb for 10% mass reduction. For situations where no mass is removed at zero cost, the committee estimates the likely cost for 10% mass reduction will be $1.18/lb.

Scenario 4 – 15 Percent Mass Reduction

This scenario analyzes a conversion to an aluminum car body for a 3,800 lb vehicle using data from the EDAG study with other estimates (Table 6.9). The steel body weighs approximately 863 lb, and the aluminum equivalent body is estimated to be 40 percent lighter than the steel body. The closure panels (hood, deck lid, and four doors) can also be

---

[5] $\frac{\$319 + \$188}{175 \text{ kg}} = \$2.90/\text{kg}$.

TABLE 6.9 Analysis of Mass Reduction and Direct Manufacturing Cost for Conversion to an Aluminum Body Vehicle

	Steel	Aluminum	Difference		Comments
Vehicle Weight	3800				Curb weight
Final Body Weight (BIW) (lbs)	863	518	345.3	40%	From EDAG study (BIW)
					Al BIW 40% lighter than steel
Offal	1.4	1.4			40% industry average for scrap
Total Required Material (lbs)	12009	725	-483.4		Pounds
Average Cost ($/lb)	$0.50	$2.00			Various Steel/Aluminum grades
Total Material Cost	$605	$1,450			
Offal (lbs)	345	207	-138.1		
Offal Value	$0.10	$1.10			Scrap value per pound
Offal Recycled (lbs)	311	186		0.9	Pounds recycled per BIW
Reclamation Value	-$31.08	-$205.12	-$174		Material less recycled scrap
Net Total Mtl. Cost	$573.61	$1,245.22	$671.61		Estimate: weld/adhesive/fasteners
Joining Cost	$250	$500	$250		Steel: $0.05 – 4000 spot welds
					BIW Cost
Total Material Assembly	$824	$1,745	$922		
Mass Reduction Analysis					
Body (lbs)	863	518	345.3		
Closures (lbs)	367	277	89.9		From EDAG Study (doors, hood, lid)
Decompounding (lbs)	0	-174	174		40% of total mass reduction
Total	1230	621	608.9		
Total			16%		
Cost Analysis					
Body	$824	$1,745	$921		
Closures			$141		Costs from EDAG Study
Decompounding			-$174	$1.00	Assume $ value per pound saved
Total			$888		
Cost per Pound Mass Reduction			$1.46		

made from aluminum. Recognizing increases in the average material cost (about $2.00/lb for aluminum versus $0.50/lb for steel), recycling of waste material (recycling value of $1.10/lb for aluminum and $0.10/lb for steel), and the additional costs for joining (aluminum joining estimated to cost twice as much as steel), a $921 cost increase was estimated for the aluminum body-in-white (BIW). The cost increase for aluminum closures (hood, deck lid and four doors), modified from values in the EDAG study (not shown in the EDAG chart above, but available in the study), was $141.10 for a 89.9 lb weight reduction. Mass decompounding is estimated at 40 percent of the weight savings and returned the value of $1.00/lb saved (see EDAG Cost Study–2011 Honda Accord; aluminum closure cost estimation in Singh 2012).

In the analysis below, the conversion from steel to aluminum for the BIW and the closures is estimated to cost $888 to save 609 pounds, or $1.46/lb. Hence the committee estimates the cost of 15 percent mass reduction at $1.46/lb, assuming no mass reduction is available at zero cost. Alternatively, if 6.25 percent weight reduction is available at zero cost and the next 8.75 percent costs $1.46/lb, the cost estimate for 15 percent lightweighting is $0.86/lb.

Scenario 5 – 20 Percent Mass Reduction

This scenario approaches the most aggressive scenarios on the EDAG chart (similar to Option 3). In addition to an aluminum-intensive body and 40 percent mass decompounding, the aggressive use of magnesium components and composites is needed. The hood and roof will be composite, and the doors may be a combination of high-strength steel, composite, aluminum, and magnesium. Cost estimates (per pound) increase over the 15 percent scenario above, but by how much is difficult to estimate. While possible to implement, high-volume manufacturers struggle more with plastic/composite body panels because of quality (dimensional stability and surface finish) and paintability of nonmetals. The EDAG report adequately points out a number of these risks and manufacturing trade-offs. Twenty percent mass reduction (rather than 25 percent) may be more achievable for volume manufacturers through 2025 as they compromise on some of the options outlined in the EDAG report. Assuming all mass removal will have a cost, 20 percent total mass reduction is estimated at \$2.03/lb. Allowing instead for 6.25 percent of the mass to be removed at zero cost, the committee estimates that a 20 percent reduction is achievable at \$1.40/lb.

Scenario 6 – 25 Percent Mass Reduction

This is the most aggressive scenario, with composite body panels (carbon fiber) and aggressive use of aluminum and magnesium, along with aggressive decompounding. Expect both cost and mass reduction opportunities to be somewhat less than the EDAG chart due to risks and trade-offs. Limitations arise from quality (dimensional stability), joining complexity (extensive use of adhesives with greater complexity), production cycle times (composites process much slower than metal), long-term reliability, and recycling. Although this mixed-material pathway is the most promising, it also needs the greatest amount of development and supply chain advancement. Significant progress will be made with this technology by 2030 and beyond. For scenarios that do not allow any percentage of free mass removal, the cost is likely to be \$3.28/lb. Assuming that 6.25 percent of the vehicle mass is removed at no cost, removing 25 percent of the mass from a vehicle design is likely to cost \$2.46/lb.

Learning

Each of the six scenarios above is progressively more complex in terms of the design, development, and manufacturing of the vehicle. The amount of learning, and associated cost reduction, should be considered. For the most part, Scenarios 1 – 3 will have minimal learning associated with them. The major technology changes for Scenarios 1 – 3 rely mainly on material substitution: high-strength steel for mild steel and aluminum for steel. One of the positive attributes of these lightweighting pathways is that the metals are similar in many ways and use much of the existing steel infrastructure (predictive design, fabrication, and assembly). Furthermore, all auto manufacturers have experience with these materials and have been working with them for several years. The learning curve for aluminum is largely confined to the development phase for launching a new plant (which takes months, not years). The cost estimates for Scenarios 1 – 4 are essentially mature and are not expected to achieve any significant cost reductions.

Scenarios 5 and 6, however, have significant opportunity for learning and cost reduction. Most manufacturers have limited experience with mixed-material vehicles, especially those involving composites in high-volume production (over 100,000 units/year). Composites are the least standardized material relative to the metals used in automotive applications. More reinforced composites (glass and carbon), such as those proposed in Scenario 6, are often referred to as "engineered solutions" because their chemistries are uniquely developed for a specific application. Scenario 5 entails broader use of plastic and composites, along with adhesives and fasteners, which all have learning opportunity at mass production. Scenario 6 is similar, but with even greater complexity due to the more sophisticated engineered materials. Estimates for the learning potential for the six scenarios are listed in Table 6.10.

Fuel Consumption

Ricardo conducted a modeling study for fuel economy effectiveness; the results are summarized in Table 6.11. The fuel economy improvement is converted to fuel consumption improvement using the following formula:

$$\text{Fuel consumption } \% = 1 - \frac{1}{(1 + \text{Fuel Economy } \%)}$$

This formula was used to convert the fuel economy improvement estimates in Table 6.11 to fuel consumption reduction estimates in Table 6.12 by using a value of 10 percent mass reduction as a midpoint for the range of mass reduction levels considered in this study.

TABLE 6.10 Learning Factors for Levels of Mass Reduction

Mass Reduction (%)	Scenario	2012	2017	2020	2025
2.5	1	1.00	1.00	1.00	1.00
5	2	1.00	0.994	0.991	0.985
10	3	1.00	0.985	0.975	0.960
15	4	1.00	0.969	0.951	0.920
20	5	1.00	0.962	0.938	0.900
25	6	1.00	0.942	0.908	0.850

TABLE 6.11 Ricardo Estimates for Fuel Economy Improvements by Percent Weight Reduction over the EPA Combined Drive Cycle

	Passenger Vehicle		Truck	
	Base Engine	Downsized Engine	Base Engine	Downsized Engine
Gasoline	0.33%	0.65%	0.35%	0.47%
Diesel	0.39%	0.63%	0.36%	0.46%

SOURCE: Ricardo (2007).

TABLE 6.12 Reduction in Fuel Consumption per Percent Mass Reduction (percent)

	% Improvement in Fuel Economy per % Weight Reduction, EPA Combined Drive Cycle (Fuel Consumption Equivalent in Brackets)			
	Passenger Vehicle		Truck	
	Base Engine	Downsized Engine	Base Engine	Downsized Engine
Gasoline	0.33% (0.32%)	0.65% (0.61%)	0.35% (0.338)	0.47% (0.449)
Diesel	0.39% (0.375%)	0.63% (0.592%)	0.36% (0.348)	0.46% (0.440)

Mass Reduction Effectiveness

A measure of technology effectiveness (TE) is to divide the technology cost by the fuel consumption benefit. The smaller the TE, the more appealing the technology is for its cost effectiveness in reducing fuel consumption. A cost-effective pathway to reduce fuel consumption can use TE to prioritize the most cost-effective technologies to achieve a fuel consumption target. There are reasons, however, why companies might not always follow the TE ranking.

In addition to TE, two factors that affect the relative appeal of technology selection are availability and performance. A technology may be unavailable or have risk associated with it (e.g., supply chain or technology risk) that a manufacturer wishes to avoid. In lightweighting, for example, there is a potential for a material shortage if there is a major change in the market affecting supply or demand, as may be the case for aluminum today. In the case of performance, most technologies impact the driver in ways other than fuel consumption. Manufacturers vigorously compete across car models, focusing on vehicle performance experienced by the driver, for example, in these ways:

- Safety (crashworthiness),
- Steering feel,
- Driving responsiveness,
- Ride comfort,
- Noise from wind or the road,
- Vibrations, and
- Acceleration and stopping.

Noise and vibration concerns have been raised by the manufacturers in connection with the lightweighting studies, and they have indicated that countermeasures for these attributes are necessary. Crashworthiness is maintained or improved with all lightweighting designs (or the designs are modified to be safe). The other attributes (steering feel, driving responsiveness, acceleration, and stopping) can all be improved with lighter vehicles. As discussed earlier in this chapter, these attributes are important competitive differentiators that might favor lightweighting, even with a less competitive fuel economy cost effectiveness than other technologies that might degrade one or more performance attributes.

Table 6.13 summarizes the committee's estimates for the costs and effectiveness of mass reduction. As with its other cost estimates, committee members held a range of views on the best estimates of cost and effectiveness, and Table 6.13 reports the range of most likely values based on expert views of the committee. It is also important to repeat that these values are not meant to represent the full range of possible values for technology cost and effectiveness. This is especially true for mass reduction. The committee concluded that the uncertainty surrounding the cost of mass reduction is particularly large due to the wide array of approaches for reducing mass and the observation that each particular vehicle model is at a different starting point in terms of mass reduction opportunities. This is in contrast to SI technologies, where the step from naturally aspirated engines to turbocharged-downsized engines is a fairly distinctive step and much more common across OEMs.

Safety

When determining the maximum potential CAFE standards, NHTSA must assess whether a new technology or change in vehicle design to save fuel will have implications

TABLE 6.13 Committee Estimates of Direct Manufacturing Costs and Effectiveness for Mass Reduction for Midsized Cars, Large Cars, and Light-Duty Trucks

Midsized and Large Cars (3,500 lbs and 4,500 lbs)

Mass Reduction (%)	Cost Estimates Include Decompound	Percent Reduction in Fuel Consumption (%)	Most Likely Cost Estimates ($ per lb) 2017	2020	2025	TSD Estimates for 2017
2.5	No	0.80	0.00 - 0.25	0.00 - 0.25	0.00 - 0.25	$0.11
5	No	1.60	0.00 - 0.50	0.00 - 0.50	0.00 - 0.50	$0.22
10	Yes	6.10	0.44 - 1.18	0.43 - 1.17	0.43 - 1.15	$0.44
15	Yes	9.15	0.86 - 1.46	0.84 - 1.43	0.82 - 1.39	$0.65
20	Yes	12.21	1.40 - 2.03	1.37 - 1.98	1.31 - 1.90	$0.87
25	Yes	15.26	2.46 - 3.28	2.37 - 3.16	2.22 - 2.96	$1.09

Light-Duty Trucks (5,500 lb)

Mass Reduction (%)	Cost Estimates Include Decompound	Percent Reduction in Fuel Consumption (%)	Most Likely Cost Estimates ($ per lb) 2017	2020	2025	TSD Estimates for 2017
2.5	No	0.85	0.00 - 0.28	0.00 - 0.28	0.00 - 0.28	$0.11
5	No	1.69	0.00 - 0.56	0.00 - 0.56	0.00 - 0.55	$0.22
10	Yes	4.49	0.49 - 1.32	0.49 - 1.31	0.48 - 1.29	$0.44
15	Yes	6.73	0.96 - 1.64	0.94 - 1.60	0.91 - 1.55	$0.65
20	Yes	8.98	1.56 - 2.27	1.53 - 2.22	1.47 - 2.13	$0.87
25	Yes	11.22	2.76 - 3.67	2.65 - 3.54	2.49 - 3.31	$1.09

for the safety of the vehicle's passengers. It is the Agency's goal to draft rules that encourage manufacturers to develop solutions that maintain, or increase, safety while improving fleet fuel economy.

In order to discourage OEMs from improving fuel economy by simply making vehicle models smaller and lighter, NHTSA fuel economy regulations are based on vehicle models' length and width (or "footprint") rather than their weight or mass (this approach was recommended in the NRC [2011] Phase One report). In this section, the committee assesses recent evidence about the effects of reducing vehicle mass while maintaining footprint. At the time this report is being written, however, even the most comprehensive analyses and studies are challenged to isolate the effects of design from the multiple causes of crashes and their severity. Over the past 10-15 years, the understanding of the relationship between mass and safety has been enhanced by consideration of the role of vehicle footprint (as opposed to the role of mass) in occupant protection. However, even these analyses are confounded by driver and environmental influences on safety outcomes, as described below.

The biggest contributors to a fatal collision are commonly viewed as: (1) the driver, (2) the environment, and (3) the vehicle(s). For the purposes of this report, the committee focuses on the vehicle's role in providing occupant protection in the event of a crash and, more briefly, vehicle characteristics that may assist in avoiding crashes. However, the committee acknowledges the importance of the driver and the environment to safety and touches on both to give perspective on the safety implications of reducing mass in vehicles.

Driver behavior has long been recognized as the single biggest factor in the cause, or avoidance, of a fatal collision (Evans 1991). However, driver behavior in crashes (risky driving, distraction) and driver judgment and skills are extremely difficult to measure. Instead, researchers attempting to estimate the roles of mass and footprint on vehicle safety in statistical analysis rely on proxy measures of driver behavior associated with crashes, such as gender and age. Available proxy measures of driver characteristics, however, are crude and therefore imperfect for separating driver from vehicle characteristics in isolating the effects of mass reduction on vehicle occupant protection. A NHTSA study (Singh 2012) demonstrated that drivers under 30 or over 70 are more likely to be at fault in a two-vehicle crash than drivers between 30 and 70. An Insurance Institute for Highway Safety (IIHS) study (McCartt and Teoh 2014) for the years 2008 to 2012 stated that 28.5 percent of fatalities of drivers between the ages of 15 and 17 years old occurred in small vehicles. Midsized and large cars accounted for 23.4 percent and 11.7 percent of fatalities, respectively. While lighter-weight vehicles are capable of being more nimble and therefore may be better able to avoid crashes, younger male drivers also tend to drive vehicles at higher speeds, resulting in higher crash rates and, presumably, higher impact speeds (Wenzel

2012b). The speed at impact in crashes, however, is not measured, which further illustrates the difficulty and complexity of any analyses attempting to isolate the role of mass reduction and vehicle size from other causal factors (Singh 2012).

Environmental factors, such as highway design and traffic patterns, are understood to be the second biggest determinants of whether a crash will be fatal (Evans 1991). Although designed for higher speeds than other roads, rural interstates have the lowest fatal crash rates per mile driven of all highway classes; design minimizes opportunities for vehicles to conflict by eliminating crossing intersections and reduces the severity of crashes by having wide medians, shoulders and crash barriers that reduce opportunities for vehicle impacts and reduce the severity of the impacts that occur (GAO 2004). Rural two-lane roads, in contrast, often post high speeds relative to design, offer opportunities for vehicles to conflict at driveways and intersections, and provide less protection to avoid vehicles striking fixed objects off the roadway. Characterizing the specific environmental conditions under which each accident occurred is difficult, however, because minimal details are provided about roadway design and operating conditions in the police crash reports that researchers rely on in statistical analyses of vehicle safety. Crash reports provide basic information about time of day, visibility, and type of roadway, but not about the specific traffic characteristics that existed at the time of a crash or local design features that may have contributed to the cause and severity of the crash.

Vehicle Size, Vehicle Mass, and Crash Physics

One of the most common generalizations about mass and safety, often referred to as "the simple physics argument," is that all else being equal, the passenger in the lighter vehicle is at more risk than the passenger in the heavier vehicle. Other vehicle attributes, driver characteristics, and crash circumstances have a much greater effect on fatality risk than a reduction in vehicle mass or footprint (Wenzel 2012a). Even so, occupants of smaller vehicles are at greater risk of fatality in crashes, particularly in a crash with a vehicle of greater mass. When discussing the vehicle itself, the most comprehensive statistical analyses to date suggest that vehicle footprint has a greater influence on fatality risk than vehicle mass (Wenzel 2012a). The 2012 NHTSA study (Singh 2012) and the Dynamic Research, Inc. (DRI) analysis (Kebschull and Sekiguchi 2008) are in agreement that, in cars, reducing vehicle mass and footprint is associated with a higher risk of fatality than if mass is reduced while holding footprint constant. By reducing footprint, a vehicle is more likely to have less crush distance[6] to absorb crash forces, and this may increase the propensity for a vehicle with a high center of gravity to roll over. The DRI studies estimated the effect of mass reduction while holding footprint constant and identified the effects of the separate components of footprint, track width, and wheelbase on fatality risk. One of the major findings from the DRI study is that holding mass constant and increasing vehicle footprint improved societal safety. Second, holding vehicle size constant while increasing mass increased societal risk.[7] A reduction in crush distance resulted in a passenger's body absorbing more of the kinetic (impact) energy over a shorter period of time. Unlike if mass is increased, if size is increased while also improving safety design, safety will improve by reducing rollovers by providing a wider track width, or lowering the vehicle's center of gravity, and/or adding crush distance and improved design of the occupant compartment to better absorb impact forces. Additionally, the most recent analyses done by NHTSA, Lawrence Berkeley National Laboratory (LBNL), and DRI all seem to suggest that fatalities per vehicle miles traveled (VMT) will increase if footprint were reduced, holding mass constant. However, these studies are in agreement that if mass is reduced from the heaviest light trucks, societal fatality risk will decrease slightly (Wenzel 2012a; Kahane 2012). According to these studies, there appears to be a greater increase in risk when 100 lb is removed from lighter-than-average cars (< 3,106 lb) than when 100 lb is removed from heavier cars. In addition, based on the results of these studies, removing a greater percentage of mass from heavier vehicles and a smaller percentage of mass, if any, from lighter vehicles appears to maintain societal safety while improving fuel economy, which is in agreement with the TSD (EPA/NHTSA 2012b).

The DRI studies also separated the two components of fatality risk per VMT—crash probability (crashes per VMT) and crash outcome (fatalities per crash)—and found that, for lighter-than-average cars, mass reduction is associated with a large increase in crashes per VMT but a small decrease in fatalities per crash, leading to a net increase in fatalities per VMT. This is important because there is no obvious explanation for this result except that it might be a spurious correlation between mass and driver behavior (DRI and NHTSA seem to agree on this). It is possible that male or younger drivers are more robust than female or older drivers and therefore more likely to survive a crash. Another possible explanation is that the time of day at which the crash occurred was not taken into account in the analysis. Additionally, since heavier or larger vehicles may suffer less damage, these crashes may not be reported as regularly as crashes involving smaller or lighter vehicles. The DRI studies addressed other considerations left out, such as NHTSA's use of a single exposure measure (choice of exposure measure from among reasonable alternatives changes the significance and signs of

[6] Crush distance is the distance over which forces are absorbed during a crash.

[7] In this report, societal risk is used to describe the statistical probability of a fatality occurring for the occupants of the subject vehicle, the occupants of any involved vehicle(s), and any pedestrians or cyclists involved in a given crash. Personal risk, or occupant risk, is the statistical probability of a fatality occurring for only the occupants of the subject vehicle.

some key coefficients, indicating that results are not robust). NHTSA concludes that the DRI two-stage regression model does not cleanly separate the effects of crash avoidance and crashworthiness (Kahane 2012, Section 4.6). However, NHTSA concludes that this problem does not affect the baseline model of fatality risk per VMT since it combines the effects of mass reduction on both crash frequency and fatality risk per crash.

Identifying the effects of vehicle attributes on vehicle safety is challenging because driver behavior is many times as important as vehicle attributes in determining crash probability and severity, and vehicle crashes have complex causal factors that are often difficult to observe and measure. As a consequence, even small correlations between relevant driver behaviors and vehicle attributes, given that behavior is less than perfectly represented in a model, will result in spurious correlations that can easily dominate the results. Therefore, it may not be surprising to find that statistical analyses of the effects of vehicle mass and size on safety do not produce robust inferences for every vehicle class and size. However, in nearly every case that mass was removed from lighter-than-average cars, there was a slight statistically significant increase in fatality risk. Under such circumstances, willingness to explore alternative hypotheses and to test the robustness of results is critically important. DRI and LBNL suggested and conducted 19 sensitivity analyses to test the robustness of the baseline NHTSA regression model. For the most part, the effects estimated by the NHTSA baseline regression model are in the middle of the effects estimated by the 19 alternative regression models, and the effects of the alternative models are within the level of uncertainty of the baseline NHTSA regression model. In the NHTSA baseline model, the (positive) mass reduction coefficient is statistically significant for lighter-than-average cars only; the estimated mass reduction coefficient is positive for 18 and is statistically significant for 17 of the 19 alternative regressions. While the magnitude of the effect of mass reduction on fatality risk in lighter-than-average cars varies substantially depending on the choice of the measure of exposure and the data and control variables used, in virtually every case mass reduction is associated with a small increase in fatality risk in lighter-than-average cars.

Specifically, the DRI studies have made three crucially important contributions to understanding the relationship between vehicle size and weight and safety (Van Auken and Zellner 2013a, 2013b, 2013c):

1. In its 2003 study, DRI separated the effects of footprint and weight on fatality risk. In a 2010 study, NHTSA updated its 2003 results, including both mass and footprint in the same regression model. In the 2012 studies, NHTSA, DRI, and LBNL used analysis of variance inflation factors to demonstrate that including both footprint and weight in the same regression model would not produce inaccurate results. The 2012 studies indicate that mass reduction while holding footprint constant is associated with a small increase in risk for lighter-than-average cars only; the estimated effect on other vehicle types is not statistically significant.

   The DRI analyses separated the effects of size and weight on crash probability as distinct from crash outcome using a two-stage regression model.
2. The DRI analysis' separation of the effects of size and weight on crash probability resulted in the inference that mass does not operate to reduce fatalities and injuries given a crash but rather appears to reduce the probability of a crash in certain cases. There is the possibility that the empirical relationship between mass and crash outcome might be a spurious correlation rather than a genuine causal relationship. One of the things that supports this view is the way that the relationship changes depending on the exposure measure used. However, another explanation suggests that the many years of new safety regulations and crash testing have resulted in vehicle designs that mitigate the theoretical safety penalty in crash outcomes in lighter vehicles.
3. DRI also showed that the exposure measure chosen to normalize fatalities or crashes has an important effect on the inferences one might draw from the analysis. Exposure measures are in fact explanatory variables whose mathematical relationship to the dependent variable (e.g., fatalities, serious injuries) and coefficient are constrained by assumption. An exposure measure is by definition the factor or variable that the analyst asserts would have a unitary relationship to the dependent variable except for the effects of the other explanatory variables and a random error term. If this is not true, then coefficient estimates will be biased if either the erroneous portion of the exposure measure is correlated with the explanatory variables or there are omitted variables that are correlated with the explanatory variables. The fact that alternative exposure measures lead to quite different coefficient estimates suggests that one or both of these phenomena are present.

The crash outcome regression coefficients were not strongly affected by the change in exposure measure. Those coefficients tend to show consistent patterns across model formulations and data sets. In contrast, the crash probability regressions exhibit greater variability across data sets and model formulations in the relationships among mass, size, and crash probability, particularly for mass reduction in lighter cars showing a change from a 1.96 percent increase to a 1.45 percent increase in crash probability; a footprint reduction in lighter cars showing a change from a 1.36 percent increase to a 1.82 percent increase in crash probability; and a footprint reduction in heavier cars yielding a change from a 1.32 percent increase to a 1.82 percent increase in crash probability. Both the NHTSA 2012 safety study (Kahane 2012) and DRI's analysis reasonably surmise that the cause is deficiencies in the model, particularly imprecise or omitted explanatory variables. The implication is that the coefficients may be biased. While both studies make rigorous attempts to account for the effects of multicollinearity on the variance of coefficient estimates, there is no attempt to account for potential bias. This should be a priority in future research. In the future, studies in this area may benefit from including information on crash severity. This could potentially be performed by using currently available information, such as whether or not a vehicle was towed from the crash scene.

The regression models described above are being used to draw general conclusions about the effect of vehicle mass reduction on societal risk. However, it should be noted that they are estimating the recent historical relationship between mass and risk, after accounting as carefully as possible for differences in vehicles, drivers, crash times, and locations. There are likely to be other factors that have not been accounted for in reducing mass from any one vehicle, and conditions, vehicles, and technologies are likely to be different in the future.

Mitigating Mass Reductions Through Design

In the context of mass reduction, safety is primarily a design issue. Advanced designs that disperse crash forces and optimize crush space and energy management can allow weight reduction while maintaining or even improving safety. In a crash, occupant protection is provided by designing the vehicle structure to absorb energy and prevent intrusion into the occupant compartment. For instance, in single-vehicle collisions, the NHTSA 2012 study concludes that removing mass from a single vehicle while maintaining footprint would allow for less energy to be absorbed over a fixed distance, which should reduce the risk to passengers (Singh 2012). This would require that the structural strength of the vehicle be maintained or improved. The report recommended that the most effective methods of reducing mass and maintaining footprint are (1) substitution of lighter weight materials; (2) substitution of stronger materials while using less of them; (3) downsizing the engine and powertrain; (4) use of lightweighted features; and (5) reduction of body overhang outside the wheel dimensions.

Advanced materials such as high-strength steel, aluminum, and polymer-matrix composites (PMC) have significant advantages in terms of strength versus weight. For example, pound for pound, aluminum absorbs two times the energy in a crash compared to steel and can be up to two and a half times stronger. The high strength-to-weight ratio of advanced materials allows a vehicle to maintain, or even increase, the size and strength of critical front and back crumple zones without increasing vehicle weight and to maintain a manageable deceleration profile. And, given that all light-duty vehicles likely will be downweighted, vehicle-to-vehicle crashes should also be mitigated due to the reductions in kinetic energy of the vehicles. Lastly, assuming mass reduction without size reduction, vehicle handling (exacerbated by smaller wheelbases, for instance) is not an issue.

Fleet Mix and Transition

During the transition period, when masses of heavier vehicles are being reduced, there are concerns that there might be a negative impact on safety due to variance in the distribution of vehicle masses across the vehicle fleet. Simulation work has been done for four fleetwide mass reduction scenarios (Kahane 2012):

1. 100-lb reduction in all vehicles;
2. proportionate (2.6%) mass reduction in all vehicles;
3. mass reduction of 5.2% in heavier light trucks, 2.6% in all other vehicle types except lighter cars, whose mass is kept constant; and
4. a safety-neutral scenario, where mass is reduced 0.5% in lighter cars, 2.1% in heavier cars, 3.1% in CUVs/minivans, 2.6% in lighter light trucks, and 4.6% in heavier light trucks.

The most aggressive of these scenarios (reducing mass 5.2% in heavier light trucks and 2.6% in all other vehicles types except lighter cars) is estimated to result in a small reduction in societal risk.

Pedestrian Safety

In addition to occupant protection, vehicle design and material selection can influence the safety of pedestrians in the event of a collision. The primary focus of NHTSA has been to prevent a vehicle-to-pedestrian collision in the first place (e.g., Safe Routes to School). However, NHTSA is taking into consideration Global Technical Regulation No. 9 (GTR 9), which would affect the hood and bumper design of vehicles. For example, the 2013 Ford Fusion utilizes an aluminum hood in the United States (as a mass reduction measure) while the same vehicle in Europe requires a steel hood design due to the pedestrian injury laws in Europe (Ramesh et al. 2012). NHTSA is continuing to evaluate the potential of more stringent requirements (GTR 9) but had some concerns about whether the regulations would be relevant to the mix of vehicles in the United States (i.e., SUVs and trucks).

ROLLING RESISTANCE

In addition to aerodynamic drag and inertial force due to vehicle mass, tire rolling resistance is one of many forces that must be overcome in order for a vehicle to move.

$$F = C_{rr} N$$

where F is the force of the rolling resistance, N is the normal force, and C_{rr} is the rolling resistant coefficient.

Tires

When rolling, a tire is continuously deformed by the load exerted on it from the vehicle's weight. The repeated deformation during rotation causes energy loss known as rolling resistance. Rolling resistance is affected by tire design: materials, shape and tread design, and inflation. Underinflated tires increase rolling resistance. The opportunity to improve fuel economy by reducing rolling resistance is already used by OEMs to deliver lower fuel consumption.

There are performance trade-offs involving tires that tire manufacturers consider during design and manufacturing. These trade-off variables include, for example, tread compound, tread design, bead/sidewall, belts, casing, and tire mass. Important tire performance criteria affected by design and manufacturing include rolling resistance, tire wear, stopping distance with respect to road surface conditions, and cornering grip. Wear and grip are closely correlated to tread pattern, softer-gripping tread compounds, and footprint shape. A typical low-rolling-resistance tire's attributes could include increased specified tire inflation pressure, material changes, tire construction with less hysteresis, geometry changes (e.g., reduced aspect ratios), and reduction in sidewall and tread deflection. These changes would generally be accompanied by additional changes to vehicle suspension tuning and/or suspension design.

The impact of emphasizing one performance objective, such as low rolling resistance, over other performance criteria is inconclusive. While tires with low rolling resistance do not appear to compromise traction, they may wear out tread faster than conventional tires. A 2008 study by *Consumer Reports* summarized by *Automotive News* (Snyder 2008) concludes that there may be a reduction in traction of low-rolling-resistance tires that could increase the vehicle's stopping distance. However, the study was not rigorously controlled, and other influences may have confounded the results. The response by one tire manufacturer, Michelin (Barrand and Bokar 2008) argued that low-rolling-resistance tires can be achieved without sacrificing performance factors by balancing the design and manufacturing process variables. Tire makers continue to research how to get optimal performance (including fuel economy) without sacrificing other criteria such as safety or wear. Goodyear points out that performance trade-offs between rolling resistance, traction, and tread wear can be made based on materials and process adjustments, which also affect cost (Goodyear Tire and Rubber Company 2009).

Rolling resistance can also be affected by brakes. Low drag brakes reduce the sliding friction of disc brake pads on rotors when the brakes are not engaged because the brake pads are pulled away from the rotating rotor. The benefit compared to conventional brakes may be about a 1 percent reduction in fuel consumption.

Rolling resistance is also affected by tire inflation, so any technology that affects inflation levels can also affect fuel economy. Reducing tire inflation levels increases rolling resistance, which in turn increases fuel consumption. A tire pressure monitoring system (TPMS), required by federal regulation for all light-duty vehicles, can be set to different pressure thresholds, and the TPMS must alert the operator when "one or more of the vehicle's tires is 25 percent or more below the manufacturer's recommended inflation pressure" (NHTSA 2005). Goodyear has been developing a tire inflation monitoring system capable of self-pumping a tire when it falls below the recommended pressure. This technology is currently being deployed on a small scale, specifically in heavy-duty vehicles, but it is possible the trend will extend in to the light-duty vehicle industry.

Fuel Consumption

Reducing tractive energy does not translate into a directly proportional reduction of fuel consumption because of (1) the accessory load and (2) the possibility that the powertrain may then operate at worse efficiency points. Ensuring powertrain efficiency requires downsizing the engine and/or changing transmission shift points at the same time because powertrain efficiency is reduced with a lighter load, especially with SI engines that will then operate with more throttling. An OEM designing a vehicle with low-resistance tires can fully take advantage of rolling resistance changes by optimizing the powertrain.

A report on tires and fuel economy estimates that a 10 percent reduction in rolling resistance will reduce fuel consumption by 1 to 2 percent (NRC 2006). This reduction, however, is without changes in the powertrain. If the powertrain could be adjusted to give the same performance, then the benefit of a 10 percent reduction might be as much as 3 percent. Goodyear and Michelin supported these estimates in dialogue with this committee.

In 2005, measured rolling-resistance coefficients, C_{rr}, ranged from 0.00615 to 0.01328, with a mean of 0.0102. The best is 40 percent lower than the mean, equivalent to a fuel consumption reduction of 4-8 percent. Some tire companies have reduced their rolling-resistance coefficient by about 2 percent per year for at least 30 years. OEMs have an incentive to provide their cars with low-rolling-resistance tires to maximize fuel economy during certification.

In the 2050 Transitions report (NRC 2013), scenario projections of reductions in light-duty new-vehicle fleet rolling resistance by 2030 for the midrange case was 26 percent for passenger cars to 15 percent for light trucks. The optimistic-case rolling-resistance reductions were projected to be 40 percent for passenger cars to 30 percent for light trucks. Estimates of up to 40 percent were provided by certain tire manufacturers, but were put in the context of a high-end (cost premium, unquantified) tire targeted at low-rolling-resistance performance.

The TSD considered two levels of rolling resistance, one targeting a 10 percent reduction and the other targeting a 20 percent reduction in rolling resistance. The first level,LRR1 was defined as a 10 percent reduction in rolling resistance from a base tire, which was estimated to show a 1 to 2 percent effectiveness improvement in the Agencies' final rule for MYs 2017-2025. The 2011 Ricardo study used by the Agencies used a 1.9 percent fuel consumption reduction for LRR1 for all vehicle classes. LRR1 tires are widely available today and appear to constitute a larger and larger portion of tire manufacturers' product lines as the technology continues to improve and mature. The second level,LRR2 is defined as a 20 percent reduction in rolling resistance from a base tire, yielding an estimated 3.9 percent fuel consumption reduction for all vehicle classes. In the Agencies' CAFE model, this resulted in a 2.0 percent incremental effectiveness increase from LRR1. Tire industry input endorsed these numbers.

Future improvements in tire-pressure-monitoring technology and other innovative strategies include tire pressure self-pumping designs within the tire to maintain correct tire pressure in real-time; taller and narrower tires with higher inflation pressures; and other sophisticated tire-pressure-monitoring systems.

Timing

Low-rolling-resistance tires are already used by OEMs. Vehicle manufacturers have an incentive to provide their cars with low-rolling-resistance tires to maximize fuel economy during certification. In fact, some OEMs have been recommending one specific tire for each vehicle and urging vehicle owners not to use different tires. The 2050 Transitions report estimated about a 2 percent reduction in rolling resistance per year (NRC 2013). All vehicles today are being offered with low-tire-pressure monitors to warn the driver of underinflated tires for safety and fuel economy.

The discussion in the NHTSA/EPA rulemaking support documents concluded that tire technologies that enable improvements of 10 and 20 percent have been in existence for many years. Achieving improvements up to 20 percent involves optimizing and integrating multiple technologies, with a primary contributor being the adoption of a silica tread technology. This approach was based on the use of new silica along with a specific polymer and coupling agent combination. Tire suppliers have indicated there are one or more innovations that they expect to occur in order to move the industry to the next quantum reduction of rolling resistance.

Costs

The Phase 1 Report estimated the incremental cost for low-rolling-resistance tires to be $2 to $5 per tire (NRC 2011). One tire manufacturer suggested that tires that do not compromise stopping distance or tread wear could cost 10 to 20 percent more than conventional tires. (Note: The uncertainty surrounding low-rolling-resistance tires with respect to increased tread wear and stopping distance is the reason for increasing the estimated cost beyond the $1.00 per tire cost cited in NRC 2006). The NRC (2006) study recognized that an acceptable increase in tread wear and stopping distance might occur. An additional cost can be expected to minimize this increase in tread wear.

The 2050 Transitions report estimated that average future improvements by 2030 are estimated to provide 20-28 percent reduction in rolling resistance relative to 2010, for a fuel consumption reduction of 5-8 percent at a cost of $6.25 per tire (NRC 2013).

The TSD shows 2017 DMC estimates for LRR1 as $5 relative to baseline tires and LRR2 as $63 relative to the baseline tires (Table 6.14). This agrees with the 2012-2016 light-duty vehicle rule, since NHTSA/EPA estimated the incremental DMC at an increase of $5 (2007 dollars) per vehicle for the LRR1 10 percent reduction in rolling resistance. This included the costs associated with five tires. The Agencies used MY 2017 as the starting point for market entry for LRR2 and took into account the advances in industry knowledge and an assumed increase in demand for improvements in this technology, arriving at an interpolated DMC for LRR2 of $63 (2010 dollars) per vehicle relative to the baseline tire. The Agencies did not include a cost for the spare tire because they believed manufacturers would not include a LRR2 as a spare given the $63 DMC. At this time, data are not available to differentiate LRR2 costs for different size vehicles. Tire manufacturers endorsed the numbers used by the Agencies in their rulemaking support documents. However, they said that the manufacturing costs did not appear to take into account R&D costs, which are about 3 percent of total sales. The committee's summary of current studies on low-rolling-resistance tires can be found in Table 6.15.

TABLE 6.14 EPA/NHTSA Technical Support Document Direct Manufacturing Cost Estimates for Low-Rolling-Resistance Tires: Levels 1 and 2 in the 2017-2025 Time Frame (2010 dollars)

Cost Type	Lower Rolling- Resistance Tire Technology	Incremental to	2017	2018	2019	2020	2021	2022	2023	2024	2025
DMC	Level 1	Baseline	5	5	5	5	5	5	5	5	5
DMC	Level 2	Baseline	63	63	51	51	40	39	38	37	36

SOURCE: EPA/NHTSA (2012b).

TABLE 6.15 Summary of Estimated Direct Manufacturing Costs for Low-Rolling-Resistance Tires from Various Studies

Study	Year Published	DMC ($)	Rolling Resistance Reduction (%)	FC Reduc-tion (%)	Comments
NRC 2011 Report	2008	30-40	5-10	1-2	Stop distance and durability rely on quality of materials which cost more. Needs regulation.
2050 Transitions Report	2011	25	21-28	4-6	No degradation.
EPA/NHTSA TSD	2012a and 2012b	LRR1: 5 LRR2: 63	LRR1: 10 LRR2: 20	1.9 3.9	No degradation. Incremental to baseline.
Committee Estimates	2015	LRR1: 5 LRR2: 63	LRR1: 10 LRR2: 20	1.9 3.9	No degradation. Incremental to baseline.

Barriers

If performance trade-offs such as stopping distance, durability, or NVH are associated with lowering the rolling resistance of tires, then there would be significant barriers to its marketplace acceptance. The TSD indicated that the use of improved polymers, coupling agent, and silica was known to reduce tire rolling resistance at the expense of tread wear, but new approaches using novel silica reduced the tread wear trade-off. Recent research has indicated that reductions in rolling resistance can occur without adversely affecting wear and traction (Pike Research and ICCT 2011).

Low Drag Brakes

Low drag brakes reduce the sliding friction of disc brake pads on rotors when brakes are not engaged because the brake pads are pulled away from the rotating disc either by mechanical or electric methods.

Fuel Consumption

NHTSA and EPA estimated the fuel consumption reduction to be 0.8 percent based on the 2011 Ricardo study. The committee believes this is a reasonable estimate.

Costs

NHTSA/EPA estimated the DMC cost at $59 (2010 dollars). This estimate appears reasonable to the committee.

VEHICLE ACCESSORIES

As discussed in the Phase 1 Report, automakers are beginning to introduce electric devices (such as motors and actuators) that can reduce the mechanical load on the engine, reduce weight, and optimize performance, resulting in reduced fuel consumption. The most advantageous opportunities for converting mechanical devices to electrical are devices that operate only intermittently, such as power steering and air conditioning (AC) compressor. With the new EPA test procedures, some of the benefits from accessory electrification will be reflected in the fuel economy labels, and improvements in these areas will be pursued by auto manufacturers.

Power Steering

In the past, most power steering systems used a hydraulic system to steer the vehicle's wheels. The hydraulic pressure typically comes from a gerotor (sometimes referred to as a rotary vane pump) driven by the vehicle's engine. A double-acting hydraulic cylinder applies a force to the steering gear, which in turn steers the road wheels. Sensors detect the position and torque of the steering column, and a computer module signals a motor that provides assisting torque via the motor, which connects to either the steering gear or steering column. This allows varying amounts of assistance to be applied depending on driving conditions. The steering-gear response can be tailored to variable-rate and variable-damping suspension systems, optimizing ride, handling, and steering for each vehicle.

Electric power steering (EPS) gives more assistance as the vehicle slows down and less at faster speeds. EPS eliminates a belt-driven engine accessory and several high-pressure hydraulic hoses between the hydraulic pump, mounted on the engine, and the steering gear, mounted on the chassis. This greatly simplifies manufacturing and maintenance. By incorporating electronic stability control, electric power steering systems can instantly vary torque assist levels to aid the driver in corrective maneuvers.

Fuel Consumption

EPS and electrohydraulic power steering (EHPS) provide a potential reduction in CO_2 emissions and fuel consumption over hydraulic power steering because of reduced overall accessory loads. The systems eliminate the parasitic losses associated with belt-driven power steering pumps that consistently draw load from the engine to pump hydraulic fluid through the steering actuation systems even when the wheels are not being turned. The Phase 1 Report stated that Ricardo found that EPS reduced combined fuel consumption by about 3 percent based on full system simulation calculations. From this and the estimates provided in recent regulatory activities by NHTSA and EPA, the committee estimated that EPS reduces combined fuel consumption by about 1 to 3 percent on the FTP. However, the committee recognized that the fuel consumption reduction could be as high as 5 percent under in-use driving conditions. The 2050 Transitions Report estimated that EPS consumes 2-3 percent less fuel (NRC 2013). Some weight reduction is realized, and costs are similar to hydraulic systems.

EPA and NHTSA estimated a 1.5/1.3/1.1 percent reduction in fuel consumption for small/compact/full-sized passenger cars and a 1.2/1.0/0.8 percent reduction for light trucks of varying size (EPA/NHTSA 2012b). The 2010 Ricardo study confirmed this estimate. For large pickup trucks the Agencies used EHPS due to the utility requirement of these vehicles. The effectiveness of EHPS is estimated to be 0.8 percent.

Timing

EPS is an enabler for all vehicle hybridization technologies since it provides power steering when the engine is off. EPS also may be implemented on most vehicles with a standard 12 V system. Some heavier vehicles may require a higher voltage system or EHPS, which may add cost and complexity. At least one OEM said that electric power steering would be on all vehicles by 2014 MY.

Costs

The DMC for EPS was estimated to be $87 for the 2017 MY (Table 6.15). The Agencies use the same DMC for EPS as for EHPS. The Agencies consider EPS/EHPS technology to be on the flat portion of the learning curve and have applied a low complexity in their indirect cost multiplier (ICM) analysis. The committee agrees with their estimates for the DMCs for EPS (Table 6.16).

High-Efficiency Alternator

Alternators charge the battery and power the electrical systems when the engine is running. Typical alternator efficiency is 65 percent. Typical losses include electrical, magnetic, and mechanical losses. Clearly as the alternator becomes more efficient in the process of converting mechanical energy into electrical power, less fuel is consumed. Another issue is that the efficiency of an alternator varies with load; therefore, at certification loads, the efficiency is low.

Fuel Consumption

The NHTSA/EPA final rule considered two levels of improved accessories. For level one, IACC1, NHTSA incorporated a high-efficiency (70 percent) alternator, an electric water pump, and electric cooling fans. The second level of improved accessories, IACC2, added the higher efficiency alternator and incorporated a mild regenerative alternator strategy, as well as improved cooling. NHTSA estimated of 1.2/1.0 percent reduction in fuel consumption for small/large cars and 1.01 to 1.61% reduction for small/large light-duty trucks for IACC1 and an incremental effectiveness for IACC2 relative to IACC1 ranging from 1.85 to 2.55 percent reduction in fuel consumption for small/large cars and 1.74/2.15 percent reduction for small/large light trucks.

Costs

In the TSD, the Agencies estimated the DMC of IACC1 at $71 (2007 dollars) for the 2012-2016 rule. Converting to 2010 dollars and applying the appropriate learning factor, the DMC becomes $71 for this analysis, applicable in MY 2017 and consistent with the heavy-duty rule. The Agencies consider IACC1 technology to be on the flat portion of the learning curve and have applied a low complexity ICM of 1.24 through 2018 then a long-term ICM of 1.19 thereafter. The assumed cost is higher for IACC2 due to the inclusion of a higher efficiency alternator and a mild level of regeneration. The Agencies estimate the DMC of the higher efficiency alternator and the regeneration strategy at $43 (2010 dollars) incremental to IACC1, applicable in MY 2017. Including the costs for IACC1 results in a DMC for IACC2 of $114 (2010 dollars) relative to the baseline case and applicable in MY 2017. The Agencies consider the IACC2 technology to be on the flat portion of the learning curve. They have applied a low complexity ICM of through 2018, then a long-term ICM of 1.19.

Heating, Ventilating, and Air-Conditioning

Air-conditioning (AC) is standard equipment in nearly all new cars and trucks. According to the Agencies' TSD, over

TABLE 6.16 Direct Manufacturing Costs of Electric/Electrohydraulic Power Steering (2010 dollars)

Cost type	2017	2018	2019	2020	2021	2022	2023	2024	2025
DMC	87	86	84	82	80	79	77	76	74

SOURCE: EPA/NHTSA (2012b).

95 percent of the new cars and light trucks in the United States are equipped with mobile air conditioning systems. Three recent studies have estimated the impact of AC use on the fuel consumption of motor vehicles in the United States. Based on a combination of the results from these studies, EPA and NHTSA estimated that AC use accounts for 3.9 percent of the car and light truck fuel consumption in the United States (EPA/NHTSA 2012b).

There are two mechanisms by which vehicle AC systems contribute to increased fuel consumption and emission of greenhouse gases (GHGs). The first is direct leakage of the refrigerant into the air. The hydrofluorocarbon refrigerant compound currently used in all recent model year vehicles is R-134a. The second mechanism by which AC systems contribute to GHG emissions is the consumption of additional fuel required to provide power to the AC system and from carrying the weight of the additional fuel. This section will focus on the second mechanism related to fuel consumption by the AC system.

The fuel economy values obtained on the two-cycle (i.e., city and highway) fuel economy test do not reflect potential improvements in air-conditioning system efficiency, refrigerant leakage, or refrigerant global warming potential (GWP), termed off-cycle benefits. NHTSA and EPA allow auto manufacturers to count such decreases in fuel consumption and GHG emissions from HVAC improvements through credits. Credits can be earned for other technologies as noted in this chapter and Chapter 10, and are also earned for general compliance with the standards, as noted in the credit trading section in Chapter 10. Since EPA and NHTSA recognized that cost-effective air-conditioning system improvements will be available in the 2017-2025 time frame, the Agencies increased the stringency of the target curves based on their assessment of the ability of manufacturers to implement these changes. For the CAFE standards, an offset was included based on air-conditioning system efficiency improvements. For the GHG standards, a stringency increase was included based on air-conditioning system efficiency, leakage, and refrigerant improvements

For MYs 2017-2019, the new AC17 test will provide the means for manufacturers to demonstrate eligibility for AC efficiency credits. The AC17 test is replacing the previously required AC Idle test which did not capture the majority of the driving or ambient conditions when the AC is in operation. Results from the AC17 test allow manufacturers access to the credits from a menu based on the design of their AC systems. In MYs 2020 and thereafter, the AC17 test will be used not only to demonstrate eligibility for efficiency credits, but also to partially quantify the amount of the credit. AC17 test results with AC on and off ("A" to "B" comparison) equal to or greater than the menu value will allow manufacturers to claim the full menu value for the credit. A test result less than the menu value will limit the amount of credit to that demonstrated on the AC17 test.

Fuel Consumption

Air conditioning contributes significantly to the on-road efficiency gap between CAFE

certification values and real-world fuel consumption. The air conditioner is turned off during the CAFE tests consisting of the FTP and the highway fuel economy test (HFET) drive cycles, but in the real world, drivers tend to use air conditioning in warm, humid conditions and in cooler conditions for defrost operations. In the 2012-2016 MY rulemaking, the Agencies estimated the average impact of an air conditioning system at approximately 14.3 g CO_2 over an SCO3 test for an average vehicle without any of the improved air conditioning technologies discussed in that rulemaking.[8] For a 27 mpg (330 g CO_2/mi) vehicle, this is approximately 20 percent of the total estimated on-road gap.

Most of the excess load from AC systems on the engine is due to the compressor, which pumps the refrigerant around the system loop. Additional loads on the engine come from electrical or hydraulic fans used to facilitate the exchange of heat across the condenser and radiator. The technologies that manufacturers are expected to use to generate credits for improved AC efficiency and to improve fuel efficiency are discussed below. These technologies focus on the compressor, electric motor controls, and system controls, which reduce the overall load on the AC system. The Agencies' goal is to improve efficiency of the AC system without sacrificing passenger comfort.

- *Reduced reheat using an externally-controlled, variable-displacement compressor.*
  "External control" of a variable-displacement compressor is defined as control of the displacement of the compressor based on the temperature set point or cooling demand of the AC system inside the passenger compartment. In contrast, conventional internal controls adjust the displacement of the compressor based on conditions within the AC system, such as head pressure, suction pressure, or evaporator outlet temperature. With external control, the compressor load is matched to the cooling demand of the cabin. With internal controls, the amount of cooling delivered by the system may be greater than desired, at which point the cooled cabin air is "reheated" to achieve the desired cabin comfort. This reheating of the air reduces the efficiency of the AC system. The SAE Improved Mobile Air Conditioning Cooperative Research Program (IMAC) program determined that by reducing reheat through external control of the compressor, an efficiency improvement of 24.1 percent was possible with this technology alone. The Agencies estimated that additional improvements to this technology are

[8] SC03 refers to the EPA Supplemental Federal Test Procedure (SFTP) with Air Conditioning. The test runs for approximately 10 minutes with an average speed of 21.55 mph, resulting in approximately 3.5 miles covered.

possible and that reducing reheat can provide a 30 percent reduction in fuel consumption of the AC system.

- *Reduced reheat using an externally-controlled, fixed-displacement or pneumatic variable-displacement compressor.*
  When using a fixed-displacement or pneumatic variable-displacement compressor (which controls the stroke, or displacement, of the compressor based on system suction pressure), reduced reheat can be realized by disengaging the compressor clutch momentarily to achieve the desired evaporator air temperature. This disengaging, or cycling, of the compressor clutch must be externally-controlled in a manner similar to that described above. The Agencies believe that a reduced reheat strategy for fixed-displacement and pneumatic variable-displacement compressors can result in an efficiency improvement of 20 percent. This lower efficiency improvement estimate (compared to an externally-controlled variable displacement compressor) is due to the thermal and kinetic energy losses resulting from cycling a compressor clutch off-and-on repeatedly.

- *Defaulting to recirculated cabin air.*
  In ambient conditions where air temperature outside the vehicle is much higher than the air inside the passenger compartment, most AC systems draw air from outside the vehicle and cool it to the desired comfort level inside the vehicle. This approach wastes energy because the system is continuously cooling the hotter outside air instead of having the AC system draw its supply air from the cooler air inside the vehicle (also known as recirculated air, or "recirc"). By cooling only this inside air (i.e., air that has been previously cooled by the AC system), less energy is required, and AC idle Tests conducted by EPA indicate that an efficiency improvement of 35 to 40 percent is possible under idle conditions. Ongoing testing on the new AC17 test, described below, is expected to provide data on the overall effectiveness of this technology during other driving conditions. To maintain freshness and humidity inside the cabin, EPA believes some manufacturers will control the air supply in a "closed-loop" manner, equipping their AC systems with humidity sensors or fog sensors (which detect condensation on the inside glass), allowing them to adjust the blend of fresh-to-recirculated air and optimize the controls for maximum efficiency. Vehicles with closed-loop control of the air supply (i.e., sensor feedback used to control the interior air quality) will provide a 30 percent reduction in fuel consumption of the AC system. Vehicles with open-loop control (where sensor feedback is not used to control interior air quality) will provide a 20 percent reduction in fuel consumption of the AC system.

- *Improved blower and fan motor controls.*
  To control the speed of the direct current (dc) electric motors in an air conditioning system, resistive elements are often used to reduce the voltage supplied to the motor. However, these resistive elements produce heat, which is typically dissipated into the air ducts of the AC system. Not only does this consume electrical energy, but it also contributes to the heat load on the AC system. Controlling dc voltage with a pulsewidth modulated controller on the motor can reduce the amount of energy wasted. EPA and NHTSA believe that when more efficient speed controls are applied to either the blower or fan motors, an overall improvement in AC system efficiency of 15 percent is possible.

- *Internal heat exchanger.*
  An internal heat exchanger (IHX), which is a suction line heat exchanger, transfers heat from the high-pressure liquid entering the evaporator to the gas exiting the evaporator. An IHX will reduce compressor power consumption and improve the efficiency of the AC system. In the MY 2012-2016 rule, the Agencies indicated that, with the changeover to an alternative refrigerant such as HFO-1234yf, the different expansion characteristics of that refrigerant (compared to R-134a) would necessitate an IHX. The Agencies believed that a 20 percent improvement in efficiency relative to the baseline configuration can be realized with an IHX.

- *Improved-efficiency evaporators and condensers.*
  The evaporators and condensers in an AC system are designed to transfer heat to and from the refrigerant. The evaporator absorbs heat from the passenger compartment air and transfers it to the refrigerant, while the condenser transfers heat from the refrigerant to the outside ambient air. The efficiency, or effectiveness, of this heat transfer process directly effects the efficiency of the overall system, as more work is required if the process is inefficient. A method for measuring the heat transfer effectiveness of these components is to determine the Coefficient of Performance (COP) for the system using the method described in SAE standard J2765 – Procedure for Measuring SystemCOP of a Mobile Air Conditioning System on a Test Bench. The COP is the ratio of the cooling at the evaporator to the work supplied to the compressor. The manufacturer must submit bench-test-based engineering analysis at the time of certification. The Agencies will consider the baseline component to be the version that a manufacturer most recently had in production on the same vehicle or a vehicle in a similar EPA vehicle classification. The design characteristics of the baseline component will be documented in an engineering analysis and compared to the improved components, along with data demonstrating the COP improvement.

If these components can demonstrate a 10 percent improvement in COP versus the baseline components, EPA and NHTSA estimate that a 20 percent improvement in overall system efficiency is possible.

- *Oil separator.*
In a typical AC system, oil circulates throughout the system for the purpose of lubricating the compressor. Because the oil is in contact with inner surfaces of the evaporator and condenser, and the coating of oil reduces the heat transfer effectiveness of these devices, the overall system efficiency is reduced. Inefficiency also results from "pushing around and cooling" an extraneous fluid that results in a dilution of the thermodynamic properties of the refrigerant. By containing the oil within the part of the compressor where it is needed, the heat transfer effectiveness of the evaporator and condenser will improve. The overall COP will also improve due to a reduction in the flow of diluent. The SAE IMAC program estimated that overall system COP could be improved by 8 percent if an oil separator was used. EPA and NHTSA believe that if oil is prevented from circulating throughout the AC system, an overall system efficiency improvement of 10 percent can be realized. Manufacturers will need to submit an engineering analysis to demonstrate the effectiveness of their oil separation technology.

The Phase 1 Report estimated that fuel consumption (in gallons per 100 miles driven) could be reduced by about 3 to 4 percent with a variable-stroke HVAC compressor and better control of the amount of cooling and heating used to reduce humidity. Other technologies that can yield incremental reductions in fuel consumption are UV filtering glazing and cool/reflecting paints, but these technologies are not currently pursued very aggressively because they are not taken account of in the official CAFE certification tests

The 2050 Transitions Report estimated that improved HVAC design would reduce air conditioning related fuel consumption by 40 percent by 2030. Better cabin thermal energy management through use of solar-reflective paints, solar-reflective glazing, and parked car ventilation was projected to reduce air conditioner related fuel consumption by 26 percent (Rugh et al. 2007). This study estimates that 2030 fuel consumption reduction for improved air conditioning of 50 percent would yield a 2 percent reduction in fuel consumption overall.

The cooperative industry and government SAE IMAC program demonstrated that average AC efficiency can be improved by 36.4 percent (compared to an average MY 2008 baseline AC system) when utilizing "best-of-best" technologies (EPA/NHTSA 2012b). EPA and NHTSA consider a baseline AC system to contain the following components and technologies: internally-controlled fixed-displacement compressor (in which the compressor clutch is controlled based on internal system parameters, such as head pressure, suction pressure, and/or evaporator outlet temperature); blower and fan motor controls that create waste heat (energy) when running at lower speeds; thermostatic expansion valves; standard efficiency evaporators and condensers; and systems that circulate compressor oil throughout the AC system. These baseline systems are inefficient in their energy consumption because they add heat to the cooled air out of the evaporator in order to control the temperature inside the passenger compartment. In addition, many systems default to a fresh air setting, which brings hot outside air into the cabin rather than recirculating the already-cooled air within the cabin.

A summary of the efficiency-improving AC technologies and the associated credits are listed in Table 6.17. As indicated earlier, for MYs 2020 and thereafter, the AC17 test will be used to qualify the amount of the credit. AC17 test results ("A" to "B" comparison) must be equal or greater than the credits shown in Table 6.17 to qualify for the full credit listed.

Based on vehicle simulation research by EPA, the impact of AC usage on average CO_2 emissions and fuel consumption is 11.9 g/mi (0.001339 gal/mi) for cars and 17.2 g/mi (0.001935 gal/mi) for trucks, as shown in Table 6.18. The final CAFE rule will encourage the reduction of CO_2 emissions from AC usage from cars and trucks by up to 42 percent from current baseline levels. Applying the 42 percent reduction to the average CO_2 emissions yields the maximum AC CO_2 credit opportunity of 5 g/mi (0.000563 gal/mi) for cars and 7.2 g/mi (0.000810 gal/mi) for trucks, as shown in Table 6.18.

Timing

EPA and NHTSA believe that the efficiency-improving technologies discussed in the previous sections are available to manufacturers today, and their feasibility and effectiveness have been demonstrated by the SAE IMAC program and various industry sources. The Agencies also believe that when these individual components and technologies are fully designed, developed, and integrated into AC system designs, manufacturers will be able to achieve the estimated reductions in CO_2 emissions and to generate appropriate AC efficiency credits, which are discussed in the following section. The NRC committee did not receive any comments from vehicle manufacturers that were contrary to this assessment by EPA and NHTSA.

Mercedes noted an electric air conditioner compressor and water pump would be viable with a 48 V electric system. Variable stroke compressors and reduction of subcooling are been developed and should appear in vehicles in the next 3 to 5 years.

Costs

The direct manufacturing costs for AC efficiency improvements are shown in Table 6.19. These costs are for

TABLE 6.17 Efficiency-Improving Air Conditioning Technologies and CO_2 and Fuel Consumption Reduction Credits

Technology Description	AC CO_2 Emission and Fuel Consumption Reduction	Car AC Credit and Adjustment (g/mi CO_2/ gal/mi)	Truck AC Credit and Improvement (g/mi CO_2/ gal/mi)
Reduced reheat, with externally-controlled, variable-displacement compressor	30%	1.5 (*30% of 5.0 g/mi impact*) / 0.000169	2.2 (*30% of 7.2 g/mi impact*) / 0.000248
Reduced reheat, with externally-controlled, fixed-displacement or pneumatic variable displacement compressor	20%	1.0 / 0.000113	1.4 / 0.000158
Default to recirculated air with closed-loop control of the air supply (sensor feedback to control interior air quality) whenever the outside ambient temperature is 75°F or higher (although deviations from this temperature are allowed if accompanied by an engineering analysis)	30%	1.5 / 0.000169	2.2 / 0.000248
Default to recirculated air with open-loop control of the air supply (no sensor feedback) whenever the outside ambient temperature is 75°F or higher (although deviations from this temperature are allowed if accompanied by an engineering analysis)	20%	1.0 / 0.000113	1.4 / 0.000158
Blower motor control which limit wasted electrical energy (e.g. pulsewidth modulated power controller)	15%	0.8 / 0.000090	1.1 / 0.000124
Internal heat exchanger (or suction line heat exchanger)	20%	1.0 / 0.000113	1.4 / 0.000158
Improved evaporators and condensers (with engineering analysis on each component indicating a COP improvement greater than 10%, when compared to previous design)	20%	1.0 / 0.000113	1.4 / 0.000158
Oil Separator (internal or external to compressor)	10%	0.5 / 0.000056	0.7 / 0.000079

SOURCE: EPA/NHTSA (2012b).

TABLE 6.18 Derivation of Maximum CO_2 Credits for Air Conditioning Efficiency Improvements

	Cars	Trucks
CO_2 Emissions for AC Usage on SC03 Cycle Based on EPA Simulation, assuming: - U.S. typical AC on-times: 23.9% manual AC, 35% automatic AC - Market Penetration: 62% manual, 38% automatic - Average AC compressor loads based on environmental conditions in the U.S. (from NREL)	11.9 g CO_2/mi (0.001339 gal/mi)	17.2 g CO_2/mi (0.001935 gal/mi)
Maximum AC CO_2 Credit - Equal to a 42% reduction encouraged by the final CAFE rule	5 g CO_2/mi (0.000563 gal/mi)	7.2 g CO_2/mi (0.000810 gal/mi)

NOTE: Factor to convert from CO_2 to gal/mi is 8,887 g CO_2/gal.

TABLE 6.19 Costs for Air Conditioning Efficiency Improvements

		Estimated Direct Manufacturing Costs ($)		
Technology		2017	2020	2025
Car	2012 - 2016 Efficiency Improvements	46	43	39
	2017 - 2025 Efficiency Improvements	1	1	1
	Total	47	44	40
Truck	2012 - 2016 Efficiency Improvements	32	30	27
	2017 - 2025 Efficiency Improvements	1	15	13
	Total	33	45	40

SOURCE: EPA/NHTSA (2012b).

improved compressors, expansion valves, heat exchangers, and the control of these components for the purposes of reducing tailpipe CO_2 emissions and fuel consumption as a result of AC use. The 2012-2016 rule technologies represent the reference case in terms of controls and costs. However, additional costs are included for indirect efficiency improvements as the 2012-2016 MY vintage systems penetrate to the entire fleet. The Agencies expect the AC efficiency costs to be incurred consistent with their estimated ramp up of manufacturers' use of AC credits (as shown in Table 5-13 of the TSD). The ramp up of credits is factored in the costs shown in Table 6.19. The Agencies received no public comments on these AC costs. Likewise, no vehicle manufacturer provided comments on these AC costs to the NRC committee. The Agencies consider technologies for most of the AC system improvements to be on the flat portion of the learning curve.

As indicated earlier, for MYs 2020 and thereafter, the AC17 test will be used to qualify the amount of the credit. AC17 test results ("A" to "B" comparison) must be equal to or greater than the credits shown in Table 6.18 to qualify for the full credit listed.

Direct manufacturing costs estimated by EPA and NHTSA for the AC efficiency improvements listed in Table 6.17 are shown in Table 6.19. EPA and NHTSA imply that the costs shown in the table would be associated with obtaining the maximum AC CO_2 credits shown in Table 6.18.

In addition to the foregoing discussion of indirect CO_2 and fuel consumption reduction credits for AC efficiency improvements, the final GHG/CAFE rule expands provisions for manufacturers to generate credits for reduced AC leakage and alternative low GWP[9] refrigerants. However, unlike the AC efficiency improvements, reductions in AC leakage and alternative low GWP refrigerants do not count toward the CAFE calculations since these improvements do not improve fuel economy. The reduced AC leakage hardware includes improved hoses, connectors, and seals. The low GWP refrigerants require additional refrigerant hardware. The CO_2 credits and costs for reduced AC leakage and GWP refrigerants are described in the TSD (EPA/NHTSA 2012b). EPA estimated that there would be significant penetration of AC technologies for leakage reduction and efficiency improvements to gain credits, and this was reflected in the stringency of the standards.

Crediting Off-Cycle Efficiency Technologies

The combined city/highway, or "two-cycle," certification test for fuel economy is known to produce results substantially higher than the average fuel economy in real-world driving. Furthermore, certain technologies deliver real-world fuel savings that are not reflected, or are not fully reflected, in two-cycle fuel economy test values. These include, for instance, air conditioner efficiency improvements, which do not provide fuel savings on the test cycle because the air conditioner remains off throughout the two-cycle test. In order to incentivize the development and deployment of such technologies, the Agencies have defined "off-cycle" credits that manufacturers apply toward the fuel economy and CO_2 emissions averages for their vehicles. The 2017-2025 rule was the first use of off-cycle credits in the CAFE program.

Due to the anticipated difficulty of having manufacturers' demonstrate the off-cycle savings for individual technologies, the Agencies estimated fuel economy credit values for several technologies that they judged could be reasonably quantified in a generic fashion. Default fuel savings values, separate for cars and for trucks, for these "preapproved" technologies appear in two menus of off-cycle credits in the rule and are shown in the final rule (EPA/NHTSA 2012a, Tables II-21 and II-22). Aside from air conditioning efficiency improvements, manufacturers may claim these credits simply by providing the specifications of their equipment and stating the number of their cars and trucks on which the technology appears. However, there is a cap of 10 g/mi CO_2 (about 0.001 gal/mi), in total and on average over cars and trucks, on non-AC credits obtained through this menu approach (EPA/NHTSA 2012a, 62727).[10] Manufacturers can also obtain off-cycle credits by directly demonstrating the fuel savings of the technology, and credits so obtained are not subject to the 10 g/mi cap.

For air conditioning efficiency improvement and two other off-cycle technologies that they believed would be widely used, the Agencies incorporated the expected benefits into the stringency of the standards. The two others are stop-start systems and active aerodynamic improvements (EPA/NHTSA 2012a, 62720). Stop-start systems provide fuel savings over the certification cycle but are eligible for off-cycle credits based on the notion that the real-world idle fraction on average is significantly larger than the idle fraction in the two-cycle test. Active aerodynamics refers to aerodynamic technologies that are activated only at certain speeds, including active grille shutters and active ride height control.

In modeling compliance with the 2017-2025 standards, the Agencies assumed that these three categories of technologies would be used by manufacturers to achieve the standards and were thus incorporated into the stringency of the standards. The committee's assessment of the effectiveness of AC improvements can be found in Table 6.20.

[9] Global warming potential (GWP) is a relative measure of heat trapped in the atmosphere by a greenhouse gas. GWP is calculated over a certain time period and is expressed as a factor of carbon dioxide (whose GWP is standardized to 1).

[10] These credit amounts are significant from a compliance perspective. For example, for a 45 mpg vehicle, a fuel savings credit equivalent to 10 g/mi would increase nominal fuel economy by 2.4 mpg.

TABLE 6.20 Compilation of Effectiveness of Improved Accessories from Various Studies and Organizations

Potential Reduction of Fuel Consumption with the Use of Vehicle Accessories	Reduction in Fuel Consumption (%)	Comments
NRC Phase 1 Study (NRC 2011)		
Variable stroke HVAC compressor	3 - 4	Improved cooling, heating, and humidity control.
Low transmissivity glazing, "cool" paint, and parked vehicle ventilation	~ 1	Lower heat buildup in vehicle decreases AC load.
Electrohydraulic power steering	4	Combined electric and hydraulic power for midsize to larger vehicles. Reduces continuous load on engine.
Electric power steering	1 - 5	
NRC 2050 Transitions Study (NRC 2013)		
HVAC and thermal management	2	Electric power steering for smaller vehicles reduces continuous load on engine – smaller benefits (1-3%) estimated for the FTP.
Electric power steering	2 - 3	
NHTSA/EPA Final Rule (NHTSA 2005)		
High-efficiency alternator	1.2 - 1.8	
Improved cooling	1.74 - 1.55	

Barriers

The technologies discussed above focused on the compressor, electric motor controls, and system controls, which reduce the overall load on the AC system. The goal is to improve efficiency of the AC system without sacrificing passenger comfort.

AUTOMATED AND CONNECTED VEHICLES

Automated and connected vehicles are attracting increasing attention. They can function under current highway conditions, as has been demonstrated by the vehicles put on the road in California by Google that traveled well over 200,000 miles. They are touted as a means to improve driving safety, reduce travel times, enable otherwise-incapable people to operate cars, and reduce fuel consumption. NHTSA believes that automated and connected vehicles represent "a historic turning point for automotive travel" (NHTSA Preliminary Policy on Automated Vehicles 2015). In 2013, the Agency released a policy statement that describes the technologies available, a summary of the research it has been pursuing, and its recommendations on a safe implementation of these technologies. The technologies deliver capabilities that range from increasing the available information to a driver to a vehicle being capable of operating under complete autonomy. NHTSA has classified these technologies into categories defined by five levels of autonomy:

- *No automation (Level 0).* The driver is in complete and sole control of the primary vehicle controls—brake, steering, throttle, and motive power—at all times.
- *Function-specific automation (Level 1).* Automation at this level involves one or more specific control functions. Examples include electronic stability control or precharged brakes, where the vehicle automatically assists with braking to enable the driver to regain control of the vehicle or stop faster than would be possible by acting alone.
- *Combined function automation (Level 2).* This level involves automation of at least two primary control functions designed to work in unison to relieve the driver of control of those functions. An example of combined functions enabling a Level 2 system is adaptive cruise control in combination with lane centering.
- *Limited self-driving automation (Level 3).* Vehicles at this level of automation enable the driver to cede full control of all safety-critical functions under certain traffic or environmental conditions and in those conditions to rely heavily on the vehicle to monitor for changes requiring transition back to driver control. The driver is expected to be available for occasional control, but with sufficiently comfortable transition time. The Google car has mostly operated under limited self-driving automation.
- *Full self-driving automation (Level 4).* The vehicle is designed to perform all safety-critical driving functions and monitor roadway conditions for an entire trip. Such a design anticipates that the driver will provide destination or navigation input but is not expected to be available for control at any time during the trip. This includes both occupied and unoccupied vehicles.

The following section attempts to characterize these technologies by level and describes potential effects on fuel consumption and passenger safety.

Level 1: Function-Specific Autonomy

In the near term, it is likely that the technologies that will be commercially available will remain in Level 1 through Level 3. Some of these technologies have been commercially available on certain models and, in some situations, NHTSA has found the technologies to be essential to improving vehicle safety. Function-specific autonomy can describe technologies such as maintaining speed, assisting in stopping, and warning of action that would increase the risk to the vehicle occupants.

Conventional cruise controls maintain a constant (or nearly constant) speed, effectively relieving the driver of one task. The newer, autonomous cruise control addresses two things: proximity to vehicles ahead of the subject car and lane boundaries. The autonomous (or "adaptive") cruise control(ACC) keeps the speed of the vehicle at a level that maintains a preset distance behind the vehicle ahead of the controlled vehicle. Such controls are currently available, for example, on four models of Ford cars and three Lincoln cars. The current market price for these systems ranges between $1,200 and $3,000. The sensing systems in these controls are now radar; formerly, some were laser light beams. Regarding fuel consumption reduction, advanced cruise control systems have the potential to keep the vehicle at a more constant speed, which uses less fuel than having the driver attempt to maintain vehicle speed. However, the reduction in fuel consumption would be highly dependent on the driver's behavior and is difficult to quantify. It is unknown what impacts this technology may have on safety since it controls only one task for the driver. Assuming that the driver is attentive to controlling the vehicle, it is likely that maintaining a constant speed will increase safety for other drivers on the road and therefore reduce risk.

Electronic stability control (ESC) is one area of research that NHTSA has identified as a promising approach to increasing safety. Often referred to as "traction control," ESC is a digital technology that works by its ability to automatically apply braking to individual wheels in order to avoid dangerous changes in vehicle heading. In other words, it supports the driver's behavior, such as driving along a curved road, by noting the intended change in direction, calculating whether or not braking is required to maintain vehicle stability for each wheel, and then executing the braking required to steer the vehicle in the intended direction. The entire process usually occurs so quickly that the driver is often ignorant of the fact that it is even functioning. NHTSA has found the technology to be so promising that every light-duty vehicle after MY 2011 has been required to include it.

Level 2: Combined Function Autonomy

Many of the technologies in Level 2 are also available today. These technologies are capable of controlling the vehicle in two functions, thereby removing the tasks from the driver. A more elaborate cruise control system, in a testing stage now, provides an audible and visual warning signal when vehicles ahead reduce speeds, and then prepares and applies brakes, presumably faster than the human driver could move a foot from accelerator to brake pedal. A version of cruise control under development now monitors the deviations of the vehicle's path from any strictly in-lane motion. It goes by the name "Lane Keep Assist" (LKA). A simpler device is the "Lane Departure Warning" (LDW), which simply alerts the driver to the beginnings of a deviation. The LDW control puts in a resistance to any steering that would take the car from a linear, in-lane path, literally nudging it back into its lane. Fuel savings from widespread implementation of cruise controls might result if simulations represent their effects accurately. The travel times from simulations were 20 to 37.5 percent shorter than those without autonomous cruise controls, so the total fuel consumption would be lower than at present (Treiber and Kesting 2012; Suzuki and Nakatsuji 2003). With respect to maintaining lane control, this approach does not seem to have any effect on fuel performance. Assuming that these technologies function as intended, it is likely that societal and personal safety will improve as they are included in more and more vehicles. However, the degree of improvement on safety is an unknown as drivers are unlikely to report when a collision is avoided because of the technology functioning properly.

Longer-Term Fully Automated and Connected Vehicles (Levels 3 and Up)

Fully automated vehicles require sensors to recognize all the relevant characteristics of their surroundings and their current operation, but they also must have both links to the mechanical controls of the vehicle and a control system, presumably based in a computer, to determine the appropriate response to the sensors' signals for adjusting the mechanical controls.

There are two levels of sensor technology associated with self-driving vehicles. One is the set of sensors and controls for an individual vehicle that could operate under existing conditions on streets and highways. The second level has communication between autonomous vehicles and between vehicles and their surroundings. While there are many benefits to making individual, independent vehicles self-driving, still more benefits could be achieved if there were, as the advisory firm KPMG Consulting describes it, "a convergence of sensor-based technologies and connected-vehicle communication."

Sensor-based technologies that recognize other nearby vehicles, signals, signs, pedestrians, and cyclists come at various levels of sophistication and with a choice of sensor technology. Sensor technology is typically based in either radar (radio frequency or microwave radiation) or lidar (optical radiation). At present, the cost of any adequate sensor is far too high to make the technology widely available. For example, the lidar system used by Google for its autonomous vehicles costs $70,000 (Bunkley 2012). Obviously such costs would drop significantly with mass application, but by how much is unknown to the committee.

Fuel Consumption

The potential of autonomous vehicles for reducing fuel consumption is considerable indeed, provided that large numbers of cars have that capability. Maintaining smooth traffic flow would reduce mean travel times by reducing congestion and fuel consumption. "Platooning," or having significant numbers of vehicles move in synchrony, would reduce air resistance much as "drafting" does when a cyclist rides close behind other cyclists or behind a vehicle. Estimates of fuel savings associated with having significant numbers of autonomous cars vary, but a conservative minimum is about 20 percent. However, the full implementation of high levels of autonomy in vehicles may produce unexpected results in terms of fuel consumption due to an increase in vehicle miles traveled.

Timing

Estimates of the time it will take until autonomous vehicles constitute a significant fraction of the car market vary widely. The most optimistic say that this could happen by 2022; others say 2050. There is a growing view that at some time in the next half-century, we will see many self-driving cars in use. One aspect is of course establishing a manufacturing structure, but a complementary one is creating the means to retrofit existing vehicles both for sensing and communicating with other vehicles and stationary signaling centers.

Costs

The costs of fully automated and connected vehicles have yet to be determined. While studies are currently being performed on this topic, the range of estimates on the costs of implementing autonomous technologies more advanced than Level 4 have significant ranges of uncertainty associated with them.

Barriers

Given the experimental status of autonomous vehicles, the National Highway Traffic Safety Administration (NHTSA) has established a requirement that a human driver must be able to take control at any time of an otherwise autonomous vehicle. This requirement will presumably remain in effect until such a time when these vehicles are fully tested and have become acceptable on the basis of their reliability and safety. There will also be a very long period of time when fully automated and connected vehicles share the road with conventional vehicles.

The technology could have net negative consequences in terms of fuel consumption and traffic flow. The autonomy enables people to operate cars who would be unable to drive conventional vehicles. For example, Google made a video of a full round-trip in an autonomous vehicle with a blind man in the driver's seat. This might lead to an increase in the number of vehicles on the roads and an increase in the fleet's vehicle miles traveled.

FINDINGS AND RECOMMENDATIONS

Finding 6.1 The committee's estimates of fuel consumption reduction effectiveness and direct manufacturing costs for aerodynamic improvements, low rolling resistance tires, low drag brakes, electric power steering, and improved accessories are shown in Table 6.21 and are in agreement with NHTSA's estimates.

Finding 6.2 The mass reduction targets identified in the NHTSA/EPA TSD are conservative. Automakers are more likely to implement the more aggressive levels of mass reduction estimated in Table 6.22.

Finding 6.3 As more mass is removed from a vehicle, incremental costs tend to increase. The initial reductions in mass come from easier and less-complex alternatives than later alternatives, particularly with mass substitution options. For example, the progression in the industry to lightweighting shifts applications from mild steel to high-strength steel, or from high-strength steel to aluminum, or aluminum to composites and magnesium. Each increasingly aggressive step removes additional mass and comes at a higher cost than the preceding step. Manufacturers will be constrained by high costs rather than by options in reducing mass between now and 2030.

Finding 6.4 The mass reductions cited above, in the 15-20 percent range for larger vehicles, are expected, especially when they are accomplished with a complete vehicle design and consequent mass decompounding and drivetrain optimization. Such mass reductions could be cost effective, especially for electric- and hybrid-powered vehicles, because savings associated with mass reductions are more significant for these powertrains than for conventional spark-ignition powertrains.

Finding 6.5 It is the committee's view that mass will be reduced across all vehicle sizes, with proportionately more mass removed from heavier vehicles. The most current studies that analyze the relationships between vehicle foot-

TABLE 6.21 Estimates of Technology Effectiveness and Costs for 2017, 2020, and 2025

Technology	Fuel Consumption Reduction (%)	Direct Manufacturing Costs[a] (in 2010$)		
		2017	2020	2025
Aerodynamic Improvement 1(10% C_d)				
Small and Large Car	2.3	39	37	33
Light Truck	2.3	39	37	33
Aerodynamic Improvement 2(20% C_d)[b]				
Small and Large Car	2.5	117	110	100
Light Truck	2.5	117	110	100
Low Rolling Resistance Tires Level 1	1.9	5	5	5
Low Rolling Resistance Tires Level 2[b]	2.0	58	46	31
Low Drag Brakes	0.8	59	59	59
Electric Power Steering				
Small Car	1.3	87	82	74
Large Car	1.1	87	82	74
Light Truck	0.8	87	82	74
Improved accessories Level 1				
Small Car	1.2	71	67	60
Large Car	1.0	71	67	60
Light Truck	1.6	71	67	60
Improved accessories Level 2[b]				
Small Car	2.4	43	40	37
Large Car	2.6	43	40	37
Light Truck	2.2	43	40	37

[a] Relative to baseline except as noted.
[b] Relative to Level 1.

TABLE 6.22 Mass Reductions Foreseen by NHTSA/EPA and by the Committee (percent)

Vehicle	NHTSA/EPA TSD Estimate	Committee Estimate
Small car	0	5
Midsize car	3.5	10
Large car	10	15
Minivan	20	20
Light duty truck	20	20

print, mass, and safety support the argument that removing mass across the fleet in this manner while keeping vehicle footprints constant will have a beneficial effect on societal safety risk. Additionally, with the introduction of improved crash simulation and vehicle design techniques, new materials, and crash avoidance technology (such as lane change warning and autonomous frontal braking), crashworthiness and crash avoidance should be improved. During the transition period when vehicle masses are being reduced, there could be an increase in safety risk due to variance in the distribution of the mass across the vehicle fleet.

Recommendation 6.1 NHTSA should carefully consider and, if necessary, take steps it believes could mitigate the possible threats to safety during the transition period, as the fleet moves from current vehicle designs to a more light-weighted fleet.

Finding 6.6 The cost to repair and insure lighter weight vehicles will increase as manufacturers employ material substitution to lightweight cars. Material substitution with high-strength steel, aluminum, magnesium, and composites add complexity to vehicles, making them more expensive to insure and repair. The service industry will have to increase training and upgrade equipment to be able to evaluate and repair vehicles with a broader mix of materials and joining technologies.

Recommendation 6.2 Automobile manufacturers, the insurance industry, and the repair industry should continue coordination so that appropriate procedures and technology are ready for the introduction of vehicles with nontraditional materials.

Finding 6.7 The evolution of the materials' industries, especially steel and aluminum, is significant and warrants monitoring. The availability of aluminum or other aspects of the aluminum supply chain (e.g., annealing, rolling, recycling, tooling/forming, etc.) may limit the industry's move toward lighter weight cars. The economic impact of transitioning to higher aluminum content may have a significant impact on the associated workforce.

Finding 6.8 There have been several teardown studies to help assess the opportunity and cost for reducing mass in vehicles, but little attention has been given to interpreting how best to use the results. The committee feels that the studies are hard to generalize and apply to other vehicles due to the wide variation across vehicle models and because extrapolating results from one or two studies to the entire fleet would be problematic. Future teardown studies would benefit from careful selection of vehicles that are representative of their class.

Finding 6.9 The committee finds the use of the lightweight optimization studies combining computer-aided engineering and teardown analyses to be an improvement over the current method used for the 2017-2025 rulemaking. These types of analyses can be helpful in identifying components where lightweighting is possible, illustrating examples of material substitution, and taking an integrated approach to mass reduction over the entire vehicle. However, the committee recognizes the limitations of vehicle-specific studies when used to estimate costs across the entire fleet, or even across a vehicle class. The vehicle model selected for the analyses will have a large impact on the opportunities for mass reduction. Factors such as the substantial differences in the starting point of vehicle models, the varied materials in current designs, and individual business considerations—such as global platforms and maintaining vehicle NVH—mean that such studies must be supplemented with other analysis. There is high potential for misinterpretation of the cost estimates resulting from these vehicle-specific studies if they are applied to other vehicle designs in a general fashion, and this potential is much greater for mass reduction techniques than it is for other types of technologies.

Recommendation 6.3 The committee recommends that the Agencies augment their current work on vehicle design optimization with a materials-based approach that looks across the fleet to better define opportunities and costs for implementing lightweighting techniques, especially in the area of decompounding. Such an approach might include assessing opportunities for well-defined substitution, such as replacing a hood or door with a lighter material across the light-duty fleet. A characterization of current vehicles in terms of materials content is a prerequisite for such a materials-based approach and for quantifying the opportunities to incorporate different lightweighting materials in the fleet. The committee recommends that the Agencies consider undertaking such a characterization.

Finding 6.10 An estimated 20 percent reduction in rolling resistance appears reasonable, with a 4 percent incremental fuel consumption reduction in the 2020 to 2025 time frame. However, there are engineering challenges associated with tire design with respect to rolling resistance, tread wear, and traction, because these attributes affect tire costs.

Finding 6.11 The NHTSA/EPA TSD estimates that electric power steering could provide an incremental fuel consumption reduction of 1.3 percent for small cars, 1.0 percent for large cars, and 0.8 percent for small light trucks, 1.0 percent for medium light trucks and 0.8 percent for large light trucks at a cost of $109 appear reasonable.

Recommendation 6.4 NHTSA/EPA should consider the contributions of tire pressure management systems and autonomous tire inflation technology to reducing fuel consumption and improving vehicle safety.

Recommendation 6.5 NHTSA should continue to maintain the current tire safety regulatory structure, which will not allow safety performance to be traded off for improved fuel economy performance. And, in this vein, NHTSA should maintain a regulatory structure—including a rating system, especially fuel economy ratings, for tire consumers,—to support marketplace decisions that could result in aftermarket tire performance that does not significantly differ from new vehicle tire performance.

Finding 6.12 EPA and NHTSA have estimated that the final CAFE rule will encourage a reduction in AC CO_2 emissions for cars and trucks of up to 42 percent from current baseline levels. Since the AC is not turned on as a part of the fuel economy and GHG emissions standards compliance test drive cycles, the effect of AC efficiency improvements is an off-cycle effect. AC credits for efficiency improvements are applicable to both GHG emissions and fuel consumption. For MYs 2017 to 2019, the AC17 test will be used to demonstrate eligibility for AC credits using a menu based on the AC design. For MYs 2020 and thereafter, the AC17 test will be used to qualify the amount of the credit. AC17 test results must demonstrate reductions in fuel consumption equal to or greater than the allowable credits to qualify for the full credits listed by EPA and NHTSA. AC efficiency improvements are estimated by the Agencies to have a direct manufacturing cost of $40 (2010 dollars) by MY 2025.

Finding 6.13 While there has been much publicity over the potential benefits of connected vehicle technologies, significant uncertainty remains about the impact these technologies will have in the 2025 timeframe. In addition to the uncertainty of how these technologies will affect fuel consumption, there is even greater uncertainty about how these technologies will be implemented with regard to laws and regulations, how they will affect vehicle miles traveled, and safety considerations for more advanced connected vehicle technologies.

Recommendation 6.6 NHTSA/EPA should continue to evaluate the potential contribution of automated and connected vehicle technologies for improving fuel economy. The Agencies should consider the desirability and feasibility of providing CAFE-related credits to incentivize the adoption of appropriate technologies.

REFERENCES

3M Company. n.d. Auto Part Targets for Lightweight Plastics and Rubber.

Alonso, E., T.M. Lee, C. Bjelkengren, R. Roth, and R. Kirchain. 2012. Evaluating the potential for secondary mass savings in vehicle lightweighting. Environmental Science & Technology.

Barrand, J., and J. Bokar. 2008. Reducing tire rolling resistance to save fuel and lower emissions. SAE Paper 2008-01-0154. SAE International, Warrendale, PA.

Bunkley, K. 2012. The car without a driver. Automotive News, July 16. http://www.autonews.com/article/20120716/OEM06/307169920/the-car-without-a-driver.

California Air Resources Board (CARB). 2011. Appendix Q Proposed LEVIII GHG Technical Support Document. December 7. Sacramento, Calif.

Cheah, L.W. 2010. Cars on a Diet: The Material and Energy Impacts of Passenger Vehicle Weight Reduction in the U.S. PhD. Dissertation, Massachusetts Institute of Technology.

DOE (U.S. Department of Energy). 2010. Material Content per Light Vehicle, 1995 and 2008. http://energy.gov/eere/vehicles/fact-642-september-27-2010-material-content-light-vehicle-1995-and-2008.

DOE. 2013. Workshop Report: Light-Duty Vehicles Technical Requirements and Gaps for Lightweight and Propulsion Materials, February.

Ducker Worldwide. 2014. 2015 North American Light Vehicle Aluminum Content Study.

Ducker Worldwide. 2011. Aluminum in 2012 North American Light Vehicles: Executive Summary. September 7.

EEA (Energy and Environmental Analysis, Inc.). 2007. Update for Advanced Technologies to Improve Fuel Economy of Light Duty Vehicles—Final Report. Prepared for U.S Department of Energy. October.

EPA (Environmental Protection Agency). 2015. Cost-Effectiveness of a Lightweight Design for 2020-2025: An Assessment of a Light-Duty Pickup Truck. SAE World Congress. April 2015. Detroit, Mich.

EPA. 2008. EPA Staff Technical Report: Cost and Effectiveness Estimates of Technologies Used to Reduce Light-Duty Vehicle Carbon Dioxide Emissions. EPA420-R-08-008. Ann Arbor, MI.

EPA/NHTSA (Environmental Protection Agency/National Highway Traffic Safety Administration). 2012a. 2017 and Later Model Year Light-Duty Vehicle Greenhouse Gas Emissions and Corporate Average Fuel Economy Standards. EPA 40 CFR 85, 86, 600; NHTSA 49 CFR 523, 531, 533, 536, 537. August 28.

EPA/NHTSA. 2012b. Joint Technical Support Document: Final Rulemaking for the 2017-2025 Light-Duty Vehicle Greenhouse Gas Emission Standards and Corporate Average Fuel Economy Standards. EPA-420-R-12-901. http://www.epa.gov/otaq/climate/documents/420r12901.pdf.

Evans, L. 1991. Traffic Safety and the Driver. New York: Van Nostrand Reinhold Publishing.

Forbes. 2008. Automakers' Natural Appeal. Forbes Magazine, September 26.

GAO (U.S. Government Accountability Office). 2004. Highway Safety: Federal and State Efforts to Address Rural Road Safety Challenges. U.S. General Accounting Office Report to Congressional Committees. http://www.gao.gov/assets/250/242663.pdf.

Gessat, J. 2007. Electrically Powered Hydraulic Steering Systems for Light Commercial Vehicles. SAE 2007-01-4197.

Goodyear Tire and Rubber Company. 2009. Letter from Donald Stanley, Vice President, Product Quality and Plant Technology, April 13, 2009; email exchanges with Donald Stnaley, January 28, 2009, and February 25, 2009.

Hucho, H.W. 2005. Aerodynamik des Automobils, Wieweg+Tuebner, 5th ed.

Kahane, C.J. 2012. Relationships Between Fatality Risk, Mass, and Footprint in Model Year 2000-2007 Passenger Cars and LTVs – Final Report. DOT HS 811 665. August. National Highway Traffic Safety Administration.

Kebschull, S.A., and M. Sekiguchi. 2008. Brief Summary of a Furthur Analysis of the effects of SUV Weight and Length Changes on SUV Crashworthiness and Compatibility Using Fleet Modeling and Risk-Benefit Analysis. Dynamic Research, Inc., DRI-TM-08-24.

Kolwich, G. 2014. Silverado 1500 Mass-Reduction and Cost Analysis Project Review. Follow-up to NAS/NRC Committee visit to NVFEL, July 31st, 2014. Presentation to the Committee on Assessment of Technologies for Improving the Fuel Economy of Light-Duty Vehicles, Phase 2, September 4.

Lotus Engineering. 2010. An Assessment of Mass Reduction Opportunities for a 2017-2020 Model Year Vehicle Program. Washington, D.C.: The International Council on Clean Transportation.

McCartt, A.T. and E.R. Teoh. 2014. Type, Size and Age of Vehicles Driven by Teenage Drivers Killed in Crashes During 2008-2012. Arlington, Virginia: Insurance Institute for Highway Safety.

NHTSA (National Highway Traffic Safety Administration). 2005. Federal Motor Vehicle Safety Standards; Tire Pressure Monitoring Systems; Controls and Displays. Docket No. NHTSA 2005-20586. http://www.nhtsa.gov/cars/rules/rulings/tpmsfinalrule.6/tpmsfinalrule.6.html

NRC (National Research Council). 2006. Tire and Passenger Vehicle Fuel Economy: Informing Consumers, Improving Performance-Special Report 286. Washington, D.C.: The National Academies Press.

NRC. 2011. Assessment of Fuel Economy Technologies for Light-Duty Vehicles. Washington, D.C.: The National Academies Press.

NRC. 2013. Transitions to Alternative Vehicles and Fuels. Washington, D.C.: The National Academies Press.

Pike Research and International Council Clean Transportation. 2011. Opportunities to Improve Tire Energy Efficiency. http://theicct.org/tire-energy-efficiency.

Ramesh C.K. et al. 2012. Design of Hood Stiffener of a Sedan Car for Pedestrian Safety. M.S. Ramaiah School of Advanced Studies. Bangalore, India.

Rugh, J.P. et al. 2007. Reduction in Vehicle Temperatures and Fuel Use from Cabin Ventilation, Solar-Reflective Paint, and a New Solar-Reflective Glazing. NREL/CP-540-40986. Presented at the 2007 SAE World Congress. Detroit, MI, April 16-19.

Schutte and Joost. 2012. Energy, Materials, and Vehicle Weight Reduction. U.S. Department of Energy Presentation. Washington, D.C., May 6.

Singh, H. 2012. Mass Reduction for Light-Duty Vehicles for Model Years 2017-2025: Final Report. DOT HS 811 666. August. National Highway Traffic Safety Administration.

Snyder, J. 2008. A big fuel saver: Easy-rolling tires (but watch braking). Automotive News, July 21. http://www.autonews.com/article/20080721/OEM01/307219960.

Suzuki, H., and T. Nakatsuji. 2003. Effect of adaptive cruise control (ACC) on traffic throughput: numerical example on actual freeway corridor. JSAE Review 24(4): 403-410. doi: 10.1016/S0389-4304(03)00074-2.

Treiber, M., and A. Kesting. 2012. Validation of traffic flow models with respect to the spatiotemporal evolution of congested traffic patterns. Transportation Research Part C: Emerging Technologies 21: 31-41.

Van Auken, R.M., and J.W. Zellner. 2013a. Updated Analysis of the Effects of Passenger Vehicle Size and Weight on Safety, Phase I; Updated Analysis Based on 1995 to 2000 Calendar Year Data for 1991 to 1999 Model Year Light Passenger Vehicles. DRI-TR-11-01-1. Torrance, CA: Dynamic Research, Inc. May.

Van Auken, R.M., and J.W. Zellner. 2013b. Updated Analysis of the Effects of Passenger Vehicle Size and Weight on Safety, Phase II; Preliminary Analysis Based on 2002 to 2008 Calendar Year Data for 2000 to 2007 Model Year Light Passenger Vehicles to Induced-Exposure and Vehicle Size Variables. DRI-TR-13-02. Torrance, CA: Dynamic Research, Inc. May.

Van Auken, R.M., and J.W. Zellner. 2013c. Updated Analysis of the Effects of Passenger Vehicle Size and Weight on Safety; Sensitivity of the Estimates for 2002 to 2008 Calendar Year Data for 2000 to 2007 Model Year Light Passenger Vehicles to Induced-Exposure and Vehicle Size Variables. DRI-TR-13-04. Torrance, CA: Dynamic Research, Inc. May.

Wenzel, T. 2010. Analysis of the Relationship Between Vehicle Weight/Size and Safety, and Implications for Federal Fuel Economy Regulation, Final Report prepared for the Office of Energy Efficiency and Renewable Energy, U.S. Department of Energy. http://energy.lbl.gov/ea/teepa/pdf/lbnl-3143e.pdf.

Wenzel, T. 2012a. Assessment of NHTSA's Report "Relationships Between Fatality Risk, Mass, and Footprint in Model Year 2000-2007 Passenger Cars and LTVs". Final report prepared for the Office of Energy Efficiency and Renewable Energy, US Department of Energy. Lawrence Berkeley National Laboratory. August. LBNL-5698E. http://energy.lbl.gov/ea/teepa/pdf/lbnl-5698e.pdf.

Wenzel, T. 2012b. An Analysis of the Relationship between Casualty Risk per Crash and Vehicle Mass and Footprint for Model Year 2000-2007 Light-Duty Vehicles. Final report prepared for the Office of Energy Efficiency and Renewable Energy, US Department of Energy. Lawrence Berkeley National Laboratory. August. LBNL-5697E. http://energy.lbl.gov/ea/teepa/pdf/lbnl-5697e.pdf.

Wenzel, T. 2013. Assessment of DRI's Two-Stage Logistic Regression Model Used to Simultaneously Estimate the Relationship between Vehicle Mass or Size Reduction and U.S. Fatality Risk, Crashworthiness/Compatibility, and Crash Avoidance. Draft report prepared for the Office of Energy Efficiency and Renewable Energy, U.S. Department of Energy. Lawrence Berkeley National Laboratory. January.

Wernle, B. 2014. Ford Dealers Throng Training on How to Fix Aluminum F-150. Automotive News, May 5. http://www.autonews.com/article/20140505/RETAIL05/305059973/ford-dealers-throng-training-on-how-to-fix-aluminum-f-150.

7

Cost and Manufacturing Considerations for Meeting Fuel Economy Standards

ESTIMATING THE COSTS OF MEETING THE FUEL ECONOMY STANDARDS

Technologies' costs and fuel economy impacts are both important for setting fuel economy and greenhouse gas (GHG) emissions standards. The direct manufacturing costs of component technologies or technology packages are the most important pieces of information, but the degree to which the technology and design changes will affect the indirect costs of firms also matters. The rate of technological and design changes required by the standards can also affect costs by making capital investments prematurely obsolete or requiring greater than normal engineering effort. Technology and design changes can also have secondary impacts on consumers' satisfaction and corporate profits.

Uncertainty about future costs is inescapable because of the uncertain rate and direction of future technological progress, as well as uncertainties about the future prices of materials, energy, labor, and capital. Although technological change is certain, its direction, magnitude, and impacts on cost are difficult to predict. For most components, manufacturing costs tend to decrease with increased production volumes and with the accumulation of experience. However, there are no exact methods for predicting future rates of learning by doing or technological progress. Assuming no technological progress or cost reduction via learning will likely overestimate the costs of compliance. On the other hand, overly optimistic assumptions will result in underestimation of costs. In this chapter the committee discusses methodological issues with estimating the costs of fuel economy technologies and manufacturing issues with deploying these technologies.

Defining Costs

Rigorously estimating costs requires carefully defining terms. The most important concepts are *direct manufacturing costs, indirect costs, and total costs.*[1]

The elementary cost components of manufacturing are materials, energy, labor, and capital. Cost estimation begins with an understanding of the quantities of each component required to produce a certain number of units per year: Unit cost[2] is calculated by multiplying the quantities by their prices, summing to yield total cost, and dividing by the volume of the production. In general, total cost is not a linear function of production volume. There are *fixed costs* that are required to produce any units and *variable costs* that do increase more linearly with production volume, but even this is an oversimplification. In general, total costs increase less than linearly with production volume due to *economies of scale*. Thus, estimates of unit costs require specification of a production volume. Fixed costs are typically amortized over assumed levels of production for a certain number of years. Thus, unit cost estimates also depend on production volumes and the usable lifetimes of fixed investments. In addition, prices of all of the components tend to vary over time and location, to different degrees. Prices may vary with the degree of competition in the supply chain. Price variability is not the only source of uncertainty. The quantities of labor, capital, and energy required may also vary over time and with improvements in manufacturing processes.

Direct manufacturing costs (DMC) are defined alternatively as the price an original equipment manufacturer (OEM) would pay a supplier for a fully manufactured part ready for assembly in a vehicle, or the OEM's total cost of internally manufacturing the same part. Thus, if the part is made by the OEM, the DMC is typically considered to be

[1] As discussed in Chapter 1, the committee defines total costs for a technology as direct manufacturing costs plus indirect costs. Total cost does not represent the cost of ownership to the consumer, which is discussed in Chapter 9 as the private cost of ownership.

[2] Unit cost is the cost of a specified unit of a product or service.

the materials, labor, and energy required by the OEM to manufacture the part, similar to the definitions provided by Helfand and Sherwood (2013) and Blincoe (2013). If the part is made by a supplier, the DMC includes the research and development and indirect costs of the supplier. This would not include integration and other indirect costs borne by the OEM, however. It includes all the suppliers' fixed costs necessary to manufacture the part, as well as a normal rate of return for the supplier on capital. DMCs are the most important element of cost estimation because they typically make up the majority of total costs and because indirect costs tend to vary with DMC.

Indirect costs (IC) include expenditures not directly required for manufacturing a component technology but necessary for the operation of an automobile manufacturing firm. Direct manufacturing costs plus indirect costs equal what economists call the long-run average cost of manufacturing, or, simply, total costs. In an ideal competitive market in equilibrium, price would equal long-run average cost. In the less-than-perfect dynamic markets of the real world, the two frequently differ. Indirect cost estimates may also have a large impact on estimated total costs. For a typical OEM, indirect costs (including a typical rate of profit) average about 50 percent of direct manufacturing costs (RTI/UMTRI 2009).

There are many ways to describe indirect costs. In the following section the committee discusses the two principal approaches to representing indirect costs for a technology: (1) the retail price equivalent (RPE) markup and (2) the indirect cost multiplier (ICM). Though the methods for estimating these multipliers are different, both the RPE and ICM represent the costs for producing a technology that are not included in the estimate of the direct manufacturing costs. Table 7.1 shows a breakdown of the components included in indirect costs based on information available for publicly traded firms (RTI/UMTRI 2009, Table 3-3).

TABLE 7.1 Breakdown of Indirect Costs for an OEM

OEM vehicle production overhead	Warranty
	Research and development
	Capital depreciation and amortization
	Maintenance, repair, and operations
Corporate overhead	Administration
	Employee benefits (e.g. health insurance, retirement plans)
Sales including dealer costs	Transportation
	Marketing
	Dealer selling costs (inventory, advertising, labor, etc.)
	Dealer profit
Corporate profit	

Estimating Direct Manufacturing Costs

There are a variety of sources of information about current manufacturing costs. Teardown studies performed by competent third parties are the most reliable sources of cost estimates but are also the most expensive. Other sources are useful but may have biases or inadequacies (e.g., basing direct manufacturing costs on market prices that also reflect dynamic effects of supply and demand) and so must be interpreted with caution. Cost quotes for fully manufactured components may be provided by OEMs or suppliers. The different sources, as well as their strengths and weaknesses, were described in the Phase 1 report (NRC 2011).

The 2025 rulemaking makes use of a small number of detailed teardown cost analyses. The NRC Phase 1 report recognized the need for cost estimates based on teardown studies and concluded that increasing the use of this approach would increase confidence in the accuracy of the costs (NRC 2011). In 2009, the Environmental Protection Agency (EPA) contracted with FEV, Inc., to perform a cost analysis of converting a conventional naturally aspirated, fuel-injected I4 engine to a stoichiometric, direct-injected, turbocharged and downsized I4 engine (FEV 2009). This appears to be the first instance of the Agencies' use of cost estimation based on tearing down vehicles, itemizing each part, estimating the purchase or direct manufacturing cost of each part and subsystem, and constructing detailed cost estimation models for the complete system change. This level of detail, together with documentation of the costs of every item and process, makes possible a more complete understanding of the reasons for differences in cost among technologies. Since the Phase 1 report, additional teardown cost studies have been carried out for mild (FEV 2011a) and full hybrid (FEV 2012), valve-train technologies (FEV 2011a), advanced transmissions (FEV 2011b), and mass reduction (FEV 2012). Key assumptions include high-volume manufacturing (450,000 units/year) in North America. The volume assumptions made by the Agencies in order to scale costs are further discussed in the following section on economies of scale.

Estimating Indirect Costs

Rogozhin et al. (2010) calculated indirect costs by category for eight major automobile manufacturers using publicly available annual reports from the OEMs for 2007. Total indirect costs as a ratio to direct manufacturing costs were similar for all manufacturers (Figure 7.1): The minimum ratio was 0.45, the maximum was 0.49, and the average was 0.46. The indirect cost calculations excluded a category entitled "Other costs (not included as contributors)," shown in Figure 7.1 as "Other." Had the "Other" costs been included, it would have raised the average to 0.50. The distribution of indirect costs by category is more variable than the total, however. Costs classified as "Corporate overhead" vary from 0.04 to 0.14,

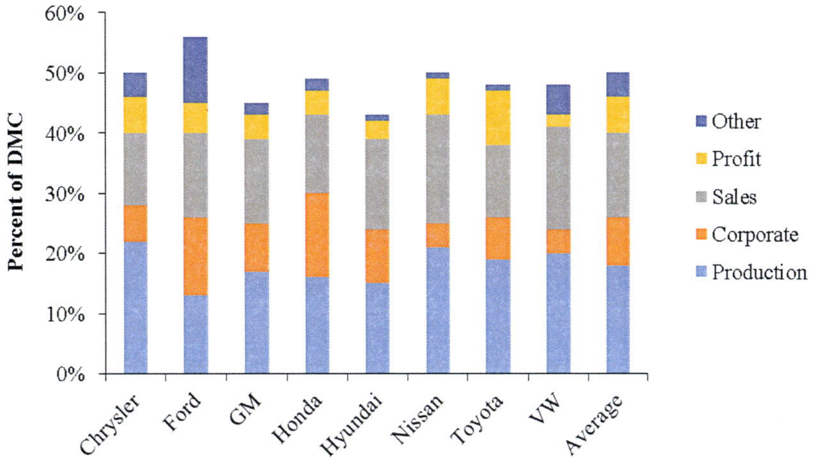

FIGURE 7.1 Indirect costs as a percent of direct manufacturing costs by OEM, 2007.
SOURCE: Rogozhin et al. (2010).

and "Production overhead" ranges from 0.13 to 0.22. The rate of manufacturer profit (excluding dealer profit) in 2007 also varied in a relatively narrow range across OEMs from 3 percent to 9 percent.

The small variations in total indirect costs among firms may reflect differences in accounting conventions as much as real cost differences. There is no strict definition of what must be included in direct manufacturing costs. If the standard is the price an OEM would pay a supplier for a fully manufactured part ready for assembly, then direct manufacturing costs should include amortization of capital required for the subassembly, maintenance of facilities used, and profit on the operations. It is not clear that internal definitions of direct manufacturing costs always include all of these components.

The industry average ratio of indirect costs to direct manufacturing costs appears to fluctuate within a range of +/− 0.1 over time. An investigation of the ratio of total costs to direct manufacturing costs by the National Highway Traffic Safety Administration (NHTSA) for 1972-1997 found that the ratios fluctuated between 1.4 and 1.6 but without any apparent trend (Figure 7.2). In this regard, the ratio of total to direct costs represents the average markup above direct costs for all technologies produced in a given year. Figure 7.2 indicates that to cover costs, provide a return to investors, and remain competitive in the marketplace, OEMs have typically set prices that average 1.5 times direct costs (Blincoe 2013). The consistency of the 1.5, or 50 percent, markup from manufacturing costs to retail price is noteworthy and suggests that it is reasonable to assume that the relative share of indirect costs

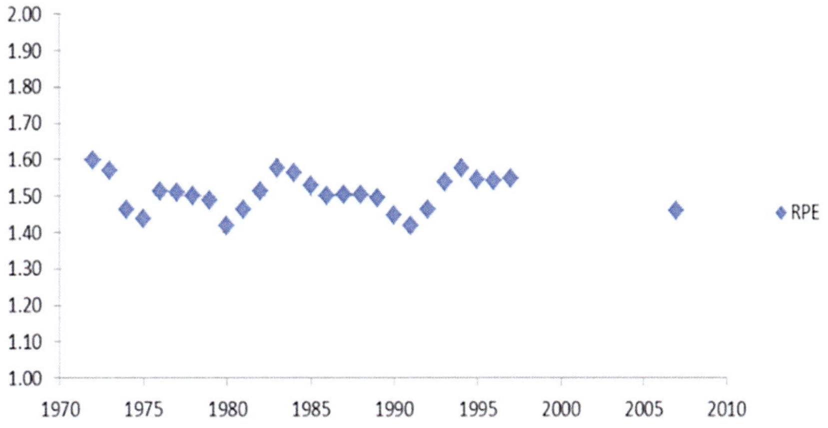

FIGURE 7.2 Total costs as a ratio to direct manufacturing costs (RPE), 1972-1997 and 2007.
SOURCE: EPA/NHTSA (2012).

may not change much in the future. For this reason, the Final Rule increased the estimated 46 percent markup to 50 percent for use in estimating the indirect cost multipliers used in calculating the total costs of technologies used to meet the standards (EPA/NHTSA 2012). The similarity of the indirect cost shares across manufacturers and over time is also consistent with the view that competition among manufacturers is robust in that a manufacturer with substantially above-average indirect costs would be unsuccessful in the market.

RPE and ICM Methods

Retail Price Equivalent

The RPE makes no distinctions among technologies with respect to their impacts on indirect costs. It assigns an average indirect cost percentage to all technologies, thereby avoiding the question of attributing changes in indirect costs to specific technologies or design changes. This approach maintains the typical markup rate because fuel economy technologies add to the price of the vehicle. Based on the available data, a reasonable RPE multiplier would be 1.5.

The RPE method was used in previous rulemakings by NHTSA (DOT/NHTSA 2009, 173) and previous NRC fuel economy studies. NRC (2002) used a value of 1.4, while the Phase 1 study and NHTSA (DOT/NHTSA 2009) determined that a value of 1.5 was appropriate (NRC 2011, 36). A study comparing estimates of RPE multipliers concluded that a value of 1.56 was appropriate for outsourced parts purchased from suppliers (Vyas et al. 2000). Duleep (2008) recommended RPEs ranging from 1.65 to 1.73 depending on the complexity of the part; In 2003, McKinsey (as quoted in Bussmann 2008) produced a 1.7 RPE multiplier; Bussmann (2008) calculated an RPE of 2.0 for data obtained from Chrysler for 2003-2004. The Phase 1 report committee commissioned a study by IBIS Associates (2008) that costed out all components of a Honda Accord sedan and a Ford F-150 pickup truck. For the Honda, the RPE multiplier based on average transaction price was 1.39, while the markup to manufacturer's suggested retail price (MSRP) was 1.49. For the F-150 the corresponding markups were 1.52 for transaction price and 1.54 for MSRP (NRC 2011, 33).

Indirect Cost Multiplier

The alternative is to estimate the impact of each technology on each component of indirect cost. In theory, this approach seems clearly superior to assuming identical impacts for all technologies regardless of their nature. However, attribution can be ambiguous, especially for future costs. Whether or not a specific technology will add to warranty or advertising costs is highly uncertain, for example. Does selling cars with a higher value added require additional corporate staff? Does it add to dealers' inventory costs? Unfortunately, there is a general lack of empirical data on which to base such attributions.

The indirect cost multiplier (ICM) method is an application of activity based costing (ABC) methods to fuel economy technologies. ABC accounting attempts to assign costs to products based on the activities they require. In the case of ICMs, however, the costs are not assigned to the final product (an automobile) but rather to specific components of the final product. The difficulty is that the accounting data for automobile manufacturers that is publicly available in annual reports is organized according to standard financial accounting principles. Financial accounting is intended to give creditors, investors, and the government a fair and accurate representation of the firm's transactions, revenues, profits, and losses. Costs are generally classified by function rather than by attributing activities to specific products or components of products. Therefore, additional analysis is required to assign costs to components.

The ICM method attempts to estimate the specific impacts of technologies and technology packages on indirect costs. While the ICM method is logically appealing, rigorous implementation is very difficult because it requires extensive knowledge of a firm's operations and involves uncertainty about components such as warranty costs. RTI International and UMTRI (2009) and Rogozhin et al. (2010) provide descriptions and examples of how EPA estimates ICMs. As described by Helfand and Sherwood (2013), EPA relied on its engineers and scientists with expertise in automotive product development and production to develop the ICMs used in the Corporate Average Fuel Economy (CAFE)/GHG standards. The agency used two different experiments, a consensus approach (Rogozhin et al. 2010) and a Delphi-based approach (Helfand and Sherwood 2009) to develop the ICM.

The ICM estimation process applied by the Agency begins by associating each technology with one of four degrees of innovation and complexity (Rogozhin et al. 2010). The different levels of innovation and complexity are judged to affect research and development, corporate and dealer labor, and warranty components of the ICM differently. The levels of innovation used by the Agency were these:

- Incremental: compatible with existing core components of the automobile;
- Modular: changes core components but not their interaction;
- Architectural: changes interactions of components but not their fundamental function; and
- Differential: establishes new functions for components and changes their interaction.

The technologies were further classified by complexity. The panel of Agency experts was given four examples:

- Low complexity: passive aerodynamic drag reductions;
- Medium complexity: turbocharging with downsizing;

- High 1 complexity: hybrid electric vehicle; and
- High 2 complexity: plug-in hybrid electric vehicle.

In the final rule, the Agencies point out the limitations of both the RPE and ICM methods. Because the accounting methods of manufacturers differ and costs are generally not classified as direct or indirect, the estimation of RPEs requires judgment to allocate costs between the two categories. As described above and in the related references, the ICM method requires grouping technologies into different categories based on levels of innovation and complexity, which requires judgment and assumes that all technologies within a group have identical impacts on indirect costs. Expert judgment is also relied upon to estimate the impact values for each technology and cost component. Finally, ICM estimates have not been validated by directly measuring the indirect cost impacts of specific technologies.

The current rulemaking attempts to estimate transitional as well as long-run average indirect cost impacts. Higher short-run costs are represented by higher ICMs. The higher short-run ICM impact factors are intended to represent additional engineering effort that is initially required to integrate a new technology component into the overall vehicle system. Engine changes, for example, will require adjustments to transmissions, computer control algorithms, and other elements of the vehicle. Once a new technology has been integrated into a first-generation vehicle, the indirect cost multipliers assume the long-run values. The short run is defined as one production cycle lasting 4-5 years, and >5 years is the long run (Rogozhin et al. 2010). Short-run ICMs range from 4 percent to 18 percent higher than the long-run ICMs, with greater differences as the degree of complexity/innovation increases (Table 7.2).

Undoubtedly, different technologies and different design changes to vehicles affect indirect costs differently. It is therefore appropriate for the Agencies to work toward a methodology that assigns different indirect cost multipliers to different fuel economy and emissions technologies. As described above, the empirical basis for such multipliers is still lacking, and, since their application depends on expert judgment, it is not possible to determine whether the Agencies' ICMs are accurate or not. In a presentation to the committee, the EPA presented evidence that on average the ICM method resulted in a ratio of total costs to direct manufacturing costs of approximately 1.5 (EPA 2014). The committee notes the seeming incongruity in the result where, on average, EPA estimates that the ratio of total costs to direct costs is about 1.5, but almost all of the individual ICMs are less than 1.5. The committee encourages the Agencies to continue research on ICMs with the goal of developing a sound empirical basis for their estimation. One possible method for developing such an empirical basis is through case studies to break down the costs associated with the specific steps required to integrate a new technology and trace its impacts on indirect costs. The committee notes that the specific values for the ICMs are critical since they may affect the overall estimates of cost and benefits for the overall standards and the cost effectiveness of the individual technologies.

TABLE 7.2 Indirect Cost Multipliers Used in the 2025 Rule

Complexity/Innovation	Near Term	Long Term
Low	1.24	1.19
Medium	1.39	1.29
High 1	1.56	1.35
High 2	1.77	1.50

SOURCE: EPA/NHTSA (2012, Table 3-1).

Economies of Scale

Scale economies are an important determinant of cost in the automobile industry. The NRC 2011 report asserted that scale economies would generally be reached at between 100,000 and 500,000 units per year, citing evidence from Martec (2008) and Honda (DOT/NHTSA 2009, 185). Husan (1997) cites estimates of scale economies for different production processes that range from 120,000 to 240,000 for powertrain manufacturing to 2,000,000 for foundry/forging and pressing. Even within a category, estimates of maximum scale economies vary widely, for example, from 120,000 to 1,000,000 for powertrain manufacturing. The exception is final assembly, where the range of estimates for volumes at which maximum scale economies are achieved is narrower: 100,000 to 300,000.

Economies of scale are often summarized by a scale elasticity that represents the relative reduction in cost with a 1 percent increase in scale as the optimum is approached. Data from Husan (1997) suggest a scale elasticity of approximately −0.1 for final assembly (Figure 7.3). The evidence also suggests that both optimal production volumes and scale elasticities will vary by manufacturing process. It should be noted that the Agencies do not separate cost reductions from increasing scale and cost reductions from learning by doing.

The Agencies use economies of scale to develop cost estimates of direct manufacturing costs. The incremental DMC of a technology is developed as discussed previously, assuming a North American production of 450,000 units. This is justified based on an FEV turbocharging and downsizing pilot study (FEV 2009) which used MY 2007 WardsAuto data to estimate U.S. domestic light-duty engine production volumes of 350,000 to 480,000 for moderate- to high- volume applications. The assumption of 450,000 units therefore is not production of a given technology by each manufacturer but by the entire United States or North American market (it is not clear which). The scale assumption of 450,000 units is used in such areas as the teardown studies supporting the rule and as an input to the BatPac model for battery costing, for example.

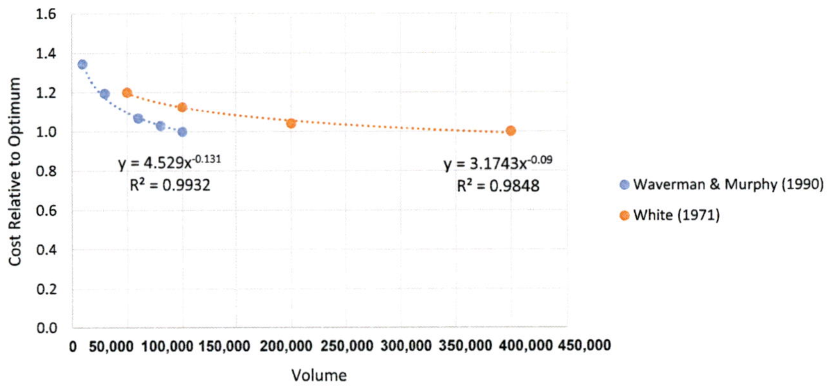

FIGURE 7.3 Estimates of scale economies in automobile manufacturing.
SOURCE: Data from Husan (1997).

As discussed further below, assuming such high volumes across the board assumes that all technologies will be implemented at large scales, and that economies of scale operate across multiple manufacturers, not simply within a single manufacturer. This volume assumption is problematic for technologies at low volume, like hybrid technologies, and for technologies with significant proprietary issues preventing suppliers from producing components at scale, such as for transmissions. The Agencies recognized that if their assumption of high volume is wrong, then actual costs will be higher than those predicted. Economies of scale are also connected to learning as defined by the Agencies and described below.

Learning by Doing

Numerous retrospective studies have documented how the price of a novel technology declines with cumulative production (Wene 2000). For many technologies price, p, has been shown to decrease as a function of the logarithm of cumulative production, X, known as a learning curve.

$$p_X = p_0 X^{-a}$$

The progress ratio (equal to 2^{-a}) measures the relative cost for each doubling of production. For example, if $a = 0.074$, the progress ratio is 0.95, which implies a 5 percent reduction in manufacturing costs for every doubling of cumulative production. In theory, p_0 is the price of the first unit of sales. However, in practice it is very difficult to know the price of the very first unit of any commodity, and the calibration of the learning curve can be very sensitive to the assumed initial conditions.

While numerous studies have estimated learning curves using historical data, there is no rigorous method for predicting learning curves for novel technologies in the absence of empirical data. Further complicating matters, observed price reductions are typically due to a combination of scale economies, exogenous technological change, and learning by doing in manufacturing processes. It is useful to separate the three since scale economies are a function of the current volume of production, rather than cumulative production, and exogenous technological change represents general advances in science and technology over time. Mathematically, it is straightforward to model separate effects for the three components, yet few empirical analyses have done so. If the overall combined effect is attributed to learning by doing, the result will be an overestimate of the potential for cost reduction via cumulative production (Nordhaus 2009).

The NRC Phase 1 report noted that cost estimates were usually based on three key assumptions (NRC 2011, 25):

- High-volume production (100,000 to 500,000 units);
- Learned production costs; and
- Competition in the supply chain from at least three global suppliers.

Under these assumptions, the Phase 1 committee concluded that it was not appropriate to use traditional learning curves to predict future cost reductions for technologies already in mass production. On the other hand, that committee also concluded that it would be appropriate to apply learning curves to cost estimates for truly novel technologies that did not reflect learning by doing. The Technical Support Document (TSD) for the 2017-2025 standards explains that the Agencies employ a nontraditional learning curve method using time- instead of volume-based learning and distinguishing between novel and established technologies (EPA/NHTSA 2012).

The Agencies' rulemaking distinguishes two types of learning by doing, "steep" and "flat," as well as the case of no learning (EPA/NHTSA 2012). Only newer technologies are subject to steep learning. The list of newer technologies includes only six items, subject to steep learning during the following periods:

- Air conditioner alternative refrigerant, MY 2016-2020;
- P2 hybrid vehicle battery pack components, MY 2012-2016;
- Electric/plug-in vehicle battery pack components, MY 2012-2025;
- Electric/plug-in vehicle battery charger components, MY 2012-2025;
- Stop-start, MY 2012-2015; and
- Lower rolling resistance tires, level 2, MY 2017-2021.

After their periods of steep learning, each of the above technologies converts to flat learning until 2025. Twenty-one technologies are subject to only flat learning, which begins in 2012 and continues through 2025. Only five technologies are considered to have no opportunities for future cost reduction via learning:

- Engine modification to accommodate low friction lubes;
- Engine friction reduction – levels 1 and 2;
- Lower rolling resistance tires – level 1;
- Low drag brakes; and
- Electric/plug-in vehicle battery charger installation labor.

Chapter 8 shows the various learning curves used by the Agencies and adopted by the committee in most cases to estimate future costs. The Agencies use a base year within their model of learning by doing. This base year is when the technology is considered "mature" and is when the Agencies assume volumes of 450,000, discussed previously. This base year is also used as the year off which the indirect cost is calculated for each technology. As discussed in Chapter 8, the concept of negative learning is used for low-volume technologies.

Steep learning follows a step function that produces a 20 percent reduction in cost after the first 2 years of implementation and another 20 percent reduction after another 2 years of production. Once two fast learning cycles are complete, new technologies follow the flat learning cycle. Flat learning begins with 5 years of cost reduction at 3 percent per year, followed by 5 years at 2 percent per year, and finally 5 years at 1 percent per year. For a technology subject to steep learning, this results in a more than 50 percent (52.8 percent) cost reduction after 19 years. Flat learning results in more than a 25 percent (26.2 percent) reduction in cost after 15 years. It is common practice in the automotive industry for OEMs to negotiate contracts with suppliers that stipulate annual cost reductions in the range of 1 percent to 3 percent, depending on the technology. Since such contracts are generally considered proprietary, it is difficult to document and measure this phenomenon. The Agencies apparently assign learning functions to technologies based on their expert judgment.

The Agencies' use of the learning-by-doing concept is unconventional in that it is strictly a function of time rather than cumulative production. This formulation avoids the complication of endogeneity, the simultaneous determination of the cost of a technology and its adoption by manufacturers. It also sidesteps the problems of choosing a learning rate and an initial level of cumulative production. Within the TSD, the Agencies provide a detailed discussion of how learning is applied (EPA/NHTSA 2012). For example, as described above, when steep learning is in effect, then a 20 percent decrease in cost is assumed for every doubling in production volume. This is implemented by time rather than volume, however, so it implicitly assumes that a doubling of volume occurs every 2 years for technologies subject to steep learning. They also describe the difficulties of implementing volume-based learning within the models used by NHTSA (the Volpe model) and EPA (the OMEGA model), which would require the models to endogenously estimate a production volume and then apply a volume-based learning curve for any specific time period through an iterative feedback loop (EPA/NHTSA 2012). The committee appreciates this difficulty. On the other hand, this approach to learning allows a technology to accomplish very significant cost reductions even if its production volumes remain very low. Further, while the Agencies' description of how they apply learning and how they are not separating learning into volume-based and time-based learning is clear, the figure displaying their conceptual model of learning (Figure 3-1 in the TSD) titles the curve as the volume-based learning curve with an x-axis labeled "cumulative production." This conveys to the reader that learning is estimated from production volume. It would be helpful to rectify this situation in order to eliminate ambiguity about what was actually done.

The Agencies decided that learning should affect only direct manufacturing costs and not indirect costs, except for warranty costs. Their reasoning is that learning affects only direct manufacturing costs, except that warranty costs involve replacement of parts whose costs should also decrease with learning. It is difficult to evaluate this assertion due to the lack of empirical data.

Stranded Capital

If the rate of fuel economy improvement or GHG reduction required by the standards necessitates replacing capital investments before their normal depreciated lifetime, it may be appropriate to attribute the remaining amortized cost of the capital equipment, the "stranded capital," to the replacement technology.[3] The NRC Phase 1 report also noted that accelerated rates of redesign and technology adoption could demand more engineering resources than are available, potentially driving up labor costs.

---

[3] Even this attribution is somewhat arbitrary in that there are generally many technologies available with which to improve fuel economy. The attribution of stranded capital cost to a specific "replacement" technology may make it a less economical choice than alternatives.

An estimate of the potential cost impact of stranded capital was made by FEV, Inc., (2011c) for the EPA. The study is not based on historical data nor is it an analysis of how the standards are likely to cause capital investments to be stranded in the future. Rather it is a "what if" assessment of the potential impact of stranded capital costs under a variety of different assumptions. The study estimates "the potential saddling of cost onto a new technology configuration as a result of the production equipment and/or tooling for the baseline configuration being abandoned before the planned fully depreciated life" (FEV, Inc. 2011a, 1-1). Six case studies were carried out: two conventional engines replaced by downsized, turbocharged, direct-injection engines; three upgraded transmissions; and one conventional V6 powertrain replaced by a power-split hybrid system. The additional cost of truncating the useful life of the productive capital after 3, 5, and 8 years was estimated. The results indicate relatively modest yet nontrivial cost impacts relative to the total cost of the technologies in question (Table 7.3). EPA and NHTSA (2012) provide descriptions of how the regulatory models apply costs representing the stranded capital cost of the replaced technology to the cost for producing the newly applied technology.

The FEV analysis does not address whether premature retirement is likely for specific technologies and how prevalent it might be. Although more analysis on this topic would improve the estimate of costs of the standards, the automobile industry in general is moving toward more rapid model updates due to consumer and competitive pressures (Oliver Wyman 2013; IHS Automotive 2013; Bloomberg 2014; Finlay 2014).

There are alternative approaches to using the engineering costs of prospective technologies for estimating the costs of the standards. Empirical economic analyses have used models of manufacturer and consumer behavior to infer the costs of CAFE standards. Jacobsen (2013) estimated the costs to those automakers that appeared to be substantially constrained by the standards during the late 1990s, when gasoline prices were low. He estimates the added costs of the last miles per gallon improvement to meet the standard will vary a good deal across the affected manufacturers, with direct costs to the domestic manufacturers ranging from $52 to $438 per car, and from $157 to $264 per truck. Using a different approach, Anderson and Sallee (2011) take the decision by companies to produce additional flexible fuel vehicles (FFVs) in lieu of reducing emissions in other ways as indicative of the cost of the fuel economy regulations at the margin. Their analysis covers the period from 1996 to 2006, and they find that the implied direct costs of the standards at the margin are quite low, between about $9 and $27 per vehicle. These studies examine a period when standards were relatively flat, which may not be as relevant to the future regulations, but the approach to estimating costs of regulatory compliance ex post is valuable. To date, there has been no careful study comparing ex-post economic studies to the ex-ante forecasted engineering costs for the same period.

There are also economic models that attempt to forecast the future effects, costs, and benefits of variations in the fuel economy regulations. These include Liu et al. (2014), Skerlos and Whitefoot (2012), and Jacobsen (2013). These studies use underlying engineering cost estimates but incorporate economic decisions in response to the regulations and thus the ability to capture more complete measures of the costs of the regulations. Such models and approaches can serve as important additional input in the regulatory analysis of the rules.

MANUFACTURING ISSUES—TIMING CONSIDERATIONS FOR NEW TECHNOLOGIES

Technology Introduction in Vehicle Manufacturing

The timing for introducing new vehicle features and technologies is a significant strategic and competitive issue for automotive manufacturers. More rapid technology introduction will keep products current and more likely to be recognized as new and state of the art, resulting in higher consumer demand. Accelerating new product development provides

TABLE 7.3 Estimated Added Cost of Stranded Capital

		Potential Stranded Capital Cost per Vehicle (Productive Capital Stranded After *X* Years)		
Replaced Technology	New Technology	3 Years	5 Years	8 Years
Conventional V6	DSTGDI I4	$56	$40	$16
Conventional V8	DSTGDI V6	$60	$43	$17
Six-speed AT	6-speed DCT	$55	$39	$16
Six-speed AT	8-speed AT	$48	$34	$14
Six-speed DCT	8-speed DCT	$28	$20	$8
Conventional V6	Power-split HEV	$111	$79	$32

SOURCE: FEV (2011c).

a market advantage to the manufacturer and, as discussed earlier, is a general industry trend in response to consumer and competitive pressures. In the context of fuel economy, new products are likely to have more recently developed technologies that reduce fuel consumption. Adopting new technologies, however, must be considered in the broader context of vehicle design and manufacturing. More substantial technological advancements, such as lightweighting or downsizing powertrains, may require major reengineering efforts and their introduction would be, therefore, timed with a new vehicle or new powertrain program. Other technologies that may require less integration engineering (such as low rolling resistance tires, low global warming potential (GWP), A/C systems, or other technologies) can be implemented during minor upgrades, which occur more frequently throughout the overall model life cycle.

Resource constraints that limit rapid implementation of new technology are costs, development and validation lead time, engineering resource availability, and financial risk. These same resource constraints also apply to suppliers who may be cooperating in new product development with the manufacturer. Technologies requiring high capital expenditures (e.g., tooling for powertrain or body components) tend to have longer product life cycles to help lower amortization rates. Much newer technologies, such as forming parts out of aluminum instead of steel, may require greater engineering lead time and resources than simply changing the shape of a part made of the same material. Additionally, the greater the change in the technology the greater the challenge of achieving both production and product validation, which can result in higher risk. Stop/start, safety issues with batteries, and utilization of die cast structural components instead of formed panels are examples of technological changes that potentially pose such risks.

As an example, a body structure is shown in Figure 7.4, composed of conventional and ultra-high-strength steel panels, both traditionally stamped and hot formed, as well as aluminum panels, extruded and rolled profiles, and die cast components.

Lightweighting Body—Near Term

As discussed in Chapter 6, while the majority of the car body today is made from steel, the body is also made from a mixture of materials that includes many grades of steel, grades of aluminum, a variety of composites, magnesium, and other materials. The industry, in general, is trending toward a broader distribution of materials that will bring about a significant transformation in the manufacturing process. The evolution of steel implementation in vehicles has been significant, from the frequent use of mild-strength steel in the early 1980s to today's vehicles having over 50 percent of their body made from high-strength steel (HSS). It is not unusual to have over 15 grades of steel in a single vehicle body. Today, the implementation of steel in vehicles utilizes the following design tools:

- Engineering modeling software to simulate the forming of parts and assembly (welding) of the body. Modeling expertise and algorithms have to be modified as new material grades get introduced.
- Extensive cold-stamping infrastructure with stamping plants that in some cases have 30 or more press lines. Today's state-of-the-art press lines (e.g., a servo press)

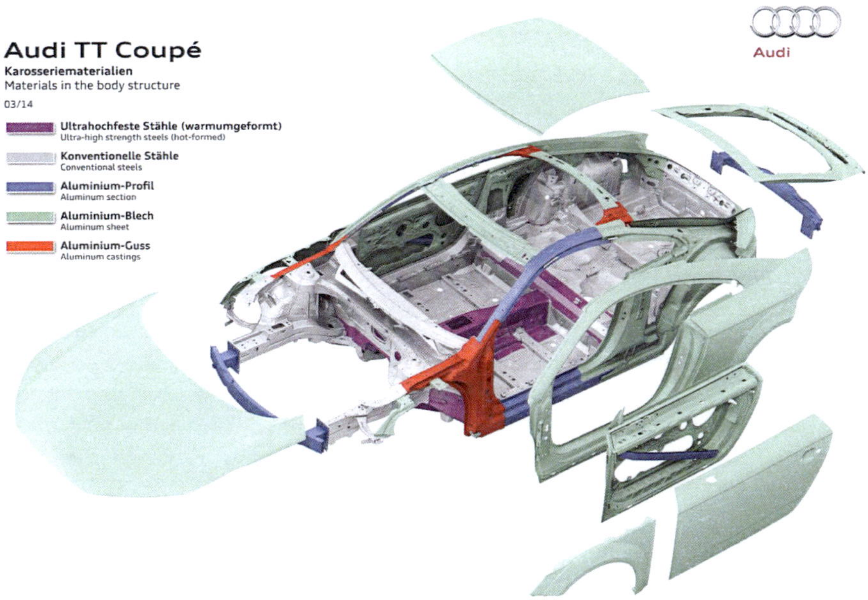

FIGURE 7.4 Example vehicle incorporating a combination of steel and aluminum types.
SOURCE: Audi MediaServices (n.d.).

can cost upwards of $70 million, whereas a tandem or transfer press line might cost $20 to $40 million apiece. A single high-volume vehicle may require 8 to 10 press lines to support it with stamped steel parts. Automation between the presses typically uses magnets, designed for steel components, to pick up parts.
- Tooling industry that produces hundreds of dies to fabricate parts for a single vehicle body. The automotive tool and die industry is steel-centric, and the die cost for an all-new vehicle body will cost $200 million to $400 million. There are also assembly tools (to weld) and checking tools (to measure parts) that are designed specifically for steel. Assembly makes occasional use of adhesive and other joining technologies but is dominated by spot welding, where the typical car has 4,000 or more spot welds in it. Most spot welding is performed by robots, and a new body shop will have 300 to 1,000 robots in it. The number of new joining technologies has been increasing in the body shop—for example, adhesive gluing, laser welding, and fastening (e.g., rivets). With adjustments, this general infrastructure can be used with new materials such as high-strength steel or even aluminum. However while the traditional steel body shop is dominated by spot welding, a body shop with other materials will likely be dominated by other joining technologies.
- The body-in-white is sent to the paint shop, where it is sealed, cleaned, and painted. Different materials paint differently and many paint shops rely on electrostatic paint, thus requiring a conductive metal surface. Paint shops last 20 or 30 years or longer and can cost over $500 million. They are subject to strict environmental laws, and upgrades are expensive and time consuming. Their design often influences the type of vehicle that can be made at the assembly plant.

As the materials being used in a vehicle's body change, there can be significant impact on the design tools described above. Today, the industry has less experience designing, tooling, and joining aluminum components of a vehicle than it has with those of steel. A task as menial as moving a finished component around the factory floor requires a different engineering approach since the magnets used to move steel pieces do not function on aluminum pieces. However, some of these design and engineering challenges may be mitigated by OEMs moving towards an all-aluminum vehicle design. This can be seen today, as design, tooling, and stamping processes for aluminum are migrated from steel with a transitory learning curve.

While OEMs are already looking toward aluminum as the next step from steel, vehicle design further into the future may yield vehicle bodies that use multiple materials to achieve the fuel consumption and safety requirements of future standards. However, the industry's limited experience base with and supply infrastructure for composites for the body presents a bigger challenge than the those associated with aluminum. As the number of materials in a body increases, joining and painting them becomes more difficult. Joining with adhesive has made significant advancements over the past 10 years, and it now provides a better joint than spot welding, although at higher cost. The paint shop can then be tuned to this new material, even if the paint shop was initially designed for steel. Other materials such as magnesium and composites are used only in selective locations while the body is still dominated by steel and aluminum.

Lightweighting Body—Long Term

The next migration to more mixed materials will occur as new assembly plants and paint shops get upgraded in the next 10 to 20 years. It is important to note in this context that the changeover to new materials almost never is a 1:1 substitution of a part or component. It entails, instead, a concept change for the functional design as well as for the manufacturing processes and systems used, as all the recent major changeovers reflect. While the opportunity for reducing mass with a mixed material vehicle is much better than with a monolithic design, mixing materials into the car poses new challenges not confronted by a steel-intensive vehicle.

The material sectors (steel, aluminum, magnesium, composites) are largely autonomous, with little to no collaboration between them. They are, individually, fierce competitors attempting to promote the use of their material in the car, often at the expense of another material. Consequently, there has been little to no cooperation between the material sectors. Challenges to mixing materials into the car, though significant, are not insurmountable; they just have not received the attention that individual materials have received. The principal challenges include the following:

- *Coefficient of thermal expansion*. Different materials expand and contract differently, and this can distort the body, especially when it is exposed to a heated paint station, for example.
- *Joining different materials*. Different materials require different joining methods. The methods of primary interest include spot welding, laser welding, friction stir welding, weld-bond adhesive, riveting, and fasteners. Each of these joining methods has an extensive research base, but except for spot welding, most experience is outside the auto industry. Organizations like EWI (Columbus, Ohio) have been researching these methods for years.
- *Predictive modeling*. Two aspects of CAE modeling include static and dynamic analysis. The static analysis principally looks at individual component forming and strength properties. Dynamic modeling attempts to evaluate a structure during a simulated crash. Both modeling methods become increasingly challenging with different materials and different joining methods.

- *Supply chain*. The maturity level of each supply chain is different. Steel is well established with many commodity materials. Aluminum is positioned to expand, but price volatility is a concern. Composites have a unique challenge: Many suppliers and extensive branding by company for their products makes material standardization, specification, and testing difficult.

The current U.S. knowledge base for automotive lightweighting materials, material properties, designing, forming, and joining is distributed throughout the vehicle manufacturers and Tier 1 and Tier 2 suppliers. Ford has pioneered the move to mass-produced aluminum structures with annual volumes of about 650,000 units for the 2015 MY aluminum body F-150, which is about 10 times the volume of niche-market luxury vehicles made of aluminum (SAE 2014). Ford worked with suppliers, including Alcoa and Novelis, to fine-tune the compositional specifications of the aluminum alloys for the F-150. In 1993, Ford developed the experimental aluminum-intensive Mercury Sable and introduced the first aluminum-intensive Jaguar XJ in 2003.

Changes to Powertrain (Adding Technologies, Downsizing, and Electrification)

Traditionally, development cycles in powertrains have been disengaged from vehicle model cycles. In part, this practice was due to the strategy of continuously making smaller incremental upgrades to the powertrains. In addition, manufacturers also prefer to limit the risk of combining the launch of a new vehicle with a new powertrain at the same time. Periods to completely retool for machines and assembly lines for engines and transmissions have generally been in the 10+ years range.

While powertrain design improvements have mainly been driven by performance improvements, a recent focus on fuel economy improvements, GHG emission reduction, and vehicle lightweighting has prompted radical changes in system and component designs as well as more dynamic and quicker-paced design implementation. This trend has had a significant impact on the manufacturing process of powertrain components, both in the initial steps of casting or forging and in the highly complex and highly automated machining and assembly processes.

In order to understand and assess these changes, a more differentiated view of the components of the powertrain makes sense. Essentially the powertrain comprises the engine, turbochargers, transmission, one or more drive shafts, differentials, and the related axle drives. Engines are traditionally manufactured in-house by the OEMs and very rarely shared by multiple OEMs in the form of collaborative projects. This indicates a lesser desire for standardization of engines among the different OEMs.

For example, downsizing the engine does not change the casting process of engine blocks and cylinder heads itself, but does necessitate a changeover of casting molds, cores, risers, and other components. Advanced engine concepts usually lead to more intricate and complex shapes and surfaces of the castings, however. These concepts require a more controlled flow of air and/or fuel, which leads to more complex and intricate casting geometries on cylinder heads. Other challenges in casting technologies can include mechanical core stability, riser geometries, and cycle times.

Within each OEM, considerable efforts are being made to standardize engine features to accommodate increased manufacturing flexibility. Designs having three, four, or six in-line cylinders with identical cylinder specifications can be machined and assembled on the same lines; this is the case for V6 and V8 engines as well. This standardization of engine features can even extend to the changeover capability between gasoline and diesel engines and will be facilitated by the foreseeable introduction of aluminum blocks for diesel engines. Flexible automation has made the youngest generation of production lines more capable of introducing even substantial product innovations at a quicker pace than before without a substantial cost penalty.

Attached to the engine, but fully separate from a manufacturing point of view, turbochargers have become important elements of powertrains with the use of sometimes up to three units on a single engine. Casting aside, manufacturing challenges arise when balancing the machined parts and the precision assembly of ball bearing systems for high rpm and low friction performance, especially with such advanced concepts as twin scroll or variable geometry systems. Turbochargers, or other supercharging systems, are generally manufactured by suppliers aiming to benefit from economies of scale and then delivered to the engine assembly sites.

Transmissions are often manufactured by suppliers, but many OEMs maintain in-house production capability for various volumes. For vehicle designs ranging from more basic and inexpensive to the more luxurious and sophisticated, transmissions with six or more speeds have become the standard. As discussed in Chapter 5, transmissions with eight, nine, or ten speeds will become increasingly prevalent, and continuously variable transmissions (CVTs) and dual-clutch transmissions will continue to play important roles. From a manufacturing viewpoint, all of these designs share a much higher package density than their predecessors. Consequently, machining and assembly tolerances have tightened significantly, leading to more process monitoring and control efforts on the production side and on overall integration.

The manufacture of the remaining components of the powertrain, drive shafts, differentials, and axles, which are mostly produced by suppliers instead of the OEMs, is impacted by fuel consumption and lightweighting targets. Issues here are predominantly caused by the need to reduce friction, which results in tighter machining, balancing, assembly tolerances, and bearing specifications, which all have a potential to increase manufacturing cost.

The Product Development Process

The timing for implementing change on the automobile has trended to different product life cycles for different subsystems on the car. Table 7.4 approximates the product development process (PDP) for different changes.

While automakers wish to keep the vehicle "fresh" for consumers, cost, lead time, resource availability, and risk limit the rate of change. Although accelerating the PDP will challenge these constraints, the availability of new technologies (e.g., safety technologies, electronics, and software) may warrant faster introduction than initially planned, and this typically increases cost. Traditionally, the rate of introducing new technology was embedded in an OEM's overall strategic planning over several model life cycles and incorporated into the PDP and the manufacturing process. However, automakers have indicated that the steadily increasing mandates to improve fuel economy will necessitate more frequent product updates than their PDPs are designed to accommodate. Technologies that are ready for deployment cannot wait until the next upgrade, which might be as long as a 4-8 year cycle for upgrading the powertrains or models shown in Table 7.4, or the automaker risks not meeting the fuel economy target and consumer expectations. Therefore, a shorter deployment cycle is needed and a similar speeding up of the PDP is also necessary for the complete vehicle assembly. This also reduces the length of time available for engineering and predeployment testing, which means that automakers have relied more on accelerated laboratory testing and environmental test chambers and less on testing in the field. Introducing technology faster than planned will lead to issues related to stranded capital, which is discussed earlier in this chapter, and higher product development costs.

Standardized product design and process design principles are a hallmark of mass production to ensure competitive cost and quality performance. Major automakers have standardized procedures for both designing and producing vehicle subsystems. As with all standards, there has to be a balance between constraining innovation and achieving cost and quality objectives. For example, changing from steel to aluminum for a body part will probably require a change in standard practice as to how the part is designed (since aluminum cannot always bend into the same shape as steel), tooled, fabricated, and assembled. Deviations from standard practice result in higher development costs and risk.

Another industry trend inhibiting the rate of technology deployment is the globalization of platforms, discussed below. The reasons for designing different vehicles on a single platform are to reduce cost by keeping the overall volume of the part/component at high volumes, to decrease the need for engineering development resources, to improve quality, and to reduce risk. However, one consequence of this level of global standardization is that the cost to modify the platform now impacts the global vehicle design, not just those vehicles sold in a particular region. Another concern is that when a problem occurs, it occurs to a very large, potentially worldwide, population of vehicles, resulting in higher warranty or recall costs. Global platforms and standardized product and process development by the automakers will add complexity and cost to introducing new technology, and these impacts will weigh against the scale economies of having higher volumes on a smaller number of platforms. Another potential cost savings is that fewer platforms means that fewer redesigns will be needed to add technology across the fleet.

Table 7.5 summarizes the timing, cost, and integration into manufacturing of a variety of fuel economy technologies as estimated by this committee.

TABLE 7.4 Product Lifetimes and Development Cycles

Type of Change	Typical Frequency	Description (Engineering and Tooling)	Investment (Approximate Scale) (billion $)
New vehicle platform (clean sheet)	7 - 10 years	Total engineering for chassis and body and trim. Sporadically in conjunction with new powertrain development.	1.0 - 2.5
Major vehicle upgrade (on established platform)	6 - 8 years	Most of chassis may be carried over. Major body changes (with some carryover).	0.5 – 0.75
Minor vehicle upgrade (re-skin)	2 - 4 years	Minimal engineering with mostly cosmetic changes such as trim. Changes may be implemented that affect aerodynamics, rolling resistance or vehicle accessories,	0.25
New powertrain/transmission	10 - 15 years	All new engine or transmission design. Little to no carryover from previous generation.	0.75 - 1.5
Upgrade powertrain and transmission	4 - 8 years	Technology advancement such as changing 6-speed AT to 6-speed DCT or converting V6 to an I4 with turbo resulting in modifications of the assembly process and tooling	0.2 to 0.4

SOURCE: FEV (2011c, Table 4-1).

TABLE 7.5 Manufacturing Considerations with Associated Timelines and Costs Estimated for the Introduction of Various Vehicle Fuel Economy Technologies

Technology	Description	Integration with Manufacturing	Time	Cost
Powertrain				
Downsize and turbocharge	Engine downsize and addition of boosting	New engine re-design – requires new product development. (Turbo charger available from outsource)	Engine PDP – 2 to 3 years for engine development, 4 to 5 years including emissions certification	Expensive re-tooling requirements
Electrification				
Stop-start	Modifications to ICE	Minor modification of existing manufacturing line	Not significant, but best planned with new model launch	Not difficult to integrate into existing facility
Mild hybrid	Modifications to ICE, new batteries and power electronics	Minor modification of existing manufacturing line	Not significant, but best planned with new model launch	Not difficult to integrate into existing facility- relatively high development costs if volumes remain low
P2 hybrid	New chassis with existing or common body architecture, new batteries and power electronics	Unique chassis, torque converter removed and electric motor/ generator installed in its place without changing the engine or transmission	All new engine and powertrain assembly	New and unique hybrid powertrain line operating outside of standard PDP - relatively high development costs if volumes remain low
PS hybrid	New chassis and powertrain with existing or common body architecture, including batteries and power electronics	Unique chassis and powertrain system to be integrated with traditional body assembly	All new engine and powertrain assembly	New and unique ICE and hybrid powertrain line operating outside of standard PDP - relatively high development costs if volumes remain low
PHEV 40 mi electric range	New chassis and powertrain with existing or common body architecture, including batteries and power electronics	Unique chassis and powertrain system to be integrated with traditional body assembly	All new engine and powertrain assembly	New and unique electrified powertrain line, perhaps with new ICEs, operating outside of standard PDP - relatively high development costs if volumes remain low
EV 75 mi range	New chassis and powertrain with existing or common body architecture, including batteries and power electronics	Unique chassis and powertrain system to be integrated with traditional body assembly	All new engine and powertrain assembly	New and unique electrified powertrain line operating outside of standard PDPP - relatively high development costs if volumes remain low
Body – Lightweighting				
Steel to High-Strength Steel	HSS substitution for individual parts	HSS up to ~1000 MPa has minor issues and minimal increase in cost. Over ~1000 MPa requires change in forming process that will increase cost 50% or more.	Not a significant time impact in most cases. Supply chain availability concern.	Generally, cost premium increases as the steel strength increases. Significant cost increase for steel over 1000MPa due to hot forming. A hot formed part may cost two or more times a cold stamped part.
Steel to Aluminum (Closures)	Hood, deck lid, doors (aluminum hood and deck lids are common today. Aluminum doors and roof panels will be increasing	The challenges for converting to aluminum closure panels are more in design than manufacturing. Not a major manufacturing concern, but some changes with handling, dust/dirt and joining complexity (e.g., fasteners and adhesives) will add cost.	Minimal	Aluminum closures will cost about 25% more than steel equivalents. The cost premium is over $3.00 per pound saved.

continued

TABLE 7.5 Continued

Technology	Description	Integration with Manufacturing	Time	Cost
Steel to Aluminum (Body)	Aluminum body-in-white (traditional for lower volume and premium vehicles)	Significant changes required. The technology is known but the execution at scale production will be a challenge in stamping and body assembly.	Increased launch efforts as the industry learns to do this at volume. There are no significant timing hurdles for this except for the supply chain. Aluminum supply requires at least 30 months lead-time for a high volume vehicle.	Cost increase for converting from a steel to aluminum body-in-white (and closures) is in the order of $1.50 to $2.00 per pound weight reduction (high volume) – over $500 per vehicle.
Steel or Aluminum to Composites	Semi-structural components can be made with composites. A high-volume composite body-in-white is not considered viable due to cost and complexity until sometime past 2025/2030.	Complex. Integration hurdles in assembly (using adhesive) and new painting processes will be required that take a long time and large expense to convert.	Not expected until past 2030 for high volume. Will see composite panels for premium vehicles, usually with a metal subframe.	Expensive – several thousand dollars. New supply chain needed and existing infrastructure (presses, welders and tool making) require overhaul.
Non-Powertrain Technologies				
Aerodynamics	Passive and active technologies. Passive technologies can be readily implemented. Active ones generally require more time. Impact on styling will constrain options.	Minimal barriers	Can generally be implemented with a minor facelift.	Not significant
Low Rolling Resistance Tires	Integration with vehicle, road noise, etc. required.	Not an issue	Not an issue	Not significant
Electric Power Steering	Integration with vehicle	Not an issue	Not an issue	Not significant
Improved Accessories	Integration with vehicle	Not an issue	Not an issue	Not significant
Smart vehicle technology	Sensors and computers and communication devices require design into structure. System design for redundancy.	Electronics primarily from external supply chain with minimal constraints. Integration into vehicle requires some effort.	Not a significant issue (assuming electronics are available with necessary performance)	Not significant

Growing Impact of Global Platforms on Vehicle Design Optimization

The automotive industry is experiencing significant growth in the use of global platforms as a way to reduce cost and increase engineering efficiency. It's estimated that 30 percent of the vehicles produced in 2013 will be made on global platforms (Sedgwick 2014), and this number is continuing to increase. Recently, Ford announced that it will reduce its number of platforms from 15 down to 9 by 2015, and that those 9 platforms will account for 99 percent of the vehicles they manufacture. General Motors announced even more aggressive plans to reduce its number of platforms from 26 to 4, and Volkswagen has indicated that it plans to produce 40 different vehicle models globally on a single vehicle platform. Other companies such as Volvo, Nissan-Renault, BMW, and Toyota are in the process of executing their own versions of global platforms. Each manufacturer is developing its own methods and focusing on different areas of the vehicle to standardize. While potential benefits may be realized by increasing economies of scale and reducing cost and time to develop new model variations, there are also potential drawbacks such as the reduction of design flexibility and suboptimal vehicle design. Vehicle platforms are designed to accommodate certain component modifications but generally not new major technology advances. Despite these possible limitations, it is estimated that the top 10 global platforms will account for over 200 vehicle models by 2017.

There are a number of methods by which vehicle manufacturers are developing global platforms. For example, the modular transverse matrix (MQB), developed by Volkswagen, established a uniform mounting position for all engines and a standardized front carriage structure, which allows it to produce models with different wheelbases and track widths on the same assembly line. Likewise, Nissan-Renault has developed its common module family (CMF), an architecture based on the assembly of compatible modules for the engine bay, cockpit, front underbody, rear underbody, and electrical/electronic architecture. Although these new methods do not necessarily fit the definition of a typical platform, they share the common goal of increasing commonality and standardization across vehicle models (increasing economies of scale and standardizing supply chains). The resulting reduction in unique engineering content and components across different models reduces cost while maintaining product choices for the consumer. In some cases, this can also result in over-engineered parts (designed for the greatest application load).

Global platforms are engineered to anticipate the introduction of various future modifications to the platform as technology advances or other changes are desired. The platform design might limit the ability to implement some changes but expedite the implementation of others that fit within the standard design, reducing development costs. However, since global platforms produce vehicles for different countries, they are designed to accommodate the most stringent requirements for powertrain performance, emission controls, and safety.

Vehicle manufacturers see global platforms as a way to maximize efficiency and reduce cost over a wide range of vehicle models. Nissan-Renault estimated that it will reduce its engineering cost by 30 to 40 percent and part cost by 20 to 30 percent by moving to its CMF system. Volkswagen has also estimated that its MQB could cut production cost by as much as 20 percent. The primary objective of the global platform is to reduce costs through economies of scale. Some regulatory technologies may benefit from platforms, whereas some may not because of differences in different countries. A vehicle platform is essentially the basic building block of components and systems from which a vehicle can be built. Increasing the number of vehicles shared on a single platform—which accounts for nearly half of the product development cost—can significantly reduce engineering cost. Similarly, purchasing and tooling cost can be reduced through economies of scale of component sharing and single sourcing of equipment.

There are several risks and potential limitations that vehicle manufacturers must manage when developing global platforms. With common systems and components shared across many vehicles, design flaws can significantly increase the exposure of a manufacturer's vehicle fleet to recalls. Manufacturers will have to commit significant upfront investment, which may limit the flexibility to modify components and manufacturing processes over time. This could lead to stranded capital if the OEM is unable to amortize the initial investment due to the increased frequency of new design implementations. At the same time, platforms must have enough flexibility to differentiate the product from model to model for consumers to feel as if each product is different. This differentiation may be achieved by standardizing only systems that do not significantly affect styling. An important challenge is that components and structures may be overengineered for some vehicles such that they meet the requirements of all vehicles shared under the same platform. It is probable that shared components will be specified based on the most demanding and more expensive vehicles within the platform. Under such a condition some vehicles may incur increased cost and suboptimal design to meet the specifications of the platform. Each manufacturer must find a balance between the desire to increase economies of scale and the risk of overengineering its vehicles.

FINDINGS AND RECOMMENDATIONS

Finding 7.1 The committee conceptually agrees with the Agencies' method of using an indirect cost multiplier instead of a retail price equivalent to estimate the costs of each technology since ICM takes into account design challenges and the activities required to implement each technology. In the absence of empirical data, however, the committee was unable to determine the accuracy of the Agencies' ICMs. Due to this lack of empirical information, the committee generally assessed only the direct manufacturing costs for each technology. Historically, many studies have concurred on an average markup factor of 1.5.

Recommendation 7.1 The Agencies should continue research on indirect cost multipliers with the goal of developing a sound empirical basis for their estimation. One possible method for developing such an empirical basis is through case studies to break down the costs associated with the specific steps required to integrate a new technology and trace the impacts of a new technology on indirect costs. The committee provides an example earlier in the report where the committee used its knowledge to attempt to construct an ICM that demonstrates some of the insights that can be gathered from an empirical approach.

Finding 7.2 The Agencies' use of the learning-by-doing concept is unconventional in that it is strictly a function of time rather than cumulative production. It allows a technology to accomplish significant cost reductions even if its production volumes remain very low. And in some of its presentations on how they approach learning, the Agencies convey the notion that cumulative production volume is used in the estimates. However, the committee appreciates the difficulties of implementing volume-based learning in the compliance models used by the Agencies.

Recommendation 7.2 The Agencies should make clear the terminology associated with learning and should assess whether and how volume-based learning might be better incorporated into their cost estimates, especially for low-volume technologies. The Agencies should also continue to conduct and review empirical evidence for the cost reductions that occur in the automobile industry with volume, especially for large-volume technologies that will be relied on to meet the CAFE/GHG standards. The committee also recommends that, once the Agencies have decided on an implementation scenario, they should regard their production volumes as fixed and look for inconsistencies in their scenario with respect to cost reductions from learning (i.e., they have assigned a large cost reduction from learning for technologies with very low market penetrations).

Finding 7.3 The committee disagrees with the methodology of assigning direct manufacturing costs based on a 450,000 unit production volume since some technologies, especially those related to electric and hybrid vehicles, may take many years, if ever, to reach this market penetration. Additionally, since the 450,000 unit production volume applies to the entire United States, the smaller production volumes of each manufacturer would not bring the expected cost reductions.

Recommendation 7.3 For technologies such as electric vehicles, which may not reach 450,000 unit production volumes in North America in the time frame of the standards, the Agencies should use an appropriate, lower production volume to project direct manufacturing costs.

Finding 7.4 The product development process of auto manufacturers is accelerating for several reasons, one of which is to implement new technologies faster. Manufacturers traditionally bundle technologies and implement them in predetermined cycles. New regulations now encourage more rapid deployment as soon as a technology is ready to avoid falling behind on satisfying the steadily increasing regulations. More rapid deployment, although better for meeting regulations and responding to consumer demands, will increase stranded capital and incur higher product deployment costs.

Finding 7.5 The growth of global platforms used by automakers supports scale economies with shared components and engineering content. However, shared content may inhibit the use of lightweighting materials and fuel-economy technologies if they are not readily available in local markets, or because these technologies are not appropriate for the local market. Global platforms thus can be considered a constraint, especially in the short term, whereby supply chains are not fully developed, as well as an opportunity, especially in the long term, whereby scale economics can provide cost reductions. Since attributes that are unique to a local market (such as emission/fuel economy regulations or crashworthiness) may call for unique technologies, there is a risk that these may not be compatible with the global platform. In some cases, where regionally unique attributes are embedded in the global platform, a less-than-optimal solution (such as a subsystem engineered for the greatest load case) is likely to exist in the global platform that prevents an "optimal" design in every region.

REFERENCES

Audi. 2014. Audi TT Coupe. Audi MediaServices. https://www.audi-mediaservices.com/publish/ms/content/de/public/grafiken/2014/03/20/TTC140091.html.

Blincoe, L. 2013. NHTSA's Application of Indirect Costs. Presentation to the National Research Council Committee on Assessment of Technologies for Improving Light-Duty Vehicle Fuel Economy, Phase 2, March 27.

Bloomberg. 2014. VW plans Apple-like model updates. Automotive News, July 28. http://www.autonews.com/article/20140728/OEM03/140729897/vw-plans-apple-like-model-updates#.

Bussmann, W.V. 2008. Study of industry-average markup factors used to estimate retail price equivalents (RPE). Presentation to the National Research Council Committee on Assessment of Technologies for Improving Light-Duty Vehicle Fuel Economy, January 24.

DOT (Department of Transportation)/NHTSA (National Highway Traffic Safety Administration). 2009. Average fuel economy standards, passenger cars and light trucks, model year 2001: Final rule. 49 CFR Parts 523, 531, 533, 534, 536, and 537, Docket No. NHTSA-2009-0062, RIN 2127-AK29. Washington D.C., March 23.

Duleep, K.G. 2008. Analysis of technology cost and retail price. Presentation to the National Research Council Committee on Assessment of Technologies for Improving Light-Duty Vehicle Fuel Economy, January 25.

EPA (Environmental Protection Agency). 2012. Regulatory Impact Analysis: Final Rulemaking for 2017-2025 Light-Duty Vehicle Greenhouse Gas Emission Standards and Corporate Average Fuel Economy Standards. PA-420-R-12-016.

EPA. 2014. MTE Technologies and Costs: 2022-2025 GHG Emissions Standards. Presentation to the National Research Council Committee on Assessment of Technologies for Improving Fuel Economy of Light Duty Vehicles, Phase 2, July 31.

EPA (Environmental Protection Agency)/NHTSA (National Highway Traffic Safety Administration). 2012. Joint Technical Support Document: Final Rulemaking for 2017-2025 Light-Duty Vehicle Greenhouse Gas Emission Standards and Corporate Average Fuel Economy Standards, EPA-420-R-12-901, U.S. Environmental Protection Agency, Ann Arbor, MI.

FEV, Inc. 2009. Light-Duty Technology Cost Analysis Pilot Study. EPA-420-R-09-020, U.S. Environmental Protection Agency, Ann Arbor, MI.

FEV, Inc. 2011a. Light-Duty Vehicle Technology Cost Analysis, Mild Hybrid and Valvetrain Technology. EPA-420-R-11-023, U.S. Environmental Protection Agency, Ann Arbor, MI.

FEV, Inc. 2011b. Light-Duty Vehicle Technology Cost Analysis, Advanced 8-Speed Transmissions. EP-C-07-069, U.S. Environmental Protection Agency, Ann Arbor, MI.

FEV, Inc. 2011c. Potential Stranded Capital Analysis on EPA Light-Duty Technology Cost Analysis. EPA-420-R-11-019, U.S. Environmental Protection Agency, Ann Arbor, MI, November.

FEV, Inc. 2012. Light-Duty Vehicle Mass Reduction and Cost Analysis – Midsize Crossover Utility Vehicle. EPA-420-R-12-026, U.S. Environmental Protection Agency, Ann Arbor, MI, August.

Finlay, S. 2014. '15 Toyota Camry Reflects Shorter Product Cycle. WardsAuto, October 1. http://wardsauto.com/vehicles-technology/15-toyota-camry-reflects-shorter-product-cycle.

Helfand, G., and T. Sherwood. 2013. Automobile industry retail price equivalent and indirect cost multipliers studies. Presentation to the National Research Council Committee on Assessment of Technologies for Improving Light-Duty Vehicle Fuel Economy, Phase 2, March 27.

Husan, R. 1997. The continuing importance of economies of scale in the automotive industry. European Business Review 97(1): 38-42.

IBIS Associates, Inc. 2008. Data Collection and Analysis: Vehicle Systems Costs. Report to the National Research Council Committee on Assessment of Technologies for Improving Light-Duty Vehicle Fuel Economy, December.

IHS Automotive. 2013. Ford Earns Top Marks in Polk Automotive Loyalty Awards; Volkswagen Named Most Improved. Polk, January 15. https://www.polk.com/company/news/ford_earns_top_marks_in_polk_automotive_loyalty_awards_volkswagen_named_mos.

Martec Group, Inc. 2008. Variable costs of fuel economy technologies. Presentation to the National Research Council Committee on Assessment of Technologies for Improving Light-duty Vehicle Fuel Economy, January 24.

NHTSA. 2012. Final Regulatory Impact Analysis: Corporate Average Fuel Economy for MY 2017-MY2025 Passenger Cars and Light Trucks. Office of Regulatory Analysis and Evaluation, National Center for Statistics and Analysis.

Nordhaus, W.D. 2009. The Perils of the Learning Model for Modeling Endogenous Technologies Change. Working Paper 14638, National Bureau of Economic Research, Cambridge, MA, January.

NRC (National Research Council). 2002. Effectiveness and Impact of Corporate Average Fuel Economy Standards. Washington, D.C.: The National Academies Press.

NRC. 2011. Assessment of Fuel Economy Technologies for Light-Duty Vehicles. Washington, D.C.: The National Academies Press.

Oliver Wyman. 2013. Automotive Value-Chain Structure Changes Massively. Automotive Manager: Trends, Opportunities and Solutions along the Entire Value Chain. Oliver Wyman, January. http://www.oliverwyman.com/content/dam/oliver-wyman/global/en/files/archive/2013/Oliver_Wyman_Automotive_Manager_I_2013.pdf.

Rogozhin, A., M. Gallaher, G. Helfand, and W. McManus. 2010. Using indirect cost multipliers to estimate the total cost of adding new technology in the automotive industry. International Journal of Production Economics 124: 360-368.

RTI International/UMTRI (University of Michigan Transportation Research Institute). 2009. Automobile Industry Retail Price Equivalent and Indirect Cost Multipliers. EPA-420-R-09-003, U.S. Environmental Protection Agency, Ann Arbor, MI.

SAE. 2014. Automotive Engineering, November 4.

Sedgwick, D. 2014. Carmakers bet on big global platforms to cut costs. Autonews, August 4. http://www.autonews.com/article/20140804/OEM10/308049988/carmakers-bet-on-big-global-platforms-to-cut-costs.

Vyas, A., D. Santini, and R. Cuenca. 2000. Comparison of Indirect Cost Multipliers for Vehicle Manufacturing. Center for Transportation Research, Argonne National Laboratory, Argonne, IL.

Wene, C.O. 2000. Experience Curves for Energy Technology Policy. International Energy Agency, OECD, Paris.

8

Estimates of Technology Costs and Fuel Consumption Reduction Effectiveness

INTRODUCTION

The committee's estimates of fuel consumption reduction effectiveness and costs of the technologies discussed in previous chapters are examined from an application perspective in this chapter. The previous chapters identified current and future technologies that are effective for reducing fuel consumption. Although many of these technologies, such as engine friction reduction, are applicable to most vehicles, others are specific to particular classes of vehicles. Secondary axle disconnect, for example, would be applicable only to four-wheel-drive vehicles. Some of the technologies discussed in previous chapters have already been incorporated in current vehicles, and additional technologies are expected to be applied by the 2016 MY. Vehicle classes and baselines are discussed in this chapter followed by the central topics of fuel consumption reduction effectiveness and cost estimates of technologies and the implementation of the technologies in vehicles.

FUEL CONSUMPTION REDUCTION EFFECTIVENESS AND COST OF TECHNOLOGIES

Vehicle Classes

The National Highway Traffic Safety Administration (NHTSA) used twelve vehicle classes in its support documentation for the final CAFE rule: Subcompact Car, Compact Car, Midsize Car, and Large Car; performance versions of these four classes of cars; Small sport utility vehicle (SUV)/Pickup/Van, Midsize SUV/Pickup/Van, Large SUV/Pickup/Van, and Minivan. The NRC Phase 1 study considered ten vehicle classes, although these classes were consolidated into five classes for evaluating costs and effectiveness (NRC 2011). From this evaluation, the relative costs and effectiveness values were found to be primarily influenced by engine type, specified as I4, V6, or V8, rather than by vehicle type. Based on the results from the NRC Phase 1 study, the following three classifications of vehicles and associated engine types were selected to be appropriate for the analysis of overall costs and effectiveness for the current study:

- Midsize Car with I4 dual overhead camshaft (DOHC) engine,
- Large Car with V6 DOHC engine, and
- Large Light Truck with V8 overhead valve (OHV) engine.

The committee used these classifications in evaluating the Agencies' estimates since comparability with the Agencies' classifications was important. The Environmental Protection Agency (EPA)/NHTSA joint Technical Support Document (TSD) (EPA/NHTSA 2012a), which was used to provide baseline information in this section, uses classifications that are slightly different from those used by the NHTSA Regulatory Impact Analysis (RIA) (NHTSA 2012). Generally, the Compact/Midsize car classes of the RIA were aligned with the Midsize Car class used in the TSD. Likewise, the Large Car class in the RIA was aligned with the Standard/Large Car class used in the TSD. The Large Light Truck classification was consistent in the RIA and the TSD.

Baselines

The selection of a baseline is important in assessing the overall costs and effectiveness of technologies. EPA and NHTSA defined a null[1] or baseline vehicle as consisting of the following features:

[1] The null vehicle concept was developed by EPA and NHTSA as a reference point against which effectiveness and cost can be consistently measured (Olechiw 2014). It is defined as a vehicle having the lowest level of technology in the 2008 MY. Technologies are first added to bring the null vehicle into compliance with the 2016 standards, followed by compliance with the 2021 and 2025 standards. The concept is particularly important because, even though NHTSA and EPA use different compliance models, the effectiveness values determined by both Agencies are relative to the same null package; each compliance model uses the same base data. This committee applied the null vehicle concept to illustrate effectiveness and cost in an example pathway.

- Spark ignition (SI) engine,
- Naturally aspirated,
- Four valves per cylinder (except two valves per cylinder for OHV engines),
- Port fuel injection (PFI),
- Fixed valve timing and lift, and
- Four-speed automatic transmission.

Although the Agencies' analysis began with the 2008 MY, very few vehicles in the 2008 MY had the limited content of this null vehicle. Many, however, contained some of the EPA and NHTSA technologies defined in the TSD. EPA's and NHTSA's analysis for the final rule began by identifying technologies in the 2008 MY vehicles that were in addition to those in the null vehicle. Although the Agencies correctly ascribe technologies already applied in their compliance models, a revised null vehicle and better discussion of the concept is appropriate for the mid-term review. In this chapter, the baseline for most of the initial technologies within a category is the baseline or null vehicle. The effectiveness and cost of the technologies that follow the initial technologies are specified as "relative to" one of the following: (1) the baseline or null vehicle, (2) the previously applied technology, or (3) another defined reference condition as discussed in a later section of this chapter.

Effectiveness of Technologies

Many of the technologies identified in previous chapters are broadly applicable across most light-duty vehicle classes, although some limitations must be considered. As discussed in Chapter 5, dual clutch transmissions (DCTs) may not be acceptable to customers for midsize and larger cars due to launch and gear shifting quality concerns that contrast with the smooth performance provided by a conventional automatic transmission with a hydraulic torque converter. Likewise, continuously variable transmissions (CVTs) have torque limitations that preclude applications requiring high torque loads in vehicles with larger engines, where towing is an important functional attribute.

Table 8A.1 provides a compilation of the committee's low and high most likely estimates of fuel consumption reduction effectiveness for the technologies discussed in the previous chapters (see Table 8A.1). The derivations of the low and high most likely estimates, which are discussed in previous chapters, relied on (1) fundamental technical analyses, (2) literature reviews, including the Phase 1 NRC study, (3) full system simulation, (4) EPA certification test data, (5) inputs received from vehicle manufacturers and suppliers, (6) comparisons with extensive EPA and NHTSA evaluations using full system simulations, including the lumped parameter model, and (7) the committee's expert opinion. For reference, EPA and NHTSA estimates of fuel consumption reductions, which are provided in the TSD (EPA/NHTSA 2012a), are shown in Appendix S, Table S.1.

The committee's most likely estimates of fuel consumption reduction effectiveness are comparable to NHTSA's estimates for many of the technologies defined by NHTSA. The committee estimated higher most likely effectiveness values for several technologies, including mass reduction (12.2 percent compared to NHTSA's estimate of 10.2 percent for a 20 percent mass reduction) and high-efficiency gearbox technology (4.9 to 5.4 percent compared to NHTSA's estimate of 2.7 percent when applied to an eight-speed automatic transmission, although NHTSA's eight speed transmission is assumed to include some benefits of efficiency improvements not included in the 2.7 percent improvement). For some other technologies, including several of the turbocharged, downsized engine technologies and P2 hybrids, the committee extended the range of most likely estimates of effectiveness to include lower values. For several other technologies, including shift optimization (0.5 to 1.0 percent compared to NHTSA's 3.9 to 4.1 percent) and eight-speed automatic transmissions (1.5 to 2.0 percent compared to NHTSA's 4.6 to 5.3 percent), the committee's low and high range of most likely estimates were lower than NHTSA's estimates.

In addition to listing technologies defined by EPA and NHTSA, Table 8A.1 also lists the effectiveness estimates of other technologies not considered by EPA and NHTSA that may be available either by the 2025 MY or later, extending to the 2030 MY. The technologies that might be available by the 2025 MY could provide additional reductions in fuel consumption or, possibly, alternative approaches at lower cost. In addition, the committee has identified several technologies that might be available after 2025, although these technologies are generally in the research phase of development. Technologies using alternative fuels may also provide some opportunities for reductions in fuel consumption.

Costs of Technologies

The direct manufacturing costs of technologies for reducing fuel consumption were estimated by (1) developing cost estimates for key subsystems and components for each technology, (2) using the detailed cost teardown studies conducted by EPA with appropriate updates, (3) considering input from the vehicle manufacturers and suppliers, (4) referring to the Phase 1 NRC Study, and (5) evaluating estimates provided by experts through presentations and publications. These low and high most likely cost estimates for the technologies, discussed in the earlier chapters, are shown in Tables 8A.2a, b, and c for the 2017, 2020, and 2025 MYs, respectively. Tables 8A.2a, b, and c also show costs estimates for technologies not considered by EPA and NHTSA that may be available either by the 2025 MY or later, extending to the 2030 MY. For reference, EPA and NHTSA cost estimates contained in the TSD (EPA/NHTSA 2012a) are shown in Appendix S, Tables S.2a, b, and c.

The committee's estimates of direct manufacturing costs are comparable to NHTSA's estimates for some of the technologies defined by NHTSA. The committee extended the range to include higher estimates of direct manufacturing costs for some of these technologies, including several SI engine technologies, several transmission technologies, and electrified powertrain technologies. The ranges of most likely direct manufacturing costs for several technologies, including advanced diesel engines (with an estimated cost of $2,572 for an I4 advanced diesel engine for a midsize car compared to NHTSA's estimate of $1,752 in 2025), several transmission technologies (including a high-efficiency gearbox with an estimated cost of $267 compared to NHTSA's estimate of $163), and mass reduction (ranging from $0.43 to $1.15 per pound for cars compared to NHTSA's estimate of $0.35 per pound for 10 percent mass reduction in 2025), were higher than NHTSA's estimates.

Cost analyses made by EPA and NHTSA were generally based on a production assumption of 450,000 units per year. Although this production volume may be valid for very high-volume vehicles, typical vehicles from one manufacturer will have significantly lower annual production volumes. In addition, one vehicle line will tend to have several engines and transmissions, which further lowers the production volume of engine- and vehicle-specific components.

For newer technologies, the assumption of 450,000 units per year appears to be optimistic. For example, electric vehicles are assumed to be 2 percent of the U.S. fleet in 2025. Assuming a total fleet of 16 million vehicles, 2 percent is 320,000 units per year. However, this volume may not be concentrated in industry common components but distributed among many manufacturers, which could reduce the volume for a particular manufacturer to 32,000 units per year or less and negatively impact the costs relative to the assumption of 450,000 units per year industry volume. EPA and NHTSA have recognized the need to represent low volume introductions with costs that exceed these estimates based on mature production volumes by applying the concept of negative learning, which is described later in the chapter.

Relative Effectiveness and Cost

The effectiveness and cost of technologies listed in Tables 8A.1 and 8A.2a, b, and c are dependent on the application of the specific technology. Since the technologies may be applied differently, Tables 8A.1 and 8A.2a, b, and c contain the column labeled "Relative To" to define the specific application. The initial technologies within a category are generally shown relative to the baseline, which is considered to be the baseline, or null vehicle, discussed previously. Subsequent technologies may be shown as "Relative To" one of the following: (1) the baseline, or null vehicle, (2) the previously applied technology listed in the table, or (3) another defined reference condition. For example, all of the "Other Technologies" in the diesel engine category are shown relative to the Advanced Diesel technology since they depend on this technology having been previously implemented.

For the mass reduction sections of Tables 8A.1 and 8A.2a, b, and c, effectiveness and costs are shown in two formats; one is for the mass reduction relative to the baseline vehicle, such as 0-10 percent mass reduction and the other is for the mass reduction relative to the previous mass reduction, such as for the 5-10 percent mass reduction increment. A transition occurs at 15 percent mass reduction, which will likely involve a change from a vehicle with high-strength steel to one with an aluminum body. This transition is shown in the tables as follows:

- Table 8A.1 shows fuel consumption reductions for the aluminum body vehicle for two cases. For the 0-15 percent mass reduction case, 9.15 percent reduction in fuel consumption is shown relative to the original baseline. For the 10-15 percent mass reduction case, only a 3.25 percent reduction in fuel consumption is shown relative to the previous mass reduction, which would be the case of having already achieved 10 percent mass reduction with the high-strength steel vehicle. This case is considered the most likely assumption for the transition from high-strength steel to aluminum-body vehicles.
- Tables 8A.2a, b, and c show the costs for the aluminum-body vehicle as relative to the baseline for both the incremental 10-15 percent and for the absolute 0-15 percent mass reduction cases. The reason for this is that the previous high-strength steel body vehicle, which achieved 10 percent mass reduction, cannot be reused for the aluminum-body vehicle, so the costs are reset back to the original baseline vehicle. The table shows that the cost of the aluminum body vehicle is the same whether the starting point is the original baseline vehicle or the high-strength steel body vehicle that has already achieved a 10 percent mass reduction.

Learning Curves

EPA and NHTSA developed learning curves that provide learning factors as a function of the model year. Examples of these learning curves are shown in Figure 8.1. An important feature of the learning curve is the basis, which is the year in which the learning factor equals 1.00, indicating that the technology is mature. NHTSA defines a mature technology as one that has reached a production volume of 450,000 units per year in North America. The learning factor is applied to the direct manufacturing cost for the base year to determine the direct manufacturing costs for the other years of interest. The effects of learning curves are reflected in the estimated direct manufacturing costs shown in Tables 8A.2a, b, and c. Generally the committee applied the same learning curves used by NHTSA, although a learning curve different from NHTSA's assumption was used for mass reduction, as discussed in Chapter 6. A variety of learning curves is shown in Figure 8.1. Learning curve 6 is flat with no learning,

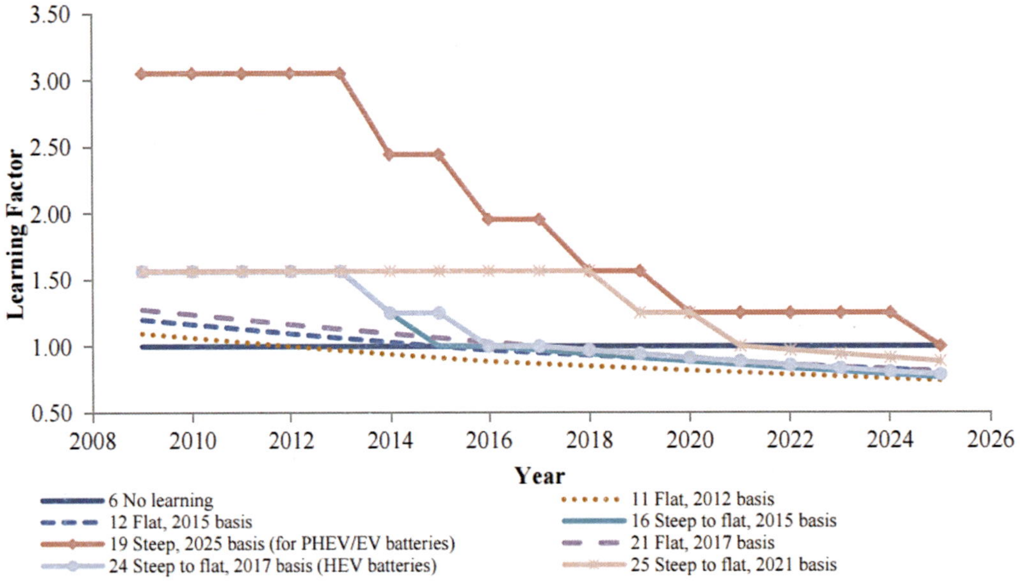

FIGURE 8.1 Learning factors for several different learning curves.
SOURCE: NHTSA (2012).

which, for example, was applied to low friction lubricants. Typical learning curves have a basis in 2012, 2015, or 2017. However, learning curves for newer technologies have their basis as late as 2025. The base year of these learning curves tends to be preceded by steep learning schedules, which is the concept of applying negative learning to estimate the costs of new technologies during early, low-volume introductions into production. Steep learning curves assume 20 percent decreases in the learning factor every 2 years during the initial years of production, for a maximum of two learning cycles, before converting to the flatter learning curves.

Interaction of Technologies

EPA and NHTSA discussed technologies in their joint TSD relative to a null vehicle. NHTSA structured its analysis in the RIA so that each successive technology is added to the preceding technology and the fuel consumption reduction effectiveness values are dependent on and incremental to each of the previous technologies that have already been applied (NHTSA 2012). In many cases, this means accounting for synergies among technologies.[2]

NHTSA used decision trees to illustrate the order of application of technologies and the effectiveness of a technology relative to previous technologies. An excerpt of a decision tree for a midsize car is shown in Figure 8.2. In this decision tree, turbocharging and downsizing—level 1 is shown to have an incremental effectiveness of 8.3 percent relative to the previous technologies, which included friction reduction, variable valve timing and lift, and stoichiometric gasoline direct injection. The relative effectiveness shown in the decision tree is consistent with the 12.9 to 14.9 percent effectiveness relative to the baseline null vehicle shown in the TSD. The lower effectiveness shown in the decision tree results from the application of a technology that reduces friction and pumping losses and improves thermodynamic efficiency after many other technologies have already been applied that provided similar improvements. This example illustrates the significant reduction in effectiveness that depends on the order in which a technology is applied. Effectiveness values for SI engine technologies shown in Table 8A.1 are relative to the previously applied technologies. The order of application of the technologies listed in the table follow the order developed by NHTSA in the decision trees.

Accounting for interaction of multiple technologies was important when combining technologies in the order presented in Table 8A.1. The effects of potential positive and negative synergies were considered. EPA and NHTSA identified the effects of interactions using the lumped parameter model, which was validated using the Ricardo full system simulations (Ricardo Inc. 2011). Many of the committee's interactions were directly scaled from interaction effects that had been defined by NHTSA in the RIA and in the decision trees. To confirm that the interactions of technologies were

[2] Two or more technologies applied together might be negatively synergistic, meaning that the sum of their effects is less than the impact of the individual technologies (contributes less to reducing fuel consumption, in this case). Or, they might be positively synergistic, meaning that the sum of the technologies' effects is greater than the impact of the individual technologies (in this case, contributes more to reducing fuel consumption) (EPA/NHTSA 2009).

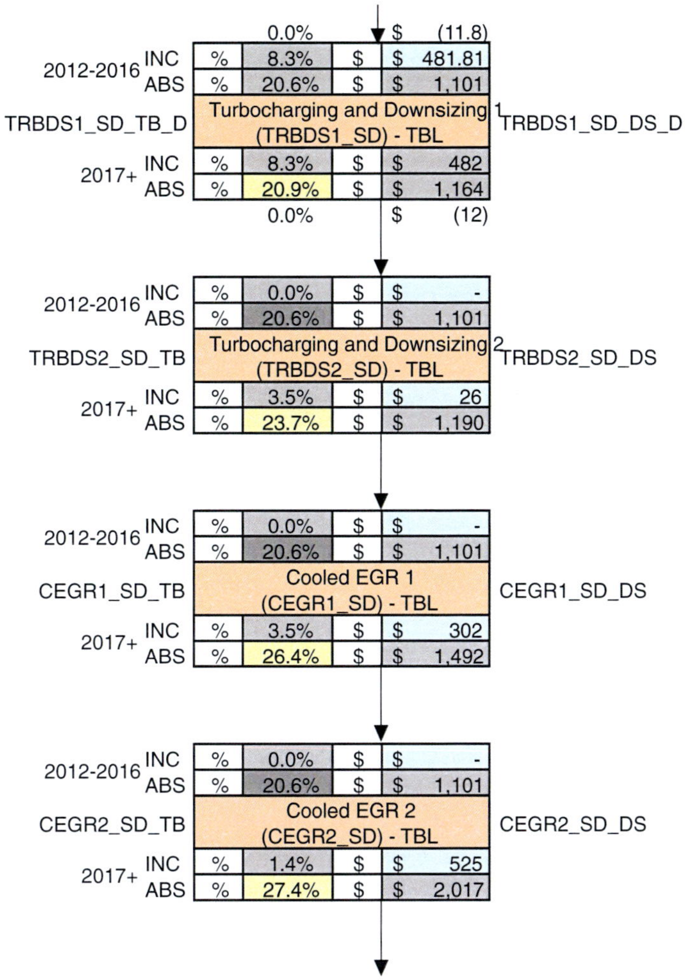

FIGURE 8.2 Excerpt from NHTSA's decision tree for a midsize car.
SOURCE: NHTSA (2012).

appropriately accounted for, the committee contracted with the University of Michigan to conduct full system simulations. The results of the full system simulations, described later in this chapter, generally agreed with the interactions developed by EPA and NHTSA.

Effect of Engine Downsizing on Costs

An important factor affecting costs of turbocharged, downsized engines is downsizing displacement. In some downsizing cases, the number of cylinders is reduced instead of continuing to proportionally downsize the displacement of each cylinder. NHTSA recognized that there are limits to reducing cylinder size since heat losses increase with smaller cylinder displacements. As shown in Table 8.1, NHTSA specified the cases in which displacement reduction requires a reduction in the number of cylinders (NHTSA 2012). The committee followed the same schedule shown in Table 8.1 for reducing the number of cylinders.

Costs for turbocharging and downsizing are shown in the TSD relative to the null vehicle. However, since NHTSA assumes that turbocharging and downsizing occur after the application of many other engine technologies, turbocharging and downsizing costs need to be adjusted, as shown for the example of an I4 engine downsized to an I3 engine in a midsize car in Table 8.2. All of the previously applied technologies for the four cylinders of the baseline engine need to be reduced to only three cylinders to provide cost savings

TABLE 8.1 Changes in Number of Cylinders as Engines Are Downsized

Base Engine	18-bar Engine	24-bar Engine	27-bar Engine
I4	I4	I3	I3
V6	I4	I4	I4
V8+	V6	V6	I4

TABLE 8.2 Effects of Reducing Number of Cylinders on Direct Manufacturing Cost When Changing from Level 1 to Level 2 Turbocharged, Downsized Engine

Baseline:	I4 Engine	Incremental Costs ($)
Previously Added Features (Cylinder Number Dependent)	LUB2 x 4	51
	EFR1 x 4	49
	DVVL x 4	116
	SGDI x 4	186
Total Deleted Costs		402
Turbocharged Downsized Engine	I3 Engine	Incremental Costs ($)
Downsizing I4-I3 (TSD Table 3-32)	TSD Table 3-32	-174
Turbocharging (TSD Table 3-31)	TSD Table 3-31	182
Previously Added Features (Cylinder Number Dependent)	LUB2 x 3	38
	EFR1 x 3	37
	DVVL x 3	87
	SGDI x 3	140
Total Added Costs		310
Net Cost		-92

NOTE: Direct manufacturing costs (2010$) based on NHTSA decision trees, cost files, and TSD.

that are in addition to the savings from reducing the number of cylinders. Similar adjustments are applied to the costs for a V8 engine downsized to a V6 engine and a V6 engine downsized to an I4 engine. The resulting revised costs are noted with asterisks and are shown on the shaded rows in Tables 8A.2a, b, and c for cases where the number of cylinders is reduced. These costs on the shaded rows are shown below the costs for turbocharged, downsized engines without a change in the number of cylinders. A complete description of the derivation of turbocharged, downsized engine costs shown in Table 8A.2 is provided in Appendix T.

Synergies

The effectiveness values of technologies for reducing fuel consumption are generally defined in the TSD (EPA/NHTSA 2012a) relative to a null, or baseline, vehicle. However, when adding a new technology to a vehicle that already contains other technologies for reduced fuel consumption, NHTSA developed a method for accounting for positive and negative synergies. The method is briefly described in this section and subsequently applied in several of the committee's estimates, shown in Table 8A.2 and in an example pathway described later in this chapter.

NHTSA defined decision trees that consist of separate paths for SI engines, diesel engines, transmissions, accessories, hybrids, mass reduction, low rolling resistance tires, aerodynamic drag reduction, and low drag brakes and secondary axle disconnect. These decision trees will also be discussed in a later section of this chapter. Within each decision tree path, successive technologies are applied and their effectiveness values are shown relative to the preceding technology, rather than to the null, or baseline, vehicle. NHTSA generally determined these application-specific effectiveness values by applying the lumped parameter model, which was previously validated by the full system simulations developed by Ricardo (Ricardo Inc. 2011).

NHTSA developed another method for accounting for synergies when crossing over to another decision tree path, such as adding technologies from the transmission path after the applicable technologies in the SI engine path had been added. For the case of crossing over to other decision tree paths, NHTSA developed Tables V-30a-f in its RIA (NHTSA 2012). These tables list technology pairings and incremental synergy factors associated with those pairings. The incremental synergy factors for all instances of a technology in the incremental synergy tables that match technologies already applied to the vehicle are summed and applied to the percent reduction in fuel consumption of the technology being applied.

Examples of applying the synergy factors for technologies from the transmission decision tree path to an engine that already has all of the technologies in the SI engine decision tree path are shown in Table 8.3. As shown in the table, the adjusted percent reductions in fuel consumption of several transmission technologies are significantly reduced relative to the baseline engine when applied to an engine containing all of the fuel consumption reduction technologies. The relatively close agreement of the adjusted percent reductions in fuel consumption with estimates using the lumped parameter model is shown in the table for reference. The ratios of the adjusted percent fuel consumption reduction to the base

TABLE 8.3 Synergy Factors for Application of Transmission Technologies to 27 bar BMEP (CEGR2) Engines

Applying NHTSA Method Using NHTSA RIA Tables V-30 a-c for a Midsize Car						
Technology	% FC Impr. (Relative to Base [Null] Engine)	Synergy Factor Pairs	Synergy Factor	Adjusted % FC Impr.	Notes	Ref: Lumped Parameter Model % FC Impr.
Engine Decision Tree Path						
ICP						
DCP						
CVVL						
SGDI						
TRBDS1						
TRBDS2						
CEGR1						
CEGR2						
Transmission Decision Tree Path						
IATC	3.0	Sum of Synergy Factors	-1.4	1.6	3.0 - 1.4 = 1.6	0.8
		IATC - ICP	-1.6			
		IATC - CVVL	-0.6			
		TRBDS1 - IATC	-0.8			
		IATC - TRBDS1	1.6			
NUATO	2.0	Sum of Synergy Factors	-2.0	0.0	2.0 - 2.0 = 0	0.3
(6 sp AT)		NUATO - ICP	-1.2			
		TRBDS2 - NUATO	-1.2			
		CEGR2 - NUATO	-0.8			
		NUATO - TRBDS1	1.2			
8 sp AT	4.6	Sum of Synergy Factors	-0.7	3.9	4.6 - 0.7 = 3.9	3.8
		8 sp AT - ICP	-2.5			
		8 sp - CVVL	-0.7			
		8 sp AT - TRBDS	2.5			
SHFTOPT	4.1	Sum of Synergy Factors	-1.3	2.8	4.1 - 1.3 = 2.8	3.1
		DCP - SHFTOPT	-0.6			
		TRBDS2 - SHFTOPT	-0.7			

fuel consumption reduction values shown in Table 8.3 were applied in the committee's estimates of transmission technologies in Tables 8A.1 and 8A.2 for example pathways discussed later in this chapter.

Cost Effectiveness of Technologies

The cost effectiveness of the technologies defined by NHTSA for SI engines is illustrated in Figure 8.3. The NHTSA-defined technologies are shown on the plot of NRC estimated incremental 2025 MY direct manufacturing cost in 2010 dollars versus percent reduction in fuel consumption. Lines of constant cost per percent reduction in fuel consumption are overlaid on the plot to illustrate the cost effectiveness of the technologies. The costs per percent reduction in fuel consumption for the SI engine technologies range from less than $25 per percent to more than $100 per percent.

The cost effectiveness values of NHTSA's technologies for overall SI engine technologies leading to a 27 bar BMEP engine as well as for advanced diesel engine and strong hybrid technologies are shown in Figure 8.4. The 2025 MY direct manufacturing cost per percent reduction in fuel consumption of an SI engine with all of NHTSA's technologies included is less than $50 per percent, which is lower than advanced diesel engines and strong hybrids, which are in the range of $75 to $100 per percent reduction in fuel consumption. The cost per percent reduction in fuel consumption of a mild hybrid exceeds $100 per percent.

The cost of some of the SI engine technologies, especially cooled exhaust gas recirculation (EGR)—level 2 at over

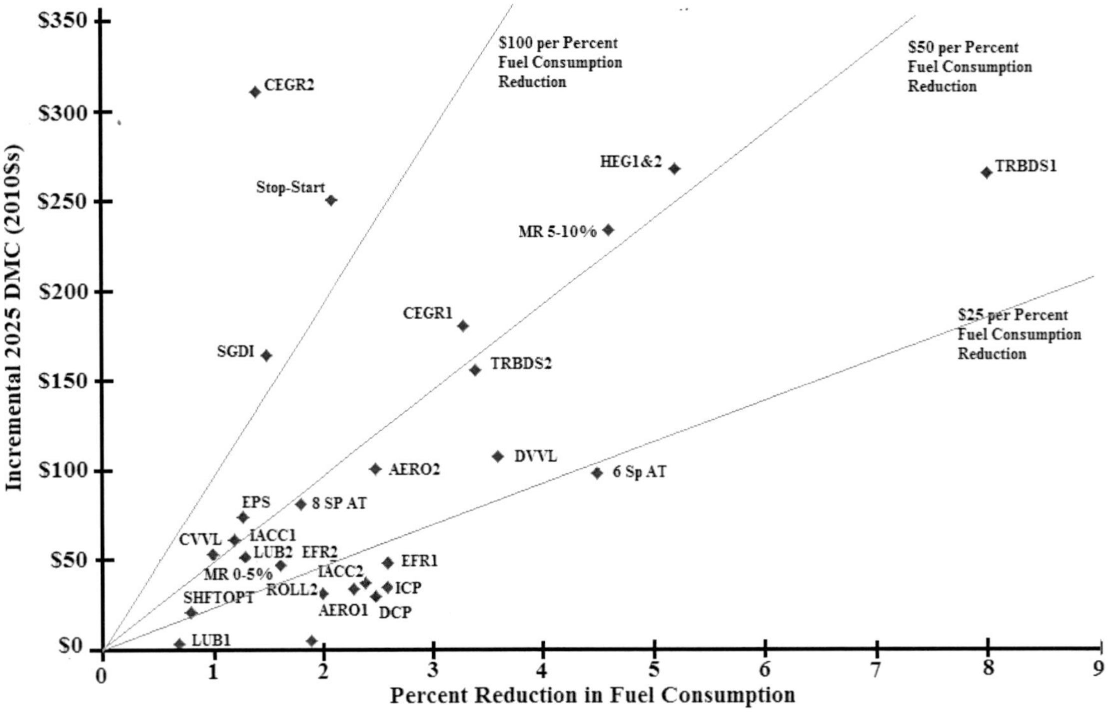

FIGURE 8.3 NHTSA technologies for spark ignition I4 engines in midsize cars shown on a plot of NRC-estimated incremental 2025 MY direct manufacturing cost in 2010 dollars versus percent reduction in fuel consumption.

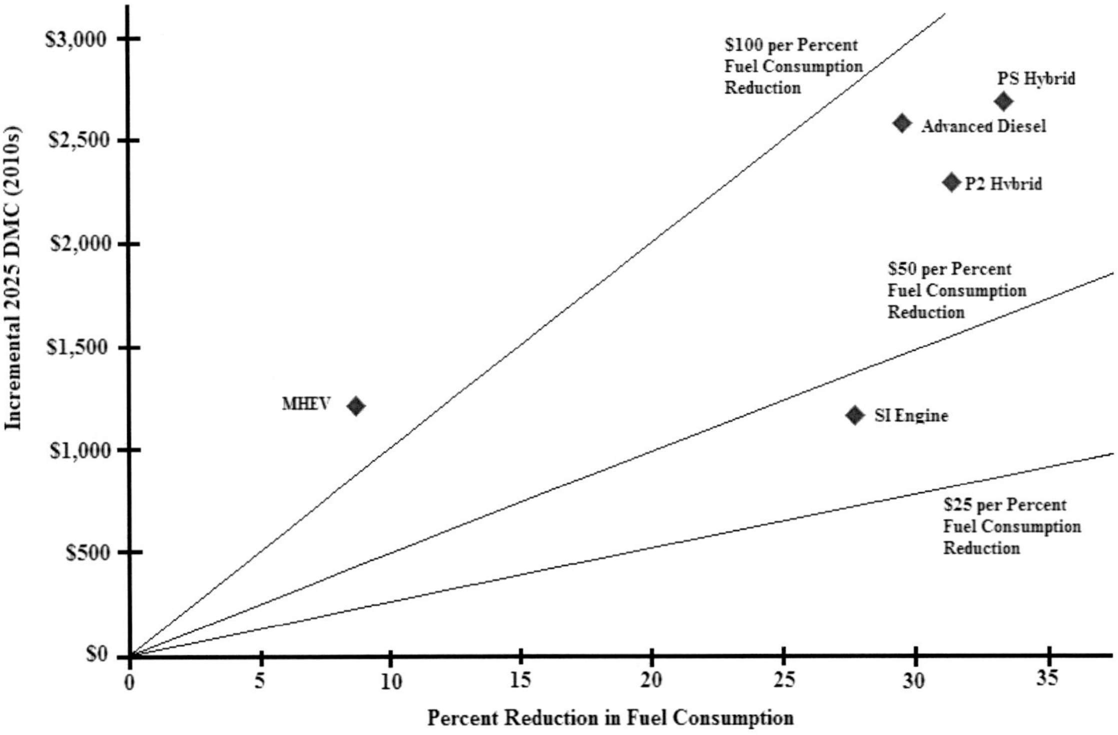

FIGURE 8.4 SI engine technologies, hybrid, and advanced diesel technologies in midsize cars shown on a plot of NRC-estimated incremental 2025 MY direct manufacturing cost in 2010 dollars versus percent reduction in fuel consumption.

$200 per percent reduction in fuel consumption, significantly exceed the cost of hybrids and diesels at less than $100 per percent, as shown in Figure 8.4. This suggests that the most cost-effective approach for a manufacturer may include the selective application of hybrids and diesels before considering the more expensive SI engine technologies, with cost exceeding $100 per percent reduction in fuel consumption.

TECHNOLOGY PATHWAY EXAMPLE

The committee developed a technology pathway example to illustrate the overall effectiveness and cost of applying many of the technologies discussed in the previous technology chapters to a specific vehicle. Important factors in developing a technology pathway include sequencing of the technologies and synergies of the technologies within a decision tree path and across such paths.

It is critical that the results of the committee's technology pathway examples not be interpreted as assessments of the compliance costs for the 2017-2025 standards. Assessing compliance costs was not part of the charge to the committee, and given limitations on the committee's resources to model fleet and vehicle models in more detail, it did not estimate such costs. As discussed in Chapter 10, the models used by NHTSA and EPA for estimating the cost of compliance track technology additions for approximately 1,300 separate vehicle models through the compliance period (2017-2025). These models also take into account the various crediting provisions described in Chapter 10 that the Agencies are permitted to use in determining the stringency of the standards. The committee notes that a simple "roll-up" of the NRC's cost and effectiveness estimates for the technologies in the Agencies' compliance demonstration path for a sample vehicle cannot be used to estimate future compliance costs. An estimate of compliance costs would require similar roll-ups for all vehicles together with consideration of flexibilities (including credits for air conditioning, off-cycle technologies, alternative fuel and advanced technology vehicles, and the banking and trading of credits) that reduce compliance costs. Such analysis was well beyond the committee's resources and capabilities. Instead, the committee looked at costs and technology benefits for three representative vehicles and did not estimate the full impacts of the various flexibilities available to OEMs. Nevertheless, technology roll-ups are a convenient device for illustrating the aggregate fuel consumption reductions and costs of technology packages, and the analysis in the following sections provides such examples for that reason. Such an approach was used in earlier NRC reports on fuel economy technologies (NRC 2002, 2011).

Sequencing of Technologies

The NRC Phase 1 (NRC 2011) report identified the following factors that a vehicle manufacturer will consider, at a minimum, when implementing technologies to reduce fuel consumption:

- Cost effectiveness, which is defined as the incremental cost per percent reduction in fuel consumption ($/% FC);
- Ability to integrate the technology into the vehicle and engine cycle plans;
- Impact on vehicle performance characteristics and other functional characteristics;
- Applicability to the specific product or vehicle class; and
- Customer acceptance.

The following considerations were applied in ranking the technologies for the example pathway:

- EPA and NHTSA defined a null, or baseline, vehicle, which was used as the starting point for the pathway.
- The technologies were ranked in the order of cost effectiveness wherever possible. Exceptions to this ranking of technologies include the following:
  - —A less cost-effective technology will precede a more cost-effective technology if the less cost-effective technology is required prior to implementation of the more cost-effective technology. For example, stoichiometric gasoline direct injection does not rank high on the basis of cost effectiveness but is considered to be a requirement before turbocharging and downsizing can be applied.
  - —Some technologies require more development time before being available for production implementation. The technology must be implementation-ready 3 to 5 years before production implementation, which is in contrast to a future development being explored in the research laboratory. Implementation readiness implies that all aspects of the technology have been proven, including function, durability, reliability, cost, and supplier readiness. Some potentially attractive technologies cannot be considered because they are not implementation-ready.
- Some technologies are considered by NHTSA to be applicable anytime, such as improved accessories, but are generally ranked in the order of cost effectiveness.

Since NHTSA uses decision trees to determine the order in which technologies are applied to a vehicle, the committee followed a similar approach in developing pathways. The following decision trees were utilized in developing the pathways: Engine Technology, Transmission, Mass Reduction, Low Rolling Resistance Tires, Low Drag Brakes, Aerodynamic Drag Reductions, Electrification/Accessory, and Hybrid Technology. These decision trees are shown in Figures 8.5, 8.6, and 8.7.

The decision trees provide the sequence for applying individual technologies to vehicles in NHTSA's Volpe model and

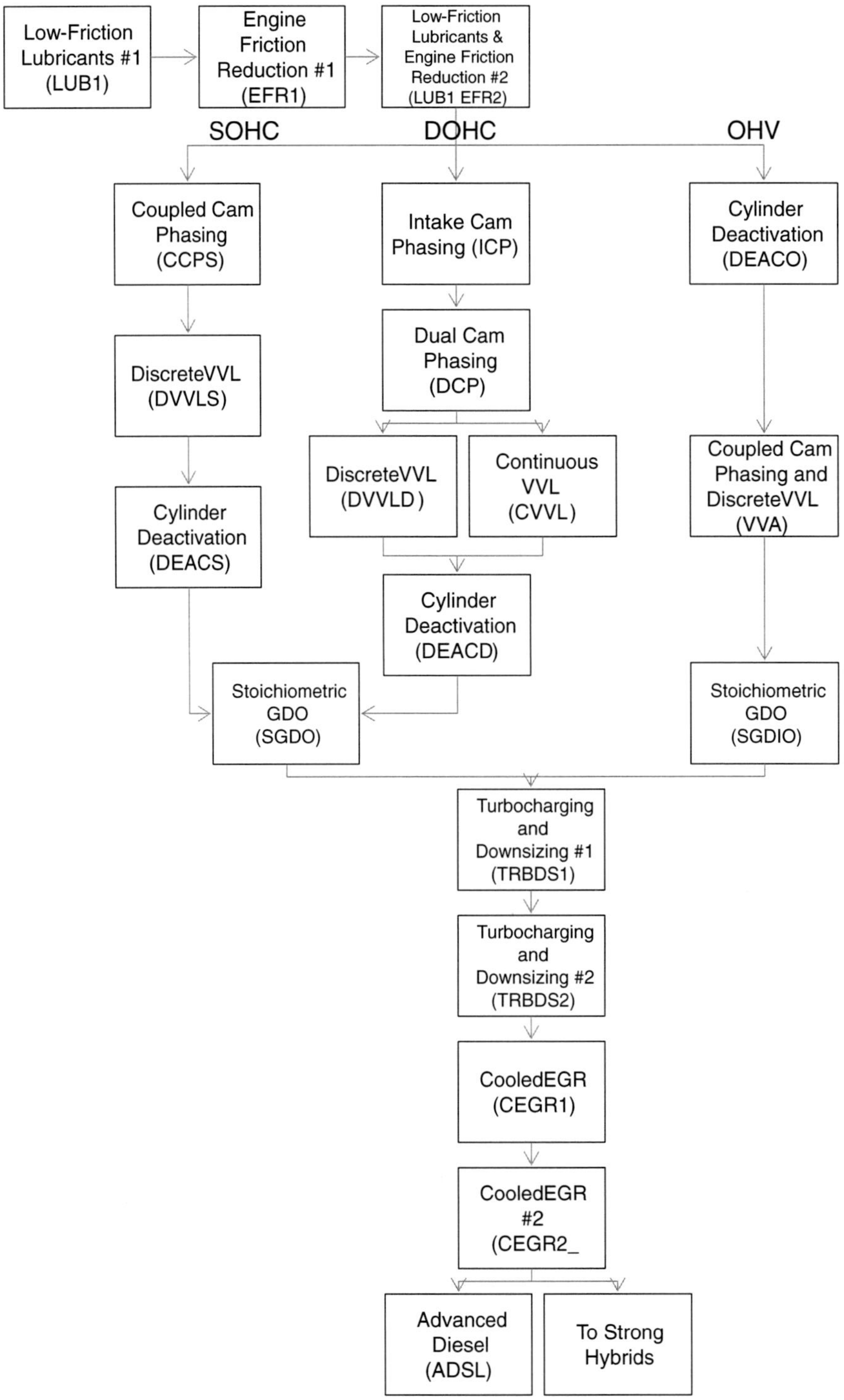

FIGURE 8.5 Engine technology decision tree.
SOURCE: NHTSA (2012).

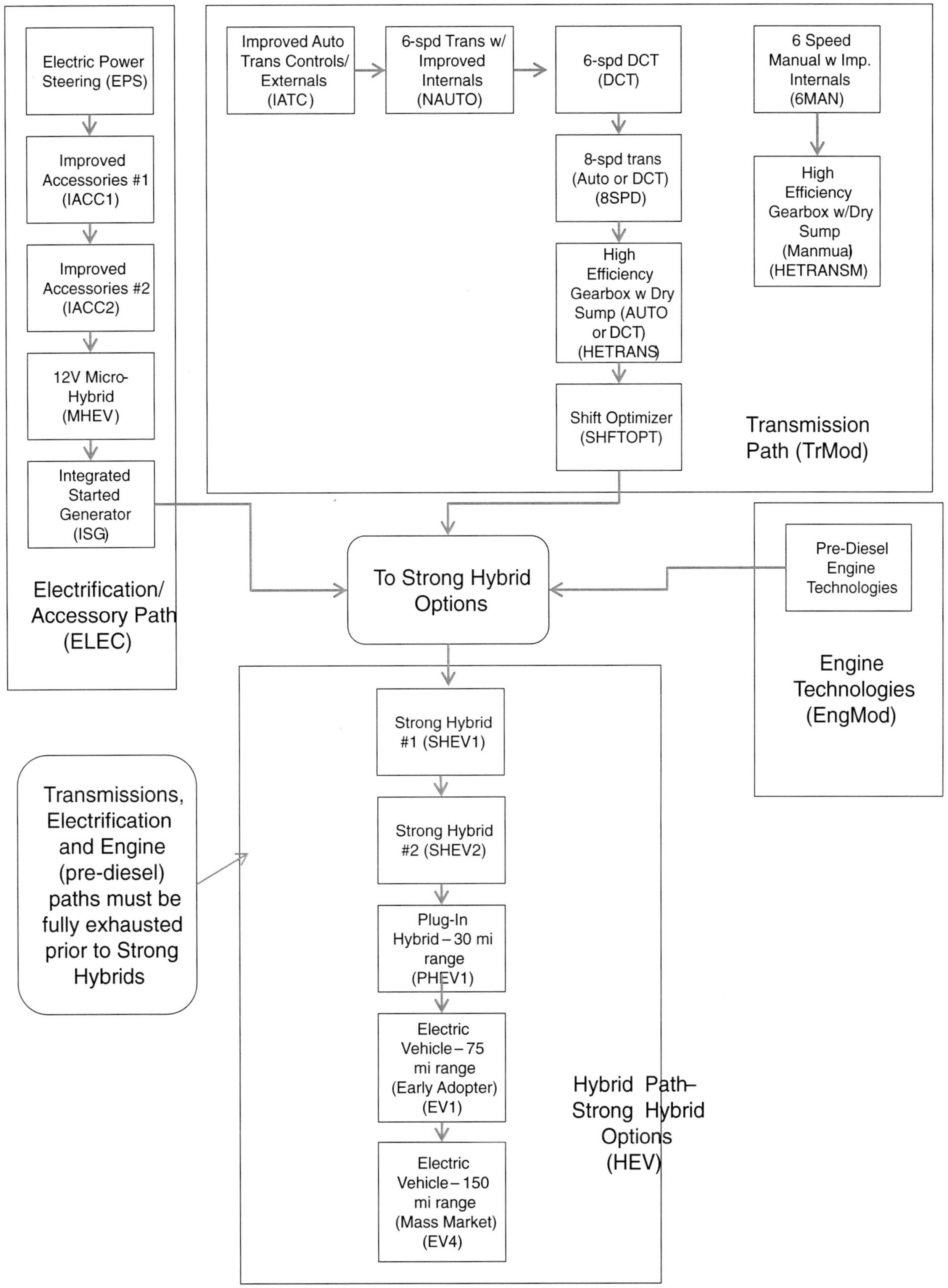

FIGURE 8.6 Electrification/accessory, transmission, and hybrid technology decision tree.
SOURCE: NHTSA (2012).

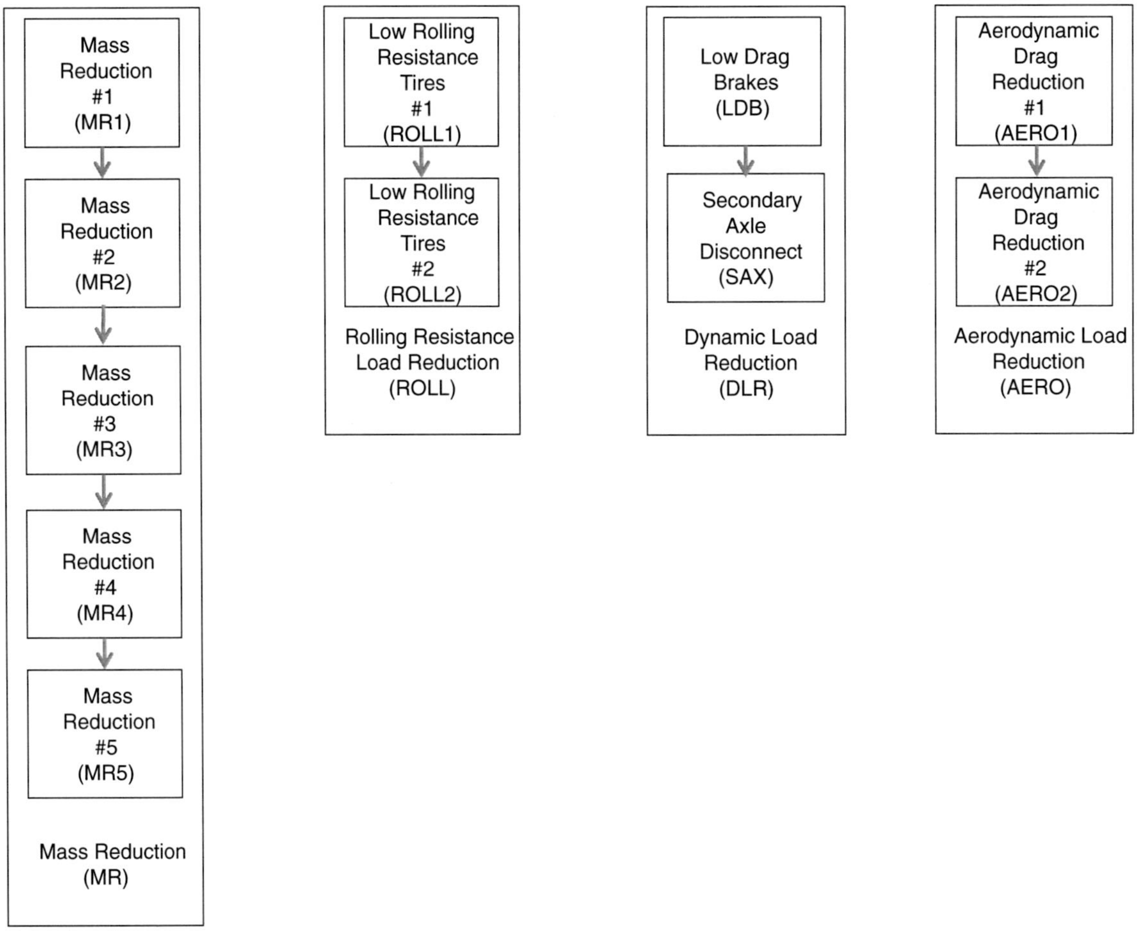

FIGURE 8.7 Vehicle technology decision tree.
SOURCE: NHTSA (2012).

have generally been followed in developing the committee's pathways. For the engine decision trees, several pathways are provided for different valve configurations, including DOHC, single overhead camshaft (SOHC), and OHV. For the first level of a turbocharged, downsized engine (TRBDS1), all engines are converted to DOHC configurations so that there are no longer any path-dependent variations. After all of the available SI engine technologies have been applied, the decision tree splits either to the advanced diesel or to the hybrid pathway.

The transmission decision tree follows the general pathway that includes a six-speed automatic transmission with improved controls and external features, a possible transition to a DCT, followed by an eight-speed transmission and a high-efficiency gearbox. The hybrid decision tree begins with electrified accessories, followed by stop-start, integrated starter generator followed by strong hybrids, followed by plug-in hybrids and electric vehicles. The vehicle technology decision trees provide a progression of more advanced technologies for mass reductions, low rolling resistance tires, low drag brakes and other vehicle driveline technologies, and aerodynamic drag reduction.

The pathways developed by the committee followed the fuel consumption reduction and cost methodologies that are used by NHTSA to ensure that synergies are properly included within each pathway. Detailed decision trees that include NHTSA's accounting of effectiveness and cost for each technology in the pathways are provided at the NHTSA fuel economy website.

Committee Example of Technology Pathway

The example technology pathway developed by the committee illustrates the process of combining the technologies discussed in this study. As described in the preceding section, the criteria for adding technologies in the pathway consisted of (1) cost effectiveness, (2) prerequisite technical requirements, (3) applicability to the specific product, and (4) implementation readiness. Technologies applied in the example here include only NHTSA-defined technologies, as these were the technologies for which the committee had the most complete information on effectiveness, costs, and interaction with other technologies. However, the committee applied mass reduction up to 10 percent, in keeping with Finding 6.2 in Chapter 6; this was in contrast to the Agencies, which limited mass reduction for midsize cars to 3.5 percent in their compliance scenario. The technology pathway example for a midsize car with an I4 DOHC SI engine is shown in Tables 8.4a and b using the committee's low and high most likely estimates. Cost effectiveness values are shown in the right column of the example pathway. The pathway begins with the null vehicle identified above the "Possible Technologies" column. As noted earlier, the null vehicle was defined by EPA and NHTSA as a vehicle having a naturally aspirated engine, four valves per cylinder, fixed valve timing and lift and a four-speed automatic transmission. Although the Agencies' analysis began with the 2008 MY, very few vehicles in 2008 had only the content of the null vehicle. Many 2008 vehicles contained some of the early EPA and NHTSA technologies. The committee reviewed the best-selling vehicles in the midsize vehicle classification. An example vehicle was selected that had the CAFE fuel economy closest to the average of the best-selling vehicles. The additional technologies included in this specific vehicle that were additions to the null vehicle were identified. These technologies were applied first in the pathways so that the example vehicle could be aligned with the 2008 MY, as shown in Table 8.4.

As additional technologies are applied in the pathway, the fuel consumption reductions are derived from a multiplicative combination of one minus the individual estimated fuel consumption reduction fraction (percent reduction divided by 100), while the cumulative costs are derived from an addition of the individual costs. This approach was used in earlier NRC reports to represent the fuel consumption benefits of multiple technologies (NRC 2002, 2011). These fuel consumption reductions are converted to miles per gallon, based on the EPA certification CAFE fuel economy of the example vehicle. The 2016 and 2025 CAFE targets for the example vehicle, based on its footprint, are indicated at the appropriate locations along the pathway. The direct manufacturing costs for 2017, 2020, and 2025 are listed for each technology and are then added to provide a cumulative direct manufacturing cost for the pathway. The pathway shows the cumulative direct manufacturing costs of the technologies applied to the null vehicle to the 2016 time frame, the technologies applied in the 2017 to 2025 time frame, and the technologies that may be available beyond 2025.

The results from the example pathway for a midsize car with an I4 SI engine using the committee's low and high most likely estimates are summarized in Figure 8.8. Applying technologies in the order of cost effectiveness results in the increasing incremental direct manufacturing costs per percent reduction in fuel consumption, as shown in the figure. As shown in Tables 8.4a and b and illustrated in Figure 8.8, the cost effectiveness of technologies range from under $10 per percent reduction in fuel consumption to a high of $260 per percent reduction in fuel consumption for cooled EGR—level 2. This example pathway shows the low and high most likely estimates of the direct manufacturing costs to reach the 2016 CAFE target, which becomes the baseline for achieving the 2025 CAFE target for this example vehicle. As shown in Figure 8.8, both the lower pathway, which uses the low cost and high effectiveness combinations, and the higher pathway, which uses the high cost and low effectiveness combinations, reach the 2025 target without exhausting the available NHTSA-defined technologies. As noted above, both pathways include 10 percent mass reduction, unlike NHTSA's compliance scenario. Pathways were not developed for other vehicle classifications to determine the ability of the NHTSA-defined technologies to reach the 2025 MY CAFE target.

A similar pathway using NHTSA estimates for both direct manufacturing cost and total costs is provided in Appendix U for reference and summarized in Figure 8.9. The committee's estimates are compared with NHTSA's estimates in Table 8.5. To achieve the CAFE target for the 2025 MY from the 2016 MY baseline, the committee's example calculation of cumulative direct manufacturing cost estimates exceeded the estimate using NHTSA's technology cost and effectiveness estimates by 11 percent in the lower pathway and by 56 percent in the higher pathway. This was due to lower committee effectiveness estimates for some technologies and higher cost estimates for other technologies. It is important to note that these calculations did not include full CAFE/GHG program flexibilities so are not intended to be an estimate of actual compliance costs. In this example, technologies were applied to achieve the CAFE targets without consideration of other vehicles in a manufacturer's fleet and without consideration of credits. The results for other vehicle classifications may vary considerably from this example.

Alternative Pathways

The pathways shown in Figure 8.8 were developed by applying technologies that were defined by NHTSA for SI engines, transmissions, and vehicle technologies in the TSD together with 10 percent mass reduction. As shown in Tables 8A.1 and 8A.2, the committee also identified other technologies with the potential for additional reductions in fuel consumption or possibly lower cost alternatives to the technologies defined by NHTSA. Alternative pathways were developed using several of these technologies applied individually in addition to the NHTSA-defined technologies or in place of several NHTSA technologies. These pathways are provided in Appendix V and a summary of the results is shown in Table 8.6 and compared to the previously discussed example pathway using the committee's effectiveness and cost estimates.

The first alternative technology was a high compression ratio with exhaust scavenging, followed by turbocharging

TABLE 8.4a Midsize Car SI Engine Pathway Showing NRC Low Most Likely Estimates for 2017, 2020, and 2025

Midsize Car with SI Engine Pathway with 10% MR - NRC Low Most Likely Estimates - Direct Manufacturing Costs (2010$)

Low Most Likely Cost Estimates Paired with High Most Likely Effectiveness Estimates

Possible Technologies	% FC Reduction	FC Reduction Multiplier	Cumulative FC Reduction Multiplier	Fuel Consumption (gal/100 mi)	Cumulative Percent FC Reduction	Unadjusted Combined FE (mpg)	2017 Cost Estimates	2020 Cost Estimates	2025 Cost Estimates	2017 Cost/ Percent FC ($/%)
Null Vehicle[a]		1.000	1.000	3.240	0.0%	30.9				
Intake Cam Phasing ICP	2.6%	0.974	0.974	3.156	2.6%	31.7	$37	$35	$31	$14.23
Dual Cam Phasing DCP (vs. ICP)	2.5%	0.975	0.950	3.077	5.0%	32.5[b]	$31	$29	$27	$12.40
2008 Example Vehicle										
Low Rolling Resistance Tires - 1 ROLL1	1.9%	0.981	0.932	3.018	6.8%	33.1	$5	$5	$5	$2.63
Low Friction Lubricants - 1 LUB1	0.7%	0.993	0.925	2.997	7.5%	33.4	$3	$3	$3	$4.29
6 Speed Automatic Transmission[c] 6 SP AT with Improved Internals IATC	1.6%	0.984	0.910	2.949	9.0%	33.9	$37	$34	$31	$23.13
Aero Drag Reduction - 1 AERO1	2.3%	0.977	0.889	2.882	11.1%	34.7	$39	$37	$33	$16.96
Engine Friction Reduction - 1 EFR1	2.6%	0.974	0.866	2.807	13.4%	35.6	$48	$48	$48	$18.46
Improved Accessories - 1 IACC1	1.2%	0.988	0.856	2.773	14.4%	36.1	$71	$69	$60	$59.17
Electric Power Steering EPS	1.3%	0.987	0.845	2.737	15.5%	36.5	$87	$82	$74	$66.92
Mass Reduction - 2.5% MR2.5 (-87.5 lbs)	0.8%	0.992	0.838	2.715	16.2%	36.8	$0	$0	$0	$0.00
2016 Target 36.6 mpg										
Discrete Variable Valve Lift DVVL	3.6%	0.964	0.808	2.617	19.2%	38.2	$116	$109	$99	$32.22
Mass Reduction - 2.5%-5.0% MR5-MR2.5 (-87.5 lbs)	0.8%	0.992	0.801	2.596	19.9%	38.5	$0	$0	$0	$0.00
Stoichiometric Gasoline Direct Injection SGDI (Required for TRBDS)	1.5%	0.985	0.789	2.557	21.1%	39.1	$192	$181	$164	$128.00
Turbocharging & Downsizing - 1 (I-4 to I-4) TRBDS1 33% DS 18 bar BMEP	8.3%	0.917	0.724	2.345	27.6%	42.6	$288	$271	$245	$34.70
Turbocharging & Downsizing - 2 (I-4 to I-3) TRBDS2 50% DS 24 bar BMEP	3.5%	0.965	0.698	2.263	30.2%	44.2	-$92	-$89	-$82	-$26.29
8 Speed Automatic Transmission[c] 8 SP AT	1.7%	0.983	0.687	2.225	31.3%	45.0	$56	$52	$47	$32.94
Shift Optimizer[c] SHFTOPT	0.7%	0.993	0.682	2.209	31.8%	45.3	$26	$24	$22	$37.14
Improved Accessories - 2 IAAC2	2.4%	0.976	0.665	2.156	33.5%	46.4	$43	$40	$37	$17.92
Low Rolling Resistance Tires ROLL2	2.0%	0.980	0.652	2.113	34.8%	47.3	$58	$46	$31	$29.00
Aero Drag Reduction - 2 AERO2	2.5%	0.975	0.636	2.060	36.4%	48.5	$117	$110	$100	$46.80
Mass Reduction - 5.0%-10.0% MR10-MR5 (-175 lbs)	4.6%	0.954	0.607	1.965	39.3%	50.9	$154	$151	$151	$33.48
Low Friction Lub - 2 & Engine Friction Red - 2 LUB2_EFR2	1.3%	0.987	0.599	1.940	40.1%	51.6	$51	$51	$51	$39.23
Continuously Variable Valve Lift CVVL (vs. DVVL)	1.0%	0.990	0.593	1.920	40.7%	52.1	$58	$55	$49	$58.00
High Efficiency Transmission HEG1 & 2	5.4%	0.946	0.561	1.817	43.9%	55.0	$314	$296	$267	$58.15
2025 Target 54.2 mpg										
Cooled EGR - 1 CEGR1 50% DS 24 bar BMEP	3.5%	0.965	0.541	1.753	45.9%	57.0	$212	$199	$180	$60.57
Cylinder Deactivation DEACD	0.0%	1.000	0.541	1.753	45.9%	57.0				
Cooled EGR - 2 (I-3 to I-3) CEGR2 56% DS 27 bar BMEP	1.4%	0.986	0.533	1.729	46.7%	57.9	$364	$343	$310	$260.00
Totals										
Relative to Null Vehicle	46.7%	0.533					$2,315	$2,181	$1,983	$49.62
Null Vehicle - 2008 MY Vehicle	5.0%	0.950					$68	$64	$58	$13.51
2008 MY Vehicle - 2016 MY	11.8%	0.882					$290	$278	$254	
2017 MY- 2025 MY	33.1%	0.669					$1,381	$1,297	$1,181	$41.74
Beyond 2025 MY	4.9%	0.951					$576	$542	$490	$118.74

[a] Null vehicle: I4, DOHC, naturally aspirated, 4 valves/cylinder PFI fixed valve timing and 4 speed AT.

[b] An example midsize car in 2008 was 46.64 sq ft and had a fuel economy of 32.5 mpg. Its standard for MY2016 would be 36.6 mpg and for MY2025 would be 54.2 mpg.

[c] These technologies have transmission synergies included. Green highlighting indicates a technology order different than the NHTSA pathway, shown in Appendix S.

TABLE 8.4b Midsize Car with SI Engine Pathway Showing NRC High Most Likely Estimates for 2017, 2020, and 2025

Midsize Car with SI Engine Pathway with 10% MR - NRC High Most Likely Estimates - Direct Manufacturing Costs (2010$)

High Most Likely Cost Estimates Paired with Low Most Likely Effectiveness Estimates

Possible Technologies	% FC Reduc-tion (%)	FC Reduction Multiplier	Cumulative FC Reduction Multiplier	Fuel Consumption (gal/100 mi)	Cumulative Percent FC Reduction	Unadjusted Combined FE (mpg)	2017 Cost Estimates	2020 Cost Estimates	2025 Cost Estimates	2017 Cost/ Percent FC ($/%)
Null Vehicle[a]		1.000	1.000	3.240	0.0%	30.9				
Intake Cam Phasing ICP	2.6%	0.974	0.974	3.156	2.6%	31.7	$43	$41	$36	$16.54
Dual Cam Phasing DCP (vs. ICP)	2.5%	0.975	0.950	3.077	5.0%	32.5[b]	$35	$33	$31	$14.00
					2008 Example Vehicle					
Low Rolling Resistance Tires - 1 ROLL1	1.9%	0.981	0.932	3.018	6.8%	33.1	$5	$5	$5	$2.63
Low Friction Lubricants - 1 LUB1	0.7%	0.993	0.925	2.997	7.5%	33.4	$3	$3	$3	$4.29
6 Speed Automatic Transmission[c] 6 SP AT with Improved Internals IATC	1.3%	0.987	0.913	2.958	8.7%	33.8	$37	$34	$31	$28.46
Aero Drag Reduction - 1 AERO1	2.3%	0.977	0.892	2.890	10.8%	34.6	$39	$37	$33	$16.96
Engine Friction Reduction - 1 EFR1	2.6%	0.974	0.869	2.815	13.1%	35.5	$48	$48	$48	$18.46
Improved Accessories - 1 IACC1	1.2%	0.988	0.858	2.781	14.2%	36.0	$71	$67	$60	$59.17
Electric Power Steering EPS	1.3%	0.987	0.847	2.745	15.3%	36.4	$87	$82	$74	$66.92
Mass Reduction - 2.5% MR2.5 (-87.5 lbs)	0.8%	0.992	0.841	2.723	15.9%	36.7	$22	$22	$22	$27.50
					2016 Target 36.6 mpg					
Discrete Variable Valve Lift DVVL	3.6%	0.964	0.810	2.625	19.0%	38.1	$133	$125	$114	$36.94
Mass Reduction - 2.5%-5.0% MR5-MR2.5 (-87.5 lbs)	0.8%	0.992	0.804	2.604	19.6%	38.4	$66	$66	$66	$82.50
Stoichiometric Gasoline Direct Injection SGDI (Required for TRBDS)	1.5%	0.985	0.792	2.565	20.8%	39.0	$192	$181	$164	$128.00
Turbocharging & Downsizing - 1 TRBDS1 33% DS 18 bar BMEP	7.7%	0.923	0.731	2.368	26.9%	42.2	$331	$312	$282	$42.99
Turbocharging & Downsizing - 2 TRBDS2 50% DS 24 bar BMEP	3.2%	0.968	0.707	2.292	29.3%	43.6	-$96	-$92	-$86	-$30.00
8 Speed Automatic Transmission[c] 8 SP AT	1.3%	0.987	0.698	2.262	30.2%	44.2	$151	$126	$115	$116.15
Shift Optimizer[c] SHFTOPT	0.3%	0.997	0.696	2.255	30.4%	44.3	$26	$24	$22	$86.67
Improved Accessories - 2 IAAC2	2.4%	0.976	0.679	2.201	32.1%	45.4	$43	$40	$37	$17.92
Low Rolling Resistance Tires ROLL2	2.0%	0.980	0.666	2.157	33.4%	46.4	$58	$46	$31	$29.00
Aero Drag Reduction - 2 AERO2	2.5%	0.975	0.649	2.103	35.1%	47.5	$117	$110	$100	$46.80
Mass Reduction - 5%-10% MR10-MR5 (-175 lbs)	4.6%	0.954	0.619	2.006	38.1%	49.8	$325	$322	$315	$70.65
Low Friction Lub - 2 & Engine Friction Red - 2 LUB2_EFR2	1.3%	0.987	0.611	1.980	38.9%	50.5	$51	$51	$51	$39.23
Cooled EGR - 1 CEGR1 50% DS 24 bar BMEP	3.0%	0.970	0.593	1.921	40.7%	52.1	$212	$199	$180	$70.67
High Efficiency Transmission HEG1 & 2	4.9%	0.951	0.564	1.827	43.6%	54.7	$314	$296	$267	$64.08
					2025 Target 54.2 mpg					
Continuously Variable Valve Lift CVVL (vs. DVVL)	1.0%	0.990	0.558	1.809	44.2%	55.3	$67	$63	$56	$67.00
Cylinder Deactivation DEACD	0.0%	1.000	0.558	1.809	44.2%	55.3				
Cooled EGR - 2 CEGR2 56% DS 27 bar BMEP	1.4%	0.986	0.550	1.783	45.0%	56.1	$364	$343	$310	$260.00
					Totals					
Relative to Null Vehicle	45.0%	0.550					$2,744	$2,584	$2,367	$61.03
Null Vehicle - 2008 MY Vehicle	5.0%	0.950					$78	$74	$67	$15.49
2008 MY Vehicle - 2016 MY	11.5%	0.885					$312	$298	$276	$27.15
2017 MY- 2025 MY	32.9%	0.671					$1,923	$1,806	$1,658	$58.42
Beyond 2025 MY	2.4%	0.976					$431	$406	$366	$180.64

[a] Null vehicle: I4, DOHC, naturally aspirated, 4 valves/cylinder PFI fixed valve timing and 4 speed AT.
[b] An example midsize car in 2008 was 46.64 sq ft and had a fuel economy of 32.5 mpg. Its standard for MY2016 would be 36.6 mpg and for MY2025 would be 54.2 mpg.
[c] These technologies have transmission synergies included. Green highlighting indicates a technology order different than the NHTSA pathway, shown in Appendix S.

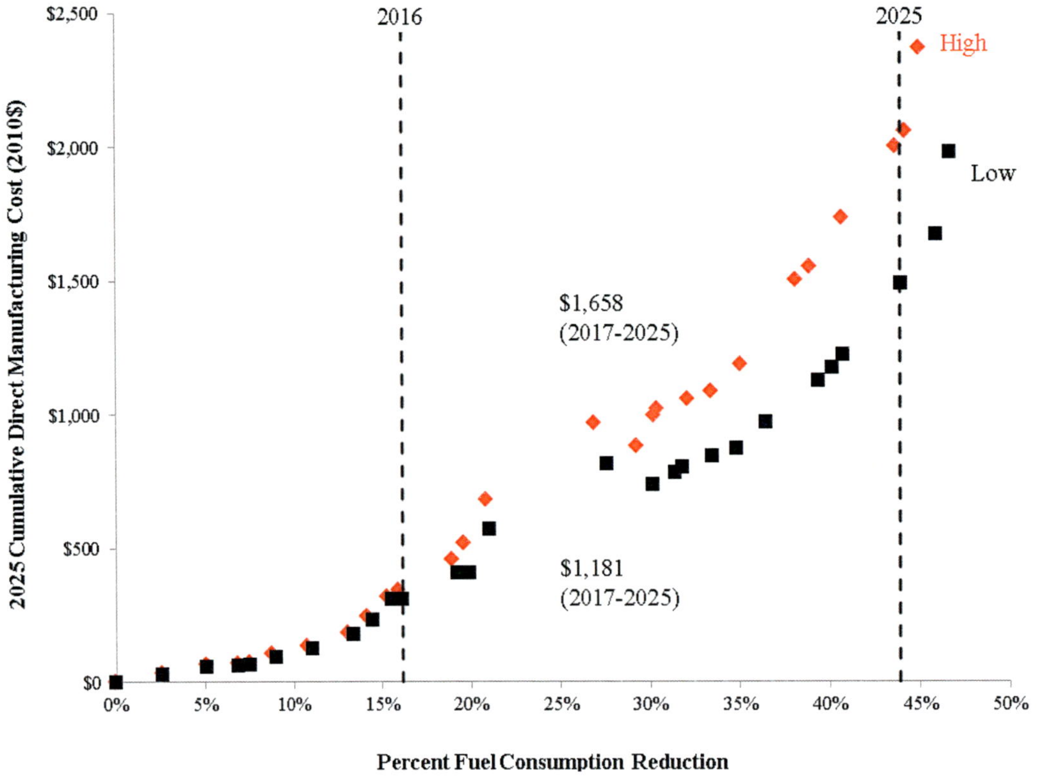

FIGURE 8.8 Pathway example for midsize car with I4 SI engine showing NRC low and high most likely estimates of 2025 MY direct manufacturing costs and fuel consumption reduction effectiveness.

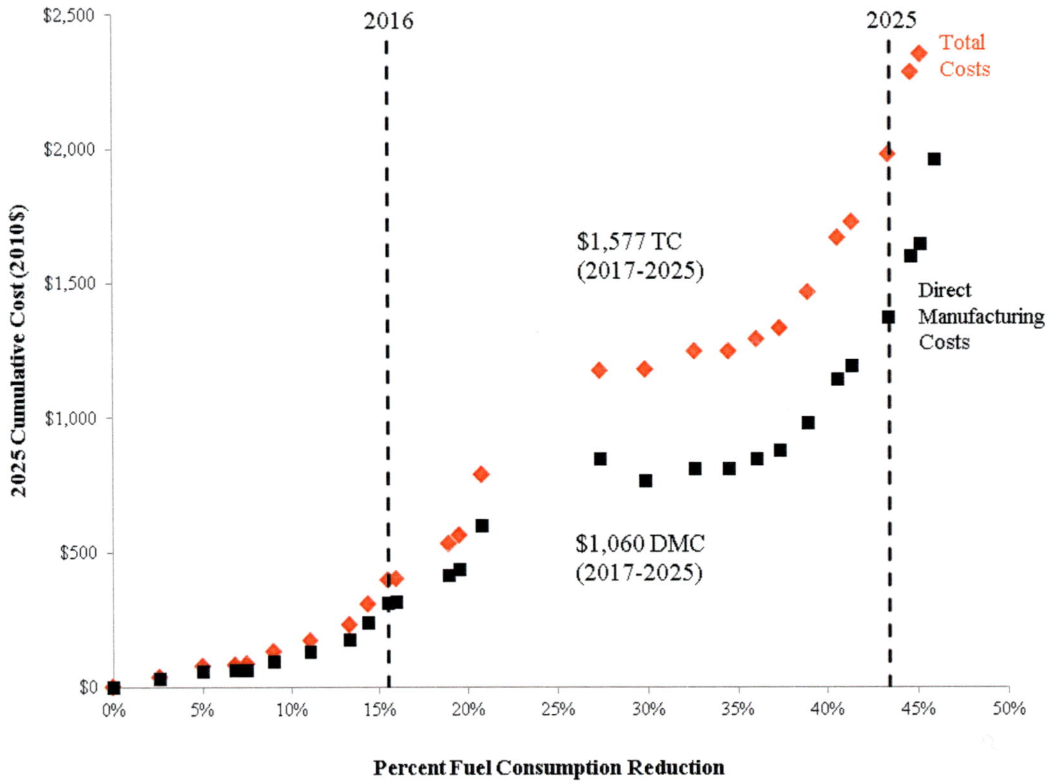

FIGURE 8.9 Midsize car with I4 SI engine pathway example using NHTSA's estimates of 2025 MY cumulative direct manufacturing and total costs.

TABLE 8.5 Illustrative Incremental Direct Manufacturing Costs for the 2017-2025 Target for an Example Midsize Car with an I4 SI Engine (2010 dollars)

	NRC Most Likely Estimates 2025 MY*	NHTSA Estimates 2025 MY
Direct manufacturing costs in 2017-2025 time frame	1,181 - 1,658	1,060
Reference: Total cost		1,577

* Successive technologies were generally added to the pathways in Table 8.4 according to cost effectiveness. Adding the technology, High Efficiency Transmission, provided an incremental effectiveness which was larger than required to exactly meet the 2025 CAFE target. As a result, the Low Friction Lub and Engine Friction Reduction – Level 2 technology, which had been added earlier, could have been deleted from the pathway, thereby reducing the 2017-2025 direct manufacturing costs shown in this table by $51.

TABLE 8.6 Alternative Pathways for a Midsize Car with an I4 Gasoline Engine

Pathway	Fuel Consumption Reduction (%)		2025 MY Direct Manufacturing Cost (2010 dollars)	
	Overall (From Null Vehicle)	2017-2025 MY Time Frame	Overall (From Null Vehicle)	2017-2025 Time Frame
NHTSA Technologies with 10% Mass Reduction (Figure 8.9a and b)	45.0 - 46.7	32.9 - 33.1	1,983 - 2,367	1,181 - 1,658
High Compression Ratio with Exhaust Scavenging (In addition to TRBDS 1 and 2)	48.3 - 49.9	32.8 - 33.7	2,233 - 2,617	1,115 - 1,641 (−17 to −66)
EAVS-Supercharger with partial MHEV Function (Replacing TRBDS1 and 2, SS, IACC1, IACC2)	52.7 - 53.7	30.8 - 32.0	3,025 - 3,376	1,566 - 1,800 (+142 to +385)

and downsizing. This technology was effective in reducing the direct manufacturing cost in the 2017 to 2025 MY time frame by $17 to $66 relative to the pathway using NHTSA-defined technologies with 10 percent mass reduction, as shown in Table 8.6. However, as indicated in Chapter 2, the future path for high compression ratio with exhaust scavenging is not clear with respect to the applicability of turbocharging and downsizing to this concept.

The next alternative technology was the electrically assisted variable speed (EAVS) supercharger system as a replacement for turbocharging to level 2, and improved accessories levels 1 and 2, as described in Chapter 2. The EAVS supercharger system also has the potential to provide stop-start and mild hybrid functions, although these features were not included in the pathways shown in Figure 8.8. The EAVS supercharger system increased overall effectiveness by 7 to 8 percent relative to the pathways shown in Figure 8.8. However, direct manufacturing cost in the 2017 to 2025 time frame increased by $142 to $385 relative to the pathways using NHTSA-defined technologies with 10 percent mass reduction, as shown in Table 8.6. This increase occurred because the EAVS supercharger system with cost effectiveness of $50 per percent reduction in fuel consumption replaced turbocharging and downsizing technology, which has lower cost effectiveness values. Although a reduction in cost was not shown for the 2017 to 2025 time frame, the additional reduction in overall fuel consumption provided by the EAVS supercharger system might provide either longer-term advantages or other opportunities compared to technologies with lower overall fuel consumption reduction. For example manufacturers mayoverachieve in a particular vehicle line by applying this technology while saving costs in another vehicle line, which then may require the application of fewer fuel consumption reduction technologies.

Application of Credits

EPA and NHTSA provide manufacturers with preapproved technologies that qualify for off-cycle credits. For the first time, NHTSA also is providing indirect credits in the 2017 to 2025 final CAFE rule for improvements in air conditioning efficiency. The air conditioning efficiency indirect credits and off-cycle credits are summarized in Table 8.7. The table also shows air conditioning direct leakage and low GWP credits for CO_2, but these improvements do not have associated CAFE credits. For the analysis supporting the final CAFE rule, NHTSA assumed that off-cycle credits for active aerodynamics and stop-start technologies would be available to manufacturers for compliance with the CAFE targets, similar to the other available fuel-economy-improving technologies. Therefore, NHTSA included the assessment of off-cycle credits in the assessment of maximum feasible standards.

TABLE 8.7 Air Conditioning Efficiency and Off-Cycle Credits

	Car		Truck	
	CO_2 g/mi	gal/mi	CO_2 g/mi	gal/mi
A/C Credits (Projected estimated use of credits)				
Direct Leakage Credit (R-134a) (Not applicable to CAFE)	6.3		7.8	
Direct Credit for Low GWP A/C (Not applicable to CAFE)	13.8		17.2	
Indirect Credit (AC Efficiency)	5	0.000563	7.2	0.00081
Active Aerodynamic Improvements				
3% Reduction	0.6	0.000068	1.0	0.000113
Stop-Start (with heater circulation system)	2.5	0.000282	4.4	0.000496
Off Cycle Electrical Load Reduction (Lighting)				
100 W Reduction with high efficiency exterior lights	1	0.000113	1.0	0.000113
Solar Panels (75 watt)	3.3	0.000372	3.3	0.000372
Battery Charging Only				
Active Transmission Warm-up	1.5	0.000169	3.2	0.000361
Active Engine Warm-up	1.5	0.000169	3.2	0.000361
Exhaust Heat				
Secondary coolant loop				
Solar/Thermal Control	Up to 3	0.000338	Up to 4.3	0.000484
Total		0.002052		0.003105

SOURCE: EPA/NHTSA (2012a).

The cost savings from air conditioning efficiency indirect credits and active aerodynamics off-cycle credits were evaluated using the example pathways for a midsize car. The two technologies provide a total credit of 0.000631gal/mi, as shown in Table 8.7, which lists the credits for the applicable technologies. The results from using this credit for the midsize car pathway are shown in Tables 8.8a and b for the most likely low and high cost estimates. The 2017 to 2025 MY costs for the example pathway with and without credits are compared in Table 8.9. By using the credit, the costs of the eight-speed automatic transmission and low friction lubricant with engine friction reduction—level 2 technologies used in the original pathways to reach the 2025 MY CAFE targets would be saved. Without these technologies, direct manufacturing cost savings of $98 to $166, or approximately 8 to 10 percent of the cumulative costs from 2017 to 2025 MY for this example pathway, would be realized. Although the cost savings of 8 to 10 percent are realized, the fuel consumption reduction was diminished by approximately 6 percent, since the technologies with the higher cost per percent fuel consumption reduction were selected for replacement by the credits. The savings from the application of credits were evaluated only for this example pathway, but the benefits of credits are expected to be directionally similar for other pathways. Although credits for only two technologies were applied in this example, the total possible credits listed in Table 8.9 exceed twice the sum of these two credits. However, most of the additional credits require additional costs to implement the technologies associated with the credits.

FULL SYSTEM SIMULATION MODELING OF FUEL CONSUMPTION REDUCTIONS

In order to further understand fuel consumption benefits, the committee contracted with experts at the University of Michigan's Department of Mechanical Engineering (referred to as U of M throughout this section) to use full system simulation modeling to analyze the effects of technologies (Middleton et al. 2015). The committee recognizes that as more technologies are added to vehicles that are aimed at reducing the same type of losses, the possibility of overestimating fuel consumption reduction becomes greater. The results of the simulations assisted the committee in evaluating the aggregated fuel consumption reduction values provided by these technologies. However, it is important to note that full system simulation modeling for powertrains and vehicles requires a great deal of financial and human resources as well as specific engine and other vehicle data, which were beyond the scope of this study. Thus, the committee engaged U of M to look at a combination of critical technologies for an SI engine and automatic transmission powertrain for a single vehicle class in order to provide some additional evidence for the committee's estimates. Although other vehicle efficiency technologies are also important, a review of NHTSA's analysis of synergies revealed that the largest synergies resulted from adding various combinations of powertrain technologies. Therefore, the limited scope of this full system simulation modeling was focused on the powertrain technologies. The committee did not use the results of the full system simulation modeling directly in its estimates of fuel consumption benefits, but it did use them as a way to understand the potential magnitude of the interactions among individual technologies.

Methodology

The approach used by U of M applies GT-Power, a widely used engine and powertrain simulation tool (Gamma Technologies n.d.), to analyze a series of engine and powertrain modifications that are expected to be available to engine manufacturers in the 2017-2025 time frame and to estimate their impact on overall vehicle fuel economy. The committee focused on benefits for a midsize passenger car. Because the models are physics-based and integrated, the effects of each change can be investigated step-by-step in a consistent manner to more accurately take into effect nonlinear influences and avoid double counting. The study focused solely on powertrain changes; accordingly, vehicle parameters such as test weight, drag coefficient, and rolling resistance were held constant. In addition, rear axle ratios, gear ratios, and engine sizes were chosen to ensure similar 0-60 mph acceleration for each configuration. With each vehicle configuration, drive cycle simulations were carried out over the standard fuel economy compliance test (the Federal Test Procedure [FTP] city cycle and highway cycle [HWY]) and a 0-60 acceleration mode. Each engine configuration was modeled to maintain, as closely as possible, the torque curve of the baseline naturally aspirated engine so that equal performance, as measured by 0-60 mph acceleration time, would be maintained.

Engine and Powertrain Model

A schematic of the model architecture is shown in Figure 8.10. The engine model consists of a number of physics-based submodels, including a standard entrainment combustion model, an autoignition integral knock model, and normal breathing, friction, and heat transfer models included within GT-Power. A turbocharged boost system was modeled using a generic turbocharger map, scaled where necessary depending on engine size. The transmission was modeled as a multispeed automatic configuration with a representative loss map specified as a function of speed, load, and drive ratio together with a torque converter map. A generic continuously variable transmission (CVT) was similarly modeled using a separate loss map derived from data for several modern CVT designs. In all cases, representative shift schedules and rear axle ratios were used and provided relatively constant 0-60 performance.

Powertrain Technologies

The powertrain technologies that were modeled relate to both engine and transmission and reflect several of the key technologies defined by NHTSA. The engine technologies include (in the order of application) valve train improvements using dual cam phasing (DCP), which allows independent adjustments to valve timing (included in the baseline 2012 vehicle); engine friction reduction and lubricant improvements; discrete variable valve lift (DVVL) or cam profile switching, which provides reductions in pumping work; gasoline direct injection (GDI), which provides better fuel control than port fuel injection (PFI) and cooling of the intake charge; boosted operation with a turbocharger (TC) and reduced engine displacement to reduce the relative contribution of friction and pumping losses; and cooled EGR (CEGR), which has been reported to reduce knock, especially with boosted engines, and provide additional benefits of dilute combustion. Transmission technologies included six- and eight-speed automatic transmissions (6 AT, 8 AT) as well as CVTs. In order to simulate in-use transmission behavior that would be acceptable to consumers, the time to execute a gear shift was set at 0.5 seconds, and the minimum time in any given gear before up or down shifting was set at 2 seconds.

Model Results

The results of the simulations are summarized in Table 8.10. Included in the table are the FTP, HWY, and combined cycle fuel economy results as well as the corresponding combined fuel consumption values. Also shown are the incremental and cumulative changes in fuel consumption. The incremental changes are relative to the previous powertrain configurations in the table.

As expected the simulations show that the current baseline (Task 3) is more efficient than the previous reference (Task 2). This change is due to a broadening of the optimal engine operating range resulting from the use of DCP. Beginning with the baseline Task 3 and progressing to Task 6, the results show a steady improvement in fuel economy with friction reduction, DVVL, and direct injection. The downsizing and boosting of Task 7 (33 percent downsizing) shows major improvement, while further downsizing in Task 8 (50 percent downsizing) shows additional gain. Interestingly, replacing the six-speed transmission with an eight-speed in Task 9 shows a small (0.4 percent) decrease in fuel economy, similar to what happened in a recent design of experiments simulation study examining a number of transmission features, which found only a small improvement (0.2 percent) in fuel economy when going from a six- to eight-speed transmission (Robinette 2014). Further optimization of the gear and final drive ratios, shift strategy,

TABLE 8.8a Extract of Midsize Car Pathway Showing the Effect of A/C Efficiency Credits and Active Aerodynamics Off-Cycle Credits on the NRC Low Most Likely Cost Estimates

Midsize Car with SI Engine Pathway with 10% MR - NRC Low Most Likely Estimates - Direct Manufacturing Costs (2010$)
Low Most Likely Cost Estimates Paired with High Most Likely Effectiveness Estimates

Possible Technologies	% FC Reduction	FC Reduction Multiplier	Cumulative FC Reduction Multiplier	FC (gal/100 mi)	Cumulative Percent FC Reduction	Unadjusted Combined FE (mpg)	2017 Cost Estimates	2020 Cost Estimates	2025 Cost Estimates	2017 Cost/ Percent FC ($/%)
Null Vehicle[a]		1.000	1.000	3.240	0.0%	30.9				
Intake Cam Phasing (ICP)	2.6%	0.974	0.974	3.156	2.6%	31.7	$37	$35	$31	$14.23
Dual Cam Phasing DCP (vs. ICP)	2.5%	0.975	0.950	3.077	5.0%	32.5[b]	$31	$29	$27	$12.40
2008 Example Vehicle										
Low Rolling Resistance Tires - 1 ROLL1	1.9%	0.981	0.932	3.018	6.8%	33.1	$5	$5	$5	$2.63
Low Friction Lubricants - 1 LUB1	0.7%	0.993	0.925	2.997	7.5%	33.4	$3	$3	$3	$4.29
6 Speed Automatic Transmission[c] 6 SP AT with Improved Internals IATC	1.6%	0.984	0.910	2.949	9.0%	33.9	$37	$34	$31	$23.13
Aero Drag Reduction - 1 AERO1	2.3%	0.977	0.889	2.882	11.1%	34.7	$39	$37	$33	$16.96
Engine Friction Reduction - 1 EFR1	2.6%	0.974	0.866	2.807	13.4%	35.6	$48	$48	$48	$18.46
Improved Accessories - 1 IACC1	1.2%	0.988	0.856	2.773	14.4%	36.1	$71	$69	$60	$59.17
Electric Power Steering EPS	1.3%	0.987	0.845	2.737	15.5%	36.5	$87	$82	$74	$66.92
Mass Reduction - 2.5% MR2.5 (-87.5 lbs)	0.8%	0.992	0.838	2.715	16.2%	36.8	$0	$0	$0	$0.00
2016 Target 36.6 mpg										
Discrete Variable Valve Lift DVVL	3.6%	0.964	0.808	2.617	19.2%	38.2	$116	$109	$99	$32.22
Mass Reduction - 2.5%-5.0% MR5-MR2.5 (-87.5 lbs)	0.8%	0.992	0.801	2.596	19.9%	38.5	$0	$0	$0	$0.00
Stoichiometric Gasoline Direct Injection SGDI (Required for TRBDS)	1.5%	0.985	0.789	2.557	21.1%	39.1	$192	$181	$164	$128.00
Turbocharging & Downsizing - 1 (I-4 to I-4) TRBDS1 33% DS 18 bar BMEP	8.3%	0.917	0.724	2.345	27.6%	42.6	$288	$271	$245	$34.70
Turbocharging & Downsizing - 2 (I-4 to I-3) TRBDS2 50% DS 24 bar BMEP	3.5%	0.965	0.698	2.263	30.2%	44.2	-$92	-$89	-$82	-$26.29
8 Speed Automatic Transmission[c] 8 SP AT	1.7%	0.983	0.687	2.225	31.3%	45.0	$56	$52	$47	$32.94
Shift Optimizer[c] SHFTOPT	0.7%	0.993	0.682	2.209	31.8%	45.3	$26	$24	$22	$37.14
Improved Accessories - 2 IAAC2	2.4%	0.976	0.665	2.156	33.5%	46.4	$43	$40	$37	$17.92
Low Rolling Resistance Tires ROLL2	2.0%	0.980	0.652	2.113	34.8%	47.3	$58	$46	$31	$29.00
Aero Drag Reduction – 2 (AERO2)	2.5%	0.975	0.636	2.060	36.4%	48.5	$117	$110	$100	$46.80
Mass Reduction - 5.0%-10.0% MR10-MR5 (-175 lbs)	4.6%	0.954	0.607	1.965	39.3%	50.9	$154	$151	$151	$33.48
Low Friction Lub - 2 & Engine Friction Red - 2 LUB2_EFR2	1.3%	0.987	0.599	1.940	40.1%	51.6	$51	$51	$51	$39.23
Continuously Variable Valve Lift CVVL (vs. DVVL)	1.0%	0.990	0.593	1.920	40.7%	52.1	$58	$55	$49	$58.00
High Efficiency Transmission HEG1 & 2	5.4%	0.946	0.561	1.817	43.9%	55.0	$314	$296	$267	$58.15
2025 Target 54.2 mpg										
Cooled EGR - 1 CEGR1 50% DS 24 bar BMEP	3.5%	0.965	0.541	1.753	45.9%	57.0	$212	$199	$180	$60.57
Cylinder Deactivation DEACD	0.0%	1.000	0.541	1.753	45.9%	57.0				
Cooled EGR - 2 (I-3 to I-3) CEGR2 56% DS 27 bar BMEP	1.4%	0.986	0.533	1.729	46.7%	57.9	$364	$343	$310	$260.00
Totals										
Relative to Null Vehicle	46.7%	0.533					$2,315	$2,181	$1,983	$49.62
Null Vehicle - 2008 MY Vehicle	5.0%	0.950					$68	$64	$58	$13.51
2008 MY Vehicle - 2016 MY	11.8%	0.882					$290	$278	$254	
2017 MY- 2025 MY	33.1%	0.669					$1,381	$1,297	$1,181	$41.74
Beyond 2025 MY	4.9%	0.951					$576	$542	$490	$118.74
Credits to replace 8 sp AT:					2.263	-2.225	0.038			
Credits to replace LUB2_EFR2:					1.965	-1.940	0.026			
Total to be replaced with credits							0.064			
Technology Not Required with Credits (2017 MY - 2025 MY)							$107	$103	$98	
Reduced 2017 MY - 2025 MY Costs with Credits							$1,274	$1,194	$1,083	
Percent Cost Savings with Credits (2017 MY - 2025 MY)							7.7%	7.9%	8.3%	
Credits (gal/100 mi)										
AC Efficiency					0.0563					
Active Aerodynamics					0.0068					
Stop-start (N/A w/ o SS)					0					
Total =					0.0631					

[a] Null vehicle: I4, DOHC, naturally aspirated, 4 valves/cylinder PFI fixed valve timing and 4 speed AT.
[b] An example midsize car in 2008 was 46.64 sq ft and had a fuel economy of 32.5 mpg. Its standard for MY2016 would be 36.6 mpg and for MY2025 would be 54.2 mpg.
[c] These technologies have transmission synergies included. Green highlighting indicates a technology order different than the NHTSA pathway, shown in Appendix S.

TABLE 8.8b Extract of Midsize Car Pathway Showing the Effect of A/C Efficiency Credits and Active Aerodynamics Off-Cycle Credits on the NRC High Most Likely Cost Estimates

Midsize Car with SI Engine Pathway with 10% MR - NRC High Most Likely Estimates - Direct Manufacturing Costs (2010$)
High Most Likely Cost Estimates Paired with Low Most Likely Effectiveness Estimates

Possible Technologies	% FC Reduction (%)	FC Reduction Multiplier	Cumulative FC Reduction Multiplier	FC (gal/100 mi)	Cumulative Percent FC Reduction	Unadjusted Combined (mpg)	2017 Cost Estimates	2020 Cost Estimates	2025 Cost Estimates	2017 Cost/ Percent FC ($/%)
Null Vehicle[a]		1.000	1.000	3.240	0.0%	30.9				
Intake Cam Phasing ICP	2.6%	0.974	0.974	3.156	2.6%	31.7	$43	$41	$36	$16.54
Dual Cam Phasing DCP (vs. ICP)	2.5%	0.975	0.950	3.077	5.0%	32.5[b]	$35	$33	$31	$14.00
					2008 Example Vehicle					
Low Rolling Resistance Tires - 1 ROLL1	1.9%	0.981	0.932	3.018	6.8%	33.1	$5	$5	$5	$2.63
Low Friction Lubricants - 1 LUB1	0.7%	0.993	0.925	2.997	7.5%	33.4	$3	$3	$3	$4.29
6 Speed Automatic Transmission[c] 6 SP AT with Improved Internals IATC	1.3%	0.987	0.913	2.958	8.7%	33.8	$37	$34	$31	$28.46
Aero Drag Reduction - 1 AERO1	2.3%	0.977	0.892	2.890	10.8%	34.6	$39	$37	$33	$16.96
Engine Friction Reduction - 1 EFR1	2.6%	0.974	0.869	2.815	13.1%	35.5	$48	$48	$48	$18.46
Improved Accessories - 1 IACC1	1.2%	0.988	0.858	2.781	14.2%	36.0	$71	$67	$60	$59.17
Electric Power Steering EPS	1.3%	0.987	0.847	2.745	15.3%	36.4	$87	$82	$74	$66.92
Mass Reduction - 2.5% MR2.5 (-87.5 lbs)	0.8%	0.992	0.841	2.723	15.9%	36.7	$22	$22	$22	$27.50
					2016 Target 36.6 mpg					
Discrete Variable Valve Lift DVVL	3.6%	0.964	0.810	2.625	19.0%	38.1	$133	$125	$114	$36.94
Mass Reduction - 2.5%-5.0% MR5-MR2.5 (-87.5 lbs)	0.8%	0.992	0.804	2.604	19.6%	38.4	$66	$66	$66	$82.50
Stoichiometric Gasoline Direct Injection SGDI (Required for TRBDS)	1.5%	0.985	0.792	2.565	20.8%	39.0	$192	$181	$164	$128.00
Turbocharging & Downsizing - 1 TRBDS1 33% DS 18 bar BMEP	7.7%	0.923	0.731	2.368	26.9%	42.2	$331	$312	$282	$42.99
Turbocharging & Downsizing - 2 TRBDS2 50% DS 24 bar BMEP	3.2%	0.968	0.707	2.292	29.3%	43.6	-$96	-$92	-$86	-$30.00
8 Speed Automatic Transmission[c] 8 SP AT	1.3%	0.987	0.698	2.262	30.2%	44.2	$151	$126	$115	$116.15
Shift Optimizer[c] SHFTOPT	0.3%	0.997	0.696	2.255	30.4%	44.3	$26	$24	$22	$86.67
Improved Accessories - 2 IAAC2	2.4%	0.976	0.679	2.201	32.1%	45.4	$43	$40	$37	$17.92
Low Rolling Resistance Tires ROLL2	2.0%	0.980	0.666	2.157	33.4%	46.4	$58	$46	$31	$29.00
Aero Drag Reduction - 2 AERO2	2.5%	0.975	0.649	2.103	35.1%	47.5	$117	$110	$100	$46.80
Mass Reduction - 5%-10% MR10-MR5 (-175 lbs)	4.6%	0.954	0.619	2.006	38.1%	49.8	$325	$322	$315	$70.65
Low Friction Lub - 2 & Engine Friction Red - 2 LUB2_EFR2	1.3%	0.987	0.611	1.980	38.9%	50.5	$51	$51	$51	$39.23
Cooled EGR - 1 CEGR1 50% DS 24 bar BMEP	3.0%	0.970	0.593	1.921	40.7%	52.1	$212	$199	$180	$70.67
High Efficiency Transmission[c] HEG1 & 2	4.9%	0.951	0.564	1.827	43.6%	54.7	$314	$296	$267	$64.08
					2025 Target 54.2 mpg					
Continuously Variable Valve Lift CVVL (vs. DVVL)	1.0%	0.990	0.558	1.809	44.2%	55.3	$67	$63	$56	$67.00
Cylinder Deactivation DEACD	0.0%	1.000	0.558	1.809	44.2%	55.3				
Cooled EGR - 2 CEGR2 56% DS 27 bar BMEP	1.4%	0.986	0.550	1.783	45.0%	56.1	$364	$343	$310	$260.00
					Totals					
Relative to Null Vehicle	45.0%	0.550					$2,744	$2,584	$2,367	$61.03
Null Vehicle - 2008 MY Vehicle	5.0%	0.950					$78	$74	$67	$15.49
2008 MY Vehicle - 2016 MY	11.5%	0.885					$312	$298	$276	$27.15
2017 MY- 2025 MY	32.9%	0.671					$1,923	$1,806	$1,658	$58.42
Beyond 2025 MY	2.4%	0.976					$431	$406	$366	$180.64
	Credits to replace 8 sp AT				2.292	-2.262	0.030			
	Credits to replace LUB2_EFR2				2.006	-1.980	0.026			
	Total to be replaced with credits						0.056			
	Technology Not Required with Credits (2017 MY - 2025 MY)						$202	$177	$166	
	Reduced 2017 MY - 2025 MY Costs with Credits						$1,721	$1,629	$1,492	
	Percent Cost Savings with Credits (2017 MY - 2025 MY)						10.5%	9.8%	10.0%	

Credits (gal/100 mi)	
AC Efficiency	0.0563
Active Aerodynamics	0.0068
Stop-start (N/A w/ o SS)	0
Total =	0.0631

[a] Null vehicle: I4, DOHC, naturally aspirated, 4 valves/cylinder PFI fixed valve timing and 4 speed AT.
[b] An example midsize car in 2008 was 46.64 sq ft and had a fuel economy of 32.5 mpg. Its standard for MY2016 would be 36.6 mpg and for MY2025 would be 54.2 mpg.
[c] These technologies have transmission synergies included. Green highlighting indicates a technology order different than the NHTSA pathway, shown in Appendix S.

TABLE 8.9 Example of the Effect of Credits on 2017 MY to 2025 MY Direct Manufacturing Costs for a Midsize Car with an I4 Gasoline Engine

Pathway for Midsize Car with I4 Gasoline Engine	FC Reduction 2017 MY-2025 MY (%)	2025 Direct Manufacturing Costs (2010 dollars)
Without Credits	32.9 - 33.1	1,181 - 1,658
With Credits for A/C Efficiency and Active Aerodynamics	31.0 - 31.1	1,083 - 1,492
Savings		98 - 166

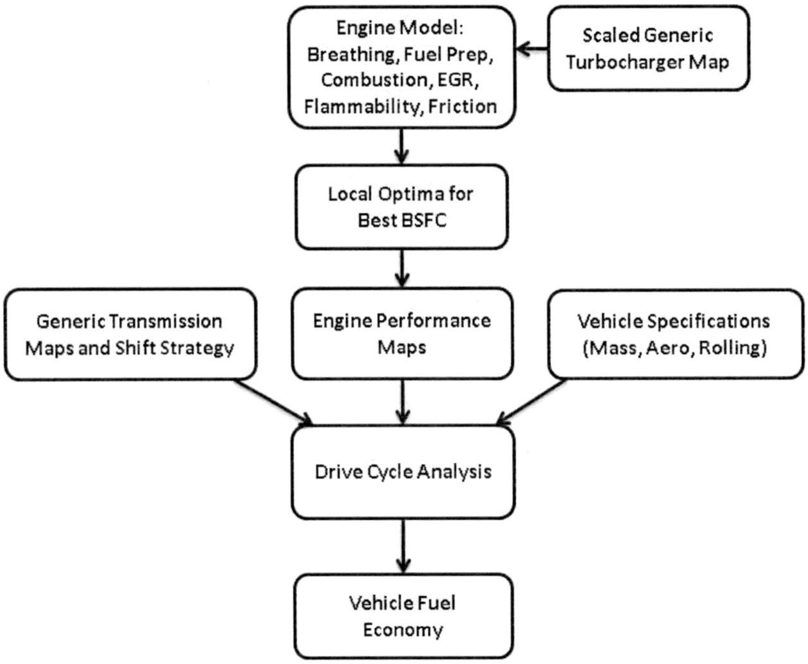

FIGURE 8.10 Schematic of engine–vehicle model.
SOURCE: Middleton et al. (2015).

deceleration fuel shut-off, and minimum time in gear and shift execution might yield fuel economy improvements for the eight-speed transmission. Scherer et al. (2009) described a new eight-speed transmission design with significant friction reductions. The effects of such reductions are simulated in Tasks 9B through 9E, where arbitrary levels of friction reduction up to 60 percent are introduced. Approximately 1.8 to 1.9 percent fuel consumption reductions were shown for each 15 percent reduction in losses. The introduction of the CVT with losses representative of current production CVTs in Task 10-A showed fuel economy comparable to the 6 AT in Task 8. Analysis showed a significant tightening of the visitation points on the engine map toward the optimal region; however, this effect was apparently not enough to compensate for the CVT's higher losses compared to the six-speed automatic transmission. To examine this effect further, the CVT loss map was replaced with the loss map of the eight -speed automatic transmission and rerun as Task 10-B. The result showed only a minor improvement in fuel economy, since moderate speed and load losses were not improved significantly.

Comparison with NHTSA RIA Estimates and EPA Lumped Parameter Model

The results from the U of M full system simulation are compared with NHTSA's estimates in the RIA and estimates provided by EPA's lumped parameter model in Table 8.11. The incremental fuel consumption reductions are similar for all three estimation methods, although some differences appear for individual technologies. However, the U of M full system simulation shows significantly lower improvements for the eight-speed automatic transmission and the high-efficiency gear box, with a 30 percent reduction in losses relative to the other two estimation methods. As suggested in Chapter 5, NHTSA should investigate the benefits of trans-

TABLE 8.10 Detailed Fuel Economy Results from the U of M Full System Simulation Study

	Engine												Fuel Consumption	Incr. Fuel Consumption	Cumul. Fuel
					Cams										
Task	Disp (L)	Air	Fuel	Fric	Fixed	DCP	DVVL	CEGR	Trans	FTP (mpg)	HWY (mpg)	Combined (mpg)	(FC) (gal/100)	Change (%)	Consumption change (%)
1	2	TC	GDI						----					Model validation engine	
2	2.5	NA	PFI		Fixed				6 AT	22.9	40.4	28.4	3.52	Engine reconfigured to 2.5L NA	
3	2.5	NA	PFI			DCP			6 AT	26.2	42.9	31.8	3.15	2012 Ford Fusion simulation	
4	2.5	NA	PFI	EFR-LUB		DCP			6 AT	27.5	44.5	33.2	3.01	-4.4	-4.4
5	2.5	NA	PFI	EFR-LUB		DCP	DVVL		6 AT	29.3	46.7	35.2	2.84	-5.7	-9.8
6	2.5	NA	GDI	EFR-LUB		DCP	DVVL		6 AT	29.7	47.0	35.6	2.81	-1.1	-10.8
7	1.68	TC	GDI	EFR-LUB		DCP	DVVL		6 AT	33.3	50.7	39.4	2.54	-9.6	-19.4
8	1.25	TC	GDI	EFR-LUB		DCP	DVVL	CEGR	6 AT	35.1	52.7	41.3	2.42	-4.6	-23.0
9-A	1.25	TC	GDI	EFR-LUB		DCP	DVVL	CEGR	8 AT	34.7	53.1	41.1	2.43	0.4	-22.8
9-B[a]	1.25	TC	GDI	EFR-LUB		DCP	DVVL	CEGR	8AT -15%	35.4	54.2	41.9	2.39	-1.9	-24.2
9-C[a]	1.25	TC	GDI	EFR-LUB		DCP	DVVL	CEGR	8AT -30%	36.0	55.4	42.8	2.34	-1.9	-25.7
9-D[a]	1.25	TC	GDI	EFR-LUB		DCP	DVVL	CEGR	8AT -45%	36.7	56.4	43.5	2.30	-1.8	-27.0
9-E[a]	1.25	TC	GDI	EFR-LUB		DCP	DVVL	CEGR	8AT -60%	37.3	57.4	44.3	2.26	-1.8	-28.3
10-A[b]	1.25	TC	GDI	EFR-LUB		DCP	DVVL	CEGR	CVT	35.5	52.6	41.6	2.41	6.6	-23.6
10-B[b]	1.25	TC	GDI	EFR-LUB		DCP	DVVL	CEGR	CVT	35.6	52.8	41.8	2.40	-0.4	-23.9

[a] Tasks 9-B – 9-E have reduced transmission torque losses relative to the 6 AT and 8 AT transmissions in Tasks 8 and 9-A, as indicated by the percentage in the table under Trans.
[b] Task 10-B is the same as Task 10-A except that CVT-2 uses the more efficient loss map of the 6 AT automatic transmissions.
SOURCE: Middleton et al. (2015).

TABLE 8.11 Comparison of U of M Full System Simulation Results with NHTSA RIA and EPA Lumped Parameter Model Estimates

	U of M Full System Simulation			From NHTSA RIA Table V-126			From Lumped Parameter Model			
Technology	U of M[a] FSS Results Incremental % FC Reduction	U of M FSS Results Incremental % FC Reduction	U of M FSS Results Cumulative % FC Reduction	NHTSA RIA Incremental % FC Reduction	NHTSA RIA Incremental % FC Reduction	NHTSA RIA Cumulative % FC Reduction	EPA LPM Incremental % FC Reduction	EPA LPM Incremental % FC Reduction	EPA LPM Cumulative % FC Reduction	Adjusted EPA LPM Cumulative % FC Reduction
DCP and 6AT-IATC	Baseline	1.000	0			0.0		0.929	7.1	0.0
LUB1				0.7	0.993		0.5	0.995	7.6	0.5
EFR1				2.6	0.974		1.8	0.982	9.3	2.4
EFR2				1.3	0.987		1.4	0.986	10.6	3.8
LUB and EFR	4.4	0.956	4.4	4.5	0.955	4.5	3.8	0.962		
DVVL	5.7	0.943	9.8	4.6	0.954	8.9	2.2	0.978	12.6	5.9
SGDI	1.1	0.989	10.8	1.5	0.985	10.3	1.5	0.985	13.9	7.3
TRBDS1	9.6	0.904	19.4	8.3	0.917	17.7	6.4	0.936	19.4	13.2
TRBDS2				3.5	0.965		2.6	0.974	21.5	15.5
CEGR1				3.5	0.965		3.6	0.964	24.3	18.5
TRBDS2 and CEGR1	4.6	0.954	23.0	6.9	0.931	23.4	6.1	0.939		
8 AT	-0.4	1.004		3.9	0.961	26.4[b]	3.7	0.963	27.1	21.5
8 AT-HETRANS[a]	3.8	0.962	25.7	2.7	0.973	28.4[b]	4.0	0.960	30.0	24.7
Summary for Turbocharged Downsized Engines										
SGDI		0.989			0.987			0.985		
TRBDS1		0.904			0.917			0.936		
TRBDS2 and CEGR2		0.954			0.931			0.939		
		0.853	14.7%		0.843	15.7%		0.866	13.4%	

[a]30% loss reduction.
[b]Synergies applied.

missions with an additional number of gears when applied to advanced SI engines that already have significantly reduced pumping and friction losses.

Agencies' Full System Simulation Programs

Recently, EPA and NHTSA initiated full system simulation programs. These simulation programs will enhance the Agencies' capability for analyzing fuel consumption reduction technologies and are in response to the recommendation of the Phase 1 NRC study (2011). The Advanced Light-Duty Powertrain and Hybrid Analysis (ALPHA) tool was created by EPA to evaluate the GHG emissions and fuel efficiency of light-duty vehicles (Lee et al. 2013). NHTSA is investigating simulating all technology combinations for all vehicle classes using the Autonomie vehicle simulation model developed by Argonne National Laboratories (NHTSA 2014). Autonomie deployment and support are now handled by LMS International. Engine maps will be developed in GT-Power by IAV. NHTSA anticipates that this simulation model will replace the synergy factors described earlier in this chapter.

IMPLEMENTATION STATUS OF FUEL CONSUMPTION REDUCTION TECHNOLOGIES

Fuel Consumption Reductions from EPA Certification Data Compared to NHTSA Estimates

Some of the engine, transmission, and vehicle technologies anticipated by EPA and NHTSA for the 2017 to 2025 MY CAFE targets have already been introduced in production vehicles. Vehicles with these technologies provided an opportunity for the committee to examine the status of fuel consumption reductions achieved by current 2014 MY vehicles. The committee analyzed the fuel consumption reductions achieved by several vehicles, based on EPA certification test data, and compared the results to the NHTSA-estimated fuel consumption reductions. NHTSA estimated fuel consumption reductions for the vehicles by applying its estimated fuel consumption reductions for each of the technologies that had been applied to four high volume midsize cars. The baseline for this comparison used the 2008 MY vehicles for two reasons. First, NHTSA used the 2008 MY as a baseline in the TSD (EPA/NHTSA 2012a). Second, the vehicles that were reviewed, which are listed in Table 8.12, approximated, with minor exceptions, the NHTSA null vehicle configuration consisting of a naturally aspirated engine, port fuel injection, fixed valve timing and lift, and a four- or five-speed automatic transmission. The engine technologies included in the 2014 MY vehicles were determined from the 2014 EPA Fuel Economy Datafile. However, the vehicles' fuel consumption reduction technologies were less well defined.

The results from comparing the 2014 MY vehicles to the 2008 MY baseline vehicles are shown in Table 8.12, and they indicate that the actual fuel consumption reductions based on EPA certification test data meet, and in some cases exceed, the aggregation of NHTSA technology effectiveness estimates. During this time frame, a combination of engine, transmission, and vehicle technologies have been applied to these vehicles, providing fuel consumption reductions ranging from 14 percent to 21 percent. Since these technologies have already been applied to the 2014 MY vehicles shown in the table in order to comply with the current CAFE standards, they will be included in the baseline vehicles for the beginning of the 2017 to 2025 MY CAFE standards.

Table 8.12 uses EPA uncorrected FTP75 and HWFET combined fuel economy data that are used for CAFE compliance and are obtained from the EPA Fuel Economy Datafile. The left-hand group of columns in the table, which lists the technologies generally available in the 2008-2014 MY time frame, is shown for reference. Not all of these technologies were utilized by all of the example vehicles listed since the OEMs were able to meet their overall 2014 MY CAFE targets without incorporating all of them.

EPA Certification Fuel Economy Compared to CAFE Targets

The fuel economy values of selected vehicles of interest, which have already incorporated some of the technologies identified by NHTSA, were compared to the current and future CAFE targets. The EPA certification fuel economy values for these vehicles (two-cycle CAFE certification test, see Chapter 10) together with other pertinent characteristics, including the fuel economy improvement technologies and footprint, are provided in Appendix W, Tables W.1 and 2. For reference, the label fuel economy values (five-cycle CAFE label test, see Chapter 10) are also provided for comparison with the two-cycle CAFE fuel economy values. Similar information for several hybrid vehicles is also provided in Appendix W, Table W.3.

The CAFE fuel economy values of these vehicles are plotted on the NHTSA fuel economy target curves for the 2012 MY through 2025 MY shown in Figure 8.11 for cars and Figure 8.12 for light trucks. The fuel economy values cluster around the 2016 MY targets and between the 2019 MY and 2021 MY targets. In particular, the fuel economy values of several cars with SI engines are notably above the 2016 MY targets. This includes a vehicle with turbocharging and downsizing; one with variable valve lift and a CVT transmission; one with Multi-Air variable valve timing and lift and a DCT transmission; and one with a three-cylinder naturally aspirated engine and a CVT transmission. The vehicle with the three-cylinder engine with a CVT currently has the highest EPA fuel economy for the 2015 MY. Notable on this figure is the BMW 740Li. This vehicle incorporates many of the technologies used for improving fuel economy in a high-performance vehicle. This example illustrates that implementing these technologies may provide incremental improvements but does not ensure that high fuel economy will be achieved.

TABLE 8.12 Fuel Consumption Reductions for 2014 MY Compared to 2008 MY Midsize Cars, Based on EPA Certification Test Data Compared to Aggregation of NHTSA Technology Effectiveness Estimates

				Actual EPA Fuel Consumption Reductions Compared to Aggregation of NHTSA Technology Effectiveness Estimates							
				Example Vehicles from EPA Fuel Economy Guide							
Baseline Vehicle				Ford Fusion		Chevrolet Malibu		Toyota Camry		Honda Accord	
NA, 4 valves/cyl				2.3L I4 NA		2.4L I4 NA		2.4L NA		2.4L NA	
Fixed valve timing				ICP - VVT		DCP - VVT		ICP - VVT		ICP - VVT	
AT-4				AT-5		AT-4		AT-5		I-VTEC (DVVL)	
										AT-5	
Possible Technologies	NHTSA % FC Reductions			EPA Data	NHTSA Estimates	EPA Data	NHTSA Estimates	EPA Data	NHTSA Estimates	EPA Data	NHTSA Estimates
			2008 MY EPA Combined FE (uncorrected)	30.57	Base	32.48	Base	32.89	Base	31.6	Base
LUB1	0.7	0.993		X	0.993			X	0.993	X	0.993
				5W-20 GF5		5W-30		0W-20		0W-20	
ROLL1	1.9	0.981		Assumed	0.981	Assumed	0.981	Assumed	0.981	Assumed	0.981
EFR1	2.6	0.974		50% Assumed	0.987	50% Assumed	0.987	50% Assumed	0.987	50% Assumed	0.987
EPS	1.3	0.987		X	0.987	X	0.987	X	0.987	X	0.987
IACC1	1.22	0.988		50% Assumed	0.994	50% Assumed	0.994	50% Assumed	0.994	50% Assumed	0.994
6 SP AT	2.04	0.980		X	0.980	X	0.980	X	0.980	X	0.934
										(CVT - Assumed = 8 spd)	
ICP	2.62	0.974									
DCP	2.47	0.975		X	0.975			X	0.975	X	0.975
Sub-Total	13.9%	0.861									
DVVL	3.64	0.964				X	0.964				
SGDI	1.5	0.985		X	0.985	X	0.985	MFI		X	0.985
TRBDS1	7.49	0.925		X	0.925						
33%											
AERO1	2.3	0.977		Assumed	0.977	Assumed	0.977	Assumed	0.977	Assumed	0.977
SS	2.1	0.979		X	0.979	X	0.979				

			Ford Fusion		Chevrolet Malibu		Toyota Camry		Honda Accord	
			1.5L TC		2.5L NA		2.5L NA		2.4L NA	
			I & E VVT		I & E VVT		I & E VVT		I & E VVT	
			AT-6		AT-6		AT-6		CVT	
			GDI		GDI		MFI		GDI	
			SS		DVVL				DVVL	
	0.7228	2014 MY EPA Combined FE (uncorrected)	38.92	0.785	39.00	0.845	38.19	0.880	40.22	0.827
Total- Possible Technologies	-27.7%	FC Reduction %	-21.5%	-21.5%	-16.7%	-15.5%	-13.9%	-12.0%	-21.4%	-17.3%
Differences: EPA data vs. calculation using NHTSA estimates				0.0%		1.2%		1.9%		4.1%

SOURCE: EPA (2008, 2014); EPA/NHTSA (2012b); Ford Parts; GM Parts; Toyota Parts; Honda Parts.

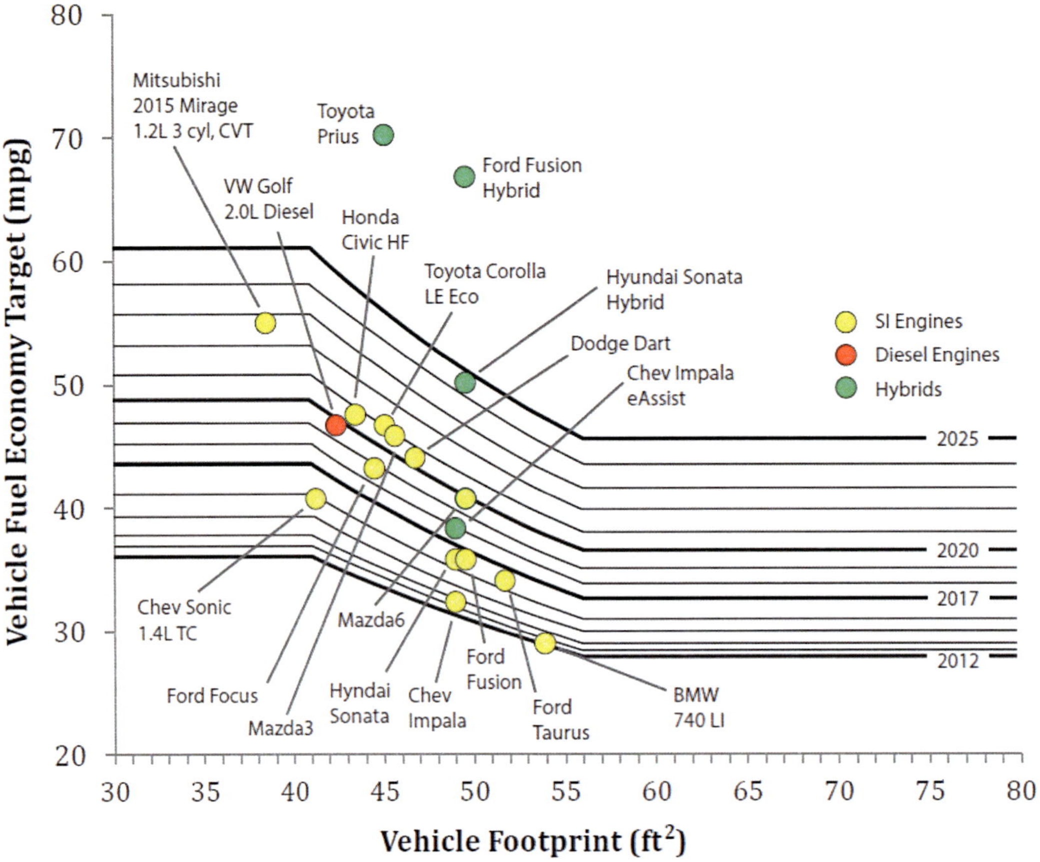

FIGURE 8.11 Fuel economy values of 2013 and 2014 MY cars incorporating many CAFE technologies plotted on NHTSA CAFE target curves.
SOURCE: EPA/NHTSA (2012b); EPA (2008, 2014); Cars.com.

The CAFE fuel economy values of several hybrid vehicles are also plotted on Figure 8.11. The hybrid vehicles are outliers on this plot since they achieve fuel economy values well above their conventional SI engine counterparts. The high levels of fuel economy for hybrid vehicles illustrates why many OEMs are pursuing hybrid technology as part of a broad CAFE/GHG compliance plan. The fuel economy values of two examples of 2014 MY powersplit hybrid vehicles currently exceed the 2025 MY targets, while the fuel economy of an example of a 2014 MY P2 hybrid closely approaches the 2025 MY targets. This figure illustrates the potential for a manufacturer to use hybrid powertrains to offset vehicles with conventional SI engines with fewer fuel consumption reduction technologies than might be required without the offsetting hybrid vehicles.

The CAFE fuel economy values for the pickup trucks cluster around the 2016-2017 targets with two exceptions. The Ram pickup truck with a 3.0L diesel engine shows a 19 percent improvement in fuel economy over a similar truck with a V6 gasoline engine. The 2015 F150 pickup truck with an aluminum body with a reported 700 lb weight reduction and a nearly 50 percent downsized and turbocharged V6 engine also shows a 19 percent improvement over a similar truck with a V6 turbocharged engine. As shown in Appendix W, Table W.2, the 22 mpg combined fuel economy label for the aluminum F150 is within 1 mpg of the 23 mpg combined fuel economy label of the Ram diesel pickup truck. Lower fuel costs for gasoline compared to diesel could eliminate the operating cost differences between these vehicles and possibly favor the gasoline engine, depending on relative fuel cost differences.

FINDINGS AND RECOMMENDATIONS

Finding 8.1 (Partitioning technologies by time frame) EPA and NHTSA have defined many technologies with the potential to reduce fuel consumption. Costs and benefits of the CAFE final rule were assessed from a baseline fleet of vehicle makes and models as they existed in 2008 and in 2010. To assemble this baseline fleet, technologies were

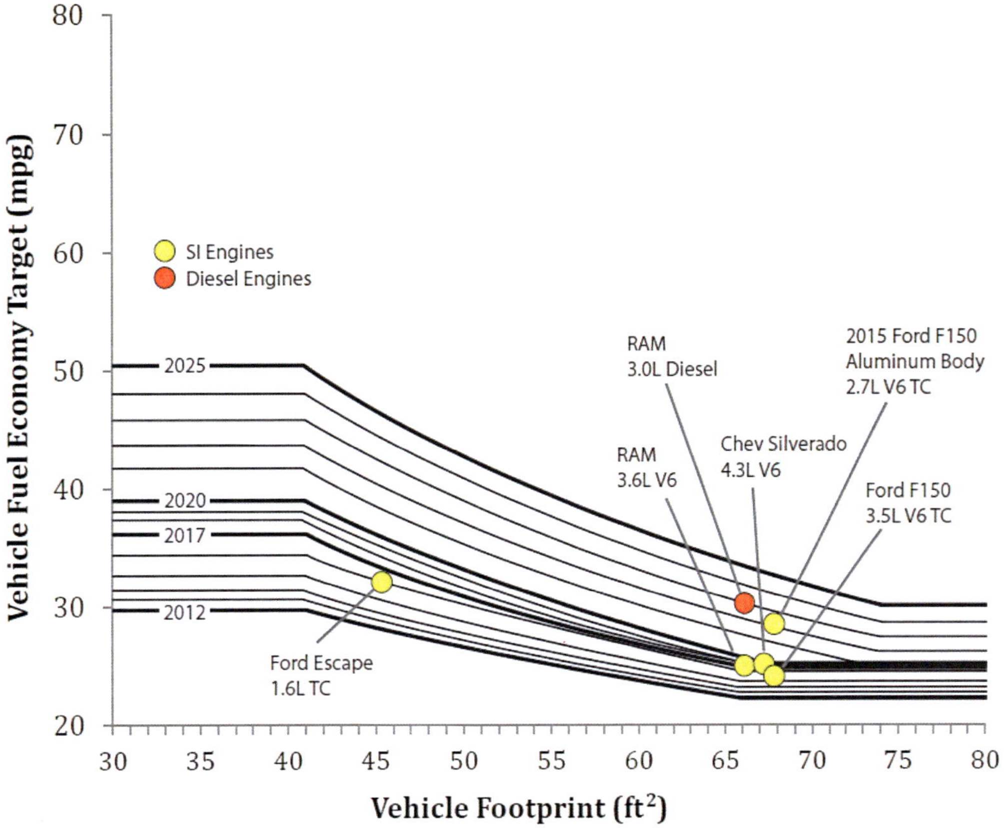

FIGURE 8.12 Fuel economy values of 2013, 2014, and 2015 MY trucks incorporating many CAFE technologies plotted on NHTSA CAFE target curves.
SOURCE: EPA/NHTSA (2012b); EPA (2008, 2014); Cars.com

added to a "null" vehicle, defined as one having an engine with four valves per cylinder, fixed valve timing and lift, port fuel injection, and a four-speed automatic transmission. In EPA's and NHTSA's compliance models, technologies were added to baseline fleet vehicles to reach compliance with the fleet average, footprint-based standards. This was done for the 2012-2016 MY standards, and then, using the modeled compliance paths to 2016, new technologies were added to further comply with the 2017-2025 standards. EPA and NHTSA used different compliance models with slightly different baseline fleets.

Recommendation 8.1 (Define new 2016 null vehicle) The committee compliments EPA and NHTSA on their plans to determine the actual technology penetration rates for the 2016 MY fleet, as data becomes available. The committee recommends the Agencies establish a new definition of a "null" vehicle, representative of the most basic vehicle in the 2016 MY time frame as well as a baseline 2016 MY fleet reflecting actual technology penetration rates. The vehicles in the 2016 MY fleet should be assigned EPA certification fuel economy values and reasons for any differences between actual and estimated effectiveness of the technologies applied to the 2008 MY vehicles, derived from the original null vehicle, should be determined. This updated baseline should consider changes in performance of these 2016 MY vehicles relative to the 2008 MY vehicles when estimating the effectiveness of the technologies applied to the 2016 MY vehicles. Updated null vehicles with technologies applied for the 2016 MY will assist in distinguishing between technologies that can be applied for the 2017 to 2025 MY CAFE targets from technologies that have already been applied to achieve the 2016 MY CAFE targets.

Finding 8.2 (Effectiveness) The committee's most likely estimates of fuel consumption reduction effectiveness are comparable to NHTSA's estimates for many of the technologies defined by NHTSA. However, the committee estimated higher most likely effectiveness values for several technologies, including mass reduction and high-efficiency gearbox technology. For some other technologies, including several of the turbocharged, downsized engine technologies and P2 hybrids, the committee extended the range of most likely estimates of effectiveness to include lower values. For several other technologies, including eight-speed automatic transmissions and shift optimization, the committee's low and high ranges of most likely estimates were lower than NHTSA's estimates.

Finding 8.3 (Costs) The committee's estimates of direct manufacturing costs are comparable to NHTSA's estimates for some of the technologies defined by NHTSA. The committee extended the range to include higher estimates of direct manufacturing costs for some technologies, including several SI engine technologies, several transmission technologies, and electrified powertrain technologies. The ranges of most likely direct manufacturing costs for several other technologies, including diesel engines, several transmission technologies, and mass reduction, were estimated to be higher than NHTSA's estimates.

Recommendation 8.2 (Updating cost and effectiveness) While the committee concurred with the Agencies' costs and effectiveness values for a wide array of technologies, in some cases the committee developed estimates that significantly differed from the Agencies' values, so the committee recommends that the Agencies pay particular attention to the reanalysis of these technologies in the mid-term review.

Finding 8.4 (Cost effectiveness) The cost effectiveness of individual technologies defined as the cost per percent fuel consumption reduction for the technologies for SI engine, transmission, and vehicle technologies ranges from less than $25 to significantly over $100 per percent fuel consumption reduction. The cost effectiveness of a spark ignition engine with all of NHTSA's technologies included is less than $50 per percent reduction in fuel consumption, which is lower than advanced diesel engines and strong hybrids, which are in the range of $75 to $100 per percent reduction in fuel consumption.

Finding 8.5 (Effectiveness depends on the prior technologies) Some of the technologies defined by NHTSA have already been incorporated in current vehicles, and additional technologies are expected to be applied to achieve the 2016 MY CAFE targets. By the 2016 MY, many vehicles will include variable valve timing and lift, stoichiometric gasoline direct injection, six- or eight-speed automatic transmissions, and some will have turbocharged, downsized engines at level 1 with 18 bar BMEP or higher. Although these technologies are included in the complete list of NHTSA's technologies relative to the baseline null vehicle that approximates a vehicle prior to the 2008 MY, only the additional technologies beyond those applied by the 2016 MY will be available in the 2017-2025 time frame and beyond to provide additional reductions in fuel consumption. The effectiveness of a technology depends on the technologies that have already been applied to a vehicle. For example, although the relative effectiveness of a turbocharged, downsized engine at level 1 was estimated by NHTSA to provide 12.9 to 14.9 percent reduction in fuel consumption relative to the null vehicle, the effectiveness of this technology is reduced to 7.7 to 8.3 percent when applied to a vehicle already having friction reduction, variable valve timing and lift, and stoichiometric gasoline direct injection technologies.

Finding 8.6 (Other technologies) In addition to the technologies defined by NHTSA, the committee has identified other technologies that might be available by the 2025 MY that could provide additional reductions in fuel consumption or provide alternative approaches at lower cost. In addition, the committee has identified several technologies that might be available after the 2025 MY, although these technologies are generally still in the research phase of development. As discussed in Chapter 2, alternative fuels combined with SI technologies (e.g., flex-fuel vehicles, ethanol-boosted direct injection systems) also may provide some opportunity for petroleum reductions. For each of the technologies listed in the Alternative Fuels section of Table 8A.1, energy consumption reduction (as gasoline gallons equivalent, gge) is shown, followed by the CAFE petroleum reduction in brackets. The application of indirect credits for air conditioning efficiency together with active aerodynamics and stop-start off-cycle credits provide opportunities for cost savings in achieving the CAFE targets.

Finding 8.7 (Full system simulation) Full system simulations provide estimates of effectiveness of technologies applied either singularly or in combination with other technologies, as in the case of applying multiple technologies to achieve future CAFE targets. Full system simulations can provide these estimates before experimental test data are available. The committee contracted with the University of Michigan to develop a full system simulation, which confirmed the effectiveness trends provided by the EPA lumped parameter model and incorporated in NHTSA's decision tree paths together with their synergy tables. For projections of technologies without test data, full system simulations must include detailed models that are correlated to baseline hardware with available test data. The correlation of models ensures that the results will reflect only the effectiveness of the technology of interest.

Recommendation 8.3 (Full system simulation and teardown cost analysis) The committee notes that the use of full vehicle simulation modeling in combination with lumped parameter modeling and teardown studies contributed substantially to the value of the Agencies' estimates of fuel consumption and costs, and it therefore recommends they continue to increase the use of these methods to improve their analysis. The committee recognizes that such methods are expensive but believes that the added cost is well justified because it produces more reliable assessments.

REFERENCES

EPA (Environmental Protection Agency). 2008. Fuel Economy Data Files. http://www.fueleconomy.gov/feg/download.shtml.

EPA. 2012. Regulatory Impact Analysis: Final Rulemaking for 2017-2025 Light-Duty Vehicle Greenhouse Gas Emission Standards and Corporate Average Fuel Economy Standards. EPA-420-R-12-016, August.

EPA. 2014. Fuel Economy Data Files. http://www.fueleconomy.gov/feg/download.shtml.

EPA/NHTSA (National Highway Traffic Safety Administration). 2009. Draft Joint Technical Support Document: Proposed Rulemaking to Establish Light-Duty Vehicle Greenhouse Gas Emission Standards and Corporate Average Fuel Economy Standards. EPA-420-D-09-901, September.

EPA/NHTSA. 2012a. Joint Technical Support Document, Final Rulemaking 2017-2025 Light-Duty Greenhouse Gas Emission Standards and Corporate Average Fuel Economy Standards. EPA-420-R-12-901, August.

EPA/NHTSA. 2012b. 2017 and Later Model Year Light-Duty Vehicle Greenhouse Gas Emissions and Corporate Average Fuel Economy Standards. EPA 40 CFR Parts 85, 86 and 600, NHTSA 49 CFR Parts 523, 531, 533, 536 and 537, August 28.

Gamma Technologies. n.d. GT-POWER Engine Simulation Software: Engine Performance Analysis Modeling. http://www.gtisoft.com/upload/Power.pdf.

Goodman, L.A. 1962. The variance of the product of K random variables. Journal of the American Statistical Association. 57(297).

Heard, D.C. 1987. A Simple Formula for Calculation the Variance of Products and Dividends. Department of Renewable Resources, Government of the NWT, Yellowknife, NWT.

Lee, B., S. Lee, J. Cherry, A. Neam, J. Sanchez, and E. Nam. 2013. Development of Advanced Light-Duty Powertrain and Hybrid Analysis Tool. SAE Technical Paper 2013-01-0808. doi: 10.4271/2013-01-0808.

Middleton, R., O. Gupta, H-Y Chang, G.A. Lavoie, and J. Martz. 2015. Fuel Economy Estimates for Future Light Duty Vehicles. University of Michigan Report. Ann Arbor, Michigan.

NHTSA. 2012. Final Regulatory Impact Analysis: Corporate Average Fuel Economy for MY 2017-MY2025 Passenger Cars and Light Trucks. Office of Regulatory Analysis and Evaluation, National Center for Statistics and Analysis.

NHTSA. 2014. NHTSA's Recent Activities on Light-Duty Fuel Economy. Presentation to the National Research Council Committee on Assessment of Technologies for Improving Fuel Economy of Light Duty Vehicles, Phase 2, June 23.

NRC (National Research Council). 2002. Effectiveness and Impact of Corporate Average Fuel Economy (CAFE) Standards. Washington, D.C.: The National Academies Press.

NRC. 2011. Assessment of Fuel Economy Technologies for Light-Duty Vehicles. Washington, D.C.: The National Academies Press.

Olechiw, M. 2014. Baseline Vehicles. Email communication to the committee, June 27.

Ricardo, Inc. 2011. Draft Project Report: Computer Simulation of Light-Duty Vehicle Technologies for Greenhouse Gas Emission Reduction in the 2020-2025 Timeframe. Report to EPA Office of Transportation and Air Quality, EP-W0-07-064, Ann Arbor, Michigan, April 6.

Robinette, D. 2014. A DFSS Approach to determine automatic transmission gearing content for powertrain-vehicle system integration. SAE Int. J. Passeng. Cars - Mech. Syst. 7(3): 1138-1154. doi:10.4271/2014-01-1774.

Scherer, H., M. Bek, and S. Kilian. 2009. ZF New 8-speed Automatic Transmission 8HP70 - Basic design and hybridization. SAE Int. J. Engines 2(1): 314-326. doi:10.4271/2009-01-0510.

ANNEX TABLES

TABLE 8A.1 NRC Committee's Estimated Fuel Consumption Reduction Effectiveness of Technologies

Percent Incremental Fuel Consumption Reductions: NRC Estimates

		Midsize Car I4 DOHC	Large Car V6 DOHC	Large Light Truck V8 OHV	
Spark Ignition Engine Technologies	Abbreviation	Most Likely	Most Likely	Most Likely	Relative To
NHTSA Technologies					
Low Friction Lubricants - Level 1	LUB1	0.7	0.8	0.7	Baseline
Engine Friction Reduction - Level 1	EFR1	2.6	2.7	2.4	Baseline
Low Friction Lubricants and Engine Friction Reduction - Level 2	LUB2_EFR2	1.3	1.4	1.2	Previous Tech
VVT- Intake Cam Phasing (CCP - Coupled Cam Phasing - OHV)	ICP	2.6	2.7	2.5	Baseline for DOHC
VVT- Dual Cam Phasing	DCP	2.5	2.7	2.4	Previous Tech
Discrete Variable Valve Lift	DVVL	3.6	3.9	3.4	Previous Tech
Continuously Variable Valve Lift	CVVL	1.0	1.0	0.9	Previous Tech
Cylinder Deactivation	DEACD	N/A	0.7	5.5	Previous Tech
Variable Valve Actuation (CCP + DVVL)	VVA	N/A	N/A	3.2	Baseline for OHV
Stoichiometric Gasoline Direct Injection	SGDI	1.5	1.5	1.5	Previous Tech
Turbocharging and Downsizing Level 1 - 18 bar BMEP 33%DS	TRBDS1	7.7 - 8.3	7.3 - 7.8	6.8 - 7.3	Previous Tech
Turbocharging and Downsizing Level 2 - 24 bar BMEP 50%DS	TRBDS2	3.2 - 3.5	3.3 - 3.7	3.1 - 3.4	Previous Tech
Cooled EGR Level 1 - 24 bar BMEP, 50% DS	CEGR1	3.0 - 3.5	3.1 - 3.5	3.1 - 3.6	Previous Tech
Cooled EGR Level 2 - 27 bar BMEP, 56% DS	CEGR2	1.4	1.4	1.2	Previous Tech
Other Technologies					
By 2025:					
Compression Ratio Increase (with regular fuel)	CRI-REG	3.0	3.0	3.0	Baseline
Compression Ratio Increase (with higher octane regular fuel)	CRI-HO	5.0	5.0	5.0	Baseline
Compression Ratio Increase (CR~13:1, exh. scavenging, DI (aka Skyactiv, Atkinson Cycle))	CRI-EXS	10.0	10.0	10.0	Baseline
Electrically Assisted Variable Speed Supercharger[a]	EAVS-SC	26.0	26.0	26.0	Baseline
Lean Burn (with low sulfur fuel)	LBRN	5.0	5.0	5.0	Baseline
After 2025:					
Variable Compression Ratio	VCR	Up to 5.0	Up to 5.0	Up to 5.0	Baseline
D-EGR	DEGR	10.0	10.0	10.0	TRBDS1
Homogeneous Charge Compression Ignition (HCCI) + Spark Assisted CI[b]	SA-HCCI	Up to 5.0	Up to 5.0	Up to 5.0	TRBDS1
Gasoline Direct Injection Compression Ignition (GDCI)	GDCI	Up to 5.0	Up to 5.0	Up to 5.0	TRBDS1
Waste Heat Recovery	WHR	Up to 3.0	Up to 3.0	Up to 3.0	Baseline
Alternative Fuels[c]:					
CNG-Gasoline Bi-Fuel Vehicle (default UF = 0.5)	BCNG	Up to 5 Incr [42]	Up to 5 Incr [42]	Up to 5 Incr [42]	Baseline
Flexible Fuel Vehicle (UF dependent, UF = 0.5 thru 2019)	FFV	0 [40 thru 2019, then UF TBD]	0 [40 thru 2019, then UF TBD]	0 [40 thru 2019, then UF TBD]	Baseline
Ethanol Boosted Direct Injection (CR = 14:1, 43% downsizing) (UF~0.05)	EBDI	20 [24]	20 [24]	20 [24]	Baseline

continued

TABLE 8A.1 Continued

Percent Incremental Fuel Consumption Reductions: NRC Estimates

		Midsize Car I4 DOHC	Large Car V6 DOHC	Large Light Truck V8 OHV	
Diesel Engine Technologies	Abbreviation	Most Likely	Most Likely	Most Likely	Relative To
NHTSA Technologies					
Advanced Diesel	ADSL	29.4	30.5	29.0	Baseline
Other Technologies					
Low Pressure EGR	LPEGR	3.5	3.5	3.5	ADSL
Closed Loop Combustion Control	CLCC	2.5	2.5	2.5	ADSL
Injection Pressures Increased to 2,500 to 3,000 bar	INJ	2.5	2.5	2.5	ADSL
Downspeeding with Increased Boost Pressure	DS	2.5	2.5	2.5	ADSL
Friction Reduction	FR	2.5	2.5	2.5	ADSL
Waste Heat Recovery	WHR	2.5	2.5	2.5	ADSL
Transmission Technologies	Abbreviation	Most Likely	Most Likely	Most Likely	Relative To
NHTSA Technologies					
Improved Auto. Trans. Controls/Externals (ASL-1 & Early TC Lockup)	IATC	2.5 - 3.0	2.5 - 3.0	2.5 - 3.0	4 sp AT
6-speed AT with Improved Internals - Lepelletier (Rel to 4 sp AT)	NUATO-L	2.0 - 2.5	2.0 - 2.5	2.0 - 2.5	IATC
6-speed AT with Improved Internals - Non-Lepelletier (Rel to 4 sp AT)	NUATO-NL	2.0 - 2.5	2.0 - 2.5	2.0 - 2.5	IATC
6-speed Dry DCT (Rel to 6 sp AT - Lepelletier)	6DCT-D	3.5 - 4.5	3.5 - 4.5	N/A	6 sp AT
6-speed Wet DCT (Rel to 6 sp AT - Lepelletier) (0.5% less than Dry Clutch)	6DCT-W	3.0 - 4.0	3.0 - 4.0	3.0 - 4.0	6 sp AT
8-speed AT (Rel to 6 sp AT - Lepelletier)	8AT	1.5 - 2.0	1.5 - 2.0	1.5 - 2.0	Previous Tech
8-speed DCT (Rel to 6 sp DCT)	8DCT	1.5 - 2.0	1.5 - 2.0	1.5 - 2.0	Previous Tech
High Efficiency Gearbox Level 1 (Auto) (HETRANS)	HEG1	2.3 - 2.7	2.3 - 2.7	2.3 - 2.7	Previous Tech
High Efficiency Gearbox Level 2 (Auto, 2017 and Beyond)	HEG2	2.6 - 2.7	2.6 - 2.7	2.6 - 2.7	Previous Tech
Shift Optimizer (ASL-2)	SHFTOPT	0.5 - 1.0	0.5 - 1.0	0.5 - 1.0	Previous Tech
Secondary Axle Disconnect	SAX	1.4 - 3.0	1.4 - 3.0	1.4 - 3.0	Baseline
Other Technologies					
Continuously Variable Transmission with Improved internals (Rel to 6 sp AT)	CVT	3.5 - 4.5	3.5 - 4.5	N/A	Previous Tech
High Efficiency Gearbox (CVT)	CVT-HEG	3.0	3.0	N/A	Previous Tech
High Efficiency Gearbox (DCT)	DCT-HEG	2.0	2.0	2.0	Previous Tech
High Efficiency Gearbox Level 3 (Auto, 2020 and beyond)	HEG3	1.6	1.6	1.6	Previous Tech
9-10 speed Transmission (Auto, Rel to 8 sp AT)	10SPD	0.3	0.3	0.3	Previous Tech
Electrified Accessories Technologies	Abbreviation	Most Likely	Most Likely	Most Likely	Relative To
NHTSA Technologies					
Electric Power Steering	EPS	1.3	1.1	0.8	Baseline
Improved Accessories - Level 1 (70% Eff Alt, Elec. Water Pump and Fan)	IACC1	1.2	1.0	1.6	Baseline
Improved Accessories - Level 2 (Mild regen alt strategy, Intelligent cooling)	IACC2	2.4	2.6	2.2	Previous Tech
Hybrid Technologies	Abbreviation	Most Likely	Most Likely	Most Likely	Relative To
NHTSA Technologies					
Stop-Start (12V Micro-Hybrid) (Retain NHTSA Estimates)	SS	2.1	2.2	2.1	Baseline

continued

TABLE 8A.1 Continued

Percent Incremental Fuel Consumption Reductions: NRC Estimates					
		Midsize Car I4 DOHC	Large Car V6 DOHC	Large Light Truck V8 OHV	
Integrated Starter Generator	MHEV	6.5	6.4	3.0	Previous Tech
Strong Hybrid - P2 - Level 2 (Parallel 2 Clutch System)	SHEV2-P2	28.9 - 33.6	29.4 - 34.5	26.9 - 30.1	Baseline
Strong Hybrid - PS - Level 2 (Power Split System)	SHEV2-PS	33.0 - 33.5	32.0 - 34.1	N/A	Baseline
Plug-in Hybrid - 40 mile range	PHEV40	N/A	N/A	N/A	Baseline
Electric Vehicle - 75 mile	EV75	N/A	N/A	N/A	Baseline
Electric Vehicle - 100 mile	EV100	N/A	N/A	N/a	Baseline
Electric Vehicle - 150 mile	EV150	N/A	N/A	N/A	Baseline
Other Technologies					
Fuel Cell Electric Vehicle	FCEV	N/A	N/A	N/A	Baseline
Vehicle Technologies	Abbreviation	Most Likely	Most Likely	Most Likely	Relative To
NHTSA Technologies					
Without Engine Downsizing[d]					
0 - 2.5% Mass Reduction (Design Optimization)	MR2.5	0.80	0.80	0.85	Baseline
2.5 - 5% Mass Reduction		0.81	0.81	0.85	Previous MR
0 - 5% Mass Reduction (Material Substitution)	MR5	1.60	1.60	1.69	Baseline
With Engine Downsizing (Same Architecture)[d]					
5 - 10% Mass Reduction		4.57	4.57	2.85	Previous MR
0 - 10% Mass Reduction (HSLA Steel and Aluminum Closures)	MR10	6.10	6.10	4.49	Baseline
10 - 15% Mass Reduction (Aluminum Body)		3.25	3.25	2.35	Previous MR
0 - 15% Mass Reduction (Aluminum Body)	MR15	9.15	9.15	6.73	Baseline
15 - 20% Mass Reduction		3.37	3.37	2.41	Previous MR
0 - 20% Mass Reduction (Aluminum Body, Magnesium, Composites)	MR20	12.21	12.21	8.98	Baseline
20 - 25% Mass Reduction		3.47	3.47	2.46	Previous MR
0 - 25% Mass Reduction (Carbon Fiber Composite Body)	MR25	15.26	15.26	11.22	Baseline
Summary - Mass Reduction Relative to Baseline					
0 - 2.5% Mass Reduction	MR2.5	0.80	0.80	0.85	Baseline
0 - 5% Mass Reduction	MR5	1.60	1.60	1.69	Baseline
0 - 10% Mass Reduction	MR10	6.10	6.10	4.49	Baseline
0 - 15% Mass Reduction	MR15	9.15	9.15	6.73	Baseline
0 - 20% Mass Reduction	MR20	12.21	12.21	8.98	Baseline
0 - 25% Mass Reduction	MR25	15.26	15.26	11.22	Baseline
Low Rolling Resistance Tires - Level 1 (10% Reduction)	ROLL1	1.9	1.9	1.9	Baseline
Low Rolling Resistance Tires - Level 2 (20% Reduction)	ROLL2	2.0	2.0	2.0	Previous Tech
Low Drag Brakes	LDB	0.8	0.8	0.8	Baseline
Aerodynamic Drag Reduction - Level 1 (10% Reduction)	AERO1	2.3	2.3	2.3	Baseline
Aerodynamic Drag Reduction - Level 2 (20% Reduction)	AERO2	2.5	2.5	2.5	Previous Tech

[a] Comparable to TRBDS1, TRBDS2, SS, MHEV, IACC1, IACC2
[b] With TWC aftertreatment. Costs will increase with lean NO_x aftertreatment.
[c] Fuel consumption reduction in gge (gasoline gallons equivalent) [CAFE fuel consumption reduction]
[d] FC Reductions – Ricardo 2007. Car without engine downsizing: +3.3% mpg/10% MR = -3.2% FC/10% MR. Car with engine downsizing (for MR > 10%): +6.5% mpg/10%MR = -6.1% FC/10% MR. Truck without engine downsizing: +3.5% mpg/10% MR = -3.4% FC/10% MR. Truck with engine downsizing (for MR > 10%): +4.7% mpg/10%MR = 4.5% FC/10% MR.
NOTE: Midsize car: 3,500 lbs, large car: 4,500 lbs, large light truck: 5,500 lbs.

TABLE 8A.2a NRC Committee's Estimated 2017 Direct Manufacturing Costs of Technologies

2017 MY Incremental Direct Manufacturing Costs (2010$): NRC Estimates

		Midsize Car I4 DOHC	Large Car V6 DOHC	Large Light Truck V8 OHV	
Spark Ignition Engine Technologies	Abbreviation	Most Likely	Most Likely	Most Likely	Relative To
NHTSA Technologies					
Low Friction Lubricants - Level 1	LUB1	3	3	3	Baseline
Engine Friction Reduction - Level 1	EFR1	48	71	95	Baseline
Low Friction Lubricants and Engine Friction Reduction - Level 2	LUB2_EFR2	51	75	99	Previous Tech
VVT- Intake Cam Phasing (CCP - Coupled Cam Phasing - OHV)	ICP	37 - 43	74 - 86	37	Baseline for DOHC
VVT- Dual Cam Phasing	DCP	31 - 35	72 - 82	37 - 43	Previous Tech
Discrete Variable Valve Lift	DVVL	116 - 133	168 - 193	37 - 43	Previous Tech
Continuously Variable Valve Lift	CVVL	58 - 67	151 - 174	N/A	Previous Tech
Cylinder Deactivation	DEACD	N/A	139	N/A	Previous Tech
Variable Valve Actuation (CCP + DVVL)	VVA	N/A	N/A	157	Baseline for OHV
Stoichiometric Gasoline Direct Injection	SGDI	192	290	277 - 320	Previous Tech
Turbocharging and Downsizing Level 1 - 18 bar BMEP 33%DS	TRBDS1	288 - 331	-129 to -86	942 - 1,028	Previous Tech
V6 to I4 and V8 to V6			-455* to -369*	841* to 962*	
Turbocharging and Downsizing Level 2 - 24 bar BMEP 50%DS	TRBDS2	182	182	308	Previous Tech
I4 to I3		-92* to -96*			
Cooled EGR Level 1 - 24 bar BMEP, 50% DS	CEGR1	212	212	212	Previous Tech
Cooled EGR Level 2 - 27 bar BMEP, 56% DS	CEGR2	364	364	614	Previous Tech
V6 to I4				-524* to -545*	
Other Technologies					
By 2025:					
Compression Ratio Increase (with regular fuel)	CRI-REG	50	75	100	Baseline
Compression Ratio Increase (with higher octane regular fuel)	CRI-HO	75	113	150	Baseline
Compression Ratio Increase (CR~13:1, exh. scavenging, DI (aka Skyactiv, Atkinson Cycle))	CRI-EXS	250	375	500	Baseline
Electrically Assisted Variable Speed Supercharger	EAVS-SC	1,302	998	N/A	Baseline
Lean Burn (with low sulfur fuel)	LBRN	800	920	1,040	Baseline
After 2025:					
Variable Compression Ratio	VCR				Baseline
D-EGR	DEGR				TRBDS1
Homogeneous Charge Compression Ignition (HCCI) + Spark Assisted CI[a]	SA-HCCI				TRBDS1
Gasoline Direct Injection Compression Ignition	GDCI				Baseline
Waste Heat Recovery	WHR				Baseline
Alternative Fuels:					
CNG-Gasoline Bi-Fuel Vehicle	BCNG	6,000	6,900	7,800	Baseline
Flexible Fuel Vehicle	FFV	75	100	125	Baseline
Ethanol Boosted Direct Injection (incr CR to 14:1, 43% downsizing)	EBDI	740	870	1,000	Baseline

continued

TABLE 8A.2a Continued

2017 MY Incremental Direct Manufacturing Costs (2010$): NRC Estimates

		Midsize Car I4 DOHC	Large Car V6 DOHC	Large Light Truck V8 OHV	
Diesel Engine Technologies	Abbreviation	Most Likely	Most Likely	Most Likely	Relative To
NHTSA Technologies					
Advanced Diesel	ADSL	3,023	3,565	3,795	Baseline
Other Technologies					
Low Pressure EGR	LPEGR	133	166	166	ADSL
Closed Loop Combustion Control	CLCC	68	102	102	ADSL
Injection Pressures Increased to 2,500 to 3,000 bar	INJ	24	26	26	ADSL
Downspeeding with Increased Boost Pressure	DS	28	28	28	ADSL
Friction Reduction	FR	64	96	96	ADSL
Waste Heat Recovery	WHR	N/A	N/A	N/A	
Transmission Technologies	Abbreviation	Most Likely	Most Likely	Most Likely	Relative To
NHTSA Technologies					
Improved Auto. Trans. Controls/Externals (ASL-1 & Early TC Lockup)	IATC	50	50	50	Baseline 4 sp AT
6-speed AT with Improved Internals - Lepelletier (Rel to 4 sp AT)	NUATO-L	-13	-13	-13	IATC
6-speed AT with Improved Internals - Non-Lepelletier (Rel to 4 sp AT)	NUATO-NL	195	195	195	IATC
6-speed Dry DCT (Rel to 6 sp AT - Lepelletier)	6DCT-D	-149 to 31	-149 to 31	N/A	6 sp AT
6-speed Wet DCT (Rel to 6 sp AT - Lepelletier)	6DCT-W	-88 to 88	-88 to 88	-88 to 88	6 sp AT
8-speed AT (Rel to 6 sp AT - Lepelletier)	8AT	56 - 151	56 - 151	56 - 151	Previous Tech
8-speed DCT (Rel to 6 sp DCT)	8DCT	179	179	179	Previous Tech
High Efficiency Gearbox Level 1 (Auto) (HETRANS)	HEG1	120	120	120	Previous Tech
High Efficiency Gearbox Level 2 (Auto, 2017 and Beyond)	HEG2	194	194	194	Previous Tech
Shift Optimizer (ASL-2)	SHFTOPT	26	26	26	Previous Tech
Secondary Axle Disconnect	SAX	100	100	100	Baseline
Other Technologies					
Continuously Variable Transmission with Improved internals (Rel to 6 sp AT)	CVT	179	179	N/A	Baseline
High Efficiency Gearbox (CVT)	CVT-HEG	125	125	N/A	Baseline
High Efficiency Gearbox (DCT)	DCT-HEG	150	150	150	Baseline
High Efficiency Gearbox Level 3 (Auto, 2020 and Beyond)	HEG3	150	150	150	Baseline
9-10 speed Transmission (Auto, Rel to 8 sp AT)	10SPD	75	75	75	Baseline
Electrified Accessories Technologies	Abbreviation	Most Likely	Most Likely	Most Likely	Relative To
NHTSA Technologies					
Electric Power Steering	EPS	87	87	87	Baseline
Improved Accessories - Level 1 (70% Eff Alt, Elec. Water Pump and Fan)	IACC1	71	71	71	Baseline
Improved Accessories - Level 2 (Mild regen alt strategy, Intelligent cooling)	IACC2	43	43	43	Previous Tech
Hybrid Technologies	Abbreviation	Most Likely	Most Likely	Most Likely	Relative To
NHTSA Technologies					
Stop-Start (12V Micro-Hybrid)	SS	287 - 387	325 - 425	356 - 456	Baseline

continued

TABLE 8A.2a Continued

2017 MY Incremental Direct Manufacturing Costs (2010$): NRC Estimates

		Midsize Car I4 DOHC	Large Car V6 DOHC	Large Light Truck V8 OHV	
Integrated Starter Generator	MHEV	1,087 - 1,253	1,087 - 1,377	1,087 - 1,438	Previous Tech
Strong Hybrid - P2 - Level 2 (Parallel 2 Clutch System)	SHEV2-P2	2,463 - 3,126	2,908 - 3,726	2,947 - 3,762	Baseline
Strong Hybrid - PS - Level 2 (Power Split System)	SHEV2-PS	3,139	3,396	N/A	Baseline
Plug-in Hybrid - 40 mile range	PHEV40	13,193 - 14,776	17,854 - 20,141	N/A	Baseline
Electric Vehicle - 75 mile	EV75	14,812 - 15,446	19,275 - 20,393	N/A	Baseline
Electric Vehicle - 100 mile	EV100	16,831	21,123	N/A	Baseline
Electric Vehicle - 150 mile	EV150	22,257	26,193	N/A	Baseline
Other Technologies					
Fuel Cell Electric Vehicle	FCEV	N/A	N/A	N/A	Baseline
Vehicle Technologies	Abbreviation	Most Likely	Most Likely	Most Likely	Relative To
NHTSA Technologies					
Without Engine Downsizing					
0 - 2.5% Mass Reduction (Design Optimization)	MR2.5	0 - 22	0 - 28	0 - 39	Baseline
2.5 - 5% Mass Reduction		0 - 66	0 - 84	0 - 116	Previous MR
0 - 5% Mass Reduction (Material Substitution)	MR5	0 - 88	0 - 113	0 - 154	Baseline
With Engine Downsizing (Same Architecture)[b]					
5 - 10% Mass Reduction		154 - 325	198 - 419	270 - 572	Previous MR
0 - 10% Mass Reduction (HSLA Steel and Aluminum Closures)	MR10	154 - 413	198 - 531	270 - 726	Baseline
10 - 15% Mass Reduction (Aluminum Body)		452 - 767	581 - 986	792 - 1,353	Baseline
0 - 15% Mass Reduction (Aluminum Body)	MR15	452 - 767	581 - 986	792 - 1,353	Baseline
15 - 20% Mass Reduction		528 - 654	679 - 841	924 - 1,144	Previous MR
0 - 20% Mass Reduction (Aluminum Body, Magnesium, Composites)	MR20	980 - 1,421	1,260 - 1,827	1,716 - 2,497	Baseline
20 - 25% Mass Reduction		1,173 - 1,449	1,508 - 1,863	2,079 - 2,549	Previous MR
0 - 25% Mass Reduction (Carbon Fiber Composite Body)	MR25	2,153 - 2,870	2,768 - 3,690	3,795 - 5,046	Baseline
Mass Reduction Cost ($ per lb.)					
0 - 2.5% Mass Reduction	MR2.5	0.00 to 0.25	0.00 to 0.25	0.00 - 0.28	Baseline
0 - 5% Mass Reduction	MR5	0.00 to 0.50	0.00 to 0.50	0.00 - 0.56	Baseline
0 - 10% Mass Reduction	MR10	0.44 to 1.18	0.44 to 1.18	0.49 - 1.32	Baseline
0 - 15% Mass Reduction	MR15	0.86 - 1.46	0.86 - 1.46	0.96 - 1.64	Baseline
0 - 20% Mass Reduction	MR20	1.40 - 2.03	1.40 - 2.03	1.56 - 2.27	Baseline
0 - 25% Mass Reduction	MR25	2.46 - 3.28	2.46 - 3.28	2.76 - 3.67	Baseline
Low Rolling Resistance Tires - Level 1 (10% Reduction)	ROLL1	5	5	5	Baseline
Low Rolling Resistance Tires - Level 2 (20% Reduction)	ROLL2	58	58	58	Previous Tech
Low Drag Brakes	LDB	59	59	59	Baseline
Aerodynamic Drag Reduction - Level 1 (10% Reduction)	AERO1	39	39	39	Baseline
Aerodynamic Drag Reduction - Level 2 (20% Reduction)	AERO2	117	117	117	Previous Tech

*Costs with reduced number of cylinders, adjusted for previously added technologies – see Appendix T for the derivation of the turbocharged, downsized engine costs.

[a] With TWC aftertreatment. Costs will increase with lean NO_x aftertreatment.

[b] Includes mass decompounding: 40% for cars, 25% for trucks.

NOTE: Midsize car: 3,500 lbs, large car: 4,500 lbs, large light truck: 5,500 lbs.

TABLE 8A.2b NRC Committee's Estimated 2020 MY Direct Manufacturing Costs of Technologies

2020 MY Incremental Direct Manufacturing Costs (2010$): NRC Estimates

		Midsize Car I4 DOHC	Large Car V6 DOHC	Large Light Truck V8 OHV	
Spark Ignition Engine Technologies	Abbreviation	Most Likely	Most Likely	Most Likely	Relative To
NHTSA Technologies					
Low Friction Lubricants - Level 1	LUB1	3	3	3	Baseline
Engine Friction Reduction - Level 1	EFR1	48	71	95	Baseline
Low Friction Lubricants and Engine Friction Reduction - Level 2	LUB2_EFR2	51	75	99	Previous Tech
VVT- Intake Cam Phasing (CCP - Coupled Cam Phasing - OHV)	ICP	35 - 41	70 - 81	35- 41	Baseline for DOHC
VVT- Dual Cam Phasing	DCP	29 - 33	67 - 76	35 - 41	Previous Tech
Discrete Variable Valve Lift	DVVL	109 - 125	158 - 182	N/A	Previous Tech
Continuously Variable Valve Lift	CVVL	55 - 63	142 - 163	N/A	Previous Tech
Cylinder Deactivation	DEACD	N/A	131	147	Previous Tech
Variable Valve Actuation (CCP + DVVL)	VVA	N/A	N/A	261 - 301	Baseline for OHV
Stoichiometric Gasoline Direct Injection	SGDI	181	273	328	Previous Tech
Turbocharging and Downsizing Level 1 - 18 bar BMEP 33%DS	TRBDS1	271 - 312	-122 to -81	877 - 958	Previous Tech
V6 to I4 and V8 to V6			-432* to -349*	779* - 891*	
Turbocharging and Downsizing Level 2 - 24 bar BMEP 50%DS	TRBDS2	172	172	289	Previous Tech
I4 to I3		-89* to -92*			
Cooled EGR Level 1 - 24 bar BMEP, 50% DS	CEGR1	199	199	199	Previous Tech
Cooled EGR Level 2 - 27 bar BMEP, 56% DS	CEGR2	343	343	579	Previous Tech
V6 to I4				-522* to -514*	
Other Technologies					
By 2025:					
Compression Ratio Increase (with regular fuel)	CRI-REG	50	75	100	Baseline
Compression Ratio Increase (with higher octane regular fuel)	CRI-HO	75	113	150	Baseline
Compression Ratio Increase (CR~13:1, exh. scavenging, DI (aka Skyactiv, Atkinson Cycle))	CRI EXS	250	375	500	Baseline
Electrically Assisted Variable Speed Supercharger	EAVS-SC	1,302	998	N/A	Baseline
Lean Burn (with low sulfur fuel)	LBRN	800	920	1,040	Baseline
After 2025:					
Variable Compression Ratio	VCR				Baseline
D-EGR	DEGR				TRBDS1
Homogeneous Charge Compression Ignition (HCCI) + Spark Assisted CI[a]	SA-HCCI				TRBDS1
Gasoline Direct Injection Compression Ignition	GDCI				Baseline
Waste Heat Recovery	WHR				Baseline
Alternative Fuels:					
CNG-Gasoline Bi-Fuel Vehicle	BCNG	6,000	6,900	7,800	Baseline
Flexible Fuel Vehicle	FFV	75	100	125	Baseline
Ethanol Boosted Direct Injection (incr CR to 14:1, 43% downsizing)	EBDI	740	870	1,000	Baseline

continued

TABLE 8A.2b Continued

2020 MY Incremental Direct Manufacturing Costs (2010$): NRC Estimates

		Midsize Car I4 DOHC	Large Car V6 DOHC	Large Light Truck V8 OHV	
Diesel Engine Technologies	Abbreviation	Most Likely	Most Likely	Most Likely	Relative To
NHTSA Technologies					
Advanced Diesel	ADSL	2,845	3,356	3,571	Baseline
Other Technologies					
Low Pressure EGR	LPEGR	125	157	157	ADSL
Closed Loop Combustion Control	CLCC	64	96	96	ADSL
Injection Pressures Increased to 2,500 to 3,000 bar	INJ	23	25	25	ADSL
Downspeeding with Increased Boost Pressure	DS	26	26	26	ADSL
Friction Reduction	FR	60	91	91	ADSL
Waste Heat Recovery	WHR	N/A	N/A	N/A	
Transmission Technologies	Abbreviation	Most Likely	Most Likely	Most Likely	Relative To
NHTSA Technologies					
Improved Auto. Trans. Controls/Externals (ASL-1 & Early TC Lockup)	IATC	46	46	46	Baseline 4 sp AT
6-speed AT with Improved Internals - Lepelletier (Rel to 4 sp AT)	NUATO-L	-12	-12	-12	IATC
6-speed AT with Improved Internals - Non-Lepelletier (Rel to 4 sp AT)	NUATO-NL	181	181	181	IATC
6-speed Dry DCT (Rel to 6 sp AT - Lepelletier)	6DCT-D	-138 to 28	-138 to 28	N/A	6 sp AT
6-speed Wet DCT (Rel to 6 sp AT - Lepelletier)	6DCT-W	-82 to 82	-82 to 82	-82 to 82	6 sp AT
8-speed AT (Rel to 6 sp AT - Lepelletier)	8AT	52 - 126	52 - 126	52 - 126	Previous Tech
8-speed DCT (Rel to 6 sp DCT)	8DCT	167	167	167	Previous Tech
High Efficiency Gearbox Level 1 (Auto) (HETRANS)	HEG1	113	113	113	Previous Tech
High Efficiency Gearbox Level 2 (Auto, 2017 and Beyond)	HEG2	183	183	183	Previous Tech
Shift Optimizer (ASL-2)	SHFTOPT	24	24	24	Previous Tech
Secondary Axle Disconnect	SAX	94	94	94	Baseline
Other Technologies					
Continuously Variable Transmission with Improved internals (Rel to 6 sp AT)	CVT	168	168	NA	Baseline
High Efficiency Gearbox (CVT)	CVT-HEG	117	117	NA	Baseline
High Efficiency Gearbox (DCT)	DCT-HEG	141	141	141	Baseline
High Efficiency Gearbox Level 3 (Auto, 2020 and beyond)	HEG3	141	141	141	Baseline
9-10 speed Transmission (Auto, Rel to 8 sp AT)	10SPD	71	71	71	Baseline
Electrified Accessories Technologies	Abbreviation	Most Likely	Most Likely	Most Likely	Relative To
NHTSA Technologies					
Electric Power Steering	EPS	82	82	82	Baseline
Improved Accessories - Level 1 (70% Eff Alt, Elec. Water Pump and Fan)	IACC1	67	67	67	Baseline
Improved Accessories - Level 2 (Mild regen alt strategy, Intelligent cooling)	IACC2	40	40	40	Previous Tech
Hybrid Technologies	Abbreviation	Most Likely	Most Likely	Most Likely	Relative To
NHTSA Technologies					
Stop-Start (12V Micro-Hybrid)	SS	261 - 336	296 - 371	325 - 400	Baseline

continued

TABLE 8A.2b Continued

2020 MY Incremental Direct Manufacturing Costs (2010$): NRC Estimates

		Midsize Car I4 DOHC	Large Car V6 DOHC	Large Light Truck V8 OHV	
Integrated Starter Generator	MHEV	1,008 - 1,160	1,008 - 1,274	1,008 - 1,329	Previous Tech
Strong Hybrid - P2 - Level 2 (Parallel 2 Clutch System)	SHEV2-P2	2,295 - 2,912	2,410 - 3,472	2,744 - 3,503	Baseline
Strong Hybrid - PS - Level 2 (Power Split System)	SHEV2-PS	2,954	3,196	N/A	Baseline
Plug-in Hybrid - 40 mile range	PHEV40	9,763 - 11,253	13,172 - 15,325	N/A	Baseline
Electric Vehicle - 75 mile	EV75	10,189 - 10,768	13,310 - 14,331	N/A	Baseline
Electric Vehicle - 100 mile	EV100	11,482	14,492	N/A	Baseline
Electric Vehicle - 150 mile	EV150	14,954	17,737	N/A	Baseline
Other Technologies					
Fuel Cell Electric Vehicle	FCEV	N/A	N/A	N/A	
Vehicle Technologies	Abbreviation	Most Likely	Most Likely	Most Likely	Relative To
NHTSA Technologies					
Without Engine Downsizing					
0 - 2.5% Mass Reduction (Design Optimization)	MR2.5	0 - 22	0 - 28	0 - 39	
2.5 - 5% Mass Reduction		0 - 66	0 - 84	0 - 116	
0 - 5% Mass Reduction (Material Substitution)	MR5	0 - 88	0 - 113	0 - 154	
With Engine Downsizing (Same Architecture)[b]					
5 - 10% Mass Reduction		151 - 322	194 - 414	270 - 567	Previous MR
0 - 10% Mass Reduction (HSLA Steel and Aluminum Closures)	MR10	151 - 410	194 - 527	270 - 721	Baseline
10 - 15% Mass Reduction (Aluminum Body)		441 - 751	567 - 965	776 - 1,320	Baseline
0 - 15% Mass Reduction (Aluminum Body)	MR15	441 - 751	567 - 965	776 - 1,320	Baseline
15 - 20% Mass Reduction		518 - 635	666 - 817	907 - 1,122	Previous MR
0 - 20% Mass Reduction (Aluminum Body, Magnesium, Composites)	MR20	959 - 1,386	1,233 - 1,782	1,683 - 2,442	Baseline
20 - 25% Mass Reduction		1,115 - 1,379	1,433 - 1,773	1,961 - 2,426	Previous MR
0 - 25% Mass Reduction (Carbon Fiber Composite Body)	MR25	2,074 - 2,765	2,666 - 3,555	3,644 - 4,868	Baseline
Mass Reduction Cost ($ per lb.)					
0 - 2.5% Mass Reduction	MR2.5	0.00 - 0.25	0.00 - 0.25	0.00 - 0.28	
0 - 5% Mass Reduction	MR5	0.00 - 0.50	0.00 - 0.50	0.00 - 0.56	Baseline
0 - 10% Mass Reduction	MR10	0.43 - 1.17	0.43 - 1.17	0.49 - 1.31	Baseline
0 - 15% Mass Reduction	MR15	0.84 - 1.43	0.84 - 1.43	0.94 - 1.60	Baseline
0 - 20% Mass Reduction	MR20	1.37 - 1.98	1.37 - 1.98	1.53 - 2.22	Baseline
0 - 25% Mass Reduction	MR25	2.37 - 3.16	2.37 - 3.16	2.65 - 3.54	Baseline
Low Rolling Resistance Tires - Level 1 (10% Reduction)	ROLL1	5	5	5	Baseline
Low Rolling Resistance Tires - Level 2 (20% Reduction)	ROLL2	46	46	46	Previous Tech
Low Drag Brakes	LDB	59	59	59	Baseline
Aerodynamic Drag Reduction - Level 1 (10% Reduction)	AERO1	37	37	37	Baseline
Aerodynamic Drag Reduction - Level 2 (20% Reduction)	AERO2	110	110	110	Previous Tech

*Costs with reduced number of cylinders, adjusted for previously added technologies – see Appendix T for the derivation of the turbocharged, downsized engine costs.

[a] With TWC aftertreatment. Costs will increase with lean NO_x aftertreatment.

[b] Includes mass decompounding: 40% for cars, 25% for trucks.

NOTE: Midsize car: 3,500 lbs, large car: 4,500 lbs, large light truck: 5,500 lbs.

TABLE 8A.2c NRC Committee's Estimated 2025 MY Direct Manufacturing Costs of Technologies

2025 MY Incremental Direct Manufacturing Costs (2010$): NRC Estimates					
		Midsize Car I4 DOHC	Large Car V6 DOHC	Large Light Truck V8 OHV	
Spark Ignition Engine Technologies	Abbreviation	Most Likely	Most Likely	Most Likely	Relative To
NHTSA Technologies					
Low Friction Lubricants - Level 1	LUB1	3	3	3	Baseline
Engine Friction Reduction - Level 1	EFR1	48	71	95	Baseline
Low Friction Lubricants and Engine Friction Reduction - Level 2	LUB2_EFR2	51	75	99	Previous Tech
VVT- Intake Cam Phasing (CCP - Coupled Cam Phasing - OHV)	ICP	31 - 36	63 - 73	31 - 36	Baseline for DOHC
VVT- Dual Cam Phasing	DCP	27 - 31	61 - 69	31 - 36	Previous Tech
Discrete Variable Valve Lift	DVVL	99 - 114	143 - 164	N/A	Previous Tech
Continuously Variable Valve Lift	CVVL	49 - 56	128 - 147	N/A	Previous Tech
Cylinder Deactivation	DEACD	N/A	118	133	Previous Tech
Variable Valve Actuation (CCP + DVVL)	VVA	N/A	N/A	235 - 271	Baseline for OHV
Stoichiometric Gasoline Direct Injection	SGDI	164	246	296	Previous Tech
Turbocharging and Downsizing Level 1 - 18 bar BMEP 33%DS	TRBDS1	245 - 282	-110 to -73	788 - 862	Previous Tech
V6 to I4 and V8 to V6			-396* to -316*	700* - 800*	
Turbocharging and Downsizing Level 2 - 24 bar BMEP 50%DS	TRBDS2	155	155	261	Previous Tech
I4 to I3		-82* to -86*			
Cooled EGR Level 1 - 24 bar BMEP, 50% DS	CEGR1	180	180	180	Previous Tech
Cooled EGR Level 2 - 27 bar BMEP, 56% DS	CEGR2	310	310	523	Previous Tech
V6 to I4				-453* to -469*	
Other Technologies					
By 2025:					
Compression Ratio Increase (with regular fuel)	CRI-REG	50	75	100	Baseline
Compression Ratio Increase (with higher octane regular fuel)	CRI-HO	75	113	150	Baseline
Compression Ratio Increase (CR~13:1, exh. scavenging, DI (aka Skyactiv, Atkinson Cycle))	CRI-EXS	250	375	500	Baseline
Electrically Assisted Variable Speed Supercharger	EAVS-SC	1,302	998	N/A	Baseline
Lean Burn (with low sulfur fuel)	LBRN	800	920	1,040	Baseline
After 2025:					
Variable Compression Ratio	VCR	597	687	896	Baseline
D-EGR	DEGR	667	667	667	TRBDS1
Homogeneous Charge Compression Ignition (HCCI) + Spark Assisted CI[a]	SA-HCCI	450	500	550	TRBDS1
Gasoline Direct Injection Compression Ignition	GDCI	2,500	2,875	3,750	Baseline
Waste Heat Recovery	WHR	700	805	1,050	Baseline
Alternative Fuels:					
CNG-Gasoline Bi-Fuel Vehicle	BCNG	6,000	6,900	7,800	Baseline
Flexible Fuel Vehicle	FFV	75	100	125	Baseline
Ethanol Boosted Direct Injection (incr CR to 14:1, 43% downsizing)	EBDI	740	870	1,000	Baseline

continued

TABLE 8A.2c Continued

2025 MY Incremental Direct Manufacturing Costs (2010$): NRC Estimates

		Midsize Car I4 DOHC	Large Car V6 DOHC	Large Light Truck V8 OHV	
Diesel Engine Technologies	Abbreviation	Most Likely	Most Likely	Most Likely	Relative To
NHTSA Technologies					
Advanced Diesel	ADSL	2,572	3,034	3,228	Baseline
Other Technologies					
Low Pressure EGR	LPEGR	113	141	141	ADSL
Closed Loop Combustion Control	CLCC	58	87	87	ADSL
Injection Pressures Increased to 2,500 to 3,000 bar	INJ	20	22	22	ADSL
Downspeeding with Increased Boost Pressure	DS	24	24	24	ADSL
Friction Reduction	FR	54	82	82	ADSL
Waste Heat Recovery	WHR	700	805	1,050	ADSL
Transmission Technologies	Abbreviation	Most Likely	Most Likely	Most Likely	Relative To
NHTSA Technologies					
Improved Auto. Trans. Controls/Externals (ASL-1 & Early TC Lockup)	IATC	42	42	42	Baseline 4 sp AT
6-speed AT with Improved Internals - Lepelletier (Rel to 4 sp AT)	NUATO-L	-11	-11	-11	IATC
6-speed AT with Improved Internals - Non-Lepelletier (Rel to 4 sp AT)	NUATO-NL	165	165	165	IATC
6-speed Dry DCT (Rel to 6 sp AT - Lepelletier)	6DCT-D	-127 to 26	-127 to 26	N/A	6 sp AT
6-speed Wet DCT (Rel to 6 sp AT - Lepelletier)	6DCT-W	-75 to 75	-75 to 75	-75 to 75	6 sp AT
8-speed AT (Rel to 6 sp AT - Lepelletier)	8AT	47 - 115	47 - 115	47 - 115	Previous Tech
8-speed DCT (Rel to 6 sp DCT)	8DCT	152	152	152	Previous Tech
High Efficiency Gearbox Level 1 (Auto) (HETRANS)	HEG1	102	102	102	Previous Tech
High Efficiency Gearbox Level 2 (Auto, 2017 and Beyond)	HEG2	165	165	165	Previous Tech
Shift Optimizer (ASL-2)	SHFTOPT	22	22	22	Previous Tech
Secondary Axle Disconnect	SAX	86	86	86	Baseline
Other Technologies					
Continuously Variable Transmission with Improved internals (Rel to 6 sp AT)	CVT	154	154	NA	Baseline
High Efficiency Gearbox (CVT)	CVT-HEG	107	107	NA	Baseline
High Efficiency Gearbox (DCT)	DCT-HEG	127	127	127	Baseline
High Efficiency Gearbox Level 3 (Auto, 2020 and beyond)	HEG3	128	128	128	Baseline
9-10 speed Transmission (Auto, Rel to 8 sp AT)	10SPD	65	65	65	Baseline
Electrified Accessories Technologies	Abbreviation	Most Likely	Most Likely	Most Likely	Relative To
NHTSA Technologies					
Electric Power Steering	EPS	74	74	74	Baseline
Improved Accessories - Level 1 (70% Eff Alt, Elec. Water Pump and Fan)	IACC1	60	60	60	Baseline
Improved Accessories - Level 2 (Mild regen alt strategy, Intelligent cooling)	IACC2	37	37	37	Previous Tech
Hybrid Technologies	Abbreviation	Most Likely	Most Likely	Most Likely	Relative To
NHTSA Technologies					
Stop-Start (12V Micro-Hybrid)	SS	225 - 275	255 - 305	279 - 329	Baseline

continued

TABLE 8A.2c Continued

2025 MY Incremental Direct Manufacturing Costs (2010$): NRC Estimates					
		Midsize Car I4 DOHC	Large Car V6 DOHC	Large Light Truck V8 OHV	
Integrated Starter Generator	MHEV	888 - 1,018	888 - 1,115	888 - 1,164	Previous Tech
Strong Hybrid - P2 - Level 2 (Parallel 2 Clutch System)	SHEV2-P2	2,041 - 2,588	2,410 - 3,086	2,438 - 3,111	Baseline
Strong Hybrid - PS - Level 2 (Power Split System)	SHEV2-PS	2,671	2,889	N/A	Baseline
Plug-in Hybrid - 40 mile range	PHEV40	8,325 - 9,672	11,189 - 13,135	N/A	Baseline
Electric Vehicle - 75 mile	EV75	8,451 - 8,963	11,025 - 11,929	N/A	Baseline
Electric Vehicle - 100 mile	EV100	9,486	11,971	N/A	Baseline
Electric Vehicle - 150 mile	EV150	12,264	14,567	N/A	Baseline
Other Technologies					
Fuel Cell Electric Vehicle	FCEV	N/A	N/A	N/A	
Vehicle Technologies	Abbreviation	Most Likely	Most Likely	Most Likely	Relative To
NHTSA Technologies					
Without Engine Downsizing					
0 - 2.5% Mass Reduction (Design Optimization)	MR2.5	0 - 22	0 - 28	0 - 39	Baseline
2.5 - 5% Mass Reduction		0 - 66	0 - 85	0 - 112	Previous MR
0 - 5% Mass Reduction (Material Substitution)	MR5	0 - 88	0 - 113	0 - 151	Baseline
With Engine Downsizing (Same Architecture)[b]					
5 - 10% Mass Reduction		151 - 315	194 - 405	264 - 558	Previous MR
0 - 10% Mass Reduction (HSLA Steel and Aluminum Closures)	MR10	151 - 403	194 - 518	264 - 710	Baseline
10 - 15% Mass Reduction (Aluminum Body)		431 - 730	554 - 938	751 - 1,279	Baseline
0 - 15% Mass Reduction (Aluminum Body)	MR15	431 - 730	554 - 938	751 - 1,279	Baseline
15 - 20% Mass Reduction		486 - 600	626 - 772	866 - 1,064	Previous MR
0 - 20% Mass Reduction (Aluminum Body, Magnesium, Composites)	MR20	917 - 1,330	1,179 - 1,710	1,617 - 2,343	Baseline
20 - 25% Mass Reduction		1,026 - 1,260	1,319 - 1,620	1,807 - 1,947	Previous MR
0 - 25% Mass Reduction (Carbon Fiber Composite Body)	MR25	1,943 - 2,590	2,498 - 3,330	3,424 - 4,290	Baseline
Mass Reduction Cost ($ per lb.)					
0 - 2.5% Mass Reduction	MR2.5	0.00 - 0 .25	0.00 - 0 .25	0.00 - 0.28	Baseline
0 - 5% Mass Reduction	MR5	0.00 - 0.49	0.00 - 0.49	0.00 - 0.55	Baseline
0 - 10% Mass Reduction	MR10	0.43 - 1.15	0.43 - 1.15	0.48 - 1.29	Baseline
0 - 15% Mass Reduction	MR15	0.82 - 1.39	0.82 - 1.39	0.91 - 1.55	Baseline
0 - 20% Mass Reduction	MR20	1.31 - 1.90	1.31 - 1.90	1.47 - 2.13	Baseline
0 - 25% Mass Reduction	MR25	2.22 - 2.96	2.22 - 2.96	2.49 - 3.12	Baseline
Low Rolling Resistance Tires - Level 1 (10% reduction in rolling resistance)	ROLL1	5	5	5	Baseline
Low Rolling Resistance Tires - Level 2 (20% reduction in rolling resistance)	ROLL2	31	31	31	Previous Tech
Low Drag Brakes	LDB	59	59	59	Baseline
Aerodynamic Drag Reduction - Level 1	AERO1	33	33	33	Baseline
Aerodynamic Drag Reduction - Level 2	AERO2	100	100	100	Previous Tech

* Costs with reduced number of cylinders, adjusted for previously added technologies – see Appendix T for the derivation of the turbocharged, downsized engine costs.

[a] With TWC aftertreatment. Costs will increase with lean NO_x aftertreatment.

[b] Includes mass decompounding: 40% for cars, 25% for trucks.

NOTE: Midsize car: 3,500 lbs, large car: 4,500 lbs, large light truck: 5,500 lbs.

9

Consumer Impacts and Acceptance Issues

INTRODUCTION

The success of the Corporate Average Fuel Economy/Greenhouse Gas (CAFE/GHG) national program depends in important ways on how consumers respond to the more fuel-efficient vehicles of the future. The vehicles that will be needed for compliance with the rules will implement fuel economy and other vehicle changes in a variety of ways that the consumer may or may not perceive, and they will have higher initial costs. How much will consumers value and be willing to pay for better fuel economy? Will consumers accept vehicle models with new fuel economy technologies? How will they trade off fuel economy improvements with other attributes? What are the prospects for alternative fuel vehicles (AFVs) and advanced technology vehicles (ATVs) in terms of cost, performance, and overall sales compared with conventional vehicles? This chapter addresses these and related questions of consumer impacts and acceptance of fuel economy technologies.

The chapter begins by looking at trends in the new vehicle characteristics over time. This provides context for the fuel economy standards and improvements in technology that have allowed for increases in both fuel economy and performance. The chapter then explores the issue of how consumers value fuel economy. A key premise of the regulation and a central part of the analysis of the costs and benefits of the new rules is that consumers undervalue fuel economy when they are considering new car purchases. This is evidenced by the net private savings for consumers in fuel costs relative to vehicle purchase price attributed to the rule (EPA/NHTSA 2012a, 62627). Despite a large body of literature on this issue, there appears to be no consensus on the extent to which consumers undervalue fuel economy or on how attitudes on this issue vary across the population. The chapter then turns to look in detail at evidence about consumer reaction to new technologies for fuel economy or to vehicles currently in the market. The chapter concludes with an assessment of the effect of the rules on affordability and sales.

TRENDS IN VEHICLE CHARACTERISTICS

Vehicle technology has been improving over time, and it is useful to look at the relative changes in characteristics during periods when fuel economy standards were increasing and when they were constant.

Evidence on Past Changes in Technology and Vehicle Characteristics

There is much data on average characteristics of the vehicle fleet over the past 40 years. This historical evidence shows trends in vehicle characteristics and technological change and indicates how those might be affected by such events as changes in fuel prices and fuel economy standards. Figure 9.1 shows the relative change in label fuel economy, the horsepower/weight ratio, and time to accelerate from 0 to 60 mph since 1975, using sales-weighted averages across the fleet. The average horsepower/weight ratio stayed relatively constant during the late 1970s while fuel economy increased rapidly. Both horsepower and weight were declining, but the ratio remained relatively constant. This was during the time when gasoline prices were relatively high and rising and the first CAFE standards were put in place.

The period from about 1985 to 2005 was a time with relatively low gasoline prices and constant CAFE standards for vehicles. There was substantial technological progress for vehicles during this period, and it appeared to go almost exclusively into better performance, such as horsepower and acceleration (EPA 2013; Knittel 2012). During this time there was a shift to larger vehicles in the on-road fleet—toward trucks and SUVs and away from cars. Average weight of vehicles increased, but Figure 9.1 shows that horsepower increased even more rapidly, so the horsepower/weight ratio increased while fuel economy stayed the same or fell slightly. Time to accelerate from 0 to 60 has been continually decreasing since 1982. According to MacKenzie and Heywood, "Ninety-five percent of vehicles sold today achieve a level of acceleration performance that beats the average from 1992,

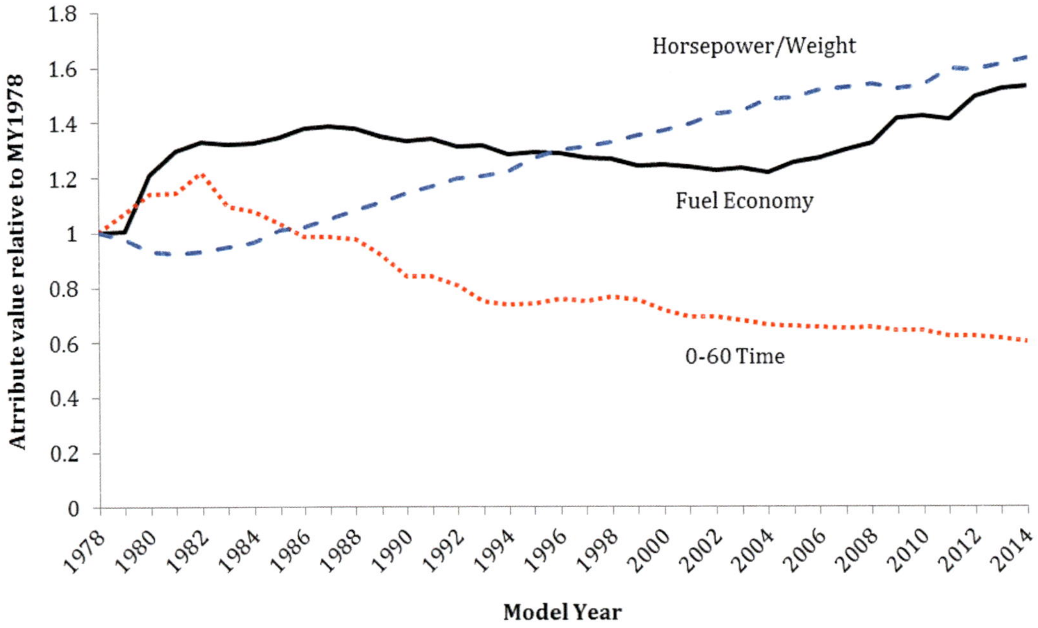

FIGURE 9.1 Label fuel economy, horsepower/weight ratio, and 0-60 mph acceleration time for MY 1978-2014 light-duty vehicles. SOURCE: EPA (2014).

and would have put them in the top 5% in 1985" (2012). However, the rate of improvement in acceleration time has been decreasing in recent years.

The final period, from 2005, reflects important changes. Real gasoline prices started to increase rapidly around 2004, and new fuel economy standards were put in place: for light-duty trucks starting in 2005, further tightened in 2008, and for cars in the 2011 MY. Both the higher gas prices and stricter fuel economy standards tend to push fuel economy of the fleet higher, and fuel economy trended upward over this period.

The difficulty in drawing conclusions from Figure 9.1 about the relationship between performance, standards, and fuel economy is that a number of different factors are changing at the same time—gasoline prices, vehicle mix, CAFE standards, and the technologies. The variables may also not be independent of each other. For example, CAFE standards themselves have the effect of accelerating technological change in the vehicle market. Some have argued that the standards spur the rate of technology change or encourage the development of new technologies that would not otherwise have been developed (Jaffe et al. 2003). Those technologies have impacts not only on fuel economy but on other vehicle attributes as well. Klier and Linn (2013) find some statistical evidence that the light truck standards the U.S. adopted in 2007 contributed to technical change during the period from 2008 to 2012. It is unclear how much of the recent trend in greater fuel economy is due to new technology going to improved fuel economy of specific vehicles and how much is due to vehicle mix (more of the more fuel-efficient vehicles being purchased). Khanna and Linn (2013) look at this issue for 2000 to 2012 and find that about half of the improvement of roughly 4 mpg in the overall fleet is due to change in vehicle mix (market shares) and the other half to changes in technologies applied to specific vehicles.

It is clear from Figure 9.1 that vehicle technology has been continually improving over time. Figure 9.2 further illustrates this point. Knittel (2011) estimated the trade-off between fuel economy and performance (measured in horsepower) in 1980 (blue dots) and 2006 (grey squares) and found that technological change has allowed for improvements in both fuel economy and horsepower over this period. The rate of improvement is approximately 2 percent per year.

Figure 9.2 shows the general trade-off between performance and fuel economy as well as the improvements in both vehicle characteristics between 1980 and 2006. Knittel (2012) extended his original analysis through MY 2011 and found continued technological progress, consistent with improvements in fuel economy of 1.97 and 1.51 percent per year from 2006 to 2011 for passenger cars and light-duty trucks, respectively, holding other attributes constant. These technology improvements provided opportunities for increases in both horsepower and fuel economy. A combination of market choices and regulations is likely to continue to determine the allocation of future technology change between performance and fuel economy.

Improvements in technology have also led to greater reliability. Figure 9.3 shows the results of a vehicle depend-

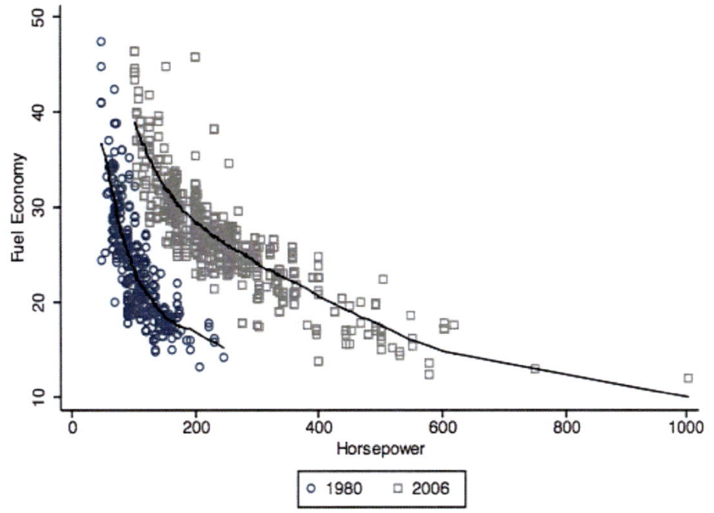

FIGURE 9.2 Fuel economy in miles per gallon (two-cycle certification CAFE) vs. horsepower, passenger cars in 1980 (blue dots) and in 2006 (grey squares).
SOURCE: Knittel (2011).

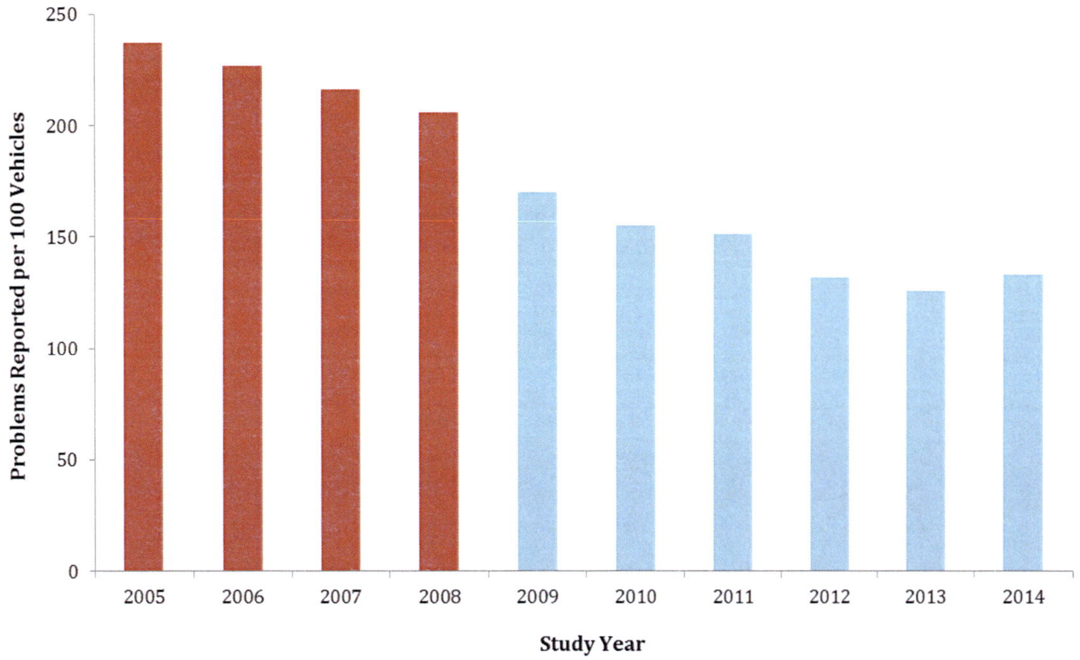

FIGURE 9.3 Results of J.D. Power U.S. Vehicle Dependability Study, in which owners of 3-year-old vehicles report problems they have experienced with their vehicles. Reported problems range from safety issues that required repair to consumer dissatisfaction with vehicle attributes. The figure colors change in 2008 to indicate a change in methodology in that year, preventing comparisons between the two periods, though the trend remains consistently downward.
SOURCE: Data from J.D. Power (2013).

ability study from 2013 that covers the years 2005-2013. Over this time period, owners of 3-year-old vehicles reported fewer problems in each subsequent year. Some of this increased reliability may be due to changes made to meet the longer warranty requirements for emissions control equipment and enhanced systems robustness with tightened onboard diagnostic monitoring requirements. In large part because vehicles are more dependable, they last longer in the fleet. The average age of both cars and trucks increased over the last 20 years, with the average age of vehicles increasing from 8.4 years in 1995 to 11.4 years in 2013.

Despite the many improvements in vehicle attributes and performance, the price of a new vehicle has risen relatively slowly over the last 20 years, increasing only about 5 percent since 1996 according to the Bureau of Labor Statistics (BLS 2014). In addition to the small increase in purchase price for new cars, there has been a shift in the variable versus fixed cost for driving new cars. Figure 9.4 compares variable and fixed costs of driving 15,000 miles, which is roughly a year of driving. Variable costs include fuel and maintenance costs and fixed costs include insurance, license, registration, taxes, depreciation, and finance charges. In the 1990s, declining gas prices and increases in fixed costs led to a declining share of total cost of driving attributed to fuel costs (or variable costs). However, in recent years, variable costs as a share of the total have been rising as a result of higher gasoline prices. Gas prices will continue to vary, increasing or decreasing the impact of fuel costs on the variable costs of driving.

Gasoline costs are likely to play an important role in how smoothly the CAFE standards are implemented in the coming years. There is evidence that consumers are more likely to purchase fuel-efficient vehicles when gasoline prices are high and less likely when gasoline prices are low. Figure 9.5 shows the trends in real gasoline prices, actual fuel economy, and the fuel economy standards for cars and trucks over time. Gasoline prices varied over the period but were generally high in the early 1980s, decreasing beginning in 1984, and were flat in real terms between 1988 and 2004. Since 2004, gasoline prices have trended upward except during the recession starting in 2008 and in the recent period starting at the end of 2014. Figure 9.5 also shows that the fuel economy standards themselves were relatively unchanged during the period from 1985 to 2011 (2004 for trucks) when real gasoline prices remained relatively low.

The actual fuel economy data in Figure 9.5 start in 1990 and show that fuel economy remains fairly flat through the 90s, starting to trend up when real gasoline prices begin to rise around 2004. Actual fuel economy continues to increase through the rest of the decade both because gas prices were sharply increasing until 2009 and because of increasing fuel economy standards for trucks starting in 2005 and cars starting in 2011. Some of the increases in fuel economy before 2008 may have been in anticipation of the standards as automakers overcomplied to bank credits for later use. Actual fuel economy is above the standards in nearly all of the years for this period except for trucks beginning in 2013. If real gasoline prices fall and remain low in the future and consumers choose larger vehicles or less-efficient used vehicles, the new, stronger CAFE/GHG standards may not deliver as much benefit as originally predicted, due to reduced sales of the more costly low-fuel-consuming vehicles.

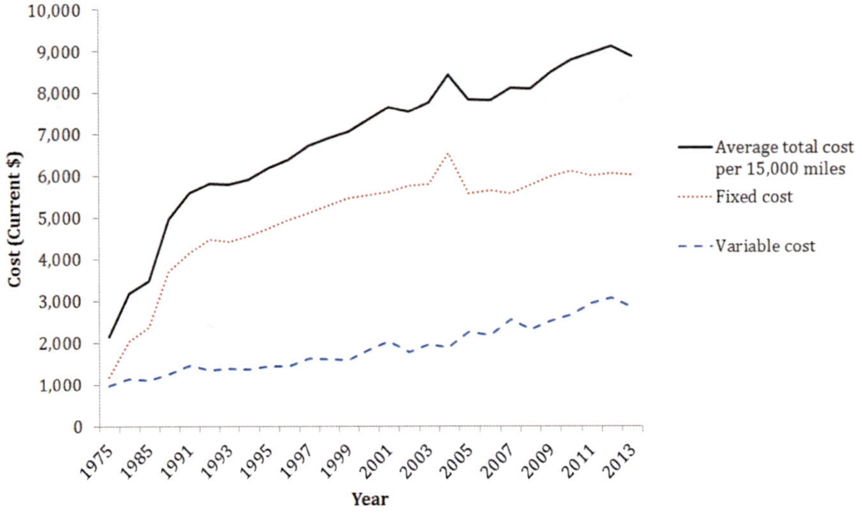

FIGURE 9.4 Average annual cost of driving a new car. The costs are an annual average over the first 5 years, assuming 15,000 miles driven per year, in current dollars. Variable costs include gasoline, maintenance, and tire costs. Fixed costs include insurance, taxes, depreciation, finance charges, license, and registration.
SOURCE: American Automobile Association (2013).

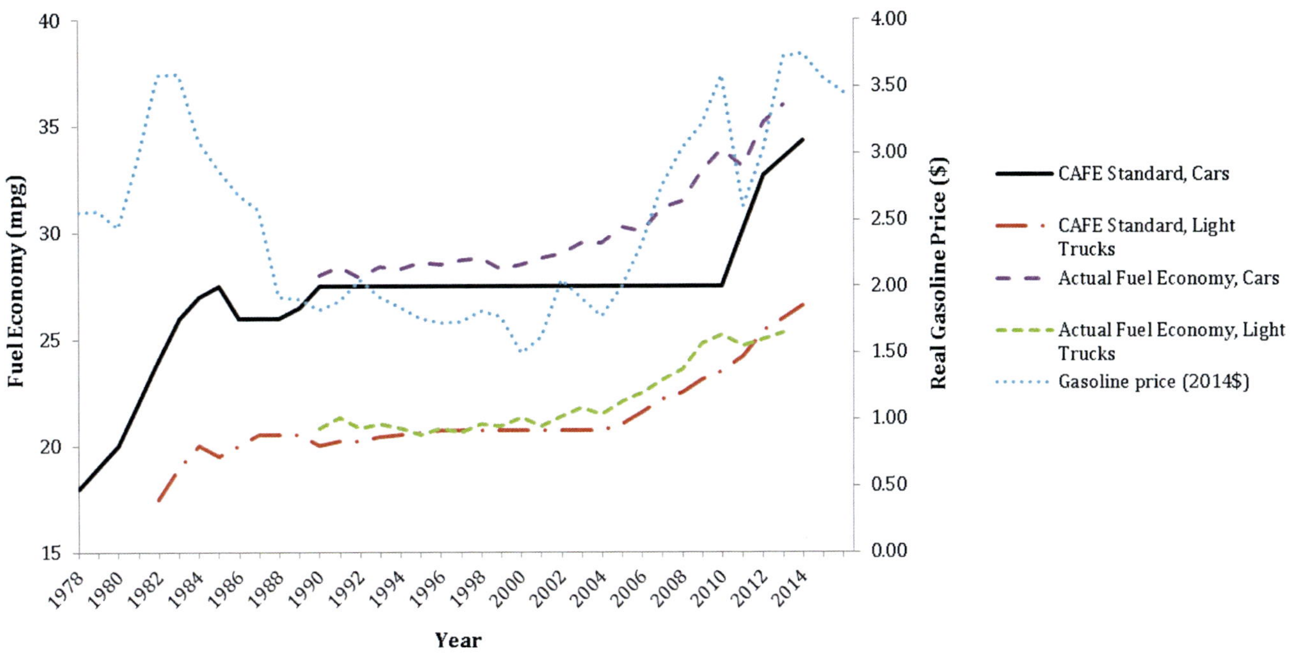

FIGURE 9.5 Fuel economy standards and actual fuel economy of cars and trucks by year plotted against real gasoline prices in 2014 dollars. All fuel economy standard and actual values are the certification fuel economy values.
SOURCE: DOT (2014); BTS (2014); EIA (2014).

Recent Changes in Fuel Economy

As shown in Figure 9.5, year-over-year improvements in fuel economy have occurred since about 2004 and will continue to occur at an accelerated rate due to stricter standards. Figure 9.6 illustrates some of the improvement that began in earnest from 2007 to 2010, focusing on the percent of vehicles in different miles per gallon categories. The percent of vehicles above 23 mpg increased from 14 percent to 50% percent in those 12 years.

Improvements in fuel economy were achieved across segments from 2009 to 2014, shown in Figure 9.7. Each segment showed a shift to a greater number of more fuel-efficient vehicle models in that period, though the study does not recognize the difference between SUVs and CUVs, nor does it describe how the fuel economy improvements were achieved.

In summary, technology used in vehicles has dramatically improved over the last three decades. The potential to trade-off fuel economy, horsepower, and other attributes has not remained fixed over time; innovation has pushed out the technology frontier for improvements to both fuel economy and other attributes. In recent years, some of this improvement has gone into better fuel economy or reliability, but much has gone to improving other vehicle attributes such as horsepower and acceleration. Fuel economy standards have impacted vehicle characteristics, but so have consumer preferences and gasoline prices. How consumers value fuel economy and other attributes is a critical question for the Agencies in setting standards and for vehicle manufacturers in selling vehicles. The economic theories attempting to understand consumer behavior are explained in the following sections, as are consumer responses to changing vehicle characteristics.

CONSUMER VALUATION OF FUEL ECONOMY: THE ENERGY PARADOX?

How consumers value fuel economy when they purchase a new or used car is critically important to evaluating the benefits and costs of fuel economy standards. The standards are essentially set based on the Agencies' assessment of what technologies will be available and technically feasible, as long as the benefit of the entire rule exceeds the costs. Standard setting is further discussed in Chapter 10. Key considerations in understanding the standards' impacts on consumers, automakers, and the country are whether or not the market undervalues future fuel savings relative to their expected full-lifetime discounted economic value, how that impacts the sales of vehicles with varying fuel economy, and how that bears on the need for standards, as well as the costs and benefits of these standards.

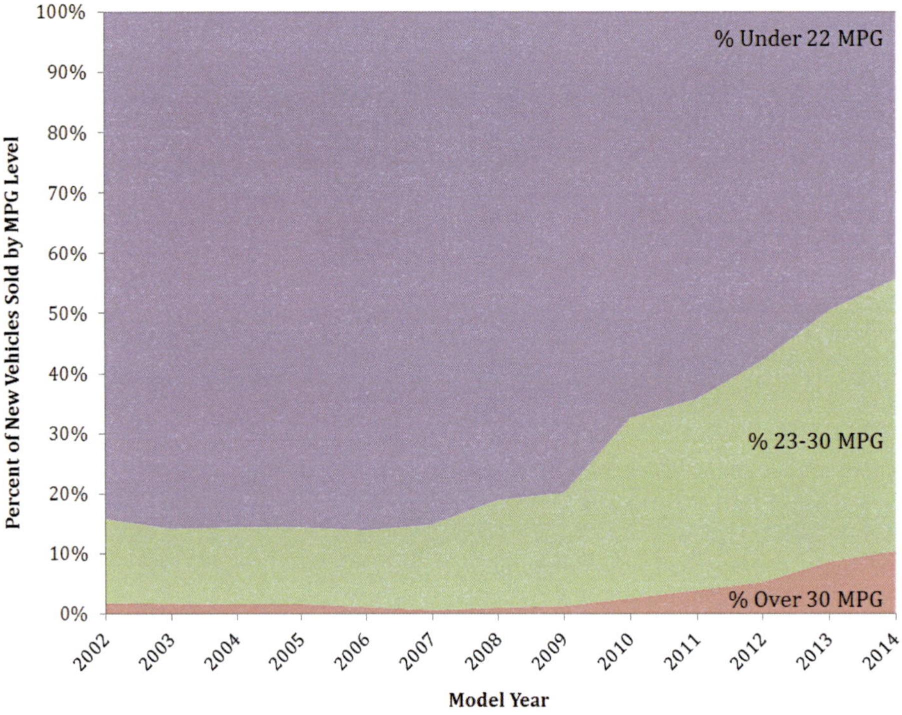

FIGURE 9.6 Percent of passenger vehicles in different label fuel economy categories by year. The percent of vehicles over 23 mpg increased from 14 percent to over 50 percent between 2002 and 2014.
SOURCE: Committee-generated, using data from Consumer Federation of America (2014).

What Is the Energy Paradox?

The "energy paradox" describes the seeming failure of markets for energy-using durable goods to adopt apparently cost-effective, energy-efficient technology (see Gillingham and Palmer 2013; Allcott and Greenstone 2012; Jaffe and Stavins 1994; Sanstad and Howarth 1994). In well-functioning markets, energy-efficient technologies should be applied up to the point where the cost of the last technology adopted equals the present value of the energy it will save. The existence of a substantial technological potential for cost-effective energy efficiency improvement that has not been implemented in the market contradicts the premise of a well-functioning market, hence the paradox. If the energy paradox exists, there are net energy savings for consumers, savings they would value, that are not being realized. Whether or not an energy paradox actually exists is the subject of continuing debate.

The Final Rule for the 2017 and later model year standards asserts that the rule will achieve cost-effective energy savings for car-buyers, implying the existence of an energy paradox:

> Although the agencies estimate that technologies used to meet the standards will add, on average, about $1,800 to the cost of a new light duty vehicle in MY 2025, consumers who drive their MY 2025 vehicle for its entire lifetime will save, on average, $5,700 to $7,400 (7 and 3 percent discount rates, respectively) in fuel, for a net lifetime savings of $3,400 to $5,000. (EPA/NHTSA 2012a, 62627)

Whether such favorable results can be achieved by the standards not only affects their costs and benefits as reflected in consumers' satisfaction with new vehicles but also affects manufacturers' revenues, profits, and employment by virtue of the demand for new vehicles. If the Agencies' estimates are correct and consumers perceive an increase in the value of an average new car of more than $5,000 at a cost of only $1,800, sales should increase. On the other hand, if consumers substantially undervalue future fuel savings (or if the Agencies' estimates of the costs are incorrect), a price increase for fuel economy technologies would lead to decreased sales. If the market for fuel economy is already functioning optimally, consumers will be less satisfied, and sales and industry profits will be reduced. If, however, there are inefficiencies in the fuel economy market, public policy could save consumers money and increase industry profits (see Fischer et al. 2007; Allcott, Mullainathan, and Taubinsky 2012).

When comparing vehicles, the rational consumer of economic theory equates the additional cost of a more fuel-efficient vehicle to the present value of the future fuel

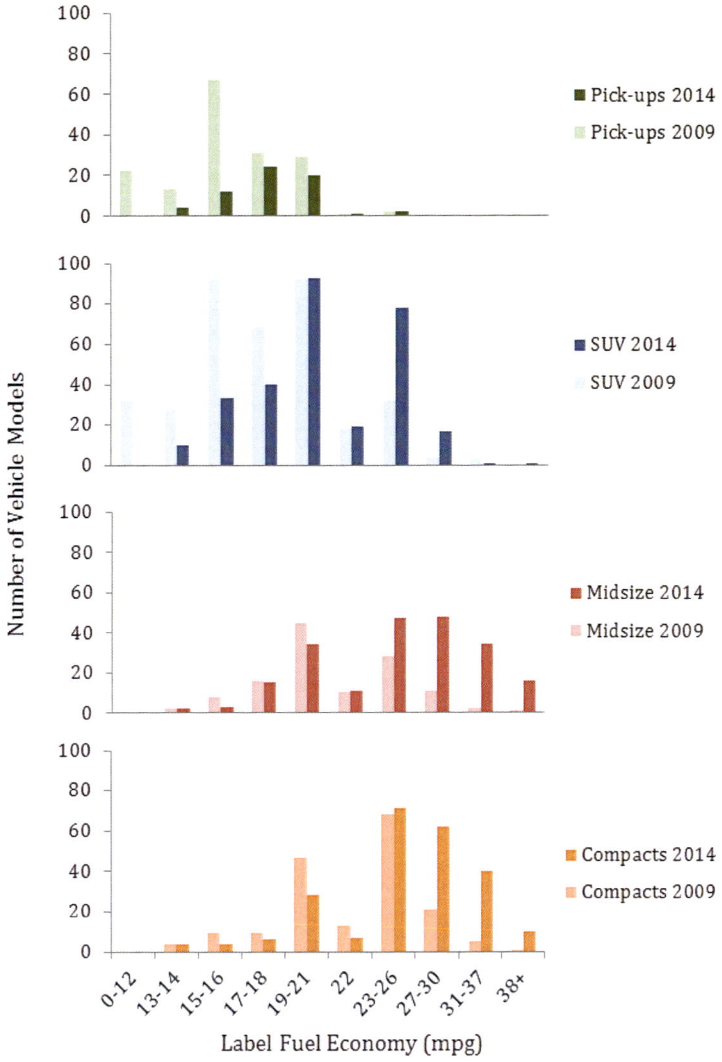

FIGURE 9.7 Number of models in several categories of label fuel economy. Between 2009 and 2014, there was a greater choice of vehicle models with better fuel economy across a variety of vehicle classes. Shown is the number of vehicle models available in 2014 and in 2009 by fuel economy range for both midsize vehicles and SUVs as an example.
NOTE: Compacts include small station wagons, and midsize includes midsize station wagons.
SOURCE: Data from Consumer Federation of America (2014).

savings it would provide, other things being equal. If consumers acted based on this model, then they should obtain the private benefits of fuel economy technologies without regulations. Mathematically, the present value, V, of future savings is the sum over the vehicle's useful lifetime, L, of the product of the price of fuel in year i, P_i, miles traveled in year i, M_i, and the difference in fuel consumption per mile (1/E = 1/miles per gallon) between the lower (1) and the higher (2) fuel economy vehicle, $(1/E_1 - 1/E_2)$, converted to present value at discount rate r.

$$V = \sum_{i=1}^{L} \frac{P_i M_i \left(\frac{1}{E_1} - \frac{1}{E_2} \right)}{(1+r)^i}$$

There is no doubt that typical consumers do not conform precisely to the fully informed, economically rational model. Obtaining accurate information about fuel economy improvements and costs, especially within vehicle packages, may be difficult if not impossible. Also, in calculating costs and benefits, the CAFE/GHG standards assume technologies are applied to fuel economy rather than performance, though in the real market, consumers will value and choose between many different attributes, some of which may be competing. While the above calculation may be beyond the capabilities of many vehicle buyers, they might develop accurate intuition based on their experiences or obtain the estimate from another source, such as a website or a mobile application. The question is how great the differences are and whether they cause important deviations from the ideal model in ways that fuel economy standards might either correct or exacerbate.

Undervaluation of fuel economy improvement creates what Herrnstein et al. (1993) termed an "internality," a welfare loss that consumers impose on themselves when they fail to make optimal decisions.[1] Internality inefficiencies affect only the efficiency of energy-using durable goods and not their utilization.[2] This is in contrast to externalities associated with energy use, such as carbon dioxide emissions, which imply both underinvestment in energy efficiency and excessive consumption of energy services (e.g., vehicle travel). If both energy efficiency investment inefficiency internalities and energy use externalities are present, then it would be appropriate to implement a combination of policies targeting vehicle fuel economy, such as the CAFE/GHG standards and externality taxes. The distinction between externalities and energy efficiency investment inefficiencies is important because it provides a theoretical basis for economically efficient fuel economy regulations.

A fuel tax could be a policy means to correct a market externality, but it might not correct market inefficiencies in investments in fuel economy. If consumers undervalue fuel economy, then a tax on fuel equal to the marginal externality costs of fuel use would not lead to a social-welfare-maximizing market outcome for fuel economy. This is because consumers would continue to undervalue the savings (fuel cost plus tax) due to improved fuel economy. Under such conditions, Allcott et al. (2012) show that a combination of an externality tax on fuel and subsidies to efficient vehicles can maximize social welfare. If properly designed, a "feebate" consisting of a fee on inefficient vehicles with rebates to efficient vehicles can be an efficient internality tax. Likewise, an efficiency regulation that induces an appropriate shadow price on energy efficiency combined with a tax on petroleum can also be an economically efficient solution. If markets systematically underinvest in fuel economy, regulatory standards could be part of an economically efficient solution.

[1] Under the loss aversion model presented subsequently, the failure to optimize is in fact only a failure to optimize based on the expected value of fuel savings.

[2] Energy efficiency indirectly affects vehicle utilization via the energy cost per mile driven, but an undervaluation of fuel economy internality does not directly affect vehicle utilization.

Why an Energy Paradox?

Numerous potential explanations for the energy paradox have been proposed. Market inefficiencies can be caused by existing regulations, asymmetric information, transaction costs, inefficient capital markets, bounded rationality,[3] and deviations from perfect market assumptions (Sanstad and Howarth 1994). Some explanations focus on the difficulty of obtaining accurate information and limitations on consumers' cognitive skills. More recently, behavioral psychologists have added behavioral anomalies to the list of possible explanations causing deviations from the economic model of the rational, utility-maximizing consumer (see Kahneman 2011). Researchers have identified several systematic behaviors that cause decision utility (the satisfaction consumers maximize when they make a choice) to differ from experienced utility (the satisfaction consumers receive when experiencing the consequences of their choice) (Gillingham and Palmer 2013).

Loss aversion, the most firmly established theory of behavioral economics, has also been proposed as a possible explanation for the energy paradox (Greene et al. 2010). In theory, the rational economic consumer possesses all the information and computational skills necessary to make an optimal choice. Yet, as Jaffe and Stavins (1994) pointed out, markets may inadequately provide information to consumers creating uncertainty about the values of different products, about a vehicle's fuel economy, for example. Behavioral psychologists have demonstrated that human beings faced with a risky bet typically consider potential losses to be twice as important as potential gains (Kahneman 2011). Paying a higher purchase price to obtain future fuel savings is, in fact, a risky bet (Greene et al. 2010). What a vehicle's fuel economy will be in actual use,[4] what the price of fuel will be in the future, how long a vehicle will last, and how many miles it will be driven are all uncertain at the time of purchase. These uncertainties make the value of future fuel savings a probability distribution rather than a fixed quantity. By quantifying the uncertainty about future fuel savings, Greene (2011) showed that loss aversion alone could explain undervaluing future fuel savings by a factor of two or more relative to expected value.

A key property of loss aversion is that it is context-dependent. That is, if consumers do not perceive their choices as risky bets, loss aversion does not come into play. This could be relevant to evaluating fuel economy regulations, where the context will be a choice between two vehicles,

[3] The concept of bounded rationality acknowledges that not all consumers will have the technical skills or the time needed to make optimal decisions even if all the necessary information is available to them.

[4] Rated fuel economy values are estimates of the average fuel economy for all drivers. The fuel economy an individual driver gets will vary with driving style, traffic conditions, temperature, terrain, and more. Lin and Greene (2011) found that while EPA label miles per gallon estimates were relatively unbiased predictors of fuel economy estimates reported by individuals, the variance was quite large: a 2 standard deviation interval was approximately +/– 30 percent.

both of which are more fuel efficient and more expensive, rather than a choice between two vehicles, one of which has lower fuel economy and a lower price and the other of which has higher fuel economy and a higher price. If a consumer faces the choice of paying a few thousand dollars more for a hybrid vehicle that promises much better fuel economy than a nearly identical conventional vehicle, the choice may be seen as a risky bet. On the other hand, if fuel economy regulation raises the price and fuel economy of every vehicle in the market, consumers might not perceive the marketwide increase in fuel economy as a risky bet, so loss aversion would not apply. As noted above, standards will not remove all uncertainty from the consumer decision, but the uncertainty surrounding whether to pay for fuel economy technology will be reduced as there will only be higher fuel economy, higher priced choices. Uncertainty associated with being in the new vehicle market as opposed to the used vehicle market will also decrease as used vehicles improve in fuel economy. Scientific research in this area is at a very early stage, however, and there is as yet no consensus on how the theories of behavioral economics should be applied to consumers' decisions about fuel economy. There are many more statistical and econometric analyses of consumers' fuel economy choices, yet the evidence on the existence of the energy paradox is inconclusive.

Empirical Evidence of How Consumers Value Fuel Economy

Much of the prima facie evidence for the existence of an energy paradox comes from engineering–economic assessments of the cost and efficiency potential of energy technologies. Past NHTSA rulemakings (e.g., EPA/NHTSA 2012a), past NRC evaluations (e.g., NRC 2002, 2011), and other studies (see, e.g., Greene and DeCicco 2000, who review older studies) found substantial potential for fuel economy increases for which future fuel savings exceed the upfront cost but which the market had not adopted. However, none of these studies account for the opportunity cost of using technologies to improve fuel economy rather than to improve the acceleration performance or increase the size of vehicles. The potential significance of opportunity cost to the consumer is discussed later in this chapter. The impact of the possible opportunity cost on the overall cost and benefit of the regulation is discussed in Chapter 10.

During its information-gathering process, the committee found that auto manufacturers perceive that typical consumers would pay upfront for only one to four years of fuel savings, a fraction of the lifetime-discounted present value. Short payback periods imply high discount rates for fuel economy, which may indicate undervaluation of fuel economy relative to its net present value, though, as discussed below, there are other explanations for the apparent short payback periods. The academic literature, on the other hand, is nearly evenly divided between supporting and contradicting the existence of an energy paradox for motor vehicle fuel economy.

Despite the economic importance of the fuel economy standards, there has been relatively little research aimed at understanding how consumers use fuel economy information when making vehicle purchase decisions. In apparently the only study of its kind, Turrentine and Kurani (2007) conducted in-depth interviews with 57 California households across nine "life-style sectors" about their entire histories of car-buying decisions and the role that fuel economy played in their choices. They began by compiling a history of the households' car ownership decisions and asking about the reasons for buying and selling the vehicles, eventually guiding the respondents to talk about fuel economy. They found that none of the 57 California households they interviewed had ever analyzed their fuel costs in a systematic way in their vehicle purchases. Very few kept track of their gasoline costs over time or factored them into household budgets. The researchers concluded that

> One effect of this lack of knowledge and information is that when consumers buy a vehicle, they do not have the basic building blocks of knowledge assumed by the model of economically rational decision making, and they make large errors estimating gasoline cost savings over time. (Turrentine and Kurani 2007, 1213)

Lack of knowledge about fuel savings could be overcome by expert services if consumers' willingness to pay for those services exceeded their costs (or if the service were free). Yet Turrentine and Kurani's (2007) research found that the consumers in their survey did not take advantage of such services when making purchase decisions. Sallee (2013) concluded that average consumer losses due to making fuel economy choices based on incomplete information would be $100 to $300 per vehicle. Although this is not large enough to explain all of the undervaluation of fuel economy technologies, it could be one of several contributing factors.

Consumer ability to intuitively estimate fuel cost savings when comparing automobiles was tested by Allcott (2013) using a nationally representative computer-aided survey of 2,100 U.S. households. Consumers' estimates of the fuel cost differences between the vehicle they chose to purchase and a second best alternative were divided by the "true" fuel cost difference to produce a "valuation ratio." However, the "true" cost differences were calculated based on EPA label ratings. This raises two difficulties for interpreting the results. First, any given consumer's actual on-road fuel economy will differ from the EPA rating due to driving style, traffic conditions, trip lengths, speed, temperature, terrain, and other factors. Second, although there is empirical evidence concerning the bias of the EPA label ratings used until 2008 (see Lin and Greene 2010; Hellman and Murrell et al. 1984), there has been inadequate empirical measurement of the bias of the ratings revised in 2008 versus real-world experiences of motorists. Thus, the "true" fuel cost estimate may or may not be a biased estimate of the mean fuel costs for all consumers and will, in general, not be an accurate estimate of the fuel costs of a particular consumer. The results from Allcott (2013) indicate that consumers undervalue fuel economy, on average by 12 percent, and that 50 percent of consumers undervalue it by 30 percent or more. A complication is that the fuel economy and fuel cost differences between consumers' first and second choice vehicles were often small. Also, because the EPA ratings are not likely to be the actual fuel economy achieved by any given respondent, the accuracy of consumers' estimates is confounded with the accuracy of Allcott's measure of "true" fuel costs.

Larrick and Soll (2008) found that consumers significantly undervalued fuel economy improvements for low-mpg vehicles and overvalued improvements for high-mpg vehicles. This "mpg illusion" caused consumers to value miles per gallon differences equally, regardless of the level of fuel economy. Thus, the 10 mpg difference between a 10 mpg and a 20 mpg vehicle tended to be considered equivalent to the 10 mpg difference between a 30 mpg and a 40 mpg vehicle, even though the fuel savings would be six times as great for the 10 mpg to 20 mpg difference. The mpg illusion implies that car buyers will overvalue fuel economy increases for high-mpg vehicles relative to low-mpg vehicles, but it does not necessarily imply a general underutilization of fuel economy technologies. Allcott (2013) also found that his survey results supported the existence of an mpg illusion, though the magnitude of the effect was small.

A recent survey of the econometric literature from 1980 to 2009 found it nearly evenly split between studies that support the hypothesis of rational economic behavior toward automotive fuel economy and those that support a significant undervaluing of fuel economy by car buyers (Greene 2010). Summarizing the evidence from econometric studies of vehicle choice, Helfand and Wolverton (2011) concluded that 12 studies found significant undervaluing of fuel economy relative to its expected value, 8 studies concluded that consumers were close to the expected value, and 5 studies found consumers significantly overvalued fuel economy (Table 9.1). The authors attributed the widely varying results to difficulties in statistical inference in the complex, multiattribute vehicle choice decision and to the lack of a clear theoretical consensus about how consumers actually do evaluate fuel economy when comparing vehicles. Further complicating the issue is the fact that consumers are heterogeneous in their willingness to pay for fuel economy due to differing rates of vehicle use and differing discount rates among other factors (Allcott and Wozny 2014).

Studies published since the literature reviews have also reached mixed conclusions. The effects of gasoline price on consumer willingness to pay for vehicles of different fuel economies were analyzed by Busse et al. (2013). As the authors state, their analysis estimates the short-run effect of gasoline prices on consumer willingness to pay in that it is based on consumer comparisons among existing makes and models. It did not consider the long-run effect of manufacturers redesigning vehicles and adding fuel economy technologies. Testing a variety of assumptions about vehicle usage, lifetime, and price elasticities of supply and demand, the study inferred implicit discount rates for future fuel costs ranging from 20.9 percent per year to 6.8 percent per year, with an overall average of 6.8 percent, and averages for new and used vehicles of 2.6 percent and 8.8 percent, respectively. The mean values imply discount rates consistent with consumers' borrowing costs over the period of the study. The authors interpreted the estimates as not supporting undervaluation of future fuel costs by car buyers.

Whether consumers value future fuel savings at more or less than their discounted present value was evaluated by Allcott and Wozny (2014) using data on transaction prices of 83 million used-vehicle sales from 1999 to 2008. Discounted present value was calculated based on each vehicle's estimated remaining lifetime and miles. A discount rate of 6 percent was used, derived from a weighted average of actual automobile loan interest rates and the opportunity cost

TABLE 9.1 Studies on Consumer Valuation of Fuel Economy

Undervaluation of Fuel Economy	"About Right" Valuation of Fuel Economy	Overvaluation of Fuel Economy
Allcott and Wozny (2009)[a]	Brownstone et al. (1996)	Brownstone et al. (2000)
Arguea et al. (1994)	Dasgupta et al. (2007)	Cambridge Econometrics (2008)
Berry et al. (1995)	Espey and Nair (2005)	Gramlich (2008)
Bhat and Sen (2006)	Goldberg (1995)	Sawhill (2008)
Busse et al. (2009)	Goldberg (1998)	Vance and Mehlin (2009)
Eftec (2008)	Klier and Linn (2008)	
Fan and Rubin (2010)	McManus (2007)	
Feng et al. (2005)	Sallee et al. (2010)	
Fifer and Bunn (2009)		
Kilian and Sims (2006)		
Langer and Miller (2008)		
Train and Winston (2007)		

[a] This is an earlier version of Allcott and Wozny 2014.
SOURCE: Helfand and Wolverton (2011).

of capital (for purchases made with cash). How consumers value future fuel costs was inferred based on the effects of changes in gasoline price on vehicle transaction prices. A key factor is consumers' expectations regarding future fuel prices. Two alternatives were considered: (1) estimates based on oil market futures prices and (2) the assumption that current price was the best predictor of future prices. The latter "no change" assumption is consistent with the findings of an extensive analysis of national survey data on consumers' expectations about future gasoline prices from 1993 to 2010 (Anderson, Kellogg, and Sallee 2011). It is also consistent with the observation that oil prices from 1947 to 2008 are not statistically different from a random walk without drift (Hamilton 2009; Alquist and Killian 2010; Alquist et al. 2010). Allcott and Wozny found that, on average, consumers undervalued future fuel costs by about 45 percent using the better supported assumption and 24 percent if price forecasts based on futures markets were used. Consumer response to fuel price changes may be more complicated. Kilian and Sims (2006) found that consumers appear to fully value future fuel costs when gasoline prices increase but virtually ignore them when gasoline prices decrease.

Implicit consumer discount rates for the fuel savings of hybrid vehicles were estimated for the United States by Gallagher and Muehlegger (2011). The estimation assumed static fuel price expectations (consistent with fuel prices following a random walk). It also assumed that hybrid buyers would not count fuel savings over the entire expected life of the vehicle. The latter assumption was justified on the basis of either consumer short-sightedness or the failure of used car markets to attach any value to the hybrids' higher fuel economy. Of course, neither assumption is consistent with the model of the economically rational consumer. Five different model formulations were tested as well. The estimated discount rates ranged from a low of 13.0 percent for an assumed 5-year vehicle life to 41.8 percent for an 8-year life. Since U.S. light-duty vehicles typically last 14 years or more, these results would imply very high discounting of future fuel savings over the full expected life of a typical vehicle (NHTSA 2006; Davis et al. 2013, Table 3.12).

There is also empirical evidence supporting loss aversion as a possible cause of the energy paradox. Greene (2011) showed that if consumers accurately perceived the upfront cost of fuel economy improvements and the uncertainty of fuel economy estimates, the future price of fuel, and other factors affecting the present value of fuel savings, the loss-averse consumers among them would appear to act as if they had very high discount rates or required payback periods of about 3 years. Weighing potential losses (the possibility that upfront cost would exceed uncertain future savings) twice as much as potential gains results in a similar undervaluing of future fuel savings relative to their expected value. Four nationwide random sample surveys of 1,000 respondents each, conducted between 2004 and 2013, showed that consumers considered fuel economy ratings and future fuel prices to be very uncertain (Greene, Evans, and Hiestand 2013). The surveys also produced consistent evidence that consumer willingness to pay for fuel savings implies average payback periods of 2-3 years. Some respondents were asked about their willingness to pay for a given annual fuel savings while others were asked about the fuel savings that would justify a given price increase. The distribution of payback periods was very similar regardless of which way the question was posed. Nearly identical distributions of payback periods were observed for respondents who typically purchased new vehicles and those who purchased used, indicating no difference in the way fuel economy is valued in the two markets. The responses showed a wide dispersion of implied payback periods that were not strongly correlated with the attributes of the respondents, a result that is consistent with evidence about loss aversion from behavioral psychology studies. The payback periods calculated in these studies are also consistent with automobile manufacturers' statements to the committee about consumer willingness to pay for fuel economy improvements.

Public perception of the standards should be weighed along with other evidence concerning the nature of the market for fuel economy. If fuel economy standards force manufacturers to produce vehicles with too much fuel economy, too little performance and size, and too high of a price, one might expect the public to disapprove of the standards. On the other hand, if consumers are satisfied with the new, higher fuel economy vehicles (as, for example, the theory of loss aversion would imply) one would expect them to approve of the standards. There is a great deal of survey research on consumers' opinions of the desirability of raising fuel economy standards and the results are remarkably consistent over time.

Public Perception of Fuel Economy Standards

The public's perception of the CAFE standards and support for raising the standards has been highly positive for the past 25 years. Seven surveys conducted between 1988 and 1997 by groups including the American Automobile Association and the Alliance to Save Energy reported public support for raising fuel economy standards ranging from 72 percent to 95 percent (Greene and Liu 1998, Table 3). Most surveys did not make statements about the pros or cons of the fuel economy standards before asking respondents' opinions of them; however, those that did showed decreases in support when costs of the standards were asserted in framing the question. More recent polls continue to reflect high levels of public support for fuel economy standards (Table 9.2).

The consistently high support for raising fuel economy standards is noteworthy. It is possible that after almost 35 years of experience with fuel economy standards consumers do not understand that they pay more for higher fuel economy vehicles or sacrifice other vehicle attributes. Manufacturers generally expressed concern that consumers would not be willing to pay for the price increases necessary

TABLE 9.2 Recent Surveys Show High Public Support for Fuel Economy Standards

Study	Result
Knowledge Networks for Program on International Policy Attitudes, January 2005	77% supported higher fuel economy standards even when they were told, ". . . it would cost more to buy or lease a car."
Pew Research Center, 2006	86% favored requiring better fuel efficiency for cars, trucks and SUVs while only 12% opposed
The Mellman Group (for Public Opinion Strategies), November 2007	86% answered "favored" or "strongly favored" to the following question: "Do you favor or oppose requiring the auto industry to increase fuel efficiency, that is, increase the average miles per gallon of gasoline that cars, trucks and SUVs get? 71% answered strongly favored
Gallup, March 2009	80% of Americans said they favored higher fuel efficiency standards for automobiles, with 19% opposed.
Consumer Federation of America, May 2011	"Do you support or oppose the federal government requiring auto companies to increase the fuel economy of the vehicles they manufacture?" Among Republicans, 70% chose somewhat or strongly support; support by Independents and Independents leaning Republican was 69%; support among Democratic voters was 81% and 85% of Independents leaning Democratic supported fuel economy requirements.
Pew Clean Energy Program, July 2011	82% of respondents supported increasing fuel economy standards to 56 mpg by 2025
Consumer Reports, October 2011	77% of the public supported increased fuel economy standards in a U.S. nationally representative probability poll

to cover the increased costs of vehicles that could meet the 2025 standards. It is also possible that consumers do understand that standards increase the price of vehicles as well as their fuel economy and prefer the higher fuel economy outcome. In either case the overwhelmingly positive and consistent public support for fuel economy standards casts further doubt on the fully informed, economically rational model as applied to consumers' fuel economy choices.

Summary of Consumer Valuation of Fuel Economy

How markets actually value increases in new vehicle fuel economy is critical to evaluating the costs and benefits of fuel economy and GHG standards. Unfortunately, the scientific literature does not provide a definitive answer at present. Academic studies that have analyzed the evidence on consumer willingness to pay for increased fuel economy are mixed, with some studies finding little evidence of undervaluation and others finding evidence of significant undervaluation. A range of theories and explanations is put forward for why consumers may undervalue fuel economy, and some have argued that what appears to be undervaluation may in some cases be differences in preferences and circumstances among consumers. Automobile manufacturers' statements and survey evidence tend to support the view that consumers expect a quick payback for a vehicle with higher fuel economy, all else being equal. Survey evidence also indicates broad and consistent public support for raising fuel economy standards over the past 30 years.

In the committee's judgment, there is a good deal of evidence that the market appears to undervalue fuel economy relative to its expected present value, but recent work suggests that there could be many reasons underlying this, and that it may not be true for all consumers. Given the importance of this question to the rationale for regulatory standards and their costs and benefits, an improved understanding of consumer behavior about this issue would be of great value.

AUTOMAKERS' RISK AVERSION TO SUPPLYING GREATER FUEL ECONOMY

In addition to inefficiencies in how consumers value fuel economy, the Agencies in their Final Rule also raise a supply-side problem—that automakers may be risk-averse to investing in fuel efficiency and therefore undersupplying fuel economy to the marketplace in the absence of regulation (EPA/NHTSA 2012a, 2012b). The Agencies posit two reasons why this could be true: uncertainty of future consumer demand for improved fuel economy and irreversibility of the large capital investments required. While the Agencies note that risk aversion by itself does not necessarily indicate a market failure, they state that manufacturer risk aversion would lead to an underprovision of fuel economy and that increasing fuel economy standards can lead to a more optimal solution by reducing the risk for manufacturers of investing in fuel economy.

Besides poor understanding of how consumers currently value fuel efficiency, the Agencies point out that automakers face additional uncertainty in predicting future consumer demand because it appears to evolve based on a number of factors that are difficult to forecast, such as future fuel prices, economic cycles, especially recessions, and the impact of marketing efforts. The Agencies point out that consumer valuation of fuel efficiency can change more rapidly than the industry is able to change its product offerings. Long lead times for production decisions exacerbate the uncertainty since automakers must make decisions on the level of fuel efficiency years in advance, but consumer demand can change in much shorter time frames in reaction to external events such as fuel price increases and recessions.

The Agencies also raise the possibility that in the absence of higher standards, automakers are risk-averse due to the "irreversibility" of the large capital investments necessary to develop and market fuel-efficiency technologies. According to the Agencies, the effect of this irreversibility is that for a risk adverse company, being a first mover may appear to have a greater downside risk than upside risk; that is, there is a "first mover disadvantage." If the Agencies are correct, the risk of oversupplying the market is greater than the risk of undersupplying the market. The risk of oversupplying the market is that a manufacturer will not be able to recoup its investment. The risk of undersupplying the market is a loss of market share in the short run, but the manufacturer still has the option of investing in fuel efficiency and regaining some if not all of its lost market share. If the industry as a whole is risk-averse, then large-scale adoption of fuel-efficiency technologies may not occur. Sunding and Zilberman (2001) showed that for a risk-averse company, delaying adoption of a new technology in order to gain more information may be more profitable than adopting a new technology. Blumstein and Taylor (2013) note that firms may choose not to produce a more efficient product because the product will increase costs without creating a long-term competitive advantage. There is some evidence that the auto industry exhibits risk-averse behavior. In explaining the lack of innovation in the highly concentrated auto industry in the mid-1980s, Kwoka (1984) noted that quality was considered too uncertain and risky to create a competitive advantage and safer bets were comfort, convenience, power, and style.

As noted by the Agencies, requiring all manufacturers to increase fuel economy can reduce the manufacturer's perceived or actual risk of investing in a fuel economy strategy and potentially lead to a more optimal provision of fuel economy in the marketplace. Furthermore, since consumers tend to be risk-averse in adopting new technologies, having more widespread penetration of the new technologies reduces the perceived consumer riskiness of that new technology. Finally, higher volumes will bring down the costs of the fuel efficiency technologies through economies of scale, learning curves, and more rapid innovation. Reducing the costs will reduce the riskiness of the investment by accelerating consumer adoption, allowing for faster cost recovery.

As noted by Blumenstein and Taylor (2013), economic theory says that the rate of innovation is likely to be sub-optimal when the returns to society from innovation are greater than the returns to the innovator. Innovation to improve fuel economy has important social benefits from reduced oil use and lower GHG emissions, so subsidizing R&D efforts for fuel economy, such as through tax credits or grants for companies, could have important spillover social benefits. Blumstein and Taylor 2013 also note that in the literature there is anecdotal evidence that suggests the existence of supply-side problems in the energy efficiency markets, including principal/agent problems, first mover disadvantage, price discrimination, and suppression of new technology. Principal/agent problems are a type of market failure where the person making the investment, in this case in fuel economy, is not the person who will reap the benefits. Fischer (2010) finds that market power gives manufacturers a strategic incentive to price-discriminate by overproviding fuel economy in vehicle classes whose consumers, on average, value it more than consumers of vehicles of other classes, and underproviding it in classes whose consumers value it less.

The recent era of higher gasoline prices (2005 to 2014) provides some anecdotal evidence that tighter standards can play an important role in providing longer-term planning certainty for automakers. The 2008-2011 MY light-truck fuel economy standards and the first National Program standards for 2012-2016 MY helped create a predictable, stable regulatory environment that provided greater certainty for the automakers' investment plans. These standards may have also motivated some manufacturers to overcomply with the CAFE/GHG standards in these years to earn credits to be used in future years. The Alliance of Automobile Manufacturers, one of the primary industry associations for automakers, voiced its members' support for the 2009 National Program agreement, in large part due to this long-term planning certainty (Alliance of Automobile Manufacturers 2009). Domestic manufacturers of large vehicles have particularly benefited from the new footprint standards since it provides them an incentive to improve all their vehicles rather than shift to smaller cars. According to an *Automotive News* article from 2011

> Many automakers believe that the work they've done since the last big [gas] price surge, and in anticipation of higher government fuel-economy standards, leaves them better prepared this time, with stables of more competitive small cars and crossovers. . . . It could be a fairer fight this time. GM and Ford not only have more competitive small cars, but hot-selling crossovers such as the Chevrolet Equinox and Ford Edge that could benefit if consumers abandon big SUVs. (Colias 2011)

Another way of viewing the auto industry is that there is substantial competition between original equipment manufacturers (OEMs) that prompts them to take innovative risks to see what customers will buy. In order to mitigate risk while pursuing innovation, an OEM is likely to start implementation on a small scale and gauge consumer reaction over a period of time. Even in the presence of risk aversion, this innovation could eventually lead to efficient provision of fuel economy if consumers demonstrate willingness to pay full value for a technology. If a manufacturer found that market share (and maybe higher-than-average profits) could be gained by providing an efficient level of fuel economy, others would follow. Once the industry learned that consumers were willing to pay the full expected lifetime-discounted value for fuel savings, essentially all manufacturers would provide it.

EVIDENCE ON CONSUMER VALUE FOR VEHICLE ATTRIBUTES

To better understand the impact of fuel economy standards on consumers, it is important to look not just at how consumers value fuel economy, but also at how they value other attributes in the bundle of vehicle characteristics. This can suggest how they might trade-off characteristics as a result of the standards.

It is clear from past trends in vehicle performance that buyers value attributes such as horsepower and acceleration. There have been some attempts to infer the value that consumers place on different vehicle attributes (Greene and Liu 1988). Whitefoot and Skerlos (2012) summarize the ranges of values that can be inferred from this literature (Table 9.3).

The estimates vary a great deal for all of the attributes. There are a number of reasons for this. One is that there are very few studies that have attempted to carefully estimate the value of these attributes to consumers. A major difficulty with empirical estimates is that the attributes tend to be related (i.e., not independent of each other). This makes statistical analyses that attempt to value each separately very challenging. Though the Whitefoot and Skerlos study shows that the value to consumers of fuel economy is greater than the value of acceleration (Table 9.3), Klier and Linn (2012) recently attempted to carefully address the interdependence issue and found that, on average, the value of a proportionate improvement in acceleration is greater than a similar percentage improvement in fuel economy. Another reason for the variation in estimates is that consumers are different, and the range of estimates may simply reflect these differences in value for a variety of vehicle attributes.

Given the value of different attributes, it is important to ask how the characteristics of the vehicle fleet will change in the future: first in the absence of the new standards and then with the standards. The Agencies' assessments rely on the central Annual Energy Outlook forecasts of gasoline prices, which have them rising only slightly above 2012 levels in real terms by 2025 ($3.87/gal) (EPA 2012, 491). With relatively constant gasoline prices, the Agencies assume that manufacturers in the reference case without standards will not adopt more fuel-efficient technologies on their own and cite the evidence discussed above in the section Trends and Vehicle Characteristics for 1985-2005 as justification.

However, technological progress is likely to continue and other characteristics are likely to improve as they have in the past. If technology in the absence of the standards were to increase horsepower, for example, the horsepower function would continue to rise while fuel economy would remain flat (weight could increase if the fleet mix shifts toward larger vehicles). With the standards in place, fuel economy would improve and the Agencies would hold horsepower and other characteristics constant in evaluating the cost and effectiveness of technologies that could be used to meet the standards. The Agencies do acknowledge that maintaining a reference or baseline case that shows no change in other attributes in the absence of standards is a potential problem with their analysis but do not attempt to address it (NHTSA 2012, 813). The technology frontier is likely to continue to shift out, so there will be opportunity costs that are not being considered. This issue is addressed more fully in the assessment of costs and benefits in Chapter 10.

The Agencies suggest a different argument in the Final Rule: that the market for the other attributes that the technologies could provide—greater power, acceleration, and size—may not have been bundled optimally for consumers in the past. One example of this is that consumers have been switching from the less-fuel-efficient, truck-based SUVs to more fuel-efficient, car-based crossover utility vehicles (CUVs) which have less capability for towing and off-road uses. It turns out that manufacturers may have been oversupplying the market with vehicles having greater heavy-duty capabilities compared to what consumers might otherwise have chosen.

It is important to assess the value consumers place on fuel economy and other vehicle attributes and how those values may change over time. The next section reviews consumer responses to recent technologies available on vehicles in the market today.

TABLE 9.3 Inferred Willingness of Consumers to Pay for Vehicle Attributes

Vehicle Attribute	Range of Willingness to Pay for Attribute ($)
Additional square foot of vehicle size	366-2,150
Increase of 0.01 hp/lb in acceleration	97-3,345
Reduction in fuel consumption of 1 gal/100 miles	468-3,826

SOURCE: Whitefoot and Skerlos (2012).

Consumer Responses to Recent Changes in Technology

New technologies will be introduced to improve performance and fuel economy, some of which elicit consumer responses that could affect overall vehicle sales. Called "consumer-facing" technologies, these new technologies include hybrids (HEVs), plug-in electric vehicles (PEVs), turbocharged engines, new types of transmissions, such as dual clutch transmissions (DCTs), and diesel vehicles, among others. In addition to the greater fuel economy accompanying new fuel efficiency technologies, consumers may have other positive user experiences:

- Improved low-speed torque (turbocharged engines, HEVs, PEVs);

- Improved acceleration, shift quality, and reduced engine noise while cruising (transmissions with more gears and larger ratio spreads);
- Smoother acceleration and a quiet ride (HEVs);
- Greater convenience of refueling at home (PEVs); and
- Lower and more certain fuel costs (electricity vs. gasoline for PEVs).

There may also be a reduction in attributes:

- Shorter range between refueling (limited-range battery electric vehicles [BEVs]);
- Loss of trunk space (compressed natural gas vehicles, CNGVs, and hydrogen fuel cell vehicles [FCEVs]);
- Performance issues (e.g., shift optimization resulting in noise, vibration, and harshness (NVH) and shift busyness, start-stop hesitation, turbo lag, and drivability concerns with dual-clutch transmissions (DCTs)); and
- Consumer dissatisfaction with and distrust of label fuel economy relative to in-use fuel economy, particularly for high-fuel-economy vehicles.

When the committee was formed in early 2012, concerns existed about the evolving utility of vehicles due to new technologies such as BEV range anxiety and the NVH of driving a vehicle equipped with a dual-clutch transmission or stop-start. Now, more than 2 years later, manufacturers have made tremendous strides in addressing flaws and limitations of some new gearboxes and other technologies. Concerns and issues still exist, but the speed with which manufacturers can and have addressed problems is impressive. These rapid improvements are due in large part to the fact that so many of the powertrains and technologies of today are electronic or computer-assisted, allowing for minute adjustments and calibrations to very complex machinery.

General Motors (GM) recently employed a short time frame for changes to its Chevrolet Malibu, for example. The vehicle was completely redone for MY 2013 but GM announced a change to its 2.5L four-cylinder engine, adding stop-start technology for MY 2014. "The days of waiting to make changes based on traditional timing cycles is a thing of the past," said John Hanh, brand marketing manager for the Chevrolet Malibu. "We're going to react to customers as quickly as we can when we develop new technology and the customer needs it" (Mateja 2013; Pund 2013). GM stop-start implementation has been well received in the marketplace (Nagy 2014).

Not all consumer dissatisfaction with new technologies is as easily fixed, however. Some manufacturers have received such negative responses about the performance of some implementations of stop-start technology, for example, that they provide the option of disabling stop-start at each vehicle start. Other consumers have requested adjustments to the transmission shift calibration because they felt the shifts from gear to gear were too obvious or frequent, particularly on seven-, eight-, and nine-gear boxes. Some technologies require new behavior from the driver, as happened when antilock brakes were introduced, while others involve a consumer learning the new feel of driving a car with, for example, more gears.

A look at history shows examples of American consumers knowingly sacrificing some fuel economy for convenience and performance. For example, until recently, standard transmissions typically experienced better fuel economy than an automatic when used in like-for-like products, but so few Americans purchased standard transmissions that only a few OEMs even offered them on a handful of products. This trade-off is being altered by technology improvement. New technologies developed for automatic transmissions such as dual-clutch provide both the convenience of an automatic and better fuel economy than standard versions. Consumer wishes remain paramount, however, and dry DCTs came under harsh criticism from consumers and enthusiasts alike when introduced. Wet DCTs have not faced the same criticism, however, and even dry clutch versions are being improved to enhance drivability.

The diesel engine is an example of the complexity of the decisions customers must make when purchasing a new vehicle. The value to the consumer is arrived at differently than for a gasoline vehicle: the diesel-powered vehicle is more expensive than its gasoline equivalent, the fuel consumption is considerably lower, diesel fuel costs more than gasoline, and the residual value is higher (see Chapter 3). The market is also different, with fewer vehicle manufacturers offering fewer models from which to choose. Through November 2014, the German brands of VW, Audi, Mercedes, BMW, and Porsche combined made up 82 percent of the diesel car market in the United States, excluding pickup trucks from GM, Ford, and Chrysler. Newcomers like the Chevrolet Cruze and Chrysler Jeep Grand Cherokee provide more diversity for the consumer, and these new offerings could result in increased diesel penetration. The higher residual value of a diesel-powered vehicle and the lower net cost of fuel results in a diesel vehicle having a total cost of ownership lower than equivalent gasoline vehicles over a period of 3 years and 5 years. This fact, if realized by the purchaser, will have a positive effect on sales, but other negative considerations also prevail, such as fewer fueling stations, costs of maintenance of the aftertreatment system, and the net effect on the environment, such as perceived greater criteria emissions from diesel engines. EPA notes a projection for diesel penetration of 1.5 percent for passenger cars and light trucks in 2014 (EPA 2014), however sales of diesels in 2014 were 0.84 percent of all light- duty vehicle sales (Cobb 2015). This compares to over 50 percent penetration in Europe and even a 50 percent penetration in the light-heavy pickup trucks in the United States (Class 2b and 3).

Concerns about new technology implementation are not unique to fuel economy technologies. In today's high-technology world, consumers judge the utility of vehicles in ways unimaginable just a few years ago. Starting in 2006, after BMW, Audi, and Mercedes-Benz introduced new in-cabin vehicle system interface technologies, J.D. Power changed its initial quality survey (IQS) criteria to evaluate these types of systems. Vehicle owners had significant difficulties adjusting to the new, centralized functions. A manufacturer saw reliability scores for two brands plummet in new car surveys—not because of anything related to whether the engine turned on every morning, but instead related to the telematics and human–machine interface (HMI) installed in the vehicle (Bowman 2011; Consumer Reports 2012b). "It is not technology per se that generates new problems, but rather its integration and execution," Neal Oddes, Director of Product Research and Analysis at J.D. Power, noted (Janes 2013), an observation that could be made for some of the fuel-saving technologies being launched today. The reliability ratings show that the perceived value and utility of a vehicle declines precipitously if the technology does not enhance the consumer's ownership experience. This is true of consumer-facing fuel economy technologies as well.

Market Trends That Have Led to Greater Fuel Economy

Some consumer choices result in improved fuel economy. These include reduction in the number of cylinders as well as the transition from body-on-frame to unibody. With the considerable improvements in fuel economy and performance in internal combustion engines (ICEs), particularly when mated to a high-efficiency transmission, consumers have migrated away from V8 engines to more fuel efficient six- and four-cylinder engines. The average number of cylinders for light-duty vehicles dropped from 5.90 in MY 2004 to 5.27 in MY 2010 (EPA 2013). This is one of the other reasons fuel economy has improved 22 percent (5.5 mpg) since 2005 (EPA 2013). In addition, six- and four-cylinder engines benefited from significant investment in maximizing the fuel efficiency of these engines and mated transmissions. These improvements in fuel economy in ICE vehicles with familiar technologies provide the consumer with less incentive to invest in new technologies and new vehicles.

The shift from SUVs to CUVs is another example of a consumer choice that resulted in improved fuel economy. In the last 12 years, the U.S. new car buyer has moved away from traditional, truck-based SUVs like the Chevrolet Suburban or the Toyota 4-Runner toward more fuel-efficient, car-based CUVs such as the Nissan Rogue and BMW X5. According to IHS Automotive, in 2000, just 3 percent of all vehicles sold in the United States were CUVs and 17 percent were SUVs. By the end of 2012, the numbers had changed dramatically, with 26 percent of new vehicles designated as CUVs and 7.5 percent as SUVs according to WardsAuto.

This migration toward lighter, more fuel-efficient vehicles preceded the 2008 spike in fuel prices, although gas prices were rising at a slow but steady pace for the previous few years. The 2008 fuel price spike only spurred more demand for, development of, and introduction of CUVs. The sharp increase in CUV demand initially occurred outside of any direct policy mandates (see Figure 9.8).

CUVs are lighter and more fuel efficient than their SUV counterparts. Manufacturers began introducing CUVs into the United States in a trickle, with the Toyota RAV-4 arguably the first CUV to debut, in 1997, followed closely by the Honda CR-V and the Subaru Forester the next year. Lexus rolled out the venerable RX for premium buyers in 1998. As shown in Figure 9.8, the rise in CUV sales corresponded with a fall in market share for both minivans and SUVs, reflecting changing consumer preferences as new buyers, such as Generation Xers, entered the market. CUVs provide the consumer with many valued attributes such as commanding ride height, seating for 5-8 passengers, and all-wheel drive. These attributes come in a lighter, more fuel-efficient package than that of SUVs along with faster acceleration, easier handling, and a softer, more comfortable ride. Due to unibody versus body-on-frame construction and to the choices manufacturers have made about which vehicle traits to offer, most CUV models have reduced off-road capabilities, along with a lower tow rating and minimal low gear range. Most consumers do not need such rugged attributes but do value the cargo room, high seating position, comfort, and fuel efficiency the crossover affords. In a 2012 report from the Mintel Group, 60 percent of respondents said "off-road and snow capabilities" were "important" and 51 percent said "towing ability" was "important," while 93 percent said "fuel efficiency" and "steering and handling" were important. Reliability, safety, and comfort all scored even higher for importance.

As consumer acceptance of CUVs spread, more manufacturers traded out truck-based vehicle architecture for car-based, including in iconic vehicles like the Ford Explorer and the long-running Nissan Pathfinder. When Ford discontinued the truck-based Explorer in MY 2010, the label fuel economy for the XLT four-door with a V6 and five-speed automatic transmission was 14 mpg city, 20 mpg highway. The car-based 2014 XLT Explorer has a new, more fuel- efficient V6 engine and a more efficient six-speed automatic transmission and is labeled at 17 mpg city and 24 mpg highway, a full 20 percent improvement in fuel economy.

The transition from SUVs to CUVs over the last decade is one reason that average certification fuel economy for new light-duty vehicles has risen from 24.8 mpg in MY 2005 to 30.3 mpg in MY 2014 (EPA 2013). During that time, fuel prices rose from just \$2.39/gal in 2007 to \$3.37/gal as of February 3, 2014 (EIA 2014). During this gas price rise, customers continued to purchase what they perceived to be "trucks"—CUVs, SUVs, pickups, and vans—with the market consistently split 50/50 between cars and CUVs/SUVs/pickups/vans (Experian Automotive 2014). In reality, at least from a vehicle architecture perspective, the market shifted

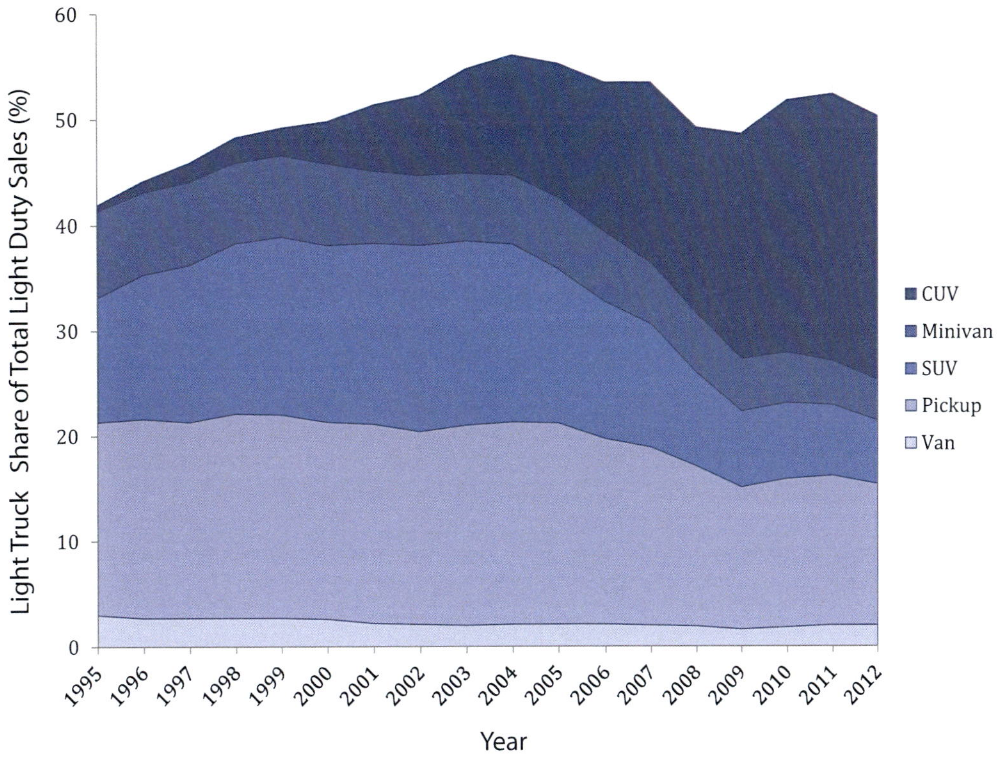

FIGURE 9.8 Light truck share of vehicle fleet by type of vehicle, showing the breakdown of the light-truck segment into CUVs, SUVs, minivans, pickups, and vans. CUVs rose in market share over the period.
SOURCE: Autodata (June 2013).

substantially in that time to more fuel-efficient car-based vehicles, which include CUVs. Shares of car-based vehicles (cars, minivans, and CUVs) increased from 66.4 percent in 2005 to 78.9 percent in 2012 (Baum 2014). The transition from V8 to V6 engines and from SUVs to CUVs indicates that consumers will make trade-offs of attributes they don't value in order to gain better fuel economy.

Alternative Technology Vehicles

Manufacturers are also introducing a host of alternative technology vehicles (ATVs) that encompass an array of fuel-saving technologies, including HEVs and PEVs as well as AFVs such as biodiesel, compressed natural gas (CNG), propane, and hydrogen fuel cell vehicles. These vehicles are helping the United States to reduce its dependence on petroleum and to reduce vehicle emissions, but availability often varies by state, and there are barriers to deployment including cost, infrastructure, technological advancements, and consumer acceptance.

Hybrid and Electric Vehicles

Sales of hybrid vehicles have gained ground since their introduction. Recently, many models have been offered in a range of vehicle classes. Market share of hybrid vehicles grew steadily from 2011 to 2013, although sales were down 8.8 percent in 2014 vs. 2013 (Autodata September 2014; Cobb 2015). In MY 2013, 19 different brands offered a total of 41 different hybrid models, featuring everything from the lowest-priced Honda Insight Hatchback ($18,725), the segment-leading Toyota Prius Hatchback ($24,200), up to the priciest model, the $120,060 Lexus LS 600h L Sedan, and the biggest, the 19-foot long Chevrolet Silverado 1500 Crew Cab pickup, now discontinued (all prices from Edmunds.com 2014). Among the 41 individual models available in 2013, 11 offered all-wheel-drive and 6 showed four-wheel-drive, including two pickup trucks from GM. The 2014 Toyota Prius offered the best label fuel economy at 51 mpg city, 48 mpg highway, while the pricey 2014 Lexus LS 600h offered the worst at just 19 mpg city, 23 mpg highway, a point below GM's five full-size truck offerings, which get 20 mpg city, 23 mpg highway.

Despite all of these models displaying an array of utility, price, fuel economy, and brands, hybrid sales accounted for just 3.4 percent of new vehicles sales in 2013, up a fraction from 3.3 percent in 2012, according to Autodata (December 2013). In 2014, hybrids represented only 2.75 percent of new light-duty vehicle sales, probably as a result of falling gasoline prices (Cobb 2015). The Toyota Prius lineup is the most popular hybrid vehicle model series, recording a 42 percent hybrid market share in 2014, down from a 45 percent market share in 2013 and 49 percent in 2012.

Consumer surveys show that there has been no upward trend in interest in purchasing an HEV or PEV in recent years (Strategic Vision, Annual surveys), although there is very wide regional variation in hybrid sales. In California, for example, the market share for conventional hybrids in 2013 was 6.8 percent of all light-duty sales. The Toyota Prius was the single best-selling model of all passenger cars, including both ICEs and ATVs, in California in 2013 (California New Dealer Association 2014). According to Experian Automotive, California accounted for 25.3 percent of all hybrids and PEVs sold in the United States in 2013. Florida, with 6.2 percent share of hybrid and PEV sales, was a very distant second (see Figure 9.9). This variation can often be attributed to the demographics and values of the consumers in each state but also localized incentive programs such as rebates, tax credits, access to HOV lanes, and availability of supporting infrastructure.

The modest size of the hybrid market does not necessarily reflect a lack of consumer demand for fuel efficiency overall but may be more a reflection of consumer perception of the lack of value for the hybrid drivetrain in particular. Furthermore, due to higher, size-based fuel economy standards in recent years, there is increasing availability of more fuel-efficient vehicles in every market segment, including pickup trucks. According to *Automotive News*, "improvements in the fuel economy of many mid-sized cars mean consumers now can save money without drastically reducing the size of their vehicles." (Bunkley 2013)

Other Alternative Fuel Vehicles

As mentioned above, other vehicles using alternative fuels such as biodiesel, CNG, propane, and hydrogen are in the market, each with its own set of challenges and opportunities. Unlike hybrid vehicles, which run on gasoline, alternative fuel vehicle challenges include establishing a fueling infrastructure. Hydrogen fuel cell vehicles in particular have the potential to be two to three times more efficient than an ICE and offer the opportunity to dramatically decrease petroleum consumption and GHG emissions, though they will require an entirely new fueling infrastructure. Currently, the infrastructure is very limited, with just 10 stations in the United States at the end of 2013 (see Chapter 4). Hyundai just released its Tucson fuel cell vehicle for lease in Southern California, where nearly all of the hydrogen fueling stations

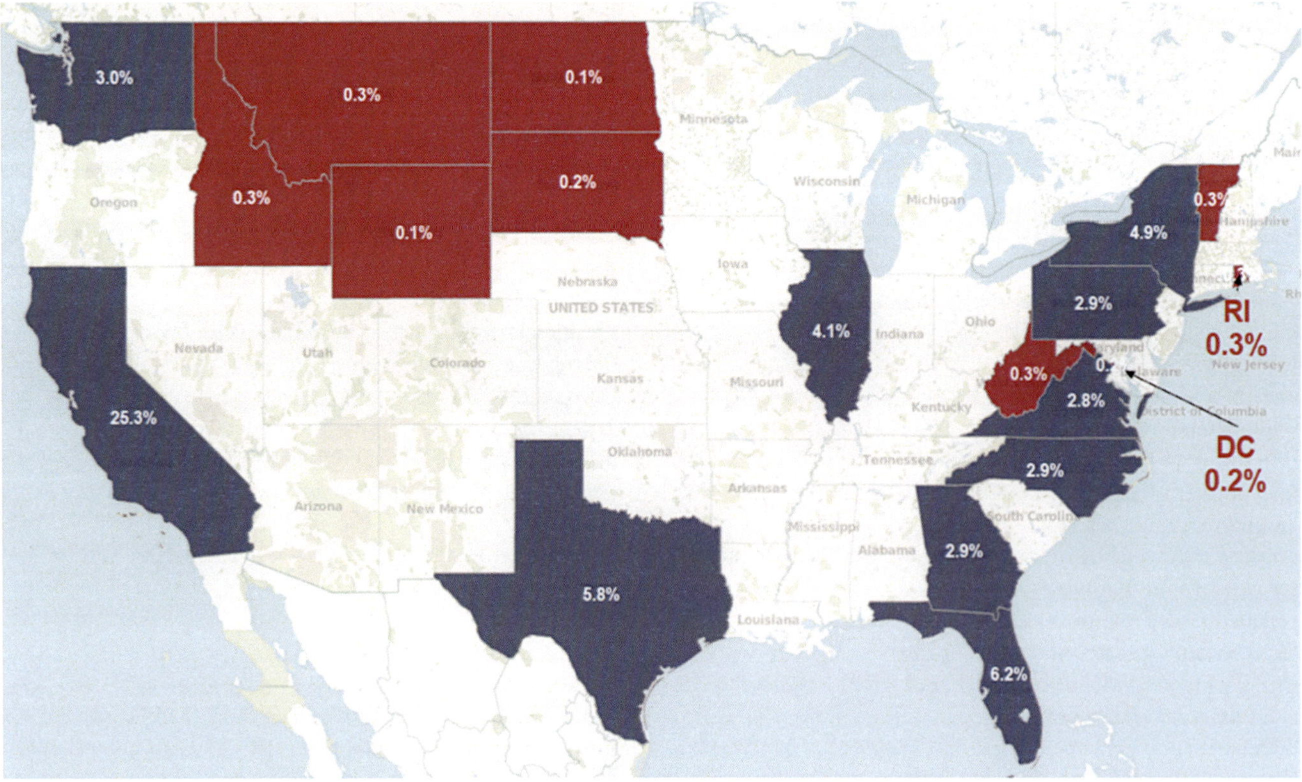

FIGURE 9.9 Uptake of hybrid and electric drivetrains is highly regional. States with the highest percent registrations of HEVs and PEVs are shown in blue, while those with the lowest percent registrations are shown in red.
SOURCE: Experian Automotive as of December 31, 2013 (U.S. light-duty vehicles only).

are currently located. The midsize crossover has a driving range up to 265 miles and takes just 10 minutes to refuel.

Natural gas vehicles come in two varieties: CNG and liquefied natural gas (LNG). Both fuels are commercially available, relatively low in price, and burn with lower emissions of criteria pollutants than gasoline or diesel, though all must meet the same tailpipe emissions standards. Vehicles equipped with CNG get about the same fuel efficiency as gasoline-powered vehicles. LNG, currently used mostly by medium- and heavy-duty trucks, has volumetric energy density about 65 percent that of gasoline, requiring about 1.5 gallons of LNG to provide the same energy as a gallon of gasoline. There are currently 737 CNG stations in various parts of the United States but only 58 LNG stations. The only car that can currently fuel with natural gas is the 2014 Honda Civic Natural Gas, but the all-new 2015 Chevrolet Impala can run on either gasoline or natural gas, using bi-fuel technology.

Propane (liquefied petroleum gas) powered vehicles have been around for decades, resulting in widespread infrastructure, with 2,714 refueling stations, mostly for medium- and heavy-duty vehicles. Similar to the natural gas vehicles, there are two different kinds of propane vehicles: dedicated, using only propane, and bi-fuel, with two systems for the two fuels. However, also similar to natural gas, there is limited product availability for light-duty vehicles. Currently there are no propane light-duty vehicles available on showroom lots. Instead, the consumer can order a truck or van from a dealership with a "prep-ready engine package" and convert the vehicle to propane use. Existing conventional vehicles can be converted to propane as well by certified installers, but the procediure can cost anywhere from $4,000 to $12,000 (DOE n.d.).

All AFV technologies address the issues of petroleum demand, emissions, and security of petroleum, but they all require significant infrastructure investment in order to become mainstream. Some automakers expressed concern that consumer familiarity with alternative fuel vehicles was a major market barrier to sales of these vehicles. Consumers are not accustomed to fueling at home or at the workplace with electricity, for example. Knowledge of and infrastructure for these vehicles may be localized due to the nature of new technology market diffusion, as discussed previously and reported in a recent NRC report (NRC 2015). These are not new challenges—these fuels were all mentioned as "alternative fuels" in the 1992 Energy Policy Act—but migrating away from the gasoline ICE continues to be a challenge in most parts of the United States. The regulatory treatment of AFVs is discussed in Chapter 10.

A challenge of increasing CAFE/GHG national program standards in 2020-2030 lies in how much improvement in fuel economy can be wrung from ICE and, beyond that, in overcoming consumers' perceived risk of new, consumer-facing technologies. Consumers are not risk-oriented when making a large purchase such as a car (Indiana University 2011, 5). If the fuel economy standards require widespread application of new technologies, will consumers be willing to pay more for these vehicles, or will they keep their existing vehicles longer and delay the greening of the fleet?

Survey Results on Fuel Economy and Other Attributes

The extent to which consumers value fuel economy will affect their willingness to purchase more fuel-efficient vehicles, especially if they perceive trade-offs with other attributes they value. When making a purchase decision, consumers may consider vehicles based on class, brand, features, or other specific considerations. Within the subset of vehicles that meet their basic needs, consumers consider a variety of broader vehicle attributes, including fuel economy but also reliability, durability, quality of workmanship, and value for money. Some attributes may be competing (such as horsepower and fuel economy), while others may be complementary (such as durability and reliability). Individual consumers value attributes differently, as shown by the wide variety of classes, makes, and models of vehicles available on the market. There is a vast amount of consumer survey data available for analysis on U.S. consumer attitudes toward fuel economy. These data indicate that many consumers value fuel economy, though it is not their only consideration.

The most reliable information about consumer preferences comes from surveys of drivers who have made a recent new car purchase. One such survey, conducted by Strategic Vision in 2013 of over 300,000 new vehicle buyers, found that while consumers value fuel economy, most do not consider it the most important attribute. In this survey, after a consumer bought or leased a new vehicle, he/she rated the importance of 49 possible reasons. In response to the question "Why did you decide to buy the particular model you did rather than some other model? How important was each of the following in your decision?" 45 percent of consumers said fuel economy was "extremely important" when selecting the vehicle they purchased or leased. The most highly rated attribute was reliability, which 68 percent of respondents said was "extremely important" for their purchase decision. Consumers could rate multiple characteristics as "extremely important," showing that they value a variety of attributes and must balance sometimes competing priorities (see Table 9.4).

Instead of rating all attributes, consumers are also often asked to select the most important attribute in purchasing a vehicle. In the Strategic Vision survey, fuel economy ranked fourth, with 8.27 percent of consumers choosing it as the most important. Value for money, reliability, and previous experience with brand were the top three reasons ranked "most important" by consumers in that survey.

Another approach to surveys asks the general public what they view as important attributes in considering a vehicle purchase. These surveys can provide information about trends in public attitudes over time. They show that consumers have a

TABLE 9.4 Survey of New Car Buyers Showing the Percent of Survey Respondents Rating a Particular Attribute as an "Extremely Important" Reason For Purchase

Rank (of 54)	Reason for Purchase	%
1	Reliability	70
2	Durability	60
3	Value for money	59
4	Quality of workmanship	59
5	Manufacturer's reputation	54
6	Safety design features in case of accident	53
7	Ease of handling/maneuverability	52
8	Warranty/guarantee	52
9	Seating comfort	49
10	Engine performance	49
11	Fuel economy	45

Survey data from Strategic Vision, October 2012-June 2013.

strong interest in fuel economy when considering a purchase, although stated interests may differ from actual purchase decisions (NADA 2014; Consumer Reports 2013; Morpace Inc. 2013). These surveys show mixed evidence about how important fuel economy is relative to other vehicle attributes for consumers intent on purchasing a vehicle. The National Automobile Dealers Association (NADA) 2013 survey asked respondents to identify the most important factor in determining which new vehicle to purchase. "Quality and dependability" was the selected factor by the most respondents (23%), with fuel economy chosen as most important by 14 percent of respondents, the second largest group. For truck buyers, 25 percent said "quality and dependability" was the most important factor, with nearly 11 percent citing brand and 9.6 percent indicating fuel economy as the most important consideration. In the 2014 New Car Buyer Survey from NADA, shown in Table 9.5, the survey question was changed to ask consumers to rank in order 10 vehicle attributes, rather than to provide the most important factor. While fuel economy was only the fourth highest rated attribute when all attributes could be rated as most important, when forced to choose the most important attribute, fuel economy was chosen as most important factor by the most respondents (NADA 2014). While fuel economy ranked first and cost of ownership second, environmental impact was ranked in seventh place, a sign that consumer interest in fuel economy might be related mostly to financial motivation.

There is much diversity in the attributes consumers value, and those values may impact their choice of desired vehicle type. In their purchase decision, consumers may select a vehicle class based on their top priorities, and then search within that class for other attributes they value. For example, consumers who value fuel economy or handling highly may search for compact vehicles, while consumers who value towing and hauling may search for SUVs or trucks. The Fuels Institute notes that shoppers interested in different vehicle classes have different priorities:

> Consumers shopping primarily for a specific class of vehicle showed a diversity of preferred vehicle attributes. Those considering smaller vehicles (coupes and sedans) were more likely to cite performance or technological attributes as influential, while those considering larger vehicles (minivans, SUVs and crossovers) found cost, capacity and safety to be more important. (Fuels Institute 2014)

In addition to variation between consumers in their valuation of vehicle attributes, there is also variation in how they value vehicle attributes at different times during the purchase

TABLE 9.5 NADA New Car and SUV/Truck Preference Surveys, August 2014

Ranking of Factors Considered during New Car Purchase			Ranking of Factors Considered during New SUV/Truck Purchase		
Factor	Average Rank	Rank[a]	Factor	Average Rank	Rank[a]
Fuel Economy	3.0	1	Fuel Economy	3.4	1
Cost of Ownership	3.8	2	Cost of Ownership	3.6	2
Power and Performance	4.2	3	Power and Performance	4.2	3
Advanced Safety Systems	4.8	4	Versatility and Utility	4.9	4
Versatility and Utility	5.2	5	Advanced Safety Systems	5.0	5
Build Quality and Reliability	5.8	6	Build Quality and Reliability	5.5	6
Vehicle Design	6.1	7	Vehicle Design	6.1	7
Environmental Impact	6.6	8	Environmental Impact	6.8	8
Brand	7.4	9	Brand	7.3	9
Technology	8.2	10	Technology	8.3	10

[a] 1 – Most Important, 10 – Least Important.

decision process: "Early on, it is about brands that excite and have stronger foundational security. Closer to purchase it is about product (mpg is a part) and price" (Alexander Edwards, Strategic Vision, pers. comm.).

At first there appears to be a contradiction between, on the one hand, many consumers saying they highly value fuel economy and, on the other, market trends that indicate large numbers of consumers continue to purchase less fuel efficient vehicles, such as trucks and SUVs. This apparent discrepancy shows that while consumers value fuel economy, they do so in the context of other attributes they also value. Consumers, therefore, may choose the most fuel- efficient vehicle that meets their needs and is in their price range, rather than the most fuel- efficient vehicle on the market. A recent survey by the Fuels Institute also highlighted this seeming contradiction between stated and revealed preference: While 84 percent of consumers say they are financially driven by fuel economy, they continue to buy larger, less fuel-efficient vehicles, indicating that they look for the most fuel-efficient version of a vehicle they already want to purchase. "An SUV customer might compare cost and fuel economy between several available SUVs, but might not be willing to consider shifting to a smaller vehicle that delivers superior economic value" (Fuels Institute 2014).

Consumers are buying fuel-efficient versions of vehicles that suit their wants and needs. This incentivizes manufacturers to provide best-in-class fuel economy—but not at the expense of other core attributes—and reinforces the decision of regulators to require different classes of vehicles to meet different standards via the different passenger car and light truck footprint standards. This accounts for why the market is seeing the success of crossovers with smaller displacement engines, for instance, but not a significant movement away from crossovers as a segment overall. It may also explain the reluctance of mainstream consumers to adopt alternative technology vehicles that may be perceived to sacrifice reliability and durability for improved fuel economy.

There is a great deal of variation in consumers' attitudes toward different types of vehicles, vehicles attributes, and fuel economy. More study to understand consumer valuation of fuel economy and indeed consumer motivation for vehicle purchases overall is critical to resolving some of these discrepancies and further understanding the buying process and its implications for the costs and benefits of the fuel economy standards.

COSTS AND BENEFITS OF THE NEW RULES TO INDIVIDUAL CONSUMERS

The Agencies assess the costs and benefits of the fuel economy rules for individual consumers and for society as a whole. The structure of the regulations and the societal impacts are discussed in Chapter 10. Consumer response to the new CAFE/GHG standards has important impacts on the costs and benefits of the rule, as well as on sales of vehicles and employment in the automotive industry. The Agencies have attempted to assess how consumers will respond to the new CAFE/GHG standards by looking at the private costs and benefits to the average consumer. They examine the costs and benefits to the consumer of owning the vehicle over the first 5 years, which is a common way of measuring cost of vehicle ownership. This approach to measuring the costs of ownership may not be the best way to understand the impact of the standards on vehicle sales. A cost of ownership is the cost once the vehicle has been purchased. More relevant for vehicle sales is the cost from the consumer's perspective before purchase, including undervaluation of fuel economy. The Agencies examine this undervaluation of fuel economy in some of their sensitivity analyses, but these analyses do not factor into the costs and benefits of the Final Rule (NHTSA 2012). The implications of the up-front perceptions of the value of fuel savings need to be included in an analysis of private consumer costs and benefits for consistency in the costs and benefits. For example, if consumers on average do not perceive much value from fuel economy, they are also unlikely to account for any residual value from a higher fuel economy vehicle. If the explanation for consumer undervaluation of fuel economy is instead that consumers are loss-averse, then the analysis should instead be based on whether consumers perceive the purchase of a more fuel efficient vehicle as a risky bet when the context of the choice is that all vehicles have been made more fuel efficient.

Agency Analysis of the Standards' Impact on Consumers Costs

The benefits to individual consumers are primarily the fuel savings they will get from driving more fuel-efficient vehicles (see Figure 10.8). As noted above and acknowledged by the Agencies, there is great variation in the value of fuel savings across individuals in the population depending on, for example, how much they drive, how much uncertainty they perceive in the fuel savings they will get, and how attentive they are to estimating the savings.

The costs to consumers are made up of the higher vehicle price due to the added components to raise fuel economy (discussed extensively in Chapters 2-6) and a range of other changes in cost, including higher sales taxes, insurance costs, financing costs, and maintenance and repair. In calculating costs and benefits, the Agencies assume no opportunity cost to the loss of attributes forgone to provide fuel economy (EPA/NHTSA 2012a, 62714). Each of these costs and the Agencies' approaches to estimating them are reviewed below. The committee's approach is necessarily qualitative, as to perform a quantitative assessment would have required significant committee resources. The committee also notes that the Agencies have undertaken such assessments in their regulatory analysis (EPA/NHTSA 2012b; EPA 2012; NHTSA 2012).

Fuel Costs

The Agencies, for their central estimates of consumers' buying responses, assume that consumers will consider 5 years of fuel savings. They based this estimate on the approximate average term of consumers' loans to finance the purchase of new vehicles (EPA/NHTSA 2012a, 62991). The savings in fuel expenditures will depend on fuel prices as well as on the type of fuel. One important uncertainty for consumers is the price of gasoline. To reflect the uncertainty of gasoline prices in the future, the Agencies' analysis is based on estimates from the Energy Information Agency (EIA) Annual Energy Outlook (AEO) 2012 of $3.87/gallon, and the range of high and low prices is drawn from the AEO 2012 high and low price of oil scenarios (EPA/NHTSA 2012a, 62617). In the most recent EIA AEO (2015), estimates of high gasoline prices are at about $3.65 per gallon in 2015 dollars, rising to $5.05 per gallon in 2030 in 2013 dollars. The low price forecast ranges from $2.21 per gallon in 2015 to $2.45/gallon in 2030, while the midrange rises through the period from $2.31 to about $3.20 per gallon in 2030. Historically there has been much variability in the EIA's AEO forecasts, however, so these numbers are likely to change with each yearly AEO update. The Agencies also rely on EIA estimates for diesel and electricity prices. Recently, gasoline prices have hit new lows, reaching $2.04 per gallon on average for the week of January 26, 2015. In the first three weeks of March 2015, the price per gallon was higher, staying above $2.40 (EIA 2015). Updated gas price forecasts for 2022-2025 should inform the midterm review.

Sales Tax

To estimate the average cost increase due to sales tax, the Agencies took the most recent auto sales tax data by state and weighted it by state population to estimate a national weighted-average sales tax of 5.46 percent. The Agencies chose to weight by using U.S. Census data on population because new vehicle sales by state were not available. To approximate new vehicle sales by state, NHTSA analyzed the change in new vehicle registrations (using R.L. Polk data) by state across recent years and found that the national weighted-average sales tax rate was almost identical to that resulting from the use of Census population estimates as weights.

Insurance Costs

The Agencies estimated the increase in collision and comprehensive (e.g., theft) car insurance over the base 5-year period since these are the portions of insurance costs that change with vehicle value. Using data from a Quality Planning study on average collision plus comprehensive insurance costs for new vehicles in 2010 dollars and data from the Bureau of Economic Analysis on average new vehicle prices, the Agencies estimated that insurance costs were, on average, 1.86 percent of the new vehicle prices. Using data from the Quality Planning study, the Agencies estimated collision plus comprehensive insurance costs due to the decrease in vehicle value when the vehicle is 5 years old to be 1.50 percent. At a 3 percent discount rate, present value of the stream of insurance costs over 5 years would be equal to 8.0 percent of the vehicle's price.

Financing Costs

The Agencies also estimated increases in costs due to the interest paid if the purchaser takes out a loan. The Agencies assumed that 70 percent of new vehicle purchases are financed based on a news article citing CNW Marketing Research (Bird 2011).

Using proprietary forecasts from IHS Automotive, the Agencies developed an average of 48-month bank and auto finance company loan rate of 5.16 percent for years 2017 through 2025 when deflated by IHS Automotive's corresponding forecasts of the consumer price index (CPI). The average person taking a loan will pay 5.43 percent more (at a 3 percent discount rate) for their vehicle over the 5 years than a consumer paying cash for the vehicle at the time of purchase. However, there will be a great deal of variation in the financing costs to different individuals. Subprime lending rates even on new cars start at about 19 percent, while those with the best credit scores will pay a good deal less than 5.16 percent.

Residual Value

To estimate residual value (or resale value) of the vehicle after 5 years, the Agencies apply the same average resale value rate for today's vehicle, 35 percent, to the increase in incremental costs that are a result of fuel-economy-improving technologies. Discounting the residual value back 5 years using a 3 percent discount rate (= 35% × 0.8755) yields an effective residual value of 30.6 percent. There is some evidence from the used car market that higher-fuel-efficiency vehicles such as hybrid and diesel vehicles can retain higher value than comparable conventional vehicles. Gilmore and Lave (2011), using auction data from Manheim Auctions, showed that after 3 years of ownership, the Toyota Prius and the Volkswagen Jetta TDI diesel vehicle retained a greater percentage of their initial purchase price than the conventional gasoline vehicles.

Maintenance and Repair Costs

The Technical Support Document (TSD) provides the most thorough discussion of maintenance and repair costs of the standards (EPA/NHTSA 2012b, 3-260). The Agencies evaluated existing data for 20 different technologies to determine if there were costs that could reasonably be attributed to the new standards. The Agencies concluded that 9 of the technologies identified could lead to a change in maintenance costs. The Agencies could not identify any other technologies that would have significantly different maintenance costs. (For further detail, see EPA/NHTSA 2012b, 3-261 to 3-264.)

The only non-EV technologies identified are two levels of low rolling resistance tires and diesel fuel filters. Tires, regardless of technology type, must be replaced multiple times over the life of the vehicle. The higher maintenance costs associated with low rolling resistance is due to the higher replacement cost. Diesel fuel filters are estimated to need replacing every 20,000 miles. Gasoline vehicles appear not to need fuel filter changes attributed to the rule.

Six of the 20 technologies with different maintenance costs due to the rule are associated with electric vehicles and four of those technologies decrease maintenance costs. The four that decrease maintenance—oil change, air filter replacement, engine coolant, and spark plug replacement—are intuitively obvious since EVs do not have an ICE. The two maintenance costs that increase are battery pack-related—battery coolant replacement and battery health check. Additionally, PEV batteries have a limited warranty of 5 or 8 years. Battery life is not known for this first generation of PEVs, but if the battery needs to be replaced in the 15 year life of a vehicle its cost would overwhelm any maintenance savings. Today's price for a Nissan Leaf battery replacement is $5,500 while a Volt battery replacement costs $2,300 (Blanco 2014; Voelcker 2014). The Nissan Leaf battery replacement price is less than the cost and Nissan admits to selling the replacement at a loss (Voelcker 2014). Battery cost estimates as evaluated by the committee are presented in Chapter 4. Battery replacement cost, should replacement need to occur, will be a large expense and will be larger than the maintenance cost of an ICE vehicle, and much larger than any of these additional ownership costs considered by the Agencies.

For repair costs, the Agencies account for costs of repairs covered by manufacturers' warranties (EPA/NHTSA 2011, 74925-74927) through the indirect cost multiplier methodology, which has a manufacturer warranty component. For repairs out of warranty, the Agencies choose not to compute an estimate due to insufficient data on frequency of different types of repairs and the immaturity of the new technologies just entering the fleet. The Agencies did not discuss possible changes in maintenance or repair costs for aluminum body vehicles, which may be higher than for steel vehicles, as noted in Chapter 6.

Agencies' Analysis of the Standards' Impact on Affordability, Sales, and Employment

As part of estimating the cost and benefit of the rule to the nation, the Agencies estimated the impact of the fuel economy standards on customer purchase decisions, vehicle sales, vehicle affordability, and employment in the automotive industry.

Private Cost of Vehicle Ownership

To evaluate sales and employment impacts, the Agencies evaluate consumer purchase decisions by estimating the private costs of ownership that consumers might include when making purchase decisions. The Agencies considered four factors discussed above: sales tax, insurance costs, financing costs, and residual values over the first 5 years of the vehicle's life (see Table 9.6). In addition, the Agencies did some sensitivity analyses, including varying assumptions of how consumers and manufacturers value years of fuel saved (NHTSA 2012, 830).

Overall, NHTSA finds that the residual value outweighs the increase in cost of ownership due to higher sales tax, insurance, and financing costs (NHTSA 2012, 832). At a 3 percent discount rate, the Agency finds that consumer valuation of the incremental purchase cost (before fuel savings are considered) is reduced by 12 percent. For the average MY 2025 incremental purchase cost of $1,836, this is equivalent to a reduction of $221 on the net consumer valuation (before fuel savings) of $1,615. This decrease in net consumer valua-

TABLE 9.6 Private Cost of Ownership at Purchase Decision Born By the Individual Consumer

	Cost as a percent of purchase price	Cost
Incremental Purchase Cost MY2025		$1,836
Discount rate	3.0%	
Increase due to higher sales tax	+5.5%	+$101
Increase due to higher insurance costs	+8.0%	+$147
Increase due to higher financing costs	+5.10%	+$94
Residual value recovered at resale	–30.64%	–$563
Change in value	–12.04%	–$221
Net consumer valuation (without fuel savings)		$1,615

NOTE: In addition to the 3 percent discount rate case presented above, the agencies also examine a 7 percent discount rate case.

tion is used in estimating the impact on sales. The cumulative impact of these additional cost considerations is to improve consumer payback, and increase sales and employment, compared to a case where these additional cost factors are not considered. The residual value does assume fuel savings after the initial 5-year period. The impact on sales should consider how consumers value fuel economy at purchase as well as any residual value recovered at resale for improved fuel efficiency.

Vehicle Demand

In addition to the impact of fuel economy regulations on individual consumer choices, the Agencies' analysis of the effects of the CAFE/GHG national program recognizes the importance of the underlying demand for vehicles. Consumer demand for vehicles may be changing in the coming years for a number of reasons. There may be a growing number of households who choose not to own a car. The number of households without a car (Madigan 2014) has increased nearly every year since 2005, when it was 8.87 percent and up to 9.22 percent in 2012, according to the University of Michigan (Sivak 2014; DeGroat 2014) and the U.S. Census Bureau. But most relevant is that this percent without a vehicle varies widely by geographical area, similar to alternative vehicle sales but in entirely different ways. Many more East Coast households are carless than in California, for example. New York City (56.5%), Washington D.C. (37.9%), Boston (36.9%), and Philadelphia (32.6%) are the top four cities for carless households, with San Francisco (31.4%) in fifth place and the only West Coast city until Seattle, in tenth place, with 16.6 percent of households. Increasing rates of urban density as well as access to public transportation could lead to decreasing rates of overall vehicle ownership over time.

Another evolving factor influencing car buying is household formation. In 2000, 68.1 percent of households were defined as "family," either married couples with children, married couples without children, single parents with children, or other family (U.S. Census Bureau 2000). In 2010, that figure dropped to 66.4 percent because single-person households climbed from 25.8 percent to 26.7 percent (U.S. Census Bureau 2012). In addition, new reports from Pew Research Center reports 42 million adults 25 and older have never been married, about 20 percent of the population. The Bureau of Labor Statistics recently reported that 50.2 percent of Americans over 16 are single, up from 37.4 percent in 1976 and the first time more adults are single than married (Madigan 2014; Miller 2014). If urbanization continues to rise, consumers may migrate away from full ownership of larger, less-fuel efficient vehicles and into smaller vehicles, they may use vehicles through car share, or they may potentially abandon vehicles altogether. At the same time, as single-person households and urbanization and density increase, EV sales could be hurt by more complicated access to charging at home as consumers move into apartments and multi-family dwellings and away from single-family homes with garages. These issues are just another example of the fluidity, complexity, and diversity of forecasting consumer demand and fuel economy.

Urbanization and household formation are also influencing the growth of alternative transportation methods such as car sharing programs, for example Car2Go and Zipcar, and on-demand transport services such as Uber and Lyft. These programs could influence vehicle ownership as well. What is clear is that there is a great deal of variation across consumers in attitude toward different types of vehicles, vehicle attributes, and fuel economy. Certainly, more study needs to be conducted on the changing demographics in the marketplace and its influence on vehicle demand.

Sales

If consumers consider higher fuel economy vehicles to be too expensive, they will purchase fewer new vehicles and retain used vehicles for a longer period, slowing down the rate of stock turnover and thereby the rate of improvement in the fuel economy of the on-road vehicle stock. On the other hand, if new, higher fuel economy vehicles are seen to be more desirable, used vehicle prices will decline relative to new vehicles, and scrappage rates will increase, as will the rate of improvement in the fuel economy of the on-road vehicle stock.

The effect of the standards on vehicle sales will depend on (1) the actual cost increase, (2) the actual fuel savings achieved, (3) how consumers perceive the value of future fuel savings at the time of purchase, including loss aversion, and (4) the price elasticity of demand for new vehicles.[5] There is reasonable evidence that the price elasticity of new vehicle demand in the United States is approximately –0.8 to –1.0 (e.g., Goldberg 1998; McCarthy 1996; Bordley 1993; Levinsohn 1988). If the price elasticity of vehicle demand is –1, then an increase in price of 1 percent causes a decrease in sales of 1 percent. If consumers assigned no value ($0) to the increased fuel economy brought about by the regulations, the regulations would unambiguously reduce vehicle sales. The Final Rule estimates that the average price of a new vehicle will increase by approximately $1,800. Prepurchase-perceived fuel savings are assumed to be about 1-4 years by the auto companies and by a number of other studies, or about $500 to $2,000, in contrast to full fuel savings over the life of a vehicle of more than $5,000, discounted to present value. If consumers valued the full fuel savings estimated by the Agencies of 5,000 present value dollars, the effect of the rule on sales would be the same as a price decrease of $3,200, and new vehicle sales would increase. If costs are higher than estimated and if the perceived fuel savings are low, as

[5] The price elasticity of demand is a unitless quantity that economists use to measure the sensitivity of demand to changes in price. It is the percent change in quantity demanded that would result from a 1 percent change in price, all else being equal.

most auto companies argue they are, the result would be an overall price increase, which would decrease vehicle sales.

If consumers undervalue fuel economy due to loss aversion, then the impact on sales will depend on the context in which consumers make their purchase. If all vehicles have higher fuel economy and higher cost due to the standards, then they will not face a risky bet and loss aversion will not come into play. If, on the other hand, they do perceive the purchase as risky, such as if they look to the used car market for a cheaper, less fuel efficient vehicle, then loss aversion will still apply. Also, views on fuel economy could change over time if there is learning about the advantages of more fuel-efficient vehicles, resulting in consumers' greater valuation of fuel economy over time. In general, new car sales impacts due to the standards will depend critically on how buyers perceive the value of the fuel savings and alternative attributes available in the vehicle market.

In addition to the effects on overall new vehicle sales, the costs and benefits of the rule may impact the mix of vehicles sold. If the costs of the rule are relatively high and the benefits relatively low, then price-sensitive consumers may opt to shift to less expensive new vehicles, such as smaller vehicles from larger vehicles. They may also purchase used vehicles. Currently the fleet mix is not being tracked. The Agencies monitor the mix of vehicles in the fleet and should track the effect of the rule on the sales mix.

The Agencies acknowledge that it is not clear which way sales will go:

> Since consumers are different and use different reasoning in purchasing vehicles, and we do not yet have an account of the distribution of their preferences or how that may change over time as a result of this rulemaking, the answer is quite ambiguous. Some may be induced by better fuel economy to purchase vehicles more often to keep up with technology, some may purchase no new vehicles because of the increase in vehicle price, and some may purchase fewer vehicles and hold onto their vehicles longer. There is great uncertainty about how consumers value fuel economy, and for this reason, the impact of this fuel economy proposal on sales is uncertain. (NHTSA 2012, 838)

As discussed in the section Trends in Vehicle Characteristics, evidence from the most recent era of higher fuel prices (2005 to present) suggests that there is no inherent conflict between higher fuel economy and increased vehicles sales. Since 2009, when the market hit a recent low of 10.4 million units, fuel economy and total vehicles sales have been steadily increasing, driven by a combination of a recovering economy, sustained high fuel prices, and more attractive vehicle offerings, including the availability of more fuel-efficient vehicle models in every market segment. The availability of new vehicle models, especially in a period of high fuel prices, can enhance vehicle sales by providing a compelling reason for consumers to upgrade their existing vehicle (Isidore 2012). Recently, fuel prices have dropped somewhat, ranging in 2014 from $3.60/gal to $2.25/gal. Low fuel prices may have uncertain effects on sales and the types of vehicles consumers want to purchase in combination with the new regulations. These effects will need to be examined if fuel prices remain low.

Affordability

If the cost and fuel savings estimates of the Final Rule are approximately correct, the standards appear to make vehicles more affordable for both new and used car buyers. This is primarily due to the fuel savings they will realize over the lifetime of the vehicle as compared to the increase in the initial purchase price of the vehicle. The committee's views on the standards' costs and benefits are presented for individual technologies in Chapters 2-6 and for the rule as a whole in Chapter 10. To the extent there are net benefits to consumers, they are likely to be greater for low-income households, which predominantly own used cars. In the United States, only the highest income quintile spends more each year on new vehicles than used vehicles (BLS 2011, Table 1). While the market values of vehicles depreciate at approximately 10 percent per year,[6] vehicle use decreases by only about 3-4 percent and fuel economy remains nearly constant over a vehicle's lifetime (see Lin and Greene 2011). As a consequence, expenditures on vehicles decrease more rapidly with vehicle age and therefore with the owner's income than expenditures on fuel (Figure 9.10). U.S. households in the lowest two income quintiles spent almost 50 percent more on fuel than on vehicles in 2011. Households in the highest income quintile spent about 25 percent less on fuel than on vehicles. The Final Rule assumes the standards will increase vehicle prices by about 6 percent but reduce fuel consumption by one-third relative to the 2016 standards. If these estimates are correct and the price of fuel is close to that assumed in the rule's analysis, all income groups would benefit.[7] Across all income groups, U.S. households spent nearly equal amounts on fuel ($2,655) and on new and used vehicles ($2,669). These are income group averages. Households that do not own vehicles would be affected only indirectly by the changes in vehicle prices and fuel economy.

Despite the fact that a cost of ownership calculation indicates that consumers will benefit from the new, higher fuel economy vehicles, if consumers are myopic when it comes to fuel savings in purchasing new or used cars, many more fuel-efficient cars may be perceived as more expensive. A new fuel-efficient car that costs even 6 percent more may appear to be less affordable than an alternative used car or than no vehicle purchase at all, if the fuel saving are only valued at

[6] Edmunds.com, Inc. (2010), for example, estimates that the market value of a 5-year-old new car will be about 40 percent of its purchase price, with higher depreciation over the first two years of 18 and 31 percent total depreciation after year 1 and year 2.

[7] This assumes that an average light-duty vehicle costs about $30,000, as reported by the Federal Trade Commission (2014).

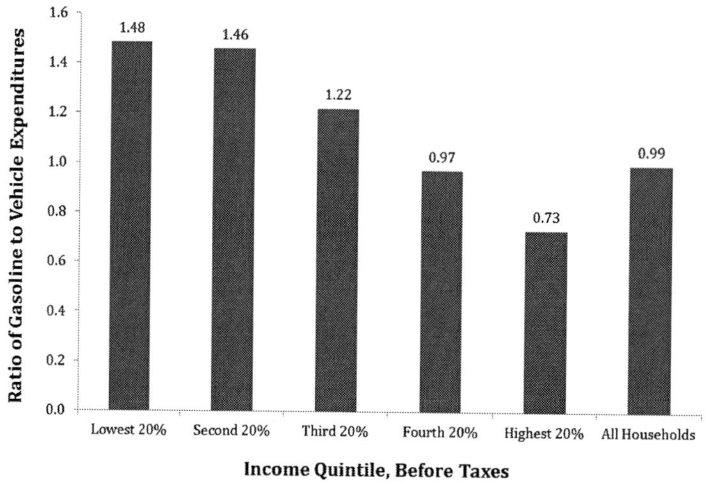

FIGURE 9.10 Ratio of household expenditures on gasoline + motor oil to expenditures on vehicles by income quintile.
SOURCE: BLS, CES (2011), Table 1.

2 or 3 years. This may be particularly true if entry-level car buyers shop between the used and new vehicle markets. As a result, people may hold onto current cars longer or may shift from the new to the used car market. This can affect the mix of new and used cars and affect the estimated GHG emissions and overall fuel savings. Used car prices may increase in this scenario and fuel economy improvements in the used car fleet may be slowed, which will affect the cost and benefits of the rule to consumers as well as to society.

Another aspect of vehicle affordability is the ability of the customer to finance his/her purchase. Though few low-income consumers purchase new vehicles, to the extent that some do, financing and liquidity constraints may be important. When making a purchase, some new car buyers may not be able to take account of long-term fuel savings because they cannot get lending terms that reflect those fuel savings. As a general rule, it appears that lenders do not consider fuel costs when qualifying applicants for auto loans, although there are some exceptions as noted in the Final Rule (EPA/NHTSA 2012a, 62950). The failure to take into account fuel savings could lead to some potential car buyers on the margin being denied loans even though the annual fuel savings could offset the higher loan payment. More investigation is needed to understand lender behavior and how the problem can be addressed.

FINDINGS AND RECOMMENDATIONS

Finding 9.1 There have been continual improvements in vehicle technology over time, enabling improvements in many vehicle attributes at relatively low cost. Much of the technology change during the period from 1985 to 2005 went to improve attributes such as horsepower and acceleration, and not to fuel economy. Since 2004, both average fuel economy and performance have trended upward. Rising gasoline prices are likely to have contributed to higher fuel economy during this latter period.

Finding 9.2 Fuel economy and GHG regulations themselves are likely to create additional incentives for innovation to reduce cost and enhance effectiveness of fuel savings approaches, beyond what would have occurred in their absence, in both known and unanticipated new technologies.

Finding 9.3 The energy paradox implies that consumers do not fully account for the expected present discounted value of fuel-saving technologies when they purchase new vehicles. Manufacturers perceive that consumers require relatively short payback periods of 1 to 4 years for fuel economy improvements. A large amount of literature in the economics and policy community attempts to understand and measure the extent and magnitude of consumer undervaluation of fuel economy, but the empirical evidence is still mixed. The results of recent studies find that consumers' responses vary from requiring payback in only 2 to 3 years to almost full lifetime valuation of fuel savings. The energy paradox is an important argument for fuel economy regulations and has a bearing on the private costs and benefits of the standards.

Finding 9.4 There are a range of theories and explanations for why consumers may or may not fully value lifetime fuel savings, all of which may have some validity as responses across the population are likely quite diverse. Some consumers may be loss-averse or risk-averse, not wanting to take on what may be perceived as a risky investment in fuel economy, while others may lack understanding of the amount and value of the future stream of fuel savings, especially as technologies are bundled into vehicle packages. Some of the apparent undervaluation may simply reflect the fact that consumers have higher value for other vehicle attributes than for fuel economy. It is important to resolve these issues because they will affect the costs and benefits of the rule.

Recommendation 9.1 The Agencies should do more research on the existence and extent of the energy paradox in fuel economy, the reasons for consumers' undervaluation of fuel economy relative to its discounted expected present value, and differences in consumers' perceptions across the population.

Finding 9.5 In the absence of increasingly stringent fuel economy standards, vehicle manufacturers may be risk-averse to long-term investments in fuel economy technologies. There has been much less analysis of supply-side considerations than of demand-side considerations. Better understanding of manufacturer risk aversion and other supply-side barriers would improve the assessment of the benefits and costs of the CAFE and GHG rules.

Recommendation 9.2 The Agencies should conduct more research on the existence and extent of supply-side barriers to long-term investments in fuel economy technologies.

Finding 9.6 The value of fuel economy is related to the value of other vehicle attributes to consumers. Therefore, understanding the value to consumers of vehicle attributes other than fuel economy is important for the assessment and implementation of the fuel economy rules. If consumers value other attributes, and such attributes are forgone to obtain fuel economy in order to meet the regulations, the benefits and costs of these attributes will need to be accounted for in an assessment of the rule. These attributes could include any technological progress to vehicles that could be made in lieu of fuel economy, including horsepower, acceleration, or accessories that add weight, for example. The few existing studies have found a wide range of values for different vehicle attributes, including for fuel economy, but new survey and statistical approaches may offer promise for improving estimation of the value of these different attributes.

Finding 9.7 Consumers consistently cite quality, reliability, and dependability as the most important factors in vehicle purchase decisions while simultaneously assigning high value to fuel economy. In balancing reliability, fuel economy, and other attributes they value, customers buy fuel-efficient versions of vehicles that suit their other wants and needs. There is a great deal of variation in consumer attitudes toward different types of vehicles, vehicles attributes, and fuel economy.

Finding 9.8 Markets are segmenting in ways that may allow for better fuel economy without a loss in other valued characteristics. For example, there has been a shift in recent years for some consumers away from SUVs and to CUVs. CUVs have similar and even sometimes greater interior size and comfort but trade the towing, off-road, and low-gear-range capabilities of the SUV for the greater fuel economy of a car platform. For a large and growing segment of buyers, this trade-off offers improved value.

Finding 9.9 There is evidence that consumers will not widely adopt technologies to improve fuel economy that interfere with driver experience or comfort. For example, some stop-start and dual-clutch transmission applications have been strongly criticized by consumers. However, over time, implementations of new technologies can improve and potentially find consumer acceptance, and some fuel economy technologies improve drivability or performance, which may be valued by consumers. A challenge in meeting the CAFE/GHG standards in 2020-2025 lies in what further improvements can be gained from the internal combustion engine. Failing that, vehicle manufacturers may encounter significant barriers in marketing new, consumer-facing technologies, especially those with perceived or real trade-offs in vehicle utility, such as driving range for limited-range BEVs.

Finding 9.10 The opportunity to finance a more fuel-efficient vehicle with higher upfront price could be improved if lenders were able to factor in the projected benefits of fuel cost savings for that vehicle model over the life of the loan.

Recommendation 9.3 The Agencies should study the value of vehicle attributes to consumers, consumer willingness to trade off other attributes for fuel economy, and the likelihood of consumer adoption of new, unfamiliar technologies in the vehicle market. This will enable the Agencies to better understand consumer response to the CAFE rules and better assess the rules' costs and benefits.

REFERENCES

Allcott, H., S. Mullainathan, and D. Taubinsky. 2014. Energy policy with externalities and internalities. Journal of Public Economics, accepted for publication and available online as of January 30.

Allcott, H. 2012. The welfare effects of misperceived product costs: Data and calibrations from the automobile market. American Economic Journal: Economic Policy 5(3): 30-66.

Allcott, H., and M. Greenstone. 2012. Is There an Energy Efficiency Gap? Working Paper 228, Energy Institute at Haas, University of California at Berkeley, January.

Allcott, H., and N. Wozny. 2014. Gasoline prices, fuel economy, and the energy paradox. The Review of Economics and Statistics, December, 96(5): 779-795.

Alliance of Automobile Manufacturers. 2009. Automakers Support President in Development of National Program for Autos. Press Release, May 18.

Alquist, R., and L. Kilian. 2010. What do we learn from the price of crude oil futures? Journal of Applied Econometrics 25(4): 539-573.

Alquist, R., L. Kilian, and R.J. Vigfusson. 2010. Forecasting the Price of Crude Oil. in Elliot and Timmerman, eds., Handbook of Economic Forecasting, forthcoming, Elsevier.

Anderson, S.T., R. Kellogg, and J.M. Sallee. 2011. What Do Consumers Believe about Future Gasoline Prices? NBER Working Paper 16974, National Bureau of Economic Research, Cambridge, MA, April.

Baum, A. 2014. Analysis of Wards Sales Data. Baum & Associates. Ann Arbor, MI. March.

Bird, C. 2011. Should I Pay Cash, Lease or Finance My New Car? CNW Marketing Research. http://www.cars.com/go/advice/Story.jsp?section=fin&story=should-i-paycash&subject=loan-quick-start&referer=advice&aff=sacbee.

Blanco, S. 2014. Chevy Volt replacement battery cost varies wildly, up to $34,000. AutoblogGreen, January 10. http://green.autoblog.com/2014/01/10/chevy-volt-battery-replacement-cost-34000/.

BLS (Bureau of Labor Statistics). 2011. Consumer Expenditures. http://www.bls.gov/opub/reports/cex/consumer_expenditures2011.pdf

BLS. 2014. Historical Consumer Price Index for Automobiles. http://www.bls.gov/cpi/cpi_dr.htm.

Blumenstain, C., and M. Taylor. 2013. Rethinking the Energy Efficiency Gap: Producers, Intermediaries, and Innovation. Working Paper. Energy Institute at Haas, Berkeley, CA.

Bordley, R.F. 1993. Estimating Automotive Elasticities from Segment Elasticities and First Choice/Second Choice Data. The Review of Economics and Statistics 75(3): 455-462.

Bowman, Z. 2011. Consumer Reports pulls Recommended rating on Ford Edge, Lincoln MKX - MyFord Touch to blame? Autoblog, January 4. http://www.autoblog.com/2011/01/04/consumer-reports-pulls-recommended-rating-on-ford-edge/.

Bradsher, K. 2002. High and Mighty: SUVs—The World's Most Dangerous Vehicles and How They Got There. PublicAffairs, a member of the Perseus Book Group.

BTS (Bureau of Transportation Statistics). 2013. Table 3-17: Average Cost of Owning and Operating an Automobile(a) (Assuming 15,000 Vehicle-Miles per Year). U.S. Department of Transportation. http://www.rita.dot.gov/bts/sites/rita.dot.gov.bts/files/publications/national_transportation_statistics/html/table_03_17.html.

BTS. 2014. Table 4-23: Average Fuel Efficiency of U.S. Light Duty Vehicle. U.S. Department of Transportation. http://www.rita.dot.gov/bts/sites/rita.dot.gov.bts/files/publications/national_transportation_statistics/html/table_04_23.html.

Bunkley, N. 2013. mpg improvements aid mid-sized sales. Automotive News, March 11.

Busse, M., C. Knittel, and F. Zettelmeyer. 2013. Are consumers myopic? Evidence from new and used car purchases. American Economic Review. February.

California New Dealer Association. 2014. California Auto Outlook 10(1), released February.

Choi, J. NADA 2013 Car Shopper Preference Survey: Quality and Dependability Trump Fuel Economy. NADA Blog. http://www.nada.com/b2b/NADAOutlook/UsedCarTruckBlog/tabid/96/entryid/298/NADA-Spring-2013-Car-Shopper-Preference-Survey-Quality-and-Dependability-Trump-Fuel-Economy.aspx.

Cobb, J. 2015. December 2014 Dashboard. Hybridcars.com, January 6. http://www.hybridcars.com/december-2014-dashboard/.

Colias, M. 2011. Buyers move toward better fuel economy. Automotive News, March 14. http://www.autonews.com/apps/pbcs.dll/article?AID=/20110314/RETAIL07/303149972/1135.

Competitive Enterprise Institute. 1998. National Environmental Survey. Prepared by the polling company for the Competitive Enterprise Institute. http://cei.org/sites/default/files/CEI%20Staff%20-%20National%20Environmental%20Survey.pdf.

Consumer Federation of America. 2014. For First Time Over 50 Percent of Current Year Models Get More than 23 MPG; Over 11 Percent Get 30 MPG! CFA, April 29. Figures 1-3. http://www.consumerfed.org/news/778.

Consumer Reports. 2012a. High gas prices motivate buyers to change direction: Nearly three-quarters of surveyed motorists would consider an alternative-fuel vehicle for their next car. Published online May 2012. http://www.consumerreports.org/cro/2012/05/high-gas-prices-motivate-drivers-to-change-direction/index.htm.

Consumer Reports. 2012b. Ford offers MyFord Touch upgrades for current owners, 2013 models. Consumer Reports News, March 5. http://www.consumerreports.org/cro/news/2012/03/ford-offers-myford-touch-upgrades-for-current-owners-2013-models/index.htm.

Consumer Reports. 2013. Survey: Car shoppers want better fuel economy, here's why. Consumer Reports News, February 4. http://www.consumerreports.org/cro/news/2013/02/survey-car-shoppers-want-better-fuel-economy-here-s-why/index.htm.

DeGroat, B. 2014. Hitchin' a ride: Fewer Americans have their own vehicle. University of Michigan Regents. Michigan News, January 23. http://ns.umich.edu/new/releases/21923-hitchin-a-ride-fewer-americans-have-their-own-vehicle.

DOE (Department of Energy). n.d. Propane Vehicle Conversions, Alternative Fuels Data Center. http://www.afdc.energy.gov/vehicles/propane_conversions.html.

DOT (Department of Transportation). 2014. National Transportation Statistics. Table 4-23: Average Fuel Efficiency of U.S. Light Duty Vehicles. http://www.rita.dot.gov/bts/sites/rita.dot.gov.bts/files/publications/national_transportation_statistics/html/table_04_23.html.

Edmunds.com, Inc. 2010. How fast does my new car lose value? http://www.edmunds.com/car-buying/how-fast-does-my-new-car-lose-value-infographic.html.

Edmunds.com. 2014. Data Center. http://www.edmunds.com/industry-center/data/.

Edwards, A. 2014. Strategic Vision. E-mail message to author, August 19.

EIA (Energy Information Administration. 2014. Weekly Retail Gasoline and Diesel Prices, Petroleum and Other Liquids. http://www.eia.gov/dnav/pet/pet_pri_gnd_dcus_nus_w.htm.

EIA. 2015. Annual Energy Outlook 2015. http://www.eia.gov/forecasts/aeo/section_prices.cfm. Accessed April 24, 2015.

EPA (Environmental Protection Agency). 2012. Regulatory Impact Analysis: Final Rulemaking for 2017-2025 Light-Duty Vehicle Greenhouse Gas Emission Standards and Corporate Average Fuel Economy Standards. PA-420-R-12-016.

EPA. 2014. Light-Duty Automotive Technology, Carbon Dioxide Emissions, and Fuel Economy Trends: 1975 through 2014. EPA-420-S-14-001.

Fischer, C., W. Harrington, and I.W.H. Parry. 2007. Should automobile fuel economy standards be tightened? The Energy Journal 28: 1-30.

Fischer, C. 2010. Imperfect Competition, Consumer Behavior, and the Provision of Fuel Efficiency in Light-Duty Vehicles, RFF Discussion paper 10-60, December 2010.

EPA/NHTSA. 2011. 2017 and Later Model Year Light-Duty Vehicle Greenhouse Gas Emissions and Corporate Average Fuel Economy Standards, Proposed Rules. EPA 40 CFR Parts 85, 86 and 600; 49 CFR Parts 523, 531, 533, 536 and 537. Federal Register 76(231):74854-75420.

EPA/NHTSA. 2012a. 2017 and Later Model Year Light-Duty Vehicle Greenhouse Gas Emissions and Corporate Average Fuel Economy Standards; Final Rule 77(199) October 15.

EPA/NHTSA. 2012b. Joint Technical Support Document, Final Rulemaking 2017-2025 Light-Duty Greenhouse Gas Emission Standards and Corporate Average Fuel Economy Standards. EPA-420-R-12-901.

Fuels Institute. 2014. Consumers and Alternative Fuels: Economics are Top of Mind. http://www.fuelsinstitute.org/ResearchArticles/ConsumersandAlternativeFuels.pdf.

Gallagher, K.S., and E. Muehlegger. 2011. Giving green to get green? Incentives and consumer adoption of hybrid vehicle technology. Journal of Environmental Economics and Management 61: 1-15.

Gallup. 2009. Americans Green-Light Higher Fuel Efficiency Standards. Accessed September 20, 2013. http://www.gallup.com/poll/118543/americans-green-light-higher-fuel-efficiency-standards.aspx.

Gillingham, K., and K. Palmer. 2013. Bridging the Energy Efficiency Gap: Insights for Policy from Economic Theory and Empirical Analysis, RFF DP 13-02, Resources for the Future, Washington, D.C., January.

Gillis, J., and M. Cooper. 2013. On the Road to 54 mpg: A Progress Report on Achievability. Consumer Federation of America. April.

Gilmore, E., and L. Lave. 2011. Comparing Resale Prices and Total Cost of Ownership for Gasoline, Hybrid and Diesel Passenger Cars and Trucks. Working Paper. Carnegie Mellion University.

Goldberg, P.K. 1998. The effects of Corporate Average Fuel Efficiency standards in the U.S. Journal of Industrial Economics XVLI(1): 1-33.

Greene, D.L. 2010. How consumers value fuel economy: a literature review. EPA 420-R-10-008, Environmental Protection Agency, Office of Transportation and Air Quality, March.

Greene, D.L. 2011. Uncertainty, loss aversion and markets for energy efficiency. Energy Economics 33: 608-616.

Greene, D.L., D. Evans, and J. Hiestand. 2013. Survey evidence on the willingness of U.S. consumers to pay for automotive fuel economy. Energy Policy 61: 1539-1550.

Greene, D.L., and J. Liu. 1988. Automotive Fuel Economy Improvements and Consumers' Surplus. Transportation Research Part A 22A(3): 203-218.

Hamilton, J.D. 2009. Understanding crude oil prices. The Energy Journal 30(2): 179-206.

Helfand, G., and A. Wolverton. 2011. Evaluating the consumer response to fuel economy: a review of the literature. International Review of Environmental and Resource Economics 5(2): 103-146.

Hellman, K.H., and J.D. Murrell. 1984. Development of Adjustment Factors for the EPA City and Highway mpg Values. SAE Technical Paper Series No. 840496, Society of Automotive Engineers, Warrendale, Pennsylvania.

Herrnstein, R.J., G. Lowenstein, D. Prelec, and W. Vaughan, Jr. 1993. Utility maximization and melioration: Internalities in individual choice. Journal of Behavioral Decision Making 6: 149-185.

Hyundai. n.d. We've Reimagined the Idea of an Electric Vehicle. https://www.hyundaiusa.com/tucsonfuelcell.

Indiana University. 2011. Plug-in Electric Vehicles: A Practical Plan for Progress. The Report of an Expert Panel. School of Public and Environmental Affairs at Indiana University. http://www.indiana.edu/~spea/pubs/TEP_combined.pdf.

Isidore, C. 2012. Fuel efficiency drives car sales. CNNMoney, April 2. http://money.cnn.com/2012/04/02/autos/fuel-efficiency-car-sales/.

Jaffe, A.B., R.G. Newell, and R.N. Stavins. 2003 Technological change and the environment. in K-G. Mäler and J. Vincent (Eds.) Handbook of Environmental Economics, North-Holland.

Jaffe, A.B., and R.N. Stavins. 1994. The energy-efficiency gap: What does it mean? Energy Policy 22(10): 804-810.

Janes, G. 2013. The Increasing Importance of User Experience. Visual Logic, September 2. http://vlgux.com/2013/09/importance-of-user-experience/.

Jensen, C. 2011. Consumer Reports Poll Shows Support for Stronger Fuel-Economy Standards. New York Times. http://wheels.blogs.nytimes.com/2011/11/14/consumer-reports-poll-shows-support-for-stronger-fuel-economy-standards/?_r=0.

Kahn, J. 1986. Gasoline prices and the used automobile market: A rational expectations asses price approach. Quarterly Journal of Economics 101(2): 323-340.

Kahneman, D. 2011. Thinking Fast and Slow. Farrar, Straus and Giroux, New York.

Khanna, S., and J. Linn. 2013. Do Market Shares or Fuel Economy Explain Rising Shares of New Vehicle Fuel Economy? Resources for the Future Discussion Paper, 13-29. September.

Kilian, L., and E.R. Sims. 2006. The Effects of Real Gasoline Prices on Automobile Demand: A Structural Analysis Using Micro Data. University of Michigan.

Klier, T., and J. Linn. 2013. Technological Change, Vehicle Characteristics, and the Opportunity Costs of Fuel Economy Standards, RFF Discussion paper, Washington, D.C.

Klier T., and J. Linn. 2012. New vehicle characteristics and the cost of the Corporate Average Fuel Economy standards. Rand Journal of Economics 43(1):186-213.

Knittel, C. 2011. Automobiles on Steroids: Product Attribute Trade-offs and Technological Progress in the Automobile Sector. American Economic Review, December, pp. 3368-3399.

Knittel, C. 2012. Reducing petroleum consumption from transportation. Journal of Economic Perspectives. 26(1): 93-118.

Knittel, C. 2013. Are consumers myopic? Evidence from new and used car purchases. The American Economic Review 103(1): 1-41.

Krupnick, A., I.W.H. Parry, M. Walls, T. Knowles, and K. Hayes. 2010. Toward a New National Energy Policy: Assessing the Options. Resources for the Future, Washington, D.C. www.Energypolicyoptions.org.

Kwoka Jr., J.E. 1984. Market power and market change in the U.S. automobile industry. Journal of Industrial Economics 32(4): 509-22.

Levinsohn, J. 1988. Empirics of Taxes on Differentiated Products: The Case of Tariffs in the U.S. Automobile Industry. Trade Policy Issues and Empirical Analysis, National Bureau of Economic Research, pp. 9-44.

Lin, Z., and D.L. Greene. 2011. Predicting Individual On-road Fuel Economy Using Simple Consumer and Vehicle Attributes. SAE Technical Paper 11SDP-0014, Society of Automotive Engineers, Warrendale, PA, April 12.

MacKenzie, D., and J. Heywood. 2012. Acceleration Performance Trends and the Evolving Relationship Between Power, Weight, and Acceleration in U.S. Light-Duty Vehicles: A Linear Regression Analysis. MIT Department of Mechanical Engineering. http://web.mit.edu/sloan-auto-lab/research/beforeh2/files/MacKenzie&Heywood-TRB-2012-1475.pdf.

Madigan, K. 2014. Vital signs: More households don't own a car. The Wall Street Journal, January 21. http://blogs.wsj.com/economics/2014/01/21/vital-signs-more-households-dont-own-a-car/?mg=blogs-wsj&url=http%253A%252F%252Fblogs.wsj.com%252Feconomics%252F2014%252F01%252F21%252Fvital-signs-more-households-dont-own-a-car.

Mateja, J. 2013. Chevy's Gamble on Stop/Start Paying Off, Exec Says. WardsAuto, December 23. http://wardsauto.com/vehicles-amp-technology/chevy-s-gamble-stopstart-paying-exec-says.

McCarthy, P.S. 1996. Market price and income elasticities of new vehicle demands. The Review of Economics and Statistics LXXVII(3): 543-547.

McConnell, V. 2013. The New CAFE Standards: Are they Enough on Their Own? RFF Discussion paper 13-14. Washington, D.C., May. http://www.rff.org/Publications/Pages/PublicationDetails.aspx?PublicationID=22180.

Mellman Group. 2007. Voters Strongly Support Stricter Fuel Efficiency Standards Which They See As Vital For National Security. The Mellman Group, Inc. and Public Opinion Strategies. http://www.pewtrusts.org/uploadedFiles/wwwpewtrustsorg/Reports/Global_warming/poll-energy-national-security.pdf.

Miller, R. 2014. Is Everybody Single? More Than Half the U.S. Now, Up From 37% in '76. Bloomberg. http://www.bloomberg.com/news/2014-09-09/single-americans-now-comprise-more-than-half-the-u-s-population.html.

Mintel Group. 2012. SUVs and CUVs - US - April 2012. http://reports.mintel.com/display/590296/?__cc=1.

Morpace, Inc. 2013. PACE Study: Consumers are willing to pay more and sacrifice performance for vehicles with better fuel economy. PR Newswire, November 4. http://www.prnewswire.com/news-releases/pace-study-consumers-are-willing-to-pay-more-and-sacrifice-performance-for-vehicles-with-better-fuel-economy-230529961.html.

NADA. 2014. 2014 New Car Shopper Preference Survey. NADA Used Car Guide: Perspective. August. http://automotivedigest.com/wp-content/uploads/2014/08/2014-NADA-New-Car-Shopper-Preference-Survey.pdf.

Nagy, B. 2014. 2014/2015 Chevrolet Malibu buyers embrace start/stop technology. Kelley Blue Book, July 23. http://www.kbb.com/car-news/all-the-latest/20142015-chevrolet-malibu-buyers-embrace-startstop-technology/2000010939/.

NHTSA (National Highway Traffic Safety Administration). 2012. Final Regulatory Impact Analysis: Corporate Average Fuel Economy for MY 2017-MY2025 Passenger Cars and Light Trucks. Office of Regulatory Analysis and Evaluation, National Center for Statistics and Analysis.

NRC (National Research Council). 2002. Effectiveness and Impact of Corporate Average Fuel Economy (CAFE) Standards. Washington, D.C.: The National Academies Press.

NRC. 2011. Assessment of Fuel Economy Technologies for Light-Duty Vehicles. Washington, D.C.: The National Academies Press.

NRC. 2015. Overcoming Barriers to Deployment of Plug-In Electric Vehicles. Washington, D.C.: The National Academies Press.

Pew Charitable Trusts. 2011. Pew Clean Energy Program poll finds strong support for ambitious standards. http://www.pewenvironment.org/news-room/press-releases/administration-auto-industry-in-sync-with-americans-opinion-on-fuel-economy-85899362431.

Pund, D. 2013. Marchionne Says Nine-Speed Auto Still in the Works for Dodge Dart [2013 Detroit Auto Show]. Car and Driver, January 14. http://blog.caranddriver.com/dodge-dart-to-get-nine-speed-automatic-2013-detroit-auto-show/.

Sallee, J.M. 2013. Rational Inattention and Energy Efficiency. Working Paper 19545, National Bureau of Economic Research, Cambridge, MA, October.

Sanstad, A.H., and R.B. Howarth. 1994. Normal markets, market imperfections and energy efficiency. Energy Policy 22(10): 811-818.

Sivak, M. 2014. Has Motorization in the U.S. Peaked? Part 4: Households without a Light-Duty Vehicle. University of Michigan Transportation Research Institute. January. http://deepblue.lib.umich.edu/bitstream/handle/2027.42/102535/102988.pdf.

Smith, D. H. 2011. GM Investments in Fuel Economy, Design and Quality Paying Off. Automotive Discovery, August 5. http://automotivediscovery.com/gm-investments-in-fuel-economy-design-and-quality-paying-off/929726/.

Stoklosa, A. 2013. 2013 Dodge Dart Rallye 1.4T Manual. Car and Driver, February. http://www.caranddriver.com/reviews/2013-dodge-dart-rallye-14t-manual-long-term-test-wrap-up-review-no-one-will-miss-us-page-4.

Sunding, D., and D. Zilberman. The Agricultural Innovation Process: Research and Technology Adoption in a Changing Agricultural Sector. Chapter 4 in Handbook of Agricultural Economics, Volume 1, eds. B. Gardner and G. Rausser (Elesvier 2001).

Thurlow, A. 2013. Survey: For U.S. shoppers, fuel economy trumps performance. Automotive News, November 8.

Turrentine, T.S., and K.S. Kurani. 2007. Car buyers and fuel economy? Energy Policy 35: 1213-1223.

U.S Bureau of Labor Statistics. 2011. Consumer Expenditures Survey 2011, Table 2. Income before taxes: Average annual expenditures and characteristics. http://www.bls.gov/cex/2011/Standard/income.pdf. Accessed Feb. 3, 2014.

U.S. Census Bureau. 2012. Households and Families: 2010. 2010 Census Briefs, 5. http://www.census.gov/prod/cen2010/briefs/c2010br-14.pdf.

U.S. Federal Trade Commission. 2014. Buying a New Car. http://www.consumer.ftc.gov/articles/0209-buying-new-car.

Voelcker, J. 2014. Nissan Leaf $5,500 Battery Replacement Loses Money, Company Admits. Green Car Reports, July 24. http://www.greencarreports.com/news/1093463_nissan-leaf-5500-battery-replacement-loses-money-company-admits.

WardsAuto. Data Center. http://wardsauto.com/data-center.

Whitefoot, K.S., and S.J. Skerlos. 2012. Design incentives to increase vehicle size created from the U.S. footprint-based fuel economy standards. Energy Policy 41: 402-411.

World Public Opinion (WPO). 2006. U.S. Public Favors Raising Auto Fuel Efficiency (CAFE) Standards. http://www.worldpublicopinion.org/pipa/articles/btenvironmentra/193.php?nid=&id=&pnt=193&lb=bte.

10

Overall Assessment of CAFE Program Methodology and Design

The reformed Corporate Average Fuel Economy (CAFE) standards adopted into final regulations in 2010 for model year (MY) 2012-2016 vehicles and then in 2012 for MY 2017-2021 are quite different from the earlier CAFE standards in a number of important ways. The most significant changes have been mentioned already in this report and include harmonizing the fuel economy and greenhouse gas (GHG) standards and increasing the stringency of the standards in each successive year from MY 2012 to MY 2021. The GHG standards are final to 2025, and the CAFE standards are final to 2021 and "augural" for MY 2022 to 2025.[1] Standards had to be set in each year at "maximum feasible" levels through 2030, considering "technological feasibility, economic practicability, the effect of other motor vehicle standards of the Government on fuel economy, and the need of the United States to conserve energy" (49 U.S.C. 32902 (f)). This chapter discusses the legislative mandates calling for standards, their enforcement via test cycles, design of the standards and possible societal costs and benefits. Key design changes for the 2017-2025 CAFE and GHG standards are highlighted. Regulation of fuel economy by vehicle footprint and credit banking and trading, as well as by the special provisions for alternative technology vehicles (ATVs) and alternative fuel vehicles (AFVs) are all discussed.

CHOICE OF VEHICLE ATTRIBUTES IN THE DESIGN OF CURRENT REGULATIONS

The Energy Independence and Security Act (EISA) legislation in 2007 required that new fuel economy standards be based on vehicle attributes—that is, they would vary in some way by vehicle mass, size, or other relevant characteristics. The relevant fuel economy or GHG target for a vehicle would be calculated based on a mathematical formula that related the attribute to the target. The attribute-based standards began in MY 2009 for light trucks[2] and MY 2012 for passenger cars. Compliance with the standards is assessed at the manufacturer level, so a sales-weighted average of a manufacturer's fleet must meet the sales-weighted attributed-based standard. Vehicle footprint, determined by multiplying the vehicle's wheelbase by the vehicle's average track width, was chosen as the attribute upon which to base the standards (EPA/NHTSA 2012a, 62639). The National Highway Traffic Safety Administration (NHTSA), the Environmental Protection Agency (EPA), and others argue that the footprint standard encourages more technology for improved fuel economy across all vehicle sizes, with less incentive to either upweight, as could be the case with a mass-based standard, or to downsize if the standard was the same for all vehicles. In fact, NHTSA appears to be most concerned with the potential safety implications of downsizing. The issue of how the standards and vehicle safety interact is complex and is discussed in more detail below.

Several possible attributes were considered in setting the standards before the footprint standard was chosen. European, Japanese, and Chinese standards depend on vehicle weight, with heavier vehicles allowed to have more lenient standards. However, one argument for the footprint standard over the weight standard is that the former would provide incentives for manufacturers to improve fuel economy by weight reduction, rather than size reduction, thus mitigating any adverse safety impacts. A weight-based standard would not provide the same incentive, since lighter vehicles would face a tighter standard (German and Lutsey 2010). There is also concern that weight-based standards may incentivize manufacturers to make vehicles heavier, to reach a lower standard, thereby undermining some of the fuel sav-

[1] NHSTA describes the "augural" MYs 2022-2025 standards as not final and "as representative of what levels of stringency the agency currently believes would be appropriate in those model years, based on the information before us today."

[2] CAFE standards for light trucks for MY 2008-2011 included a reform to the structure for CAFE standards for light trucks and gave manufacturers the option for MY 2008-2010 to comply with the reformed standard or to comply with the unreformed standard. The reformed standard was based on the vehicle footprint. The unreformed standard for 2008 was set to be 22.5 mpg.

ings and CO_2 emissions reductions. A recent study by Ito and Sallee (2013) of the weight-based regulations in Japan suggests that vehicle weight may have increased as a result of the standards there.

Relative to a weight-based standard, incentives to increase vehicle size under a footprint standard are less clear because some argue that moving to a larger footprint requires a significant redesign of the vehicle (German and Lutsey 2010). Incentives to increase size are still theoretically possible and will depend on the cost of meeting the standard for vehicles of different sizes and consumer willingness to pay higher prices for vehicles of different sizes.

A strong motivation for a footprint-based standard is that its cost tends to fall more evenly on all manufacturers, allowing the domestic companies who produce a larger-sized fleet to meet a less-stringent fleetwide average standard than, for example, the Asian companies. The Asian manufacturers tend to advocate a more uniform standard since they build smaller, lighter vehicles. However, the footprint standard does not give any advantages to the European manufacturers, who tend to build vehicles with higher horsepower for their size.

The committee turns next to more discussion of the effects of the footprint standards on vehicle size mix and on vehicle safety. Vehicle safety from a mass reduction standpoint is also discussed in Chapter 6.

Effects of the Footprint Standard on Vehicle Size and Size Mix in the Fleet

There is some concern that the footprint standard may create the unintended incentive for manufacturers to increase the size of any given vehicle so as to lower the applicable standard. As discussed above, many argue that this type of perverse incentive is less likely compared to a weight-based standard, but it may still be an issue and should be carefully considered. In fact, the earlier CAFE standards were also attribute-based—vehicle class in that case—with one standard for passenger cars and a less stringent one for light trucks. The lower standard for trucks may have helped to accelerate the dramatic growth in CUVs, SUVs, and minivans (most of which are classified as light trucks) in the late 1980s and 1990s, when light truck sales went from 20 percent of the light-duty fleet in 1980 to over 50 percent of the fleet in 2000. The less stringent standard for light trucks was not the only reason for this change, but some of the class shifting that occurred may have been an unintended consequence of the regulations.

The footprint curves for cars and trucks in MY 2017-2025 are shown in Figure 10.1. The curves indicate the standard a vehicle of a given footprint must meet, with a new, more stringent curve for each model year. There is one set of curves for cars and another for light trucks, with the light truck curve being less stringent for any given footprint and also with a different slope and cutpoints than for cars.

The Agencies carried out extensive analyses about how to set the slope and cutpoints of the car and truck footprint curves. To try to prevent incentives to shift the size of vehicles, the Agencies developed an empirical relationship between footprint and fuel use based on sales-weighted 2008 fuel economy and footprint data, and used this relationship to set the slope of the curve. This is a reasonable attempt to reflect a general trade-off between footprint and energy use, but it does not ensure that there are no incentives to upsize or downsize created by the regulations. The incentives will depend, for example, on the costs of increasing a vehicle's footprint compared to the savings in meeting a lower fuel economy level. This will differ across vehicles. It will also depend on the profitability of vehicles of different sizes and the ability to pass higher costs on in the marketplace (elasticities of demand for vehicles of different sizes and types).

Possible Outcomes of a Footprint-Based Standard

Three outcomes related to the size of vehicles in the fleet are possible due to the regulations: Manufacturers could change the size of individual vehicles, they could change the mix of vehicle sizes in their portfolio (i.e., more large cars relative to small cars), or they could change the mix of cars and light trucks. The questions are these: What, if any, incentives are created by the footprint standard? How important are the resulting sales outcomes for the goals of the policy, including safety, fuel consumption, and GHG emissions?

The current assumptions in the societal cost-benefit analysis of the rule are that there will be no change in vehicle size or in vehicle size mix from the reference case (no regulation) as a result of the regulation. However, size mix is assumed to change to a slightly smaller vehicle fleet between 2017 and 2025, regardless of the regulation (EIA 2014).

Shifts in the Car/Light Truck Mix

Separate car and light-truck standards might incentivize a shift to light trucks from cars. The light truck standards are less strict and do not rise with size as fast for light trucks as they do for cars. This is especially true for large light trucks. The Agencies give a number of reasons why this is the case, including the fact that many large trucks tend to have low weight relative to their size (e.g., flat beds in pickups) and have greater need for towing capabilities. Several auto companies argued that the standards favor companies with a relatively large number of trucks in their fleets, implying there may be incentives to make larger, less fuel-efficient vehicles. However, factors such as added weight and four-wheel drive (as opposed to two-wheel drive) make it more expensive to shift a vehicle from a car to a truck, by definition. The committee heard arguments that there should be a single footprint curve for all vehicles instead of separate ones for cars and for trucks. As the standards become more stringent each year,

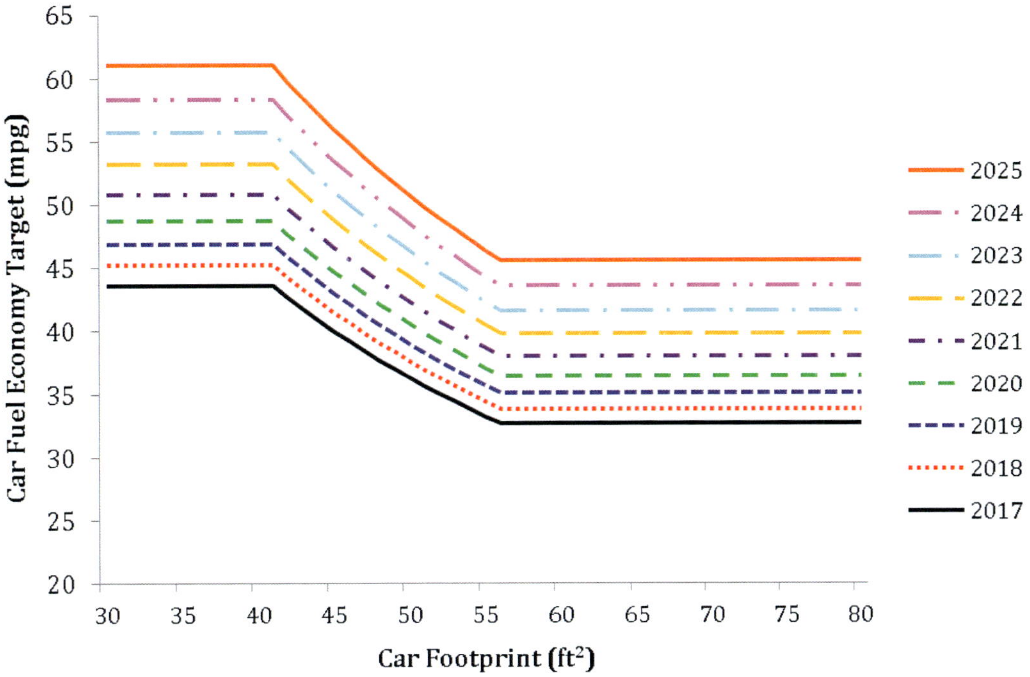

FIGURE 10.1a Fuel economy target vs. vehicle footprint for cars in each model year from 2017 to 2025. The fuel economy target increases for a given vehicle footprint as the standards become more stringent over time.

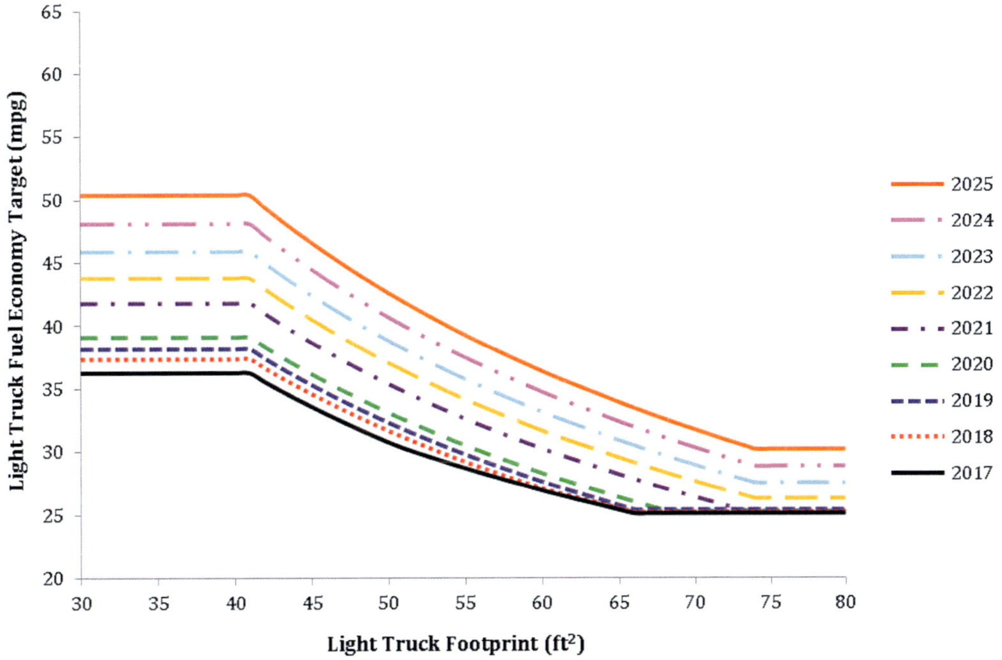

FIGURE 10.1b Fuel economy target vs. vehicle footprint for trucks in each model year from 2017 to 2025. The fuel economy target increases for a given vehicle footprint as the standards become more stringent over time.

relative costs may also change over time. In the 2017-2025 National Program, credit trading between car and truck fleets will be allowed, as discussed in more detail in this chapter. This creates a new set of incentives, depending on the cost of meeting the rules on the different types of vehicles. For example, if higher fuel economy for trucks is more costly or less profitable than for cars, the manufacturers could comply by exceeding the standards for cars, while being below the standard for trucks.

Changes in Vehicle Model Footprint

Individual vehicle footprints could also change over time. The footprint of a specific vehicle model could be increased because the cost is less than the higher cost of compliance with the standard at the lower footprint, or decreased if cost savings are greater if size is reduced. Alternatively, a specific vehicle model configuration could be dropped and another adopted because of incentives created by the rules. NHTSA and EPA have considered the issue of size shifting in setting the footprint standard and in the Regulatory Impact Assessment (RIA). The EPA RIA references the study by Whitefoot and Skerlos (2011), which uses an economic–engineering model of the vehicle fleet and changes in the fleet over time and finds that there could be some increase in vehicle size overall as a result of the footprint-based regulations, based on data available at the time of the study. As a first approach to determining if there are particular trends emerging as a result of the rules, changes in such vehicle nameplates and footprints can be monitored over time, although it is difficult to distinguish between changes occurring in the fleet due to the rule versus those that are occurring for other reasons.

Changes in Vehicle Fleet Mix

An economic behavioral model would be useful for predicting the effects of the standards on the fleet. For example, as the fuel economy standards are made more stringent over time, what is the relative shift in the marginal costs for vehicles of different sizes and how would those changes affect purchase decisions across the fleet? Are the proportionate changes in small car costs greater than large car costs, as might be expected? What is known about the elasticities of demand for vehicles of different sizes and market segments? This last question is relevant for predicting how difficult it will be to pass costs forward in different model segments. There are some estimates from the industry and from the economics literature on elasticities, but it is unclear how reliable these are. Estimates tend to show that the vehicle size/types with the lowest own-price elasticities of demand are for both large and small SUVs and large pickup trucks. If the costs of compliance are relatively lower for larger vehicles and these costs can more readily be passed on in the form of higher prices, then there could be some shift toward larger vehicles. An estimate of these impacts could help assess whether the standards create incentives that adversely affect fuel consumption, safety, and environmental outcomes.

Some recent papers have tried to incorporate these effects into models that integrate the engineering data with economic behavior of manufacturers and consumers over time. These models are potentially useful for looking at the full effects of the regulations over time, including producer and consumer responses to costs and price changes in different vehicle segments. They can also provide insight about the full costs and benefits of the regulations. The Whitefoot and Skerlos (2011) analysis is one such model, and there have been others, including one by Jacobsen (2013a) and one by Gramlich (2010). All of these models find some tendency for the size of the fleet to increase with the current footprint standards, with larger effects in the market for trucks. In particular, because of the shape of the footprint curves, the greatest incentives are for small and medium-sized trucks to get larger. Vehicle mix may also be affected by shifts from the light-duty to the medium-duty market, such as from Class 2a trucks to Class 2b trucks if the trade-off in costs is favorable. Another potential effect of the rule is to influence old vehicle scrappage rates as the price of both new and used vehicles changes over time (e.g., Jacobsen and van Benthem 2015). More attention to issues related to the overall effect of the regulations on the fleet composition is warranted. Discussion of the impact of the standards on vehicle size mix from a consumer perspective can be found in Chapter 9. The mix of vehicles needs to be tracked over time, as the Agencies are starting to do, but economic models are also important for forecasting how the mix will change.

Effect of the Footprint Standard on Vehicle Safety

The effect of attribute-based standards on vehicle safety is a complex question for a number of reasons. The mass disparity between the vehicles involved in a crash seems to be a key factor in assessing safety, with the risks for those in the lighter vehicle increasing along with the mass of the heavier vehicle (Evans 1991; NHTSA 2012a; LBNL 2012a). This implies that the greater the size disparity in the fleet, the more fatalities there are likely to be. In the past, size and mass have been highly correlated, but that has become less true recently, and the footprint standards will tend to reduce mass while attempting to keep the footprint relatively constant.

Figures 10.2a and 10.2b show the trends in distribution of vehicle weight in the fleet by car and truck from 1975 to 2007. The figures show a good deal of variation in weight for both cars and trucks, with light trucks substantially heavier than cars. Since 1991, there has been a trend toward greater weight for both cars and trucks. This likely reflects in part preferences for larger cars that consumers perceive to be safer. Figures 10.2a and 10.2b also indicate that cars have been getting heavier except for the largest passenger cars, which became much less common between 1975 and 1991, with the shift from cars to trucks. Small trucks (less than

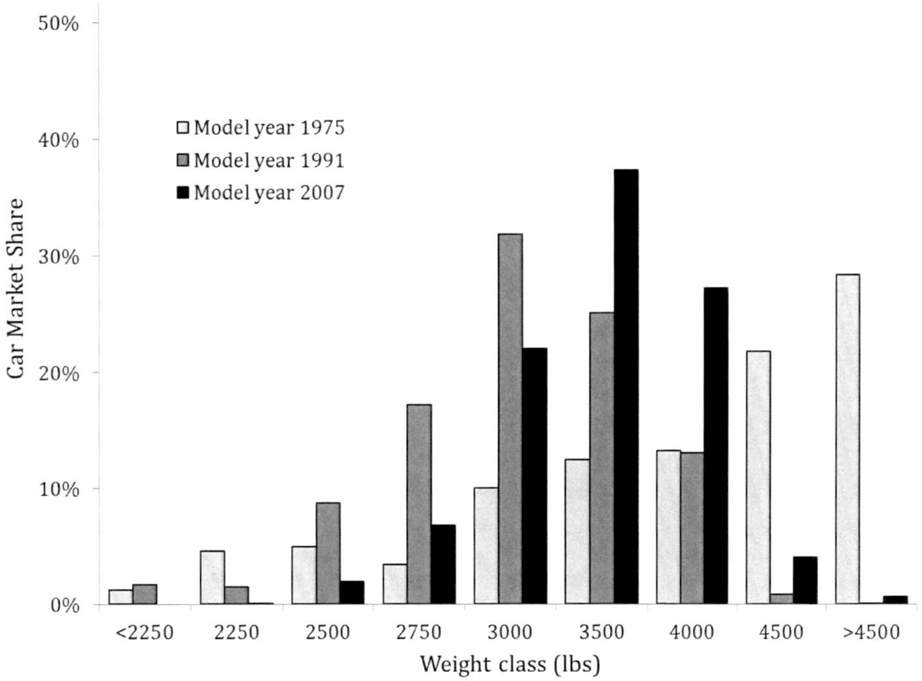

FIGURE 10.2a Changes in the distribution of car weights in MY 1975-2007.

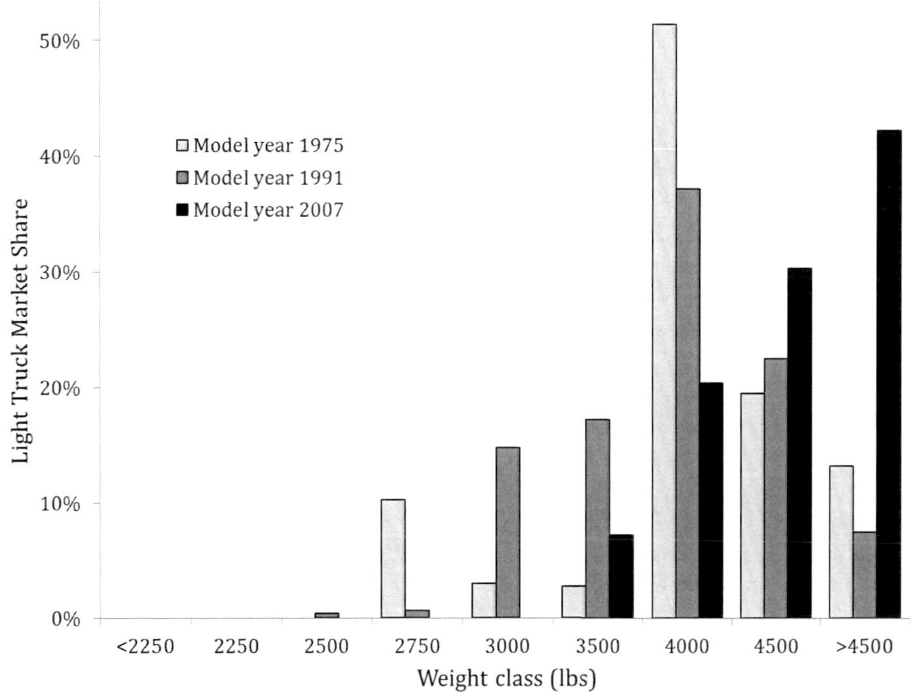

FIGURE 10.2b Changes in the distribution of light truck weights in MY 1975-2007.

3,000 lb) largely disappeared from the market, and there was a large increase in trucks greater than 4,000 lb between 1991 and 2007.

NHTSA favors the footprint-based standard in part because the Agency is concerned that any CAFE regulation that results in a downsizing of new vehicles will result in more fatalities, at least in the short term, compared to a regulation that tends to add new vehicles with the current footprint size distribution. NHTSA argues that alternative CAFE rules such as a more uniform standard would tend to result in some downsizing. There is concern that downsizing the fleet at this time will have adverse effects on safety. Much of the recent statistical evidence on safety suggests, among other things, that maintaining vehicle footprint while reducing vehicle mass may have better safety outcomes than a policy that reduces both the mass and size of the vehicles in the fleet. These studies, which are summarized in Chapter 6, attempt to isolate the effects of vehicle footprint from mass on fatalities in vehicle crashes. This is a major reason for NHTSA's implementation of the footprint standard in the recent CAFE revisions.

The Agencies may want to consider the impact on overall fleet mix and associated safety due to individual choices. When some consumers buy larger vehicles because they believe they will afford them more safety in a crash, they are not accounting for the external costs of their decision. These consumers are likely more concerned with their own safety and may not consider the societal impacts of their decision. From a social welfare perspective, this leads to a vehicle fleet that is on average heavier than is optimal and may result in more fatalities (Li 2012) than would a lighter fleet. In one study of vehicle safety, Anderson and Auffhammer (2013) attempt to estimate the magnitude of this accident-related externality and find that it is quite large.

The estimated effects of reducing mass or footprint are small compared to other vehicle attributes, driver characteristics, and crash circumstances (Figures 2.5 to 2.10 of Wenzel 2012). While, on average, mass reduction in lighter-than-average cars is associated with a small increase in fatality risk, there is a large range in risk for cars of the same mass, even after accounting for differences in vehicles, drivers, and crash circumstances (Section 4 of Wenzel 2012). It is important to note that the data used for the statistical analyses rely on historical data from recent vehicle designs and that the mass and size distribution of the fleet, and designs of vehicles, are likely to change by the time the standards become effective in MY 2017 to 2025.

NHTSA argues that alternative CAFE rules such as a more uniform standard would tend to result in some downsizing of the fleet in terms of both size and weight and that this would have an adverse effect on safety. One interesting study from the economics literature finds that this may not be the case, however. Jacobsen (2013a) analyzed different regulatory approaches for CAFE using a model of accidents that accounts for different vehicle size and safety attributes and driver behavior. His analysis suggests one standard or set of standards for vehicles, and not a separate one for cars and trucks. Though there may be some downsizing from this approach, there may also be a shift away from trucks, which makes the fleet more uniform in size. He finds that the changes in these risks offset each other.

In any case, the credit trading discussed later in this chapter should equalize the net marginal cost of fuel economy improvements across all vehicle types and manufacturers, in theory, limiting concerns about multiple standards.

Overall, evidence from available data suggests that the effect of the fuel economy rules on vehicle safety is likely to be relatively small. The selection of footprint as the attribute on which the standards are based provides a reasonable approach to a safety-neutral standard based on the information currently available. However, there should be continued study of the relationship between vehicle size, weight, and safety, and the effects of overall fleet size and mix on societal risk.

CREDIT TRADING

An important aspect of the 2017-2025 MY CAFE/GHG standard is flexibility in the means and timing of compliance offered via new opportunities in banking, borrowing, transferring, and trading "credits." Vehicle manufacturers have always had some flexibility in meeting CAFE standards, such as averaging across models in their fleet, banking credits, and paying civil penalties to comply. In the CAFE/GHG standards, credits can be earned for vehicles that have lower fuel use or GHG emissions than the target for that footprint, and can be used to offset higher fuel use or emissions of vehicles that are above the footprint-based target. Auto manufacturers have additional opportunities to earn credits, such as by producing certain AFVs or implementing technologies with off-cycle benefits (e.g., improved air conditioner efficiency). These technology-based credits are described in this chapter and in Chapter 6. The principle under fully tradable fuel economy and emissions credits is that there is a target total amount of fuel consumption and greenhouse gas emissions reductions over a period of time, but when those reductions occur and which vehicles and companies implement them are flexible. This allows the targets to be met at a lower cost.

The 2017-2025 CAFE/GHG standards allow greater flexibilities for credit trading over time, between car and truck fleets and across manufacturers. Opportunities for a manufacturer to bank and borrow credits over time will allow that manufacturer to better match product redesign cycles that are usually between 3 and 5 years, with the standard increasing in stringency every year. In addition, trading credits across companies can allow cost savings because some companies have a much greater difficulty meeting the standards than others due to differences in product types and range of vehicles offered. This increased flexibility in meeting standards is likely to be important for manufacturer compliance with the regulations. Credit trading is just beginning under the new rules, and it will be important to assess and possibly revise the provisions of the trading rules over the next few years. A key element of this assessment is whether the credit provisions of the two Agencies allow similar flexibility or whether one set of rules is more binding.

Manufacturer Averaging of Fuel Use and Emissions Across Models in Their Fleet

Each manufacturer is in compliance with the national program standards if the footprint-based, sales-weighted fleet average of fuel economy and GHG emissions is at least equivalent to the fleet-average, footprint-based standard given the actual size mix of vehicle sizes sold by the firm. A manufacturer faces two standards under both the CAFE and GHG regulations, one for cars and a more lenient one for trucks.[3] Manufacturers can average fuel consumption or emissions across all vehicles in a class (cars or light trucks), allowing substantial variation in individual model emissions and fuel consumption, even for vehicles of the same footprint. Figures 10.3 and 10.4 show the variation in fuel consumption by footprint across the entire fleet for cars and for trucks. Figure 10.3 shows the actual fuel economy for each of about 1,100 car makes and models (in red) relative to the fuel economy standard of each car based on its footprint for model year 2014. Certification fuel economy vs. vehicle footprint data for each manufacturer also shows a range of actual fuel economies relative to the standards for individual vehicle manufacturer car fleets (not shown). Figure 10.4 shows footprint and fuel economy for all trucks for the 2014 model year along with the truck standard in 2014.

It is clear from these graphs that there are a range of vehicles on the market with different fuel economies and other characteristics, even with similar footprints. Averaging within the vehicle classes (and across vehicle types and manufacturers, as we discuss below) allows manufacturers to offer a range of vehicle types and characteristics, including fuel economy. Consumers will continue to have choices about fuel economy relative to other characteristics, so the issue of how they value fuel economy remains critically important to the impact of the standards.

Defining Fuel Economy and GHG Credits

Both NHTSA and EPA allow companies to use credit surpluses or deficits in meeting the standards, but the two Agencies define credits differently, due to different regulatory mandates. This may have important consequences for automakers in meeting both standards. Because the 2017-2025 standards depend on the footprint of each vehicle, under both rules, the relevant car and light truck standards for each manufacturer will be different from those of other manufacturers and will depend on the mix of sizes of vehicles the firm sells. For EPA, the greenhouse gas standards are in terms of grams of CO_2 equivalent per mile, or in total grams of CO_2 equivalent over the lifetime of the vehicle. A manufacturer earns credits when it produces vehicles with less CO_2 per mile than its production-weighted footprint standard. Deficits are the opposite—they occur when the manufacturer's actual fleet GHG emissions exceed its production-weighted footprint standard. Credits or deficits are converted into total grams of CO_2 under or over the standard over the life of the vehicles, using an estimated car or truck lifetime vehicle miles traveled (VMT). Cars and trucks are assumed to have different lifetime VMTs, estimated by NHTSA at 195,264 miles for passenger cars and 225,865 miles for light trucks.

Under CAFE rules, a manufacturer earns credits when the vehicles it produces use less fuel per mile than the production-weighted footprint standard requires and faces credit deficits when it produces vehicles that on average have fuel use greater than the standard. A credit or deficit is earned for each 0.1 mpg difference between the standard and the actual mpg for each vehicle. Total credits earned by a manufacturer are the sum of these differences across all vehicles produced in a given year. Credits must be traded in terms of fuel consumption rather than fuel economy and so are adjusted for a vehicle's fuel consumption over the life of the vehicle.

The ability of manufacturers to earn credits is also impacted by other provisions of the NHTSA and EPA rule. For example, in the EPA program, credits are earned for the production of alternative fuel vehicles, for off-cycle emissions reductions, and for air conditioning adjustments. Many of these additional ways of earning credits are described in detail in Chapter 6. The following sections address how credits may be traded, the role they have in compliance, and whether robust markets in credits are likely to develop.

Transferring Credits Between Cars and Trucks

One new provision of the rules that went into effect starting in model year 2012 is that each manufacturer can trade credits between its own car and light truck lines for both fuel use and GHG emissions. This is referred to by the Agencies as transferring credits. So, for example, if a manufacturer's light truck fleet does not meet the light truck standard, that manufacturer can overcomply on the cars it produces and transfer the credits to the light truck fleet to make up the shortfall.

The preliminary evidence is that in the first few years of the EPA GHG program, 2012 and 2013, most of the auto companies earned many more GHG emissions credits for cars than they earned for trucks (EPA 2014a). In addition, in

[3] Since the inception of the CAFE standards, manufacturers have been allowed to average fuel economy across models within their own car and light truck fleets. An exception to this is that each manufacturer must meet a specific minimal standard for domestically produced vehicles. The domestically produced standard cannot be met with more efficient imported vehicles.

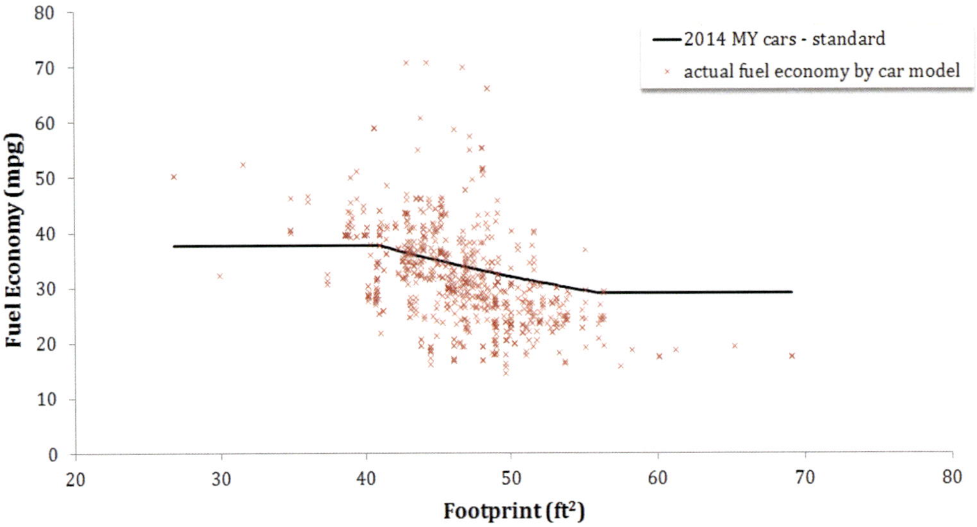

FIGURE 10.3 EPA certification fuel economy vs. vehicle footprint, plotted with the CAFE footprint standard for cars by vehicle nameplate, MY 2014.

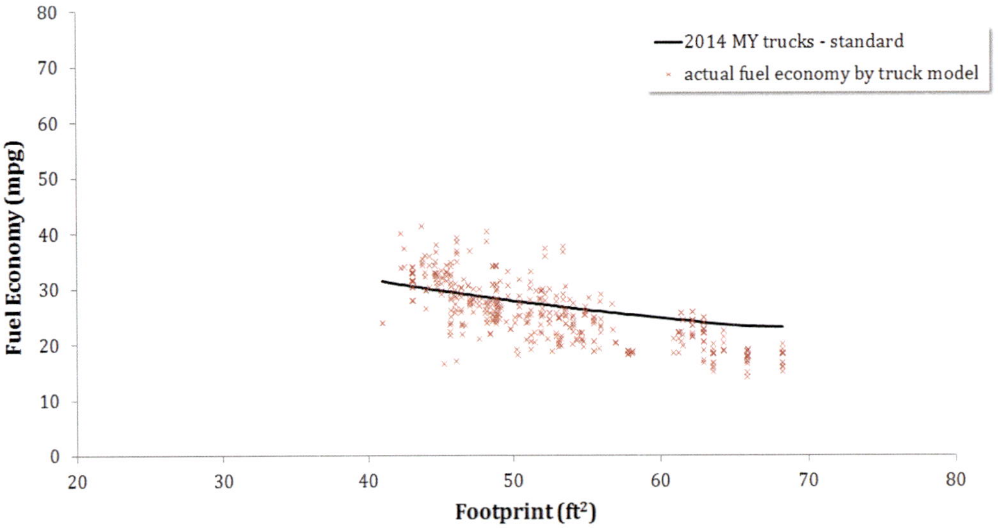

FIGURE 10.4 EPA certification fuel economy vs. vehicle footprint, plotted with the CAFE footprint standard for trucks by vehicle nameplate, MY 2014.

reporting to the EPA, about half of the roughly 20 manufacturers reported earning credits for overcompliance of their car fleets, and the other half were in deficit due to undercompliance. For trucks, only three of the companies earned credits in 2012, and overall, the industry was in a deficit with respect to the truck standards (EPA 2014a, 16). The data are not yet available from NHTSA on credits earned by cars and trucks by manufacturer, but it is likely to be similar.

This suggests that it may be more costly for manufacturers to comply with the early standards by reducing fuel use or emissions from light trucks than cars. The issue also appears to be driven by the fact that the large car market shrunk during the 1990s and was replaced by SUVs and CUVs classified as light trucks, making it easier for manufacturers to meet their car standards. It must be noted that some of the manufacturers reporting deficits in their light truck fleets used previously accumulated credits from earlier years to offset those deficits. Because credits can be banked and used in later years, drawing conclusions for any one year is difficult. It will be interesting to see if the tendency for lower compliance on the truck side persists.

Allowing manufacturers to transfer credits between its car and truck fleets will allow the automakers to meet the standards in the most profitable and cost-effective ways.

This flexibility will become increasingly important as the standards become stricter over time.

Banking Credits

Both NHTSA and EPA allow credits to be traded backward and forward over time, called banking. They allow firms to carry credits up to 5 years into the future and back in time for up to 3 years. For example, if a company cannot comply with its average standard for cars this year, it can borrow forward from its future fleets in one or more of the next 3 years, effectively making the standards they must meet more stringent in at least one of those years.

In a system with banking of credits, it is important to determine when to allow the companies to start banking. Both EPA and NHTSA allowed companies to bank credits for 3 years before the first year of the new rules (NHTSA has always allowed banking). This means companies were allowed to bank in 2009, 2010, and 2011. Not all companies overcomplied with the rules in those years and earned credits, but many did. The standards through 2011 did not depend on vehicle footprint, and the early credit accumulation will tend to favor smaller, lighter vehicle manufacturers. The number of credits earned by manufacturers in those years is quite uneven. Three companies at the end of the period in 2011 held 90 percent of all credits earned, and those companies continued to add to their credits in MY 2012. The ratio of banked GHG credits to annual production volumes in 2012 varies across the manufacturers, from about 35 down to about 5 banked credits per vehicle produced (EPA 2014a). NHTSA also reports credit holdings. A concern for vehicle manufacturers is the uncertainty about the cost and feasibility of compliance in the later years of the CAFE program, after 2016. Certain companies are in a much better position going into the 2017 regulatory phase than others.

Trading Credits Between Manufacturers

Trading between companies is now allowed under the CAFE and GHG rules and should help to address the issue of the different situations of the auto companies. Companies that have high costs or the greatest difficulties in complying can purchase credits from other companies. EPA and NHTSA both have a mechanism for companies to report trades. There have been a handful of trades between companies since trading was allowed in the beginning of 2010. For example, the EPA reports (2014a) that Mercedes, Ferrari, and Chrysler bought GHG credits, while Nissan, Tesla, and Honda sold credits in 2012. NHTSA reports credit holdings but not trades. Little is known about the prices of the transactions, as the prices are not reported, but it is likely that the prices have been lower than the fine that can be paid to comply, which is \$5.50 per 0.1 mpg per vehicle shortfall (or \$55/mpg/vehicle) under NHTSA's rules. A robust market for trading is more likely to develop if there is transparency about prices.

Differences Between NHTSA and EPA Credit Programs

There are a number of differences between NHTSA and EPA rules about credits and credit trading, illustrated in Table 10.1. In effect, two separate standards and two separate credit markets can be used to help meet those standards. Manufacturers are likely to hold, buy, or sell credits in both markets. The two credit programs are not entirely harmonized at this point. Table 10.1 shows some of these differences (Leard and McConnell 2015). First, credits are defined differently, as described above. Credits under NHTSA's rules are defined as 0.1 mpg. This means that to transfer credits over time to vehicles of different efficiencies or across vehicle classes (cars to light trucks or vice versa), an adjustment must be made to ensure that gallons of fuel used are not increased by the trade. Credits under EPA's program are in grams of CO_{2e} so they are more directly transferable. Both Agencies attempt to account for emissions or fuel used over the life of the vehicles, and they assume the VMT of cars is lower than that for light trucks. However, each Agency had different assumptions about the average number of lifetime miles of cars and of light trucks in the 2012 to 2016 rule, though VMT assumptions are now apparently the same for the 2017-2025 rule.

One of the most important differences in the two programs is that under NHTSA rules, companies can pay a fine to comply: \$5.50 per 0.1 mpg for each vehicle over the standard.[4] This is like a "safety valve" on the costs of the regulations. If the rules turn out to be more expensive than anticipated, or fall more heavily on some firms than others, then the fine sets a ceiling on the cost of additional reductions. A number of automakers have complied in this manner in the past, paying fines ranging from tens of thousands to millions of dollars per year. However, under the Clean Air Act, EPA cannot allow the auto companies to pay a fine to comply with the CO_2 standard. Instead, auto companies will be out of compliance with the Clean Air Act if they cannot demonstrate compliance by producing lower emitting vehicles, by using credits generated internally, or through trading with other manufacturers. They will have to stop offering for sale noncompliant vehicles and need to pay potentially large penalties for noncompliance, up to \$37,500 per vehicle (EPA 2009). This is likely to make the EPA rules much more binding, especially for some companies. It may also create a stronger demand for EPA credits. Credit prices could increase to high levels, depending on how difficult the standards are to meet in the later years. Some other credit markets initiated by EPA, such as in the SO_2 market, have used a safety valve mechanism to limit the increase in credit prices: Credits can be sold by the Agency at an established price and time.

Another difference in the Agencies' rules is that NHTSA puts limits on how many credits can be transferred by a

[4] Fines for compliance as described here differ from fines levied for noncompliance, such as those paid recently by automakers for incorrect testing procedures that resulted in fuel economy values too high for certain models.

TABLE 10.1 Comparison of Credit Programs under NHTSA and EPA

Provisions Related to Credits under the New Regulations	
NHTSA (fuel consumption under ECPA)	**EPA (GHG emissions under the Clean Air Act)**
Definition of a Credit	
1/10th mpg below the vehicle manufacturer's footprint-based standard	1 gram per mile CO_2 equivalent below the manufacturer's required grams per mile standard (also framed as megagrams CO_2 over life of vehicles)
FFVs accounted for as specified under EISA, assumed to have low gasoline consumption relative to other ICEs.	FFVs earn credits according to EISA provisions; but special treatment for FFVs ends in 2015.
Banking and Borrowing Credits	
Carry forward	
5 years	5 years, and credits earned between 2010 and 2016 can be carried forward through 2021
Carry back	
3 years	3 years
Transferring Credits between Car and Truck Categories	
Limits on credits that can be transferred: MY 2011- 2013, 1 mpg MY 2014 -2017, 1.5 mpg MY 2018 on, 2.0 mpg	No limits on transfers
Transfers from car to truck or vice versa must be converted from mpg to gallons of fuel.	Credits are in grams of CO_2, so grams can be traded directly between cars and trucks, and across manufacturers
Other Restrictions on Using Credits	
Credits cannot be used to meet the domestic minimum fuel economy standard (Congress established a separate minimum standard for vehicle produced in the U.S.)	No differences for vehicle produced domestically or in other countries
Exemptions	
No exemptions for manufacturers with limited product lines; fines can be paid.	Temporary Lead-time Alternative Allowance Standards (TLAAS) for manufacturers with limited product lines; also exemptions for operationally independent manufacturers
Non-compliance Penalties	
$5.50/tenth mpg over standard, per vehicle, as a fine delineated in 49 USC 32912(b), adjusted for inflation	No payment of fine to comply with the Clean Air Act. Auto manufacturers who cannot demonstrate compliance with their own fleet and accrued or acquired credits will be out of compliance with the Clean Air Act and will have to stop selling non-compliant vehicles and pay potentially large penalties, up to $37,500 per vehicle.

SOURCE: Leard and McConnell (2015).

manufacturer between its car and light truck fleets. Table 10.1 shows these limits. It is not clear why there are limits to the number of trades that can be made. EPA has no limits. Also, NHTSA does not allow credit trading from the overall car fleet to the domestic fleet for meeting the minimum domestic fleet standard.

NHTSA and EPA have differing provisions for calculating compliance fuel economy or GHG emissions and hence credits earned for production of flex-fuel vehicles (FFVs). Currently, FFVs are treated in a similar way by the two Agencies. They are allowed to be counted as having very low CO_2 emissions (discussed in more detail later in the chapter). This favorable treatment for FFVs is currently set to expire at the end 2015 under the EPA rules, but it will not expire for the CAFE rules until MY 2020, which was what the automakers agreed to when they supported the original MY 2012-2016 GHG and CAFE program. There are a handful of manufacturers that earn substantial credits by producing these vehicles. These manufacturers will be bound by EPA's more stringent FFV credit system after 2015. Described in more detail later in the chapter, beginning in 2016, the compliance GHG emissions of FFVs will assume they operate on 100 percent gasoline unless automakers choose to use national averages of E85 use, currently estimated as 14 perecnt of all fuel used in FFVs, or petition to use manufacturer-specific data on FFV fuel use (EPA 2014d). Also, no extra incentive for the alternative fuel portion of FFV compliance fuel economy, the 0.15 factor, will be used in the EPA program.

Expected changes to the EPA credit program are likely to affect the ability of some manufacturers to earn credits in the future. The EPA's Temporary Lead-Time Alternative Allowance Standards (TLAAS) for manufacturers with limited product lines is only in place through the 2015 MY. Under these provisions, manufacturers with sales of less than 400,000 in the United States in 2009 are allowed to meet a lower standard for MY 2012 to 2015. Manufacturers such as Mercedes and Porsche are eligible for this exception and have complied with a more lenient standard. When this provision expires in 2016, compliance may be difficult for many of these automakers. They have frequently paid fines to comply with CAFE standards in the past but will not be able to pay fines under the EPA rules.

Overall Assessment of Credit Provisions

Credit use within firms and across vehicle classes and trading across firms will become increasingly important for keeping costs of compliance down as the CAFE and GHG standards tighten over time. There are a number of restrictions that limit the use of credits. Some of these limits seem unnecessary, such as the NHTSA restriction on the number of credits that can be transferred between cars and light trucks. With the banking provisions, and increasingly strict standards, the committee expects that many firms will overcomply in the early years, so that they can exceed the standards in later years. It is likely to be cost effective to spread the costs of complying over time (this happened in the SO_2 trading market in the 1990s). Also, the value of a credit, whether it is transferred or traded to another company, is expected to rise over time as the standards get stricter. The use of credits conveys key information about the ease or difficulty of meeting standards, and the price of credits should reflect the cost of additional controls for meeting the standards. Both EPA and NHTSA are monitoring manufacturers' compliance with the rules. Collecting this information and making it available is key for a smoothly functioning credit program and credit market.

It is clear that manufacturers are facing very different situations today, in terms of their credit positions, due to different vehicles and the different fuel economies of vehicles in the market. Some firms have no credits or very few, and others have a great many credits already accumulated. Because of such different positions, firms would likely benefit from being able to trade with each other. Some firms appear to have very high costs per vehicle for meeting standards and some much lower costs; otherwise, they would not find it advantageous to trade credits. It is important that a robust market be allowed to develop to ensure the regulations are successful. Uncertainty about technological progress and consumer acceptance of new technologies may make firms reluctant to trade credits. The midterm review is an appropriate time to consider what the credit market barriers might be as the standards tighten over the next few years. Finally, whether the NHTSA and EPA credit markets should be more harmonized should be explored. If they are not harmonized, what are the implications for how manufacturers comply with the rules?

ASSESSING ADEQUACY OF THE CERTIFICATION TEST CYCLES

Why Is the Test Cycle Important?

Compliance with the CAFE standards is determined by testing vehicles on dynamometers in a laboratory over carefully defined test cycles under controlled conditions. This is necessary to ensure consistency of measurements across vehicles and manufacturers, and over time. Most of the testing is done by the manufacturers, who certify to the EPA that the testing has been done correctly. The EPA tests a smaller number of vehicles to monitor compliance. Certification is based on a weighted average of two test cycles.

The fuel economy tests used to certify vehicles' compliance with the CAFE standards tend to overestimate the average fuel economy motorists will typically achieve in actual driving (EPA/NHTSA 2012a, 62988). This is reflected in the systematically-adjusted lower fuel economy values the government reports on new vehicle window stickers, via the website www.fueleconomy.gov, and in the *Fuel Economy Guide*. The Agencies also adjust the certification fuel economy values downward when evaluating the future impacts of the fuel economy and GHG standards, using a 20 percent fuel economy shortfall for vehicles operating on liquid fuels and a 30 percent shortfall for hybrid vehicles (EPA/NHTSA 2012a, 62989). The difference is in part due to the greater opportunities hybrids offer to "engineer to the test." The relationship between the test values and fuel economy performance in the real world is of great importance because the primary benefits of the CAFE standards depend entirely on the in-use improvements achieved: (1) reduced petroleum consumption, (2) reduced GHG emissions, and (3) fuel cost savings to consumers. Investments in vehicle technology and design changes that do not produce real-world fuel economy or GHG benefits are wasted. However, as long as the *ratio* of real-world to test-cycle fuel economy remains constant as test-cycle fuel economy improves, the expected benefits will be realized. On the other hand, if the ratio decreases over time and the gap between real-world and test-cycle fuel economy grows, the benefits of the standards will be smaller than expected and the standards' cost/benefit ratio will likely increase.

The Energy Policy and Conservation Act of 1975 (EPCA) that established the CAFE standards limited EPA's ability to modify the certification test procedures. In particular, the EPCA stipulated that "the Administrator shall use the same procedures for passenger automobiles the Administrator used for model year 1975. . . or procedures that give comparable results" (49 U.S.C. 32904(c)). This requirement has prohibited EPA from changing the fuel economy test procedures in any way that meaningfully changes the resulting miles per gallon estimates. In the 2012 Final Rule, the EPA argued that it should be allowed to change the fuel economy test cycles in ways that do not replicate the 1975 test results. The Agency's rationale is that a new interpretation of the restriction is warranted, given the need to harmonize the CAFE standards with new GHG regulations that are not bound by the EPCA's 1975 limitation. If this interpretation is validated in court, it may create an opportunity to modify the existing two-cycle test.

Adequacy of the Two-Cycle Certification Test Procedure

EPA recognizes that the two drive cycles currently used to certify vehicles for fuel economy compliance—the FTP, or "city," cycle and the HWFET, or "highway," cycle—are not adequate representations of real-world driving behavior (EPA 2011b; Rosca n.d.). Both cycles were originally developed to measure pollutant emissions and were subsequently adopted for measuring fuel economy in 1975. Early evidence that the test cycles overestimated average real-world fuel economy led to the development and implementation of correction factors in 1984 by the EPA used to inform the public (Hellman and Murrell 1984). The correction factors discounted the city fuel economy estimates by 10 percent and the highway estimates by 22 percent (EPA 2012, A-11). Compliance with the CAFE standards remained based on the unadjusted two-cycle tests.

Although the EPCA limits EPA's authority to change the fuel economy test procedures, the Agency has latitude under the Clean Air Act to modify the emissions tests. The three additional test procedures were adopted by EPA in 1996 to better reflect criteria pollutant emissions of automobiles during real-world operating conditions (EPA/NHTSA 2012a, 62803). The Agency noted that pollutant emissions from vehicles were often substantially higher when the vehicles were operated at speeds, acceleration rates, and under other conditions not present in the two-cycle tests. Data collected in actual traffic confirmed that higher speeds and acceleration were commonplace, as was air conditioner use in warm weather. A new high-speed cycle was added to include speeds up to 80 mph and maximum acceleration rates more than 2.5 times those of the city and highway test cycles (Table 10.2). An air conditioner test was added to estimate the impacts of AC use in hot weather. Finally, a cold temperature test was added that repeats the city cycle test in an ambient temperature of 20°F.

In 2008, the adjustment factors for fuel economy labels were revised once again, adding further downward adjustments, this time based on the five different test cycles. Two of these are the FTP and HWFET cycles (on which CAFE compliance is based), while the other three reflect more aggressive and higher speed driving (US06), use of air conditioners (SC03), and a "cold temperature" version of the FTP cycle (EPA 2012, A-9). Relative to the previously adjusted city and highway numbers, the new adjusted numbers were approximately 11 percent lower for the city cycle and 8 percent lower for the highway cycle, although the degree of adjustment is higher the higher a vehicle's fuel economy numbers, suggesting an expectation that the test vs. real-world gap will increase with increasing miles per gallon. For example, the current gap in fuel economy between the certification two-cycle test and the adjusted label five-cycle fuel economy values is approximately 20 percent

TABLE 10.2 Comparison of EPA Test Cycles

Driving Schedule Attributes	Test Schedule				
	City	Highway	High Speed	AC	Cold Temp
Trip Type	Low speeds in stop-and-go urban traffic	Free-flow traffic at highway speeds	Higher speeds; harder acceleration & braking	AC use under hot ambient conditions	City test w/ colder outside temperature
Top Speed	56.7 mph	60 mph	80 mph	54.8 mph	56.7 mph
Average Speed	21.2 mph	48.3 mph	48.4 mph	21.2 mph	21.2 mph
Max. Acceleration	3.3 mph/sec	3.2 mph/sec	8.46 mph/sec	5.1 mph/sec	3.3 mph/sec
Simulated Distance	11 mi.	10.3 mi.	8 mi.	3.6 mi.	11 mi.
Time	31.2 min.	12.75 min.	9.9 min.	9.9 min.	31.2 min.
Stops	23	None	4	5	23
Idling Time	18% of time	None	7% of time	19% of time	18% of time
Engine Startup	Cold	Warm	Warm	Warm	Cold
Lab Temperature	68–86°F	68–86°F	68–86°F	95°F	20°F
Vehicle Air Conditioning	Off	Off	Off	On	Off

NOTE: Though the FTP test is run over 11.1 miles, the first cold transient portion and last hot transient portion of 3.6 mi are weighted at 0.43 and 0.57, respectively, with the middle cold stabilized portion of 3.9 mi weighted at 1.0. See CFR 40 chapter I subchapter U part 1066 Subpart 1 §1066.820.
SOURCE: DOE (2014).

for conventional vehicles and 30 percent for hybrid electric vehicles (HEVs).

An additional problem with the certification test procedures is the method currently used by EPA in setting vehicle weight for chassis dynamometer testing. For fuel economy certification tests, a loaded vehicle weight is determined from the vehicle weight plus 300 lb (two passengers). In the current test procedure, the dynamometer inertial load is not set to the actual loaded vehicle weight, but instead bins loaded vehicle weight into predetermined ranges, called equivalent test weight classes (ETWCs). ETWC ranges are narrower (125 lb) for lighter vehicles and broader for heavier vehicles (250 lb up to 500 lb if the 250 lb ETWC setting is not available). Broad ranges mean that weight reduction does not reduce the vehicle's test weight until the reduction is sufficient to move it into the next lower weight class. This limits the CAFE/GHG compliance benefits of weight reduction to automakers using the current certification procedures. Using actual vehicle weight plus 300 lb to set the chassis dynamometer would allow automakers to more fully realize the compliance benefits of implemented weight reductions. The Worldwide harmonized Light vehicles Test Procedure, adopted in Europe, will use actual vehicle weight for testing, with benefits discussed by the International Council on Clean Transportation (ICCT) (Mock 2011). One complication of using actual loaded vehicle weight rather than ETWC is that manufacturers can currently group several series of a vehicle line into a single test, reducing their compliance burden. This complication might be addressed by continuation of the practice of permitting several series of a vehicle line to be grouped within the sales-weighted, average vehicle test weight.

Need for Real-World Fuel Economy Data

The value of the test cycle estimates as predictors of real-world fuel economy can only be determined by comparing them to real-world fuel economy data. As estimates of real-world fuel economy, the test cycle and adjusted miles per gallon estimates should be evaluated on the basis of bias and accuracy. Bias measures the degree to which the estimates consistently over- or understate the mean, or average, fuel economy experienced by all drivers in actual driving. Accuracy measures the degree to which the fuel economy estimates deviate from the individual fuel economies achieved by individual drivers in actual driving. For the purposes of ensuring that the fuel economy standards achieve the goals of reducing light-duty vehicle petroleum consumption, GHG emissions, and fuel costs, unbiasedness is sufficient. For public information purposes, however, accuracy is also important.

While it is obvious that measuring outcomes in the real world is necessary to determine the real-world performance of the standards, scientifically valid real-world fuel economy data has not been collected in the United States for almost 20 years. The 1984 adjustment factors were based on an extensive statistical analysis of real-world driving data collected by the EPA from diverse sources (Hellman and Murrell 1984). The 2008 adjustments were based largely on engineering analysis and judgment, with more limited statistical analysis of real-world driving data (EPA 2006). Unfortunately, there is no current scientific survey of real-world fuel economy in the United States.

It is now possible to collect real-world fuel economy data using vehicles' On-Board Diagnostic systems (OBDII). Statistics Canada (2014) has been collecting data on vehicle use and fuel consumption via engine data loggers connected to vehicles' OBDII systems since 2013. The data loggers automatically record data on a vehicle's operation when its engine is on, as frequently as every second. In the Canadian survey, about 150 vehicles are active at any given time and a vehicle remains in the sample for three weeks. Data are collected on approximately 7,000 to 8,000 vehicles a year. There are still issues to be resolved in estimating fuel consumption from OBDII data (Posada and German 2013). The ICCT conducted a feasibility and scoping study to estimate the cost of designing and implementing a U.S. survey (ERG, Inc. 2013).

Modern information technology may also enable the estimation of more accurate, individualized fuel economy numbers. No single test cycle can represent the range of differences in driving behavior, traffic, and environmental conditions that exist in the real world. For this reason, fuel economy labels have always cautioned motorists that "your mileage may vary." To be of greatest value to car buyers, fuel economy information should be accurate for the individual driver. Developing more accurate, individualized fuel economy estimates may now be possible thanks to vehicles' computer systems, GPS, and advances in vehicle simulation modeling. By continuously recording data from a vehicle's OBDII system, it should be possible to create an individualized driving pattern specific to a driver's actual driving conditions and driving behavior that could then be used to predict a driver's individual fuel economy. Innovators have already begun to develop applications for analyzing such data to predict how changing behavior might improve fuel economy or to create individual fuel economy estimates for vehicles the consumer has never driven (see Fleetcarma 2014; Fiat 2014).

A valid understanding of the relationship between test-cycle and real-world fuel economy based on in-use data could fill three important information gaps for regulators and consumers:

- Unbiased estimates necessary for quantifying the benefits of the fuel economy and GHG standards;
- Assurance that time and money spent to increase fuel economy on the test cycles would result in real benefits to motorists and society; and
- Improved methods of estimating individual fuel economy that would increase the value of fuel economy information and perhaps reduce the tendency of consumers to undervalue future fuel savings.

The evidence to date is mixed and limited by the lack of statistically valid in-use fuel economy data. Canada may be the only country to have continued measuring on-road fuel economy through the 1990s until today. A 1999 study by Natural Resources Canada (ECMT 2005) concluded that fuel consumption of passenger cars was 23 percent higher than combined city/highway test estimates and that the comparable number for light trucks was 27.9 percent. Analyzing U.S. data, Mintz et al. (1993) found an average miles per gallon shortfall of 18.6 percent for passenger cars and 20.0 percent for light trucks in the early 1990s. Since this exceeded the approximately 15 percent adjustment adopted earlier by the EPA (Hellman and Murrell 1984), Mintz et al. concluded that the mpg gap was widening over time. However, as ECMT (2005) observed, the findings of Mintz et al. are actually consistent with estimates made earlier by McNutt et al. (1982). McNutt et al. (1982) also concluded that the mpg shortfall increased with increasing mpg. In the Final Rule (EPA/NHTSA 2012a, 62988), NHTSA calculated that actual fuel economy for passenger cars was 21-23 percent lower than the test- cycle numbers but only 16-18 percent lower for light trucks, based on a comparison with Federal Highway Administration estimates. The Federal Highway Administration estimates average national fuel economy by estimating aggregate VMT and then dividing by aggregate fuel consumption. The Agencies noted that the gap between compliance and in-use fuel economy may increase in response to the Final Rule and promised to monitor real-world fuel economy performance and improve and update their estimates of the on-road gap, as appropriate (EPA/NHTSA 2012b).

Mock et al. (2013, 2014) found that EU certification fuel consumption estimates fell short of real-world estimates by about 8 percent in 2001, increasing to 21 percent by 2011 and 38 percent by 2013. That study was based on a number of data sources, including approximately 6,000 records per year self-reported to the German website www.spritmonitor.de, and 1,200 vehicles tested by the EU auto club ADAC. The authors attributed the growing gap to increasing use of tolerances and loopholes in test settings, the inability of the test cycle to represent real-world driving conditions, and an increasing market share of vehicles equipped with air conditioning.

U.S. studies relying on 20,000 vehicle records self-reported by users of the website www.fueleconomy.gov found that the 1984 adjusted EPA estimates were almost perfectly unbiased estimators of real-world fuel economy for vehicles with spark-ignition internal combustion engines (ICEs) (Lin and Greene 2011; Greene et al. 2007). The study also found that the adjusted EPA estimates slightly underpredicted diesel vehicle fuel economy and substantially overpredicted hybrid vehicle fuel economy. The variance of the reported hybrid fuel economy numbers was also considerably greater than that of ICE-only vehicles. Neither the 1984 adjusted nor the 2008 adjusted estimates were accurate for a specific vehicle: a two-standard deviation confidence interval was estimated to be +/− 7 miles per gallon.

Whether the gap between the two-cycle certification tests and real-world fuel economy will increase in the future is important to estimating the costs and benefits of the fuel economy standards. Deviation of real-world fuel economy from EPA window sticker value, as well as from the CAFE compliance values, is expected to increase as some additional fuel economy technologies are applied to vehicles. When a vehicle is driven more aggressively, such as at higher speeds and higher acceleration rates than specified by the FTP75 and HWFET drive cycles used for CAFE compliance, more fuel will be consumed. If the vehicle has a conventional, naturally aspirated engine, the fuel consumption outside the CAFE drive cycles differs from on-cycle fuel consumption due to the gradual changes in BSFC values on the fuel consumption map of the engine and the increased power requirements at the higher speeds or acceleration rates.

For a turbocharged, downsized engine, changes in the BSFC values on the fuel consumption map outside the CAFE drive cycles can be greater than with a naturally aspirated engine. For example, with a highly turbocharged and downsized engine, higher speeds may require enrichment to limit exhaust temperature to protect the turbocharger and catalyst. This enrichment would increase fuel consumption beyond what would be experienced with a naturally aspired engine. A similar effect would occur at higher acceleration rates.

The available evidence suggests that there may be a tendency for the miles per gallon shortfall to increase over time as fuel economy increases and advanced technologies like hybrid vehicles and turbocharged, downsized engines increase their market share. This is noted in Chapters 2 and 4. However, the evidence is not conclusive. A definitive answer will likely require an effort to collect statistically representative in-use fuel economy data.

THE TREATMENT OF "ALTERNATIVE" TECHNOLOGIES IN THE CAFE/GHG PROGRAM

This section provides a review of the regulatory structure for CAFE and GHG compliance of AFVs and ATVs and assesses how the methods might align with program goals and actual performance.

CAFE Program

The goal of the CAFE program, as established by the Energy Policy and Conservation Act of 1975 (EPCA, P.L. 94-163), is to reduce U.S. dependence on oil primarily by raising the fuel efficiency of cars and trucks. In 1988, Congress modified provisions of the CAFE program through the Alternative Motor Fuels Act of 1988 (AMFA, Pub. L. 100 94). The goal of AMFA was to increase energy security and improve air quality by promoting the widespread use of alternative fuels. For ethanol, methanol, and natural gas,

AMFA is very specific on how the Agencies should provide CAFE credits. For electric vehicles, AMFA did not specify the credit but authorized the Department of Transportation (DOT) to provide additional incentives if it found it necessary to stimulate production. AMFA also limited the extent to which a manufacturer can use flex-fuel and dual-fuel vehicle credits to increase average fuel economy. For model years 1993 through 2004, the maximum increase was 1.2 mpg for each category of automobiles (domestic and imported passenger car fleets and light truck fleets). AMFA allowed the incentive program to be extended on the approval of the Secretary of Transportation for up to 4 years beyond MY 2004, but at a ceiling reduced from 1.2 mpg to 0.9 mpg. In 2004, DOT set the limit at 0.9 mpg for MY 2005-2008.

The Energy Independence and Security Act of 2007 (EISA) extended the fuel economy credits for flexible-fuel vehicles (FFVs) and dual-fuel AFVs through MY 2019 (P. L. 110-140) but phased out these credits by MY 2020. The maximum increase that may result from FFVs and dual-fuel AFVs was capped at 1.2 mpg through 2014, after which it declines in 0.2 mpg increments to 0.2 mpg by 2019 and then expires in 2020. EISA did not phase out the fuel economy credits for dedicated AFVs.

In May 2009, the Obama administration announced a new harmonized national policy for GHG emission and CAFE standards. EPA's GHG program phases out FFV and dual-fuel credits by MY 2016 for GHG compliance purposes, 4 years earlier than EISA phases them out for CAFE compliance purposes. EPA also developed new methodologies for GHG ratings for AFVs based on well-to-wheels GHG emissions and better estimates of actual alternative fuel usage. Starting with MY 2020, EPA and NHSTA will use the same methodology for FFVs and dual-fueled AFVs, which will be based on estimates on or actual data from the fraction of miles that such vehicles operate on the alternative fuel. Dedicated AFVs will continue to use the same methodology in current law for CAFE compliance and the full well-to-wheels methodology for GHG compliance purposes.

Dedicated Alternative Fuel Vehicles

For dedicated liquid alternative fuel vehicles (including methanol or ethanol high-blend fuels), 49 U.S.C. 32905, as originally specified by AMFA, requires the certification fuel economy for CAFE compliance purposes to be based on the fuel economy when tested on the alternative fuel (such as M85 or E85) adjusted by a "fuel content" factor of 15 percent by volume of petroleum-derived fuel (either gasoline or diesel).[5] The general formula is as follows (Rubin and Leiby 1998):

$$MPG_{CAFE\ dedicated\ AFV} = \frac{MPG_{measured}}{0.15}$$

For example, if a dedicated E85 vehicle was rated at 25 mpg when tested on E85, the fuel economy would be adjusted by dividing by 0.15 (equivalent to multiplying by 6.67), yielding a fuel economy for compliance purposes of 167 mpg [(1/0.15) × (25) = 167 mpg].

Dedicated natural gas-powered automobiles are treated in a similar manner to dedicated alcohol fuel vehicles. For dedicated natural gas automobiles, 49 U.S.C. 329, as originally specified by AMFA, requires that that the certification fuel economy for CAFE compliance purposes be based on the rated or measured fuel economy when running on natural gas (miles per 100 cubic feet) adjusted by the energy content conversion factor (0.823 gallons per 100 cubic feet) and divided by the same fuel content factor as used for alcohol-fueled vehicles (0.15).[6] For example, a dedicated natural gas vehicle that achieves 25 miles per 100 cubic feet of natural gas would have a CAFE value of 203 mpg [(25/100) × (100/0.823)(1/0.15) = 203 mpg]. Unlike with alcohol fuels, there is no physical justification for the 15 percent adjustment factor; the apparent intent of Congress when it adopted AMFA was to provide the identical incentive for natural gas vehicles as for E85 AFVs, a treatment that has been extended to biodiesel (B20) vehicles and electric vehicles, as discussed below.

For battery electric vehicles (BEVs), 49 U.S.C. 329 requires NHTSA to calculate the fuel economy for CAFE compliance purposes using a conversion factor, called the Petroleum Equivalency Factor (PEF), developed by the Department of Energy (DOE).[7] The fuel economy for compliance purposes (or the "petroleum-equivalent fuel economy") is simply the PEF (in Wh/gallon) divided by the rated energy efficiency (in Wh/mile). The PEF for electricity has been set by DOE at 82,049 Wh/gal. The PEF is derived by first calculating a well-to-wheels, gasoline-equivalent energy content of electricity (E_g) and then dividing it by the same 0.15 "fuel content" factor used for alcohol and natural gas-powered vehicles.[8] E_g is calculated as follows:[9]

[5] AMFA defined "alcohol" as a mixture containing 85 percent or more by volume of methanol, ethanol, or any other alcohol. AMFA recognized dedicated AFVs as those that operate exclusively on a 70 percent or greater methanol or ethanol concentration, or only on compressed or liquefied natural gas as "dedicated" AFVs. This treatment has been extended to EVs.

[6] The conversion factors for other gaseous alternative fuels (in gallons equivalent per 100 standard cubic feet): LNG = 0.823; LPG (Grade HD-5) = 0.726; hydrogen = 0.259; hythane = 0.741 (Federal Register Vol. 61 No. 64).

[7] Note EPA is tasked under EPCA to measure and calculate fuel economy for individual models. 49 U.S.C. 32904(a)(2)(B) expressly requires EPA to calculate the fuel economy of electric vehicles using the PEF developed by DOE, which contains an incentive for electric operation already.

[8] The general form of the PEF equation is: $PEF = E_g \times (1/0.15) \times AF \times DPF$, where AF is the petroleum-fueled accessory factor for EVs with auxiliary petroleum-fueled accessories such as cabin heater/defroster systems and DPF is the driving pattern factor (set to 1.0 assuming capabilities similar to conventional vehicles) (CFR 65, 113). Most PEVs do not use petroleum-fueled accessories.

[9] "Electric and Hybrid Vehicle Research, Development, and Demonstration Program; Petroleum-Equivalent Fuel Economy Calculation; Final Rule," 10 CFR Part 474, 2000-06-12.

$$E_g = (T_g \times T_t \times C) / T_p$$

where:
T_g = U.S. average fossil-fuel electricity generation efficiency = 0.328
T_t = U.S. average electricity transmission efficiency = 0.924
T_p = Petroleum refining and distribution efficiency = 0.830
C = Watt-hours of energy per gallon of gasoline conversion factor = 33,705 Wh/gal
E_g = (0.328 × 0.924 × 33705)/0.830 =12,307 Wh/gal[10]

For example, a BEV that is rated on the certification test cycle at 230 Wh/mi (roughly equivalent to a Nissan Leaf) is treated as a vehicle with a 357 mpg petroleum-equivalent fuel economy for compliance purposes [82,049 Wh/gal × (1/230 Wh/mi) = 357 mpg]. In contrast, the same vehicle would be rated at 147 mpg-equivalent on a tank-to-wheel basis (33,705 Wh/gal × 1/230 Wh/mi), and 54 mpg-equivalent on a well-to-wheels energy equivalency basis (12,307 Wh/gal × 1/230 Wh/mi).

Flex-Fuel and Dual-Fuel Vehicles

The methodology for the fuel economy of FFVs and dual-fuel vehicles for CAFE compliance purposes through 2019 MY is specified in 49 U.S.C. 32905. The basic calculation is a harmonic average of the fuel economy for the alternative fuel and the conventional fuel (a 50/50 split), regardless of the fraction of each type of fuel actually used. In addition, the fuel economy value for the alternative fuel is significantly increased by dividing by a "fuel content" factor of 0.15 (equivalent to multiplying by 6.67). The general formula is as follows (Rubin and Leiby 1998):

$$MPG_{CAFE\ FFV} = \frac{1}{\dfrac{0.5}{MPG_{measured,gas}} + \dfrac{0.5}{\dfrac{MPG_{measured,alt\ fuel}}{0.15}}}$$

For FFVs, 49 U.S.C. 32905 requires that the fuel economy be calculated similar to dual-fuel vehicles as the harmonic average of the measured fuel economy when running on petroleum fuel and the compliance fuel economy when running on alcohol fuel.[11] This is equivalent to assuming that the vehicles would operate 50 percent of the time on petroleum fuel and 50 percent of the time on alcohol fuel, and continues to adjust the alcohol fuel economy by dividing by the fuel content factor, 0.15. For example, for an FFV that is rated at 25 mpg on the petroleum fuel and 17 mpg when operating on an alcohol fuel, the resulting fuel economy for compliance purposes would be 41 mpg:

$$MPG_{CAFE\ FFV} = \frac{1}{\dfrac{0.5}{25} + \dfrac{0.5}{\dfrac{17}{0.15}}} = 41 \text{ mpg}$$

However, EPA recently finalized an E85 use weighting factor of 0.14 rather than 0.5 for the GHG standard for MY 2016-2018 that manufacturers may use for weighting CO_2 emissions, as discussed in Chapter 2. Therefore, the EPA weighting factor for CO_2 emissions is more restrictive than the 0.5 weighting factor for CAFE and will limit the application of FFVs.

For dual-fuel natural gas vehicles, 49 U.S.C. 32905, as originally specified by AMFA, requires that the certification fuel economy be calculated as the harmonic average of the tested or measured fuel economy when running on conventional fuel and that when running on natural gas using the same 0.15 volumetric conversion factor as for dedicated alcohol-powered vehicles. The calculation is the same as the FFV. PHEVs are another example of a dual-fuel vehicle. Through 2019, dual-fueled vehicles such as PHEVs are considered to operate 50 percent of the time on gasoline and 50 percent on the alternative fuel. Beginning in 2020, dual-fueled vehicle fuel economy will be weighted by modeled usage of the two types of fuel.

Beginning in MY 2020, EPA has authority under EPCA to develop measurement and fuel economy protocols for the CAFE program (49 U.S.C. 32906) for FFVs and dual-fuel vehicles. Under the MY 2017-2025 Final Rule, EPA finalized its proposal to use the same methodology to weight the alternative and conventional fuel use for both CAFE standards and GHG emissions compliance.[12] For ethanol FFVs, manufacturers have the choice of using national average E85 usage data or manufacturer-specific E85 usage data. The default is to use the gasoline fuel economy value for FFVs. For PHEVs and dual-fuel CNG vehicles, the fuel economy weightings will be determined using the Society of Automotive Engineer (SAE) utility factor methodology, SAE J1711. The SAE J1711 utility factor approach is its recommended practice for measuring the exhaust emissions and fuel economy of HEVs.[13] The SAE J1711 procedure calculates a utility factor that is based on the vehicles' electric range and assumes that drivers charge once per day and drive duty cycles similar to the average light-duty passenger vehicle. For example, based on the cycle-specific fleet utility factors, the 2012 Chevrolet Volt PHEV, which has an all-electric range of 38 miles over EPA's two-cycle certifi-

[10] Dividing E_g by 0.15 yields the PEF = 82,049 Wh/gal.

[11] Under EISA, B20 (20% biodiesel and diesel mixture) is also given the same 0.15 fuel content factor as other liquid alternative fuels such as E85 and M85.

[12] Note that while the weighting methodology is the same, the CO_2 and fuel economy ratings methodologies when operating on an alternative fuel still differ.

[13] 76 FR 39504-39505 and 40 CFR 600.116-12(b). For more detailed information on the development of this SAE utility factor approach, see http://www.SAE.org, specifically SAE J2841 "Utility Factor Definitions for Plug-In Hybrid Electric Vehicles Using Travel Survey Data," September 2010.

cation tests, has a combined city/highway cycle utility factor of 0.69, meaning that the average Volt driver is projected to drive about 69 percent of the miles on grid electricity and about 31 percent of the miles on gasoline. The following equations are the J1711 method for petroleum-only fuel economy, the method used for MY 2020 and beyond CAFE fuel economy compliance (Al-Alawi and Bradley 2014):

$$UF_{Urban} = \frac{1}{\frac{1-UF_U}{PCT_U}}$$

$$UF_{Hwy} = \frac{1}{\frac{1-UF_H}{PCT_H}}$$

$$UF_{Petroleum\ FE} = \frac{1}{\frac{0.55}{UF_{Urban}} + \frac{0.45}{UF_{Hwy}}}$$

where

UF_{Urban} is the utility factor-weighted fuel economy for the urban drive cycle;

UF_U is the urban utility factor, essentially the fraction of urban driving expected to be displaced by an AFV of certain range;

UF_{Hwy} is the utility factor-weighted fuel economy for the highway drive cycle;

UF_H is the highway utility factor, essentially the fraction of highway driving expected to be displaced by an AFV of certain range;

$UF_{Petroleum\ FE}$ is the combined city/highway petroleum only fuel economy;

PCT_U is the partially charged test fuel economy for the urban drive cycle; and

PCT_H is the partially charged test fuel economy for the highway drive cycle.

Using this method, Alawi and Bradley (2014) calculate that a compact car PHEV with 20-mile range (PHEV20) would have a compliance fuel economy of 90 mpg and a compact car PHEV with 60-mile range would have a compliance fuel economy of 226 mpg.

Dual-fueled natural gas vehicles would use the same method as PHEVs to weight the natural gas and petroleum driving portions. EPA provides specific utility factors based on the SAE methodology, which appear identical to the PHEV utility factor (a 50 mi range dedicated natural gas vehicle has a utility factor of 0.689, the same as a PHEV50). A dual-fuel CNG vehicle with a 150-mi two-cycle CNG range would result in a compliance assumption of 92.5 percent operation on CNG and 7.5 percent operation on gasoline. A dual-fuel CNG vehicle with a driving range of less than 30 miles would use a utility factor of 0.50.

CO_2 and Fuel Economy Incentives for Advanced Technologies in Full-Size Pickup Trucks

Under the authority of the Clean Air Act and EPCA, EPA provides a per-vehicle CO_2 credit in the GHG program and NHTSA provides an equivalent fuel consumption improvement value in the CAFE program for manufacturers that sell significant numbers of large pickup trucks that are mild or strong HEVs or exceed a specific CO_2 performance threshold. EPA's rationale for these incentives is that it believes that the MY 2012-2025 standards will be "challenging for large vehicles, including full-size pickup trucks often used in commercial applications." EPA's intent is to pull forward penetration of new technologies, especially hybrids, in the MY 2017-2021 time frame that will help manufacturers meet the more stringent MY 2022-2025 truck standards.

There are four different incentives for advanced technology full-size pickups: two technology-based and two performance-based. The technology-based incentives differ for mild and strong HEV pickup trucks. Mild and strong HEV pickup trucks are defined based on energy flows to the high-voltage battery. The performance-based incentives are for other promising technologies besides hybridization that can provide significant reductions in GHG emissions and fuel consumption, such as lightweight materials. To avoid double-counting, no truck will receive credit under both the HEV and the performance-based approaches.

Mild HEVs are eligible for a per-vehicle CO_2 credit of 10 g/mi and an equivalent 0.0011 gal/mi petroleum credit during MYs 2017-2021. To be eligible, at least 20 percent of a company's full-size pickup production in MY 2017 must be mild HEVs, and that ramps up to at least 80 percent in MY 2021. Strong HEV pickup trucks are eligible for a 20 g/mi credit (0.0023 gal/mi) during MY 2017-2025 if the technology is used on at least 10 percent of a company's full-size pickups in that model year.

Full-size pickup trucks certified as performing 15 percent better than their applicable CO_2 target will receive a 10 g/mi credit (0.0011 gal/mi), and those certified as performing 20 percent better than their target will receive a 20 g/mi credit (0.0023 gal/mi). The 10 g/mi performance-based credit will be available for MY 2017 to 2021 and, once qualifying, a vehicle model will continue to receive the credit through MY 2021, provided its CO_2 emissions level does not increase. The 20 g/mi performance-based credit will be provided to a vehicle model for a maximum of 5 years within the 2017 to 2025 MY period provided its CO_2 emissions level does not increase. Minimum sales penetration thresholds apply for the performance-based credits, similar to those adopted for HEV credits.

GHG Standard Program Treatment of BEVs, PHEVs and FCEVs

EPA has broader discretionary authority under the Clean Air Act for treatment of AFVs and ATVs, though the basis for the treatment must be grounded in effective reductions in air pollutants. The permanent regulatory treatment for GHG emissions compliance of plug-in hybrid electric vehicles (PHEVs), BEVs, and fuel cell electric vehicles (FCEVs) will use a well-to-wheels analysis (EPA/NHTSA 2012a, 62820). EPA's GHG standard program has two incentives for BEVs, PHEVs, and FCEVs: zero emission treatment and sales multipliers. For MY 2017-2021, the GHG emission program sets a value of 0 g/mi for the tailpipe CO_2 emissions compliance value for BEVs, FCEVs, and PHEVs (based on electricity usage). For MY 2022-2025, the program allows the 0 g/mi treatment for up to a per-company cumulative sales cap tiered as follows: (1) 600,000 BEV/PHEV/FCEVs for companies that sell 300,000 BEV/PHEV/FCEVs in MY 2019-2021; or (2) 200,000 BEV/PHEV/FCEVs for all other manufacturers. Starting with MY 2022, the compliance GHG emissions value for BEVs, FCEVs, and the electric portion of PHEVs in excess of individual automaker cumulative production caps must be based on net upstream accounting of CO_2 emissions.

The GHG standard program also provides a sales multiplier that allows a manufacturer to count each BEV/PHEV/FCEV/compressed natural gas (CNG) vehicle as more than one vehicle in the manufacturer's compliance calculation. EPA's rationale for providing multipliers is "to provide temporary regulatory incentives to promote advanced vehicle technologies" (EPA/NHTSA 2012a, 62650). EPA provides CNG vehicle multipliers since it believes that the infrastructure and technologies for CNG vehicles could serve as a bridge to use of advanced technologies such as hydrogen fuel cells. BEVs and FCEVs start with a multiplier value of 2.0 in MY 2017 and phase down to a value of 1.5 in MY 2021. PHEVs and CNG vehicles start at a multiplier value of 1.6 in MY 2017 and phase down to a value of 1.3 in MY 2021.

The impact of these zero emission treatment and multipliers is to effectively provide a CO_2 credit toward a manufacturer's fleet average compliance calculation. For every 1 percent of total passenger car production, the zero emission treatment alone is worth 2.1 g/mi starting in MY 2017, declining to 1.4 g/mi in MY 2021. When the 2.0 multiplier in MY 2017 is considered, the credit is worth 4.2 g/mi.

Appropriateness of Credits for Alternative Technologies

Battery Electric Vehicle Incentives

The CAFE incentives for BEVs provide a fuel economy credit toward a manufacturer's compliance, thereby reducing the average fuel economy required of its conventional vehicle fleet. Based on the fuel economy standards for passenger cars in the MY 2017-2025 Final Rule, the committee estimates that if a manufacturer chose to produce 1 percent BEVs, it could reduce the fuel economy of its conventional fleet by 0.35 mpg in MY 2017 and 0.47 mpg in MY 2025. The incentives effectively create a trade-off: Current petroleum consumption will be higher under the rule with the PEV incentives, in exchange for the potential for greater petroleum reductions in the future due to the deployment of PEVs. The value of the credits may be thousands of dollars per vehicle. This incentive may drive additional deployment of PEVs. But this may not be the most cost-effective way to increase the number of alternative fuel vehicles in the long run.

California's Zero Emission Vehicle (ZEV) requirements will also influence the rate of adoption of PHEVs, BEVs, and FCEVs by auto manufacturers. Large-volume manufacturers will be required to supply 15.4 percent of the vehicles they sell in California and other participating states as zero emission vehicles (i.e., as either PEV or FCEV by 2025 (CARB 2012)). As of 2014, nine states in addition to California are adopting the ZEV standards, representing a total of about 28 percent of the new vehicle market in the United States (ZEV Program Implementation Task Force 2014). The volumes of PHEVs, BEVs, and FCEVs estimated by the California Air Resources Board to be produced in compliance with the ZEV mandate in California are illustrated in Figure 10.5. The ZEV mandate does not directly impact the CAFE standards but will influence the way these manufacturers meet the federal CAFE/GHG standards. The vehicles that manufacturers sell to comply with the ZEV mandate will form a part of their compliance with the CAFE/GHG standards, meaning that they will need fewer fuel economy improvements from their conventional vehicles than would have been required without the ZEV mandate.

MY 2020 and Beyond Methodology for FFVs and Dual-Fuel Vehicles

The impact of the utility factor approach for PHEVs and dual-fueled CNG and the demonstration of actual usage for FFVs will be to increase the credits for PHEVs and dual-fuel CNG vehicles and decrease the credits for FFVs. The Agencies believe that while weighting to better reflect real-world usage is a major change, this change "orients the calculation procedure more to the real-world impact on petroleum usage, consistent with the statute's overarching purpose of petroleum conservation" (EPA/NHTSA 2012a, 62829).

Many analysts have pointed out that the current crediting system for FFVs that assumes a 50/50 split of alcohol to petroleum has led to the unintended consequence of increasing petroleum dependency since only a small fraction of FFVs actually uses E85 (DOT et al. 2002; GAO 2007; Liu and Hefland 2009). Tying FFV credits to actual use of an alternative fuel is consistent with DOT et al. (2002) and GAO (2007) recommendations. DOT et al. (2007) recommended "linking the CAFE credit to actual alternative fuel used." GAO (2007) recommended that the dual-fuel program should

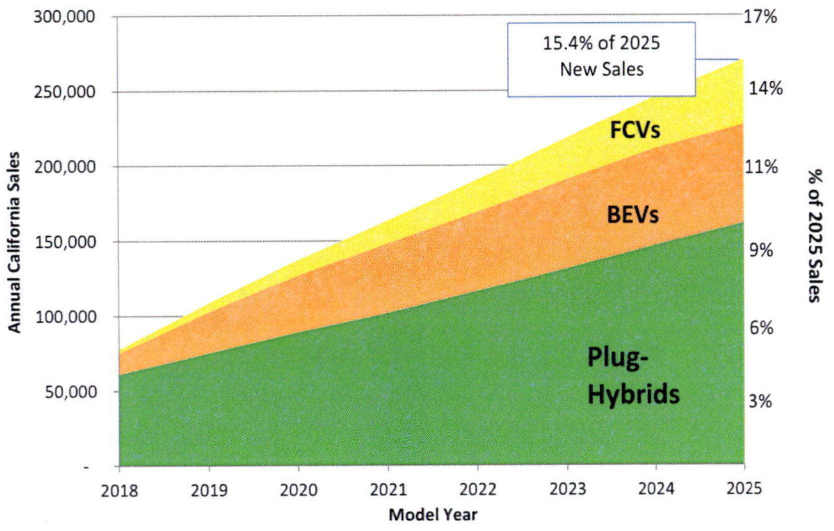

FIGURE 10.5 Annual California sales for MY 2018-2025 expected for ZEV regulation compliance, showing projected PHEV, BEV, and FCEV sales.
SOURCE: CARB (2011).

be "eliminated or revised." It recommended "lowering the credit to more accurately reflect how often these vehicles are actually run on alternative fuels could be appropriate."

Adoption of the "utility factor" method to credit PHEVs and dual-fuel CNG vehicles is an improvement over the previous method of assuming a 50/50 split. However, while the utility factors are based on actual survey data of travel behavior, this will not necessarily correlate with alternative fuel refueling behavior in the real world. The Agencies postulate that if a driver spends the extra money on PHEV or CNGV, he/she is more likely to use the alternative fuel. This assumption may hold well for PHEVs that have multiple refueling options (home charging). In fact, early adopters of the Chevrolet Volt, as studied in the EV Project, drive on average 75 percent on electricity, more than the utility factor predicts (ECOtality 2013). The assumption that utility factors accurately predict alternative fuel usage behavior may not be as reasonable for a dual-fuel CNGV if infrastructure is not readily available.

APPROACH AND METHODOLOGY USED TO SET STANDARDS AND EVALUATE COSTS AND BENEFITS

Introduction

The Agencies use a series of models to carry out the quantitative analysis necessary to estimate feasible levels of fuel economy increases and GHG reductions and their costs. Each Agency models plausible technology changes to forecasted future fleets that would result in compliance for each manufacturer. The structure of the analytical methodology is similar for both of the Agencies and is illustrated in Figure 10.6. Data on the performance of vehicle components and systems, including engine maps, aerodynamic drag coefficients, and other information, are inputs to the full vehicle simulation model. The full vehicle simulation model predicts the impacts of advanced fuel economy and GHG technologies on fuel consumption and emissions for seven base vehicles representing seven vehicle classes. These data are used to calibrate a simpler lumped parameter model that can be used to estimate impacts for millions of technology-vehicle combinations. These impacts, together with estimated technology costs, are used by the Agencies' compliance models (Volpe and OMEGA) to estimate manufacturer-specific compliance with fuel economy and emissions standards.

In the compliance models, technologies are ordered by cost effectiveness subject to engineering and manufacturing constraints. The models iterate, adding technologies to individual makes, models, and engine-drivetrain configurations with the objective of achieving cost-effective reductions in petroleum use and GHG emissions. Through the use of these models, the Agencies developed what the committee termed the EPA/NHTSA compliance demonstration path representing a cost-effective set of technologies that automobile manufacturers could adopt to meet the standards. These compliance demonstration paths are broken down by individual manufacturer and by model year. These are reported as the technology penetrations by manufacturer and year in the Final Rule and in supporting documents. Although the Agencies' analysis demonstrates a possible technology path to compliance, each OEM will plot its own future course to compliance. Thus, what the

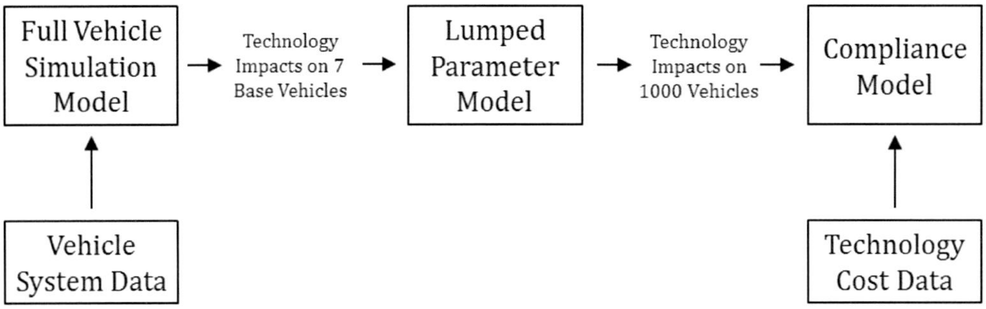

FIGURE 10.6 Simplified diagram illustrating the Agencies' methodology for setting standards.

Agencies' analysis shows is a demonstration of possibility, not a forecast of the future.

Baseline and Reference Vehicle Fleets

In addition to establishing the technologies that could be implemented for fuel economy by 2025, the Agencies also evaluated the costs and benefits of the rule. Developing both the technology paths and the costs and benefits used the concepts of the null, baseline, and reference vehicle fleets. The committee developed Figure 10.7 to aid in its understanding of the relationship among the null vehicle, baseline fleet, reference fleet, reference case, and control case.

The Agencies developed the null vehicle package as a reference point against which effectiveness and cost can be consistently measured across compliance models (Olechiw 2014). Chapter 8 defines the null vehicle concept and describes how the baseline fleet is built up from a null vehicle to an estimate of the actual fleet as it existed in either 2008 or 2010. This building up of the null vehicle to the actual fleet is shown Step 2 of Figure 10.7 using grey squares to represent the technologies added to the individual vehicle models. One complexity in building the baseline fleet for the 2017-2025 CAFE/GHG standards is the use of two baseline years. This was done in large part due to the effects of the economic recession on the 2008 sales and sales mix. The recession not only caused a drastic reduction in vehicle sales but altered the distribution of sales and led to the termination of certain vehicle makes. Thus, the Agencies developed a second baseline fleet using 2010 certification data.

Once the baseline fleet for 2008 or 2010 had been defined, technologies were added to the individual vehicle models within an OEM's lineup until, combined with projected sales volumes for each model, each manufacturer's car and truck sales met the 2016 CAFE/GHG standards. The OMEGA and Volpe models are used for this purpose. The projected fleet that reached compliance in 2016 defined the reference fleet for 2017. Figure 10.7 represents the technologies added to move from the baseline to reference fleets in Step 3. The varying numbers of technologies and lengths of time between technology additions in Figure 10.7 illustrate that each model may start at a different level of technology and may apply a different number of technologies at different times.

The 2017 reference fleet formed the basis of both the reference case and the control case. To make the reference case, the reference fleet was futured by assuming no technology improvements and no improvements in fuel economy beyond 2016, but allowing for forecasted changes in vehicle sales and class/model mix over time. The 2017 reference fleet was also used to form the control case, in which technologies were added to increase fuel economy and reduce GHG emissions from 2017 to 2025 to meet the 2025 standards. The difference between the control case and the reference case amounted to the Agencies' assessment of the costs and benefits of the standards.

The Agencies' use of the two 2008 and 2010 baseline fleets is a step toward acknowledging the uncertainty of future vehicle markets. The large differences between the 2008 and 2010 data (only 2 years apart) is a reminder of the difficulty of forecasting the vehicle market as far as 15 years into the future. Makes and models will come and go, the popularity of vehicle classes will change, new vehicle types will be created, and existing ones will fade away. The committee knows of no methods for accurately predicting the volume, composition, and technology of light-duty vehicle sales 15 years into the future; however, important economic uncertainties should be included as a sensitivity analysis in the reference case. Also, there is potential for economic–engineering models to use forecast data on cost, effectiveness, and demand elasticities to make useful uncertainty analyses of the broad effects of the rule. The price of gasoline, for example, is critically important to the costs and benefits of the rule, and various assumptions about the price of gasoline should be included among the economic uncertainties evaluated. Analyzing the nature and degree of uncertainty over similar time periods in the past may provide useful guidance about the nature and degree of uncertainty that can be expected in the future.

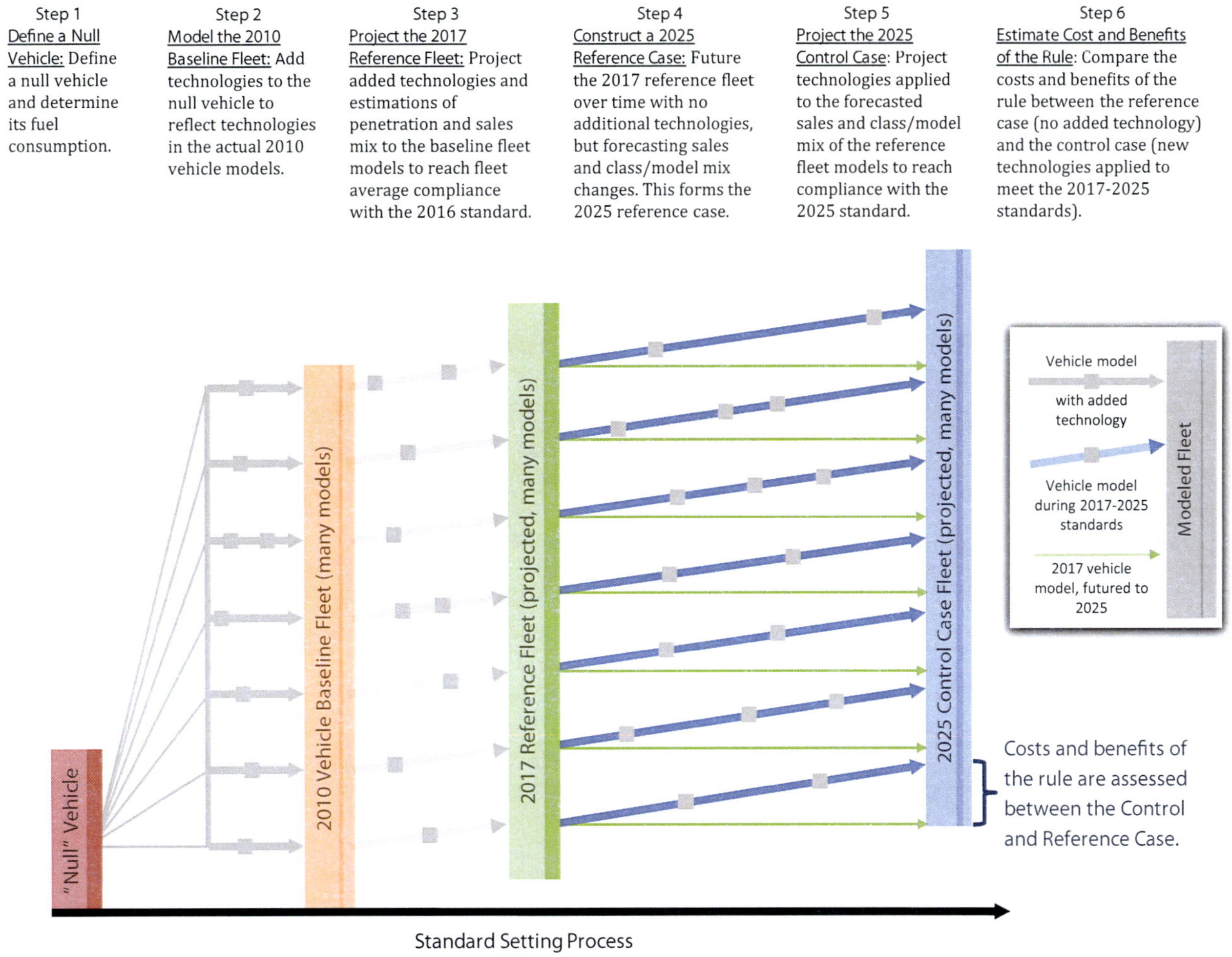
Step 1
Define a Null Vehicle: Define a null vehicle and determine its fuel consumption.
Step 2
Model the 2010 Baseline Fleet: Add technologies to the null vehicle to reflect technologies in the actual 2010 vehicle models.
Step 3
Project the 2017 Reference Fleet: Project added technologies and estimations of penetration and sales mix to the baseline fleet models to reach fleet average compliance with the 2016 standard.
Step 4
Construct a 2025 Reference Case: Future the 2017 reference fleet over time with no additional technologies, but forecasting sales and class/model mix changes. This forms the 2025 reference case.
Step 5
Project the 2025 Control Case: Project technologies applied to the forecasted sales and class/model mix of the reference fleet models to reach compliance with the 2025 standard.
Step 6
Estimate Cost and Benefits of the Rule: Compare the costs and benefits of the rule between the reference case (no added technology) and the control case (new technologies applied to meet the 2017-2025 standards).
"Null" Vehicle
2010 Vehicle Baseline Fleet (many models)
2017 Reference Fleet (projected, many models)
2025 Control Case Fleet (projected, many models)
Vehicle model with added technology
Vehicle model during 2017-2025 standards
2017 vehicle model, futured to 2025
Modeled Fleet
Costs and benefits of the rule are assessed between the Control and Reference Case.
Standard Setting Process

FIGURE 10.7 Schematic illustrating the Agencies' definition of null vehicles, 2010 baseline fleet, 2017 reference fleet, reference case, and control case used to evaluate the CAFE/GHG standards. Each arrow represents a vehicle model progressing through time. The slanting arrows represent increasing fuel economy over time. Technologies added to the models are represented by squares. In setting the standards, the agencies define a null vehicle, a 2008 or 2010 baseline fleet modeled on the real 2008 and 2010 fleets, and a 2017 reference fleet projected to comply with the 2012-2016 standards. The Agencies then form a reference case, representing the 2017 reference fleet progressing through time with no added technologies, and a control case, representing the 2017 reference fleet progressing through time with added technologies to meet the more stringent standards. The costs and benefits of the rule are determined between the reference and control cases, summed over the sales-weighted vehicle models.

Reference Case and Implications for Estimating Costs and Benefits

The costs and benefits of improved fuel economy as a result of the rule from 2017 to 2025 are estimated relative to a reference case that assumes some growth over this period in the overall vehicle fleet and a relative shift toward cars and away from trucks. It assumes, however, that there will be no changes after 2016 to the fuel economy of individual model vehicles. The implication of comparing the improvements in each year to the reference case is that in the absence of the rule, the fuel economy of the fleet would not have changed at all through the 2017 to 2025 time period—fuel economy would have remained through time at its 2016 level. The Agencies acknowledge this assumption and believe it is consistent with consumer choices from 1984 to 2004 (see Figure 9.1).

Using the 2016 vehicle as a reference also implicitly assumes that there would be no other improvement to other vehicle characteristics in the absence of the standards. This is equivalent to a reference case with no further technical change in the vehicle market from 2017 to 2025. An alternate reference case for the benefit cost analysis would account for a rate of technological progress similar to what has occurred in the past. The rate of technological progress in vehicle attributes and efficiency has been strong and continual over the past 30 years, as shown in Figure 9.1. Also, EPA (2014c) provides further evidence of past trends.

Developing a reference case that reflects technological progress over time is important for attempting to account for costs and benefits that might be left out of the analysis. The reference case with no fuel economy changes should instead include some attempt to measure improvements in other vehicle attributes likely to occur over time. Then, with the introduction of the rule, and all improvements going toward fuel economy, there will be opportunity costs in terms of the other attributes that are forgone. NHTSA acknowledges this issue in the Final Rule when they state, "the true economic costs of achieving higher fuel economy should include the opportunity costs to vehicle owners of any accompanying reductions in vehicles' performance, carrying capacity, and utility, and omitting these will cause the agency's estimated technology costs to underestimate the true economic costs of improving fuel economy" (EPA/NHTSA 2012a, 62988). The committee recognizes the difficulty of determining an appropriate reference fleet over time and of estimating the opportunity costs. But there are various approaches that could be developed to incorporate such forecasts in the reference fleet. There is past evidence about the rate of technological change that provides some guidance. There are continuing efforts to improve estimates to the value of fuel economy and other vehicle attributes to consumers that could inform the estimates of opportunity costs.

Technology Impact Estimation

Estimating the impacts of technologies with the potential to reduce fuel consumption and greenhouse gas emissions is complicated by three issues:

- Implementations of technological concepts differ across manufacturers and even across vehicles made by the same manufacturer.
- Technologies often have secondary effects on vehicle attributes that may require additional engineering or design changes that affect fuel economy.
- Dynamometer testing of two vehicles with and without the technology in question but that are otherwise identical is generally not possible.

As a consequence, estimating the impacts of technologies on fuel consumption always involves a degree of uncertainty.

Full Vehicle Simulation Modeling

The National Research Council's (NRC's) 2011 report on fuel economy technologies recommended that the Agencies make use of full system simulation modeling (a.k.a. full vehicle simulation) to estimate the impacts of fuel economy technologies:

> Full system simulation (FSS), based on empirically derived powertrain and vehicle performance and fuel consumption data maps, offers what the committee believes is the best available method to fully account for system energy losses and synergies and to analyze potential reductions in fuel consumption as technologies are introduced into the market. (NRC 2011, 155)

The Agencies have endeavored to follow this recommendation and made extensive use of full vehicle simulation modeling in their technical analyses in support of the 2017-2025 rule. Simulation analyses for the 2011-2016 rule were carried out by Ricardo, Inc. using its commercially available simulation model, EASY5. While the EASY5 model is commercially available, Ricardo used proprietary input for the engine maps, transmission efficiencies, and shift schedules for the EPA analysis. Those analyses included 26 technology packages applied to five vehicle classes. For the 2017-2025 rule, an additional 107 vehicle packages were simulated by the same consulting firm for seven vehicle classes (Ricardo 2011). In addition, a design of experiments method was used to vary input parameters to develop data for predicting the combinations of factors such as engine size and final drive ratio that would yield the greatest reduction in fuel consumption while meeting the requirement of performance equivalent to the baseline and reference fleets (EPA 2012a, 3-55). The simulations included four advanced engine concepts, five advanced transmissions, and two hybrid vehicle architectures (EPA 2012a, Tables 3-5 and 3-6).

The Agencies have made substantial progress toward the goal of full system simulation modeling for every important technology pathway and for every vehicle class. Full vehicle simulation modeling has limitations, however. First, because it is skilled-labor- and data- intensive, it is also relatively expensive. Second, the expertise and software resources have historically been found only in the OEMs and industry research and consulting firms. Some of the necessary information held by these firms is proprietary, which constrains the ability of the Agencies to obtain peer review and to accomplish full disclosure. Third, full vehicle simulation modeling, as envisioned by the 2011 NRC committee, can only be carried out for technologies that have already been competently incorporated into at least one vehicle. Only then can the performance and data maps be empirically derived. For technologies that have not been implemented, existing engine maps and other key inputs can be modified by expert judgment, which makes validation difficult. A preferred alternative to this approach is to use a detailed engine model calibrated to an existing engine map, to develop maps for engines that have not been developed in hardware (as was done in the committee's University of Michigan full system simulation discussed in Chapter 8). The Agencies should consider adopting this approach for technologies that have not been implemented. Finally, due to resource limitations, full vehicle simulation modeling is not feasible for every one of the approximately 1,000 vehicles in the baseline or reference fleets. Because the technologies present on other vehicles will differ from the configuration used in the simulation modeling (and their implementations will vary as well), other methods must be used to estimate the impacts on each and every vehicle in the baseline and reference fleets.

Validation of the full vehicle simulation model runs is difficult because advanced technologies are sometimes not available in an actual vehicle, especially in the full range of combinations considered in the modeling. The EPA conducted an external peer review of the modeling by Ricardo, Inc., in which the review panel expressed frustration with their lack of access to proprietary data and models. The EPA reported to the committee on actions it has taken to validate its simulation modeling results by comparisons with dynamometer tests on existing vehicles (EPA 2014b). The EPA also reported that it is developing its own simulation model named ALPHA, in order to allow full public disclosure of the model and its input data (EPA 2012a).

Lumped Parameter Modeling

Lumped parameter models simplify the representation of a complex system by using a smaller number of elements and associated parameters to approximate the behavior of the full system. The objective of lumped parameter modeling of vehicle fuel economy is to represent the synergies in reductions of energy losses among technologies in a model that is orders of magnitude less complex than a full vehicle simulation model. The EPA developed a lumped parameter model in order to estimate the impacts of combinations of technologies on the baseline and reference fleets. In 2012, NHTSA and EPA used outputs of the lumped parameter model to calibrate inputs to NHTSA's Volpe model and EPA's OMEGA model.

EPA's lumped parameter model represents the conversion of chemical energy in fuel to thermal and mechanical energy in the vehicle. It quantifies the losses of energy in the vehicle system: the determinants of the forces the vehicle must overcome to accomplish the dynamometer test cycles as well as energy dissipated in braking (EPA 2012a, 3-69). The baseline vehicle is described by a fixed percentage of chemical energy going to each category of energy use (loss), including thermodynamic losses, exhaust heat, pumping losses, engine friction losses, transmission losses, vehicle road load losses, and inertial losses. Fuel economy technologies reduce specific categories of energy losses by a certain percentage. This avoids double counting of benefits and helps ensure that the overall impact estimates do not violate physical laws. Because it is far less complex than full vehicle simulation models, the lumped parameter model could be used to estimate impacts on all 1,000 or so vehicles and millions of vehicle–technology combinations for the baseline and reference fleets. These results are used by the OMEGA and Volpe models in estimating compliance with the fuel economy and emissions standards.

The EPA's lumped parameter model is calibrated to the full vehicle simulation results. This is presumably done by adjusting the energy loss shares; however, the technical support document does not describe the calibration process in sufficient detail to evaluate it. EPA presented to the committee a sample of comparisons between the lumped parameter model predictions and the 2011 Ricardo simulation results (EPA 2014b). The comparisons supported EPA's assertion that the lumped parameter model predictions are within 3 percent of the full simulation modelling results for the seven baseline vehicles and "with a few exceptions" within 5 percent for advanced technology packages.

Cost Estimation

From the perspective of the costs and benefits of fuel economy and GHG standards, accurately estimating costs is as important as estimating technology impacts. The methods the Agencies use for estimating direct and indirect costs over time are discussed in Chapter 7. This section addresses the way costs are used in the methodology for setting standards. Technology costs are used to calculate one or more cost-effectiveness indices for each technology. Conceptually, cost-effectiveness is defined as the incremental cost per percent reduction in fuel consumption ($/% FC). A high cost-effectiveness can be expressed either as a high fuel consumption reduction effectiveness/cost or as the inverse, a low cost/fuel consumption effectiveness, as is reported in the EPA RIA and in the example pathways in Chapter 8. Technologies are applied to vehicles in the compliance models in order of cost-effectiveness, subject to a number of constraints. The constraints include the applicability of a technology to a specific vehicle class, its availability in the year being simulated, its compatibility with other technologies in use on the vehicles, and whether it requires that other technologies be implemented prior to its use. A technology's retail price equivalent cost includes both direct manufacturing and indirect costs.

Compliance Models

The Agencies are obligated to provide a least-cost compliance path that shows how each OEM might comply with the standards, not necessarily the path they will actually follow. Each Agency has its own model for estimating compliance with fuel economy and GHG standards and the costs and benefits thereof. Both models are available to the public. The NHTSA model (a.k.a. the Volpe model) was developed for earlier rulemakings and revised in 2012 (NHTSA 2012a). The EPA's OMEGA model is similar with respect to inputs and outputs and the logic for determining compliance (OMEGA). Both models apply technologies to the baseline and reference fleets in order of cost-effectiveness, subject to constraints to represent availability, applicability, and engineering logic. Both models calculate manufacturer-specific standards based on the footprints and sales of the vehicles in the baseline and reference fleets using the footprint versus fuel economy and GHG functions. Neither model estimates the impacts of fuel economy or GHG standards on the mix of vehicles sold, although both Agencies have research projects under way to investigate the feasibility and value of estimating such impacts. The models iteratively apply technologies to each manufacturer's vehicles until the specified standard is met or the available technologies are exhausted. The models differ in their methodology for adding technologies to a vehicle. The Volpe model applies technologies to every vehicle model within each manufacturer's fleet using decision trees until the manufacturer's fleet achieves compliance. In contrast, the OMEGA model develops "master sets of technology packages" for each vehicle class. The OMEGA model applies these packages to the vehicle classes in the entire U.S. fleet rather than by manufacturer, as in the Volpe model. The Volpe model also allows manufacturers to pay a penalty if the cost of meeting the standard exceeds the statutory fine for noncompliance. Since that feature is not an option under the GHG regulations, the Volpe model allows it to be disabled.

Both models take account of the availability of technologies in time and the normal redesign cycles for vehicles. Technologies may be designated as applicable or not applicable to each class of vehicle. Each technology is also described by an earliest year in which it becomes available for use and a later year in which it will no longer be available. For recently introduced technologies, limits can be placed on how rapidly the technology can be adopted. One difference between the OMEGA and Volpe models is that the Volpe model's algorithm calculates compliance by model year whereas the OMEGA model applies technologies for a "redesign cycle," which is assumed to be approximately 5 years.

According to EPA, "OMEGA assumes that a manufacturer has the capability to redesign any or all of its vehicles within this redesign cycle. OMEGA does not attempt to determine exactly which vehicles will be redesigned by each manufacturer in any given model year" (2012b, 6). This method does not permit evaluation of banking, borrowing, and credit trading, features that have been shown to be important to manufacturers' abilities to cost-effectively achieve compliance with fuel economy and emissions standards (Rubin et al. 2009; Bunch and Greene 2011; Liu et al. 2014). It also does not permit a year-by-year analysis of the industry's investment requirements.

Once a final level of fuel economy is achieved, the models estimate the social costs and benefits of the standards. These include both private costs and benefits to individual consumers and external costs and benefits that accrue to society as a whole. The external benefits of the rule include the value of reductions in GHG emissions, the energy security benefits of reduced petroleum consumption, and health improvements due to particulate matter reductions.[14] There are also some small external costs in terms of more congestion and accidents from more driving due to the rebound effect.[15] The private costs to consumers are the higher upfront cost of vehicles (termed program costs), and the private benefits are the fuel savings, the savings from less frequent refueling, and the value of additional miles driven due to the rebound effect. The private benefit of the fuel savings is by far the largest benefit, though it may not be considered by car buyers at the time of purchase, as discussed at length in Chapter 9. The relative sizes of the private and public (external) costs and benefits associated with the lifetimes of 2017-2025 MY light-duty vehicles, assuming a 3 percent discount rate, are shown in Figure 10.8.

The costs of petroleum dependence and greenhouse gas emissions are particularly important because they represent the primary motivation for the standards. The rulemaking lists the benefits of reduced oil consumption as "reduction in petroleum market externalities" (EPA/NHTSA 2012a, 63080), indicating a misunderstanding of the nature of oil dependence costs. It is important to recognize that the salient market failure in the case of oil dependence is imperfect competition. Imperfect competition is not an externality and should not be estimated as if it were. The economic harm done by higher than competitive market prices has two components: (1) reduced GDP due to the increased economic scarcity of petroleum and (2) a transfer of wealth from oil-importing economies to oil-exporting economies (Greene 2010). The second component is not an economic loss from a global perspective but is an economic loss from the perspective of the U.S. economy. This fact has apparently created some confusion about how to add up costs and benefits.[16] Petroleum dependence can also impose external costs associated with military expenses.

The Agencies estimate a variety of different costs to petroleum dependence, including macroeconomic costs imposed by disruptions in oil imports, higher cost of oil due to U.S. demand in the world market (termed the monopsony component, discussed below), and military costs to secure oil imports from unstable regions and maintain the Strategic Petroleum Reserve. The Final Rule notes that only the macroeconomic disruption costs are incorporated into the cost benefit analysis (EPA/NHTSA 2012a, 62717). The Agencies

[14] EPA estimated $PM_{2.5}$ reductions because the net emissions reductions from reduced fuel refining, distribution, and transport is larger than the emissions due to increased VMT and increased electricity production (EPA/NHTSA 2102a, 62899).

[15] The rebound effect is an increase in vehicle use as a consequence of the reduction in the cost of energy per mile of driving due to increased fuel economy.

[16] The OMEGA model discussion also appears to confuse the costs of oil import dependence with the costs of oil dependence. The economic costs of oil dependence are a function not only of the quantity of imports but also of the total quantity of oil consumed throughout the economy.

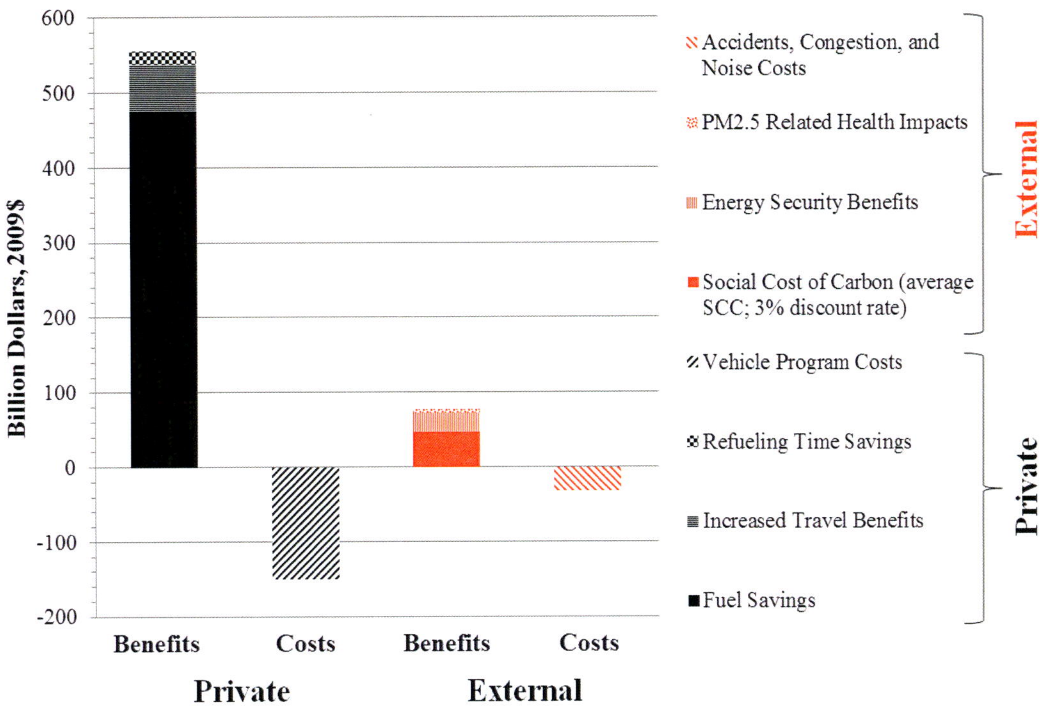

FIGURE 10.8 Distribution of lifetime private benefits and costs (black) and external benefits and costs (red) of 2017-2025 MY light-duty vehicles under the standards, using a 3 percent discount rate.
SOURCE: Data from EPA/NHTSA (2012a, Tables III-104 and III-105).

justify excluding the military security costs because they are difficult to quantify. The difficulty of estimating national defense and foreign policy costs due to oil dependence has been noted elsewhere (NRC 2010). Difficulty of estimation may lead to estimates that are uncertain but it does not imply zero cost as assumed in the Final Rule (EPA/NHTSA 2012a, 63088).[17]

In considering the standards' GHG and oil dependence benefits, the Agencies excluded what are termed "monopsony benefits" (EPA/NHTSA 2012a, 62939). Monopsony benefits measure the benefit to the U.S. economy of a reduction in the world price of oil due to reductions in U.S. oil demand brought about by the standards. Because the U.S. accounts for more than one-fifth of the world's petroleum consumption, large changes in U.S. demand can affect world prices in both the short and long run. The Agencies quantified the monopsony benefit at \$9.77/bbl, slightly larger than the \$8.26/bbl benefit attributed to reduced disruption of oil supplies (EPA/NHTSA 2012a, 62939). In accounting for costs and benefits of the rule, the oil disruption benefit was incorporated as an external energy security benefit and the monopsony benefit was not incorporated. The justification for excluding the monopsony benefit was that it is a transfer and not a benefit from the perspective of the global economy. The reasoning was that if one includes the full global benefit of reduced U.S. GHG emissions, one must also take the global perspective when it comes to the transfer of wealth due to higher than competitive oil market prices. Since the U.S. economy's gain is canceled by a corresponding loss of revenue to oil exporters, there is no monopsony benefit from the global perspective. The fallacy in this reasoning resides in insisting that the scope of the two problems, oil dependence and climate change, must be the same. In fact, oil dependence is a national concern of the United States. Like national defense, it is inherently adversarial (i.e., oil consumers against producers using monopoly power to raise prices). The problem of climate change is inherently global and requires a global solution. If each nation considered only the benefits to itself in determining what actions to take to mitigate climate change, an adequate solution could not be achieved. Likewise, if the U.S. considers the economic harm its reduced petroleum use will do to monopolistic oil producers it will not adequately address its oil dependence problem. Thus, if the United States is to solve both of these problems it must take full account of

[17] The Agencies did include an estimated military cost in their sensitivity analysis but this does not adequately address the need to consider national defense and foreign policy costs.

the costs and benefits of each, using the appropriate scope for each problem.

Uncertainty

Estimating the potential for future fuel economy improvement and its costs and benefits is complex and uncertain. As this report has repeatedly noted, estimating even the current costs and impacts of fuel economy technologies involves substantial uncertainty. The Agencies discuss these sources of uncertainty at length in Section IV of the Final Rule, as well as within the Agencies' respective RIAs (EPA/NHTSA 2012a; EPA 2012c; NHTSA 2012b). The Agencies have done sensitivity and uncertainty analysis for the 2017-2025 CAFE/GHG standards in theses RIAs. EPA conducted a sensitivity analysis regarding benefits from reducing GHG emissions and fuel savings for different assumptions of the rebound rates—that is, the increase in vehicle use that results if an increase in fuel efficiency lowers the cost per mile of driving (EPA 2012c). It also looked at the sensitivity of the regulation's benefits under varying assumptions concerning health impacts of air pollution and the global warming potential of various GHGs. NHTSA performed a sensitivity analysis on fuel prices and a probabilistic uncertainty analysis using Monte Carlo simulation. The Monte Carlo analysis included uncertainties in (1) technology costs, (2) technology effectiveness, (3) fuel prices, (4) manufacturers' decisions to produce vehicles with higher fuel economies than mandated by the CAFE standards, (5) VMT, (6) passenger car share of the new market, (7) value of oil consumption externalities, and (8) rebound effects (NHTSA 2012b). The results of the NHTSA probabilistic assessment shows that, for a range of assumed discount rates, there is a high degree of certainty (99 percent) that higher CAFE standards will produce a net societal benefit in each of the combined fleet model years covered by this rule.

A more comprehensive modeling of uncertainty would integrate all the components noted above, including uncertainty about the baseline and reference vehicle fleet size and composition. Looking ahead to 2025 and 2030 brings in additional sources of uncertainty:

- The pace and direction of future technological progress,
- Current and future consumer behavior and preferences,
- Future market conditions, including the prices of key commodities from oil to aluminum,
- Future regulatory initiatives, and
- The impacts of global climate change and the importance of GHG mitigation.

At present, it is not clear how to carry out such a comprehensive uncertainty analysis. The committee is well aware of the challenges posed by assigning probability distributions to point estimates of costs and fuel consumption impacts. Introducing a much wider array of uncertain parameters would not only magnify the challenge but create a far greater complication: representing the relationships among the factors.

The committee could not conduct a probabilistic uncertainty analysis given resource and time constraints. The committee throughout its report emphasizes where it sees important uncertainties for both the technology benefits and costs as well as for other factors that will impact the cost and implementation of the new standards. In particular, Chapter 9 emphasizes uncertainties in the estimation of how consumers value fuel economy and other vehicle improvements and how willing consumers are to purchase innovative technologies. Though the committee could not quantify these uncertainties, they are noted throughout the report as topics for follow-up analysis by the Agencies.

Agency Coordination

The Final Rule supporting documents reflect a high degree of coordination between the Agencies with respect to data, methods, and premises. Differing regulatory mandates require effort in coordination of analysis, and the Agencies retain different compliance models to allow them to represent the different requirements of the CAFE and Clean Air Act laws. Multiple support documents are required of NHTSA and EPA in the process of setting the standards due to differing regulatory mandates. The analysis and documentation in support of the rule contains redundancies and inconsistencies. Redundancies, such as development of different full vehicle simulation models, are a waste of limited resources and should be consolidated. Redundancies can also lead to inconsistencies, which should be minimized to ease understanding of and compliance with the standards.

Reconciling GHG and CAFE Treatment of Alternative Fuel Vehicles

Fuel economy and GHG emissions are regulated by two Agencies, NHTSA and EPA. The motivations for these regulations include energy efficiency, reduction of GHG emissions, and energy security. The standards have been harmonized such that for gasoline-powered vehicles excluding AC credits, the CAFE and GHG objectives are consistent based on the relationship between a given volume of gasoline and the associated mass of CO_2 produced upon combustion. Difficulties arise when other fuels are used for propulsion as the standards are not harmonized with those fuels in mind. For example, alternative fuels benefit energy security but not necessarily GHGs. The first challenge in regulating under two objectives lies in misalignment between the petroleum reduction and GHG benefits for some AFVs. For example, domestically-produced fuels such as ethanol and natural gas provide substantial energy security benefits but only modest GHG benefits. A second complication for regulating under two metrics is that to appropriately account for the GHG emissions of vehicles using a variety of fuels, upstream emissions not produced at the tailpipe must be included. The GHG benefits for BEVs and FCEVs, for example, are entirely dependent on how the electricity or hydrogen is produced. The GHG benefits of natural gas and ethanol will also be significantly impacted by their upstream emissions.

The first complication in regulating under two metrics is harmonizing standards from a petroleum consumption and GHG emissions perspective for vehicles powered by fuels other than gasoline, such as ethanol, electricity, hydrogen, or natural gas. For example, BEVs must be assigned a compliance mpg-equivalent value even though they use no petroleum onboard. Compliance fuel economy of electric vehicles and other AFVs are increased using a 0.15 divisor (equivalent to multiplying fuel economy values by 6.67), which appears to be based on providing the same incentive multiplier as E85 and is not directly related to the petroleum consumption of electricity, hydrogen, or natural gas. This is equivalent to assuming that all alternative fuels provide an 85 percent reduction in GHGs per unit of energy. For FFVs, where the 0.15 divisor is related to the petroleum content of the alternative fuel, the regulatory treatment assumed 50 percent use of E85 and 50 percent use of gasoline, when in fact very few consumers use E85. As noted previously, this treatment is appropriately being phased out. While alternative fuels can lead to major discrepancies between the CAFE and GHG benefits, diesel vehicles also present a complication. Diesel combustion results in more carbon dioxide emitted per gallon than gasoline, so a diesel vehicle that meets the CAFE target for its size would exceed its GHG target. Even gasoline vehicles rely on corrections to the GHG/CAFE relationship—for example, through air conditioning emissions credits.

The second complication with regulating alternative fuel vehicles is how to appropriately account for upstream emissions of GHGs and consumption of petroleum. A well-to-wheels analysis is appropriate to assess and compare the very different upstream GHG and petroleum impacts of fuels. Light-duty vehicles of all fuel types are assessed for fuel consumption and tailpipe GHG emissions in the compliance test cycles described earlier in this chapter. For CAFE and GHG compliance of gasoline and diesel vehicles, no direct accounting is made for upstream emissions and petroleum consumption from refining or transportation of petroleum. As described earlier in the chapter, the permanent GHG regulatory treatment of PHEVs, BEVs, and FCEVs will use a well-to-wheels analysis and corrects for the upstream GHG emissions of a comparable gasoline vehicle in order to provide equitable treatment. Up to a certain cumulative production volume by a manufacturer, however, these well-to-wheels emissions are not taken into account as a temporary regulatory incentive, as discussed previously in the chapter. The CAFE program also accounts for upstream energy consumption, but not petroleum consumption, for PHEVs, BEVs, and FCEVs. For consumer information, a well-to-wheels analysis of GHG emissions for all vehicle types, including gasoline, diesel, and alternative fuels, is currently provided at Fueleconomy.gov.

It is not clear how to permanently resolve the differences in energy security and GHG benefits of all alternative fuels. Also, although permanent regulatory treatment of PHEVs, BEVs, and FCEVs is on a well-to-wheels basis, this is not implemented at current production volumes, nor is it used for vehicles powered by other fuels such as natural gas and ethanol. Well-to-wheels analysis provides a way to compare the GHG and petroleum impacts of a variety of fuels. While the Agencies have made commendable efforts to harmonize the CAFE/GHG national program, the committee finds that having two metrics, both greenhouse gases emissions and petroleum consumption, creates conflicts that can complicate regulations and compliance. The committee notes the strong complementarity between the two objectives. It appears that reducing the total GHG emissions (well-to-wheels) from light-duty vehicles to levels that would appropriately address the problem of climate change would also adequately solve the oil dependence problem. The Agencies should study the potential benefits, costs, and risks of establishing a standard based on a single metric that achieves both GHG and petroleum reductions in addition to continued efforts to harmonize the two regulations.

FINDINGS AND RECOMMENDATIONS

Finding 10.1 In the current assessment of the effects of the new rules, the footprint standard is assumed to have no effect on vehicle size, or on the mix of vehicle size and market shares. However, the rules could well have effects on costs and revenues for vehicles of different types and sizes, which may lead to changes in vehicle design and sales. Preliminary studies of this issue show a range of results, from little effect on the vehicle sales mix to changes in design and sales mix, leading to a larger footprint. The effects of the rule on vehicle sales mix and redesign is important because larger vehicle sizes could reduce the benefits of the rule in terms of reductions in oil consumption and GHG emissions.

Recommendation 10.1 The Agencies should monitor the effects of the CAFE/GHG standards by collecting data on fuel efficiency, vehicle footprint, fleet size mix, and price of new vehicles to understand the impact of the rules on consumers' choices and manufacturers' products offered. The Agencies have already initiated this effort, and it should be continued as a first step toward understanding the overall effect of the rule on vehicle size and size mix. Without analysis of manufacturers' and consumers' choices, however, it will be difficult to isolate the effects of the rule alone. Economic-engineering models of manufacturer decision making that take into account costs and consumer responses should also be developed as part of the assessment of the rule.

Finding 10.2 The empirical evidence from historical data appears to support the argument that the new footprint-based standards are likely to have little effect on vehicle safety and overall highway safety. If the size mix of vehicles remains roughly the same, then a reduction in the weight of vehicles is not generally associated with greater societal safety risks. To the extent the size mix and design of vehicles changes substantially, the effects on safety are not known. There will need to be continuing empirical analysis of the safety outcomes as vehicle designs and size mixes change over time.

Finding 10.3 There is no scientifically valid, comprehensive source of information on the in-use fuel economy of light-duty vehicles on U.S. roads. Therefore, the average difference between test cycle fuel economy values and in-use values is not definitively known, and differences for specific technologies are also not well understood. Furthermore there are reasons to believe that the relationship between test-cycle and real-world fuel economy may change in the future as vehicle technology changes. This information is necessary to accurately estimate the benefits of the standards and is therefore also relevant to determining the levels of the standards.

Recommendation 10.2 The Agencies, perhaps in collaboration with other federal agencies (e.g., the Bureau of Transportation Statistics and the Energy Information Administration), should conduct an ongoing scientifically-designed survey of the real-world fuel economy of light-duty vehicles. The survey should also collect information on real-world driving behavior and driving cycles. This information will be useful in determining the adequacy of the current test cycle and could inform the establishment of improved, future (post-2025) test cycles, if necessary. The survey should make use of modern information technology connecting to the on-board diagnostic systems of light-duty vehicles to make data collection simultaneously comprehensive and unobtrusive to the driver on a day-to-day basis while addressing privacy concerns.

Finding 10.4 The existing two-cycle certification tests are not a sufficiently accurate representation of real-world driving behavior where the gap between the two-cycle and five-cycle tests is 20 percent for conventional vehicles and 30 percent for HEVs. The five-cycle test procedure, as used for fuel economy labels since 2008, appears to provide a better representation of the range of real-world driving conditions.

Recommendation 10.3 Making use of information gained from the survey of real-world fuel economy (see Recommendation 10.2), the Agencies should plan a transition to replace the current two-cycle procedure with a procedure that appropriately uses the five-cycle tests. Such a new set of compliance procedures could be implemented for the next CAFE standards following the 2025 MY. This requires harmonizing the test procedures specified by EPCA with the CAA procedures.

Finding 10.5 Fuel economy of a vehicle for CAFE compliance is determined by testing the vehicle on a chassis dynamometer that simulates loaded vehicle weight equal to the vehicle's curb weight plus 300 lb (to simulate 2 passengers). Under current procedures, this simulated test weight is binned within ETWCs, which were used for setting the simulated weight for chassis dynamometer testing. ETWCs have relatively broad ranges, varying from 125 lb for lower ETWCs typical of compact cars to 250 lb for ETWCs typical of larger passenger cars and full size light trucks. As a result of these incremental steps in ETWCs, a vehicle in the upper end of an ETWC that achieves a significant mass reduction nearly equal to the range of an ETWC would not realize any fuel consumption reduction for CAFE compliance since the vehicle would still be tested within the same ETWC.

Recommendation 10.4 To realize the fuel consumption reduction benefit directly associated with the mass reduction achieved in a vehicle, EPA and NHTSA should consider adopting procedures that use the actual vehicle weight plus 300 lb for setting the simulated test weight for chassis dynamometer testing of a vehicle for CAFE compliance. Since manufacturers often group different series of a vehicle line within one ETWC to reduce the burden of testing each series, EPA and NHTSA should consider continuation of this practice by permitting several series of a vehicle line to be grouped within the sales-weighted average vehicle test weight.

Finding 10.6 The current treatment of flex-fuel and dual-fuel vehicles that assumes 50 percent alternative fuel usage has led to higher emissions of GHGs and more consumption of oil than would be the case without the 50 percent fuel use assumption. It is appropriate to phase out this treatment as currently proposed to adopt instead a system based on data for actual usage of the alternative fuel.

Recommendation 10.5 The CAFE FFV treatment that assumes 50 percent alternative fuel usage should be phased out as planned within the 2017-2025 CAFE regulation.

Finding 10.7 The current CAFE program uses a 0.15 divisor for fuel economy of alternative fuel vehicles, including natural gas and electric vehicles, to incentivize reduced oil use. This factor is more consistent with the reduced petroleum use of AFVs and less consistent with GHG benefits of all alternative fuels. Generally, EPA has broader authority under federal law than NHTSA to design its regulatory treatment in a manner consistent with GHG benefits. The GHG regulatory treatment—without incentives of temporary sales multipliers and zero tailpipe emissions treatment—is generally consistent with well-to-wheels GHG benefits for alternative fuels.

Recommendation 10.6 Permanent regulatory treatment of AFVs should be commensurate with the well-to-wheels GHG and petroleum reduction benefits when operating on alternative fuels, consistent to the greatest degree possible with NHSTA and EPA's programs and, for dual-fuel or flex-fuel vehicles, should be based on data of actual usage of the alternative fuel. If sufficient data do not exist, usage should be monitored and treatment modified as appropriate.

Finding 10.8 The Agencies' analyses of benefits and costs assume a reference case for which fuel economy does not increase after the 2016 MY. Assuming there is continued technology improvement after 2016, and that it does not go to fuel economy in the reference (no additional standards) case, then the improvements would go to enhance other vehicle attributes in the reference case. Net of costs, the value of these attributes has not been considered as an opportunity cost of the regulation, meaning that costs may have been left out of the analysis of the societal costs and benefits of the rule. The extent of this opportunity cost is linked to how consumers value fuel economy and other attributes.

Recommendation 10.7 The Agencies should consider how to develop a reference case for the analysis of societal costs and benefits that includes accounting for the potential opportunity costs of the standards in terms of alternative vehicle attributes forgone.

Finding 10.9 Firms face quite different CAFE or GHG credit holdings, partly due to the allowance for early credit accumulation before the standards became effective in 2012. Some firms accumulated credits during the period while others did not. The large variation in holdings among manufacturers reflects very different costs of meeting the standards today and in the future. A small number of manufacturers hold a large share of current credits.

Finding 10.10 The credit markets established by the Agencies—EPA for meeting GHG goals, and NHTSA for meeting fuel efficiency goals—are completely separate markets with separate rules. The credit definitions and provisions of the two Agencies are not fully harmonized, so the credits generated and the use of credits will be different in each market. The rules in one market will influence how credits are used and how compliance occurs in the other market.

Recommendation 10.8 The midterm review is a time that the Agencies should consider how the credit markets are different between the CAFE and GHG rules, and what the implications of these differences are for the auto manufacturers. At the same time, it is a good time to look at what barriers there are to effective credit markets. Ensuring that credit markets work effectively and are transparent will reduce the cost of compliance and enhance the likelihood companies will be able to comply.

Finding 10.11 The committee appreciates the difficulty for NHTSA and EPA of developing a single national program for reducing LDV petroleum consumption and GHG emissions based on their different statutory authorities and commends the Agencies for delivering it. The committee also recognizes that with differing statutory authorities come different requirements that are reflected in the compliance models, the treatment of alternative fuels, and the credit systems. The committee notes that making the CAFE and GHG regulations as consistent as possible will reduce the compliance burden for the automotive industry.

Finding 10.12 NHTSA and EPA's use of improved methods and data to establish and assess the CAFE and GHG standards is well justified because it produces more accurate assessments for standards that have very large benefits and costs for the nation. The results of these studies are reviewed and findings made in other chapters of the report. The use of full vehicle simulation modeling in combination with lumped parameter modeling has improved the estimation of the effectiveness of fuel economy improvements on individual vehicles. Use of teardown studies has also improved the estimates of costs of these improvements, although there is a risk that the process of using one example of the new technology and one example of the outgoing technology may not provide estimates that are fully representative when the technology is implemented across the entire fleet.

Recommendation 10.9 The Agencies should continue to analyze the costs and benefits of the rule with teardown studies and full system simulations and should perform more ex post review of their estimates to understand how successful vehicle manufacturers were at delivering the fuel economy and the costs estimated in the rule. Finally, the committee recommends the Agencies consider developing a short summary of its regulatory analysis. The committee found the regulatory analysis produced by the Agencies to be extensive, in-depth, and invaluable, but future efforts should be directed towards avoiding redundancies and differences among the multiple support documents, recognizing the requirements for regulatory analysis and reporting.

Recommendation 10.10 The Agencies should study more thoroughly consumer and manufacturer behavior in response to the rule. The uncertainty of choices consumers and manufacturers make in response to the standards may be greater than the uncertainty related to efficiencies and costs of the technologies.

Finding 10.13 The cost/benefit analysis recognizes that GHG mitigation must be a cooperative global effort while improving U.S. energy security is a national concern. To adequately solve the problem of climate change, individual nations need actions commensurate with the global impacts of the GHGs they emit. On the other hand, solving the problem of U.S. oil dependence includes reducing the transfer of U.S. wealth to oil-exporting countries. Although this results in a loss of profit for oil-exporting economies, it is no less a real benefit to the Unites States. The problems are of different scopes.

Recommendation 10.11 The full benefits of reducing U.S. oil dependence, including monopsony benefits, should be counted along with the global benefits of GHG reduction in the Agencies' cost/benefit analyses. The scopes of the two problems are different.

Finding 10.14 The Agencies have made commendable efforts to harmonize the GHG emissions and fuel economy standards. Harmonization is important to reducing the burden of compliance on manufacturers. However, the committee finds that having two metrics, both GHG emissions and petroleum consumption, creates conflicts that can complicate regulations and compliance. The committee notes the strong complementarity between the two objectives. It appears that reducing the total GHG emissions (WTW) from light-duty vehicles to levels that would appropriately address the problem of climate change would also adequately solve the oil dependence problem.

Recommendation 10.12 The committee recommends that the Agencies study the potential benefits, costs, and risks of establishing a standard based on a single metric that achieves both GHG and petroleum reductions in addition to continuing efforts to harmonize the two regulations.

REFERENCES

Al-Alawi, B., and T. Bradley. 2014. Analysis of corporate average fuel economy regulation compliance scenarios inclusive of plug in hybrid vehicles. Applied Energy 113: 1323-1337.

Anderson, M., and M. Affhammer. 2012. Pounds that kill: The external costs of vehicle weight. Review of Economic Studies, forthcoming.

Anderson, S.T., and J.M. Sallee. 2011. Using loopholes to reveal the marginal cost of regulations: The Case of fuel-economy standards. American Economic Review 101: 1375-1409.

Bunch, D.S., and D.L. Greene, 2011. Potential Design, Implementation, and Benefits of a Feebate Program for New Passenger Vehicles in California. State of California Air Resources Board and the California Environmental Protection Agency, Sacramento, California.http://76.12.4.249/artman2/uploads/1/Feebate_Program_for_New_Passenger_Vehicles_in_California.pdf.

CARB (California Air Resources Board). 2011. Staff Report: Initial Statement of Reasons. Advanced Clean Cars. 2012 Proposed Amendments to the California Zero Emission Vehicle Program Regulations. http://www.arb.ca.gov/regact/2012/zev2012/zevisor.pdf.

CARB. 2012. The Zero Emission Vehicle (ZEV) Regulation. California Environmental Protection Agency. Fact Sheet, p. 3. http://www.arb.ca.gov/msprog/zevprog/factsheets/general_zev_2_2012.pdf.

CARB. 2013. Staff Report: Initial Statement of Reasons for Rulemaking. 2013 Minor Modifications to the Zero Emission Vehicle Regulation. http://www.arb.ca.gov/regact/2013/zev2013/zev2013isor.pdf.

DOE (Department of Energy) 2014. Detailed Test Information: Detailed Comparison. http://www.fueleconomy.gov/feg/fe_test_schedules.shtml.

DOE (Department of Energy), DOT (Department of Transportation) and EPA (Environmental Protection Agency). 2002. Report to Congress: Effects of the Alternative Motor Fuels Act CAFE Incentive Policy.

DOT NHTSA (U.S. Department of Transportation, National Highway Traffic Safety Administration). 2012. 2017-2025 Corporate Average Fuel Economy Compliance and Effects Modeling System Documentation, DOT HS 811 670, August.

ERG, Inc. (Eastern Research Group, Inc.). 2013. Light-Duty Vehicle In-Use Fuel Economy Data Collection: Pilot Study, Report Version 8. Prepared for the International Council on Clean Transportation. http://www.theicct.org/sites/default/files/ICCT-131108%20-%20ERG%20-%20In-Use%20FE%20Pilot-%20V8FInal.pdf.

ECMT (European Council of Ministers of Transport). 2005. Making Cars More Fuel Efficient, OECD/IEA, Paris.

ECOtality. 2013. What Kind of Charging Infrastructure Do Chevrolet Volts Drivers in the EV Project Use? http://avt.inl.gov/pdf/EVProj/VoltChargingInfrastructureUsageSep2013.pdf.

EIA (Energy Information Administration). 2014. Annual Energy Outlook 2014. http://www.eia.gov/forecasts/aeo/tables_ref.cfm.

EPA (Environmental Protection Agency). 2006. Fuel Economy Labeling of Motor Vehicles: Revisions to Improve Calculation of Fuel Economy Estimates. Final Rule, Federal Register, 71(21) 5425-5513, February 1.

EPA. 2009. Clean Air Act Mobile Source Civil Penalty Policy – Vehicle and Engine Certification Requirements. Memorandum from Granta Y. Nakayama to Mobile Source Enforcement Personnel. http://www2.epa.gov/sites/production/files/documents/vehicleengine-penalty-policy_0.pdf.

EPA. 2011a. Early Trading Program. http://www.epa.gov/otaq/regs/ld-hwy/greenhouse/documents/420r13005.pdf.

EPA. 2011b. Development of Emission Rates for Light-Duty Vehicles in the Motor Vehicle Emissions Simulator (MOVES2010). U.S. Environmental Protection Agency, EPA-420-R-11-011. http://www.epa.gov/otaq/models/moves/documents/420r11011.pdf.

EPA. 2012a. Light-duty Automotive Technology, Carbon Dioxide Emissions and Fuel Economy Trends: 1975 through 2011. Appendix A Database Details and Calculation Methods, EPA-420-R-12-001a, Transportation and Climate Division, March.

EPA. 2012b. EPA Optimization Model for Reducing Emissions of Greenhouse Gases from Automobiles (OMEGA), Core Model Version 1.41. Documentation. EPA-420-R-12-024, Office of Transportation and Air Quality, Ann Arbor, August.

EPA. 2012c. Regulatory Impact Analysis: Final Rulemaking for 2017-2025 Light-Duty Vehicle Greenhouse Gas Emission Standards and Corporate Average Fuel Economy Standards. PA-420-R-12-016.

EPA. 2014a. Greenhouse Gas Emission Standards for Light Duty Vehicles. Manufacturer Performance Report for the 2012 Model Year. EPA-420-R-14-011. http://www.epa.gov/otaq/climate/documents/420r14011.pdf.

EPA. 2014b. MTE Technologies and Costs: 2022-2025 GHG Emissions Standards. Presentation to the National Research Council of the National Academies Committee on Fuel Economy of Light-Duty Vehicles, Office of Transportation and Air Quality, Ann Arbor, Michigan, July 31.

EPA. 2014c. Light-Duty Automotive Technology, Carbon Dioxide Emissions, and Fuel Economy Trends: 1975 through 2014. EPA-420-S-14-001.

EPA. 2014d. Final Determination of Weighting Factor for Testing E85 Flexible Fuel Vehicles (FFV) for MYs 2016-2018 Vehicles Under the Light-duty Greenhouse Gas Emissions Program. Enclosure to CD-14-18, Memo to Manufacturer from Byron Bunker, Dir., Compliance Division, Office of Transportation and Air Quality. http://iaspub.epa.gov/otaqpub/display_file.jsp?docid=33581&flag=1.

EPA/NHTSA (National Highway Traffic Safety Administration). 2012a. 2017 and Later Model Year Light-Duty Vehicle Greenhouse Gas Emissions and Fuel Economy Standards; Final Rule. Federal Register 77(199) October.

EPA/NHTSA. 2012b. Joint Technical Support Document: Final Rulemaking for 2017-2025 Light-Duty Vehicle Greenhouse Gas Emission Standards and Corporate Average Fuel Economy Standards. EPA-420-R-12-901, August.

Fiat. 2014. Fiat Ecodrive App. http://www.fiatusa.com/en/mobile_apps/fiat-ecodrive/. Accessed February 6, 2014.

Fleetcarma. 2014. Vehicle Modeling: A Must-Have for the Fleet Toolbox. https://www.fleetcarma.com/Resources/vehicle-modeling-ebook. Accessed February 5, 2014.

General Accounting Office. 2007. Vehicle Fuel Economy: Reforming fuel economy standards could help reduce oil consumption in cars and light trucks, and other options could complement these standards. GAO-07-021. August.

German, J., and N. Lutsey. 2010. Size or Mass? The Technical Rationale for Selecting Size as an Attribute for Vehicle Efficiency Standards. ICCT White Paper, no. 9.

Gramlich, J. 2010. Gas Prices, Fuel Efficiency, and Endogenous Product Choice in the Automobile Industry. Paper presented at Second Annual Microeconomics Conference, Federal Trade Commission. http://www.ftc.gov/sites/default/files/documents/public_events/second-annual-microeconomics-conference/gramlichppr.pdf.

Greene, D.L. 1997. Why CAFE Worked. Prepared by the Oak Ridge National Laboratory Center for Transportation Analysis, U.S. Department of Energy. ORNL/CP-94482, August.

Greene, D.L. 2010. Measuring energy security: Can the United States achieve oil independence? Energy Policy 38(4): 1614-1621.

Greene, D.L., J.L. Hopson, R. Goeltz, and J. Li. 2007. Analysis of In-Use Fuel Economy Shortfall Based on Voluntarily Reported Mile-per-Gallon Estimates. Transportation Research Record No. 1983, 99-105.

He, H. Credit Trading in the US Corporate Average Fuel Economy (CAFE) Standard. ICCT. http://www.theicct.org/sites/default/files/publications/ICCTbriefing_CAFE-credits_20140307.pdf.

Hellman, K., and J.D. Murrell. 1984. Development of Adjustment Factors for the EPA City and Highway MPG Values. SAE Technical Paper Series 840496.

ICF International. 2011. Peer Review of Ricardo, Inc. Draft Report, 'Computer Simulation of Light-Duty Vehicle Technologies for Greenhouse Gas Emission Reduction in the 2020-2025 Timeframe'. Contract No. EP-C-06-094, Work Assignment 4-04, Docket EPA-HQ-OAR-2010-0799, September 30.

Jacobsen, M. 2013a. Fuel Economy and Safety: The influences of vehicle class and driver behavior. American Economic Journal: Applied Economics 5(3). http://econweb.ucsd.edu/~m3jacobsen/Jacobsen_Safety.pdf.

Jacobsen, M. 2013b. Evaluating U.S. fuel economy standards in a model with producer and household heterogeneity. American Economic Journal: Economic Policy 5(2). http://econweb.ucsd.edu/~m3jacobsen/Jacobsen_CAFE.pdf.

Kahane, C.J. 1997. Relationships between vehicle size and fatality risk in model year 1985-93 passenger car and light trucks. DOT HS 808 570. National Highway Traffic Safety Administration, Washington, D.C.

Kebschull, S.A., J. Kelly, R.M. Van Auken, and J.W. Zellner. 2004. An Analysis of the Effects of SUV Weight and Length on SUV Crashworthiness and Compatibility using Systems Modeling and Risk Benefit Analysis. DRI-TR-04-04-2, Dynamic Research, Inc., July.

Lin, Z., and D.L. Greene. 2011. Predicting Individual On-road Fuel Economy Using Simple Consumer and Vehicle Attributes. SAE Technical Paper 11SDP-0014, Society of Automotive Engineers, Warrendale, PA, April 12.

Kling, C. L. 1994. Emission trading vs. rigid regulations in the control of vehicle emissions. Land Economics 70(2): 174-188.

Leard, B., and V. McConnell. 2015. New Markets for Pollution and Energy Efficiency: Credit Trading under Automobile Greenhouse Gas and Fuel Economy Standards. Resources for the Future Discussion Paper, No. 15-16.

Li, S. 2012. Traffic safety and vehicle choice: Quantifying the effects of the 'arms race' on American roads. Journal of Applied Econometrics 27: 34–62.

Liu, C., D.L. Greene, and D.S. Bunch, 2014. Vehicle manufacturer technology adoption and pricing strategies under fuel economy/emissions standards and feebates. The Energy Journal 35(3): 71-89.

Liu, Y., and G. Helfand. 2009. The Alternative Motor Fuels Act, Alternative Fuels and Greenhouse Gases. Transportation Research Part A. 43: 755-764.

McNutt, B.D., R. Dulla, R. Crawford, and H.T. McAdams. 1982. Comparison of EPA and On-Road Fuel Economy – Analysis Approaches, Trends and Imports. SAE Technical Paper Series 820788, Society of Automotive Engineers, Warrendale, PA, June.

Mintz, M.M., A.R.D. Vyas, and L.A. Conley. 1993. Difference between EPA Test and In-Use Fuel Economy: Are the Correction Factors Correct? TRB Paper 931104, 72nd Annual Meeting of the Transportation Research Board, Washington, D.C., January.

Mock, P. 2011. Development of a Worldwide Harmonized Light Vehicles Test Procedure (WLTP). ICCT Contribution No. 3 (focus on inertia classes). Working Paper 2011-5.

Mock, P., J. German, A. Bandivadekar, I. Riemersma, N. Ligterink, and U. Lambrecht. 2013. From Laboratory to Road: A Comparison of Official and "Real-World" Fuel Consumption and CO_2 Values for Cars in Europe and the United States. White paper, International Council on Clean Transportation, Washington, DC. http://www.theicct.org/laboratory-road.

Mock, P., U. Tiegete, V. Franco, J. German, A. Bandivadekar, N. Ligterink, U. Lambrecht, J. Kühlwein, and I. Riemersma. 2014. From Laboratory to Road: A 2014 Update of Official and "Real-World" Fuel Economy and CO_2 Values for Passenger Cars in Europe. White Paper, International Council on Clean Transportation, San Francisco, September. http://theicct.org/laboratory-road-2014-update.

NHTSA (National Highway Traffic Safety Administration). 2012a. 2017-2025 Corporate Average Fuel Economy Compliance and Effects Modeling System Documentation. U.S. Department of Transportation. ftp://ftp.nhtsa.dot.gov/CAFE/2017-25_Final/CAFE_Model_Documentation_FR_2012.08.27.pdf.

NHTSA. 2012b. Final Regulatory Impact Analysis: Corporate Average Fuel Economy for MY 2017-MY2025 Passenger Cars and Light Trucks. Office of Regulatory Analysis and Evaluation, National Center for Statistics and Analysis.

NHTSA. 2012c. Mass Reduction for Light-Duty Vehicles for Model Years 2017–2025: Final Report. DOT HS 811 666. August.

NRC (National Research Council). 2010. Hidden Costs of Energy. Washington, D.C.: The National Academies Press.

Posada, F., and J. German, 2013. Measuring in-use fuel economy in Europe and the US: Summary of pilot studies. Working Paper 2013-5, The International Council on Clean Transportation. http://www.theicct.org/sites/default/files/publications/ICCT_FuelEcon_pilotstudies_20131125.pdf.

Ricardo, Inc. 2011. Draft Project Report: Computer Simulation of Light-Duty Vehicle Technologies for Greenhouse Gas Emission Reduction in the 2020-2025 Timeframe., Report to EPA Office of Transportation and Air Quality, EP-W0-07-064, Ann Arbor, Michigan, April 6.

Rosca, A. n.d. Light Duty Vehicle Test Cycle Generation Based on Real-World. Instituto Superior Técnico of University of Lisbon, Lisbon, Portugal. https://fenix.tecnico.ulisboa.pt/downloadFile/395145957557/Alexandr%20Rosca%20Master%20Thesis%20(10%20page%20abstract).pdf.

Rubin, J., and C. Kling. 1993. An emission saved is an emission earned: An empirical study of emission banking for light-duty vehicle manufacturers. Journal of Environmental Economics and Management 25(3): 257-274.

Rubin, J., and P. Leiby. 1998. CAFE Credits for Alternative Fuel Vehicles. TAFV Model Technical Memorandum. Oak Ridge National Laboratory. January 6.

Rubin, J., P.N. Leiby, and D.L. Greene. 2009. Tradable fuel economy credits: Competition and oligopoly. Journal of Environmental Economics and Management 58(3): 315-328.

Shiau, C.-S. N., J.J. Michalek, and C. T. Hendrickson. 2009. A structural analysis of vehicle design responses to Corporate Average Fuel Economy policy. Transportation Research Part A. 43:814-828.

Statistics Canada. 2014. Fuel Consumption Survey. http://www23.statcan.gc.ca/imdb/p2SV.pl?Function=getSurvey&SDDS=2749. Accessed February 6, 2014.

Sweeney, J. 2001. Analysis of Tradable Fuel Economy Credits, Chapter 5 of Impact and Effectiveness of Corporate Average Fuel Economy Standards. NAS, Washington, D.C.

Wenzel, T. 2010. Analysis of the Relationship Between Vehicle Weight/Size and Safety, and Implications for Federal Fuel Economy Regulation, Final Report prepared for the Office of Energy Efficiency and Renewable Energy, U.S. Department of Energy. http://energy.lbl.gov/ea/teepa/pdf/lbnl-3143e.pdf.

Wenzel, T. 2012. Assessment of NHTSA's Report "Relationships Between Fatality Risk, Mass, and Footprint in Model Year 2000-2007 Passenger Cars and LTVs." Final report prepared for the Office of Energy Efficiency and Renewable Energy, U.S. Department of Energy. Lawrence Berkeley National Laboratory. August. LBNL-5698E. http://energy.lbl.gov/ea/teepa/pdf/lbnl-5698e.pdf.

Wenzel, T. 2013. The Effect of Recent Trends in Vehicle Design on U.S. Societal Fatality Risk per Vehicle. Accident Analysis and Prevention 56: 71-81. July LBNL-6277E.

Whitefoot, K.S., and S.J. Skerlos. 2012. Design incentives to increase vehicle size created from the U.S. footprint-based fuel economy standards. Energy Policy 41: 402-411.

Yiu, Y., and G. Helfand. 2012. A hedonic test of the effects of the Alternative Motor Fuels Act.Transportation Research, Part A. Policy and Practice 46(10): 1707-1715.

ZEV Program Implementation Task Force. 2014. Multi-State ZEV Action Plan. http://www.ct.gov/deep/lib/deep/air/electric_vehicle/path/multi-state_zev_action_plan_may2014.pdf.

Appendix A

Statement of Task

The committee formed to carry out this study will continue the work of the National Research Council for the U.S. Department of Transportation's National Highway Traffic Safety Administration (NHTSA) in the assessment of technologies and programs for improving the fuel economy of light-duty vehicles. While the committee will need to consider the development and deployment of fuel economy technologies up to 2019, it is tasked with providing updated estimates of the cost, potential efficiency improvements, and barriers to commercial deployment of technologies that might be employed from 2020 to 2030. It will reassess the technologies analyzed in NRC reports, Impact and Effectiveness of Corporate Average Fuel Economy (CAFE) Standards (2002) and Assessment of Fuel Economy Technologies for Improving Light-Duty Vehicle Fuel Economy (2011). It will reflect developments since these reports were issued and investigate any new technologies that may become important by 2030. The committee will also examine and make recommendations for improvements to the CAFE program. In particular, the committee shall:

1. Broadly assess the methodologies and programs used to develop standards for passenger cars and light trucks under current and proposed CAFE programs and make recommendations for future programs, including recommendations concerning the attributes used for the standards, the structure of the program necessary with the introduction of alternative technology vehicles, and the assumptions and methods used in analysis of proposed regulatory activities.
2. Examine the potential for reducing mass by up to 20%, including: technologies such as materials substitution; downsizing of existing vehicle design, systems or components; and the use of new vehicle, structural, system or component designs or other mass substitution/weight reduction categories. The committee shall consider the implications of such weight reductions on vehicle safety.
3. Examine other vehicle technologies, including aerodynamic drag reduction, improved efficiency of accessories such as alternators and air conditioners, and conversion of engine-driven equipment to electricity (e.g. power steering, fans, and water pumps).
4. Examine electric power train technologies, including the capabilities of hybrids, plug-in hybrids, battery electric vehicles, and fuel cell vehicles. The committee shall include an examination of the cost, performance, range, durability (including performance degradation over time) and safety issues related to lithium ion and other possible advanced energy storage technologies that are necessary to enable plug-in and full function electric vehicles.
5. Examine advanced gasoline and diesel engine technologies that will increase fuel economy. Advanced gasoline technologies to be examined include the high Brake Mean Effective Pressure (BMEP) and Homogeneous-Charge Compression Ignition (HCCI) engines. For diesel engines, include the capabilities of emissions control systems on advanced diesel engines to meet current and possible future criteria pollutant emissions standards, impacts on fuel consumption of emissions control systems, and the fuel characteristics needed to enable low emissions diesel technologies. For all these engines, the committee shall consider their ability to meet load demands; cost; the need for after-treatment; and market acceptability of those engines.
6. Assess the assumptions, concepts, and methods used in estimating the costs of fuel economy improvements. In particular, consider the degree to which time-based cost learning for well-developed existing technologies and/or volume-based cost learning for newer technologies should apply, what the time or volume basis should be, and whether other methods of applying cost learning are practical. Also, examine the differences between Retail Price Equivalent (RPE) and Indirect Cost

Multipliers (ICM), determine appropriate values for each, and recommend which method is preferable to use for estimating indirect costs of technologies.

7. Provide an analysis of how fuel economy technologies may be practically integrated into automotive manufacturing processes and how such technologies are likely to be applied in response to requirements for improving fuel economy. Include an analysis of how technology implementation is likely to impact capital equipment and engineering, research and development (ER&D) costs, and at what rate such technologies might be implemented to meet increases in fuel economy standards.
8. Examine the costs and benefits in vehicle value that could accompany the introduction of advanced vehicle technologies. Consider the total cost of operation of these vehicles by examining potential cost impacts on fuel, maintenance, insurance, registration fees, and other factors. In addition, assess the impact on consumers of factors that may change how they use their vehicles, such as reduction of driving range and loss of utility.
9. Examine test procedures and calculations used to determine fuel economy values for purposes of determining compliance with CAFE standards, identifying potential changes to make those procedures and calculations more relevant to and neutral in their treatment of technologies considered by the committee. In considering test procedures, the committee should examine the fuel saving potential for technologies such as adaptive cruise control, real-time traffic alerts, tire pressure sensors, and real-time fuel economy information. This analysis shall evaluate the possibility of incorporating the savings produced by such technologies within CAFE test procedures.
10. To the extent possible, the committee will address uncertainties and perform sensitivity analyses of its cost estimates and provide guidance to NHTSA on improving its uncertainty analyses given the relatively long time frame for these future estimates.
11. Write a final report documenting its conclusions and recommendations.

Appendix B

Committee Biographies

JARED L. COHON, *Chair*, is a university professor of civil and environmental engineering and engineering and public policy, director of the Scott Institute for Energy Innovation, and president emeritus at Carnegie Mellon University, where he served as president from 1997 to 2013. He is the chair of the board of the Center for Sustainable Shale Development and serves on the boards of the Health Effects Institute, the Heinz Endowments, Carnegie Corporation, Ingersoll Rand, Lexmark and Unisys. Dr. Cohon has more than 40 years of technology, research, policy, and management experience. He began his teaching career at Johns Hopkins University (JHU), where he served as assistant, associate, and full professor in the Department of Geography and Environmental Engineering (1973-1992). He also served as vice provost for research (1986-1992), associate dean of engineering (1983-1986), and assistant dean of engineering (1981-1983). Following his tenure at JHU, he became dean of the School of Forestry and Environmental Studies and a professor of environmental systems analysis at Yale University. Dr. Cohon also served as legislative assistant for energy and environment on the staff of U.S. Senator Daniel Patrick Moynihan from 1977 to 1978. President George W. Bush appointed him in 2002, and President Barack Obama reappointed him in 2009 to the Homeland Security Advisory Council on which he served until 2013. Dr. Cohon was appointed by President Bill Clinton to the Nuclear Waste Technical Review Board in 1995 and as its chair in 1997, a position he held until 2002. Dr. Cohon is a member of the National Academy of Engineering (NAE), a national authority on environmental and water resource systems analysis, and the author, co-author, or editor of more than 80 professional publications. He holds a B.S. in civil engineering from the University of Pennsylvania and a master's degree and Ph.D. in civil engineering from the Massachusetts Institute of Technology (MIT).

KHALIL AMINE is an Argonne Distinguished Fellow and manager of the Battery Technology Team within Argonne's Chemical Science and Engineering Division. He received five R&D 100 awards in the past 5 years. He also received both the Electrochemical Society Battery Technology Award and International Battery Association Award in 2010 for his advanced research on cathode materials for safe, long-lasting lithium-ion batteries. Dr. Amine founded the annual International Conference on Advanced Lithium Batteries for Automotive Applications (ABAA) and chaired the inaugural meeting in 2008. He is the president of the International Lithium Battery Association and the president of International Automotive Lithium Battery Association. He received his Ph.D. in materials science from the University of Bordeaux, France, and has studied various aspects of new materials for next-generation batteries throughout his career. Before joining Argonne in 1998, Dr. Amine led a number of advanced R&D projects at the Japan Storage Battery Company, now a subsidiary of GS-Yuasa. Prior to his private sector service, he oversaw research projects in the public sector during his tenure at the Osaka National Research Institute and Kyoto University. He was the most cited scientist in the world in the field of battery technology from 1998 to 2008 and has written the largest number of papers in the world on battery related topics from 2000 to 2011.

CHRIS BAILLIE is currently the chief engineer of new product development at AxleTech International. He was formerly the supervisor of transmission and driveline design at FEV, an internationally recognized powertrain and vehicle engineering company that supplies the global transportation industry. Mr. Baillie has extensive experience in light-duty vehicle transmission and hybrid powertrain design. He previously worked at GE Aviation as a lead engineer for gearboxes on turbine engines. He has served as lead design engineer and program manager for two parallel electric hybrid transmissions, four AMTs, a CVT, and two series hybrid powertrains. He also has experience with the design of DCTs, automatics and two-mode transmissions. He served as a consultant for the National Research Council's (NRC's) Committee on Assessment of Technologies for Improving Light-Duty Vehicle Fuel Economy. Mr. Baillie has a B.S. in mechanical engineering from Boston University.

JAY BARON is president and director of the Center for Automotive Research (CAR). He is also the director of CAR's Manufacturing, Engineering and Technology Group. Dr. Baron's recent research has focused on developing new methods for the analysis and validation of sheet metal processes including die making, tool and die tryout, and sheet metal assembly processes. He also developed functional build procedures that result in lower tooling costs and shorter development lead times, while improving quality—particularly with sheet metal assemblies. He also has been researching new technologies in the auto industry, including looking at body shop design and flexibility and evaluating the manufacturing capability of evolving technologies. Dr. Baron recently completed investigations on the state of the art of tailor-welded blank technologies, economics of weld-bond adhesives, and the analysis of car door quality and construction methods. Prior to becoming the director of manufacturing systems at CAR and subsequently president, he was the manager of manufacturing systems at the Office for the Study of Automotive Transportation at the University of Michigan Transportation Research Institute. He also worked for Volkswagen of America in quality assurance and as staff engineer and project manager at the Industrial Technology Institute in Ann Arbor and at Rensselaer Polytechnic Institute's (RPI's) Center for Manufacturing Productivity. Dr. Baron holds a Ph.D. and a master's degree in industrial and operations engineering from the University of Michigan and an MBA from RPI. He served on the NRC's Committee on Assessment of Technologies for Improving Light-Duty Vehicle Fuel Economy.

R. STEPHEN BERRY is the James Franck Distinguished Service Professor Emeritus of Chemistry at the University of Chicago and holds appointments in the College, the James Franck Institute, and the Department of Chemistry. He has also held an appointment in the School of Public Policy Studies at the University of Chicago and has worked on a variety of subjects ranging from strictly scientific matters to a variety of topics in policy. He spent 1994 at the Freie Universität Berlin as an awardee of the Humboldt Prize. In 1983 he was awarded a MacArthur Fellowship. His experimental research includes studies of negative ions, chemical reactions, detection of transient molecular species, photoionization, and other laser-matter interactions. His theoretical research has included finite-time thermodynamics, electron correlation, atomic and molecular clusters, and most recently, the micro-macro boundary. Other research has involved interweaving thermodynamics with economics and resource policy, including efficient use of energy. Since the mid-1970s, Dr. Berry has worked on issues of science and the law, and with management of scientific data, activities that have brought him into the arena of electronic media for scientific information and issues of intellectual property in that context. Dr. Berry is a member of the National Academy of Sciences (NAS) and has served on a number of NRC committees, including recent service on the Committee on Review of the U.S. DRIVE Research Program, Phase 4. He attended Harvard University, where he received an A.B. and an A.M. in chemistry and a Ph.D. in physical chemistry.

L. CATE BRINSON is currently the Jerome B. Cohen Professor of Engineering at Northwestern University and a professor in the Mechanical Engineering Department with a secondary appointment in the Materials Science and Engineering Department. After receiving her Ph.D. from California Institute of Technology (Caltech), Dr. Brinson performed postdoctoral studies in Germany at the DLR and, since 1992, she has been on the faculty at Northwestern University. Current research investigations involve characterization of local polymer mechanical behavior under confinement, nanoparticle reinforced polymers, the phase transformation response of shape memory alloys, nano- and microscale response of biomaterials, and materials genome informatics research, where investigations span the range of molecular interactions, micromechanics, and macroscale behavior. Dr. Brinson has received a number of awards, including the Nadai Medal of the American Society of Mechanical Engineers (ASME), the Friedrich Wilhelm Bessel Prize of the Alexander von Humboldt Foundation, the ASME Tom JR Hughes Young Investigator Award, and a National Science Foundation (NSF) CAREER Award. Dr. Brinson is a fellow of the Society of Engineering Science, the ASME, and the American Academy of Mechanics, and she served as a member of the Defense Science Study Group. She has given many invited technical lectures on her research and has authored one book and more than 120 refereed journal publications. She has nearly 10,000 citations and an h-index of 45 in Google Scholar and more than 5,000 citations and an h-index of 40 in ISI Web of Science. Her book has had over 30,000 chapter downloads from the e-version since publication in 2008, and a second edition is being published in early 2015. She is a member of several professional societies and served 5 years on the Society of Engineering Science board of directors, including 1 year as president of the society. Dr. Brinson has also been an associate editor of the *Journal of Intelligent Material Systems and Structures* and the *Journal of Engineering Materials and Technology.* She served two terms on the NRC's National Materials Advisory Board and has chaired two NRC studies.

MATT FRONK is president of Matt Fronk & Associates, LLC. He has more than 37 years of experience leading both research and product development projects in advanced technology, fuel cells, and energy storage. He spent 20 years leading General Motors' Fuel Cell Research and Development program. During his tenure at GM, fuel cell systems were developed from laboratory-scale systems to 100 operating vehicles—the largest of any OEM (original equipment manufacturer) auto company at the time. Mr. Fronk also has extensive global supplier development experience. After GM, he served as director of the Center for Sustainable Mobility at Rochester Institute of Technology and was instrumental in developing durability and life-cycle analyses for new product designs as they moved from concept to product. He also was a founding member and first board chair for NY BEST—an energy storage consortium in New York—and continues to this day as a board member. Mr. Fronk led the design/build of the NY BEST Battery Test Center in Rochester, New York, a state-of-the-art facility that opened in 2014. He is an expert consultant to the energy storage and fuel cell fields and co-chairs the energy innovation workgroup of the Finger Lakes Regional Economic Development Council. Mr. Fronk has a B.S. in mechanical engineering from Union College.

DAVID L. GREENE is a senior fellow of the Howard H. Baker, Jr. Center for Public Policy and a research professor of civil and environmental engineering at the University of Tennessee. Previously, he was a corporate fellow of Oak Ridge National Laboratory and a research professor of economics at University of Tennessee, Knoxville. He is author of more than 275 publications on transportation, energy, and related issues, including 100 articles in refereed journals. He is an emeritus member of both the Energy and Alternative Fuels Committees of the Transportation Research Board (TRB) and a Lifetime National Associate of the National Academies. He is a recipient of the TRB's 2012 Roy W. Crum Award for Distinguished Achievement, the Society of Automotive Engineers' (SAE's) 2004 Barry D. McNutt Award for Excellence in Automotive Policy Analysis, the Department of Energy's (DOE's) 2007 Hydrogen R&D Award and 2011 Vehicle Technologies R&D Award, the International Association for Energy Economics' Award for Outstanding Paper of 1999 for his research on the rebound effect, and the Association of American Geographers' 2011 Edward L. Ullman Award. He was also recognized by the Intergovernmental Panel on Climate Change (IPCC) for contributions to the IPCC's receipt of the 2007 Nobel Peace Prize. He holds a B.A. from Columbia University, an M.A. from the University of Oregon, and a Ph.D. in geography and environmental engineering from JHU.

ROLAND HWANG is the Energy and Transportation Program director for the Natural Resources Defense Council (NRDC) and works on sustainable transportation policies. He is an expert on clean vehicle and fuels technologies and was a member of the IPCC that won the 2007 Nobel Peace Prize. Mr. Hwang serves or has served on numerous committees and advisory panels, including for the California Plug-in Electric Vehicle Collaborative, the NRC Committee on Barriers to Electric Vehicle Deployment, and the Environmental Protection Agency's (EPA's) Mobile Source Technical Review Subcommittee. Prior to joining the NRDC, he was the director of the Union of Concerned Scientists' transportation program. He has also worked for DOE at the Lawrence Berkeley National Laboratory and for the California Air Resources Board as an air pollution engineer, and he was involved in forecasting residential and industrial energy demand, hazardous waste incinerator permitting, and evaluating toxic air emissions from landfills. Mr. Hwang has an M.S. in mechanical engineering from the University of California, Davis, and a master's degree in public policy from the University of California, Berkeley.

LINOS JACOVIDES is a professor of electrical and computer engineering at Michigan State University. He retired as director of Delphi Research Laboratories, a position he held from 1998 to 2007. Dr. Jacovides joined GM Research and Development in 1967 and became department head of electrical engineering in 1985. His areas of research were the interactions between power electronics and electrical machines in electric vehicles and locomotives. He later transitioned to Delphi with a group of researchers from GM to set up the Delphi Research Laboratories. He is a member of the NAE and a fellow of the Institute of Electrical and Electronic Engineers (IEEE) and the SAE. He was president of the Industry Applications Society of IEEE in 1990. He received a B.S. degree in electrical engineering and an M.S in machine theory from the University of Glasgow, Scotland. He received his Ph.D. in generator control systems from the Imperial College, University of London.

THERESE LANGER is the Transportation Program director for the American Council for an Energy-Efficient Economy. Her program analyzes and promotes strategies to reduce energy consumption in the U.S. transportation sector and produces annual environmental ratings of new cars and light trucks. She is the author of publications on light- and heavy-duty vehicle technologies, fuel efficiency standards, feebate policies, consumer vehicle labels, efficiency of the goods movement system, and state policies to reduce greenhouse gas emissions from the transportation sector. Dr. Langer provides guidance and analytical support on transportation energy issues to environmental groups, businesses, congressional offices, and agencies. She previously worked as staff scientist at the Rutgers Environmental Law Clinic and taught undergraduate and graduate mathematics courses at the University of Minnesota, the State University of New York at Stony Brook, and Swarthmore College. She has a B.A. from Harvard University and a Ph.D. in mathematics from the University of California, Berkeley.

REBECCA LINDLAND is a senior fellow with the King Abdullah Petroleum Studies and Research Center, spearheading their work on transportation policy, technology, and consumer demand. She was formerly the director of research for IHS Automotive where she was responsible for evaluating and assessing OEMs that participate in the U.S. and Canada marketplaces. She has a particular interest in how manufacturers' decisions reflect consumer values. As a member of IHS Automotive, Ms. Lindland was frequently quoted in the media for her coverage of new product launches and the balance sheet conditions of manufacturers and brands. Prior to her work at IHS, she worked at AlliedSignal in Rumford, Rhode Island, where she forecasted products such as Bendix brakes. A life-long automotive enthusiast, she began her career as a staff accountant with Mercedes-Benz Credit Corporation in Norwalk, Connecticut. Ms. Lindland holds a double major in accounting and business administration from Gordon College. She is a former board member of the Society of Automotive Analysts, the International Motor Press Association, and the Motor Press Guild, and was accepted into Strathmore's 2001 Who's Who in American Business.

VIRGINIA McCONNELL is a senior fellow at Resources for the Future (RFF) in the Center for Energy and Climate Economics (CECE). She is also a professor of economics at the University of Maryland, Baltimore County. Dr. McConnell has worked throughout her career to examine policies to reduce motor vehicle energy use and emissions, assessing both regulatory policies and the role of pricing and other incentive-based policies. Her recent work has focused on the evaluation of the Corporate Average Fuel Economy (CAFÉ) program and on policies toward alternative vehicles and fuels. She has studied the cost-effectiveness of various policies including those designed to increase the share of hybrids and electric vehicles in the U.S. fleet and has explored a range of policies designed to reduce local air pollution. Dr. McConnell is co-editor of *Controlling Vehicle Pollution* and has published on a range of transportation policy issues. In addition, she has served on a number of EPA and state advisory committees related to transportation, energy use, and air quality. She has been a member of several NRC panels in recent years, including the Committee on Transitions to Alternative Vehicles and Fuels and the Committee for a Study of Potential Energy Savings and Greenhouse Gas Reductions from Transportation. Dr. McConnell received a B.S. degree in economics from Smith College and a Ph.D. degree in economics from the University of Maryland.

DAVID F. MERRION is the CEO of Merrion Expert Consulting, LLC. He is the retired executive vice president of engineering for Detroit Diesel Corporation (DDC), a Daimler Trucks North America subsidiary. His positions at DDC included staff engineer, Emissions and Combustion; staff engineer, Research and Development; chief engineer, Applications; director, diesel engineering; general director, Engineering (Engines and Transmissions); and senior vice president, Engineering. Mr. Merrion has extensive expertise in the research, development, and manufacturing of advanced diesel engines, including alternative-fueled engines. He is fellow of SAE and ASME and served as president of the Engine Manufacturers Association. Mr. Merrion is a member of EPA's Mobile Sources Technical Advisory Committee, the Coordinating Research Council, and the U.S. Alternate Fuels Council. He has served on a number of NRC committees, including the Standing Committee to Review the Research Program of the Partnership for a New Generation of Vehicles; the Committee on Review of the 21st Century Truck Partnership, Phase 1; and the Committee to Assess Fuel Economy Technologies for Medium- and Heavy-Duty Vehicles. He has a bachelor of mechanical engineering from General Motors Institute (Kettering University) and an M.S. in mechanical engineering from MIT.

CLEMENS SCHMITZ-JUSTEN is partner and head of strategic consulting at CSJ Schmitz-Justen & Company. Concurrently with this appointment, he was director of international programs in the College of Business and Behavioral Science at Clemson University (2008-2010). He is the former president of BMW Manufacturing, LLC, in Spartanburg (2004-2007), where according to *Automotive News*, he "led a major update of the company's manufacturing operations in Spartanburg, S.C." During Dr. Schmitz-Justen's term of executive leadership at BMW, the Spartanburg plant built its one-millionth vehicle in the United States, underwent a multi-million dollar expansion, began using landfill methane gas to supply the paint shop, and added another generation of the popular X5 Sports Activity Vehicle to its line. He joined BMW in 1991 and served in a variety of senior management assignments within the company, such as head of the Global Painted Body Division and head of Experimental Vehicles at the Research and Innovation Center of BMW Group. Prior to that, Dr. Schmitz-Justen was a research engineer at the Fraunhofer Institute for Production Technology IPT (1981-1991), where he also earned his doctorate and served as managing chief engineer. He is an honorary adjunct professor at Chemnitz Technical University in Germany. Dr. Schmitz-Justen earned the equivalent of a master's degree and a doctorate degree in manufacturing engineering from Aachen Technical University in Aachen, Germany.

ANNA G. STEFANOPOULOU is a professor in the Mechanical Engineering Department at the University of Michigan and the director of the Automotive Research Center at the university-based U.S. Army Center of Excellence in Modeling and Simulation of Ground Vehicles. She was an assistant professor at the University of California, Santa Barbara (1998-2000), and a technical specialist at Ford Motor Company (1996-1997) where she developed nonlinear and multivariable models and controllers for advanced engines. Her algorithms were implemented and tested in experimental vehicles. She is an ASME fellow and an IEEE fellow, the founding chair of the ASME Dynamic Systems and Control Division (DSCD) Energy Systems Technical Committee, and a member of the SAE Dynamic System Modeling Standards Committee. She was an elected member of the IEEE Control Systems Society board of governors. She was the chair of the transportation committee in ASME DSCD, served as an associate editor of journals, and is a member of multiple award committees in the IEEE and ASME societies. She is a recipient of the 2012 University of Michigan College of Engineering Research Award, the 2009 ASME Gustus L. Larson Memorial Award, a 2008 University of Michigan Faculty Recognition Award, the 2005 Outstanding Young Investigator award by ASME DSCD, a 2005 Henry Russel award, a 2002 Ralph Teetor SAE educational award, and a 1997 NSF CAREER award. She was selected as one of the 2002 world's most promising innovators from the *MIT Technology Review*. She co-authored *Control of Fuel Cell Power Systems* and has 11 U.S. patents, 5 best paper awards, and 250 publications on estimation and control of internal combustion engines and electrochemical processes such as fuel cells and batteries. She obtained her diploma in naval architecture and marine engineering from the National Technical University of Athens, Greece, and her Ph.D. in electrical engineering and computer science from the University of Michigan.

WALLACE R. WADE was chief engineer and technical fellow, Powertrain Systems Technology and Processes, Ford Motor Company, where he served for 32 years prior to his retirement. Mr. Wade was responsible for the development, application, and certification of emission and powertrain control system technologies for all of Ford's North American vehicles. His technical responsibilities have included low-emission technologies for internal combustion engines; analytical and laboratory-based powertrain calibration with objective measures of driveability; the first domestic production OBD II (On- Board Diagnostic) system; technology for diesel particulate filters (DPF) with active regeneration; electronic control systems for gasoline and diesel engines; low-heat-rejection and low-friction, direct-injection diesel engines; and an ultra-low-emission, gas-turbine combustion system. Today Mr. Wade is a consultant to industry and government. He was elected to the NAE in 2011 for implementation of low-emission technologies in the automotive industry. He is a fellow of the SAE and the ASME. He has received the SAE Edward N. Cole Award for automotive engineering innovation, the ASME Soichiro Honda Award for technical achievements in automotive engineering, the Henry Ford Technology Award for exceptional technical contributions, and he has been recognized as a Distinguished Corporate Inventor by the National Inventors Hall of Fame. He has received five SAE Arch T. Colwell Awards and the SAE Vincent Bendix Automotive Electronics Engineering Award. He has received 26 patents related to improvements in powertrains and has written 25 published technical papers on powertrain research and development. He has served on three previous NRC study committees, including the Committee on Low Heat Rejection Engines and the first and second committees for the Review of the 21st Century Truck Partnership. He has an M.S.M.E. degree from the University of Michigan and a B.M.E. degree from RPI, both in mechanical engineering.

WILLIAM H. WALSH, JR., is an automobile safety consultant. He consults on vehicle safety activities with several technology companies to speed the introduction of advanced life-saving technology into the automobile fleet as well as substantive involvement in CAFE rulemakings. He held several positions at the U.S. National Highway Traffic Safety Administration (NHTSA), including senior associate administrator for policy and operations; associate administrator for plans and policy; director, National Center for Statistics and Analysis; director, Office of Budget, Planning and Policy; and science advisor to the Administrator of NHTSA. He also held the position of supervisory general engineer at DOE's Appliance Efficiency Program. His expertise covers all aspects of vehicle safety performance, cost/benefit analyses, strategic planning, statistics analyses and modeling, and policy formulation. He served on the TRB's Occupant Protection Committee and is currently serving on the NRC's Committee on the Potential for Light-Duty Vehicle Technologies 2010-2050. He has a B.S. in aerospace engineering from the University of Notre Dame and an M.S. in system engineering from George Washington University.

Appendix C

Presentations and Committee Meetings

MARCH 1-2, 2012, WASHINGTON, D.C.

Christopher Bonanti and James Tamm, National Highway Traffic Safety Administration: Motivation for the Study and NHTSA's Objectives

Edward Nam, U.S. Environmental Protection Agency, Office of Transportation and Air Quality: Presentation on EPA's Light-duty Vehicle GHG Technical Activities

Michael Stanton, Association of Global Automakers: Presentation on Perspectives of the Association of Global Automakers on Future Vehicle Technologies in the 2020 to 2030 Time Frame

Doug Greenhaus and David Wagner, National Automobile Dealers Association: Presentation on Perspectives of the National Automobile Dealers Association

Neil DeKoker, Original Equipment Suppliers Association: Presentation on Perspectives of the Motor and Equipment Manufacturers Association

JUNE 20-21, 2012, WASHINGTON, D.C.

Trevor Jones, ElectroSonics Medical Inc., Phase One Committee Chair: Lessons Learned from NHTSA Phase One Study

Ryan Harrington, U.S. Department of Transportation, Volpe National Transportation Systems Center: CAFE Compliance and Effects Modeling System – Overview

Aymeric Rousseau, Argonne National Laboratory: Argonne National Laboratory Autonomie Full Vehicle Simulation Model

Edward Nam and Lee Byungho, U.S. Environmental Protection Agency: EPA Full Vehicle Simulation Model Development

Nicholas Lutsey, University of California, Davis: Estimates of Technologies and Costs for Meeting New GHG Standards Used by the California Air Resources Board

SEPTEMBER 27-28, 2012, WASHINGTON, D.C.

Klaus Land, Regulation and Certification Division, Daimler Trucks North America: Daimler's Technology Pathway to Meet New Fuel Economy/GHG Standards

Mitsuo Hitomi, Powertrain Development Division, Mazda: Mazda's Technology Pathway to Meet New Fuel Economy/GHG Standards

Hugh Blaxill, Engineering Services NAFTA, Mahle Powertrain, LLC: Future Turbocharging and Downsizing Engine Technology Opportunities

Pete Maloney, MathWorks: Powertrain Optimization Topics Related to Fuel Economy Improvement; Automotive Engine Calibration/Controls

Ralph Brodd, Kentucky/Argonne Battery Manufacturing Research and Development Center

DECEMBER 3-6, 2012, DEARBORN, MICHIGAN

Nizar Trigui, Ford Motor Company: Ford's Technology Roadmap/Fuel Economy Strategy to Meet the 2017-2025 CAFE Standards

Scott Miller, General Motors: General Motors Technology Roadmap to Meet 2017-2025 CAFE Standards

Toyota's Technology Roadmap to Meet the 2017-2025 CAFE Standards

Gary Oshnock, Chrysler Group: Chrysler's Technology Roadmap to Meet the 2017-2025 CAFE Standards

Joseph Colucci, Automotive Fuels Consulting, Inc.: Improving Auto Fuel Economy via Fuel Changes

Ben Ellies, U.S. Environmental Protection Agency: EPA's Lump Parameter Modeling Overview

John Kasab, Ricardo: Computer Simulation of Light Duty Vehicle Technologies for Greenhouse Gas Emission Reduction in the 2020-2025 Timeframe

Don Kopinski and Ed Nam, EPA; Greg Kolwich, FEV; Javier Rodriguez, EDAG: FEV Inc.'s Cost Estimation for Gasoline HEVs, 8-speed Automatic Transmissions, and 8-Speed Dual-Clutch Transmissions

FEBRUARY 12-13, 2013, ANN ARBOR, MICHIGAN

Center for Automotive Research Lightweighting Workshop

Lixin Zhao, U.S. Department of Transportation, National Highway Traffic Safety Administration, Fuel Economy Rulemaking Division: Overview of NHTSA's Mass Studies and Projects
Harry Singh, EDAG, Inc.: Presentation on NHTSA Mass Reduction Study
Nicholas Petouhoff and Thomas Gould, Johnson Controls Automotive Seating: Vehicle Interiors
Ronald P. Krupitzer, Steel Market Development Institute: Presentation on Steel's Role in Vehicle Structure, Lightweighting, Safety and Life Cycle Emissions
Blake Zuidema, ArcelorMittal Global R&D: Presentation on the Role of Weight Reduction in Meeting the U.S. 2025 EPA/NHTSA Fuel Economy Standard
Randall Scheps, Michael Bull and Doug Richman, the Aluminum Association's Transportation Group (ATG): Automotive Aluminum: Part of the Solution
Gina Oliver and Martin Christman, American Chemistry Council: Lightweighting Vehicles Using Advanced Plastics and Composites
Dave Mason, Altair: Computer-Based Simulation and Optimization for Vehicle Design
Jackie Rehkopf, Plasan Carbon Composites: Entering Mainstream Automotive Presentation
Matt Zaluzec, Ford Motor Company: U.S. Automotive Materials Partners Presentation
Gregg Peterson, Lotus Engineering: High Development Vehicle Project for Toyota Venza

FEBRUARY 27, 2013, CAMBRIDGE, MASSACHUSETTS

Volpe National Transportation Systems Center

Volpe Staff, Discussion of Volpe Model, NHTSA Cost Methodologies and Estimates

MARCH 27-28, 2013, WASHINGTON, D.C.

Costs Workshop

David Greene, Oak Ridge National Lab (Committee Member): Overview of Issues and Methods for Estimating the Costs of Increasing Light-duty Vehicle Fuel Economy
Greg Kolwich, FEV: Presentation on Issues in Estimating the Costs of Fuel Economy Improvements under Future CAFE Regulations - Teardown analysis: State of the Art and Potential for Improvements
K.G. Duleep, H-D Systems: Estimating Fuel Economy Technology Cost and Price
Kevin Gallagher, Argonne National Laboratory: Estimating Future Costs of Lithium-ion Batteries (The BatPaC Model)
Gloria Helfand and Todd Sherwood, U.S. Environmental Protection Agency: Presentation on Automobile Industry Retail Price Equivalent and Indirect Cost Multipliers (ICM) Studies
Larry Blincoe, National Highway Traffic Safety Administration (NHTSA): NHTSAs Application of Indirect Costs
Larry Burns, University of Michigan and General Motors (retired): Comments on the Allocation of Indirect Costs in the Automotive Industry
Stephen Zoepf and John Heywood, Massachusetts Institute of Technology Sloan Automotive Laboratory: Characterizations of Deployment Rates in Automotive Technology
Steven D. Levitt, John A. List, and Chad Syverson, University of Chicago: Toward an Understanding of Learning by Doing: Evidence from an Automobile Assembly Plant
Sonia Yeh, Institute of Transportation Studies, University of California-Davis; and Edward S. Rubin, Engineering and Public Policy, Carnegie Mellon University: Presentation on Learning in Energy Technologies
Robert Van Buskirk, Lawrence Berkeley National Laboratory: Incorporating Experience Curves in Standards Analysis
Joshua Linn, Resources for the Future: Presentation on Technological Change, Vehicle Characteristics, and the Opportunity Costs of Fuel Economy Standards
Larry Burns, University of Michigan and General Motors (retired): The Challenges of Optimizing Product Plans in the Face of Regulatory Requirements, Consumer Desires, and Unknown Cost and Performance from New Technologies
Robert Lempert, RAND Pardee Center for Longer Range Global Policy and the Future Human Condition: Robust Policies under Uncertainty

JUNE 25-26, 2013, WASHINGTON, D.C.

Charles Kahane, National Highway Traffic Safety Administration (NHTSA): Relationships between Fatality Risk, Mass, and Footprint
Thomas Wenzel, Lawrence Berkeley National Laboratory: An Analysis of the Relationship between Casualty Risk per Crash and Vehicle Mass and Footprint
Stephen Ridella, National Highway Traffic Safety Administration: Using Simulation Modeling to Assess Safety of Future Light Weight Vehicles
Chuck Thomas, Honda R &D Americas, Inc.: Perspectives of Vehicle Manufacturers on Safety and Lighweighting

Chuck Nolan, Insurance Institute for Highway Safety: The Relative Safety of Large and Small Passenger Vehicles

OCTOBER 14-15, 2013, IRVINE, CALIFORNIA

Alberto Ayala, California Air Resources Board: Updates on California Air Resources Board's Tailpipe and ZEV Regulations

Stephen Ellis, American Honda Motor Company: Status of Honda's Fuel Cell Vehicle Technology

William Elrick, CA Fuel Cell Partnership: Deployment of a Hydrogen Fueling Infrastructure in California

Alexander Edwards, Strategic Vision: Results from New Car Buyer Surveys for Fuel Economy Technologies

February 13-14, 2014, Washington, D.C.

Terry Alger II, Southwest Research Institute: Presentation on BMEP Engines: Challenges and Potential Solutions

APRIL 3-4, 2014, WASHINGTON, D.C.

No public session presentations.

JUNE 24-25, 2014, WASHINGTON, D.C.

James Tamm and NHTSA Staff: Recent NHTSA Activities with Relevance to Committee

Michael Olechiw and EPA Staff: Recent EPA Activities with Relevance to Committee

JULY 31, 2014, ANN ARBOR, MICHIGAN

EPA Staff, MTE Technologies and Costs: 2022-2025 GHG Emissions Standards Briefing

SEPTEMBER 3-5, 2014, WASHINGTON, D.C.

Greg Kolwich, FEV, and EPA Staff: Presentation of EPA Lightweighting Studies

Lixin Zhao and NHTSA Staff: Presentation of NHTSA Lightweighting Studies

NOVEMBER 20-21, 2014, WASHINGTON, D.C.

No public session presentations.

Appendix D

Ideal Thermodynamic Cycles for Otto, Diesel, and Atkinson Engines

End of compression pressures shown in Figure D.1 are reasonably representative of actual engines. However, combustion processes differ significantly in actual engines, compared to the idealized cycles. In the Otto and Atkinson cycle spark ignition engines, combustion does not occur at constant volume but instead extends over a significant number of crank angle degrees (as changes in cylinder volume occur). In diesel engines, combustion does not occur at constant pressure, but instead occurs with a shorter duration which results in an increase in pressure during combustion. For these reasons, in actual engines, the diesel engine will have higher peak cylinder pressures than spark ignition engines. This results in the need for a heavier engine structure to contain the higher peak cylinder pressures of diesel engines.

REFERENCE

Ronney, P. 2013. Ideal Cycle Analysis. AME 436 Lecture 8, Spring. University of Southern California School of Engineering. http://ronney.usc.edu/AME436/Lecture8files/AME436-Lecture8.pptx. Accessed February 25, 2015.

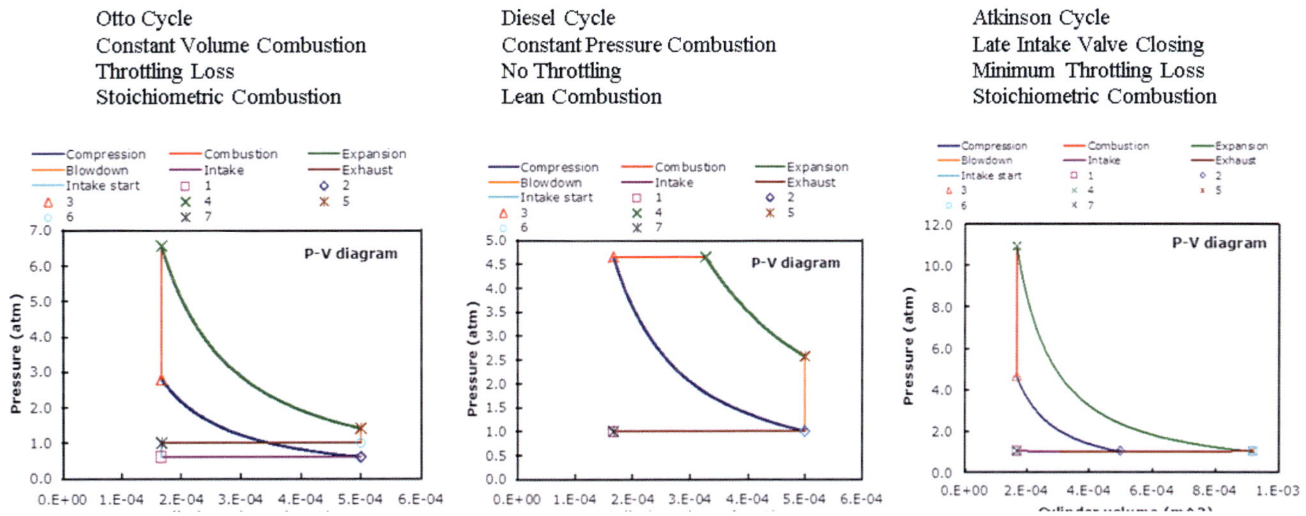

FIGURE D.1 Ideal thermodynamic cycles for Otto, Diesel, and Atkinson engines shown on pressure-volume (P-V) diagrams.
SOURCE: Paul Ronney (2013), University of Southern California, http://ronney.usc.edu.

Appendix E

SI Engine Definitions and Efficiency Fundamentals

The following definitions are helpful in discussing SI engine efficiency fundamentals (Heywood 1988):

Mean Effective Pressure (MEP) = Work per cycle/ displaced volume

Indicated Mean Effective Pressure (IMEP) = Work delivered to the piston over the compression and expansion strokes, per cycle per unit displaced volume

Friction Mean Effective Pressure (FMEP) = Total friction work per cycle per unit displaced volume

BMEP can be calculated as follows:

Brake Mean Effective Pressure (BMEP) = IMEP – FMEP (1)

FMEP consists of the following three components:

Pumping Mean Effective Pressure (PMEP) = Work per cycle done by the piston on the in-cylinder gases during the inlet and exhaust strokes. PMEP is positive for naturally aspirated engines and negative for supercharged and turbocharged engines at high loads.

Rubbing Friction Mean Effective Pressure (RFMEP) = Work per cycle dissipated per cycle in overcoming friction due to relative motion of adjacent components in the engine.

Accessory Mean Effective Pressure (AMEP) = Work per cycle required to drive engine accessories (pumps, fans, alternator, etc.) essential to engine operation.

Therefore, FMEP can be expressed as follows:

FMEP = PMEP + RFMEP + AMEP (2)

Brake thermal efficiency (BTE) is subsequently defined as the ratio of work delivered divided by the heating value of the fuel (generally lower heating value since the water in the exhaust is in vapor form):

BTE = BMEP × displaced volume / (mf × Q_{LHV}) (3)

Where: mf = mass flow rate of fuel

Q_{LHV} = Lower heating value of fuel

A similar expression is used to calculate indicated thermal efficiency (ITE).

The relationships discussed above are shown in Wade et al. (1984) for an engine operating condition representative of the FTP drive cycle.

REFERENCES

Heywood, J.B. 1988. Internal Combustion Engine Fundamentals. New York: McGraw-Hill.

Wade, W.R., J. E. White, C. M. Jones, C. E. Hunter, and S. P. Hansen. 1984. Combustion, Friction and Fuel Tolerance Improvements for the IDI Diesel Engine. SAE Technical Paper 840515.

Appendix F

Examples of Friction Reduction Opportunities for Main Engine Components

Examples of the main engine components on which vehicle manufacturers and suppliers are working to reduce friction components include the following (Truett 2013):

Smaller, low friction bearings: Smaller bearings are being designed to reduce surface area. Special coatings, such as Federal-Mogul's IROX polymer coating, have been applied to engine bearings for 2014 model year engines and these coatings can reduce friction by up to 50% compared with older, larger bearings without coatings. Coated bearings particularly help in stop-start systems, which increase wear on bearings.

Pistons: Pistons account for more than a quarter of the energy lost to friction in an engine. Piston friction is being reduced by reducing the size of the skirt and coating it with ceramic or polymer (other examples include graphite, carbon fiber, and molybdenum disulfide). Low tension piston rings currently exert about 50 percent less pressure against the bore than rings from a few years ago. Smoother, coated cylinder bore surfaces also reduce piston friction.

Valve train: Coatings, such as "Diamond-Like" on valve lifters and tappets and other engine components have been shown by Nissan to reduce friction by as much as 10 percent. Timing chains have been reduced in size and slippery guides have been applied to reduce valve train friction. Rocker arms with low friction rollers are being applied.

Seals: Low friction crankshaft seals have been developed that eliminate the spring inside the disc that squeezes the lip of the disc against the crankshaft and provide more than 50 percent reduction in friction.

Balance shaft: Ford has eliminated the balance shaft in their three-cylinder 1.0L engine by placing balance weights on the engine pulley and flywheel and using patented motor mounts. Eliminating the balance shaft reduced friction by 6 percent. Four-cylinder engines with balance shafts are being fitted with roller bearings, which reduce friction by about 2 percent.

ESTIMATION OF EFFECTIVENESS OF FUEL CONSUMPTION REDUCTION TECHNOLOGIES

Low Friction Lubricants—Level 1 (LUB1)

The effectiveness of low friction lubricants was estimated as follows. Approximately 75 percent of the friction loss is due to piston, crank, and rotating components, with approximately half of this loss due to hydrodynamic lubrication (Heywood 1988). Power loss in hydrodynamic lubrication is proportional to the lubricant viscosity. The viscosity at 100°C is reduced by approximately 25 percent by replacing the 5W-30 oil with 5W-20 oil, as indicated in Table 2.3. This viscosity reduction would be effective after the oil had fully warmed up, which is approximately a quarter of the EPA urban cycle and highway drive cycles.

Total engine friction consumes approximately 8 percent of the fuel energy. Hydrodynamic lubrication consumes half of the 75 percent of the friction loss.

8% fuel energy × 0.75 × 0.5 = 3% fuel energy consumed by hydrodynamic lubrication

A 25 percent reduction in oil viscosity, which would be effective over a quarter of the drive cycles, would result in the following reduction in engine friction:

3% fuel energy consumed by hydrodynamic lubrication × 0.25 × 0.25 = 0.19% fuel energy

The 0.19 percent reduction in fuel energy due to friction represents approximately a 2.5 percent reduction in overall engine friction (0.19 percent fuel energy/8 percent total fuel energy due to friction × 100), which would result in a 0.5 percent reduction in fuel consumption, obtained by applying 36 percent ITE.

0.19% fuel energy reduction/0.36 indicated work/fuel = 0.5% increase in indicated work

The 0.5 percent reduction in fuel consumption is within the range of EPA/NHTSA estimates in the final CAFE rule.

Low Friction Lubricants—Level 2 (LUB2)

The low friction lubricants identified for level 2 consist of 0W-20, 0W-16 or 0W-12 oils instead of 5W-20 oils. This change results in two changes. First, changing to the 0W classification reduces low temperature viscosity, as shown in Table 2.3. With an estimated 12 percent reduction in low temperature viscosity, extrapolated from Table 2.3, and assuming that this reduction would be effective over half of the drive cycles in which the oil is not fully warmed up, applying similar calculations used for level 1 for the energy consumed by hydrodynamic lubrication, above, yields a 0.5 percent reduction in fuel consumption, as follows:

3% fuel energy consumed by hydrodynamic lubrication × .12 × 0.5 = 0.18% fuel energy
0.18% fuel energy reduction / 0.36 indicated work/fuel = 0.5% increase in indicated work

The second change is the reduction in oil viscosity measured at 100°C. By changing from 5W-20 to 0W12, viscosity would be reduced by an estimated 25 percent. The calculation for level 1 indicates that this reduction in viscosity could provide a 0.5 percent reduction in fuel consumption.

Combining the 0.5 percent reduction for low temperature viscosity reduction with the 0.5 percent reduction for 100°C viscosity reduction provides an overall estimate of 1.0 percent reduction for low friction lubricants - level 2.

REFERENCES

Heywood, J.B. 1988. Internal Combustion Engine Fundamentals. New York: McGraw-Hill.

Truett, R. 2013. Chafing against engine friction. Automotive News, May 20.

Appendix G

Friction Reduction in Downsized Engines

Downsizing has a significant effect on friction reduction potential. This effect can be explained as follows with the calculation of brake specific fuel consumption (BSFC) at a typical Federal Test Procedure (FTP) cycle operating condition, assuming a constant indicated specific fuel consumption (ISFC).

$$\text{BSFC} = \text{ISFC} \times \text{IMEP/BMEP} = \text{ISFC} \times (\text{BMEP} + \text{FMEP})/\text{BMEP} \quad (1)$$

The BSFC that would result with a typical baseline friction level at a typical FTP cycle operating condition would be as follows, using Equation 1:

$$\text{BSFC} = \text{ISFC}\ (38 + 11)/38 = 1.289\ \text{ISFC}$$

Where: BMEP (brake mean effective pressure) = 38 psi (typical FTP cycle operating condition)
FMEP (friction mean effective pressure) = 11 psi (typical baseline friction at FTP operating condition)

Achieving a 10 percent reduction in friction would provide the following improvement in BSFC:

$$\text{BSFC} = \text{ISFC}\ (38 + 0.9 \text{ x } 11)/38 = 1.261\ \text{ISFC}$$

Therefore, a 10 percent reduction in friction will provide a 2.2 percent reduction in fuel consumption. Consequently, a 25 percent reduction in friction will provide a 5.6 percent reduction in fuel consumption, as discussed in the SI Efficiency Fundamentals section of Chapter 2.

For a 50 percent downsized, high BMEP engine, a BMEP level twice that of the naturally aspirated engine would be required for the same operating condition of the vehicle. The BSFC for this engine with the baseline friction is as follows:

$$\text{BSFC} = \text{ISFC}\ (2 \times 38 + 11)/(2 \times 38) = 1.145\ \text{ISFC}$$

Applying the same 10 percent reduction in friction would provide the following improvement in BSFC:

$$\text{BSFC} = \text{ISFC}\ (2 \times 38 + 0\ .9 \times 11)/2 \times 38 = 1.130\ \text{ISFC}$$

Therefore, a 10 percent reduction in friction in the downsized engine will provide only a 1.3 percent reduction in fuel consumption, which is approximately half the reduction in fuel consumption shown for the naturally aspirated engine. Consequently, a 50 percent downsized engine will require nearly twice the reduction in friction relative to that required in a naturally aspirated engine to achieve the same reduction in fuel consumption.

Even though friction reductions are not as effective in the downsized engine, the downsizing itself provides a significant reduction in friction. Friction power is calculated as follows:

$$\text{Friction Power} = K \times \text{RPM} \times \text{Displacement} \times \text{FMEP} \quad (2)$$

Equation 2 indicates that friction power would be reduced by 50 percent when the engine displacement is reduced by 50 percent, assuming constant FMEP (although FMEP would be expected to show a moderate increase due to the engine redesign to withstand higher BMEP levels).

The effect of 50 percent downsizing can be calculated by comparing the baseline conditions shown previously for both the naturally aspirated and downsized, high BMEP engines as follows:

Naturally Aspirated Engine: BSFC = 1.289 ISFC
Downsized Engine: BSFC = 1.145 ISFC

This comparison indicates that 50 percent downsizing could potentially provide a 11 percent reduction in fuel consumption, with the simplifying assuming of constant FMEP in both engines.

Appendix H

Variable Valve Timing Systems

Variable valve timing (VVT) is generally accomplished by phase shifting the camshaft relative to the crankshaft. By phase shifting the camshaft, the valve events are advanced or retarded relative to the crankshaft and the piston position within the cylinder. A cam phaser is used to rotate the camshaft relative to the timing chain sprocket driven by the crankshaft. A typical cam phaser, shown in Figure H.1, consists of the outer housing and sprocket driven by the timing chain driven from the crankshaft and an inner rotor connected to the camshaft. The inner rotor has several lobes, and the space between these lobes and similar lobes on the outer housing is filled with oil. When the oil is trapped, the inner rotor rotates with the outer housing. Adding oil from one side of the lobes and removing oil from the other side moves the inner rotor relative to the outer housing, thereby phase shifting the camshaft relative to the crankshaft.

This oil pressure actuated (OPA) system generally requires a larger oil pump to provide the additional oil flow required, which is a parasitic loss that slightly diminishes the fuel consumption reductions provided by the VVT system alone. Honda recently introduced an electrically-actuated cam phaser (from Denso) in the 1.3L engine of the 2014 MY Fit to provide the desired phase angle for cold starts and a possible reduction in parasitic losses. Electric actuation of the cam phaser replaces oil pressure actuation (OPA) used previously.

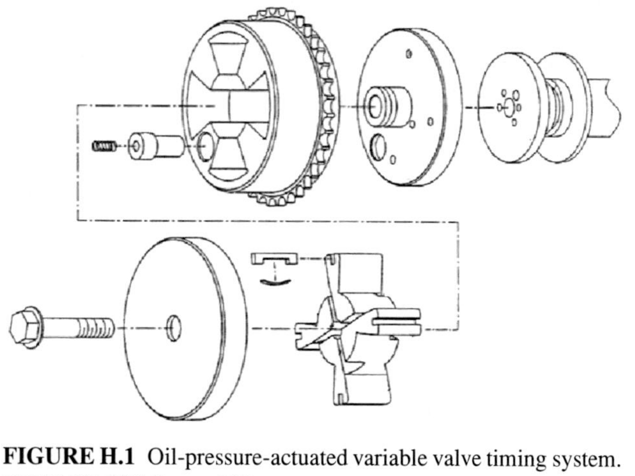

FIGURE H.1 Oil-pressure-actuated variable valve timing system.
SOURCE: U.S Patent Office.

BMW uses an alternative cam phaser design in which the end of the camshaft contains helical gear teeth. The chain-driven sprocket contains an axially moveable cap with matching helical gear teeth. By axially moving the cap, the phase of the camshaft relative to the crankshaft driven sprocket can be advanced or retarded. The moveable cap, which is attached to a double acting piston, is moved by applying oil pressure to the appropriate side of the piston.

An alternative to the oil pressure actuated (OPA) system is the cam torque actuated (CTA) system developed by BorgWarner. Rather than using oil pressure applied to the cam phaser, the CTA system relies on the reaction of the valve spring forces during valve opening or closing. During valve opening, a retarding torque is developed by the valve spring force which is used to retard the timing of the camshaft relative to the crankshaft. Conversely, during valve closing, an advancing torque is developed by the valve spring force which is used to advance the timing of the camshaft relative to the crankshaft. A solenoid-controlled spool valve directs the oil flow into the desired side of the cam phaser for advance or retard and out of the other side of the cam phaser. The advantages of the CTA system are fast response and elimination of the need for a larger oil pump. The CTA system was first introduced in the Ford 3.0L Duratec V6 engine in the 2009 MY, with later introductions in the 3.5L and 3.7L V6 and 5.0L V8 engines (Austin 2010). Ford indicated that their twin independent variable cam timing (Ti-VCT) system provides up to 4.5 percent improvement in fuel economy, with a 7 percent improvement in rated power and a 5 percent improvement in low speed torque (Ford 2010). Cost of the CTA system is estimated at 1.3 times the cost of the OPA system.

REFERENCES

Austin, M. 2010. Cam-Torque Actuated Variable Valve Timing System. Car and Driver, August. http://www.caranddriver.com/features/cam-torque-actuated-variable-valve-timing-system-feature.

Ford. 2010. Twin Independent Variable Camshaft Timing (Ti-VCT) Helps Make 2011 Ford Mustang a True Thoroughbred. Ford Motor Company, February 8. https://media.ford.com/content/fordmedia/fna/us/en/news/2010/03/24/twin-independent-variable-camshaft-timing--ti-vct--helps-make-20.html.

Weaver, S.G. 2012. The Ins and Outs of Variable Valve Timing (VVT) Systems and Their Role in Vehicle Emissions. Underhood Service, June 14. http://www.underhoodservice.com/the-ins-and-outs-of-variable-valve-timing-vvt-systems-and-their-role-in-vehicle-emissions/.

Appendix I

Variable Valve Lift Systems

A variety of two-stage, or discrete, variable valve lift (DVVL) and continuously variable valve lift (CVVL) systems have recently been incorporated in production vehicles. Several systems that have been introduced with the objective of reducing fuel consumption are described in this Appendix.

DISCRETE VARIABLE VALVE LIFT (DVVL) SYSTEMS

Two-stage, or discrete, variable valve lift (DVVL) systems generally rely on the cam profile switching (CPS) concept. CPS provides a means of switching the actuation of the intake valves between a standard high lift cam for maximum power operation and a low lift cam for efficient light load operation. The low lift cam for light load operation reduces fuel consumption by transferring air flow control from the throttle to the intake valves which results in reduced pumping losses. The incorporation of the CPS concept in numerous production applications is described in the following sections.

Honda i-VTEC

The Honda i-VTEC (intelligent-Variable Valve Timing and Lift Electronic Control) system provides continuous cam phasing (VVT) and discrete variable valve lift (DVVL) control using the cam profile switching (CPS) concept. The VTEC system consists of a camshaft with a center high lift cam and outer low lift cams and associated rocker arms, as shown in Figure I.1. For light load operation, the valves are operated by the two outer, low lift cams and associated outer rocker arms. For higher power operation, hydraulic pressure is used to insert a pin to lock the middle rocker arm to the outer rocker arms for operation on the high lift cam. Cam phasing is accomplished with a hydraulic cam phaser.

Audi Valvelift System (AVS)

The Audi Valvelift System (AVS) uses the cam profile switching concept (CPS) with a high lift cam lobe (11 mm)

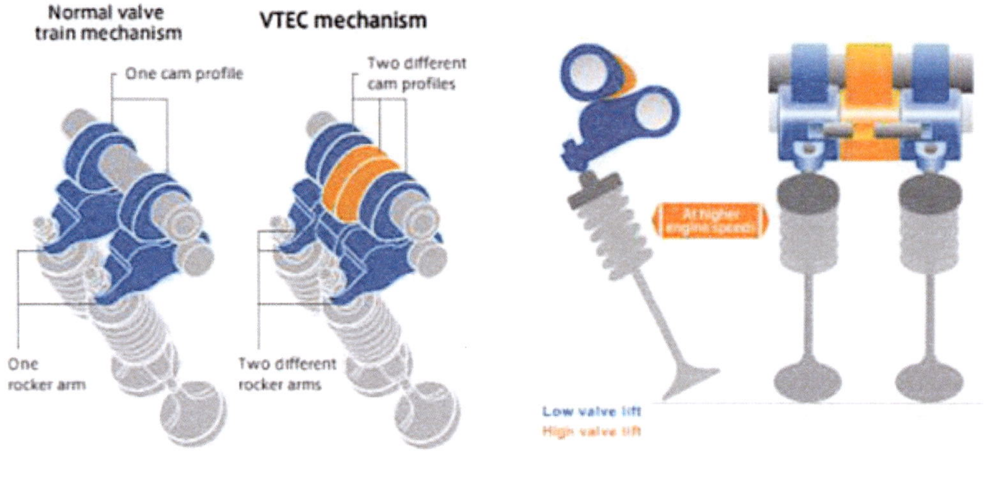

FIGURE I.1 Honda i-VTEC.
SOURCE: Courtesy of Honda (2013).

and low lift cam lobes. Two different low lift cam lobes (5.7 mm and 2 mm) are used to create swirl for improved mixing at low speeds. The two sets of the low and high lift cams for each of the two intake valves of a cylinder are mounted on a single carrier piece so that the active cams depend on the longitudinal position of the carrier piece on the camshaft. Each cam carrier piece has two spiral grooves. A pin engages one of the spiral grooves to move the carrier axially from the low lift cam to the high lift cam position. Another reverse spiral groove and pin are used to return the carrier to the low lift cam position. The pins are electrically actuated. The Audi AVS system was first introduced in the 2006 1.8L TFSI engine in the Audi A3 and subsequently applied to the 2.8L and 3.2L V6 FSI engines. Audi has indicated that the AVS system provides up to a 7 percent reduction in fuel consumption.

Mercedes Camtronic

The Mercedes Camtronic system is similar to the Audi AVS system except that it uses fewer cam carrier pieces. In the Mercedes system, each cam carrier piece serves two adjacent cylinders, instead of only one cylinder as in the Audi AVS system. Mercedes introduced the Camtronic system on the new M270 series four-cylinder engine in 2012.

Chevrolet Intake Valve Lift Control

The Chevrolet intake valve lift control (IVLC) system also uses the cam profile switching concept with low and high lift cams. The system uses rocker arms consisting of two roller followers that are electro-hydraulically latched to the rocker arm for operation on either the low or high lift cams. Chevrolet introduced the IVLC system on their 2.5L EcoTec four-cylinder engine in the 2014 MY Chevrolet Impala (SAE 2012). General Motors indicates that the IVLC system will provide a fuel savings of up to 1 mpg (approximately 4 percent (Kelly Blue Book 2013).

CONTINUOUSLY VARIABLE VALVE LIFT (CVVL) SYSTEMS

Several different principles have been applied to develop continuously variable valve lift systems (CVVL) that range from mechanical systems to hydraulic systems. Several systems that have been introduced in production with the objective of reducing fuel consumption are described here.

BMW Valvetronic

The BMW Valvetronic system was the first continuously variable valve lift mechanism, which went into production in the BMW 316ti in 2001. The goal of Valvetronic was to reduce fuel consumption. Since air flow, and thus engine output, is controlled by valve lift with the Valvetronic system, the conventional throttle valve is disabled which reduces pumping losses. Overall, BMW claims that Valvetronic can provide a 10 percent reduction in fuel consumption (Autozine 2013). Valvetronic adds an intermediate rocker arm between the camshaft and the roller finger follower that actuates the valve. The pivot location of the intermediate rocker arm is varied with an eccentric shaft controlled by an electric motor through a worm gear set. Rotating the eccentric shaft to extend the pivot location of the intermediate rocker arm results in an increase in valve lift. Although Valvetronic reduces fuel consumption at part load, maximum power is not increased since the additional components result in additional friction and inertia. The Valvetronic system adds significant height above the cylinder head so that packaging the engine is more difficult.

Toyota Valvematic (Continuous VVL)

The Toyota Valvematic system, shown in Figure I.2, inserts an intermediate rocker shaft between the camshaft and the roller rocker arm that actuates the valve (U.S. Patent Application Publication 2014). The intermediate rocker shaft contains roller followers offset from the center of the intermediate rocker shaft and additional followers. The angle between the roller follower and additional follower is controlled by an electric motor. Increasing this angle increases the resultant valve lift. Toyota implemented the Valvematic system on their 1.6L, 1.8L and 2.0L engines and subsequently introduced the system on the 2014 MY 1.8L Corolla in the U.S. market. The Valvematic system was reported to improve fuel economy by 5 percent and increase power by 6 percent in the 2014 MY Corolla (Borge 2013).

Multiair Electro-Hydraulic Valve-Timing System

In the Fiat Chrysler Multiair system, shown in Figure I.3, a piston, operated by a mechanical intake cam, is connected to the intake valve through a hydraulic chamber. The hydraulic chamber includes a solenoid valve. When the solenoid valve is closed, the oil in the line acts as a solid body and transmits the intake cam motion directly to the intake valve. By opening the solenoid valve, the hydraulic pressure is relieved and the intake valve closes under the action of the valve spring. A dedicated hydraulic brake is used to provide a soft landing of the intake valve for all engine operating conditions. A wide range of intake valve actuation modes can be obtained by controlling the solenoid valve, as illustrated in Figure I.3.

- For maximum power at high speed, the solenoid valve is always closed so that full valve opening is obtained by following the intake cam.
- For low speed torque, the solenoid valve is opened near the end of the cam profile to provide early intake valve closing to eliminate backflow into the intake manifold to maximize the air mass trapped in the cylinders.

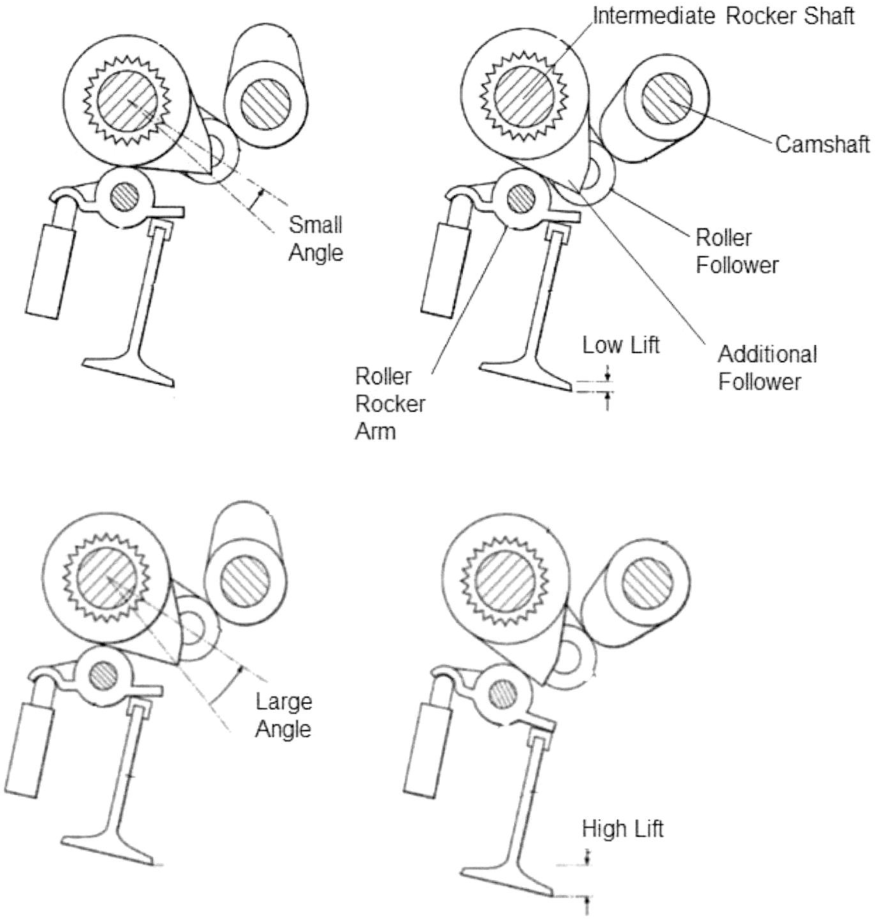

FIGURE I.2 Toyota Valvematic variable valve lift systems illustrating low and high lift configurations resulting from adjusting the angle between the intermediate roller follower and additional follower to obtain the desired valve lift.
SOURCE: Boutell (2014). Courtesy of Toyota Motor Engineering & Manufacturing North America, Inc.

- At part load conditions, the solenoid valve can be opened earlier resulting in shortened valve open times to control trapped air mass as a function of the required torque without throttling (partial valve opening).
- At idle, the solenoid valve can be closed after the mechanical cam has started and then opened early, resulting in partial valve opening so that air flows faster past the intake valve to enhance in-cylinder turbulence (late valve opening).
- The last two actuation modes can be combined to open the intake valve twice during each intake stroke to enhance turbulence and combustion rates at light loads.

Fiat Chrysler indicated that Multiair can provide a 10 percent reduction in fuel consumption with the elimination of pumping losses. Additionally, Multiair can provide up to a 10 percent increase in power when a power-oriented mechanical cam profile is used. Low speed torque can be improved by up to 15 percent (Murphy 2010). Multiair was first introduced in the 2010 MY 1.4L Chrysler Fiat 500. The next application was in the 2013 MY 2.4L Dodge Dart. Subsequent applications have included the 2014 MY Jeep Cherokee, and the 2015 MY Chrysler 200, Jeep Renegade, and Ram ProMaster City.

REFERENCES

Borge, J.L. 2013. Toyota engineers put a shine into the 2014 Corolla. SAE International, Automotive Engineering Magazine, September 9. http://articles.sae.org/12444/.

Boutell, B.T. 2014. Variable valve Operation Control Method and Apparatus. U.S. Patent Application Publication No. US 2014/0102385 A1, April 17.

Cam-changing + Cam-phasing VVT. Autozine Technical School. Accessed June 23, 2013. http://www.autozine.org/technical_school/engine/vvt_31.htm.

Continuous Variable Valve Lift (CVVL). Autozine Technical School. Accessed September 6, 2013. http://www.autozine.org/technical_school/engine/vvt_5.html.

Honda. 2013. The VTEC breakthrough: solving a century-old dilemma. http://world.honda.com/automobile-technology/VTEC/.

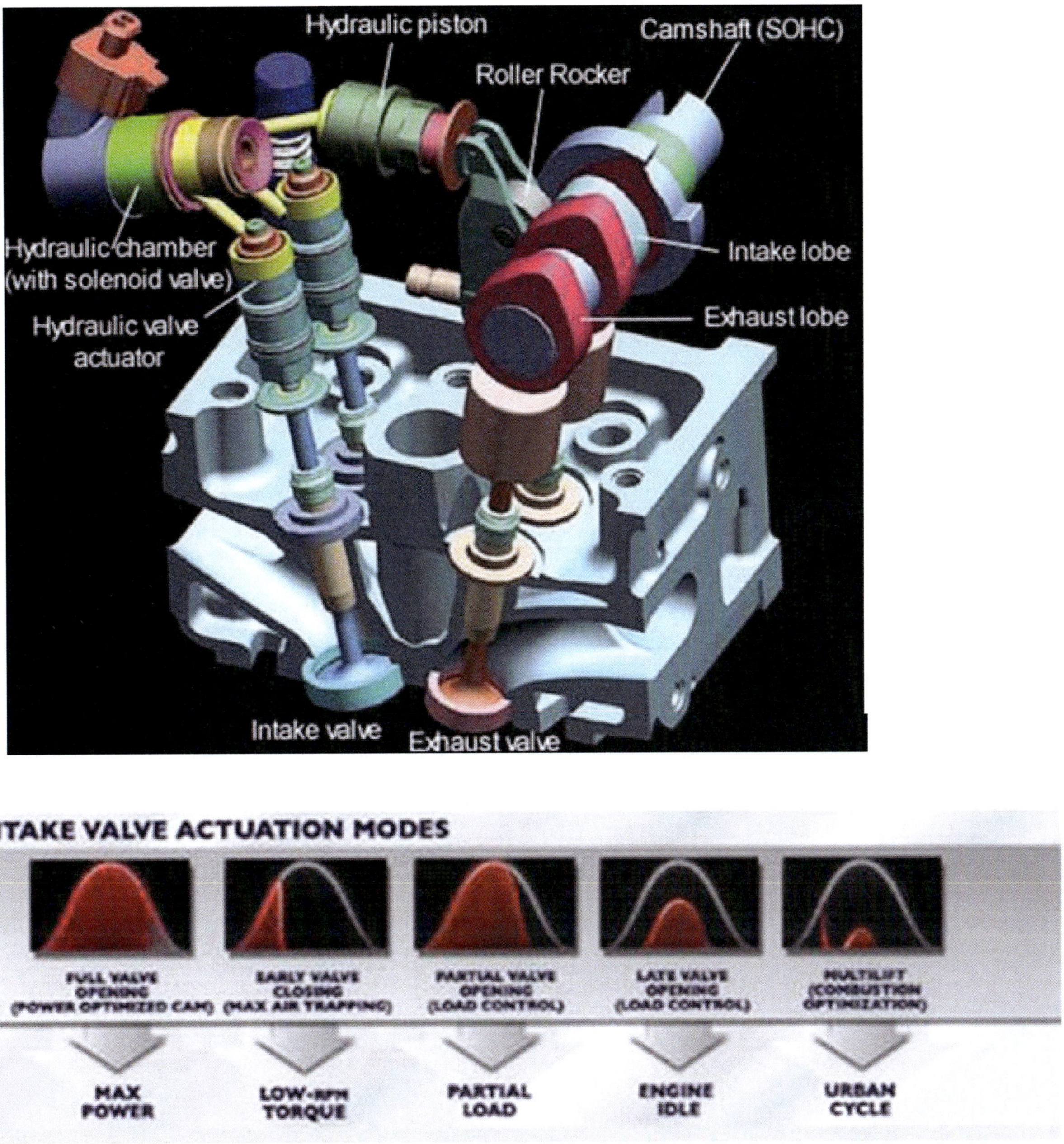

FIGURE I.3 Multiair electro-hydraulic valve-timing system.
SOURCE: Murphy (2010). Credited to FCA US LLC.

Kelly Blue Book. 2012. 2014 Chevy Impala Gets Variable Valve Lift on Ecotec 4-cylinder. Kelly Blue Book, September 17. Accessed August 6, 2013. http://www.kbb.com/car-news/all-the-latest/2014-chevy-impala-gets-variable-valve-lift-on-ecotec-4_cylinder/2000008572/.

Murphy, T. 2010. Fiat Breathing Easy With MultiAir. WardAuto, March 26. http://wardsauto.com/ar/fiat_breathing_multiair_100326.

SAE. 2012. GM's new modular Ecotec I4 architecture adds first variable intake-valve lift. October 9. http://articles.sae.org/11466/.

Appendix J

Reasons for Potential Differences from NHTSA Estimates for Fuel Consumption Reduction Effectiveness of Turbocharged, Downsized Engines

The committee estimated that the fuel consumption reduction effectiveness for a 24 bar BMEP, 50 percent downsized, turbocharged engine could be approximately 2 percentage points lower than NHTSA's estimates for the following reasons:

Reason 1: Compression ratio is likely to be reduced to accommodate 91 RON gasoline (87 AKI) in the U.S. instead of European 95 RON gasoline used in the Mahle and Ricardo analyses.

a. Compression ratios of engines cited by EPA/NHTSA:

EPA/Ricardo Report (EPA/Ricardo 2011, p. 19):
CR = 10.5:1 for full system simulation

Research engines cited in EPA/Ricardo Report (p. 19):
Sabre engine, Lotus Engineering, UK (Coltman 2008)
"Operating on regular 95 RON"
CR = 10.2:1

Mahle 30 bar BMEP engine (Lumsden 2009)
CR = 9.75
Engine was not tested on 91 RON gasoline, Mahle presentation to NRC committee (Blaxill 2012)

b. Compression ratios of U.S. turbocharged downsized engines (Mahle Wards Light Duty Engine Chart 2014):

Ford: 2.0L EcoBoost CR = 9.3:1

GM: 2.0L Turbo CR = 9.2:1
1.4L Turbo CR = 9.5:1

c. Conclusion 1: U.S. turbocharged engines operating on 91 RON gasoline compared to European engines operating on 95 RON gasoline have approximately a 1.0 lower compression ratio.

d. Effect of lower compression ratio on fuel consumption 1.0 CR results in approximately 1.5 percent FC increase (Chapter 2, High Compression Ratio with High Octane Gasoline section)

e. Conclusion 2: The 1.0 lower compression ratio of turbocharged engines in the U.S. will have approximately a 1.5 percent increase in fuel consumption relative to the cited European research engines.

Reason 2: Spark retard is likely to be required in some higher load regions encountered in the CAFE drive cycles.

a. Apply the criteria from "A review of the Effect of Engine Operating Conditions on Borderline Knock"(Russ 1996) to provide an estimate of the spark retard required for the boost pressures encountered in the CAFE drive cycles.

Criterion for RON change due with a change in compression ratio: 5 RON / 1 CR
(This criterion illustrates the reduction in CR required in the U.S. for engines operating on gasoline with 91 RON instead of 95 RON.)

Criterion for RON change with a change in intake pressure: 3-4 RON / 10 kPa (100 kPa = 14.5 psi)

Criterion for RON change with spark retard: 1 RON/ 1 degree

The pressure ratio at higher speed/load conditions in the CAFE drive cycles was estimated to be 1.5:1. The increase in pressure of 50 kPa (above 100 kPa atmospheric pressure) would result in the need for a

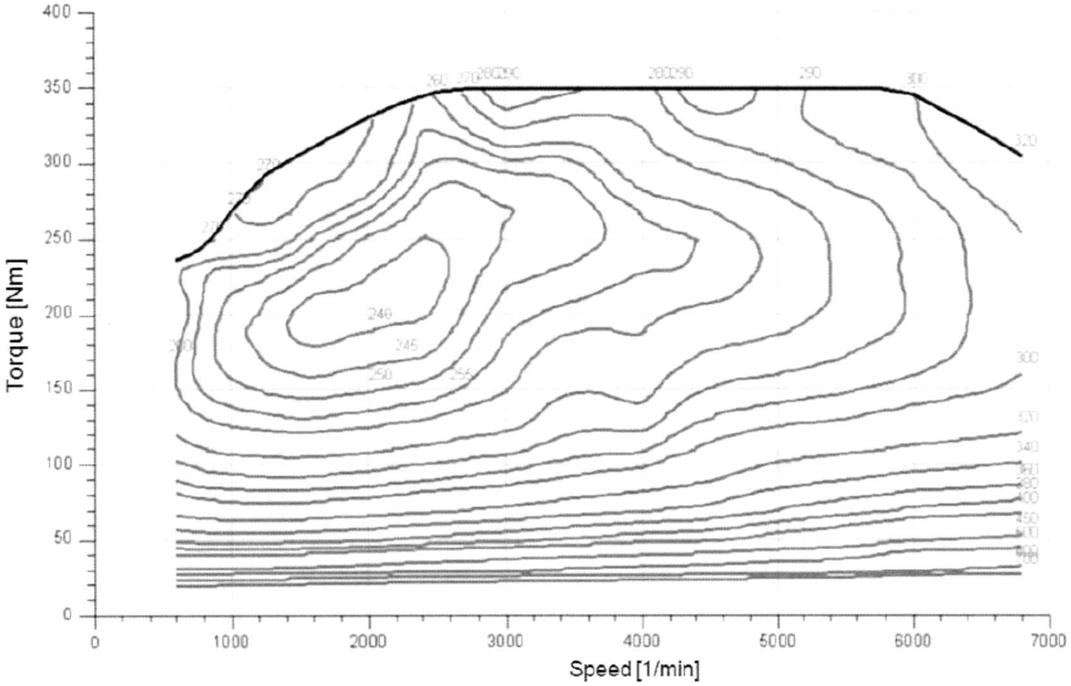

FIGURE J.1 Generic fuel island map for a 3.3L V6 gasoline engine with brake specific fuel consumption in g/kWh. SOURCE: Dick (2013). Reprinted with permission from SAE paper 2013-01-1272 Copyright © 2013 SAE International.

15 RON increase by applying the RON/kPa criteria shown above.

With the assumed 91 RON gasoline at these conditions, spark must be retarded, according to the RON/spark retard criteria shown above:

Spark Retard = 15 degrees from MBT (minimum spark advance for best torque)

b. Apply the guideline in "Study of the effects of ignition timing on gasoline engine performance and emissions" (Zareei 2013):

15 degrees spark retard: BSFC increases by 6% (at some conditions)

c. Conclusion: Spark retard to avoid knock at some conditions on the CAFE drive cycles will result in approximately 6 percent increase in fuel consumption at those conditions and an estimated 0.25 percent increase in fuel consumption on the CAFE drive cycles.

Reason 3: Wider transmission spans, or modified torque converters, may be used to overcome turbocharged lag during launch (higher engine speeds at launch)

a. The following example is based on the fuel island map shown in Figure J.1. This example is for a

	RPM	Torque lbs-ft (Nm)	Transmission 1st Gear Ratio	Driveshaft Torque (lbs-ft)	BSFC
3.3L NA Engine:	1500	133 (181)	4.0:1	532	240 g/kWh
1.65L TC Engine:[a]	1500	133 (181)	Torque not achievable at same engine speed due to turbocharger lag		
1.65L TC Engine:[a]	2000	100 (136)	5.3:1	530	255 g/kWh

[a] For the 50 percent downsized engine, torque must be multiplied by 2 to enter Figure J.1.

moderate launch acceleration rate with a 3.3L naturally aspirated engine and a 50 percent downsized, turbocharged 1.65L engine.

b. At launch, the 1.65L turbocharged (TC) engine cannot develop the torque of the 3.3L naturally aspirated (NA) engine due to turbocharger lag.
   - Higher transmission ratio is provided to TC engine for higher torque multiplication.
   - As a result, the same driveshaft torque can be provided.
   - This results in a higher engine speed for the same vehicle speed.
   - Fuel consumption will be increased by 6 percent during these periods (BSFC of 255 g/kWh for the turbocharged, downsized engine compared to 240 g/kWh for the naturally aspirated engine).

c. Conclusion: During launch from idle, which occurs 18 times during the FTP-75 drive cycle, fuel consumption of the turbocharged engine will be lower by up to approximately 6 percent and an estimated 0.25 percent increase in fuel consumption on the CAFE drive cycles.

Overall Conclusion: Fuel consumption reduction effectiveness for turbocharged, downsized engines could be approximately 2 percentage points lower than NHTSA's estimates due to 1) lower compression ratio, 2) the effects of spark retard to avoid knock at higher speed/load conditions, and 3) higher transmission ratios during launch to provide driveshaft torque comparable to the naturally aspirated engine.

REFERENCES

Blaxill, H. 2012. MAHLE Downsizing Demonstrator. Presentation to the National Research Council Committee on Assessment of Technologies for Improving Fuel Economy of Light-Duty Vehicles, Phase 2. Dearborn, Michigan, September 27.

Coltman, D., J.W.G. Turner, R. Curtis, D. Blake, B. Holland, R.J. Pearson, A. Arden and H. Nuglisch. 2008. Project Sabre: A Close-Spaced Direct Injection 3-Cylinder Engine with Synergistic Technologies to Achieve Low CO_2 Output. SAE Technical Paper 2008-01-0138.

Dick, A., J. Greiner, A. Locher, and F. Jauch. 2013. Optimization Potential for a State of the Art 8-Speed AT. SAE 2013-01-1272 .

EPA/Ricardo. 2010. Computer Simulation of Light-Duty Vehicle Technologies for Greenhouse Gas Emission Reduction in the 2020-2025 Timeframe. EPA-420-R-11-020, December.

Lumsden, G., D. OudeNijeweme, N. Fraser and H. Blaxill. 2009. Development of a Turbocharged Direct Injection Downsizing Demonstrator Engine. SAE Technical Paper 2009-01-1503.

Russ, S. 1996. A review of the effect of engine operating conditions on borderline knock. SAE Technical Paper 960497. doi:10.4271/960497.

Zareei, J., and A.H. Kakaee. 2013. Study of the effects of ignition timing on gasoline engine performance and emissions. Eur. Transp. Res. Rev. 5:109-116.

Appendix K

DOE Research Projects on Turbocharged and Downsized Engines

DOE currently has programs with Ford, General Motors, and Chrysler to demonstrate a 25 percent improvement in fuel economy while achieving Tier 2 Bin 2 emissions requirements with downsized, boosted engines and a variety of other technologies, including lean combustion, cooled EGR, advanced ignition systems, and friction reduction technologies. These programs are described here.

FORD/DOE ADVANCED GASOLINE TURBOCHARGED DIRECT INJECTION (GTDI) ENGINE DEVELOPMENT PROGRAM

DOE has a program with Ford, with support on advanced ignition concepts from Michigan Technological University, to demonstrate 25 percent fuel economy improvement in a mid-sized sedan using a downsized, advanced GTDI engine with no or limited degradation in vehicle level metrics. The vehicle is to demonstrate the capability of meeting Tier 2 Bin 2 emissions. The project includes aggressive downsizing from a large V6 engine to a small I4 engine, direct fuel injection, lean combustion with cooled EGR and advanced ignition, boosting systems with active and compounding components, cooling and aftertreatment systems, advanced friction reduction technologies, engine control strategies, and NVH countermeasures. Vehicle demonstration of greater than 25 percent weighted city/highway fuel economy and Tier 3 Bin 30 emissions on the FTP-75 test cycle is scheduled to be completed by September 30, 2015 (see Weaver 2014).

GENERAL MOTORS/DOE LEAN GASOLINE SYSTEM DEVELOPMENT FOR FUEL EFFICIENT SMALL CARS PROGRAM

DOE has a 39 month program with General Motors to demonstrate 25 percent vehicle fuel economy improvement while achieving Tier 2 Bin 2 emissions with an advanced, boosted lean gasoline combustion engine and aftertreatment system (Smith 2013). The subsystems and components that are being redesigned to support the integration of the boosted lean combustion system include new spark plugs, injector targeting changes, chamber smoothing modifications, cylinder pressure transducer provisions, split intake port cylinder head, intake port deactivation adapter assembly to provide high swirl mixture motion for increased lean dilution tolerance, and a close coupled catalyst exhaust system with cooled external EGR exhaust. General Motors has projected the following efficiency improvements for this program:

- 12.5 percent for lean dilute combustion with closely spaced multiple pulse injections and cooled EGR;
- 7.5 percent for downsizing from 2.4L PFI to 1.4L turbocharged DI engine; and
- 5 percent for vehicle integration which includes 12V stop/start and active thermal management.

The 1.4L boosted stoichiometric homogeneous engine was modified to enable lean stratified operation. A passive SCR lean aftertreatment system is being developed with efforts directed at overcoming the limitations of excess CO breakthrough during NH3 generation and insufficient NO_x reduction during high thermal operating conditions. Because of the issues with the passive SCR system, an active urea lean aftertreatment system development is continuing. The active urea dosing system initially will utilize copper zeolite SCR technology. The active and passive systems are projected to support lean and stoichiometric operation. Active thermal management is expected to provide up to an additional 1.5 percent in fuel economy benefit.

The LDB (Lean Downsized Boost) engine with the 12V stop/start and active thermal management is projected to meet the 25 percent target based on engine dynamometer test results. At the conclusion of the project in 2013, the lean downsized boosted (LDB) engine demonstrated 21 percent fuel economy improvement over the PFI baseline. The LDB engine combined with 12-volt start/stop and thermal management demonstrated the project objective of 25 percent fuel economy improvement over the PFI baseline.

CHRYSLER DOWNSIZED, TURBOCHARGED ENGINE RESEARCH PROGRAM

Chrysler has a three year, $30 million program half-funded by DOE and half-funded by Chrysler and its technology and research partners including Argonne National Laboratory, Bosch, Delphi, Ohio State University, and FEV. The three-year program began in 2010 and has since had a nine-month extension to April 2014. The objective of the program is to demonstrate a 25 percent improvement in combined city/highway fuel efficiency in a mini-van while meeting Tier 2 Bin 2 emissions and with drivability comparable to a current production vehicle. The program will downsize the base 4.0L V6 engine to a 2.4L I4 engine. Key technologies applied to this program include a purpose designed combustion chamber, 12:1 compression ratio, spray-bore cylinder liners with laser-honed surfaces for reduced mass and friction, high-energy dual-plug ignition, two-stage turbocharging, cooled EGR, secondary air injection, and a belt starter-generator (BSG). Chrysler is also using diesel micro-pilot (DMP) ignition. This system uses carefully timed pilot injection of diesel fuel to enhance and extend the gasoline burn rates at high load, while assisting the spark ignition combustion in the transition zone between high and low loads. Chrysler has found that the DMP ignition functions within very small tolerances of EGR rates that may not be controllable. As an alternative, the program is also investigating a spark-ignition dual-fuel strategy using E85 and three spark plugs per chamber.

The BSG is an enabler for iDFSO (integrated Deceleration Fuel Shut Off) functionality from 15 mph to 0 mph that is projected to provide approximately 2 percent fuel savings potential on FTP cycle. With secondary air injection, the rich combustion products react with the air to create an exotherm in the exhaust runner to provide an exhaust temperature of 842°C (compared to 455°C without air injection), within 11 seconds from a cold start for rapid catalyst warm-up.

This program has demonstrated 25 percent improvement in combined FTP city and highway fuel economy in the powertrain test cell. Vehicle results are pending as of the June 2014 DOE Annual Merit Review (Reese II 2014).

REFERENCES

Reese II, R.A. 2014. A MultiAir/MultiFuel Approach to Enhancing Engine System Efficiency. Project ID ACE062, DOE Annual Merit Review, June 19.

Smith, S. 2013. Lean Gasoline System Development for Fuel Efficient Small Car. U.S. DOE Merit Review and FY 2013 Progress Report for Advanced Combustion Engine Research and Development, Energy Efficiency and Renewable Energy. Vehicle Technologies Office.

Weaver, C.E. 2014. Advanced Gasoline Turbocharged Direct Injection (GTDI) Engine Development. Project ID ACE065, DOE Annual Merit Review, June 19.

Appendix L

Relationship between Power and Performance

The relationship between power and performance is derived in this Appendix. The propulsion or tractive force to accelerate a vehicle can be calculated from the sum of the tire rolling resistance, the aerodynamic drag, and the inertial force for the vehicle on a level road as follows (Gantt 2011):

$$T_{tr} = F_{\text{rolling resistance}} + F_{\text{aerodynamic drag}} + F_{\text{inertia}} \qquad (1)$$

Expanding this equation yields the following:

$$T_{tr} = C_{rr}\, m\, g_c + \tfrac{1}{2}\, \rho\, C_d\, A_f\, V^2/g_c + m\, dV/dt/g_c \qquad (2)$$ [1]

The following parameters were used in Equation 2 for a typical 3,500 lb midsize car:

C_{rr} = 0.0060 (tire rolling resistance)
m = 3500 lbm (mass of the vehicle)
ρ = 0.075 lbm/ft^3 (density of air)
C_d = 0.30 (aerodynamic drag coefficient)
A_f = 25 ft^2 (frontal area of vehicle)
V = 60 mph (88 ft/sec)
t_{60} = 0 to 60 mph acceleration time

During a wide-open throttle acceleration, the tractive force can be expressed as following at the 60 mph condition:

$$T_{tr}\ \text{lbf} = 0.0060 \times 3{,}500\ \cancel{\text{lbm}} \times 32.2\ \cancel{\text{ft/sec}^2}/32.2\ \cancel{\text{lbm-ft}}/\text{lbf-}\cancel{\text{sec}^2} + \tfrac{1}{2} \times 0.075\ \cancel{\text{lbm/ft}^3} \times 0.30 \times 25\ \cancel{\text{ft}^2} \times (88\ \cancel{\text{ft/sec}})^2/32.2\ \cancel{\text{lbm-ft}}/\text{lbf-}\cancel{\text{sec}^2} + 3{,}500\ \cancel{\text{lbm}} \times 88\ \cancel{\text{ft/sec}^2}/(t_{60}\ \text{sec} \times 32.2\ \cancel{\text{lbm-ft}}/\text{lbf-}\cancel{\text{sec}^2}) \qquad (3)$$

(Note the cancellation of units leaving lbf for each term of Equation 3.)

Power to propel the vehicle is obtained by multiplying the tractive effort force in Equation 3 by velocity at 60 mph (88 ft/sec) and converting the product to horsepower (hp = 550 ft lbf/sec), which yields the following equation for power:

$$Hp = 14 + 1530/t_{60} \qquad (4)$$

Applying Equation 4 yields the following relationship between 0 to 60 mph time and horsepower:

T60	% Change in 0 to 60 mph time	Hp	% Change in Hp
8 seconds	Base	205	Base
7.2 seconds	– 10%	227	+ 10.7%

These results show that approximately a 10 percent decrease in 0 to 60 mph time requires approximately a 10 percent increase in power.

REFERENCES

Allen, J. n.d. Concept Review: Unit Systems. Michigan Technological University. http://www.me.mtu.edu/~jstallen/courses/MEEM4200/lectures/energy_intro/Review_unit_systems.pdf. Accessed April 3, 2015.

Gantt, L. 2011. Energy Losses for Propelling and Braking Conditions of an Electric Vehicle. VPI MS Thesis, May.

[1] The force exerted by a mass on earth is given by the following equation, which requires the constant, g_c, which is equal to 32.2 lbm-ft/lbf-sec^2 (Allen n.d.):

$F = m \times a/gc$

Therefore, a mass of 3,500 lbm (lb mass) has a weight of 3,500 lbf (lb force) on earth, where a = 32.2 ft/sec^2, as shown by substituting in the above equation:

$F = 3{,}500\ \cancel{\text{lbm}} \times 32.2\ \cancel{\text{ft/sec}^2} / (32.2\ \cancel{\text{lbm-ft}}/\text{lbf-}\cancel{\text{sec}^2}) = 3{,}500\ \text{lbf}$

(Note the cancellation of units leaving lbf.)

This relationship is used throughout Equations 2 and 3 of this Appendix.

Appendix M

HCCI Projects

Control of auto-ignition phasing in homogeneous charge compression ignition (HCCI) combustion has been achieved with a variety of methods that affect the thermodynamic state and the chemical composition of the charge. These methods include intake air temperature modulation (Thring 1989; J. Yang et al. 2002), variable valve timing (Kaahaaina et al. 2001), exhaust throttling (Souder et al. 2004), water injection (Stanglmaier and Roberts 1999), mass flow ratio of two fuel types (Olsson et al. 2001, 2002), and variable compression ratio (Christensen et al. 1999; Drangel et al. 2002).

Investigations of the combination of ignition, multiple-injections, positive valve overlap (PVO), and boosting or supercharging are underway for extending HCCI operation (Yun et al. 2010, 2011; Kulzer et al. 2011; Mamalis et al. 2012). Specifically at high load, boosting could provide the necessary air-dilution for moderating the rate of heat release, but it also increases the pumping losses. Flexible turbocharging (variable geometry turbine and/or dual stage) or custom made supercharging will help the efficiency but add complexity and cost. Similarly, variable valve timing that can sweep the entire negative valve overlap (NVO) to positive valve overlap (PVO) space will be expensive but instrumental for low pumping losses and fast response.

Several of these projects that are currently investigating HCCI-SI mode switching and extending the gasoline HCCI load range are discussed below.

ADVANCED COMBUSTION CONTROLS - ENABLING SYSTEMS AND SOLUTIONS (ACCESS)

(AVL, Bosch, Emitech, Stanford, University of Michigan)

The Advanced Combustion Controls - Enabling Systems and Solutions (ACCESS) project, partially funded by a DOE grant, is focused on coordinating multi-mode combustion events over the engine drive cycle operating conditions. The project goal is to improve fuel economy by 25 percent with engine downsizing from a naturally aspirated 3.6L V6 engine to a turbocharged 2.0L I4 engine with part-load HCCI operation while meeting SULEV emission standards.

Figure M.1 shows the BSFC map for a multi-mode combustion 2.0L engine capable of SI (spark ignition), SACI (spark assisted-homogeneous charge compression ignition) and HCCI operation (Nüesch et al. 2014a; Yilmaz et al. 2013). The engine relies on NVO and direct injection during NVO, variable valve lift, and external EGR to cover all of the modes. On a simulated FTP75 cycle, only 44 percent of the fuel is consumed in the advanced combustion modes (20 percent of the fuel spent in the HCCI mode and 22 percent of the fuel spent in the SACI mode) which leaves a large portion (58 percent) of the fuel spent in the SI mode (low and high load and the cold part of the FTP-75 cycle).

The ACCESS project uses the combustion concept, spark assisted HCCI (SACI), to reduce the ringing intensity (RI which is related to ringing noise level) and the rate of pressure rise and to extend the HCCI high load limit as shown in Figure M.1. The RI is moderated by dividing the energy

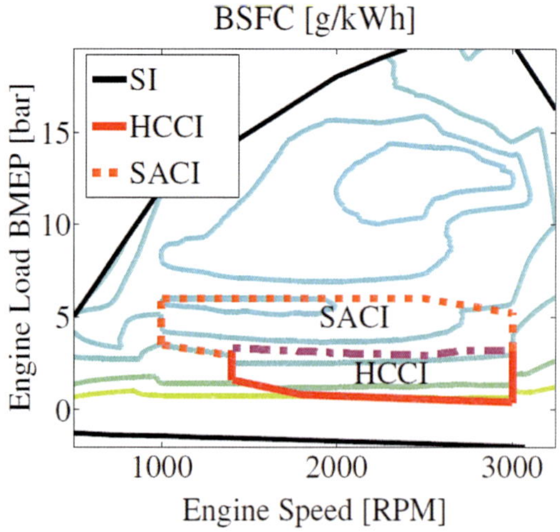

FIGURE M.1 BSFC map for a multi-mode combustion 2.0L engine showing operating regimes for lean HCCI (red solid), SACI (orange dotted), and the optimum boundary between modes (purple dotted).
SOURCE: Nüesch et al. (2014a).

release of the fuel into an SI-like heat release (slow flame propagation initiated by spark) followed by an auto-ignition, induced by the additional compression from the flame front. As a result, the pressure rise rate following the auto-ignition of the remaining compressed homogeneous fuel/air mixture is reduced. The ratio of the two combustion modes can be manipulated by the fraction of cold (external) and hot (internal) EGR trapped in the cylinder (Wagner et al. 2000), although accurate control of cooled EGR is difficult to achieve due to the transport delays in the flow paths.

Special attention to minimize the fuel penalty during mode transitions is being directed at the investigation of various mode switching strategies (Yilmaz et al. 2013; Zhu 2013). Table M.1 shows the fuel savings projected for an experimental vehicle with a 2.0L four-cylinder engine with multi-mode combustion, including SI, SACI, and HCCI modes, without fuel economy penalties for NO_x control. Fuel economy improvements approach 6 percent for the SI/SACI/HCCI combustion modes without penalties for NO_x control (Nüesch et al. 2014a).

Controlling NO_x emissions with a three-way catalyst (TWC) in multi-mode combustion incurs fuel economy penalties. These penalties were analyzed using a calibrated model for the case of SI/HCCI modes of combustion (Nüesch et al. 2014b). The results from this analysis are shown in Figure M.2 for fuel economy, engine out NO_x, and tailpipe NO_x emissions over portions of the FTP-75 cycle for several combustion modes and control cases. Case 1 is for the baseline SI mode assuming stoichiometry throughout the drive cycle to achieve maximum NO_x reduction with the TWC. Cases 2 through 5 show the results from several different SI/HCCI mode switching control strategies. Case 2 does not have any oxygen storage control and leads to the highest NO_x emissions while providing the largest improvement in fuel economy. Cases 3 through 5 apply various oxygen storage capacity (OSC) control strategies to reduce the NO_x emissions. In Case 5, rich switches are initiated if the OSC is full and the HCCI mode is vacated, and this rich depletion mode is retained until the OSC is empty. Case 5 approaches the SULEV NO_x emission requirement but negates the fuel economy benefits gained during HCCI operation (Nüesch et al. 2014a). A similar result for a drive cycle simulation was reported at the 2014 DOE Annual Merit Review (Yilmaz et al. 2014).

Based on the results shown in Figure M.2, the ACCESS project is investigating a transient lean NO_x aftertreatment system that may be required for lean HCCI operation in the multi-mode combustion concept in addition to the TWC aftertreatment. A passive SCR system is being investigated which is expected to result in an increase in cost over the initial assumption of having only a TWC aftertreatment system. A passive SCR system would likely have a lower incremental cost than the active SCR systems discussed for diesel engines in Chapter 3.

TABLE M.1 Simulated Fuel Economy Benefits for an Experimental Vehicle with a 2.0L Four-Cylinder Engine with Multi-Mode Combustion including SI, HCCI, and SACI Without Fuel Economy Penalties for NO_x Control

		FTP-75		HWFET	
Combustion Mode	Mode Switches	mpg	% Increase	mpg	% Increase
SI	None	22.8	Baseline	36.1	Baseline
SI/SACI	w/o Penalties[a]	23.5	3.2	38.0	3.5
SI/HCCI	w/o Penalties[a]	23.8	4.3	36.9	2.2
SI/SACI/HCCI	w/o Penalties[a]	24.1	5.9	38.3	6.1

Combustion Modes

SI: stoichiometric AFR, internal EGR, direct injection, TWC.

HCCI: Homogeneous highly dilute charge auto-ignites through compression; lean AFT, limited to low and intermediate speeds and loads due to high cyclic variability and very high pressure rise rates (ringing); very low NOx and reduced pumping losses.

SACI: Spark initiates a pre-flame consuming a portion of the charge, which makes the remainder of charge auto-ignite; stoichiometric AFT; internal and cooled external EGR; TWC.

Examples of Mode Switches

SI to HCCI: Cams phased from PVO (positive value overlap) to NVO (negative value overlap); cams switched to low lift; lean AFR.

SI to SACI: Cams phased from PVO (positive value overlap) to NVO (negative value overlap), cams switched to low lift; change spark timing; add external EGR; stoichiometric AFR.

[a] NOx increases relative to SI mode.

SOURCE: Derived from Nüesch et al. (2014a), Tables I and IV.

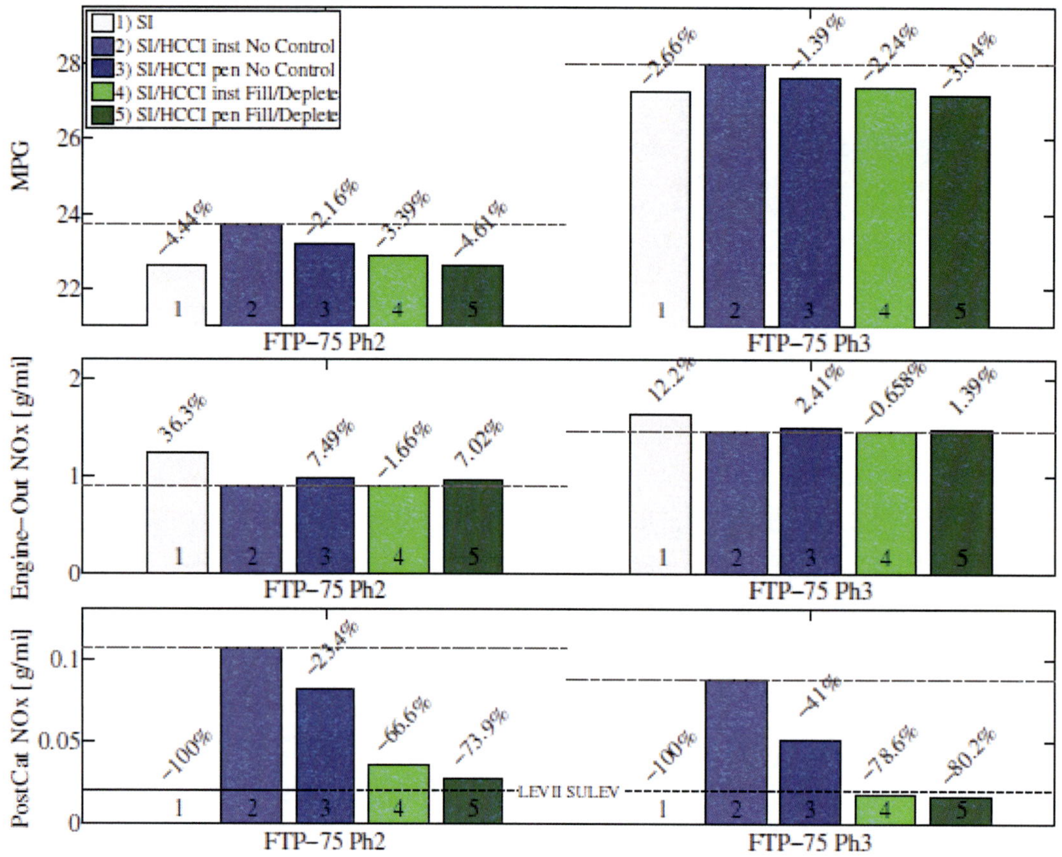

FIGURE M.2 Drive cycle simulations (FTP-75 Phase 2 and Phase 3) for fuel economy (mpg), engine out NO_x, and tailpipe NO_x with the SULEV limit of 20 mg/mi for a TWC equipped 2.0L multi-mode combustion engine.
SOURCE: Nüesch (2014b).

HOMOGENEOUS CHARGE COMPRESSION IGNITION (HCCI)

(ORNL and Delphi HCCI)

Oak Ridge National Laboratory (ORNL) and Delphi are investigating ways to expand the load where gasoline HCCI can be achieved (Szybist et al. 2013; Weall et al. 2012) on a single cylinder version of a 2.0L four-cylinder engine with the compression ratio increased to 11.85:1. Recent results have shown that the load for successful HCCI operation in a naturally aspirated engine is 3.5 bar (350 kPa) IMEP, but that this can be increased to 6.5 bar (650 kPa) IMEP with high levels of boost pressure up to 1.9 bar to provide additional air dilution as well as with the use of external EGR. Both of these methods reduce the reactivity of the systems so that more advanced fuel injection timing into the NVO is required. Under boosted conditions, NO_x emissions remained low (< 0.01 g/kWh). A peak indicated thermal efficiency of 41.5 percent (ISFC = 199 g/kWh) was achieved at a load of 6.0 bar (600 kPa) IMEP. ORNL and Delphi caution that the indicated thermal efficiency on a multi-cylinder engine may be lower due to increased pumping work when using an actual turbocharge instead of air provided by the test facility.

ORNL and Delphi have defined numerous concerns with controlling the HCCI engine. The most significant concern is with transient operation (1) within the HCCI mode and (2) during transition from the HCCI mode to the SI mode. Within the HCCI mode, boost pressures have relatively long response times due to turbocharger lag lasting at least several engine cycles. HCCI operation has been found to be extremely sensitive to NVO duration, and this sensitivity increases as engine load increases. External EGR also has a relatively long response time. During mode switching from HCCI to SI combustion, the manifold pressure for HCCI is too high for SI combustion without a large spike in torque. This issue is more severe with boosting where SI combustion requires throttled manifold pressure.

As part of their research on HCCI combustion, ORNL and Delphi have provided the following assessment of HCCI and its outlook:

- "HCCI has yet to achieve its promise of high efficiency and low NO_x emissions in a production engine despite the first published research papers appearing over 3 decades ago."

- "The load limitation is becoming even more critical for the relevance of HCCI combustion given the trends of engine downsizing.... The duty cycle of the engine is changed under normal driving conditions, resulting in a reduced amount of time spent at the conditions applicable to HCCI combustion."

GASOLINE DIRECT INJECTION COMPRESSION IGNITION (GDICI)

(University of Wisconsin and General Motors)

The University of Wisconsin and General Motors have investigated the use of 87 AKI regular grade gasoline in a high speed, direct injection, light-duty compression ignition engine to extend the low temperature combustion (LTC) regime to high loads (Ra et al. 2011). This system is termed GDICI for Gasoline Direct Injection Compression Ignition. The investigation found that GDICI operation of a light-duty engine was feasible under full load conditions of 16 bar IMEP, thereby significantly extending the low-emission combustion concept (Ra et al. 2012). The engine had a compression ratio of approximately 16.5:1 and was capable of multiple injections, specifically double and triple pulse injections. Both PM and NO_x emissions were reduced to levels of about 0.1 g/kg-f while achieving an ISFC as low as 173 g/kW-hr with a triple pulse fuel injection strategy (Ra et al. 2012). The engine was found to be very sensitive to EGR ratio, initial gas temperature, and injection pressure. The results obtained rely on high EGR rates, produced with an inlet pressure of 2.8 bar and an exhaust pressure of 3.01 bar, which might not be realizable with currently available turbocharging systems. The multiple-injection methodologies are also very difficult to control due to their sensitivity in highly varying spatial and temporal patterns of charge mixing.

GASOLINE DIRECT INJECTION COMPRESSION IGNITION (GDCI)

(Delphi, Hyundai, Wayne State University, University of Wisconsin, Wisconsin Engine Research Center)

Similar to the University of Wisconsin and General Motors GDICI engine, Delphi and Hyundai are developing, under a DOE contract, a Gasoline Direct Injection Compression Ignition (GDCI) engine with the goal of achieving full time, low temperature combustion using multiple late injections over the entire engine speed-load map from idle to full load (Sellnau 2012). Complete mixing of all the fuel in a homogeneous charge is avoided with late injections since this would lead to rapid burning of the whole mixture. Regular unleaded gasoline (90.6 RON) and unleaded gasoline with 10 percent ethanol (91.7 RON) fuels are being used in this engine. Compression ratios between 14:1 and 16.2:1 were evaluated. Low temperature combustion was demonstrated from 2 to 18 bar IMEP (Confer et al. 2012). A minimum ISFC of 181 g/kW-hr was obtained with NO_x emission less than 0.2 g/kW-hr. A multi-cylinder engine with the GDCI combustion system has been built and testing was underway as of May 2013 (Confer et al. 2013). Demonstration of this engine in a vehicle, including cold starting and transient operation, is scheduled to be completed by the end of the DOE contract in 2014. In September 2014, DOE awarded a $10M cost sharing project to Delphi to accelerate the development of the GDCI low temperature combustion technology.

This project includes more than just the development of the GDCI combustion system. In the 2012 DEER Review, Delphi and Hyundai demonstrated a 13.1 percent improvement in fuel economy due to parasitic loss reduction (crankshaft and camshaft rollerization, optimized oil pump, cooled EGR for reduced pumping losses, engine downspeeding, and exhaust heat recovery). A 13.4 percent improvement in fuel economy was achieved through improvements in the engine management system (which included GDI, VVT, cooled EGR, stop/start, and control algorithms).

REFERENCES

Christensen, M., A. Hultqvist, and B. Johansson. 1999. Demonstrating the Multi Fuel Capability of a Homogeneous Charge Compression Ignition Engine with Variable Compression Ratio. SAE World Congress. SAE 1999-01-3679.

Drangel, H., E. Olofsson, and R. Reinmann. 2002. The Variable Compression (SVC) and the Combustion Control (SCC) - Two Ways to Improve Fuel Economy and Still Comply with World-Wide Emission Requirements. SAE Technical Paper 2002-01-0996. doi:10.4271/2002-01-0996.

Kaahaaina, N., A. Simon, P. Caton, and C. Edwards. 2001. Use of Dynamic Valving to Achieve Residual-Affected Combustion. SAE Technical Paper 2001-01-0549. doi:10.4271/2001-01-0549.

Kulzer, A., T. Nier, and R. Karrelmeyer. 2011. A thermodynamic study on boosted HCCI: experimental results. SAE Technical Paper 2011-01-0905. doi:10.4271/2011-01-0905.

Mamalis, S., A. Babajimopoulos, O. Guralp, and P. Najt. 2012. Optimal Use of Boosting Configurations and Valve Strategies for High Load HCCI - A Modeling Study. SAE Technical Paper 2012-01-1101. doi:10.4271/2012-01-1101.

Nüesch, S., E. Hellstrom, L. Jiang, and A. Stefanopoulou. 2014a. Mode Switches among SI, SACI, and HCCI Combustion and their Influence on Drive Cycle Fuel Economy, 20014 American Control Conference (ACC), June.

Nüesch, S.P, A. G. Stefanopoulou, L.I. Jiang, and J. Sterniak. 2014b. Methodology to Evaluate the Fuel Economy of a Multimode Combustion Engine with Three-Way Catalytic Converter. Proceedings of the ASME 2014 Dynamic Systems and Control Conference, DSC2014-6146, October 22-24.

Olsson, J., P. Tunestål, G. Haraldsson, and B. Johansson. 2001. A Turbo Charged Dual Fuel HCCI Engine. SAE Technical Paper 2001-01-1896. doi:10.4271/2001-01-1896.

Olsson, J., P. Tunestål, B. Johansson, S. Fiveland, R. Agama, M. Willi, and D. Assanis. 2002.Compression ratio influence on maximum load of a natural gas fueled HCCI engine. SAE 2002-01-0111.

Ra, Y., P. Loeper, M. Andrie, R. Krieger, D. Foster, R. Reitz. and R. Durrett. 2012. Gasoline DICI Engine Operation in the LTC Regime Using Triple Pulse Injection. SAE Technical Paper 2012-01-1131, April.

Ra, Y., P. Loeper, R. Reitz, M. Andrie, R. Krieger, D. Foster, R. Durrett, V. Gopalakrishnan, A. Plazas, R. Peterson, and P. Szymkowicz. 2011. Study of high speed Gasoline Direct Injection Compression Ignition (GDICI) engine operation in the LTC regime. SAE Technical Paper 2011-01-1182, April.

Sellnau, M., J. Sinnamon, K. Hoyer, and H. Husted. 2012. Full Time Gasoline Direct-Injection Compression Ignition (GDCI) for High Efficiency and Low NOx and PM. SAE Paper 2012-01-0384, April.

Souder, J., P. Mehresh, J.K. Hedrick, and R.W. Dibble. 2004. A Multi-Cylinder HCCI Engine Model for Control. ASME 2004 Int. Mech. Engineering Congress and Exposition. Anaheim, California, November 13-19.

Stanglmaier, R., and C. Roberts. 1999. Homogeneous Charge Compression Ignition (HCCI): Benefits, Compromises, and Future Engine Applications. SAE Technical Paper 1999-01-3682. doi:10.4271/1999-01-3682.

Szybist, J., K. Edwards, M. Foster, K. Confer, and W. Moore. 2013. Characterization of Engine Control Authority on HCCI Combustion as the High Load Limit Is Approached. SAE Paper No. 2013-01-1665, April.

Wagner, R., J. Green Jr., J. Storey, and C. Daw. 2000. Extending Exhaust Gas Recirculation Limits in Diesel Engines. Oak Ridge National Laboratory, Internal Paper. http://web.ornl.gov/~webworks/cpr/pres/106938_.pdf.

Weall, A., J. Szybist, K. Edwards, M. Foster et al. 2012. HCCI load expansion opportunities using a fully variable HVA research engine to guide development of a production intent cam-based VVA engine: The low load limit. SAE Int. J. Engines 5(3):1149-1162. doi:10.4271/2012-01-1134.

Wheeler, J., D. Polovina, V. Frasinel, O. Miersch-Wiemers et al. 2013. Design of a 4-cylinder GTDI engine with part-load HCCI capability. SAE Int. J. Engines 6(1): 184-196. doi:10.4271/2013-01-0287.

Yang, J., T. Culp, and T. Kenney. 2002. Development of a Gasoline Engine System Using HCCI Technology - The Concept and the Test Results. SAE Technical Paper 2002-01-2832. doi:10.4271/2002-01-2832.

Yilmaz, H., O. Miersch-Wiemers, and L. Jiang. 2013. Advanced Combustion Concepts – Enabling Systems and Solutions (ACCESS) for High Efficiency Light Duty Vehicles. 2013 DOE Annual Merit Review, Arlington, Virginia, May 17. http://energy.gov/sites/prod/files/2014/03/f13/ace066_yilmaz_2013_o.pdf.

Yun, H., N. Wermuth, and P. Najt. 2010. Extending the high load operating limit of a naturally-aspirated gasoline HCCI combustion engine. SAE Int. J. Engines 3(1):681-699. doi:10.4271/2010-01-0847.

Yun, H., N. Wermuth, and P. Najt. 2011. High load HCCI operation using different valving strategies in a naturally-aspirated gasoline HCCI engine. SAE Int. J. Engines 4(1):1190-1201. doi:10.4271/2011-01-0899.

Zhu, G. 2013. Flex Fuel Optimized SI and HCCI. DOE Annual Merit Review.

Appendix N

Effect of Compression Ratio of Brake Thermal Efficiency

The effects of compression ratio on brake thermal efficiency at full load as well as part load operating conditions typical of those encountered in the CAFE test procedure are important considerations for maximizing light-duty vehicle fuel economy. Thermal efficiency of the ideal Otto cycle is given by this equation:

$$\text{Indicated Thermal Efficiency} = 1 - 1/CR^{(1-k)} \quad (1)$$

Where: CR = compression ratio

k = ratio of constant pressure to constant volume specific heats

Equation 1 shows that indicated thermal efficiency increases, but at a decreasing rate, as compression ratio is increased. However, mechanical efficiency decreases as compression ratio is increased due to higher loads on the pistons, rings, and bearings of the engine. Brake thermal efficiency is the product of indicated thermal efficiency and mechanical efficiency.

At full load conditions, mechanical efficiency can be relatively high. However, at part load conditions, mechanical efficiency will be significantly lower, even for relatively constant friction levels. The significant effect of load on mechanical efficiency is illustrated by the equation for mechanical efficiency:

$$\text{Mechanical Efficiency} = BMEP/IMEP = BMEP/(BMEP + FMEP) \quad (2)$$

Where: BMEP = brake mean effective pressure

IMEP = indicated mean effective pressure

FMEP = friction mean effective pressure

The effects of compression ratio on brake thermal efficiency together with indicated thermal efficiency and mechanical efficiency are shown in Figures 2.12(a) for full load conditions and Figure 2.12(b) for part load conditions. These figures provide the following insight into the effects of compression ratio on brake thermal efficiency:

- At full load, brake thermal efficiency increases, but at a decreasing rate, with increasing compression ratio, similar to indicated thermal efficiency.
- Up to 3 percent reduction in fuel consumption for naturally aspirated engines might be realized if compression ratio is increased from today's typical level of 10:1 to approximately 12:1. Possibly greater reductions in fuel consumption might be realized for turbocharged engines capable of operating at higher boost pressures without knock so that further downsizing could be realized. Increasing gasoline octane from 91 RON of regular grade gasoline to 95 RON has been estimated to facilitate operation at a 12:1 compression ratio.

At part load, nearly insignificant improvements in brake thermal efficiency on the CAFE test cycles are expected to be obtained by increasing compression ratio beyond approximately 12:1 due to the increasingly lower mechanical efficiency.

Appendix O

Variable Compression Ratio Engines

Harry Ricardo built the first known variable compression ratio engine as a test engine in the 1920s to study knock in aircraft engines. Variable compression ratio was achieved by raising or lowering the cylinder and cylinder head of the engine relative to the crankshaft. This work led to the development of the octane rating system that is still in use today. Variable compression ratio engines have subsequently been investigated for decades, but none have been commercially produced due to mechanical complexity, difficulty of control, and cost. Some of the variable compression ratio engines that have been investigated are shown in Figure 2.14 and include the following:

- *Adjustable piston.* An early concept for variable compression ratio was the use of an adjustable piston in the cylinder head. Adjusting the piston changes the clearance volume at top dead center, thereby altering the compression ratio. Ford and Volvo have evaluated this concept in the past.
- *Hydraulic piston.* Teledyne Continental Motors developed variable compression ratio pistons for both 1790 cu. in. and 1360 cu. in. tank engines to achieve very high output while limiting peak cylinder pressure by reducing compression ratio. The variable compression ratio piston consisted of a moveable top that used engine oil pressure to extend the piston top for the highest compression ratio. To limit peak cylinder pressure, a pressure relief valve in the piston allowed the piston top to move downward, thereby reducing compression ratio (Grundy et al. 1976). Ford patented a similar concept in the 1980s (Caswell 1984). Mercedes has also experimented with a similar concept (Joshi 2012).
- *Articulated cylinder head.* Saab introduced its system at the 2000 Geneva Motor Show. In the Saab system, one side of the cylinder head with integrated cylinders is attached to the crankcase with a hinge, and a lifting mechanism is placed on the other side of the crankcase. The cylinder head can pivot up by four degrees to increase the volume of the combustion chambers, thereby lowering the compression ratio. Saab claimed that the technology could change the compression ratio from 8:1 to 14:1 while the engine was running (Evans 2009).
- *Rocker arm.* Peugeot introduced its system at the 2009 Geneva Motor Show. In the Peugeot system, each piston is attached to a rocker arm and the center of the rocker arm is connected to the crankshaft. An intermediary gear is added to the other end of the rocker arm. Hydraulic jacks in the engine block next to the cylinders manipulate a gear rack to raise or lower the rocker arm gear in order to change the length of the piston stroke, thereby changing the volume of the combustion chamber and, consequently, the compression ratio (Evans 2009).
- *Eccentric big end rod bearing.* Waulis Motors Ltd's patented concept uses an eccentric wheel on the connecting rod big end bearing. The eccentric wheel includes a gear that meshes with a ring gear. Rotating the ring gear will adjust the position of the eccentric wheel, thereby adjusting the clearance volume at top dead center.
- *Eccentric crankshaft bearing.* Gomecsys has developed a fourth-generation variable compression ratio engine. This system houses the crankshaft bearings within eccentric wheels contained within the cylinder block. Rotating the eccentric wheels changes the clearance volume at top dead center. FEV has developed a similar variable compression ratio engine shown in Figure O.1 (Green Car Congress 2007), which is discussed later in this section.
- *Eccentric piston pin.* An eccentric piston pin has also been proposed as a means of varying the clearance volume at top dead center.
- *Multi-link rod-crank.* Nissan has developed a variable compression ratio engine in which an extra linkage is mounted to the crankshaft in place of the connecting rod. The connecting rod is connected to this linkage. The linkage is also connected to an actuator shaft so that the compression ratio can be changed (Joshi 2012). A similar concept was investigated by PSA.

An example of the compression ratios that might be used over an engine's speed/load map is shown in Figure O.1. At light loads, compression ratios as high as 15:1 are used, whereas at maximum loads, the compression ratio is reduced to 8:1. An interesting application of variable compression ratio to an FFV engine is also illustrated in Figure 2.14 (McAulay 2009). Since E85 has a significantly higher octane rating than gasoline, higher compression ratios would be used during operation on E85. As shown, at moderate loads, compression ratios about 2 ratios higher are used. At maximum loads, a compression ratio of approximately 12:1 is used.

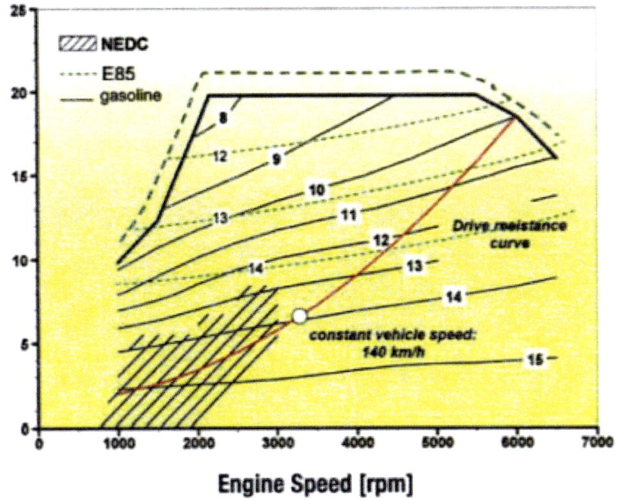

FIGURE O.1 Variable compression ratios used for gasoline and E85 for an FFV.
SOURCE: J.L. McAulay, Copyright: Massachusetts Institute of Technology, 2009. Used with permission.

REFERENCES

Evans, S. 2009. MCE-5 to Debut 220-HP .5L Engine with Variable Compression Ratio at Geneva. Motor Trend, February 25. http://wot.motortrend.com/mce5-to-debut-220hp-15l-engine-with-variable-compression-ratio-at-geneva-3891.html#ixzz2YVpc7J4G.

Green Car Congress. 2007. FEV Displays Turbocharged, Direct-Injected, E85 Variable Compression Ratio Engine. Green Car Congress, April. http://www.greencarcongress.com/2007/04/fev_displays_tu.html.

Joshi, M., and A. Kulkarni. 2012. Variable Compression Ratio (VCR) engine – A review of future powerplant for automobile. International Journal of Mechanical Engineering Research and Development 2(1), January-September.

McAulay, J.L. 2009. Assessing Deployment Strategies for Ethanol and Flex Fuel Vehicles in the U.S. Light-Duty Vehicle Fleet. MS Thesis, MIT, June.

Appendix P

Fuel Consumption Impact of Tier 3 Emission Standards

The committee estimated that the emission control technologies identified by EPA in Figure 2.17 for meeting Tier 3 emission standards for a large, light-duty truck are likely to result in less than 0.31 percent increase in fuel consumption. The increases in fuel consumption in other vehicles are expected to be less. An analysis of the reasons for this conclusion, based on three emissions control technologies that EPA identified in the Tier 3 rule, is provided below.

Technology 1: Secondary air injection with a 100 W electric air pump operating for the first 60 seconds of the cold start FTP cycle may be added.

A continuous 100 W load was estimated by EPA/NHTSA to result in an average increase of 2.5 g/mi CO_2 on the 2 cycle CAFE test (EPA/NHTSA 2012, 5-66, Table 5-18). Converting the CO_2 increase to fuel consumption yields the following:

2.5 g CO_2/mi × (gal/8887 g CO_2) = 0.00028 gal/mi

For the 2025 MY CAFE 2 cycle fleet average fuel economy requirement of 49.6 mpg, the 100 W load, operating full time, would provide the following percentage increase in fuel consumption:

(0.00028 gal/mi)/((1/49.6) gal/mi) = 1.4% increase in fuel consumption if the 100 W load were operating full time.

The 60 seconds of operation during the 505 seconds of the FTP cold start Bag 1 results in an 11.9 percent weighting. The 11.9 percent for the FTP cold start Bag 1 is weighted 43 percent when combined with the FTP hot start Bag 1. Bag 1 is weighted 36.8 percent (505 seconds/1371 seconds) over the complete 1,371 seconds of the FTP cycle. The FTP cycle is weighted 55 percent over the two CAFE test cycles. Applying these weightings to the 1.4 percent reduction in fuel consumption for a 100 W load operating full time yields the following result:

1.4% × (0.119) × (0.43) × (505/1371) × (0.55) = 0.015% increase in fuel consumption

Technology 2: An HC adsorber may be added.

A recent study of an HC adsorber showed that catalyst volume for the HC adsorber had to be added since the existing TWC catalyst volume was required for NO_x control. In this study, an adsorber with a volume of 0.67L was added to an existing 2.2L TWC, resulting in a 30 percent increase in overall catalyst volume (Gao et al. 2012). A 30 percent increase in catalyst volume due to the HC adsorber would result in the following increase in fuel consumption:

At ¼ maximum load (approximately 3 bar BMEP) and 1,500 rpm, which is a typical condition in the CAFE test cycles, an average catalyst results in an exhaust gas pressure drop of 6 mbar (Persoons 2006). This back pressure on the engine would result in a 0.2 percent increase (0.006 bar/3 bar × 100) in fuel consumption. Adding an adsorber to increase catalyst volume by 30 percent would increase fuel consumption by 0.06 percent (0.30 × 0.2%).

Technology 3: Calibration changes may consist of spark retard and increased idle speed for 30 seconds for faster catalyst warm-up.

Spark retard is likely to be required beyond the first 20-second idle period and may be needed during the next idle period for a total of 30 seconds.

By assuming that half of the maximum acceptable spark retard was used for Tier 2 emissions, the remaining half of the maximum spark retard would result in approximately a 15 percent increase in fuel consumption for 30 seconds during the FTP cold start Bag 1 (Zareei and Kakaee 2013). Applying this to the CAFE test cycle would result in 0.08 percent increase in fuel consumption as shown below.

15% × (30/505) × (0.43) × (505/1371) × (0.55) = 0.08% increase in fuel consumption

In addition to retarding spark timing, heat flux to the catalyst at idle may be increased by raising idle speed. Raising idle speed by 30 percent (from 600 rpm to 780 rpm) would result in a 30 percent increase in idle fuel consumption for the 30 seconds during the FTP cold start Bag 1. Applying this over the CAFE test cycle would result in 0.16 percent increase in fuel consumption as shown below.

30% × (30/505) × (0.43) × (505/1371) × (0.55) = 0.16% increase in fuel consumption

Therefore, spark retard and increased idle speed could result in approximately a 0.24 percent increase in fuel consumption.

Combining the above three reasons yields the following estimated increase in fuel consumption for the Tier 3 emission standards:

Percent Increases in Fuel Consumption	
100 W air pump for 60 seconds	0.01
HC Adsorber	0.06
Calibration changes for 30 seconds	0.24
Total	0.31

REFERENCES

EPA/NHTSA (Environmental Protection Agency/National Highway Traffic Safety Administration). 2012. Joint Technical Support Document, Final Rulemaking 2017-2025 Light-Duty Greenhouse Gas Emission Standards and Corporate Average Fuel Economy Standards. EPA-420-R-12-901.

Gao, Z., C. Daw, and L. Slezak. 2012. Advanced Light-Duty Engine Systems and Emissions Control Modeling and Analysis. DOE Annual Merit Review, May 16.

Persoons, T. 2006. Experimental Flow Dynamics in Automotive Exhaust Systems with Close Coupled Catalyst. Katholieke Universiteit Leuven, August.

Zareei, J., and A. H. Kakaee. 2013. Study of the effects of ignition timing on gasoline engine performance and emissions. Eur. Transp. Res. Rev. 5:109-116.

Appendix Q

Examples of EPA's Standards for Gasoline

Several examples of EPA's standards for gasoline are listed below (EPA 2013):

- The Gasoline Sulfur program was phased-in from 2004 to 2007. Refiners can produce gasoline with a range of sulfur levels as long as their annual corporate average does not exceed 30 parts per million (ppm). In addition, no individual batch can exceed 80 ppm. Sulfur can adversely affect catalysts and may also be emitted as a sulfur oxide.
- The Mobile Source Air Toxics (MSAT) rules reduce hazardous air pollutants, also known as air toxics, emitted by cars and trucks. Air toxics include benzene and other hydrocarbons such as 1,3-butadiene, formaldehyde, acetaldehyde, acrolein, and naphthalene.
- Reformulated Gasoline (RFG) was mandated for metropolitan areas with the worst smog beginning in 1995. RFG is a blended oxygenated fuel which burns cleaner than conventional gasoline, reducing emissions of smog-forming and toxic pollutants.
- EPA regulates the volatility/Reid Vapor Pressure (RVP) of conventional gasoline sold at retail stations during the summer smog season to reduce evaporative emissions that contribute to smog.
- Winter Oxygenate Fuel programs increase fuel oxygen and are mandated in certain areas for carbon monoxide control.
- E15 is a fuel containing a mixture of gasoline and ethanol, specifically 15-volume percent ethanol and 85-volume percent gasoline. EPA has granted a partial waiver to allow E15 to be introduced into commerce for use in model year 2001 and newer light-duty motor vehicles, subject to several conditions.
- Phosphorous in limited to 0.0013 g P/L since it can adversely affect exhaust catalysts.
- ASTM standards include D4814 - 13a, Standard Specification for Automotive Spark-Ignition Engine Fuel. All fuels must comply with various properties contained in this specification. Antiknock Index (AKI) is defined as the Research Octane Number + Motor Octane Number divided by two. AKI limits are not specified in the ASTM standards. AKI limits change with engine requirements and according to season and location. Fuels with an AKI of 87, 89, 91 are listed as typical for the U.S. at sea level, however higher altitudes will specify lower octane numbers.

REFERENCE

EPA (Environmental Protection Agency). 2013. Control of Air Pollution From Motor Vehicles: Tier 3 Motor Vehicle Emission and Fuel Standards. Proposed Rule, Federal Register, Vol. 78, No. 98.

Appendix R

Impact of Low Carbon Fuels to Achieve Reductions in GHG Emissions (California LCFS 2007 – Alternative Fuels and Cleaner Fossil Fuels CNG, LPG)

The low-carbon fuel standard (LCFS) is a rule that was enacted by California in 2007 and is the first low-carbon fuel standard mandate in the world. Specific eligibility criteria were defined by the California Air Resources Board (CARB) in April 2009 and became effective in January 2011. The purpose of the rule is to reduce carbon intensity in transportation fuels as compared to conventional petroleum fuels, such as gasoline and diesel. The most common low-carbon fuels are alternative fuels and cleaner fossil fuels, such as natural gas (CNG and LPG). The main purpose of a low-carbon fuel standard is to decrease carbon dioxide emissions associated with fuel-powered vehicles considering the entire life cycle ("well to wheels") in order to reduce the carbon footprint of transportation. Several bills have been proposed in the United States for similar low-carbon fuel regulation at a national level but with less stringent standards than California, but none have been approved.

The LCFS directive calls for a reduction of at least 10 percent in the carbon intensity of California's transportation fuels by 2020. These reductions include not only tailpipe emissions but also all other associated emissions from production, distribution, and use of transport fuels within the state. Therefore, the California LCFS considers the fuel's full life cycle, also known as the "well to wheels" or "seed to wheels" efficiency of transport fuels.

The carbon intensities of fuels that substitute for gasoline are shown in Table R.1. The compliance schedule for the LCFS to achieve the 10 percent reduction in carbon intensity by 2020 is shown in Figure R.1.

There are two ways to deploy alternative fuels that will help comply with the LCFS. Firstly, biofuels can be blended into conventional gasoline or diesel for consumption in the existing vehicle fleet. Secondly, advanced vehicle technologies can be deployed, which consume alternative fuels such as natural gas, electricity, or hydrogen. Compliance with the LCFS is expected to require a diverse mix of all of these alternative fuels. Due to constraints on how quickly the vehicle fleet can be turned over, however, biofuel blending is and will likely continue to be a major form of LCFS compliance until advanced vehicle technologies are deployed in higher numbers (CETC 2013).

TABLE R.1 Carbon Intensity of Fuels that Substitute for Gasoline

Fuel	Carbon Intensity[a]	Comments
California Gasoline	95.86	Gasohol with 10% ethanol
CARB LCFS for 2011	95.61	With land-use changes[b]
CARB LCFS for 2020	86.27	With land-use changes
California Ethanol	80.70	With land use changes
Cellulosic Ethanol	21.30	From farmed trees and forest ways
CNG	67.70	North American natural gas
Electricity, marginal	30.80 26.32 by 2020	Includes energy economy ratio (EER) of 3.4 for electric vehicles[c]
Hydrogen	39.42	Includes energy economy ratio (EER) of 2.5 for fuel cell vehicles[c]

[a]Grams of carbon dioxide equivalent released per megajoule of energy produced.

[b]Land use changes (i.e., land clearing) are included for biofuels and biofuel blends.

[c]Energy Economy Ratio: Distance an alternative-fueled vehicle travels divided by the distance an internal combustion engine vehicle travels using the same amount of energy.

SOURCE: CETC (2013).

In order to achieve LCFS compliance, suppliers of fuel to California are considering moving from the current 10 percent ethanol in reformulated gasoline to higher blends of ethanol, including E15 and E85. The U.S. EPA recently approved waivers for E15 consumption in model year 2001 and newer light-duty vehicles. There is considerable uncertainty today regarding the timing of E15 deployment in California.

Electricity and hydrogen used in PEVs and FCVs, respectively, promise to play significant roles in LCFS compliance, particularly in the later years of program implementation.

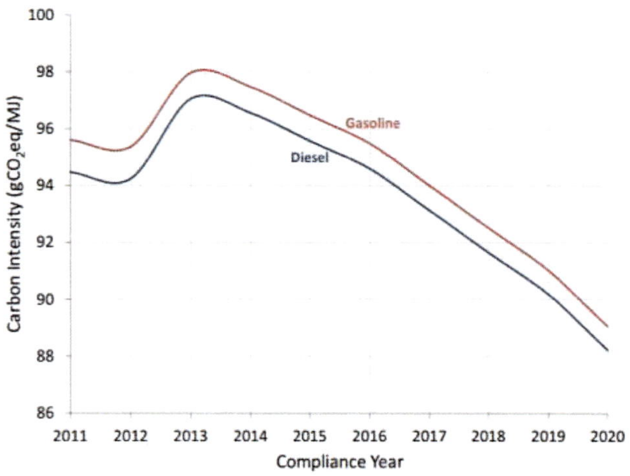

FIGURE R.1 Compliance schedule for the LCFS.
SOURCE: CETC (2013).
NOTE: CARB modified the baseline number, which was originally an average of crude oil supplied to California refineries in 2006; the values from 2013 to 2020 reflect the updated average of crude oil supplied to California refineries in 2010.

The Energy Independence and Security Act of 2007 (EISA) established new renewable fuel categories and eligibility requirements, setting mandatory life cycle greenhouse gas emissions thresholds for renewable fuel categories, as compared to those of average petroleum fuels used in 2005. EISA increased the required volume of renewable fuel produced reaching 36 billion gallons by 2022. On February 3, 2010, EPA issued its final rule regarding the expanded Renewable Fuel Standard (RFS2) for 2010 and beyond.

The use of ethanol as a blend (E10 or E15) is not expected to have significant impact on CAFE compliance (see discussion of E15 in Tier 3 Emissions section). In contrast, the expanding infrastructure for CNG may provide an increased interest in CNG vehicles with their associated CAFE benefit, particularly for full-size light trucks (as discussed in the Natural Gas and Bi-Fuel Engines section).

REFERENCE

CETC (California Electric Transportation Coalition). 2013. California's Low Carbon Fuel Standard: Compliance Outlook for 2020. Prepared for CETC by ICF International.

Appendix S

NHTSA's Estimated Fuel Consumption Reduction Effectiveness of Technologies and Estimated Costs of Technologies

TABLE S.1 NHTSA's Estimated Fuel Consumption Reduction Effectiveness of Technologies

Percent Incremental Fuel Consumption Reductions: NHTSA Estimates (TSD, RIA, Decision Trees)					
		Midsize Car I4 DOHC	Large Car V6 DOHC	Large Light Truck V8 OHV	
Spark Ignition Engine Technologies	Abbreviation	Avg	Avg	Avg	Relative To
Source: NHTSA RIA					
Low Friction Lubricants - Level 1	LUB1	0.7	0.8	0.7	Baseline
Engine Friction Reduction - Level 1	EFR1	2.6	2.7	2.4	Baseline
Low Friction Lubricants and Engine Friction Reduction - Level 2	LUB2_EFR2	1.3	1.4	1.2	Previous Tech
VVT- Intake Cam Phasing	ICP	2.6	2.7	2.5	Baseline for DOHC
VVT- Dual Cam Phasing	DCP	2.5	2.7	2.4	Previous Tech
Discrete Variable Valve Lift	DVVL	3.6	3.9	3.4	Previous Tech
Continuously Variable Valve Lift	CVVL	1.0	1.0	0.9	Previous Tech
Cylinder Deactivation (V6-DOHC, V8-OHV)	DEACD	NA	0.7	5.5	Previous Tech
Variable Valve Actuation (CCP + DVVL)	VVA	NA	NA	3.2	Baseline for OHV
Stoichiometric Gasoline Direct Injection	SGDI	1.5	1.5	1.5	Previous Tech
Turbocharging and Downsizing Level 1 - 18 bar BMEP 33%DS	TRBDS1	8.3	7.8	7.3	Previous Tech
Turbocharging and Downsizing Level 2 - 24 bar BMEP 50%DS	TRBDS2	3.5	3.7	3.4	Previous Tech
Cooled EGR Level 1 - 24 bar BMEP, 50% DS	CEGR1	3.5	3.5	3.6	Previous Tech
Cooled EGR Level 2 - 27 bar BMEP, 56% DS	CEGR2	1.4	1.4	1.2	Previous Tech
Diesel Engine Technologies					
Source: EPA/NHTSA TSD					
Advanced Diesel (Ref: Decision Trees)	ADSL	29.4	30.5	29.0	Baseline
Transmission Technologies					
Source: NHTSA RIA					
Improved Auto. Trans. Controls/Externals (ASL-1 & Early TC Lockup)	IATC	3.0	3.1	2.9	Previous Tech
6-speed Transmission with Improved Internals (Rel to 4 sp AT)	NUATO	2.0	2.0	2.1	Previous Tech
6-speed DCT (Rel to 4 sp AT) (Dry, Wet is 1% Lower)	DCT	4.1	3.8	3.8	Previous Tech
8-speed Transmission (Auto or DCT)	8SPD	4.6	4.6	5.3	Previous Tech
High Efficiency Gearbox (Auto or DCT)	HETRANS	2.7	2.6	3.7	Previous Tech
Shift Optimizer (ASL-2)	SHFTOPT	4.1	4.3	3.9	Previous Tech
Secondary Axle Disconnect	SAX	1.4	1.3	1.6	Baseline
Electrified Accessories Technologies					
Source: NHTSA RIA					
Electric Power Steering	EPS	1.3	1.1	0.8	Baseline
Improved Accessories - Level 1 (70% Eff Alt, Elec. Water Pump and Fan)	IACC1	1.2	1.0	1.6	Baseline
Improved Accessories - Level 2 (Mild regen alt strategy, Intelligent cooling)	IACC2	2.4	2.6	2.2	Previous Tech

continued

TABLE S.1 Continued

Percent Incremental Fuel Consumption Reductions: NHTSA Estimates (TSD, RIA, Decision Trees)		Midsize Car I4 DOHC	Large Car V6 DOHC	Large Light Truck V8 OHV	
Hybrid Technologies	Abbreviation	Avg	Avg	Avg	Relative To
Source: EPA/NHTSA TSD, except as noted					
Stop-Start (12V Micro-Hybrid) (RIA)	MHEV	2.1	2.2	2.1	Previous Tech
Integrated Starter Generator (RIA)	ISG	6.5	6.5	3.0	Previous Tech
Strong Hybrid - P2 - Level 2 (Parallel 2 Clutch System)	SHEV2-P2	33.6	34.5	30.1	All SI Technogies
Strong Hybrid - PS - Level 2 (Power Split System)	SHEV2-PS	33.0	32.0	33.0	Baseline for DOHC
Plug-in Hybrid - 40 mile range (w/charger & labor)	PHEV40	65.1	69.5	68.5	Baseline
Electric Vehicle - 75 miles (w/charger & labor)	EV75	87.2	87.0	NA	Baseline
Electric Vehicle - 100 mile (w/charger & labor)	EV100	87.2	87.0		Baseline
Electric Vehicle - 150 mile (w/charger & labor)	EV150	87.2	87.0		Baseline
Vehicle Technologies					
Source: NHTSA RIA					
Mass Reduction Relative to Previous Mass Reduction[a]					
Mass Reduction - Level 1 (0 - 1.5%) (Subcompact to Large LT)	MR1	0.5	0.5	0.5	Baseline
Mass Reduction - Level 2 (1.5% - 7.5%) (Subcompact to Large LT)	MR2	2.1	2.1	2.1	Previous MR
Mass Reduction - Level 3 (7.5% - 10%) (Subcompact to Large LT)	MR3	0.9	0.9	0.9	Previous MR
Mass Reduction - Level 4 (10% - 15%) (Subcompact to Large LT)	MR4	2.6	2.6	2.6	Previous MR
Mass Reduction - Level 5 (15% - 20%) (Subcompact to Large LT)	MR5	2.6	2.6	2.6	Previous MR
Mass Reduction Relative to Baseline[a]					
0 - 5% Mass Reduction	MR5	1.75	1.75	1.75	Baseline
0 - 10% Mass Reduciton	MR10	3.50	3.50	3.50	Baseline
0 - 15% Mass Reduction	MR15	7.65	7.65	7.65	Baseline
0 - 20% Mass Reduction	MR20	10.20	10.20	10.20	Baseline
Low Rolling Resistance Tires - Level 1 (10% reduction in rolling resistance)	ROLL1	1.9	1.9	1.9	Baseline
Low Rolling Resistance Tires - Level 2 (20% reduction in rolling resistance)	ROLL2	2.0	2.0	2.0	Previous Tech
Low Drag Brakes	LDB	0.8	0.8	0.8	Baseline
Aerodynamic Drag Reduction - Level 1	AERO1	2.3	2.3	2.3	Baseline
Aerodynamic Drag Reduction - Level 2	AERO2	2.5	2.5	2.5	Previous Tech

[a] 3.5% FC reduction for every 10% mass reduction - Under 10%; 5.1% FC reduction for every 10% mass reduction - Over 10%; without engine downsizing.
NOTE: Midsize car: 3500 lbs, large car: 4500 lbs, large light truck: 5500 lbs.

TABLE S.2a NHTSA's Estimated 2017 Costs of Technologies (2010 dollars)

2017 Incremental Costs (2010$): NHTSA Estimates (TSD, RIA, Decision Trees)

		Midsize Car I4 DOHC		Large Car V6 DOHC		Large Light Truck V8 OHV		
Spark Ignition Engine Technologies	Abbreviation	DMC	TC	DMC	TC	DMC	TC	Relative To
Defined by EPA and NHTSA								
Low Friction Lubricants - Level 1	LUB1	3	4	3	4	3	4	Baseline
Engine Friction Reduction - Level 1	EFR1	48	59	71	89	95	118	Baseline
Low Friction Lubricants and Engine Friction Reduction - Level 2	LUB2_EFR2	51	63	75	92	99	122	Previous Tech
VVT- Intake Cam Phasing	ICP	37	46	74	93	37	46	Baseline
VVT- Dual Cam Phasing	DCP	31	49	72	112		NA	Previous Tech
Discrete Variable Valve Lift	DVVL	116	163	168	236	240	338	Previous Tech
Continuously Variable Valve Lift	CVVL	58	81	151	212	108	151	Previous Tech
Cylinder Deactivation (V6-DOHC, V8-OHV)	DEACD		NA	139	196	157	220	Previous Tech
Variable Valve Actuation (CCP + DVVL)	VVA		NA		NA	296	416	Baseline for OHV
Stoichiometric Gasoline Direct Injection	SGDI	192	277	290	417	348	501	Previous Tech
Turbocharging and Downsizing Level 1 - 18 bar BMEP 33%DS	TRBDS1	288	482	-129	248	942	1,339	Previous Tech
V6 to I4 and V8 to V6				-455*	-120*	841*	1,212*	
Turbocharging and Downsizing Level 2 - 24 bar BMEP 50%DS	TRBDS2	182	262	182	262	308	442	Previous Tech
I4 to I3		-92*	26*					
Cooled EGR Level 1 - 24 bar BMEP, 50% DS	CEGR1	212	305	212	305	212	305	Previous Tech
Cooled EGR Level 2 - 27 bar BMEP, 56% DS	CEGR2	364	525	364	525	614	885	Previous Tech
V6 to I4						-524*	-300*	
Diesel Engine Technologies								
Defined by EPA and NHTSA								
Advanced Diesel	ADSL	2,059	2,965	2,522	3,631	2,886	4,145	Baseline
Transmission Technologies								
Defined by EPA and NHTSA								
Improved Auto. Trans. Controls/Externals (ASL-1 & Early TC Lockup)	IATC	50	63	50	63	50	63	Previous Tech
6-speed Transmission with Improved Internals (Rel to 4 sp AT)	NUATO	-13	-9	-13	-9	-13	-9	Previous Tech
6-speed DCT (Rel to 4 sp AT) (Dry, Wet is 1% Lower)	DCT	-146	-114	-146	-114	-146	-114	Previous Tech
8-speed Transmission (Auto or DCT)	8SPD	56	80	56	80	56	80	Previous Tech
High Efficiency Gearbox (Auto or DCT)	HETRANS	202	251	202	251	202	251	Previous Tech
Shift Optimizer (ASL-2)	SHFTOPT	1	2	1	2	1	2	Previous Tech
Secondary Axle Disconnect	SAX	78	98	78	98	78	98	Baseline
Electrified Accessories Technologies								
Defined by EPA and NHTSA								
Electric Power Steering	EPA	87	109	87	109	87	109	Baseline
Improved Accessories - Level 1 (70% Eff Alt, Elec. Water Pump and Fan)	IACC1	71	89	71	89	71	89	Baseline
Improved Accessories - Level 2 (Mild regen alt strategy, Intelligent cooling)	IACC2	43	54	43	54	43	54	Previous Tech

continued

TABLE S.2a Continued

2017 Incremental Costs (2010$): NHTSA Estimates (TSD, RIA, Decision Trees)		Midsize Car I4 DOHC		Large Car V6 DOHC		Large Light Truck V8 OHV		
Hybrid Technologies	Abbreviation	DMC	TC	DMC	TC	DMC	TC	Relative To
Defined by EPA and NHTSA								
Stop-Start (12V Micro-Hybrid)	SS	287	401	325	454	356	498	Previous Tech
Integrated Starter Generator	MHEV	1,087	1,634	1,087	1,634	1,087	1,634	Baseline
Strong Hybrid - P2 - Level 2 (Parallel 2 Clutch System)	SHEV2-P2	2,463	3,976	2,908	4,696	2,947	4,753	All SI Technogies
Strong Hybrid - PS - Level 2 (Power Split System) (Ref: Ricardo, 2011)	SHEV2-PS	3,139	4,990	3,396	5,398	5,023	8,146	Baseline for DOHC
Plug-in Hybrid - 40 mile range (w/charger & labor)	PHEV40	13,193	19,089	17,854	26,052		NA	Baseline
Electric Vehicle - 75 miles (w/charger & labor)	EV75	14,812	21,136	19,275	27,849		NA	Baseline
Electric Vehicle - 100 mile (w/charger & labor)	EV100	16,831	24,024	21,123	30,492		NA	Baseline
Electric Vehicle - 150 mile (w/charger & labor)	EV150	22,257	31,784	26,193	37,744		NA	Baseline
Vehicle Technologies								
Defined by EPA and NHTSA								
Mass Reduction Relative to Previous Mass Reduction								
Mass Reduction - Level 1 (0 - 1.5%) (Subcompact to Large LT)	MR1	3	4	4	5	5	6	Baseline
Mass Reduction - Level 2 (1.5% - 7.5%) (Subcompact to Large LT)	MR2	82	102	105	131	129	160	Previous Tech
Mass Reduction - Level 3 (7.5% - 10%) (Subcompact to Large LT)	MR3	67	82	86	106	105	130	Previous Tech
Mass Reduction - Level 4 (10% - 15%) (Subcompact to Large LT)	MR4	189	263	243	338	297	413	Previous Tech
Mass Reduction - Level 5 (15% - 20%) (Subcompact to Large LT)	MR5	264	367	340	472	415	577	Previous Tech
Alternative Format for Comparison to NRC Estimates								
0 - 2.5% Mass Reduction	MR2.5	4	5	5	6	6	7	
2.5 - 5% Mass Reduction		34	42	44	54	53	66	Previous MR
0 - 5% Mass Reduction	MR5	38	47	49	60	59	74	Baseline
5 - 10% Mass Reduction		113	140	146	181	178	221	Previous MR
0 - 10% Mass Reduction	MR10	151	187	194	241	237	294	Baseline
10 - 15% Mass Reduction		189	262	243	337	297	412	Previous MR
0 - 15% Mass Reduction	MR15	340	450	437	578	534	707	Baseline
15 - 20% Mass Reduction		264	367	340	472	415	577	Previous MR
0 - 20% Mass Reduction	MR20	604	817	777	1050	949	1,284	Baseline
Mass Reduction Relative to Baseline - Cost per lb.								
0 - 1% Mass Reduction	MR1	0.04	0.05	0.04	0.05	0.04	0.05	Baseline
0 - 5% Mass Reduction	MR5	0.22	0.27	0.22	0.27	0.22	0.27	Baseline
0 - 10% Mass Reduction	MR10	0.43	0.54	0.43	0.54	0.43	0.54	Baseline
0 - 15% Mass Reduction	MR15	0.65	0.86	0.65	0.86	0.65	0.86	Baseline
0 - 20% Mass Reduction	MR20	0.86	1.17	0.86	1.17	0.86	1.17	Baseline
Low Rolling Resistance Tires - Level 1 (10% reduction in rolling resistance)	ROLL1	5	7	5	7	5	7	Baseline
Low Rolling Resistance Tires - Level 2 (20% reduction in rolling resistance)	ROLL2	58	66	58	66	58	66	Previous Tech
Low Drag Brakes	LDB	59	74	59	74	59	74	Baseline
Aerodynamic Drag Reduction - Level 1	AERO1	39	49	39	49	39	49	Baseline
Aerodynamic Drag Reduction - Level 2	AERO2	117	164	117	164	117	164	Previous Tech

NOTE: Midsize car: 3500 lbs, large car: 4500 lbs, large light truck: 5500 lbs.

TABLE S.2b NHTSA's Estimated 2020 Costs of Technologies (2010 dollars)

2020 Incremental Costs (2010$): NHTSA Estimates (TSD, RIA, Decision Trees)

		Midsize Car I4 DOHC		Large Car V6 DOHC		Large Light Truck V8 OHV		
Spark Ignition Engine Technologies	Abbreviation	DMC	TC	DMC	TC	DMC	TC	Relative To
Defined by EPA and NHTSA								
Low Friction Lubricants – Level 1	LUB1	3	4	3	4	3	4	Baseline
Engine Friction Reduction – Level 1	EFR1	48	57	71	85	95	113	Baseline
Low Friction Lubricants and Engine Friction Reduction – Level 2	LUB2_EFR2	51	65	75	96	99	127	Previous Tech
VVT– Intake Cam Phasing	ICP	35	42	70	84	70	84	Baseline
VVT– Dual Cam Phasing	DCP	29	42	67	97	67	97	Previous Tech
Discrete Variable Valve Lift	DVVL	109	144	158	209	226	298	Previous Tech
Continuously Variable Valve Lift	CVVL	55	72	142	187	101	134	Previous Tech
Cylinder Deactivation (V6–DOHC, V8–OHV)	DEACD		NA	131	173	147	195	Previous Tech
Variable Valve Actuation (CCP + DVVL)	VVA		NA		NA	280	368	Baseline for OHV
Stoichiometric Gasoline Direct Injection	SGDI	181	244	273	367	328	442	Previous Tech
Turbocharging and Downsizing Level 1 – 18 bar BMEP 33%DS	TRBDS1	271	415	−122	159	877	1,172	Previous Tech
V6 to I4 and V8 to V6				–432*	–173*	779*	1,065*	
Turbocharging and Downsizing Level 2 – 24 bar BMEP 50%DS	TRBDS2	172	292	172	251	289	491	Previous Tech
I4 to I3		–89*	19*					
Cooled EGR Level 1 – 24 bar BMEP, 50% DS	CEGR1	199	292	199	292	199	292	Previous Tech
Cooled EGR Level 2 – 27 bar BMEP, 56% DS	CEGR2	343	502	343	503	579	847	Previous Tech
V6 to I4						–522*	–305*	
Diesel Engine Technologies								
Defined by EPA and NHTSA								
Advanced Diesel Technologies	ADSL	1,938	2,612	2,374	3,200	2,716	3,661	Baseline
Transmission Technologies								
Defined by EPA and NHTSA								
Improved Auto. Trans. Controls/Externals (ASL–1 & Early TC Lockup)	IATC	46	57	46	57	46	57	Previous Tech
6–speed Transmission with Improved Internals (Rel to 4 sp AT)	NUATO	–12	–9	–12	–9	–12	–9	Previous Tech
6–speed DCT (Rel to 4 sp AT) (Dry, Wet is 1% Lower)	DCT	–137	–89	–137	–89	–137	–89	Previous Tech
8–speed Transmission (Auto or DCT)	8SPD	53	71	53	71	53	71	Previous Tech
High Efficiency Gearbox (Auto or DCT)	HETRANS	184	233	184	233	184	233	Previous Tech
Shift Optimizer (ASL–2)	SHFTOPT	1	2	1	2	1	2	Previous Tech
Secondary Axle Disconnect	SAX	73	89	73	89	73	89	Baseline

continued

TABLE S.2b Continued

2020 Incremental Costs (2010$): NHTSA Estimates (TSD, RIA, Decision Trees)		Midsize Car I4 DOHC		Large Car V6 DOHC		Large Light Truck V8 OHV		
Electrified Accessories Technologies	Abbreviation	DMC	TC	DMC	TC	DMC	TC	Relative To
Defined by EPA and NHTSA								
Electric Power Steering	EPA	82	100	82	100	82	100	Baseline
Improved Accessories – Level 1 (70% Eff Alt, Elec. Water Pump and Fan)	IACC1	69	81	69	81	69	81	Baseline
Improved Accessories – Level 2 (Mild regen alt strategy, Intelligent cooling)	IACC2	40	50	40	50	40	50	Previous Tech
Hybrid Technologies								
Defined by EPA and NHTSA								
Stop–Start (12V Micro–Hybrid)	SS	261	346	296	392	325	430	Previous Tech
Integrated Starter Generator	MHEV	1,008	1,491	1,008	1,491	1,008	1,491	Baseline
Strong Hybrid – P2 – Level 2 (Parallel 2 Clutch System)	SHEV2–P2	2,295	3,394	2,410	4,008	2,744	4,068	All SI Technogies
Strong Hybrid – PS – Level 2 (Power Split System)	SHEV2–PS	2,954	4,084	3,196	4,418	4,824	6,668	Baseline for DOHC
Plug–in Hybrid – 40 mile range (w/charger & labor)	PHEV40	9,763	14,608	13,172	19,881	NA	NA	Baseline
Electric Vehicle – 75 miles (w/charger & labor)	EV75	10,189	16,175	13,310	21,446	NA	NA	Baseline
Electric Vehicle – 100 mile (w/charger & labor)	EV100	11,482	18,283	14,492	23,374	NA	NA	Baseline
Electric Vehicle – 150 mile (w/charger & labor)	EV150	14,954	23,946	17,737	28,666	NA	NA	Baseline
Vehicle Technologies								
Defined by EPA and NHTSA								
Mass Reduction Relative to Previous Mass Reduction								
Mass Reduction – Level 1 (0 – 1.5%) (Subcompact to Large LT)	MR1	3	3	3	4	4	5	Baseline
Mass Reduction – Level 2 (1.5% – 7.5%) (Subcompact to Large LT)	MR2	76	90	97	116	119	141	Previous Tech
Mass Reduction – Level 3 (7.5% – 10%) (Subcompact to Large LT)	MR3	60	72	78	92	95	113	Previous Tech
Mass Reduction – Level 4 (10% – 15%) (Subcompact to Large LT)	MR4	173	241	223	310	272	378	Previous Tech
Mass Reduction – Level 5 (15% – 20%) (Subcompact to Large LT)	MR5	242	336	311	432	380	528	Previous Tech
Alternative Format for Comparison to NRC Estimates								
0 – 1% Mass Reduction	MR1	3	4	4	5	5	6	Baseline
1 – 5% Mass Reduction		31	37	40	47	49	58	Previous MR
0 – 5% Mass Reduction	MR5	34	41	44	53	54	64	Baseline
5 – 10% Mass Reduction		103	123	133	158	162	193	Previous MR
0 – 10% Mass Reduction	MR10	137	164	177	210	216	257	Baseline
10 – 15% Mass Reduction		172	239	221	307	270	375	Previous MR

continued

TABLE S.2b Continued

2020 Incremental Costs (2010$): NHTSA Estimates (TSD, RIA, Decision Trees)

		Midsize Car I4 DOHC		Large Car V6 DOHC		Large Light Truck V8 OHV		
	Abbreviation	DMC	TC	DMC	TC	DMC	TC	Relative To
0 – 15% Mass Reduction	MR15	309	402	398	517	486	632	Baseline
15 – 20% Mass Reduction		241	334	309	430	378	525	Previous MR
0 – 20% Mass Reduction	MR20	550	737	707	947	864	1,158	Baseline
Mass Reduction Relative to Baseline – Cost per lb.								
0 – 1% Mass Reduction	MR1	0.04	0.05	0.04	0.05	0.04	0.05	
0 – 5% Mass Reduction	MR5	0.20	0.23	0.20	0.23	0.20	0.23	Baseline
0 – 10% Mass Reduction	MR10	0.39	0.47	0.39	0.47	0.39	0.47	Baseline
0 – 15% Mass Reduction	MR15	0.59	0.77	0.59	0.77	0.59	0.77	Baseline
0 – 20% Mass Reduction	MR20	0.79	1.05	0.79	1.05	0.79	1.05	Baseline
Low Rolling Resistance Tires – Level 1 (10% reduction in rolling resistance)	ROLL1	5	6	5	6	5	6	Baseline
Low Rolling Resistance Tires – Level 2 (20% reduction in rolling resistance)	ROLL2	46	54	46	54	46	54	Previous Tech
Low Drag Brakes	LDB	59	71	59	71	59	71	Baseline
Aerodynamic Drag Reduction – Level 1	AERO1	37	45	37	45	37	45	Baseline
Aerodynamic Drag Reduction – Level 2	AERO2	110	157	110	157	110	157	Previous Tech

NOTE: Midsize car: 3500 lbs, large car: 4500 lbs, large light truck: 5500 lbs.

TABLE S.2c NHTSA's Estimated 2025 Costs of Technologies (2010 dollars)

2025 Incremental Costs (2010$): NHTSA Estimates (TSD, RIA, Decision Trees)

		Midsize Car I4 DOHC		Large Car V6 DOHC		Large Light Truck V8 OHV		
Spark Ignition Engine Technologies	Abbreviation	DMC	TC	DMC	TC	DMC	TC	Relative To
Defined by EPA and NHTSA								
Low Friction Lubricants - Level 1	LUB1	3	4	3	4	3	4	Baseline
Engine Friction Reduction - Level 1	EFR1	48	57	71	85	95	113	Baseline
Low Friction Lubricants and Engine Friction Reduction - Level 2	LUB2_EFR2	51	60	75	89	99	117	Previous Tech
VVT- Intake Cam Phasing	ICP	31	39	63	78	63	78	Baseline
VVT- Dual Cam Phasing	DCP	27	39	61	90	61	90	Previous Tech
Discrete Variable Valve Lift	DVVL	99	133	143	193	204	276	Previous Tech
Continuously Variable Valve Lift	CVVL	49	67	128	174	92	124	Previous Tech
Cylinder Deactivation (V6-DOHC, V8-OHV)	DEACD		NA	118	160	133	180	Previous Tech
Variable Valve Actuation (CCP + DVVL)	VVA		NA		NA	248	336	Baseline for OHV
Stoichiometric Gasoline Direct Injection	SGDI	164	226	246	340	296	409	Previous Tech
Turbocharging and Downsizing Level 1 - 18 bar BMEP 33%DS	TRBDS1	245	388	–110	168	788	1,080	Previous Tech
V6 to I4 and V8 to V6				–396*	–142*	700*	983*	
Turbocharging and Downsizing Level 2 - 24 bar BMEP 50%DS	TRBDS2	155	214	155	214	261	361	Previous Tech
I4 to I3		–82*	5*					
Cooled EGR Level 1 - 24 bar BMEP, 50% DS	CEGR1	180	249	180	249	180	249	Previous Tech
Cooled EGR Level 2 - 27 bar BMEP, 56% DS	CEGR2	310	429	310	428	523	722	Previous Tech
V6 to I4						–453*	-289*	
Diesel Engine Technologies								
Defined by EPA and NHTSA								
Advanced Diesel	ADSL	1,752	2,420	2,146	2,954	2,455	3,392	Baseline
Transmission Technologies								
Defined by EPA and NHTSA								
Improved Auto. Trans. Controls/Externals (ASL-1 & Early TC Lockup)	IATC	42	52	42	52	42	52	Previous Tech
6-speed Transmission with Improved Internals (Rel to 4 sp AT)	NUATO	–11	–8	–11	–8	–11	-8	Previous Tech
6-speed DCT (Rel to 4 sp AT) (Dry, Wet is 1% Lower)	DCT	–124	–77	–124	–77	–124	-77	Previous Tech
8-speed Transmission (Auto or DCT)	8SPD	47	66	47	66	47	66	Previous Tech
High Efficiency Gearbox (Auto or DCT)	HETRANS	163	202	163	202	163	202	Previous Tech
Shift Optimizer (ASL-2)	SHFTOPT	0	0	0	0	0	0	Previous Tech
Secondary Axle Disconnect	SAX	66	82	66	82	66	82	Baseline

continued

TABLE S.2c Continued

2025 Incremental Costs (2010$): NHTSA Estimates (TSD, RIA, Decision Trees)		Midsize Car I4 DOHC		Large Car V6 DOHC		Large Light Truck V8 OHV		
Electrified Accessories Technologies	Abbreviation	DMC	TC	DMC	TC	DMC	TC	Relative To
Defined by EPA and NHTSA								
Electric Power Steering	EPA	74	92	74	92	74	92	Baseline
Improved Accessories - Level 1 (70% Eff Alt, Elec. Water Pump and Fan)	IACC1	64	75	64	75	64	75	Baseline
Improved Accessories - Level 2 (Mild regen alt strategy, Intelligent cooling)	IACC2	37	45	37	45	37	45	Previous Tech
Hybrid Technologies								
Defined by EPA and NHTSA								
Stop-Start (12V Micro-Hybrid)	SS	225	308	255	349	279	383	Previous Tech
Integrated Starter Generator	MHEV	888	1,249	888	1,249	888	1,249	Baseline
Strong Hybrid - P2 - Level 2 (Parallel 2 Clutch System)	SHEV2-P2	2041	2,957	2,410	3,492	2,438	3,531	All SI Technogies
Strong Hybrid - PS - Level 2 (Power Split System)	SHEV2-PS	2671	3,791	2,889	4,101	4,360	6,190	Baseline for DOHC
Plug-in Hybrid - 40 mile range (w/charger & labor)	PHEV40	8325	11,826	11,189	16,066		NA	Baseline
Electric Vehicle - 75 miles (w/charger & labor)	EV75	8451	12,226	11,025	16,159		NA	Baseline
Electric Vehicle - 100 mile (w/charger & labor)	EV100	9486	13,774	11,971	17,575		NA	Baseline
Electric Vehicle - 150 mile (w/charger & labor)	EV150	12264	17,931	14,567	21,460		NA	Baseline
Vehicle Technologies								
Defined by EPA and NHTSA								
Mass Reduction Relative to Previous Mass Reduction								
Mass Reduction - Level 1 (0 - 1.5%) (Subcompact to Large LT)	MR1	3	3	3	4	4	5	Baseline
Mass Reduction - Level 2 (1.5% - 7.5%) (Subcompact to Large LT)	MR2	67	80	86	103	106	126	Previous Tech
Mass Reduction - Level 3 (7.5% - 10%) (Subcompact to Large LT)	MR3	53	64	69	82	84	100	Previous Tech
Mass Reduction - Level 4 (10% - 15%) (Subcompact to Large LT)	MR4	152	196	196	253	239	309	Previous Tech
Mass Reduction - Level 5 (15% - 20%) (Subcompact to Large LT)	MR5	214	275	275	354	336	433	Previous Tech
Alternative Format for Comparison to NRC Estimates								
0 - 1% Mass Reduction	MR1	3	4	4	5	5	6	
1 - 5% Mass Reduction		28	33	35	42	43	51	Previous MR
0 - 5% Mass Reduction	MR5	31	36	39	47	48	57	Baseline
5 - 10% Mass Reduction		92	109	118	140	144	172	Previous MR
0 - 10% Mass Reduction	MR10	122	146	157	187	192	229	Baseline
10 - 15% Mass Reduction		153	197	197	254	240	310	Previous MR

continued

TABLE S.2c Continued

2025 Incremental Costs (2010$): NHTSA Estimates (TSD, RIA, Decision Trees)								
		Midsize Car I4 DOHC		Large Car V6 DOHC		Large Light Truck V8 OHV		
	Abbreviation	DMC	TC	DMC	TC	DMC	TC	Relative To
0 - 15% Mass Reduction	MR15	275	343	354	441	433	539	Baseline
15 - 20% Mass Reduction		214	276	275	355	336	434	Previous MR
0 - 20% Mass Reduction	MR20	489	619	629	796	769	973	Baseline
Mass Reduction Relative to Baseline - Cost per lb.								
0 - 1% Mass Reduction	MR1	0.03	0.04	0.03	0.04	0.03	0.04	Baseline
0 - 5% Mass Reduction	MR5	0.17	0.21	0.17	0.21	0.17	0.21	Baseline
0 - 10% Mass Reduction	MR10	0.35	0.42	0.35	0.42	0.35	0.42	Baseline
0 - 15% Mass Reduction	MR15	0.52	0.65	0.52	0.65	0.52	0.65	Baseline
0 - 20% Mass Reduction	MR20	0.70	0.88	0.70	0.88	0.70	0.88	Baseline
Low Rolling Resistance Tires - Level 1 (10% reduction in rolling resistance)	ROLL1	5	6	5	6	5	6	Baseline
Low Rolling Resistance Tires - Level 2 (20% reduction in rolling resistance)	ROLL2	31	38	31	38	31	38	Previous Tech
Low Drag Brakes	LDB	59	71	59	71	59	71	Baseline
Aerodynamic Drag Reduction - Level 1	AERO1	33	41	33	41	33	41	Baseline
Aerodynamic Drag Reduction - Level 2	AERO2	100	135	100	135	100	135	Previous Tech

NOTE: Midsize car: 3500 lbs, large car: 4500 lbs, large light truck: 5500 lbs.

Appendix T

Derivation of Turbocharged, Downsized Engine Direct Manufacturing Costs

The derivations of direct manufacturing costs for turbocharged, downsized engines, shown in Tables 8A.2a, b, and c (and Table S.2), are described below for an example of 2017 costs for an I4 engine. The derivation of costs for other engine types follows a similar process.

TABLE 8A.2 – WHITE ROWS (PRIMARY DOWNSIZING WITH THE SAME NUMBER OF CYLINDERS)

The turbocharged, downsized (TRBDS) engine costs for an I4 engine (downsized from a larger displacement I4 engine), shown on the white (primary) rows of Table 8A.2 are derived following NHTSA's methodology shown in the TSD (EPA/NHTSA 2012) by considering the separate costs for turbocharging and downsizing, as follows:

- Starting with TRBDS1 (18 bar BMEP with 33 percent downsizing), Table 8A.2 shows a 2017 direct manufacturing cost for this engine of $288 (using the low most likely, or NHTSA, estimate). This cost is derived from the TSD as follows:

Turbocharging for 18 bar	$365 (TSD, Table 3-31)
Downsizing I4-I4	−$77 (TSD, Table 3-32)
Net	$288 (as shown in Table 8A.2)

- The next step is TRBDS2 (24 bar BMEP, 50 percent downsizing), but the TSD only provides costs for this engine relative to the baseline engine as follows:

Turbocharging for 24 bar	$547 (TSD, Table 3-31)
Downsizing I4-I4	−$77 (TSD, Table 3-32)
Net	$470
Incremental cost (TRBDS2-TRBDS1)	$182 ($470 – $288) (as shown in Table 8A.2a)

- The next step is CEGR1 (cooled EGR added to TRBDS2), which is a standalone cost from the TSD for adding the cooled EGR system.

Cooled EGR	$212 (TSD, Table 3-34)

- The final step is CEGR2 (27 bar BMEP, 56 percent downsizing), but the TSD only provides costs (excluding the cost of cooled EGR system) relative to the baseline engine as follows:

Turbocharging for 27 bar	$911 (TSD, Table 3-31)
Downsizing I4-I4	−$77 (TSD, Table 3-32)
Net	$834
Incremental cost (CEGR2-TRBDS2)	$364 ($834 – $470) (as shown in Table 8A.2a)

NHTSA subtracts the same cost of downsizing (−$77 credit, or cost save) in each downsizing step. Therefore, this results in applying the $77 credit for downsizing only once for the turbocharged, downsized engines. The downsizing credit does not depend on the amount of downsizing (TSD, Table 3-31), as long as the downsizing is from an I4 engine to a downsized I4 engine (such as a 2.5L I4engine to a 1.68L I4 engine).

This methodology is used for all of the white entries on Table 8A.2a, b, and c for turbocharged, downsized engines (and is consistent with the TSD, Table 3-33, although this table only shows total costs).

TABLE 8A.2 – BLUE ROWS (OPTIONAL DOWNSIZING WITH REDUCED NUMBER OF CYLINDERS)

The blue rows of Table 8A.2 a, b, and c (and Table S.2) show costs for optional downsizing with a reduced number of cylinders, which is beyond the level of downsizing assumed by NHTSA in the white rows. NHTSA recognized that there are additional options for downsizing, which are shown in Table 8.1. The additional downsizing options are complicated since they occur after other technologies have already been added to the engine prior to downsizing. For example, when an I4 engine is downsized to an I3 engine, it has already received four direct fuel injectors and a number other technologies listed in Table 8A.2. When downsizing to the I3 engine by eliminating one cylinder, credit is given for eliminating one of the direct fuel injectors and for a portion of the other technologies previously added.

An example of the derivation of the costs of applying optional downsizing from an I4 engine (TRBDS1) to an I3 engine (TRBDS2) is shown in Table 8.2. The process consists of first adding the costs of all of the new technologies added to the I4 engine as shown in the upper-left side of the table. Next, the incremental costs of downsizing and turbocharging for the next BMEP level are listed, followed by the costs of the new technologies applied to the I3 engine. Notice that the turbocharging cost of $182 shown in Table 8.2 is the same cost shown on the white row of Table 8A.2a for turbocharging to TRBDS2, as described above for the white rows. Taking the costs of the added technologies minus the costs of the portion of the deleted technologies yields –$92 net cost (save) for the new I3 engine at the next BMEP level (TRBDS2). This –$92 net cost (save) is labeled with an asterisk on the blue row labeled "I4 to I3" in Table S.2.

The –$92 net incremental cost for TRBDS2 relative to TRBDS1 was also applied to the example pathways in Chapter 8 following the guidelines contained in NHTSA's decision trees and cost files for the decision trees. The –$92 (cost save) for 2017 shown in Table 8A.2a becomes –$89 for 2020 shown in Table 8A.2b and becomes –$82 for 2025 shown in Table 8A.2c (and Table S.2) by applying NHTSA's learning factors.

Tables 8A.2a, b, and c (and Table S.2) show two costs for each of the optional downsizing entries because some of the previously added technologies, which are subsequently partially deleted with a reduction in number of cylinders, had low and high most likely cost estimates. The first of the two costs on the blue row were derived using the low most likely costs of the added and subsequently partially deleted technologies, and the second of the two costs were derived using the high most likely costs of the added and subsequently partly deleted technologies.

REFERENCE

EPA/NHTSA (Environmental Protection Agency/National Highway Traffic Safety Administration). 2012. Joint Technical Support Document, Final Rulemaking 2017-2025 Light-Duty Greenhouse Gas Emission Standards and Corporate Average Fuel Economy Standards. EPA-420-R-12-901.

Appendix U

SI Engine Pathway – NHTSA Estimates – Direct Manufacturing Costs and Total Costs

TABLE U.1 Midsize Car with I4 Spark Ignition Engine Pathway Example Using NHTSA's Estimate and Showing Direct Manufacturing Costs for 2017, 2020, and 2025 MYs (2010 dollars)

SI Engine Pathway - NHTSA Estimates - Direct Manufacturing Costs

							NHTSA Cost Estimates				
Possible Technologies	NHTSA % FC Reduction	FC Reduction Multiplier	Cumulative FC Reduction Multiplier	FC (gal/100mi)	Cumulative Percent FC Reduction Multiplier	Unadj. Combined MPG	2017	2020	2025	2017 Cost/ Percent FC Reduction ($/%)	Cumulative Cost
Null Vehicle[a]		1.000	1.000	3.240	0.0%	30.9					0
Intake Cam Phasing ICP	2.6%	0.974	0.974	3.156	2.6%	31.7	$37	$35	$31	$14.23	$31
Dual Cam Phasing DCP (vs. ICP)	2.5%	0.975	0.950	3.077	5.0%	32.5[b]	$31	$29	$27	$12.40	$58
2008 Example Vehicle											
Low Rolling Resistance Tires - 1 ROLL1	1.9%	0.981	0.932	3.018	6.8%	33.1	$5	$5	$5	$2.63	$63
Low Friction Lubricants - 1 LUB1	0.7%	0.993	0.925	2.997	7.5%	33.4	$3	$3	$3	$4.29	$66
6 Speed Automatic Transmission[b] 6 SP AT with Improved Internals IATC	1.6%	0.984	0.910	2.949	9.0%	33.9	$37	$34	$31	$23.13	$97
Aero Drag Reduction - 1 AERO1	2.3%	0.977	0.889	2.882	11.1%	34.7	$39	$37	$33	$16.96	$130
Engine Friction Reduction - 1 EFR1	2.6%	0.974	0.866	2.807	13.4%	35.6	$48	$48	$48	$18.46	$178
Improved Accessories - 1 IACC1	1.2%	0.988	0.856	2.773	14.4%	36.1	$71	$69	$60	$59.17	$238
Electric Power Steering EPS	1.3%	0.987	0.845	2.737	15.5%	36.5	$87	$82	$74	$66.92	$312
2016 Target 36.6 mpg											
Mass Reduction - 1 MR1 (1.5%) (-53 lbs)	0.5%	0.995	0.840	2.723	16.0%	36.7	$3	$3	$3	$6.00	$315
Discrete Variable Valve Lift DVVL	3.6%	0.964	0.810	2.625	19.0%	38.1	$116	$109	$99	$32.22	$414
Mass Reduction - 2 MR2 (3.5%) (-70 lbs = 123 lbs-53 lbs)	0.7%	0.993	0.805	2.607	19.5%	38.4	$27	$25	$22	$38.57	$436

continued

TABLE U.1 Continued

SI Engine Pathway - NHTSA Estimates - Direct Manufacturing Costs

Possible Technologies	NHTSA % FC Reduction	FC Reduction Multiplier	Cumulative FC Reduction Multiplier	FC (gal/100mi)	Cumulative Percent FC Reduction Multiplier	Unadj. Combined MPG	NHTSA Cost Estimates 2017	2020	2025	2017 Cost/ Percent FC Reduction ($/%)	Cumulative Cost
Stoichiometric Gasoline Direct Injection SGDI (Required for TRBDS)	1.5%	0.985	0.792	2.568	20.8%	38.9	$192	$181	$164	$128.00	$600
Turbocharging & Downsizing - 1 (I-4 to I-4) TRBDS1 33% DS 18 bar BMEP	8.3%	0.917	0.727	2.355	27.3%	42.5	$288	$271	$245	$34.70	$845
Turbocharging & Downsizing - 2 (I-4 to I-3) TRBDS2 50% DS 24 bar BMEP	3.5%	0.965	0.701	2.272	29.9%	44.0	–$92	–$89	–$82	–$26.29	$763
8 Speed Automatic Transmission[b] 8 SP AT	3.9%	0.961	0.674	2.184	32.6%	45.8	$56	$53	$47	$14.36	$810
Shift Optimizer[b] SHFTOPT	2.8%	0.972	0.655	2.122	34.5%	47.1	$1	$1	$0	$0.36	$810
Improved Accessories - 2 IAAC2	2.4%	0.976	0.639	2.071	36.1%	48.3	$43	$40	$37	$17.92	$847
Low Rolling Resistance Tires ROLL2	2.0%	0.980	0.627	2.030	37.3%	49.3	$58	$46	$31	$29.00	$878
Aero Drag Reduction - 2 AERO2	2.5%	0.975	0.611	1.979	38.9%	50.5	$117	$110	$100	$46.80	$978
High Efficiency Transmission HETRANS	2.7%	0.973	0.594	1.926	40.6%	51.9	$202	$184	$163	$74.81	$1,141
Low Friction Lub - 2 & Engine Friction Red - 2 LUB2_EFR2	1.3%	0.987	0.587	1.901	41.3%	52.6	$51	$51	$51	$39.23	$1,192
Cooled EGR - 1 CEGR1 50% DS 24 bar BMEP	3.5%	0.965	0.566	1.834	43.4%	54.5	$212	$199	$180	$60.57	$1,372
2025 Target 54.2 mpg											
Stop-Start SS	2.1%	0.979	0.554	1.796	44.6%	55.7	$287	$261	$225	$136.67	$1,597
Continuously Variable Valve Lift CVVL (vs. DVVL)	1.0%	0.990	0.549	1.778	45.1%	56.2	$58	$55	$49	$58.00	$1,646
Cylinder Deactivation DEACD	0.0%	1.000	0.549	1.778	45.1%	56.2					$1,646
Cooled EGR - 2 (I-3 to I-3) CEGR2 56% DS 27 bar BMEP	1.4%	0.986	0.541	1.753	45.9%	57.0	$364	$343	$310	$260.00	$1,956
Totals											
Relative to Null Vehicle	45.9%	0.541					$2,341	$2,185	$1,956	$51.00	
Null Vehicle - 2008 MY Vehicle	5.0%	0.950					$68	$64	$58	$13.51	
2008 MY Vehicle - 2016 MY	11.1%	0.889					$290	$278	$254		
2017 MY- 2025 MY	33.0%	0.670					$1,274	$1,184	$1,060	$38.63	
Beyond 2025 MY	4.4%	0.956					$709	$659	$584	$159.83	

[a] Null vehicle: I4, DOHC, naturally aspirated, 4 valves/cylinder PFI fixed valve timing and 4 speed AT.
[b] An example midsize car in 2008 was 46.64 sq ft and had a fuel economy of 32.5 mpg. Its standard for MY2016 would be 36.6 mpg and for MY2025 would be 54.2 mpg.
[c] These technologies have transmission synergies included.

TABLE U.2 Midsize Car with I4 Spark Ignition Engine Pathway Example Using NHTSA's Estimates and Showing Total Cost Estimates for 2017, 2020, and 2025 MYs (2010 dollars)

SI Engine Pathway - NHTSA Estimates - Total Costs

							NHTSA Cost Estimates				
Possible Technologies	NHTSA % FC Reduction	FC Reduction Multiplier	Cumulative FC Reduction Multiplier	FC (gal/100mi)	Cumulative Percent FC Reduction Multiplier	Unadj. Combined MPG	2017	2020	2025	2017 Cost/ Percent FC Reduction ($/%)	Cumulative Cost
Intake Cam Phasing ICP	2.6%	0.974	0.974	3.156	2.6%	31.7	$46	$42	$39	$17.69	$39
Dual Cam Phasing DCP (vs. ICP)	2.5%	0.975	0.950	3.077	5.0%	32.5[b]	$49	$42	$39	$19.60	$78
2008 Example Vehicle											
Low Rolling Resistance Tires - 1 ROLL1	1.9%	0.981	0.932	3.018	6.8%	33.1	$7	$6	$6	$3.68	$84
Low Friction Lubricants - 1 LUB1	0.7%	0.993	0.925	2.997	7.5%	33.4	$4	$4	$4	$5.74	$88
6 Speed Automatic Transmission[c] 6 SP AT with Improved Internals IATC	1.6%	0.984	0.910	2.949	9.0%	33.9	$54	$48	$44	$33.75	$132
Aero Drag Reduction - 1 AERO1	2.3%	0.977	0.889	2.882	11.1%	34.7	$49	$45	$41	$21.27	$173
Engine Friction Reduction - 1 EFR1	2.6%	0.974	0.866	2.807	13.4%	35.6	$59	$57	$57	$22.69	$230
Improved Accessories - 1 IACC1	1.2%	0.988	0.856	2.773	14.4%	36.1	$89	$81	$75	$74.16	$305
Electric Power Steering EPS	1.3%	0.987	0.845	2.737	15.5%	36.5	$109	$100	$92	$84.17	$397
2016 Target 36.6 mpg											
Mass Reduction - 1 MR1 (1.5%) (-53 lbs)	0.5%	0.995	0.840	2.723	16.0%	36.7	$4	$3	$3	$8.00	$400
Discrete Variable Valve Lift DVVL	3.6%	0.964	0.810	2.625	19.0%	38.1	$163	$144	$133	$45.28	$533
Mass Reduction - 2 MR2 (3.5%) (-70 lbs = 123 lbs-53 lbs)	0.7%	0.993	0.805	2.607	19.5%	38.4	$34	$30	$27	$48.57	$560
Stoichiometric Gasoline Direct Injection SGDI (Required for TRBDS)	1.5%	0.985	0.792	2.568	20.8%	38.9	$277	$244	$226	$184.67	$786
Turbocharging & Downsizing - 1 (I-4 to I-4) TRBDS1 33% DS 18 bar BMEP	8.3%	0.917	0.727	2.355	27.3%	42.5	$482	$415	$388	$58.07	$1,174
Turbocharging & Downsizing - 2 (I-4 to I-3) TRBDS2 50% DS 24 bar BMEP	3.5%	0.965	0.701	2.272	29.9%	44.0	$26	$19	$5	$7.43	$1,179
8 Speed Automatic Transmission[c] 8 SP AT	3.9%	0.961	0.674	2.184	32.6%	45.8	$80	$71	$66	$20.51	$1,245
Shift Optimizer[c] SHFTOPT	2.8%	0.972	0.655	2.122	34.5%	47.1	$2	$2	$0	$0.71	$1,245
Improved Accessories - 2 IAAC2	2.4%	0.976	0.639	2.071	36.1%	48.3	$54	$50	$45	$22.57	$1,290
Low Rolling Resistance Tires ROLL2	2.0%	0.980	0.627	2.030	37.3%	49.3	$66	$54	$38	$33.00	$1,328
Aero Drag Reduction - 2 AERO2	2.5%	0.975	0.611	1.979	38.9%	50.5	$164	$157	$135	$65.78	$1,463
High Efficiency Transmission HETRANS	2.7%	0.973	0.594	1.926	40.6%	51.9	$251	$233	$202	$92.91	$1,665

continued

TABLE U.2 Continued

SI Engine Pathway - NHTSA Estimates - Total Costs

							NHTSA Cost Estimates				
Possible Technologies	NHTSA % FC Reduction	FC Reduction Multiplier	Cumulative FC Reduction Multiplier	FC (gal/100mi)	Cumulative Percent FC Reduction Multiplier	Unadj. Combined MPG	2017	2020	2025	2017 Cost/ Percent FC Reduction ($/%)	Cumulative Cost
Low Friction Lub - 2 & Engine Friction Red - 2 LUB2_EFR2	1.3%	0.987	0.587	1.901	41.3%	52.6	$63	$65	$60	$48.46	$1,725
Cooled EGR - 1 CEGR1 50% DS 24 bar BMEP	3.5%	0.965	0.566	1.834	43.4%	54.5	$305	$292	$249	$87.14	$1,974
2025 Target 54.2 mpg											
Stop-Start SS	2.1%	0.979	0.554	1.796	44.6%	55.7	$401	$346	$308	$190.95	$2,282
Continuously Variable Valve Lift CVVL (vs. DVVL)	1.0%	0.990	0.549	1.778	45.1%	56.2	$81	$72	$67	$81.00	$2,349
Cylinder Deactivation DEACD	0.0%	1.000	0.549	1.778	45.1%	56.2					$2,349
Cooled EGR - 2 (I-3 to I-3) CEGR2 56% DS 27 bar BMEP	1.4%	0.986	0.541	1.753	45.9%	57.0	$525	$503	$429	$374.82	$2,778
Totals											
Relative to Null Vehicle	45.9%	0.541					$3,445	$3,125	$2,778	$75.05	
Null Vehicle - 2008 MY Vehicle	5.0%	0.950					$95	$84	$78	$18.87	
2008 MY Vehicle - 2016 MY	11.1%	0.889					$371	$341	$319		
2017 MY- 2025 MY	33.0%	0.670					$1,971	$1,779	$1,577	$59.78	
Beyond 2025 MY	4.4%	0.956					$1,007	$921	$804	$226.96	

[a] Null vehicle: I4, DOHC, naturally aspirated, 4 valves/cylinder PFI fixed valve timing and 4 speed AT.

[b] An example midsize car in 2008 was 46.64 sq ft and had a fuel economy of 32.5 mpg. Its standard for MY2016 would be 36.6 mpg and for MY2025 would be 54.2 mpg.

[c] These technologies have transmission synergies included.

Appendix V

SI Engine Pathway – NRC Estimates – Direct Manufacturing Costs – Alternative Pathway, Alternative High CR with Exhaust Scavenging, and Alternative EVAS Supercharger

TABLE V.1 Alternative Midsize Car with SI Engine Pathway with High Compression Ratio with Exhaust Scavenging Technology Showing NRC Low Estimates for 2017, 2020, and 2025 (2010 dollars)

Midsize Car with SI Engine Pathway - NRC Low Most Likely Estimates - Direct Manufacturing Costs
Alternative Pathway - High CR with Exhaust Scavenging
Low Most Likely Cost Estimates Paired with High Most Likely Effectiveness Estimates

Possible Technologies	% FC Reduction (%)	FC Reduction Multiplier	Cumulative FC Reduction Multiplier	FC (gal/100 mi)	Cumulative Percent FC Reduction	Unadjusted Combined (mpg)	2017 Cost Estimates	2020 Cost Estimates	2025 Cost Estimates	2017 Cost/ Percent FC ($/%)
Null Vehicle[a]		1.000	1.000	3.240	0.0%	30.9				
Intake Cam Phasing ICP	2.6%	0.974	0.974	3.156	2.6%	31.7	$37	$35	$31	$14.23
Dual Cam Phasing DCP (vs. ICP)	2.5%	0.975	0.950	3.077	5.0%	32.5[b]	$31	$29	$27	$12.40
2008 Example Vehicle										
Low Rolling Resistance Tires - 1 ROLL1	1.9%	0.981	0.932	3.018	6.8%	33.1	$5	$5	$5	$2.63
Low Friction Lubricants - 1 LUB1	0.7%	0.993	0.925	2.997	7.5%	33.4	$3	$3	$3	$4.29
6 Speed Automatic Transmission[c] 6 SP AT with Improved Internals IATC	1.6%	0.984	0.910	2.949	9.0%	33.9	$37	$34	$31	$23.13
Aero Drag Reduction - 1 AERO1	2.3%	0.977	0.889	2.882	11.1%	34.7	$39	$37	$33	$16.96
Engine Friction Reduction - 1 EFR1	2.6%	0.974	0.866	2.807	13.4%	35.6	$48	$48	$48	$18.46
Improved Accessories - 1 IACC1	1.2%	0.988	0.856	2.773	14.4%	36.1	$71	$69	$60	$59.17
Electric Power Steering EPS	1.3%	0.987	0.845	2.737	15.5%	36.5	$87	$82	$74	$66.92
Mass Reduction - 2.5% MR2.5 (-87.5 lbs)	0.8%	0.992	0.838	2.715	16.2%	36.8	$0	$0	$0	$0.00
2016 Target 36.6 mpg										
Discrete Variable Valve Lift DVVL	3.6%	0.964	0.808	2.617	19.2%	38.2	$116	$109	$99	$32.22
Mass Reduction - 2.5%-5.0% MR5-MR2.5 (-87.5 lbs)	0.8%	0.992	0.801	2.596	19.9%	38.5	$0	$0	$0	$0.00
Stoichiometric Gasoline Direct Injection SGDI (Required for TRBDS)	1.5%	0.985	0.789	2.557	21.1%	39.1	$192	$181	$164	$128.00

continued

TABLE V.1 Continued

Possible Technologies	% FC Reduction (%)	FC Reduction Multiplier	Cumulative FC Reduction Multiplier	FC (gal/100 mi)	Cumulative Percent FC Reduction	Unadjusted Combined (mpg)	2017 Cost Estimates	2020 Cost Estimates	2025 Cost Estimates	2017 Cost/ Percent FC ($/%)
High Compression Ratio- Exh Scavenging EXS	6.0%	0.940	0.742	2.404	25.8%	41.6	$250	$250	$250	$41.67
Turbocharging & Downsizing - 1 (I-4 to I-4) TRBDS1 33% DS 18 bar BMEP	8.3%	0.917	0.680	2.204	32.0%	45.4	$288	$271	$245	$34.70
Turbocharging & Downsizing - 2 (I-4 to I-3) TRBDS2 50% DS 24 bar BMEP	3.5%	0.965	0.657	2.127	34.3%	47.0	-$92	-$89	-$82	-$26.29
8 Speed Automatic Transmission[c] 8 SP AT	1.7%	0.983	0.645	2.091	35.5%	47.8	$56	$52	$47	$32.94
Shift Optimizer[c] SHFTOPT	0.7%	0.993	0.641	2.076	35.9%	48.2	$26	$24	$22	$37.14
Improved Accessories - 2 IAAC2	2.4%	0.976	0.625	2.027	37.5%	49.3	$43	$40	$37	$17.92
Low Rolling Resistance Tires ROLL2	2.0%	0.980	0.613	1.986	38.7%	50.4	$58	$46	$31	$29.00
Aero Drag Reduction - 2 AERO2	2.5%	0.975	0.598	1.936	40.2%	51.6	$117	$110	$100	$46.80
Mass Reduction - 5.0%- 10.0% MR10-MR5 (-175 lbs)	4.6%	0.954	0.570	1.847	43.0%	54.1	$154	$151	$151	$33.48
Low Friction Lub - 2 & Engine Friction Red - 2 LUB2_EFR2	1.3%	0.987	0.563	1.823	43.7%	54.8	$51	$51	$51	$39.23
2025 Target 54.2 mpg										
Continuously Variable Valve Lift CVVL (vs. DVVL)	1.0%	0.990	0.557	1.805	44.3%	55.4	$58	$55	$49	$58.00
High Efficiency Transmission HEG1 & 2	5.4%	0.946	0.527	1.708	47.3%	58.6	$314	$296	$267	$58.15
Cooled EGR - 1 CEGR1 50% DS 24 bar BMEP	3.5%	0.965	0.509	1.648	49.1%	60.7	$212	$199	$180	$60.57
Cylinder Deactivation DEACD	0.0%	1.000	0.509	1.648	49.1%	60.7				
Cooled EGR - 2 (I-3 to I-3) CEGR2 56% DS 27 bar BMEP	1.4%	0.986	0.501	1.625	49.9%	61.5	$364	$343	$310	$260.00
Totals										
Relative to Null Vehicle	49.9%	0.501					$2,565	$2,431	$2,233	$51.45
Null Vehicle - 2008 MY Vehicle	5.0%	0.950					$68	$64	$58	$13.51
2008 MY Vehicle - 2016 MY	11.8%	0.882					$290	$278	$254	
2017 MY- 2025 MY	32.8%	0.672					$1,259	$1,196	$1,115	$38.34
Beyond 2025 MY	10.9%	0.891					$948	$893	$806	$87.06

[a] Null vehicle: I4, DOHC, naturally aspirated, 4 valves/cylinder PFI fixed valve timing and 4 speed AT.
[b] An example midsize car in 2008 was 46.64 sq ft and had a fuel economy of 32.5 mpg. Its standard for MY2016 would be 36.6 mpg and for MY2025 would be 54.2 mpg.
[c] These technologies have transmission synergies included. Green highlighting indicates a technology order different than the NHTSA pathway, shown in Appendix S.

TABLE V.2 Alternative Midsize Car with SI Engine Pathway with High Compression Ratio with Exhaust Scavenging Technology Showing NRC High Estimates for 2017, 2020, and 2025 (2010 dollars)

Midsize Car with SI Engine Pathway - NRC High Most Likely Estimates - Direct Manufacturing Costs
Alternative Pathway - High CR with Exhaust Scavenging
High Most Likely Cost Estimates Paired with Low Most Likely Effectiveness Estimates

Possible Technologies	% FC Reduction (%)	FC Reduction Multiplier	Cumulative FC Reduction Multiplier	FC (gal/100 mi)	Cumulative Percent FC Reduction	Unadjusted Combined (mpg)	2017 Cost Estimates	2020 Cost Estimates	2025 Cost Estimates	2017 Cost/ Percent FC ($/%)
Null Vehicle[a]		1.000	1.000	3.240	0.0%	30.9				
Intake Cam Phasing ICP	2.6%	0.974	0.974	3.156	2.6%	31.7	$43	$41	$36	$16.54
Dual Cam Phasing DCP (vs. ICP)	2.5%	0.975	0.950	3.077	5.0%	32.5[b]	$35	$33	$31	$14.00
2008 Example Vehicle										
Low Rolling Resistance Tires - 1 ROLL1	1.9%	0.981	0.932	3.018	6.8%	33.1	$5	$5	$5	$2.63
Low Friction Lubricants - 1 LUB1	0.7%	0.993	0.925	2.997	7.5%	33.4	$3	$3	$3	$4.29
6 Speed Automatic Transmission[c] 6 SP AT with Improved Internals IATC	1.3%	0.987	0.913	2.958	8.7%	33.8	$37	$34	$31	$28.46
Aero Drag Reduction - 1 AERO1	2.3%	0.977	0.892	2.890	10.8%	34.6	$39	$37	$33	$16.96
Engine Friction Reduction - 1 EFR1	2.6%	0.974	0.869	2.815	13.1%	35.5	$48	$48	$48	$18.46
Improved Accessories - 1 IACC1	1.2%	0.988	0.858	2.781	14.2%	36.0	$71	$67	$60	$59.17
Electric Power Steering EPS	1.3%	0.987	0.847	2.745	15.3%	36.4	$87	$82	$74	$66.92
Mass Reduction - 2.5% MR2.5 (-87.5 lbs)	0.8%	0.992	0.841	2.723	15.9%	36.7	$22	$22	$22	$27.50
2016 Target 36.6 mpg										
Discrete Variable Valve Lift DVVL	3.6%	0.964	0.810	2.625	19.0%	38.1	$133	$125	$114	$36.94
Mass Reduction - 2.5%-5.0% MR5-MR2.5 (-87.5 lbs)	0.8%	0.992	0.804	2.604	19.6%	38.4	$66	$66	$66	$82.50
Stoichiometric Gasoline Direct Injection SGDI (Required for TRBDS)	1.5%	0.985	0.792	2.565	20.8%	39.0	$192	$181	$164	$128.00
High Compression Ratio- Exh Scavenging EXS	6.0%	0.940	0.744	2.411	25.6%	41.5	$250	$250	$250	$41.67
Turbocharging & Downsizing - 1 TRBDS1 33% DS 18 bar BMEP	7.7%	0.923	0.687	2.226	31.3%	44.9	$331	$312	$282	$42.99
Turbocharging & Downsizing - 2 TRBDS2 50% DS 24 bar BMEP	3.2%	0.968	0.665	2.154	33.5%	46.4	-$96	-$92	-$86	-$30.00
8 Speed Automatic Transmission[c] 8 SP AT	1.3%	0.987	0.656	2.126	34.4%	47.0	$151	$126	$115	$116.15
Shift Optimizer[c] SHFTOPT	0.3%	0.997	0.654	2.120	34.6%	47.2	$26	$24	$22	$86.67
Improved Accessories - 2 IAAC2	2.4%	0.976	0.639	2.069	36.1%	48.3	$43	$40	$37	$17.92
Low Rolling Resistance Tires ROLL2	2.0%	0.980	0.626	2.028	37.4%	49.3	$58	$46	$31	$29.00

continued

TABLE V.2 Continued

Possible Technologies	% FC Reduction (%)	FC Reduction Multiplier	Cumulative FC Reduction Multiplier	FC (gal/100 mi)	Cumulative Percent FC Reduction	Unadjusted Combined (mpg)	2017 Cost Estimates	2020 Cost Estimates	2025 Cost Estimates	2017 Cost/ Percent FC ($/%)
Aero Drag Reduction - 2 AERO2	2.5%	0.975	0.610	1.977	39.0%	50.6	$117	$110	$100	$46.80
Mass Reduction - 5%-10% MR10-MR5 (-175 lbs)	4.6%	0.954	0.582	1.886	41.8%	53.0	$325	$322	$315	$70.65
Low Friction Lub - 2 & Engine Friction Red - 2 LUB2_EFR2	1.3%	0.987	0.575	1.862	42.5%	53.7	$51	$51	$51	$39.23
Cooled EGR - 1 CEGR1 50% DS 24 bar BMEP	3.0%	0.970	0.557	1.806	44.3%	55.4	$212	$199	$180	$70.67
2025 Target 54.2 mpg										
High Efficiency Transmission HEG1 & 2	4.9%	0.951	0.530	1.717	47.0%	58.2	$314	$296	$267	$64.08
Continuously Variable Valve Lift CVVL (vs. DVVL)	1.0%	0.990	0.525	1.700	47.5%	58.8	$67	$63	$56	$67.00
Cylinder Deactivation DEACD	0.0%	1.000	0.525	1.700	47.5%	58.8				
Cooled EGR - 2 CEGR2 56% DS 27 bar BMEP	1.4%	0.986	0.517	1.676	48.3%	59.7	$364	$343	$310	$260.00
Totals										
Relative to Null Vehicle	48.3%	0.517					$2,994	$2,834	$2,617	$62.03
Null Vehicle - 2008 MY Vehicle	5.0%	0.950					$78	$74	$67	$15.49
2008 MY Vehicle - 2016 MY	11.5%	0.885					$312	$298	$276	
2017 MY- 2025 MY	33.7%	0.663					$1,859	$1,760	$1,641	$55.17
Beyond 2025 MY	7.2%	0.928					$745	$702	$633	$103.92

[a] Null vehicle: I4, DOHC, naturally aspirated, 4 valves/cylinder PFI fixed valve timing and 4 speed AT.

[b] An example midsize car in 2008 was 46.64 sq ft and had a fuel economy of 32.5 mpg. Its standard for MY2016 would be 36.6 mpg and for MY2025 would be 54.2 mpg.

[c] These technologies have transmission synergies included. Green highlighting indicates a technology order different than the NHTSA pathway, shown in Appendix S.

TABLE V.3 Alternative Midsize Car with SI Engine Pathway with EAVS Supercharger Technology Showing NRC Low Estimates for 2017, 2020, and 2025 (2010 dollars)

Midsize Car with SI Engine Pathway - NRC Low Most Likely Estimates - Direct Manufacturing Costs
Alternative Pathway - EAVS Supercharger
Low Most Likely Cost Estimates Paired with High Most Likely Effectiveness Estimates

Possible Technologies	% FC Reduction	FC Reduction Multiplier	Cumulative FC Reduction Multiplier	FC (gal/100 mi)	Cumulative Percent FC Reduction	Unadjusted Combined FE (mpg)	2017 Cost Estimates	2020 Cost Estimates	2025 Cost Estimates	2017 Cost/ Percent FC ($/%)
Null Vehicle[a]		1.000	1.000	3.240	0.0%	30.9				
Intake Cam Phasing ICP	2.6%	0.974	0.974	3.156	2.6%	31.7	$37	$35	$31	$14.23
Dual Cam Phasing DCP (vs. ICP)	2.5%	0.975	0.950	3.077	5.0%	32.5[b]	$31	$29	$27	$12.40
2008 Example Vehicle										
Low Rolling Resistance Tires - 1 ROLL1	1.9%	0.981	0.932	3.018	6.8%	33.1	$5	$5	$5	$2.63
Low Friction Lubricants - 1 LUB1	0.7%	0.993	0.925	2.997	7.5%	33.4	$3	$3	$3	$4.29
6 Speed Automatic Transmission[c] 6 SP AT with Improved Internals IATC	1.6%	0.984	0.910	2.949	9.0%	33.9	$37	$34	$31	$23.13
Aero Drag Reduction - 1 AERO1	2.3%	0.977	0.889	2.882	11.1%	34.7	$39	$37	$33	$16.96
Engine Friction Reduction - 1 EFR1	2.6%	0.974	0.866	2.807	13.4%	35.6	$48	$48	$48	$18.46
Electric Power Steering EPS	1.3%	0.987	0.855	2.770	14.5%	36.1	$87	$82	$74	$66.92
Mass Reduction - 2.5% MR2.5 (-87.5 lbs)	0.8%	0.992	0.848	2.748	15.2%	36.4	$0	$0	$0	$0.00
Discrete Variable Valve Lift DVVL	3.6%	0.964	0.818	2.649	18.2%	37.7	$116	$109	$99	$32.22
2016 Target 36.6 mpg										
Mass Reduction - 2.5%-5.0% MR5-MR2.5 (-87.5 lbs)	0.8%	0.992	0.811	2.628	18.9%	38.1	$0	$0	$0	$0.00
Stoichiometric Gasoline Direct Injection SGDI (Required for TRBDS)	1.5%	0.985	0.799	2.588	20.1%	38.6	$192	$181	$164	$128.00
EAVS-Supercharger EAVS-SC	26.0%	0.740	0.591	1.915	40.9%	52.2	$1,302	$1,302	$1,302	$50.08
8 Speed Automatic Transmission[c] 8 SP AT	1.7%	0.983	0.581	1.883	41.9%	53.1	$56	$52	$47	$32.94
Shift Optimizer[c] SHFTOPT	0.7%	0.993	0.577	1.870	42.3%	53.5	$26	$24	$22	$37.14
Low Rolling Resistance Tires ROLL2	2.0%	0.980	0.566	1.832	43.4%	54.6	$58	$46	$31	$29.00
2025 Target 54.2 mpg										
Aero Drag Reduction - 2 AERO2	2.5%	0.975	0.551	1.786	44.9%	56.0	$117	$110	$100	$46.80
Mass Reduction - 5.0%-10.0% MR10-MR5 (-175 lbs)	4.6%	0.954	0.526	1.704	47.4%	58.7	$154	$151	$151	$33.48
Low Friction Lub - 2 & Engine Friction Red - 2 LUB2_EFR2	1.3%	0.987	0.519	1.682	48.1%	59.4	$51	$51	$51	$39.23
Continuously Variable Valve Lift CVVL (vs. DVVL)	1.0%	0.990	0.514	1.665	48.6%	60.0	$58	$55	$49	$58.00

continued

TABLE V.3 Continued

Possible Technologies	% FC Reduction	FC Reduction Multiplier	Cumulative FC Reduction Multiplier	FC (gal/100 mi)	Cumulative Percent FC Reduction	Unadjusted Combined FE (mpg)	2017 Cost Estimates	2020 Cost Estimates	2025 Cost Estimates	2017 Cost/ Percent FC ($/%)
High Efficiency Transmission HEG1 & 2	5.4%	0.946	0.486	1.575	51.4%	63.5	$314	$296	$267	$58.15
Cooled EGR - 1 CEGR1 50% DS 24 bar BMEP	3.5%	0.965	0.469	1.520	53.1%	65.8	$212	$199	$180	$60.57
Cylinder Deactivation DEACD	0.0%	1.000	0.469	1.520	53.1%	65.8				
Cooled EGR - 2 (I-3 to I-3) CEGR2 56% DS 27 bar BMEP	1.4%	0.986	0.463	1.499	53.7%	66.7	$364	$343	$310	$260.00
Totals										
Relative to Null Vehicle	53.7%	0.463					$3,307	$3,192	$3,025	$61.54
Null Vehicle - 2008 MY Vehicle	5.0%	0.950					$68	$64	$58	$13.51
2008 MY Vehicle - 2016 MY	13.9%	0.861					$335	$318	$293	
2017 MY- 2025 MY	30.8%	0.692					$1,634	$1,605	$1,566	$53.00
Beyond 2025 MY	18.2%	0.818					$1,270	$1,205	$1,108	$69.81

[a] Null vehicle: I4, DOHC, naturally aspirated, 4 valves/cylinder PFI fixed valve timing and 4 speed AT.

[b] An example midsize car in 2008 was 46.64 sq ft and had a fuel economy of 32.5 mpg. Its standard for MY2016 would be 36.6 mpg and for MY2025 would be 54.2 mpg.

[c] These technologies have transmission synergies included. Green highlighting indicates a technology order different than the NHTSA pathway, shown in Appendix S.

TABLE V.4 Alternative Midsize Car with SI Engine Pathway with EAVS Supercharger Technology Showing NRC High Estimates for 2017, 2020, and 2025 (2010 dollars)

Midsize Car with SI Engine Pathway - NRC High Most Likely Estimates - Direct Manufacturing Costs
Alternative Pathway - EAVS Supercharger
High Most Likely Cost Estimates Paired with Low Most Likely Effectiveness Estimates

Possible Technologies	% FC Reduction	FC Reduction Multiplier	Cumulative FC Reduction Multiplier	Fuel Consumption (gal/100 mi)	Cumulative Percent FC Reduction	Unadjusted Combined FE (mpg)	2017 Cost Estimates	2020 Cost Estimates	2025 Cost Estimates	2017 Cost/ Percent FC ($/%)
Null Vehicle[a]		1.000	1.000	3.240	0.0%	30.9				
Intake Cam Phasing ICP	2.6%	0.974	0.974	3.156	2.6%	31.7	$43	$41	$36	$16.54
Dual Cam Phasing DCP (vs. ICP)	2.5%	0.975	0.950	3.077	5.0%	32.5[b]	$35	$33	$31	$14.00
2008 Example Vehicle										
Low Rolling Resistance Tires - 1 ROLL1	1.9%	0.981	0.932	3.018	6.8%	33.1	$5	$5	$5	$2.63
Low Friction Lubricants - 1 LUB1	0.7%	0.993	0.925	2.997	7.5%	33.4	$3	$3	$3	$4.29
6 Speed Automatic Transmission[c] 6 SP AT with Improved Internals IATC	1.3%	0.987	0.913	2.958	8.7%	33.8	$37	$34	$31	$28.46
Aero Drag Reduction - 1 AERO1	2.3%	0.977	0.892	2.890	10.8%	34.6	$39	$37	$33	$16.96
Engine Friction Reduction - 1 EFR1	2.6%	0.974	0.869	2.815	13.1%	35.5	$48	$48	$48	$18.46
Electric Power Steering EPS	1.3%	0.987	0.858	2.779	14.2%	36.0	$87	$82	$74	$66.92
Mass Reduction - 2.5% MR2.5 (-87.5 lbs)	0.8%	0.992	0.851	2.756	14.9%	36.3	$22	$22	$22	$27.50
Discrete Variable Valve Lift DVVL	3.6%	0.964	0.820	2.657	18.0%	37.6	$133	$125	$114	$36.94
2016 Target 36.6 mpg										
Mass Reduction - 2.5%-5.0% MR5-MR2.5 (-87.5 lbs)	0.8%	0.992	0.814	2.636	18.6%	37.9	$66	$66	$66	$82.50
Stoichiometric Gasoline Direct Injection SGDI (Required for TRBDS)	1.5%	0.985	0.801	2.596	19.9%	38.5	$192	$181	$164	$128.00
EAVS-Supercharger EAVS-SC	26.0%	0.740	0.593	1.921	40.7%	52.0	$1,302	$1,302	$1,302	$50.08
8 Speed Automatic Transmission[c] 8 SP AT	1.3%	0.987	0.585	1.896	41.5%	52.7	$151	$126	$115	$116.15
Shift Optimizer[c] SHFTOPT	0.3%	0.997	0.584	1.891	41.6%	52.9	$26	$24	$22	$86.67
Low Rolling Resistance Tires ROLL2	2.0%	0.980	0.572	1.853	42.8%	54.0	$58	$46	$31	$29.00
Aero Drag Reduction - 2 AERO2	2.5%	0.975	0.558	1.806	44.2%	55.4	$117	$110	$100	$46.80
2025 Target 54.2 mpg										
Mass Reduction - 5%-10% MR10-MR5 (-175 lbs)	4.6%	0.954	0.532	1.723	46.8%	58.0	$325	$322	$315	$70.65
Low Friction Lub - 2 & Engine Friction Red - 2 LUB2_EFR2	1.3%	0.987	0.525	1.701	47.5%	58.8	$51	$51	$51	$39.23
Cooled EGR - 1 CEGR1 50% DS 24 bar BMEP	3.0%	0.970	0.509	1.650	49.1%	60.6	$212	$199	$180	$70.67
High Efficiency Transmission HEG1 & 2	4.9%	0.951	0.484	1.569	51.6%	63.7	$314	$296	$267	$64.08

continued

TABLE V.4 Continued

Possible Technologies	% FC Reduction	FC Reduction Multiplier	Cumulative FC Reduction Multiplier	Fuel Consumption (gal/100 mi)	Cumulative Percent FC Reduction	Unadjusted Combined FE (mpg)	2017 Cost Estimates	2020 Cost Estimates	2025 Cost Estimates	2017 Cost/ Percent FC ($/%)
Continuously Variable Valve Lift CVVL (vs. DVVL)	1.0%	0.990	0.479	1.553	52.1%	64.4	$67	$63	$56	$67.00
Cylinder Deactivation DEACD	0.0%	1.000	0.479	1.553	52.1%	64.4				
Cooled EGR - 2 CEGR2 56% DS 27 bar BMEP	1.4%	0.986	0.473	1.532	52.7%	65.3	$364	$343	$310	$260.00
Totals										
Relative to Null Vehicle	52.7%	0.473					$3,697	$3,559	$3,376	$70.12
Null Vehicle - 2008 MY Vehicle	5.0%	0.950					$78	$74	$67	$15.49
2008 MY Vehicle - 2016 MY	13.6%	0.864					$374	$356	$330	
2017 MY- 2025 MY	32.0%	0.680					$1,912	$1,855	$1,800	$59.72
Beyond 2025 MY	15.2%	0.848					$1,333	$1,274	$1,179	$87.62

[a] Null vehicle: I4, DOHC, naturally aspirated, 4 valves/cylinder PFI fixed valve timing and 4 speed AT.

[b] An example midsize car in 2008 was 46.64 sq ft and had a fuel economy of 32.5 mpg. Its standard for MY2016 would be 36.6 mpg and for MY2025 would be 54.2 mpg.

[c] These technologies have transmission synergies included. Green highlighting indicates a technology order different than the NHTSA pathway, shown in Appendix S.

Appendix W

Technologies, Footprints, and Fuel Economy for Example Passenger Cars, Trucks, and Hybrid Passenger Cars

TABLE W.1 Technologies, Footprints, and Fuel Economy for Example Passenger Cars

Technology	2014 Chev Sonic 1.4L	2014 Toyota Corolla LE Eco 1.8L	2014 Dodge Dart 1.4L	2014 Ford Fusion 1.5L	2014 Chev Impala 2.5L	2014 Ford Taurus 2.0L
Low Friction Lubricants (LUB1)	x	x		x	x	x
Variable Valve Timing ICP and DCP	x	x	Multi-air	x	x	x
Variable Valve Lift (VVL)		x	Multi-air		x	
Gasoline Direct Injection (SGDI)				x	x	x
Turbocharging and Downsizing (33%) - Level 1	x		x	x	DS only	x
Turbocharging and Downsizing (50%) - Level 2						
6 sp AT	x	CVT		x	x	x
6 sp DCT			x			
8 sp AT						
Stop-Start						
Footprint (sf)	41	44.5	45.8	48.9	48.2	51.3
Label Fuel Economy (mpg)	31	35	32	28	25	26
CAFE Fuel Economy (mpg)	40.9	46.8	43.2	36.4	32.2	34.5
2016 CAFE Target (mpg)	41.1	38.2	37.2	35.1	35.5	33.6
2025 CAFE Target (mpg)	61.1	56.6	55.1	51.8	52.5	49.5

SOURCE: EPA Fuel Economy Guides and Databases; Cars.com.

2013 BMW 740 LI 3.0L	2014 Mazda 3 2.0L	2014 Ford Focus SFE 2.0L	2014 Ford Focus 2.0L	2014 Mazda 6 2.5L	2014 Hyundai Sonata 2.4L	2015 Honda Civic HF 1.8L
x	x	x	x	x	x	x
x	x	x	x	x	x	x
x						x
x	x	x	x	x	x	
x						
	x			x	x	CVT
		x	x			
x						
x						
53.4	45.3	44	44	48.4	48.1	43.4
22	34	33	30	30	28	35
29.1	45.9	43.6	40.6	40.7	36.7	47.5
32.3	37.6	38.6	38.6	35.4	35.7	39.1
47.7	55.7	57.2	57.2	52.3	52.7	57.9

TABLE W.2 Technologies, Footprints, and Fuel Economy for Example Trucks

Technology	2013 Ford Escape 1.6L TC	2014 Chev Silverado 4.3L V6	2013 Ford F150 3.5L V6 TC	2014 RAM 3.6L V6	2014 RAM 3.0L Diesel	2015 Ford F150 2.7L V6 TC[a]
Low Friction Lubricants (LUB1)	x	x	x	x		
Variable Valve Timing ICP and DCP	x	x	x	x		x
Variable Valve Lift (VVL)						
Gasoline Direct Injection (SGDI)	x	x	x			x
Turbocharging and Downsizing (33%) - Level 1	x	x	x		x	x
Turbocharging and Downsizing (50%) - Level 2						x
6 sp AT	x	x	x			x
6 sp DCT						
8 sp AT				x	x	
Stop-Start						
Footprint (sf)	45.2	67.3	67.5	65.9	65.9	67.5
Label Fuel Economy (mpg)	26	20	18	20	23	22
CAFE Fuel Economy (mpg)	34.6	25.8	23.9	25.6	30.4	28.5
2016 CAFE Target (mpg)	32.2	24.4	24.3	24.8	24.8	24.3
2025 CAFE Target (mpg)	46.4	32.9	32.8	33.5	33.5	32.8

[a] Aluminum body.

SOURCE: EPA Fuel Economy Guides and Databases; Cars.com.

TABLE W.3 Technologies, Footprints, and Fuel Economy for Example Hybrid Passenger Cars

Technology	2014 Chev Impala eAssist	2014 Hyundai Sonata	2014 Ford Fusion Hybrid	2014 Toyota Prius Hybrid
Low Friction Lubricants (LUB1)				
Variable Valve Timing ICP and DCP				
Variable Valve Lift (VVL)				
Gasoline Direct Injection (SGDI)				
Turbocharging and Downsizing (33%) - Level 1				
Turbocharging and Downsizing (50%) - Level 2				
Hybrid Type	Belt Mounted ISG	P2	PS	PS
6 sp AT				x
6 sp DCT				
8 sp AT				
Stop-Start				
Footprint (sf)	48.2	48	48.8	44.2
Label Fuel Economy (mpg)	29	38	47	50
CAFE Fuel Economy (mpg)	38.1	51.5	66.1	70.6
2016 CAFE Target (mpg)	35.5	35.7	25.1	38.4
2025 CAFE Target (mpg)	52.5	52.7	51.9	57.0

SOURCE: EPA Fuel Economy Guides and Databases; Cars.com.

Appendix X

Full System Simulation Modeling of Fuel Consumption Reductions

This appendix provides additional details of the University of Michigan's Department of Mechanical Engineering (referred to as U of M throughout this section) project using full system simulation modeling to analyze the effects of fuel consumption reduction technologies (Middleton 2015).

VEHICLE SPECIFICATIONS

Vehicle specifications for the modeling are shown in Table X.1. For all configurations the vehicle test weight, rolling resistance, and drag characteristics were held constant at the values noted in the table. These are representative of a midsize car similar to a 2012 Ford Fusion.

COMBINATIONS OF POWERTRAIN TECHNOLOGIES

The specific combinations of powertrain technologies evaluated are shown in Table X.2.

FULL SYSTEM SIMULATION MODEL CALIBRATION

Calibration and setup of the model were the first tasks undertaken. In Task 1 the model parameters for flame speed, knock limits, and breathing were determined based on detailed dynamometer data for a 2.0L boosted GDI engine (GM- LNF). These parameters were held constant for all subsequent simulations. Engine friction values were determined by fitting the standard Chen-Flynn model to the data and adding a constant auxiliary power requirement for the fuel injection system. Validation of the baseline engine/vehicle model was accomplished in Task 3. For this, the Task 1 engine geometry was modified to simulate the baseline engine, a naturally aspirated (NA) 2.5L engine with PFI and dual cam phasing (DCP). After optimizing the cam phasing strategy, the resulting engine map was then exercised in the base vehicle over the prescribed drive cycles with appropriate transmission and rear axle ratio. With minimal parameter adjustments the results compared favorably to the EPA certification test data for a 2012 MY Ford Fusion

The remaining Tasks 3-10 include the baseline followed by successive changes in technology as noted in the table. The order chosen reflects a plausible sequence of technology adoption for decreasing fuel consumption; however, it is worth noting that the incremental gains of a given technology may not be independent of that order. The first sequence of changes are related to friction and breathing. These were evaluated in Tasks 3-5 on the naturally aspirated 2.5L engine. Task 6 added direct injection (GDI) while Task 7 introduced a turbocharger on a 33 percent downsized 1.68L engine. The boost level was chosen to maintain approximately constant peak engine power. Further downsizing by 50 percent to 1.25L was carried out in Task 8 with the addition of cooled EGR and higher boost level, again maintaining equivalent engine power. The bore and stroke sizes for both downsized engines were selected with NRC guidance to achieve a square ratio. To preclude high heat losses with exceptionally small cylinder sizes, the 50 percent downsized engines had three cylinders instead of four cylinders for the 1.68L and larger engines. Task 9 kept the same engine configuration but replaced the six-speed automatic transmission (6 AT) with an eight-speed (8 AT). Because the latest eight-speed transmissions have reduced friction losses relative to the current six-speed versions (Scherer et al. 2009), several levels of friction reduction up to 60 percent were considered in Tasks 9B through 9E. Task 10-A and 10-B both employed a CVT transmission; the former used losses representative of current

TABLE X.1 Vehicle Specifications

Vehicle Attribute	Value
Test Weight	3,625 lb (1648 kg)
Road Load Force ($F = A + BV + CV^2$)	A = 29.0 lbf (129.0 N)
	B = 0.24 lbf/mph (2.388 N/(m/s))
	C = 0.0180 lbf/mph^2 (0.4006 N/(m/s)^2)
Tire Radius	328.4 mm

TABLE X.2 Engine and Powertrain Configurations

	Engine										Transmission			
Task	Disp (L)	Air	Fuel	CR	No. Cyl.	Bore (mm)	Stroke (mm)	Friction	Cams	EGR	Design	Max Ratio	Min Ratio	Final Drive
1	2.0	TC	GDI	9.2	4	86	86	Chen-Flynn	DCP		—	—	—	—
2 Ref	2.5	NA	PFI	9.7	4	89	100	Chen-Flynn	Fixed		6 AT	4.584	0.745	3.06
3 Base	2.5	NA	PFI	9.7	4	89	100	Chen-Flynn	DCP		6 AT	4.584	0.745	3.06
4	2.5	NA	PFI	9.7	4	89	100	Red. 26%	DCP		6 AT	4.584	0.745	3.06
5	2.5	NA	PFI	9.7	4	89	100	Red. 26%	DCP + DVVL		6 AT	4.584	0.745	3.06
6	2.5	NA	GDI	9.7	4	89	100	Red. 26%	DCP + DVVL		6 AT	4.584	0.745	3.06
7	1.68	TC	GDI	9.7	4	81	81.51	Red. 26%	DCP + DVVL		6 AT	4.584	0.745	3.06
8	1.25	TC	GDI	9.7	3	81	80.86	Red. 26%	DCP + DVVL		6 AT	4.584	0.745	3.06
9-A	1.25	TC	GDI	9.7	3	81	80.86	Red. 26%	DCP + DVVL		8 AT	4.6	0.52	3.06
9-B[a]	1.25	TC	GDI	9.7	3	81	80.86	Red. 26%	DCP + DVVL		8 AT - 15%	4.6	0.52	3.06
9-C[a]	1.25	TC	GDI	9.7	3	81	80.86	Red. 26%	DCP + DVVL		8 AT - 30%	4.6	0.52	3.06
9-D[a]	1.25	TC	GDI	9.7	3	81	80.86	Red. 26%	DCP + DVVL		8 AT - 45%	4.6	0.52	3.06
9-E[a]	1.25	TC	GDI	9.7	3	81	80.86	Red. 26%	DCP + DVVL		8 AT - 60%	4.6	0.52	3.06
10-A	1.25	TC	GDI	9.7	3	81	80.86	Red. 26%	DCP + DVVL		CVT-1	2.6	0.40	5.00
10-B[b]	1.25	TC	GDI	9.7	3	81	80.86	Red. 26%	DCP + DVVL		CVT-2	2.6	0.40	5.00

[a] Cases 9-B – 9-E have reduced transmission torque losses relative to 6 AT as indicated in the table.
[b] Case 10-B is the same as case 10-A except that CVT-2 uses the more efficient loss map of the 6 AT automatic transmissions.

CVT production (CVT-1), while the latter (CVT-2) used a hypothetical CVT loss map equivalent to the more efficient map of the six-speed transmission.

FULL SYSTEM SIMULATION MODEL RESULTS

Figure X.1 graphically shows the predicted combined cycle fuel consumption results.

KNOCK MODEL RESULTS

The U of M full systems simulation uses a knock model, based on Hoepke et. al. (2012), to predict the onset of knock. The fractions of drive cycle time which required retarded spark timing to avoid knock for each of the technologies evaluated are shown in Table X.3. For the first level of turbocharging and downsizing and beyond, a significant fraction of the drive cycle time required spark retard to avoid knock. A vehicle manufacturer is likely to establish a spark timing calibration that provides significant margin to avoid knock so that complete reliance on the knock control system is not required. Such a calibration is likely to result in less than the modeled reduction in fuel consumption for the turbocharging and downsizing technologies.

FRICTION OF TURBOCHARGED DOWNSIZED ENGINES

Friction of the 2.5L naturally aspirated engine is compared with the friction of the 33 percent downsized, 1.68L turbocharged engine and the 50 percent downsized, 1.25L turbocharged engine in this section.

The Chen-Flynn friction equation used in the U of M engine simulation is shown below:

$$FMEP = FMEP_{const} + AP_{Cyl,max} + B_{cmps} + Cc^2_{mps} + L_{acc}\, D/V_d c_{rpm}$$

Where: $FMEP_{const} = 0.0$
$A = 0.004$ bar/bar
$B = 0.08$ bar/m/s
$C = 0.0$
$L_{acc} = 0.0$ (electrical loads)

Applying this equation to the 2.5L naturally aspirated engine and the 1.68L and 1.25L turbocharged engines at a typical FTP cycle engine speed of 1,500 rpm and a torque of 50 Nm yields the results shown in Table X.4. Due to the reduced stroke of the 1.68L turbocharged engine relative to the 2.5L engine, the mean piston speed decreased, which

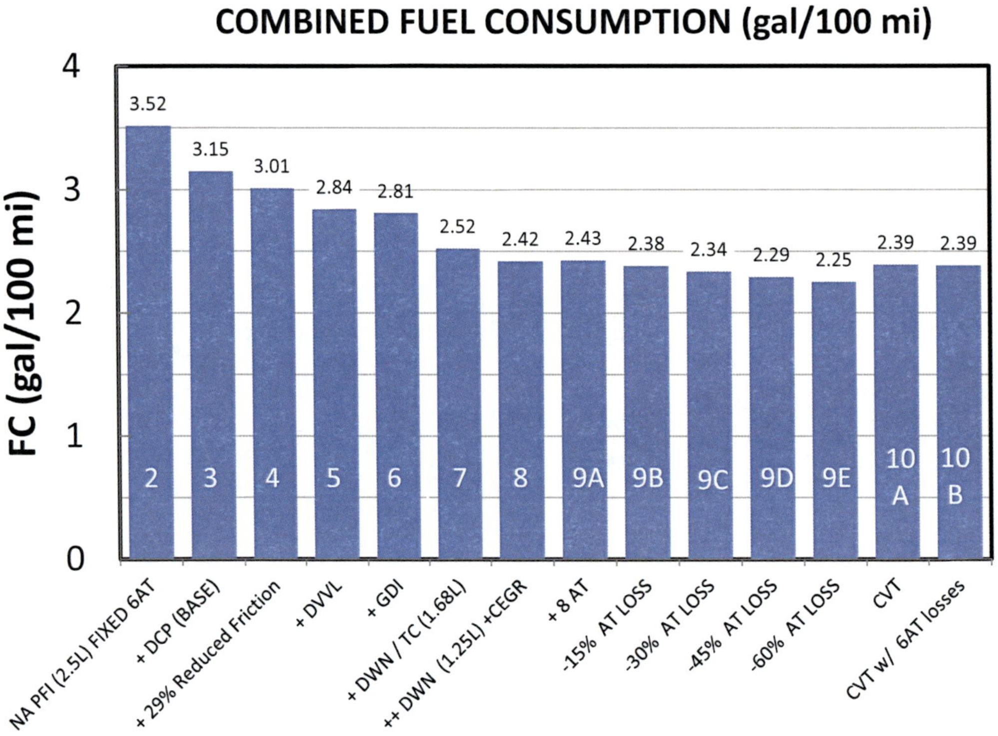

FIGURE X.1 Predicted fuel consumption (combined cycle) for various technologies. Task numbers superimposed on bars.
SOURCE: Middleton et al. (2015).

TABLE X.3 Fraction of Cycle Time with Retarded Spark Timing to Avoid Knock

Engine	FTP	HWY
Fixed Cams	0.00	0.00
DCP	0.00	0.00
EFR	0.00	0.00
DVVL	0.00	0.00
GDI 9.7:1	0.01	0.01
GDI 10.0:1	0.02	0.04
1.68L TC 6-AT	0.10	0.23
1.25L TC 6-AT	0.24	0.53
1.25L TC 8-AT	0.26	0.059

resulted in a 3 percent decrease in FMEP. However, the 1.25L turbocharged engine with nearly the same stroke as the 1.68L engine has a 7 percent increase in FMEP, due to the higher peak cylinder pressure. Even with this small increase in FMEP, the actual friction torque of the 1.25L engine will be considerably lower than the 2.5L naturally aspirated engine due to the engine downsizing.

REFERENCES

Hoepke, B., S. Jannsen, E. Kasseris, and W. Cheng. 2012. EGR effects on boosted SI engine operation and knock integral correlation. SAE Int. J. Engines 5(2):547-559. doi:10.4271/2012-01-0707.

Middleton, R., O. Gupta, H-Y. Chang, G. Lavoie, and J. Martz. 2015. Fuel Economy Estimates for Future Light-Duty Vehicles, University of Michigan Report.

Scherer, H., M. Bek, and S. Kilian. 2009. ZF new 8-speed automatic transmission 8HP70 - Basic design and hybridization. SAE Int. J. Engines 2(1): 314-326. doi:10.4271/2009-01-0510.

TABLE X.4 Comparison of Friction FMEP for 2.5L Naturally Aspirated Engine and a 50 Percent Downsized 1.25L Turbocharged Engine at 1,500 rpm Engine Speed and 50 Nm Torque

Engine Config.	Task	Torque (Nm)	BMEP (bar)	$P_{Cyl,\ max}$ (bar)	C_{mps} (m/s)	$FMEP_{PP}$ (kPa)	$FEMP_{mps}$ (kPa)	$FEMP_{acc}$ (kPa)	$FMEP_{tot}$ (kPa)
GDI 2.5L NA	6	50	2.5	16.5	5.00	4.7	28.4	5.8	38.8
1.68 L TC	7	50	3.7	21.2	4.08	6.0	23.2	8.6	37.7
1.25 L TC	8	50	5.0	24.4	4.04	6.9	23.0	11.5	41.4

Where: $P_{Cyl,\ max}$ is the peak cylinder pressure in bar, C_{mps} is the mean piston speed in m/s.
SOURCE: Derived from Middleton (2015).

Appendix Y

Acronym List

AC	air conditioning
ACCESS	Advanced Combustion Control Enabling Systems and Solutions
AFV	alternative-fuel vehicle
AKI	anti-knock index
AT	automatic transmission
BAT	binary actuation technology
BEV	battery electric vehicle
BMEP	brake mean effective pressure
BOM	bill of materials
BSFC	brake specific fuel consumption
BTE	brake thermal efficiency
CAFE	corporate average fuel economy
CAI	controlled auto-ignition
CARB	California Air Resources Board
CDPF	catalyzed diesel particulate filter
CEC	California Energy Commission
CETC	California Electric Transportation Coalition
CI	compression ignition
CNG	compressed natural gas
CO_2	carbon dioxide
CR	compression ratio
CVP	continuously variable planetary transmission
CVT	continuously variable transmission
CVVL	continuously variable valve lift
DCP	dual cam phasing
DCT	dual clutch transmission
DI	direct injection
DISI	direct injection spark ignition
DMC	direct manufacturing cost
DOC	diesel oxidation catalyst
DOE	U.S. Department of Energy
DOHC	dual overhead cam
DOT	U.S. Department of Transportation
DPF	diesel particulate filter
DVVL	discrete variable valve lift

E85	85 percent ethanol
EACC	electric accessories
ECU	engine control unit
eCVT	electronically controlled continuously variable transmission
EEA	Energy and Environmental Analysis Inc
EGR	exhaust gas recirculation
EISA	Energy Independence and Security Act of 2007
EPA	U.S. Environmental Protection Agency
EPS	electric power steering
EU	European
EVO	exhaust valve opening
FAME	fatty acid methyl ester
FC	fuel consumption
FCEV	fuel cell electric vehicle
FE	fuel economy
FSS	full system simulation
FTP	Federal Test Procedure
GDCI	gasoline direct injection compression ignition
GDI	gasoline direct injection
GDICI	gasoline direct injection compression ignition
GHG	greenhouse gas
GM	General Motors Company
HC	hydrocarbon
HCCI	homogeneous-charge compression ignition
HEV	hybrid-electric vehicle
HFET/HWFET/HWY	Highway Fuel Economy Test/ Highway Federal Emissions Test (or highway cycle)
I4	inline 4-cylinder engine
ICM	indirect cost multiplier
ICP	intake-cam phasing
IMEP	indicated mean effective pressure
ISFC	indicated specific fuel consumption
ITE	indicated thermal efficiency
IVC	intake-valve closing
LBL	low viscosity lubricants
LCFS	low-carbon fuel standard
LDV	light duty vehicle
LNT	lean NO_x traps
LP	low pressure
LPG	liquefied petroleum gas
LTC	low temperature combustion
MBT	minimum spark advance for best torque
MPFI	multi point fuel injection
MPG	miles per gallon
MPV	midsize passenger vehicle
MY	model year
NA	naturally aspirated
NA	North American
NESCCAF	Northeast States Center for a Clean Air Future

NHTSA	National Highway Traffic Safety Administration
NMHC	Non-methane hydrocarbons
NO_x	Nitrogen oxides
NRC	National Research Council
NSC	NO_x storage and reduction catalysts
NVH	noise, vibration, and harshness
NVO	negative value overlap
OBD	on-board diagnostics
OEM	original equipment manufacturer
OHV	overhead valve
ORNL	Oak Ridge National Laboratory
OSC	oxygen storage capacity
PCCI	premixed charge compression ignition
PEV	plug-in electric vehicle
PFI	port fuel injection
PGM	platinum group metals
PHEV	plug-in hybrid electric vehicle
PM	particulate matter
PVO	positive value overlap
R&D	research and development
RFG	reformulated gasoline
RIA	regulatory impact analysis
ROM	read only memory
RON	research octane number
RPE	retail price equivalent
SACI	spark assisted compression ignition
SAE	Society of Automotive Engineers
SC	supercharger
SCR	selective catalytic reduction
SGDI	stoichiometric gasoline direct injection
SI	spark ignited engine
SOC	state of charge
SOHC	single overhead cam
SULEV	super ultra-low emission vehicle
SUV	sport utility vehicle
TC	turbocharged
TSD	technical support document
TWC	three way catalyst
UDDS	urban dynamometer driving schedule
ULEV	ultra low emissions vehicle
V2G	vehicle-to-grid
V2V	vehicle-to-vehicle
V6	six-cylinder V engine
V8	eight-cylinder V engine
VEL	valve event and lift
VEM	valve-event modulation
VGT	variable geometry turbochargers
VVL	variable valve lift

VVT	variable valve timing
xEV	vehicle with an electrified powertrain, included HEV, PHEV, EV and FCEV
ZEV	zero emission vehicle